FLORA OF TROPICAL EAST AFRICA

ACANTHACEAE (Part 2)

IAIN DARBYSHIRE[1], KAJ VOLLESEN[1] & ENSERMU KELBESSA[2]

Correction to the generic key presented in Acanthaceae Part 1: 3–11.

Following the recent treatment of the genus *Brachystephanus* for the Flora, it has become clear that the generic key presented in part 1 does not adequately distinguish this genus from *Monothecium* and *Ruspolia*. The couplets covering these genera are therefore revised below:

47. Corolla limb deeply divided into five subequal lobes loosely arranged in two lips, basal tube long-linear (over 15 mm long); anthers held in throat or very shortly exserted .... 37. **Ruspolia**
    Corolla limb strongly 2-lipped, the upper lip at most shortly notched at the apex, the lower lip deeply to shallowly divided into 3 lobes, basal tube variously short to long-linear; anthers clearly exserted beyond the throat 48
48. Corolla tube up to 10 mm long; staminal filaments held within the hooded upper lip .... 39. **Monothecium**
    Corolla tube over 10 mm long or if shorter (sect. *Oreacanthus*) then staminal filaments held between the two lips, the upper lip not hooded .... 35. **Brachystephanus**

NOTE. Some important species of Acanthaceae cultivated in the F.T.E.A. region are treated on pages 730–733.

30. NEURACANTHUS

Nees in Wallich, Pl. As. Rar. 3: 76, 97 (1832) & in DC., Prodr. 11: 248 (1847); G.P. 2: 1093 (1876); C.B. Clarke in F.T.A. 5: 137 (1899); Bidgood & Brummitt in K.B. 53: 1 (1998)

Leucobarleria Lindau in Ann. R. Ist. Bot. Roma 6: 76 (1896)

Perennial herbs (incl. suffrutices with annual stems from woody rootstocks), subshrubs or shrubs; cystoliths conspicuous or not. Leaves opposite, entire to crenate. Inflorescences axillary or terminal, spicate, sometimes subtended by bract-like leaves, sometimes with sterile spinose inflorescences at base of, or surrounding, fertile inflorescences; bracts in four or six rows, imbricate, with 3–7 prominent veins, each

[1] Royal Botanic Gardens, Kew.
[2] National Herbarium, Addis Ababa University.
Iain Darbyshire: *Lepidagathis, Barleria, Brachystephanus, Isoglossa, Chlamydocardia, Rhinacanthus, Dicliptera, Hypoestes*
Kaj Vollesen: *Neuracanthus, Crabbea, Pseuderanthemum, Ruspolia, Ruttya, Monothecium, Justicia, Anisotes, Ecbolium, Megalochlamys, Cephalophis, Trichaulax*
Ensermu Kelbessa & Kaj Vollesen: *Asystasia*

subtending a single sessile or subsessile flower; bracteoles absent. Calyx 2-lipped, deeply divided between the lips, dorsal lip 3-veined, ventral 2-veined, the veins each terminating in a tooth or the teeth fused into a single apex. Corolla outside hairy in bands along middle of lobes and inside in a band on central lobe in lower lip, subactinomorphic; basal cylindric tube short; throat broadly funnel-shaped, lobes spreading to suberect, lower lip of 3 broad trianguler lobes, upper lip of 1 shortly bifid lobe. Stamens 4, didynamous, inserted at base of throat, included, ventral pair inserted below dorsal pair and smaller; ventral pair with bithecous anthers, dorsal pair with monothecous anthers, or with a second sterile theca (rarely with 2 fertile thecae), thecae oblong, glabrous or bearded at base, connective hairy. Style glabrous; stigma with a single flattened linear-oblong lobe, the upper lobe missing. Capsule 2–4-seeded, oblong or triangular-ovate, usually glabrous. Seed discoid, covered with hygroscopic hairs.

30 species in tropical Africa, Madagascar, Arabia and tropical Asia, especially in NE and East Africa.

NOTE. Recent molecular data (McDade *et al.* in Am. J. Bot. 95: 1136–1152 (2008)) place *Neuracanthus* close to both *Barlerieae* and *Whitfieldieae* but with only weak support.

1. Branched spines present below or surrounding inflorescences 2
   Branched spines absent below inflorescences 3
2. Largest leaf 4–8 cm long; fertile inflorescences with 4–10 flowers; calyces not clearly visible, 1–8 mm wide in fruit, largely concealed by the bracts; capsule 8–9 mm long . . . 6. *N. keniensis*
   Largest leaf 2–3 cm long; fertile inflorescences with 1–3(–4) flowers; calyces clearly visible, ± 10 mm wide in fruit, wider than and not concealed by the bracts; capsule ± 6 mm long .... 7. *N. polyacanthus*
3. Peduncle 3–20 mm long, with 1–3 pairs of sterile bracts; capsule 4-seeded; all inflorescences near base of stems . . 1. *N. pictus*
   Peduncle 0–4(–8) mm long, if more than 3 mm then peduncle without sterile bracts; capsule 2-seeded; inflorescences terminal or axillary in upper part of stems 4
4. Inflorescences all terminal 5
   Inflorescences all axillary or some terminal and some axillary 6
5. Suffrutescent herb with erect unbranched stems; lower and middle fertile bracts 1.5–2 cm long; corolla limb 5–8 mm in diameter when fully open .... 5. *N. decorus*
   Subshrub or shrub, stems branched; lower and middle fertile bracts 0.8–1.4 cm long; corolla limb (12–)15–22 mm in diameter when fully open .... 4. *N. ukambensis*
6. Corolla limb (12–)15–22 mm in diameter when fully open; calyces and usually bracts with stalked capitate glands; some inflorescences terminal .... 4. *N. ukambensis*
   Corolla limb 5–12 mm in diameter when fully open; bracts and calyces without capitate glands; all inflorescences axillary 7
7. Leaf-base truncate to subcordate; bracts green with pale to dark brown veins, not hard and leathery, with conspicuous shoulders to almost wings below apex .... 2. *N. africanus*
   Leaf-base cuneate to attenuate; bracts dark brown, not with darker veins, hard and leathery, narrowing gradually to apex .... 3. *N. tephrophyllus*

Bidgood & Brummitt (l.c.) divide *Neuracanthus* into three sections. The species in our area fall into these as follows:

Sect. *Neuracanthus*, species 1.
Sect. *Didymosperma* Bidgood & Brummitt in K.B. 53: 24 (1998), species 2–5.
Sect. *Leucobarleria* (Lindau) Chiov., Fl. Somala: 256 (1929), species 6–7.

1. **Neuracanthus pictus** *M.G. Gilbert* in K.B. 43: 159 (1988); Lebrun & Stork, Enum. Pl. Afr. Trop. 4: 496 (1997); Bidgood & Brummitt in K.B. 53: 22 (1998); Thulin in Fl. Somalia 3: 448 (2006). Type: Kenya, Northern Frontier District: Isiolo–Wajir road, *M.G. Gilbert & Thulin* 1093 (EA!, holo.; K!, iso.)

Perennial herb or subshrub; stems erect, to 50 cm long, densely whitish sericeous-tomentose when young with glossy silky hairs, glabrescent. Mature leaves with petiole 0–2 cm long; lamina narrowly ovate or narrowly elliptic, largest 4.5–11 × 1.2–3 cm, base attenuate, decurrent, apex subacute to rounded; beneath whitish sericeous to tomentose, above sparsely to densely sericeous-pubescent. Flowers in dense strobilate axillary spikes, all near base of the stem; peduncle 0.3–2 cm long, with 1–3 pairs of sterile bracts, sericeous to densely pubescent; spike 1.5–4 cm long; bracts densely imbricate, in 4 rows, pale green with 5 prominent crimson veins and reticulation, 1.5–2 × 1–1.7 cm, broadly elliptic, acuminate to cuspidate, minutely puberulous on veins, lamina with sparse long hairs, margins also with dense long stiff pale brown hairs to 2 mm long. Calyx lobes broadly ovate, dorsal ± 7 × 4 (± 12 × 7 in fruit) mm, with 3 acuminate teeth to 3 mm long, ventral ± 5 × 3 (± 10 × 6 in fruit) mm, with 2 acuminate teeth to 3 mm long; with long strigose hairs to 2 mm long, densest near base and on margins. Corolla very pale lilac to pale mauve with darker veins in throat sparsely sericeous-pubescent outside, 13–14 mm long of which the tube ± 5 mm, limb subactinomorphic, ± 10 mm diameter, tube ± 2 mm diameter. Larger anthers ± 1 mm long. Capsule 4-seeded, narrowly oblong, beaked, ± 10 mm long. Mature seed not seen.

KENYA. Northern Frontier District: 13 km S of Mado Gashi, 11 Dec. 1977, *Stannard & M.G. Gilbert* 935! & Isiolo–Wajir road, 114 km from turning near Isiolo, 26 April 1978, *M.G. Gilbert & Thulin* 1093!
DISTR. **K** 1, 7; Somalia
HAB. *Acacia* and *Acacia-Commiphora* bushland with *Euphorbia* and *Terminalia* on sandy soil; 300–450 m

2. **Neuracanthus africanus** *S. Moore* in J.B. 18: 37 (1880); Lindau in E. & P. Pf. IV, 3b: 315 (1895); C.B. Clarke in F.T.A. 5: 137 (1899); Fanshawe, Check List Woody Pl. Zambia Distrib.: 30 (1973); U.K.W.F., ed 2: 273 (1994); Lebrun & Stork, Enum. Pl. Afr. Trop. 4: 496 (1997); Bidgood & Brummitt in K.B. 53: 31 (1998). Type: Mozambique, Lupata, *Kirk* 10 (K!, holo.)

Perennial herb, subshrub or shrub, often forming clumps or tangles with a number of stems from large woody rootstock; stems erect to prostrate, to 1 m long, simple or branched, when young sericeous to pubescent or sparsely so with broad curly multicellular hairs (rarely glabrous). Mature leaves with petiole 0–4(–5) mm long; lamina elliptic to obovate in outline, concave or not in lower part, largest 3.7–15 × 1.8–6.5 cm, base truncate to subcordate, apex acuminate to subacute (rarely rounded); beneath glabrous to sparsely curly-pubescent on midrib, above glabrous or with sparse uniformly scattered hairs. Flowers in axillary strobilate spikes, single or 2 per axil; peduncle 0–3 mm long, apically with or without a pair of foliaceous bracts to 1(–2) cm long; spike 1–5 × 0.5–1(–1.5) cm; bracts loosely imbricate, in 4 rows, green with (3–)5–7 pale to dark brown veins and reticulation, lower and middle 0.6–1.6(–1.8) × 0.4–1(–1.2) cm, broadly elliptic to broadly

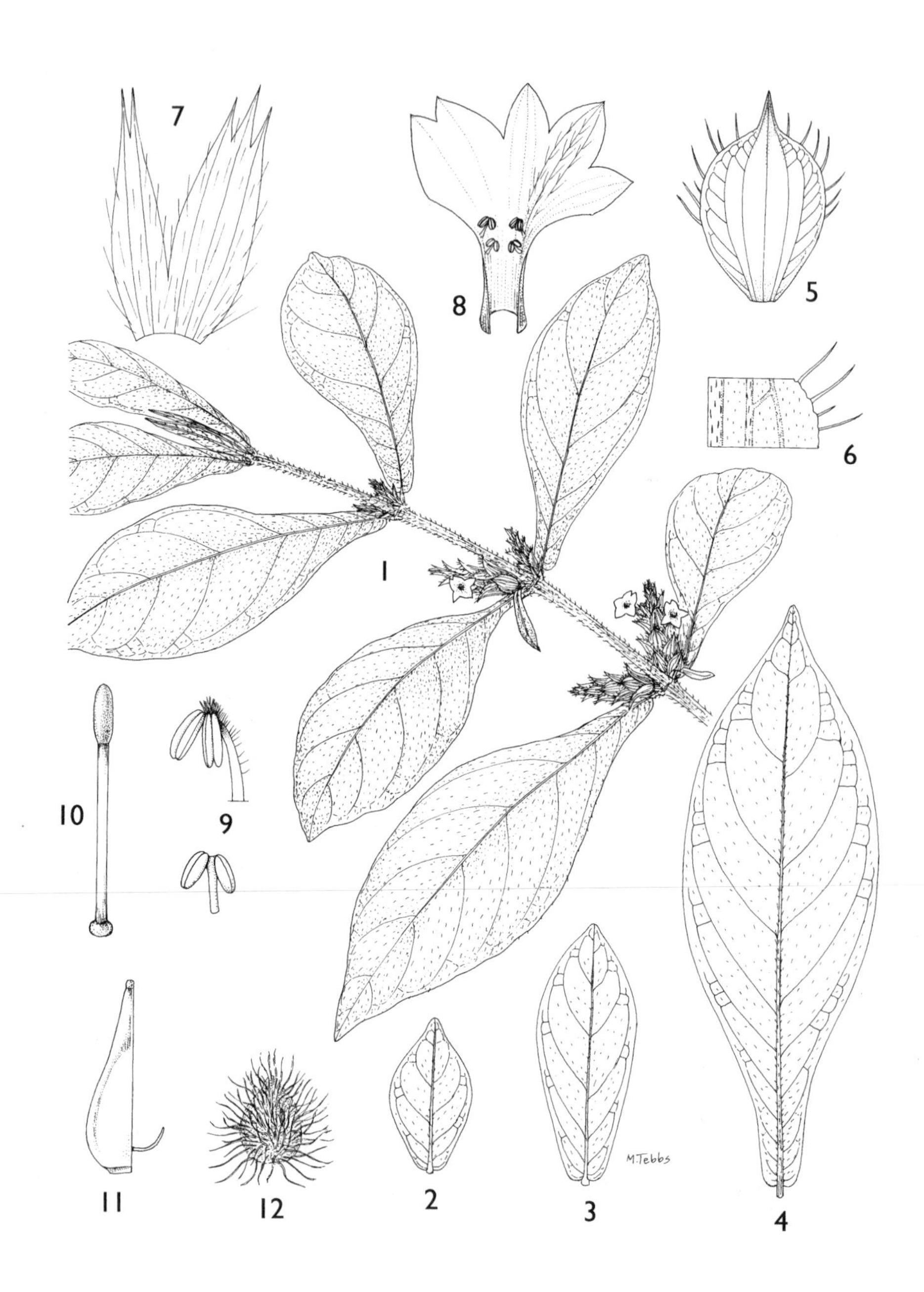

FIG. 44. *NEURACANTHUS AFRICANUS* — **1**, habit, × $^{2}/_{3}$; **2–4**, variation in leaf shape and size, × $^{2}/_{3}$; **5**, bract, × 4; **6**, detail of bract indumentum, × 8; **7**, calyx opened up, × 4; **8**, corolla opened up, × 4; **9**, stamens, × 14; **10**, style and stigma, × 14; **11**, capsule, × 4; **12**, seed, × 4. 1 & 7–10 from *Fanshawe* 8302, 2 & 5–6 from *Goodier* 855, 3 & 11 from *Fulwood* 1, 4 from *Fanshawe* 4530, 12 from *Bidgood* 1222. Drawn by Margaret Tebbs.

obovate, with conspicuous shoulders to almost wings below the spinescent acuminate tip, subglabrous to puberulous on lamina and with sparse to dense long stiff hairs on margins (rarely also on back), sometimes almost white from the dense cystoliths. Calyx lobes ovate or narrowly so, 7–11 × 2–4 (to 16 × 6 in fruit) mm, similar or dorsal slightly longer and wider, dorsal 3-toothed, ventral 2-toothed, teeth 2–3(–4) (to 6 in fruit) mm long; glabrous to sericeous-puberulous and with scattered minute stalked capitate glands also with few to many long strigose hairs, densest at base and on margins. Corolla white or white with pink to purple lines in throat or pale mauve to mauve with darker veins, sericeous to puberulous, 9–14 mm long of which the tube 4–6 mm, limb subactinomorphic, 8–12 mm diameter. Larger anthers ± 0.5 mm long. Capsule 2-seeded, ovoid, beaked, 8–11 mm long. Seed cordiform, 4–5 × 3–4 mm. Fig. 44, p. 290.

1. Leaves concave in lower part above base; calyx glabrous apart from strigose hairs and scattered capitate glands. Leaves petiolate; lower and middle fertile bracts 8–11 × 4–6 mm; corolla white with red to mauve lines in throat .... a. subsp. *africanus*
   Leaves not concave (rarely slightly so) above base; calyx puberulous or sparsely so (rarely glabrous) apart from strigose hairs and scattered capitate glands 2
2. Leaves petiolate; lower and middle fertile bracts (10–)12–16 (–18) × 5–10(–12) mm; corolla pale mauve to mauve with darker lines (rarely white or white with pink to mauve lines) .... c. subsp. *masaicus*
   Leaves sessile; lower and middle fertile bracts 6–10 × 4–6 mm; corolla white or white with pink lines ..... b. subsp. *ruahae*

a. subsp. **africanus**

Erect subshrubs or shrubs; leaves petiolate, lamina clearly concave in lower part above the base; lower and middle fertile bracts 8–11 × 4–6 mm; calyx glabrous apart from long strigose hairs and capitate glands; corolla white with red to mauve lines in throat.

TANZANIA. Mpwapwa District: 16 km N of Mwega River on Mtera track, 4 April 1988, *Bidgood et al.* 925! & 3 km S of Mayamaya on Dodoma–Singida road, 19 April 1988, *Bidgood et al.* 1222!; Kilosa District: 8 km on Malolo–Kisanga track, 3 April 1988, *Bidgood et al.* 915!
DISTR. **T** 5–7; Zambia, Malawi, Mozambique, Zimbabwe, South Africa
HAB. *Acacia*, *Acacia-Commiphora* and *Acacia-Commiphora-Euphorbia-Adansonia* bushland, usually on rocky hillsides but also on clayey loam and on calcareous soils; 500–1150 m

SYN. *Neuracanthus africanus* S. Moore var. *limpopoensis* Bidgood & Brummitt in K.B. 53: 36 (1998). Type: Zimbabwe, Chipinda Pools, *Goodier* 855 (K!, holo.; SRGH, iso.)
[*N. scaber sensu* Lindau in P.O.A. C: 369 (1895) pro parte, *non* S. Moore (1880)]

NOTE. In southern Africa subsp. *africanus* gradually gets a hairy calyx lamina (the type is glabrous) and forms referred to by Bidgood & Brummitt (l.c.) as var. *limpopoensis* have leaves very similar to those of subsp. *masaicus*. But all southern material has a white corolla with pink lines and lower and middle fertile bracts 8–12 × 5–7 mm, thus falling within the variation of subsp. *africanus* in Tanzania. Typical var. *limpopoensis* with its small leaves is quite distinct but there are – as already recognised by Bidgood & Brummitt – as many intermediate as "clean" specimens.

b. subsp. **ruahae** (*Bidgood & Brummitt*) *Vollesen* **comb. et stat. nov**. Type: Tanzania, Iringa District: Ruaha National Park, Chinaputa Escarpment, *Richards* 20957 (K!, holo.; EA!, iso.)

Erect perennial herb with unbranched stems (rarely an erect branched subshrub); leaves sessile, not (rarely slightly) concave above base; lower and middle fertile bracts 6–10 × 4–6 mm; calyx puberulous or sparsely so under the strigose hairs and capitate glands; corolla white or white with pink lines in throat.

TANZANIA. Iringa District: Ruaha National Park, 1.5 km on Msembe–Mbagi track, 26 Feb. 1961, *Greenway & Kanuri* 13961! & 2 km NE of Msembe, 17 Jan. 1972, *Björnstad* 1251! & Great Ruaha River, Mtera Dam, Ladwa, 3 Feb. 1989, *Mhoro* 6074!

DISTR. **T** 7; not known elsewhere

HAB. *Acacia-Combretum-Grewia* bushland, riverine forest, on sandy to gravelly soil or on rocky slopes; 800–900 m

SYN. *Neuracanthus africanus* S. Moore var. *ruahae* Bidgood & Brummitt in K.B. 53: 38 (1998)

c. subsp. **masaicus** (*Bidgood & Brummitt*) *Vollesen* **comb. et stat. nov**. Type: Tanzania, Dodoma District: 10 km N of Manyoni on Singida road, *Bidgood, Mwasumbi & Vollesen* 1131 (K!, holo.; DSM!, EA!, K!, MO!, iso.)

Erect to decumbent subshrub or shrub, often forming dense clumps or tangles; leaves petiolate, not concave above base (rarely slightly so); lower and middle fertile bracts (10–)12–16(–18) × 5–10(–12) mm; calyx puberulous or sparsely so under the strigose hairs and capitate glands; corolla pale mauve or mauve with darker lines in throat (rarely white or white with pink or blue lines).

KENYA. Masai District: Kajiado, between Naibor Kawadie and Torosei, 23 Feb. 1963, *Glover & Cooper* 3454! & near Lake Magadi, 24 June 1978, *Plaizier* 1413 (fide Bidgood & Brummitt).

TANZANIA. Masai District: Kitumbeine, 1 March 1969, *Richards* 24229! & 20 km on Makuyuni–Mto wa Mbu road, 23 May 1995, *Vollesen* 95/203!; Dodoma District: 30 km on Dodoma–Morogoro road, 12 April 1988, *Bidgood et al.* 1024!

DISTR. **K** 6; **T** 1, 2, 5; not known elsewhere

HAB. *Acacia* and *Acacia-Commiphora* bushland, thicket on rocky hillsides, on sandy to gravelly or clayey soils; (900–)1050–1350(–1700) m.

SYN. *Neuracanthus africanus* S. Moore var. *masaicus* Bidgood & Brummitt in K.B. 53: 37 (1998)

3. **Neuracanthus tephrophyllus** *Bidgood & Brummitt* in K.B. 53: 44 (1998); Thulin in Fl. Somalia 3: 449 (2006). Type: Somalia, 17 km W of Dinsor on Bardera road, *Gillett & Hemming* 24712 (K!, holo.; EA!, iso.)

Shrubby herb or subshrub to 0.5(–1) m tall; stems branched, when young sparsely to densely sericeous to pubescent (rarely tomentose) with broad curly multicellular hairs. Mature leaves with petiole 0–1(–1.5) cm long; lamina ovate to broadly ovate or elliptic to broadly elliptic, largest 2.5–8(–10) × 1.5–5.5(–6.5) cm, base attenuate to cuneate, decurrent or not, apex acute to rounded, usually apiculate; reticulation strongly raised; beneath pubescent to densely lanate with broad curly multicellular hairs, above subglabrous to sparsely pubescent. Flowers in axillary strobilate spikes; peduncle 1–3(–7 in fruit) mm long, with 2 pairs of green foliaceous bracts markedly different from fertile bracts, with recurved apical part, 1–3 cm long, and inside these usually with a pair of sterile bracts similar to fertile bracts; spike 0.7–3 × 0.5–1.5 (to 5 × 2 in fruit) cm; bracts densely imbricate, in 4 rows, hard and leathery, dark brown without darker veins, lower and middle 0.8–2 × 0.4–1.5 cm, ovate or elliptic to suborbicular, narrowing gradually to an acute broadly triangular apex, densely ciliate with stiff hairs, lamina subglabrous or with stiff or curly hairs, mostly near apex. Calyx lobes ovate to elliptic, 7–14 × 2–6 mm, similar or dorsal slightly longer and wider, dorsal 3-toothed, ventral 2-toothed, teeth in dorsal lobe 1–3 mm long, in ventral lobe 3–5 mm; with dense stiff hairs all over or densest along margins and towards apex and with scattered minute subsessile capitate glands. Corolla white (more rarely pale pink to pale mauve), densely sericeous, 9–13 mm long of which the tube 5–8 mm, limb subactinomorphic, 9–12 mm diameter. Larger anthers ± 0.7 mm long. Capsule 2-seeded, ovoid, beaked, 7–11 mm long; seed cordiform, 3–5 × 3–4 mm.

NOTE. This species can be divided into three reasonably well separated subspecies. As mentioned by Bidgood & Brummitt (l.c.) subsp. *conifer*, with its large cone-like spikes and densely hairy leaves, at first sight looks strikingly different from the other two. But – especially in the Mkomazi area – there are a number of collections approaching subsp. *tsavoensis*. They have a denser bract-indumentum and plants with pale pink to pale mauve corollas occur. Of the two sheets of *Vollesen* 96/49 from Mkomazi Game Reserve – both from the same population – one is typical subsp. *conifer* while the other approaches subsp. *tsavoensis* by having quite dense bract-indumentum and "white or faintly pink" corolla. *Greenway* 13032 clearly has all the characters of subsp. *tsavoensis* in flowers but the old fruiting branch on the same plant looks strikingly like subsp. *conifer*.

Bidgood and Brummitt (l.c.) mention other intermediates between the three subspecies.

1. Mature leaves usually markedly discolorous, the lower surface usually obscured by the dense pale grey indumentum; spikes (0.7–)1–3(–5) cm long; fertile bracts only hairy on edges (more rarely also a few scattered hairs on lamina), (8–)10–18(–20) × (5–)7–12(–15) mm; calyx 8–14 × 3–6 mm; corolla white (rarely with a pink tinge to pale pink or pale mauve); capsule 8–11 mm long; seed 4–5 × 4 mm . . c. subsp. *conifer*
   Mature leaves not or only slightly discolorous, the lower surface not obscured by the indumentum; spikes 0.7–2 cm long; fertile bracts hairy at least on apical part of lamina (rarely only on edges), 8–14 × 4–8 mm; calyx 7–11 × 2–3 mm; corolla white or pink; capsule 7–8 mm long; seed 3–4 × 3 mm 2
2. Fertile bracts hairy all over lamina; spikes 0.7–1.5(–2) cm long; fertile bracts 8–12(–14) × 5–6(–8) mm; corolla pale pink to pale mauve . . . . b. subsp. *tsavoensis*
   Fertile bracts only hairy on apical part of lamina (rarely only hairy on edges); spikes 0.9–2 cm long; fertile bracts 9–14 × 4–7 mm; corolla white . . . . a. subsp. *tephrophyllus*

a. subsp. **tephrophyllus**

Spikes 0.9–2 cm long; fertile bracts 9–14 × 4–7 mm, only hairy on apical part of lamina (rarely only hairy on margins); corolla white; capsule 7–8 mm long; seed 3–4 × 3 mm.

KENYA. Northern Frontier District: 27 km S of Mado Gashi, 27 Jan. 1972, *Bally & Radcliffe-Smith* 14964!; Tana River District: Malindi–Garsen road, 1 km towards Garsen from turn-off to Kibusu, 22 July 1974, *Faden* 74/1175!; Kilifi District: Baricho, 2 Dec. 1990, *Luke & Robertson* 2578!
DISTR. **K** 1, 7; Somalia
HAB. *Acacia* and *Acacia-Commiphora* bushland, often with *Dobera* and *Salvadora*, on pale to dark grey loamy to clayey often alkaline soils liable to seasonal flooding; 10–300 m

SYN. [*Neuracanthus scaber sensu* E.P.A.: 950 (1964), *quoad specim.* ex Somalia, *non* S. Moore (1880)]

b. subsp. **tsavoensis** *Bidgood & Brummitt* in K.B. 53: 50 (1998). Type: Kenya, Teita District: 6 km towards Nairobi from Maungu Station, *Faden* 74/519 (K!, holo.; EA!, MO, PRE, UPS, WAG, iso.)

Spikes 0.7–1.5(–2) cm long; fertile bracts 8–14 × 4–8 mm, hairy all over lamina; calyx 7–11 × 2–3 mm; corolla pale pink to pale mauve; capsule 7–8 mm long; seed 3–4 × 3 mm.

KENYA. Masai District: Chyulu Plains, Oleeilelu, 9 June 1991, *Luke* 2855!; Teita District: Tsavo National Park East, Voi Gate to Lugards Falls, 12 Jan. 1967, *Greenway & Kanuri* 13032! & 6 km towards Nairobi from Maungu Station, 30 April 1974, *Faden* 74/519!
TANZANIA. Pare District: Mkomazi Game Reserve, Dindira to Ngurunga, 28 May 1972, *Mbano* CAWM5790!
DISTR. **K** 6, 7; **T** 3; not known elsewhere
HAB. *Acacia* and *Acacia-Commiphora* bushland, eroded grassland, on brown sandy to stony or clayey soil; 600 –1100 m

c. subsp. **conifer** *Bidgood & Brummitt* in K.B. 53: 48 (1998); Iversen in Symb. Bot. Upsal. 29(3): 162 (1991), in obs.; U.K.W.F., ed. 2: 273 (1994), in obs. Type: Kenya, Masai District: 50 km Nairobi–Magadi road, *Bidgood & Vollesen* 1279 (K!, holo.; C!, CAS!, EA!, MO!, UPS!, WAG!, iso.)

Mature leaves usually markedly discolorous, the lower surface usually obscured by the dense pale grey indumentum; spikes (0.7–)1–3(–5) cm long; fertile bracts only hairy on edges (more rarely also a few scattered hairs on lamina), (8–)10–18(–20) × (5–)7–12(–15) mm; calyx 8–14 × 3–6 mm; corolla white (rarely with a pink tinge to pale pink or pale mauve); capsule 8–11 mm long; seed 4–5 × 4 mm.

KENYA. Masai District: 50 km on Nairobi–Magadi road, 14 Nov. 1952, *Bally* 8372! & 14 Aug. 1961, *Polhill & Verdcourt* 444! & Sinya Ondua, 25 May 1996, *Pearce & Vollesen* 948!

TANZANIA. Moshi District: Himo Station, Aug. 1965, *Beesley* 159!; Pare District: Mkomazi Game Reserve, Mbulu, 5 June 1996, *Vollesen* 96/49!; Lushoto District: 3 km on Mkomazi–Mombo road, 28 April 1988, *Bidgood & Vollesen* 1270!

DISTR. **K** 6; **T** 2, 3; not known elsewhere

HAB. *Acacia* and *Acacia-Commiphora* bushland, on gravelly to stony or clayey soil; 300–950 m in Tanzania, 1150–1500(–1700) m in Kenya

SYN. [*Neuracanthus scaber sensu* Lindau in P.O.A. C: 369 (1895) pro parte; C.B. Clarke in F.T.A. 5: 138 (1899), quoad *Volkens* 549 & 2215; T.T.C.L.: 14 (1949); Ruffo *et al.*, Cat. Lushoto Herb. Tanzania: 8 (1996), *non* S. Moore (1880)]

4. **Neuracanthus ukambensis** *C.B. Clarke* in F.T.A. 5: 138 (1899); U.K.W.F., ed. 2: 273 (1994); Lebrun & Stork, Enum. Pl. Afr. Trop. 4: 496 (1997); Bidgood & Brummitt in K.B. 53: 51 (1998). Type: Kenya, Ukamba, *Hildebrandt* 2723 (B†, holo.). Neotype: Kenya, Machakos District: Masaleni, *Napper & Jones* 1951 (K!, neo, selected by Bidgood & Brummitt (l.c.); EA!, iso.)

Subshrub or shrub to 1(–1.5) m tall; stems branched, when young sparsely pubescent to tomentose with broad curly glossy multicellular hairs to 2 mm long. Mature leaves with petiole 0–5 mm long; lamina ovate to elliptic, largest 2–10 × 1–5.5 cm, base attenuate, decurrent, apex acute to obtuse with a small apiculus; beneath sparsely curly-pubescent on midrib and veins, sparser or glabrous on lamina, above subglabrous or with sparse uniformly scattered hairs. Flowers in axillary and terminal strobilate spikes, sometimes terminal only; peduncle 0–2 mm long, with a pair of sterile bracts of same dimensions as fertile bracts; spike 1–3(–4) × 1–1.5(–2) cm; bracts densely imbricate, in 4 rows, light brown to green with 3–5 brown veins and reticulation, lower and middle 0.8–1.4 × 0.5–1 cm, broadly ovate to suborbicular, acuminate to cuspidate, without shoulders below the spinescent tip, subglabrous to puberulous on lamina and with (rarely without) stalked capitate glands, with usually dense long stiff hairs on margins (rarely also on back). Calyx lobes ovate to elliptic, 10–12 × 3–5 (to 15 × 7 in fruit) mm, dorsal slightly longer and wider, dorsal 3-toothed, ventral 2-toothed, teeth 3–5 (to 6 in fruit) mm long; with dense long stiff hairs on margins and surface and with scattered stalked capitate glands on lamina. Corolla white or white with red lines in throat (rarely very pale mauve), sericeous to puberulous, 12–16 mm long of which the cylindric tube 4–6 mm, limb subactinomorphic, (12–)15–22 mm diameter when fully open. Larger anthers ± 0.8 mm long. Capsule 2-seeded, ovoid, beaked, 8–9 mm long; seed cordiform, ± 5 × 4 mm.

KENYA. Northern Frontier District: Derisa, 28 Aug. 1947, *J. Adamson* in *Bally* 5860!; Machakos District: Bushwackers Camp, 4 June 1958, *Irwin* 412!; Kitui/Teita District: Tsavo National Park East, Lugards Falls, 31 Aug. 1969, *Bally* 13498!

DISTR. **K** 1, 4, 7; not known elsewhere

HAB. *Acacia-Commiphora* bushland with *Dobera* and *Salvadora*, on sandy to stony soil; 150–800 m

NOTE. The Wakamba rub the ashes of the burnt leaves on [skin over] spleen when ill (*J. Adamson* in *Bally* 5860).

5. **Neuracanthus decorus** *S. Moore* in J.B. 18: 307 (1880); C.B. Clarke in F.T.A. 5: 138 (1899); Benoist, Thèse Fac. Sci. Paris: 125 (1912); Lebrun & Stork, Enum. Pl. Afr. Trop. 4: 496 (1997); Bidgood & Brummitt in K.B. 53: 39 (1998). Type: Angola, *Welwitsch* 5057 (BM!, holo.; G!, K!, iso.)

Perennial herb with single (rarely 2) erect (rarely decumbent) stems from creeping rootstock with fleshy roots; stems to 35 cm long, when young densely pubescent to tomentose with broad curly multicellular hairs. Mature leaves with petiole 1–3(–6) mm long; lamina elliptic to obovate, largest 6–10.5(–16) × 2–4 cm, base cuneate to truncate, apex acute to rounded; beneath subglabrous to sparsely curly-pubescent on midrib and major veins, above with uniformly scattered sparse hairs. Flowers in a single dense strobilate quadrangular terminal spike; peduncle 1–4(–8) mm long, no sterile bracts; spike 3–7.5 × 2–3 cm; bracts densely imbricate, in 4 rows, pale green with 5–7 prominent darker veins and reticulation, 1.5–2 × 1–1.5 cm, broadly ovate to broadly obovate to almost orbicular, acute to acuminate, minutely puberulous on veins and with dense long stiff tubercle-based pale brown hairs to 2 mm long on margins (sparser on veins and lamina). Calyx lobes ovate, 9–14 × 3–4 mm, similar or dorsal slightly longer and wider, either with short (1–3 mm) teeth or divided ± halfway down; with long pale brown strigose hairs to 2 mm long, densest near base and on margins. Corolla white or white with pale pink veins, sericeous outside (puberulous on lobes), 7–10 mm long of which the tube 2–3 mm, limb subactinomorphic, 5–8 mm diameter. Anthers not seen. Capsule 2-seeded, narrowly ovoid, beaked, 9–11 mm long; seed cordiform, 4–5 × ± 3 mm.

subsp. **strobilinus** (*C.B. Clarke*) *Bidgood & Brummitt* in K.B. 53: 40 (1998). Type: Malawi, *Whyte* 138 (K!, holo.)

TANZANIA. Kigoma District: Lugufu to Uvinza, 9 Feb. 1926, *Peter* 36624!
DISTR. **T** 4; Congo-Kinshasa, Zambia, Malawi
HAB. Not recorded, but in *Brachystegia* woodland in the rest of its area; 1050 m

SYN. *Neuracanthus strobilinus* C.B. Clarke in F.T.A. 5: 138 (1899); De Wildeman, Contrib. Fl. Katanga: 200 (1921); Lebrun & Stork, Enum. Pl. Afr. Trop. 4: 496 (1997)

NOTE. Subsp. *decorus* from Angola has narrower (1.5–2(–2.5) cm) spikes and smaller (8–15 × 5–10 mm) bracts with rusty brown hairs. See discussion in Bidgood & Brummitt (l.c.) about the differences between these two taxa. The material of subsp. *decorus* is all old and poor and the prospects of getting new and good material virtually non-existent.

6. **Neuracanthus keniensis** *J.-P. Lebrun & Stork* in Adansonia 7: 155 (1985) & Enum. Pl. Afr. Trop. 4: 496 (1997); Bidgood & Brummitt in K.B. 53: 52 (1998). Type: Kenya, Northern Frontier District: 80 km on Sololo–Marsabit road, *Brémand* 17 (ALF, holo.; K!, iso.)

Shrubby herb or subshrub; stems decumbent, sometimes rooting, to 50 cm long, whitish sericeous-tomentose when young, glabrescent, old stems with corky bark. Mature leaves with petiole to 2 cm long; lamina lanceolate to elliptic, largest 4–8 × 0.5–2.5 cm, base attenuate, decurrent, apex acute or subacute; beneath densely whitish sericeous to tomentose, above sparsely to densely sericeous. Inflorescences axillary, each of 1–2 pairs of decussate sterile branched spines to 3 cm long surrounding a single fertile branch with 4–10 flowers, spines reddish brown, to 1 cm long, clustered at apex of sterile branches; peduncle, axes, basal part of spines and bracts densely tomentose with glossy silky hairs; peduncle 4–9 mm long; fertile inflorescence to 3 cm long, bracts lanceolate to narrowly ovate, 8–13 × 3–5 mm. Dorsal calyx lobe 8–13 × 3–4.5 mm, narrowly ovate, acuminate, sometimes with a pair of minute teeth below apex, ventral lobe slightly shorter and narrower; tomentose with glossy silky hairs. Corolla blue to mauve, sparsely sericeous-pubescent outside, 10–12 mm long of which the tube 3–5 mm, limb actinomorphic, 9–12 mm diameter, tube ± 2 mm diameter. Larger anthers ± 0.7 mm long, bearded at base, smaller anthers ± 0.5 mm long. Capsule 2-seeded, narrowly triangular-ovate with a long beak, 8–9 mm long; seed cordiform, ± 4 × 4 mm.

KENYA. Northern Frontier District: Isiolo, Hadado, 18 Feb. 1970, *van Swinderen* 114! & 45 km N of Marsabit, Bubbisa, 28 May 1980, *Legesse* 27! & between Mt Marsabit and Mt Kulal, Kargi, 24 Apr. 1986, *Linder* 3600!

DISTR. **K** 1, 7; not known elsewhere

HAB. *Acacia* bushland on old lava flows or sandy soil, often in semi-desert conditions; 250–1000 m

NOTE. *N. keniensis* has been collected very close to the Ethiopia–Kenya border, and the species almost certainly also occurs in the Gemu Gofa Region of SW Ethiopia.

7. **Neuracanthus polyacanthus** (*Lindau*) *C.B. Clarke* in F.T.A. 5: 139 (1899); Benoist in Notul. Syst. 2: 145 (1911); Chiovenda, Fl. Somala: 253 (1929); E.P.A.: 950 (1964); Lebrun & Stork, Enum. Pl. Afr. Trop. 4: 496 (1997); Bidgood & Brummitt in K.B. 53: 72 (1998); Thulin in Fl. Somalia 3: 454 (2006); Ensermu in Fl. Eth. 5: 434 (2006). Type: Somalia/Ethiopia, Ogaden, *Robecchi Bricchetti* 240 (FT, lecto.; selected by Bidgood & Brummitt (l.c.); K!, photo)

Dwarf shrub to 35 cm tall, often forming cushions; young branches whitish tomentellous and with scattered long stalked capitate glands, old branches with corky bark. Mature leaves with petiole to 1 cm long; lamina ovate to broadly ovate or elliptic, largest 2–3 × 1.4–1.9 cm, base cuneate to truncate, apex subacute to obtuse with a small apiculus to 1 mm long; subglabrous to densely pubescent with long glossy curly hairs. Inflorescences axillary, each of 1–2 pairs of decussate sterile branched spines 1–2.5(–3) cm long surrounding a single fertile branch with 1–3(–4) flowers, basally on each branched spine a single foliaceous bract 4–8 mm long of which ± half a terminal spine, the other bracts needle-like spines to 8 mm long, solitary or paired at each node and evenly distributed along the whole length; axes and spines sparsely pubescent and with scattered stalked capitate glands when young; peduncle 1–2 mm long; flowers solitary at lower nodes of one axis, subtended by a bract-like spine to 8 mm long. Calyx with the lobes similar, ovate to cordiform, acute, 5–7 × 2–4 (to 10 × 6 in fruit) mm, same length or dorsal lobe slightly longer; pubescent with shiny curly hairs and with scattered stalked capitate glands. Corolla blue to mauve, sparsely pubescent outside, 8–12 mm long of which the tube 1.5–3 mm, limb actinomorphic, 7–15 mm diameter, tube 1.5–2 mm diameter. Larger anthers ± 0.7 mm long, bearded at base. Capsule 2-seeded, cordiform with long beak, ± 6 mm long; seed cordiform, ± 4 × 4 mm.

KENYA. Northern Frontier District: 18 km SW of Mandera, 13 Dec. 1971, *Bally & Radcliffe-Smith* 14585! & Ramu–Banissa road, 28 km from turning to Banissa, 4 May 1978, *M.G. Gilbert & Thulin* 1405A! & Ramu–Banissa road, 29 km from turning to Banissa, 4 May 1978, *M.G. Gilbert & Thulin* 1434!

DISTR. **K** 1; Djibouti, Ethiopia, Somalia

HAB. *Acacia-Commiphora* woodland and bushland with *Delonix baccal* and *Terminalia*; 250–600 m

SYN. *Leucobarleria polyacantha* Lindau in Ann. R. Ist. Bot. Roma 6: 77 (1896); Rendle in J.B. 34: 413 (1896)

Barleria sp., Franchet in Revoil, Fl. Çomalis: 52 (1882)

31. CRABBEA

W.H. Harvey in London J. Bot. 1: 27 (1842), *nom. cons.*; Nees in DC., Prodr. 11: 162 (1847); G.P. 2: 1092 (1876); C.B. Clarke in F.T.A. 5: 118 (1899) & in Fl. Cap. 5: 38 (1901); Thulin in Nord. J. Bot. 24: 502 (2007)

Crabbea W.H. Harvey, Gen. S. Afr. Pl.: 276 (1838), *nom. rej.*
Golaea Chiov., Fl. Somala: 257 (1929)
Acanthostelma Bidgood & Brummitt in K.B. 40: 855 (1985)

Perennial or subshrubby herbs, sometimes acaulescent; cystoliths conspicuous. Leaves opposite, entire to crenate-dentate. Flowers in dense axillary sessile or pedunculate globose or subglobose heads composed of several racemoid scorpioid 3–5-flowered cymes on flattened axes, surrounded by many large spiny-bristly or entire bracts which get gradually smaller inwards, spines and margin usually thickened and yellowish, peduncle apically with two sterile bracts. Calyx glumaceous, deeply divided into 5 subequal (dorsal and ventral longer than lateral) or unequal (dorsal much larger than the rest) sepals. Corolla outside hairy and usually also glandular (rarely glabrous), subactinomorphic (the lobes fused higher up in upper lip), in bud the 3-lobed lower lip folded over the 2-lobed upper lip; basal cylindric tube widened into a long cylindric throat; lobes spreading, oblong to round and broadly rounded at apex. Stamens 4, didynamous, inserted at base of throat, included; anthers bithecous, thecae of slightly different length, rounded, with line of long white hairs on both sides of aperture. Ovary with 2–4 ovules per locule; style glabrous, articulated at base; lower stigma lobe rhomboid-rounded, the upper missing or a small tooth. Capsule 4–8-seeded, oblong (rarely ovoid). Seed discoid, covered with fine simple hygroscopic hairs.

10–15 species from Ethiopia and Somalia through eastern tropical Africa to South Africa. The species are often closely related and difficult to separate.

1. Acaulescent or sub-acaulescent herb with two pairs of leaves appressed to the ground; leaves less than twice as long as wide; heads sessile (very rarely pedunculate), bracts entire; calyx with capitate glands at base; corolla 21–32 mm long . . . 3. *C. kaessneri*
   Aerial stem present; leaves more than two pairs, usually not appressed to the ground, more than twice as long as wide; heads pedunculate (very rarely sessile); bracts on some or all heads with bristly-spinose margin; calyx without capitate glands; corolla 7–21 mm long 2
2. Corolla 7–19 mm long, white (rarely pale pink) with yellow patches in throat; peduncle (0.5–)1–9.5(–11) cm long; rhizome branched, often with several stems; largest leaf 4.5–17(–23) cm long . . . . 1. *C. velutina*
   Corolla 17–21 mm long, white with pink veins or pale mauve; peduncles 7–20 cm long; rhizome unbranched, with a single stem; largest leaf 9–25 cm long . . . . 2. *C. longipes*

1. **Crabbea velutina** *S. Moore* in J.B. 32: 135 (1894); Lindau in P.O.A. C: 368 (1894); C.B. Clarke in F.T.A. 5: 119 (1899); E.P.A.: 941 (1964); Compton, Fl. Swaziland: 551 (1976); Champluvier in Fl. Rwanda 3: 449 (1985); Blundell, Wild Fl. E. Afr.: 389, pl. 60 & 122 (1987); Iversen in Symb. Bot. Upsal. 29(3): 160 (1991); U.K.W.F., ed. 2: 270 (1994); Lebrun & Stork, Enum. Pl. Afr. Trop. 4: 475 (1997); Friis & Vollesen in Biol. Skr. 51(2): 440 (2005); Thulin in Fl. Somalia 3: 444 (2006); Ensermu in Fl. Eth. 5: 426 (2006). Types: Kenya, Mombasa, *Gregory* s.n. (BM!, syn.) & *W.E. Taylor* s.n. (BM!, syn.)

Perennial herb with 1 to several erect to decumbent or stoloniferous stems from branched creeping rhizome, roots not fleshy; stems 2–30(–50 in decumbent plants) cm long, when young puberulous to pubescent or sericeous-pubescent or densely so. Leaves with petiole 0–5(–7) cm long; lamina elliptic to obovate or narrowly so, largest 4.5–17(–23) × 1.5–6(–7) cm, more than twice as long as wide, base attenuate, often decurrent to stem, apex subacute to broadly rounded, margin crenate to dentate; puberulous to pubescent or sparsely so, beneath densest along veins (sometimes glabrous on lamina), above uniformly so. Heads 1–3.5(–4) cm across; peduncle (0.5–)1–9.5(–11) cm long, puberulous to pubescent or sericeous-pubescent or densely so; outer bracts green (rarely tinged purplish), ovate or narrowly so, 1.5–3.5 × 0.4–1.5 cm, sparsely puberulous to sparsely pubescent, at least along veins, subacute to acuminate, varying from entire to densely bristly-spinose with bristles to 12 mm long. Calyx with dorsal lobe distinctly longer and wider, 7–13(–15) × 2–3 mm, ventral and lateral 1–3 mm shorter, all subacute to acuminate; puberulous or sparsely so at base and along midribs, distinctly ciliate. Corolla white with yellow gibbous patches in throat (rarely pale pink to pale mauve), 7–19 mm long of which the cylindric tube 2–7 mm and the throat 5–12 mm long, lobes 5–10 × 4–8 mm. Filaments 1–3 and 2–4 mm long; anthers 1–2 mm long. Capsule 4–8-seeded, oblong, 8–10 mm long; seed rhomboid, 2.5–3 × 1.5–2 mm. Fig. 45, p. 299.

UGANDA. Karamoja District: Moroto, May 1956, *Wilson* 231!; Teso District: Serere, Feb. 1933, *Chandler* 1098!; Musaka Distict: Sese Islands, Bufumiro Island, July 1925, *Maitland* 775!

KENYA. Northern Frontier District: Bunduras, 20 May 1952, *Gillett* 13237!; Kitale District: NE Elgon, March 1959, *Tweedie* 1828!; Embu District: Kiambere, 25 Nov. 1951, *Kirrika* 151!

TANZANIA. Moshi District: W Kilimanjaro, Rongai Ranch, 20 April 1957, *Greenway* 9188!; Kilosa District: Ruaha Valley, 15 March 1986, *Bidgood & Lovett* 256!; Chunya District: Mbangala Village, 13 Feb. 1994, *Bidgood et al.* 2225!

DISTR. **U** 1–4; **K** 1–7; **T** 1–8; Sudan, Ethiopia, Somalia, Congo-Kinshasa, Rwanda, Burundi, Zambia, Malawi, Mozambique, Zimbabwe, Botswana, Swaziland, South Africa

HAB. *Acacia* and *Acacia-Commiphora* bushland, *Combretum* woodland and bushland, riverine forest and thicket, rocky hills, on a wide range of soils including black cotton soil, often in disturbed places and in secondary vegetation, only rarely in *Brachystegia* woodland; 150–1900 m

SYN. *Crabbea reticulata* C.B. Clarke in F.T.A. 5: 119 (1899); S. Moore in J.B. 40: 344 (1902); Chiovenda, Fl. Somala 2: 355 (1932); E.P.A.: 941 (1964). Types: Kenya, "Ukamba", *Scott Elliot* 2309 (K!, syn; BM!, iso.); Tanzania, Bukoba District: Karagwe, *Scott Elliot* 8147 (K!, syn; BM!, iso.)

NOTE. A widespread and quite variable species but always recognisable by the elongated stems and pedunculate heads.

It is quite common that the first head produced has entire bracts and the following have toothed bracts. Only rarely do the following heads have entire bracts and I have not seen specimens from the flora area where all heads have entire bracts, but there are collections from Zambia where this happens.

The corolla is usually white but in western Uganda and western Tanzania forms with pale pink or pale mauve corollas occur. These do not seem to differ in any other way from typical material but they might point to possible intergradation with *C. kaessneri*.

2. **Crabbea longipes** *Mildbr.* in N.B.G.B. 12: 718 (1935); Lebrun & Stork, Enum. Pl. Afr. Trop. 4: 475 (1997). Type: Tanzania, Masasi District: 110 km W of Lindi, Mbemkuru, *Schlieben* 6004 (B†, holo.; BM!, K!, iso.)

Perennial herb with single erect stem from unbranched creeping rhizome, roots not fleshy; stems 1–2.5(–10 fide *Mildbraed*) cm long, with scattered spreading stiff hairs, mostly at nodes. Leaves in 3–4 pairs, with petiole 0.5–1.5 cm long; lamina obovate or narrowly so, largest 9–25 × 3.5–8 cm, more than twice as long as wide, base attenuate, decurrent, apex subacute to rounded, margin subentire to crenate; beneath sparsely minutely puberulous along midrib and here also with scattered longer hairs, above with scattered puberulous hairs all over. Heads 2–3 cm across,

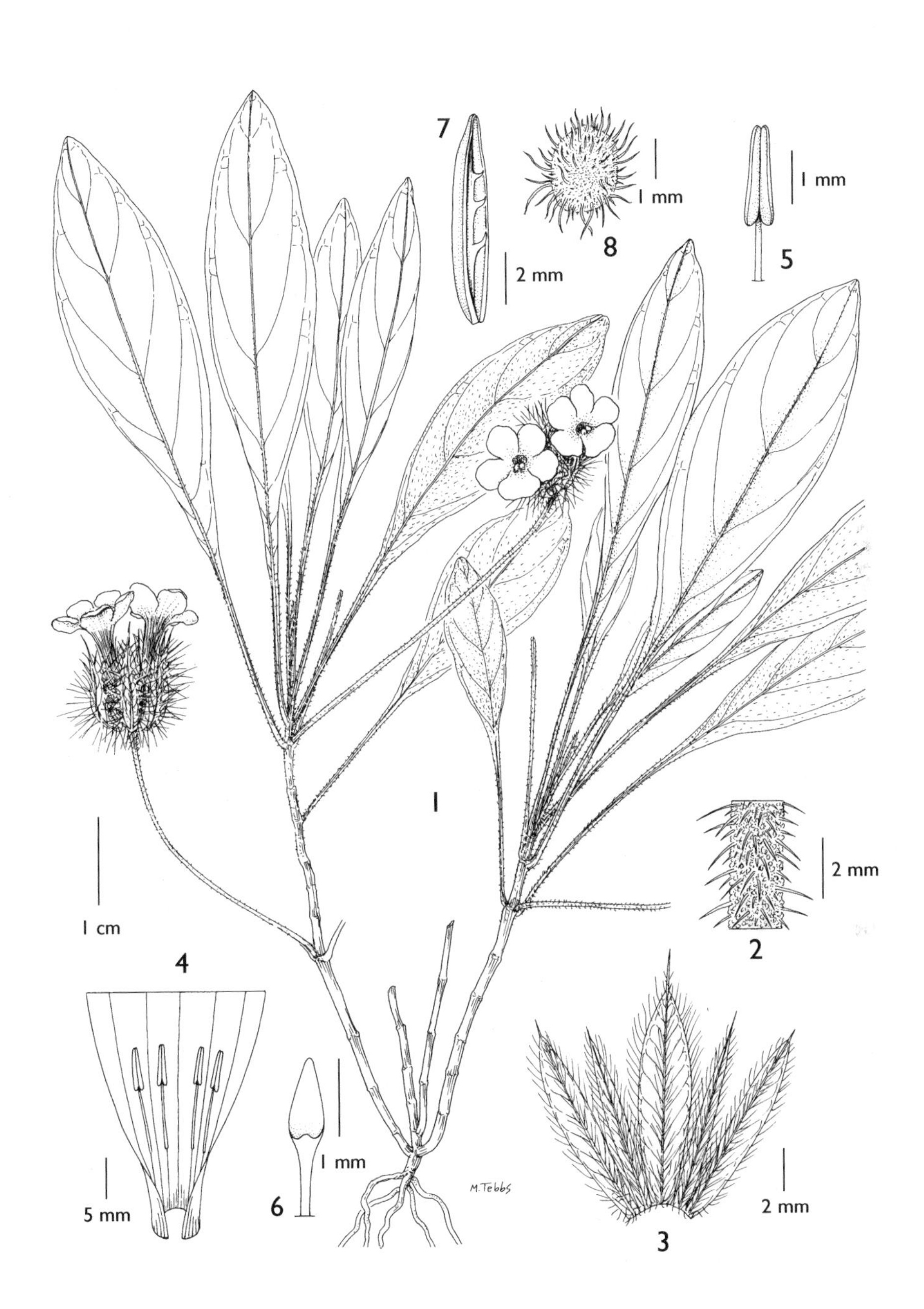

FIG. 45. *CRABBEA VELUTINA* — **1**, habit, × $^2/_3$; **2**, detail of stem indumentum, × 4; **3**, calyx opened up, × 3; **4**, corolla tube opened with stamens, × 2; **5**, anther, × 8; **6**, apical part of style and stigma, × 12; **7**, capsule, 4; **8**, seed, × 6. 1 & 3–7 from *Bidgood & Lovett* 256, 2 from *Kerfoot* 443, 8 from *Verdcourt* 815. Drawn by Margaret Tebbs.

resting on the ground; peduncle 7–20 cm long, minutely sericeous-puberulous and with scattered longer hairs; outer bracts purplish green, narrowly ovate to elliptic, 2.5–3.5 × 0.4–1 cm, sparsely puberulous along midrib and on edges, acute to rounded, varying from subentire to densely bristly-spinose with bristles to 5 mm long. Calyx with dorsal lobe distinctly longer and wider, 12–15 × 2.5–4 mm, ventral and lateral 2–5 mm shorter, all acute to acuminate; sparsely sericeous-puberulous, distinctly ciliate. Corolla white with pink veins or pale mauve and with yellow gibbous patches in throat, 17–21 mm long of which the cylindric tube 6–8 mm and the throat 11–13 mm long, lobes 7–10 × 4–7 mm. Filaments ± 3 and 4 mm long. Anthers ± 2 mm long; ovules 3 per locule. Capsule and seed not seen.

TANZANIA. Masasi District: 110 km W of Lindi, Mbemkuru, 17 Feb. 1935, *Schlieben* 6004! & Nachingwea, 16 Jan. 1962, *Anderson* 1333! & Masasi Hill, 9 March 1991, *Bidgood et al.* 1874!
DISTR. **T** 8; not known elsewhere
HAB. Open forest on rocky hills, with *Milletia*, *Dalbergia* and *Cussonia*, *Brachystegia-Julbernardia* woodland on orange-brown gritty-sandy loam; 200–500 m

NOTE. Known only from these three collections. In general facies very different from *C. velutina*, but no single character will reliably separate the two. It differs in the unbranched rhizome with a single stem, the large sparsely hairy leaves, the long slender sparsely hairy peduncle and the larger pinkish corolla.

3. **Crabbea kaessneri** *S. Moore* in J.B. 48: 252 (1910); Richards & Morony, Check List Mbala and Distr.: 229 (1969); Lebrun & Stork, Enum. Pl. Afr. Trop. 4: 475 (1997). Type: Congo-Kinshasa, Tonkoosji, *Kässner* 2337 (BM!, holo.; K!, iso.)

Acaulescent or sub-acaulescent perennial herb with unbranched (more rarely branched) creeping rhizome, roots fleshy; aerial stems absent but short subterranean stems to 2 cm long, apical part of rhizome and subterranean stems yellowish sericeous-tomentellous. Leaves in 2(–3) decussate pairs appressed to the ground; petiole 0–1 cm long, yellowish strigose pubescent; lamina ovate to elliptic or broadly so, largest 5–12 × 3.5–8 cm, less than twice as long as wide, base attenuate, decurrent, often onto stem, apex broadly rounded (rarely obtuse or retuse), margin subentire to crenate; beneath yellowish puberulous or sparsely so on midrib and lateral veins, glabrous on lamina, not ciliate, above glabrous or with scattered pale yellow hairs. Heads 1.5–4.5 cm across; peduncle 0–6(–30) mm long, densely yellowishly strigose-pubescent; outer bracts yellowish green (rarely purplish tinged), ovate to elliptic or broadly so, 2–3.5 × 0.8–2.5 cm, usually less than twice as long as wide, subglabrous to sericeous-pubescent along midrib and on veins, acute to broadly rounded, margin entire to dentate, but never bristly. Calyx with dorsal lobe longer and much wider, 12–15 × ± 2 mm, ventral and lateral 2–4 mm shorter, all acuminate to cuspidate; apically with sparse to dense strigose hairs, at base with sparse to dense short capitate glands. Corolla white with yellow gibbous patches in throat, 21–32 mm long of which the cylindric tube 7–12 mm and the throat 14–20 mm long, lobes 8–10 × 8–10 mm. Filaments 2–4 and 4–5 mm long. Anthers 1.5–2 mm long. Capsule (immature) ellipsoid, ± 13 mm long; seed not seen.

TANZANIA. Ufipa District: 15 km on Kipili–Namanyere road, 5 March 1994, *Bidgood et al.* 2659!; Iringa district: 30 km on Mafinga [Sao Hill]–Madibira road, 21 march 1988, *Bidgood et al.* 601!
DISTR. **T** 4, 7; Congo-Kinshasa, Zambia
HAB. *Brachystegia* woodland on sandy to loamy soil; 850–1650 m

NOTE. *Bidgood et al.* 2659 has a clearly peduncled inflorescence which should in theory put it into *C. velutina*, but it has the general "gestalt" of *C. kaessneri*, i.e. acaulescent, wide leaves and wide entire bracts, calyx with capitate glands. It is from an area considerably lower (850 m) than normal for the species (all Zambian material is from above 1200 m).

32. LEPIDAGATHIS

Willd. in Sp. Pl., ed. 4, 3: 400 (1800); Nees in DC., Prodr. 11: 249 (1847); C.B. Clarke in F.T.A. 5: 120 (1899)

Russeggera Endl., Nov. Stirp. Mus. Dec.: 38 (1839) & Gen. Pl., Suppl. 1: 1405 (1841)
Teliostachya Nees in Fl. Bras. 9: 71, t. 8 (1847); Nees in DC., Prodr. 11: 262, 727 (1847)
Volkensiophyton Lindau in E.J. 20: 27 (1894)
Lindauea Rendle in J.B. 34: 411, t. 362 (1896); C.B. Clarke in F.T.A. 5: 129 (1899)

Perennial herbs and subshrubs, evergreen or deciduous; cystoliths many, ± conspicuous. Stems 4-angular or subterete. Leaves opposite-decussate, blade often with minute sunken glands on lower surface. Inflorescences of unilateral (scorpioid) fasciculate or spiciform cymes, axillary, terminal and/or basal, often compounded into a dense synflorescence, rarely a dichasial thyrse or flowers solitary; bract pairs equal or dimorphic, free, sterile bracts in unilateral cymes typically imbricate and subtending the flowers, fertile bracts and bracteoles usually adpressed to calyx; bracteoles paired, (sub)equal, free. Calyx divided almost to base, unequally 5-lobed, posticous lobe broader than remaining lobes, anticous pair of lobes sometimes partially fused, lateral (inner) pair of lobes linear-lanceolate. Corolla bilabiate; tube cylindrical in lower portion, throat campanulate, base of throat ± densely hairy within; upper lip hooded, straight or arcuate, apex shortly bilobed or shallowly emarginate; lower lip 3-lobed, palate upraised with a central furrow. Stamens 4, didynamous, filaments arising from near base of corolla throat; anticous pair of stamens longer, bithecous, thecae at an equal height or offset; posticous pair shorter, monothecous or bithecous. Ovary 2-locular, 1 or 2 ovules per locule; style filiform; stigma capitate-bilobed or one lobe expanded and spoon-shaped. Capsule compressed, 2-seeded, face then ovate, or 4-seeded, face then oblong-elliptic, sometimes with a short sterile apical beak, placental base inelastic. Seeds held on retinacula, (sub)discoid, clothed in hygroscopic hairs.

A genus of ± 100 species with a tropical and subtropical distribution. Three infrageneric groups are informally recognised here on the basis of variation in inflorescence type, anther characteristics, capsule shape, number of seeds and leaf venation. Groups C and B largely correspond to C.B. Clarke's two subgenera in F.T.A. which he separated purely on whether the posticous stamens have bithecous (*Eulepidagathis*) or monothecous (*Neuracanthopsis*) anthers. *L. alopecuroides* does not conform to either of these groups due to its differing inflorescence form and is therefore placed separately in group A.

1. Inflorescence a terminal thyrse, both bracts of a pair fertile; corolla 5–7.5 mm long; capsule 3.5–4.5 mm long; slender herb of moist forest (*Group A*) ..... 1. *L. alopecuroides* p.304
   Inflorescences unilateral (scorpioid), 1 bract of each pair sterile, variously terminal, axillary and/or basal, often compounded into dense conical, globose or cushion-forming heads; corolla 7.5–28 mm long; capsule 5.5–12 mm long; plants of drier woodland and grassland 2
2. Anther thecae glabrous; posticous (shorter) pair of stamens monothecous or with vestigial second theca; capsule potentially 4-seeded but 1–2 seeds sometimes aborting, face oblong-elliptic (*Group B*) 3
   Anther thecae with short hairs and glands dorsally and with cilia along the suture; stamens all bithecous; capsule 2-seeded, face ovate (*Group C*) 11
3. Stems with a dense stellate or dendritic indumentum at least when young 4
   Stem indumentum of simple hairs or glabrescent 5

4. Synflorescences terminating well-developed branches; leafy bracts subtending the synflorescences rhombic, ovate or elliptic; corolla 13–23 mm long — 3. *L. scariosa* p.307
   Some synflorescences pseudo-axillary, terminating very short lateral branches (clearly terminal synflorescences also present on main shoots); leafy bracts subtending the synflorescences pandurate; corolla 11.5–13 mm long . . . . 4. *L. pseudoaristata* p.308
5. Stems soon woody; anticous pair of calyx lobes fused for 2–4.5 mm; inflorescence a strobilate unilateral spike with elliptic (-obovate) bracts 3–6.5 mm wide . . . . 5. *L. calycina* p.309
   Stems herbaceous or woody only at base; anticous pair of calyx lobes free or if fused for up to 2.5 mm (*L. glandulosa*) then inflorescences not strobilate and bracts (linear-) lanceolate, up to 2.5 mm wide 6
6. Bracts, bracteoles and calyx lobes aristate; corolla 7.5–14.5 mm long; anticous pair of calyx lobes fused for 1–2.5 mm beyond the short calyx tube; posticous calyx lobe margin not inrolled . . . . 2. *L. glandulosa* p.305
   Bracts, bracteoles and calyx lobes not aristate (sometimes with a flexible linear acumen); corolla 13.5–28 mm long; anticous calyx lobes free beyond the short calyx tube; posticous calyx lobe margin ± inrolled, often partially enveloping the other lobes 7
7. Inflorescences fasciculate, axillary 8
   Inflorescences congested unilateral spikes, axillary and/or terminal 9
8. Leaves pilose; inflorescences with lowermost 1–2 fertile bracts markedly larger than the remaining bracts, these pale-scarious throughout or green towards apex; sterile bracts ± falcate; corolla 13.5–20 mm long . . . . 6. *L. scabra* p.309
   Leaves glabrous or pubescent mainly on principal veins only; inflorescence with fertile bracts all ± equal in size, green in lower half and abruptly (black-) brown above at least when young (eventually turning brown-scarious); sterile bracts not falcate; corolla 20–28 mm long . . . . . . 9. *L. nemorosa* p.313
9. Night-flowering, corolla 19–28 mm long, palate of lower lip pubescent; some or all inflorescences terminal . . . . 8. *L. pallescens* p.312
   Day-flowering, corolla 13.5–20 mm long, palate of lower lip glabrous; inflorescences axillary (but can appear terminal in the rosulate *L. plantaginea*) 10
10. Plants caulescent; leaves ovate, lanceolate or elliptic, largest blade 1.8–9.5 cm long, with 3–5 pairs of lateral nerves; peduncle 0–6 mm long . . 6. *L. scabra* p.309
    Plant rosulate; leaves oblong to oblanceolate, largest blade 10–16 cm long, with 8–10 pairs of lateral nerves; peduncle 10 mm long or more . . . 7. *L. plantaginea* p.312

11. Leaves with lateral veins pinnate, ± strongly ascending and anastomosing; ovary and capsule pubescent in upper half (*L. collina* complex) 12
    Leaves prominently 3(–5)-veined from base or, if leaves linear, only midrib prominent; ovary and capsule glabrous or the former with minute sessile glands only 13
12. Stems prostrate or procumbent; inflorescences basal and cushion-forming and/or with reduced inflorescences in the lower nodes of the stems and at branching nodes; bracts, bracteoles and calyx lobes straight or weakly falcate .... 10. *L. collina* p.315
    Stems erect; inflorescences axillary in the upper half of the stems, sometimes with reduced inflorescences at the lowermost nodes; bracts, bracteoles and anticous calyx lobe strongly falcate, the latter sometimes reflexed .... 11. *L. diversa* p.316
13. Corolla 18–20 mm long, palate of lower lip lacking yellow or orange markings; unilateral cymes spiciform, 2.5–6 cm long, spreading from stems, solitary or if 2–4 at nodes then clearly separated, not forming a dense synflorescence; bracts, bracteoles and calyx lobes all caudate .... 12. *L. peniculifera* p.317
    Corolla 10–18 mm long, with yellow or orange markings on palate of lower lip; unilateral cymes fasciculate or shortly elongate, often compounded into globose or cushion-like synflorescences, or if cymes clearly separated and spiciform (*L. andersoniana*) then held ± close to stems and with bracts, bracteoles and calyx lobes more gradually narrowed (*L. andersoniana* complex) 14
14. Plants erect, decumbent or straggling; inflorescences held away from the ground at leafy axils, often in the upper half of the stems only 15
    Plants prostrate or procumbent or if leafy stems erect to decumbent then not bearing inflorescences at the leafy axils; inflorescences held against the ground, basal and/or axillary on creeping stems 17
15. Inflorescence indumentum yellow; flowering calyx lobes green throughout, lacking a darker acumen 17. *L. sp. A* p.324
    Inflorescence indumentum white or rarely pale buff-coloured; flowering calyx lobes green and usually with a conspicuous brown or purple acumen 16
16. Sterile floral bracts* lanceolate or ovate, surface ± flat, apex gradually narrowed, attenuate to long-acuminate; unilateral cymes 1(–2) per axil, alternate or opposite, often shortly spiciform, not forming dense globose heads; leaves pilose or often lanate when young .... 13. *L. andersoniana* p.318
    Sterile floral bracts broadly oblong-elliptic, rounded or somewhat obovate, surface convex, apex abruptly narrowed into an apiculum or short acumen; cymes compounded into dense, ± globose heads 3–6.5 cm in diameter, individual unilateral cymes never spiciform; leaves glabrous or sparsely pilose when young .... 14. *L. eriocephala* p.320

17. Inflorescences axillary along trailing leafy stems or if restricted to the largely leafless lower axils then "heads" only 1.5–3.7 cm in diameter, leaves 0.2–2 cm wide, sterile floral bracts 6–15 mm long — 15. *L. lanatoglabra* p.321
    Inflorescences basal (often at terminus of woody rhizome), typically compounded to form cushions, 3–11.5 cm in diameter, or more loosely clustered along the largely leafless lower axils; rarely axillary along leafy stems, then leaves over 2 cm wide and/or sterile floral bracts over 15 mm long 18
18. Synflorescences usually appearing brown or purplish in colour (sometimes green when young, then more sparsely hairy); bracteoles and calyx lobes with a prominent linear acumen, this usually conspicuously brown to purple, only this portion usually exposed beyond the sterile bracts; plants usually with mature leaves when flowering, more rarely flowering soon after burning . . . . 16. *L. fischeri* p.322
    Synflorescences appearing greyish-green in colour and densely hairy throughout, the green bracteoles and calyx lobes clearly exposed beyond the bracts, with or without a short acumen, this green throughout or the tip brown; plants often flowering before or during leaf development soon after burning . . . . 18. *L. sp. B* p.324

* The sterile floral bracts referred to in the key are the mature bracts at flowering nodes, not the reduced sterile bracts at the very base of the inflorescence or sheathing the peduncle.

Group A "Teliostachya" – species 1

Inflorescence a terminal thyrse, both bracts of a pair fertile, equal. Stamens all bithecous; filaments and thecae glabrous. Stigma capitate-bilobed. Capsule 4-seeded (or 2 by abortion), face oblong-elliptic. Seeds with hygroscopic hairs (green) brown.

NOTE. Both C.B. Clarke (F.T.A. 5: 128) and Heine (F.W.T.A. ed. 2, 2: 414) record *L. alopecuroides* as having either monothecous or bithecous anthers to the upper, shorter pair of stamens. However, all the specimens examined from our region and from central Africa were found to have four bithecous stamens.

1. **Lepidagathis alopecuroides** (*Vahl*) *Griseb.* in Fl. Brit. W. Ind.: 453 (1861); Heine in F.W.T.A. ed. 2, 2: 414, fig. 303 (1963); Akoègninou & Yédomonhan in Fl. Anal. Benin: 293, with fig. (2006). Type: Montserrat, *Ryan* s.n. (not traced)

Slender decumbent perennial herb, 5–20 cm tall, rooting at lower nodes; stems 4-angular, antrorse-pubescent or -pilose, sometimes with few patent glandular hairs when young, ± glabrescent. Leaves on petiole to 15(–23) mm long; blade ovate, elliptic or lanceolate, 2.8–6.5(–16) cm long, 0.4–2.5 cm wide, base attenuate or cuneate, apex acute, obtuse or subattenuate, surfaces glabrous or sparsely pubescent to pilose particularly on margin and veins beneath, margin sometimes with few glandular hairs; lateral veins 3–5 pairs. Thyrse ± congested, 1.5–5.5(–12) cm long, sessile or peduncle to 4 cm long; bracts green or tinged purple towards apex, ± paler towards base, scarious with age, elliptic or lanceolate, 3.5–7.5 mm long, 1.3–3 mm wide, apex subattenuate or acuminate, surface with mixed short glandular and ± longer eglandular hairs at least along margin, 3-veined from base with ± prominent reticulate tertiary venation; bracteoles as bracts but hyaline towards base, 0.5–2.3 mm wide, apex caudate, 3-veined or midrib only prominent. Calyx colour and indumentum as bracteoles, lobes caudate, anticous pair fused for up to 0.5 mm, linear-lanceolate, 3.5–5 mm long, 0.4–1 mm wide; posticous lobe (ovate-) elliptic, 4.5–7.5 mm long, 1.5–2.5 mm wide, 3-veined from base. Corolla 5–7.5 mm long, white or with limb blue or pink, often with blue to purple markings on lower lip; limb sparsely pubescent outside; tube 3–4.5 mm long; upper lip 2–3 mm long; lower lip 2–3.5 mm long including lobes 0.8–1.5 mm long. Anticous (longer) stamens with filaments 2–2.5 mm long, thecae 0.4–0.6 mm long, offset. Ovary pubescent in upper half; style glabrous or sparsely hairy in lower half. Capsule 3.5–4.5 mm long, pubescent towards apex; seeds 0.9–1.3 mm long, 0.7–0.9 mm wide.

TANZANIA. Rufiji District: Kiwengoma Forest, northern edge of the Matumbi Highlands, Nov. 1989, *Clunies-Ross & Sheil* in *Frontier Tanzania* 114!; Ulanga District: Magombera Forest, July 2003, *Luke, Luke & Arafat* 9536!

DISTR. **T** 6; widespread in W tropical Africa from Senegal to Sudan, E Congo-Kinshasa and Angola, also Neotropics

HAB. Lowland moist forest including trampled ground; 250–550 m

USES. None recorded on herbarium specimens

CONSERVATION NOTES. Although rare in the Flora region, this species is widespread and common in the forests of west Africa and the Neotropics. It appears tolerant of disturbance, often being recorded from pathsides: Least Concern (LC).

SYN. *Ruellia alopecuroidea* Vahl, Eclog. Am. 2: 49 (1798)
Teliostachya alopecuroidea (Vahl) Nees in DC., Prodr. 11: 262 (1847)
T. laguroidea Nees in DC., Prodr. 11: 264 (1847). Type: Ghana, Accra, *Vogel* s.n. (K!, holo.)
Lepidagathis laguroidea (Nees) T. Anderson in J.L.S. 7: 34 (1863); C.B. Clarke in F.T.A. 5: 128 (1899); F.P.S. 3: 183 (1956)
Teliostachya hyssopifolia Benth. in Hook., Niger Fl.: 481 (1849). Type: Sierra Leone, *Don* s.n. (K!, holo.; BM!, iso.)
Lepidagathis hyssopifolia (Benth.) T. Anderson in J.L.S. 7: 34 (1863); C.B. Clarke in F.T.A. 5: 128 (1899)
L. laurentii De Wild., Enum. Pl. Laurent.: 183, pl. 24 (1905). Type: Congo-Kinshasa, Tschopo Falls, *Laurent* s.n. (BR, holo.), **syn. nov.**

NOTE. The isolated populations in E Tanzania have a longer linear acumen to the bracteoles and calyx lobes than in most West African material and the cymes of the thyrse are more clearly separated than usual but there is some overlap in these characters.

De Wildeman records the type of *L. laurentii* as having lanceolate leaves to 15.5 cm long. I have seen leaves up to 9 cm long in other Congo material and it is likely that the long, slender leaves on the *Laurent* specimen are an ecological, rheophytic adaptation as it was collected from a waterfall.

Group B "Neuracanthopsis" – species 2–9

Inflorescences terminal and/or axillary unilateral cymes, solitary or compounded into dense synflorescences; one of each pair of bracts sterile, pairs equal or dimorphic. Corolla with upper lip straight or arcuate; lower lip with palate lacking membranous portions, glabrous or shortly pubescent on bosses. Anticous pair of stamens bithecous, posticous pair monothecous, sometimes with vestigial second theca; filaments glabrous or with few eglandular hairs at base; thecae glabrous. Stigma capitate-bilobed or with one expanded, spoon-shaped lobe. Capsule 4-seeded (or 2 by abortion), face oblong-elliptic; seeds with hygroscopic hairs (green-) brown or yellow.

2. **Lepidagathis glandulosa** *A. Rich.* in Tent. Fl. Abyss. 2: 147 (1850); Nees in DC., Prodr. 11: 243 (1847), name only; C.B. Clarke in F.T.A. 5: 128 (1899); U.K.W.F. ed. 2: 271 (1994); Friis & Vollesen in Biol. Skr. 51(2): 449 (2005); Ensermu in Fl. Eth. 5: 429, fig. 167.36 (2006) as *L. glandulosa* Hochst. ex Nees. Type: Ethiopia, Tigray, Mt Soloda [Scholoda], *Schimper* I: 44 (cited by A. Richards as 1: 41 in error) (?P, holo.; BM!, K!, iso.)

Erect, decumbent or straggling perennial herb or suffrutex, 10–90 cm tall, sometimes rooting at lower nodes; stems 4-angular, crisped-pilose or antrorse-pubescent, rarely glabrescent. Leaves sessile or petiole to 18 mm long; blade ovate, lanceolate or elliptic, (1.8–)3–10.5 cm long, (0.8–)1.2–3.5 cm wide, base attenuate to shallowly cordate, apex acute or attenuate, surfaces pilose or antrorse-pubescent at least on margin and veins beneath, rarely glabrous; lateral veins 5–8 pairs. Inflorescences terminal on main shoots and pseudo-axillary on poorly developed lateral branches, cymes fasciculate or spiciform, usually compounded into dense ovoid or globose heads, 1.3–6.5 cm long, subsessile; fertile and sterile bracts subequal, (white-) green with purple apex and main veins or brown-scarious throughout, (linear-) lanceolate, 4.5–12.5 mm long, 0.7–2.5 mm wide, apex aristate, shortly pubescent at least on margin where sometimes densely ciliate, with or without few glandular hairs, with 3–5 subparallel principal veins and ± prominent reticulate tertiary venation; bracteoles as bracts but 0.5–2 mm wide, 3-veined or only midrib prominent, glandular hairs sometimes many. Calyx as bracteoles but more tardily scarious, lobes aristate, anticous pair fused for 1–2.5 mm, (linear-) lanceolate, 6.5–12 mm long, 0.8–1.5 mm wide; posticous lobe ovate or lanceolate, 7.5–13 mm long, 1.5–3 mm wide, 3–5-veined from base and with prominent reticulate tertiary venation. Corolla 7.5–14.5 mm long, white to mauve, lower lip often with purple markings, pubescent outside mainly on limb; tube 5–9 mm long; upper lip straight, 2.5–5 mm long; lower lip 3–6 mm long including lobes 1.8–3.5 mm long. Anticous (longer) stamens with filaments 3.5–7 mm long, thecae 0.7–1.3 mm long, barely offset; posticous stamens with vestigial second theca. Ovary glabrous or apex with few hairs; style shortly pubescent and/or with minute stalked glands in lower half; stigma capitate. Capsule 6–7 mm long, largely glabrous; seeds 1.3–1.6 mm long, 1.1–1.4 mm wide.

KENYA. Trans-Nzoia District: NE Elgon, Dec. 1955, *Tweedie* 1370! & Prison Farm, Kitale, Sept. 1969, *Tweedie* 3711! & between Mt Elgon Lodge and gates of Mt Elgon National Park, Oct. 1981, *Gilbert & Mesfin* 6555!

TANZANIA. Ufipa District: 16 km on Namanyere–Kipili road, May 1997, *Bidgood et al.* 3670!; Njombe District: N slopes of Poroto Mts, above Chimala, Mar. 1991, *Bidgood, Congdon & Vollesen* 2107!; Songea District: Matengo Hills, ± 1 km SW of Lipumba, May 1956, *Milne-Redhead & Taylor* 10365!

DISTR. **K** 3; **T** 2, 4, 7, 8; Cameroon, SE Congo-Kinshasa, Sudan, Ethiopia, N Zambia

HAB. Open woodland and wooded grassland: in Kenya in *Acacia-Erythrina* woodland, in S Tanzania in *Brachystegia*, *Uapaca* and *Pterocarpus* woodland; also in forestry plantations; 850–2350 m

USES. None recorded on herbarium specimens

CONSERVATION NOTES. This species has a widespread but disjunct distribution, being absent from many apparently suitable regions, for example in Uganda. It is, however, locally common and appears adaptable to a range of habitats. It is not considered threatened: Least Concern (LC).

SYN. [*Barleria glandulosa* Hochst., *nom. nud.* in herb.]
[*L. scariosa sensu* U.K.W.F. ed. 2: pl. 118 (1994) & *sensu* Hedrén in Fl. Somalia 3: fig. 294 (2006), *non* Nees]

NOTE. Two rather distinct forms occur in our area. Plants from Kenya (matching the type from Ethiopia) are decumbent or straggling perennials, with small, fasciculate or globose heads with few to many glandular hairs, small corollas (7.5–9.5 mm long) and tardily scarious bracts and bracteoles. The common form in S Tanzania is erect-suffruticose and has larger, ± ovoid heads lacking glandular hairs, larger corollas (11.5–14.5 mm long) and soon-scarious bracts and bracteoles. This latter form, provisionally separated by D. Champluvier as *L. ciliata* ined., is also recorded in N Zambia, Congo-Kinshasa and W Ethiopia. However, a range of intermediate populations are recorded; for example *Richards* 5812 from Chilongowelo, N Zambia closely resembles the Kenyan plants in habit, inflorescence form and indumentum but has corollas ± 12 mm long, whilst *Bidgood et al.* 3460 from Tatanda, **T** 4, has a growth habit and corolla size typical of the widespread S Tanzania form but the inflorescence and indumentum closely resemble the Kenyan material. With no clear geographic trends discernable, I therefore consider this a single, polymorphic taxon.

L. glandulosa is superficially similar to *L. alopecuroides* but differs in having unilateral inflorescences which are compounded into heads (not a solitary 2-sided thyrse), the anticous pair of calyx lobes being more clearly fused, the corollas and capsules being larger, the thecae of the anticous stamens being at a subequal height (not clearly offset) and the posticous pair of stamens being monothecous (not bithecous).

3. **Lepidagathis scariosa** *Nees* in Wallich, Pl. As. Rar. 3: 95 (1832); Wight, Ic. Pl. Ind. Or. 2: t. 457 (1843); Nees in DC., Prodr. 11: 251 (1847); C.B. Clarke in Fl. Brit. Ind. 4: 520 (1885) & in F.T.A. 5: 122 (1899); F.P.S. 3: 182 (1956); U.K.W.F. ed. 1: 596 (1974) pro parte excl. fig. p. 595; U.K.W.F. ed. 2: 271 (1994) pro parte excl. pl. 118; K.T.S.L.: 604 incl. fig. (1994); Wood, Handb. Yemen Fl.: 273 (1997). Type: India, *Wallich Cat.* 2354b (K! lecto., chosen here; K-W! isolecto.)

Much-branched perennial herb or subshrub, 20–120 cm tall; stems with dense pale buff or yellow stellate or dendritic hairs, sometimes glabrescent. Leaves subsessile or petiole to 12 mm long; blade ovate or elliptic, 1–7 cm long, 0.5–3.5 cm wide, base attenuate or cuneate, apex acute or obtuse, surfaces stellate-pubescent, dense beneath particularly on principal veins, hairs often long-armed, sparse above at maturity when stellate base sometimes caducous; lateral veins 4–6 pairs. Inflorescences terminal on main shoots and well-developed lateral branches, cymes fasciculate or spiciform, compounded into globose or conical heads 1–3 cm long, sessile; leafy bracts subtending heads rhombic, ovate or elliptic, stellate pubescent; fertile and sterile floral bracts subequal, brown- or purple-scarious, rarely green, elliptic (-obovate), ovate or oblong-lanceolate, 3.5–9 mm long, 1–3 mm wide, apex mucronulate or aristate, with few to many pale long-pilose hairs at least on margin, surface often also with short glandular and/or eglandular hairs, venation parallel or inconspicuous; bracteoles (linear-) lanceolate, 4–10.5 mm long, 0.7–1.7 mm wide. Calyx usually purple-tinged, rarely green throughout, lobes mucronulate or aristate, densely long-pilose particularly on margin, surface with few to many short glandular and eglandular hairs, anticous pair fused for 2–4 mm, linear-lanceolate, 7–12.5 mm long, 1–1.5 mm wide; posticous lobe elliptic (-obovate) to oblong-elliptic, 2.5–5.5 mm wide, with 5 or 7 subparallel principal veins. Corolla 13–23 mm long, pale blue, mauve or white, often darker on lower lip, lobes with many long hairs outside, often also shortly pubescent on throat and limb; tube 9–13 mm long; upper lip arcuate, 4–9.5 mm long; lower lip 4.5–10.5 mm long including lobes 3–5.5 mm long. Anticous (longer) stamens with filaments 7.5–13 mm long, declinate, thecae 1.2–1.5 mm long, barely offset. Ovary pubescent in upper half; style pubescent in lower half; stigma capitate. Capsule 6–7.5 mm long, shortly pubescent towards apex; seeds 1.7–2.2 mm long, 1.5–2 mm wide.

UGANDA. Karamoja District: Turkana Escarpment, Mar. 1959, *J. Wilson* 707!

KENYA. Northern Frontier District: Amaya, Sept. 1976, *Powys* 225!; Machakos District: Kiboko area, June 1971, *Muriithi* 206!; Teita District: Tsavo East National Park, road entrance, Feb. 1961, *Greenway* 9828!

TANZANIA. Pare District: Kiruru, May 1927, *Haarer* 495! & Kisiwani, Feb. 1936, *Greenway* 4576!; Lushoto District: Kivingo, Jan. 1930, *Greenway* 2038!

DISTR. **U** 1; **K** 1–4, 6, 7; **T** 2–4; Mali, Benin, Nigeria, Cameroon, Congo-Kinshasa, Sudan, Eritrea, Ethiopia, Somalia, Angola, Zambia, Mozambique, Zimbabwe, Namibia; Arabia, India

HAB. Dry *Acacia-Commiphora*, *Grewia*, *Combretum*, *Euphorbia* etc. bushland and bushed grassland, on rocky hillsides or on sandy soils; 30–1650 m

USES. None recorded on herbarium specimens

CONSERVATION NOTES. A widespread and common species, favouring dry rocky or sandy habitats with limited agricultural potential: Least Concern (LC).

SYN. *Ruellia aristata* Vahl, Symb. Bot. 2: 73 (1791). Type: Yemen, *Forsskål* 372 (C microfiche 116: III. 3–4, holo., K! photo.)

Lepidagathis aristata (Vahl) Nees in DC., Prodr. 11: 251 (1847); Hepper & Friis, Pl. Pehr Forsskål's Fl. Aegypt.-Arab.: 67 (1994); Hedrén in Fl. Somalia 3: 442 (2006) pro parte excl. fig. 294; Ensermu in Fl. Eth. 5: 428 (2006); *nom. illegit., non L. aristata* Nees in Wallich, Pl. As. Rar. 3: 95 (1832) (see note)

L. terminalis Nees in DC., Prodr. 11: 251 (1847). Types: Ethiopia, Tigray, below Sessaquilla, *Schimper* II: 815 (BM!, BR!, K!, M!, WAG!, syn.) & Sudan, Upper Senaar, Fazokl [Fazohel], *Kotschy* 482 (K!, syn.)

Volkensiophyton neuracanthoides Lindau in E.J. 20: 27 (1894). Type: Tanzania, Moshi District, Kilimanjaro, Lake Chala, *Volkens* 318 (B†, holo.; BM!, K!, iso.)

Lepidagathis sciaphila S. Moore in J.B. 51: 215 (1913). Types: Congo-Kinshasa, W Kundelungu, *Kässner* 2800 (BM!, syn.; K! isosyn.) & Tanzania, [Uvira] ?Uruwira, *Kässner* 3070 (BM!, syn.), **syn. nov.**

NOTE. This is a widespread and variable taxon with several geographic variants potentially recognisable following a review of the material from across its full range; the above description covers only the variation within the Flora region. A rather distinctive form occurs in N Kenya (e.g. *Mathew* 6843), NE Uganda (*Wilson* 707) and S Ethiopia, with small leaves less than 2.5 cm long and the synflorescence comprising few unilateral cymes, one often becoming spiciform. Elsewhere the leaves are over 2.5 cm long and the synflorescence comprises many, usually fasciculate cymes. Plants from **T** 4, N and W Zambia and adjacent Congo-Kinshasa (*L. sciaphila*) have oblong-elliptic to oblong-lanceolate leaves with ± prominent reticulate venation beneath and a golden-yellow stem indumentum. Elsewhere the leaves are ovate to ovate-elliptic, usually with inconspicuous tertiary venation and the stem hairs are grey, buff or pale yellowish.

Clarke (l.c.) incorrectly placed this species and the related *L. angustifolia* C.B. Clarke from Angola in his sect. *Eulepidagathis* in which the anthers of all four stamens are bithecous; here the shorter pair of stamens are monothecous.

Nees (1832) originally applied the name *Lepidagathis aristata* to Wallich cat. no. 7163 which was later synonymised within *L. hyalina* Nees var. *riparia* C.B. Clarke in Fl. Brit. Ind. 4: 521 (1885). In De Candolle's Prodromus, Nees (1847) made the new combination *L. aristata* (Vahl) Nees based upon *Forsskål* 372, making no reference to his earlier application of this name. As the two type specimens represent different taxa, *L. aristata* (Vahl) Nees is a later homonym and illegitimate, hence *L. scariosa* is the correct name for this taxon.

4. **Lepidagathis pseudoaristata** *Ensermu* in K.B. 64: 57, fig. 1 (2009). Type: Ethiopia, Sidamo Region, Jamjam Awr., 5–10 km from Genale Doria R. towards Bitata on Dalo Mana, *Puff & Ensermu* 821229-3/5 (ETH!, holo.; WU, iso.)

Subshrub, 30–150 cm tall; stems with dense pale stellate or dendritic hairs. Leaves sometimes immature at flowering; petiole 5–20 mm long; blade ovate (-elliptic), 2.5–6 cm long, 1.5–3.5 cm wide, base attenuate to obtuse, apex acute or obtuse, surfaces stellate-pubescent, hairs often long-armed, dense beneath, sparse above at maturity when stellate base sometimes caducous; lateral veins 4–6 pairs. Inflorescences terminal on main shoots and on poorly developed lateral branches, the latter often pseudo-axillary, cymes compounded into small subglobose heads 0.8–2 cm in diameter; leafy bracts subtending heads pandurate, 10–30 mm long, stellate-pubescent; fertile and sterile floral bracts subequal, brown-scarious, obovate, spatulate or oblong, 5–7 mm long, 1–3.5 mm wide, apiculate or mucronate, long-pilose at least on margin and in upper half, surface often also with short glandular and/or eglandular hairs, venation of 3–7 parallel veins; bracteoles as bracts but linear or oblanceolate, 0.5–1.5 mm wide, apex mucronate. Calyx brown-scarious, lobes mucronate, indumentum as floral bracts, anticous pair fused for 2–3.5 mm, linear-lanceolate, 5.5–8 mm long, 0.7–1.5 mm wide; posticous lobe elliptic or obovate, 7–8.5 mm long, 3–4 mm wide, with 5 subparallel principal veins. Corolla 11.5–13 mm long, white with pink or mauve lower lip or mauve-blue throughout, lobes with many long hairs, throat and limb with shorter hairs outside; tube 7–7.5 mm long; upper lip arcuate, 3.5–4.5 mm long; lower lip 4–5 mm long including lobes 1.8–2.5 mm long. Anticous (longer) stamens with filaments 5.5–7 mm long, thecae ± 1 mm long, parallel or slightly offset. Ovary pubescent in upper half; style pubescent in lower half and with minute subsessile glands; stigma capitate. Capsule 6–7.5 mm long, shortly pubescent towards apex; seeds ± 1.8 mm long, 1.5 mm wide.

KENYA. Northern Frontier District: Moyale, July 1952, *Gillett* 13572!; Meru District: Isiolo–Wajir road just S of Shaptiga Hill, July 1974, *Faden & Faden* 74/965!
DISTR. **K** 1, 4; S Ethiopia
HAB. Dry scrub of *Acacia, Commiphora, Cussonia, Ficus, Dichrostachys* etc. on rocky hillsides; 1000–1100 m
USES. None recorded on herbarium specimens
CONSERVATION NOTES. Ensermu (l.c.) assessed this species as Near Threatened (NT) in the protologue; this assessment is maintained here.

SYN. [*L.* sp. (= *Puff & Ensermu K.* 821229–3/5) *sensu* Ensermu in Fl. Eth. 5: 429 (2006)]

NOTE. *L. pseudoaristata* is closely related to *L. scariosa* and could be considered an extreme form of that polymorphic species but is maintained here as the pseudoaxillary heads with pandurate subtending bracts give these plants a very distinctive appearance.

5. **Lepidagathis calycina** *Nees* in DC., Prodr. 11: 252 (1847); C.B. Clarke in F.T.A. 5: 127 (1899); Wood, Handb. Yemen Fl.: 273 (1997); Hedrén in Fl. Somalia 3: 442 (2006); Ensermu in Fl. Eth. 5: 430 (2006). Type: Ethiopia, Tigray, *Schimper* 1044 (?TUB, holo., n.v.)

Subshrub, 30–100 cm tall; stems soon woody, shortly white antrorse-pubescent when young. Leaves sometimes immature at flowering; petiole to 6 mm long; blade ovate or lanceolate, 2.3–4.3 cm long, 0.8–1.6 cm wide, base attenuate, apex acute or obtuse, sparsely pubescent particularly on veins and margin, with early-caducous pale stellate hairs beneath; lateral veins 5(–6) pairs. Inflorescences axillary and terminal, strobilate unilateral spikes 1–4 cm long, sessile; fertile and sterile bracts subequal, hyaline, pale green or brown with age, sometimes tinged purple towards apex, elliptic (-obovate), 8–12.5 mm long, 3–6.5 mm wide, apex attenuate or acuminate, often falcate particularly on sterile bracts, margin ciliate or sparsely so, surface with few to many minute hairs often including some glandular hairs, prominently 3-veined from base; bracteoles linear-lanceolate, 4.5–9.5 mm long, 0.3–1.5 mm wide. Calyx hyaline, indumentum as bracts but often more conspicuously ciliate; anticous lobes fused for 2–4.5 mm, lanceolate, 8.5–11.5 mm long, 1.5–3 mm wide; posticous lobe elliptic, 10–13 mm long, 3.5–6 mm wide, apex mucronulate, 3-veined from base. Corolla 13–17 mm long, white or cream, often with purple markings on lower lip and throat, pubescent outside including few glandular hairs on limb; tube 8.5–11.5 mm long; upper lip straight, 3.5–6 mm long; lower lip 4.5–6 mm long including lobes 2.5–3.5 mm long. Anticous (longer) stamens with filaments 4.5–7 mm long, thecae (0.6–)0.9–1.25 mm long, slightly offset. Ovary minutely pubescent in upper half; style with short-stalked or subsessile glands in lower half, sometimes with interspersed spreading eglandular hairs; stigma capitate. Capsule 8–9 mm long, minutely pubescent towards apex; only immature seeds seen.

KENYA. Northern Frontier District: Dandu, May 1952, *Gillett* 13210!
DISTR. **K** 1; Ethiopia, Somalia; Arabia, Pakistan
HAB. *Acacia-Commiphora* scrub on red sandy loam over granite; elsewhere also recorded on black cotton soil and over limestone; 800 m
USES. None recorded on herbarium specimens
CONSERVATION NOTES. Although widely distributed, this species appears localised, with rather few records from most countries within its range. However, suitable habitat remains widespread, with overgrazing the only likely threat; currently considered of Least Concern (LC).

SYN. *L. aparine* Chiov. in Atti Soc. Nat. Modena 44: 65 (1933). Type: Ethiopia, Bokol, between Bug-Berde and El-Bar, *Guidotti* s.n. (not traced), Syn. fide Hedrén (l.c.)

6. **Lepidagathis scabra** *C.B. Clarke* in F.T.A. 5: 129 (1899); U.K.W.F. ed. 1: 596 (1974); U.K.W.F. ed. 2: 271 (1994), both as *L. scabra* S. Moore; Ensermu in Fl. Eth. 5: 432, fig. 166.37 (2006). Type: Tanzania, Moshi District, Lake Chala, *Volkens* 320 (K!, lecto., chosen here; BM!, isolecto.)

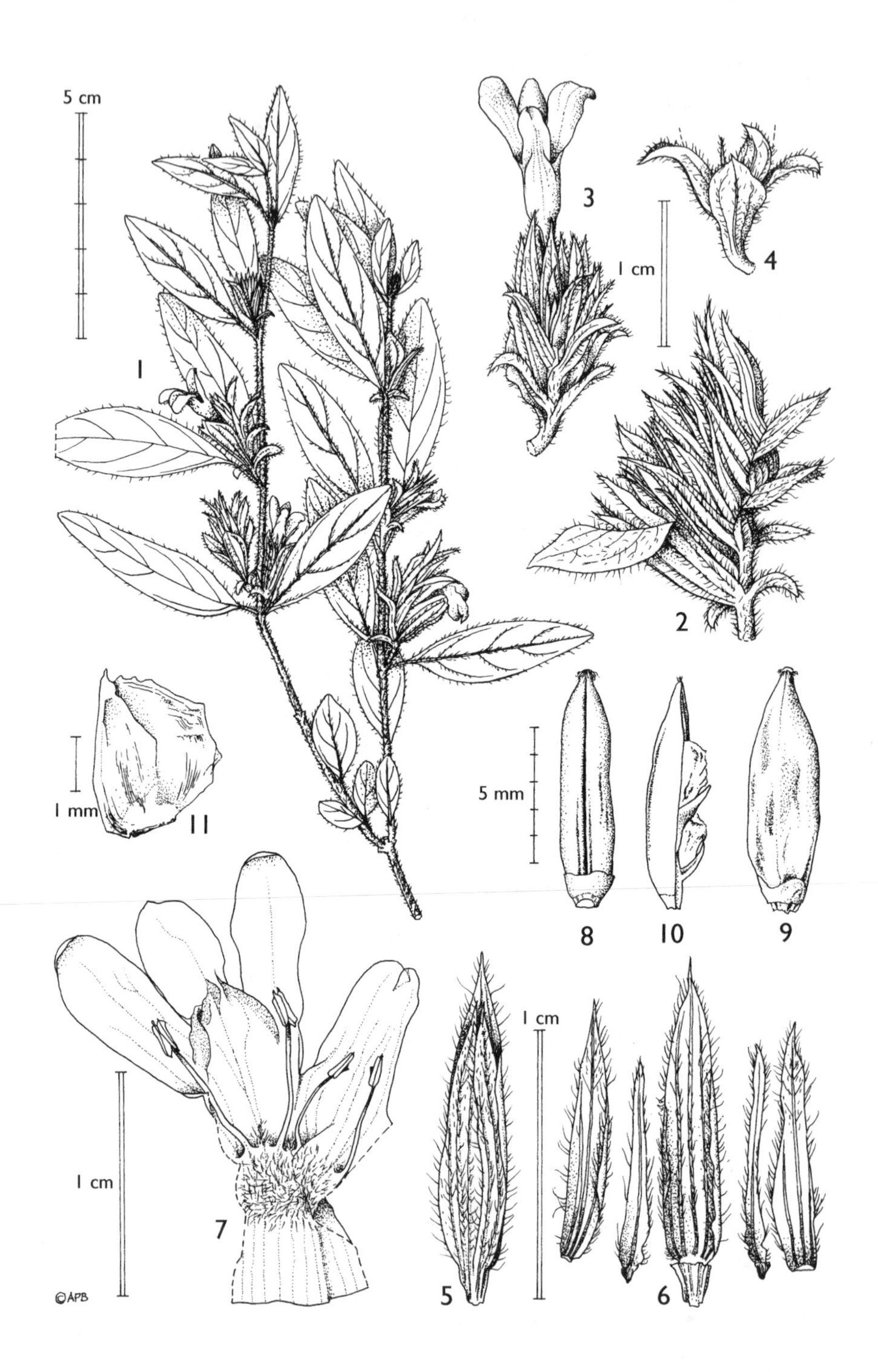
5 cm
1
3
1 cm
4
2
1 mm
11
5 mm
8
10
9
1 cm
7
©APB
5
1 cm
6

Erect, decumbent or prostrate perennial herb, typically branching mainly from base, 10–50(–120) cm tall; stems pale pilose, hairs spreading and/or antrorse. Leaves subsessile or petiole to 15 mm long; blade ovate, lanceolate or elliptic, 1.8–9.5 cm long, 0.8–3.6 cm wide, base cuneate or attenuate, apex acute to rounded or subattenuate, surfaces pale pilose; lateral veins 3–5 pairs. Inflorescences axillary, solitary, congested unilateral spikes or fascicles, 1–4.5 cm long, subsessile or peduncle to 6 mm long; sterile bracts held in two ± divergent rows, green, subulate and ± falcate, 5–13 mm long, 1–3 mm wide, with 3–5 prominent parallel veins; fertile bracts pale-scarious at least towards base, oblong-lanceolate, 7–13.5 mm long, 1.5–4 mm wide, apex acuminate, indumentum of long ascending or spreading hairs at least on veins and margin, often also with minute spreading hairs, venation parallel; lowermost 1–2 fertile bracts of each spike markedly enlarged and green, ovate to obovate, with palmate venation; bracteoles (linear-) lanceolate, 6.5–12.5 mm long, 1–2.5 mm wide. Calyx pale-scarious, indumentum as bracts; anticous lobes free, subulate, 8.5–11.5 mm long, 1.5–2.5 mm wide, margin somewhat inrolled, apex acute to acuminate; posticous lobe oblong-lanceolate, 9–13.5 mm long, 2.5–3.5 mm wide, margin inrolled and often partially enveloping other lobes. Corolla 13.5–20 mm long, pale blue, mauve or white, with purple markings on lower lip and throat, pubescent outside mainly on lower lip, upper portion of tube sometimes with few short-stalked glands outside; tube 8–10.5 mm long; upper lip straight, 4–8 mm long; lower lip 6–10 mm long including lobes 5.5–7 mm long. Anticous (longer) stamens with filaments 4–6.5 mm long, thecae 1.5–2 mm long, parallel; posticous stamens with vestigial second theca sometimes present. Ovary with or without few appressed hairs towards apex, with a ring of short hairs at style attachment; style with spreading hairs and/or short-stalked glands in lower half; stigma with spoon-shaped lobe. Capsule 7–10 mm long, glabrous or with few appressed or antrorse hairs towards apex; seeds 2.5–3 mm long, 2–2.7 mm wide. Fig. 46, p. 310.

KENYA. Northern Frontier District: Dandu, May 1952, *Gillett* 13003!; Machakos District: Bushwhackers Safari Camp, Mar. 1969, *Napper* 1968!; Teita District: Tsavo National Park (East), Feb. 1969, *Hucks* 1112!

TANZANIA. Pare District: Mkomazi Game Reserve, SE of Ndea Hill, Apr. 1995, *Abdallah & Vollesen* 95/22!; Kondoa District: Great North Road, 24 km S of Kondoa, Jan. 1962, *Polhill & Paulo* 1219!; Dodoma District: 6 km on Manyoni–Kilimatinde road, Apr. 1988, *Bidgood, Mwasumbi & Vollesen* 1088!

DISTR. **K** 1, 4, 7; **T** 1–3, 5, 7; Ghana (see note), S Ethiopia, S Zambia to South Africa & Swaziland

HAB. *Acacia-Commiphora* and/or *Combretum* woodland and grazed open bushland, open *Brachystegia-Julbernardia* woodland; 700–1700 m

USES. None recorded on herbarium specimens

CONSERVATION NOTES. This species is widely though disjunctly distributed in eastern and southern Africa and can be locally abundant. It is tolerant of or even favours limited habitat disturbance: Least Concern (LC).

NOTE. Populations from **K** 1 and S Ethiopia are rather distinctive in having few-flowered, fasciculate cymes, these elsewhere being clearly spiciform and with more flowers (see fig. 46, p. 310). The **K** 1 plants also tend to be more stunted with smaller leaves; this is probably an ecological adaptation.

FIG. 46. *LEPIDAGATHIS SCABRA* — **1**, habit; **2**, typical elongate inflorescence; **3**, inflorescence, more fasciculate variant; **4**, base of inflorescence showing enlarged basal fertile bract; **5**, calyx in situ, anterior aspect; **6**, dissected calyx showing the inner surfaces of each lobe, anticous lobes outermost, posticuous lobe central, lateral lobes in between; **7**, dissected corolla with stamens; **8**, capsule, outer surface of single valve; **9**, capsule, face view prior to dehiscence; **10**, capsule valve after dehiscence, with mature seeds; **11**, detail of mature seed. 1 from *Richards* 25542; 2 & 7–11 from *Bidgood et al.* 2212; 3–6 from *Gillett* 13003. Drawn by Andrew Brown.

Clarke (l.c.) also listed *Welwitsch* 5104 (BM!) from Angola in the protologue of *L. scabra* but this specimen clearly differs in being suffruticose and having much shorter sterile bracts with more prominent venation. This specimen is therefore rejected from *L. scabra* and *Volkens* 320 is consequently chosen as the lectotype.

Specimens from Ghana assigned to this species by Heine in F.W.T.A. ed. 2, 2: 416 (1963) have broader, more lanceolate sterile bracts and should be treated as at least subspecifically distinct.

7. **Lepidagathis plantaginea** *Mildbr.* in N.B.G.B. 13: 286 (1936). Type: Tanzania, Masasi District, Lukuledi, *Schlieben* 6339 (B†, holo.; BM!, BR!, K!, M!, S!, iso.)

Rosulate herb with short creeping rhizome. Leaves appressed to ground, 2–5 pairs; petiole to 20 mm long; blade oblong or oblanceolate, 10–16 cm long, 3.5–5.5 cm wide, base cuneate-attenuate, apex rounded or shallowly emarginate, glabrous except for sparse hairs on midrib and main veins beneath; lateral veins 8–10 pairs. Inflorescences axillary but appearing terminal above rosulate leaves, congested unilateral spikes 2–6 cm long, sometimes single-branched at base; peduncle 10–42 mm long, crisped-pilose; sterile bracts held in two divergent rows, green, (oblong-) lanceolate, ± falcate, 10–15 mm long, 1.5–4 mm wide, glabrous or margin with short-ascending and/or long-crisped hairs; fertile bracts pale brown-scarious, 10–11.5 mm long, 2.5–3.5 mm wide, apex acuminate, margin with long straight and crisped white hairs, often many towards apex, elsewhere glabrous; lowermost fertile bract of each spike somewhat larger and green; bracteoles as fertile bracts but 8–9 mm long, 1.5–2 mm wide, enveloping the calyx. Calyx scarious, indumentum as bracts; anticous lobes free, subulate, 10–11 mm long, 2–2.5 mm wide, margin somewhat inrolled, apex acuminate; posticous lobe oblong-elliptic, 10.5–12 mm long, 3.5–5 mm wide, margin strongly inrolled and enveloping the other lobes. Corolla 16–20 mm long, white to pale pink with darker markings on lower lip; throat and limb pubescent outside; tube 10.5–12.5 mm long; upper lip 3.5–4.5 mm long; lower lip 5.5–7.5 mm long including lobes 4–4.5 mm long. Anticous (longer) stamens with filaments 4.5–5 mm long, thecae 1.2–1.5 mm long, parallel; posticous stamens with vestigial second theca. Ovary glabrous except for a ring of minute hairs at style attachment; style shortly and sparsely hairy; stigma not seen. Capsule and seeds not seen.

TANZANIA. Masasi District: Lukuledi, ± 150 km W of Lindi, Apr. 1935, *Schlieben* 6339! (type) & 30 km NW of Masasi, Chiwale village, Mar. 1991, *Bidgood, Abdallah & Vollesen* 1950!

DISTR. **T** 8; not known elsewhere

HAB. *Brachystegia-Pericopsis* woodland with *Oxytenanthera* (bamboo) thicket, including on termite mounds; 300–450 m

USES. None recorded on herbarium specimens

CONSERVATION NOTES. *L. plantaginea* appears highly restricted in range within the dry woodlands of SE Tanzania, being known from only two collections. Although this area is not very well botanised, this species is clearly scarce and is apparently absent from the better-studied Lindi District and Selous National Park. Threats to its habitat are unknown but it is currently considered Vulnerable (VU D2) in view of its extremely restricted range, rendering even small-scale disturbance to its populations, or stochastic events, a significant threat.

NOTE. The inflorescence in this species is close to that of *L. scabra* but, in addition to the obvious differences in habit, leaf shape and peduncle length, they differ in *L. plantaginea* having hairs restricted to the margin and apex of the bracts and calyx lobes, a broader posticous calyx lobe and smaller anther thecae.

8. **Lepidagathis pallescens** *S. Moore* in J.B. 18: 308 (1880); C.B. Clarke in F.T.A. 5: 127 (1899). Type: Angola, Pungo Andongo, *Welwitsch* 5084 (BM!, holo.; K!, iso.)

Suffruticose perennial, 1–few erect stems from a woody base, 5–60 cm tall; stems antrorse-pubescent and/or spreading-pilose. Leaves subsessile or petiole to 7 mm long; blade ovate, lanceolate or elliptic, 3–10.5 cm long, 1.2–3.7 cm wide, base (cuneate-) attenuate, apex acute or obtuse, surfaces pubescent or pilose, sometimes restricted to margin and veins beneath; lateral veins 4–7 pairs. Inflorescences terminal and often also axillary, strobilate unilateral spikes 1.5–6 cm long, sessile or peduncle to 7(–18) mm long; bracts brown-scarious, often with a green portion at base, rarely sterile bracts pale green throughout; sterile bracts ovate or lanceolate, 10–18 mm long, 2.5–5.5 mm wide, apex attenuate or acuminate, not or rarely somewhat falcate, surface minutely pubescent and/or with long ascending or spreading white hairs sometimes restricted to margin, venation prominent, parallel towards centre and usually with prominent reticulation towards margin; fertile bracts similar but variously ovate, lanceolate, elliptic or somewhat obovate, 3.5–7 mm wide; bracteoles (oblong-) lanceolate, 8.5–19 mm long, 1.5–4 mm wide, apex narrowed into a linear acumen. Calyx brown-scarious in upper half, pale (yellow-) green below, venation and indumentum as bracts; anticous lobes free, subulate or narrowly oblong-elliptic, 11.5–16 mm long, 1.5–3.7 mm wide, apex attenuate; posticous lobe elliptic to lanceolate, 12–17 mm long, 2.5–6 mm wide, margins ± strongly inrolled and partially enveloping the other lobes. Night-flowering; corolla 19–28 mm long, white or cream, with mauve markings on lower lip; throat and limb pubescent outside; tube 11–15 mm long; upper lip 6–8.5 mm long; lower lip 8–13 mm long including lobes 6.5–10 mm long, palate shortly pubescent on bosses. Anticous stamens with filaments 4–6.5 mm long, thecae 1.4–1.7 mm long, parallel; posticous stamens with vestigial second theca. Ovary with few minute hairs at apex; style glabrous or shortly hairy; stigma with spoon-shaped lobe. Capsule 10–12 mm long, glabrous or with few hairs towards apex; seeds ± 3 mm long, 2.5 mm wide.

TANZANIA. Ufipa District: km 7, Kalambo Falls [Kapozwa] – Kwela [Kawala] road, June 1996, *Faden et al.* 96/317!; Iringa District: 42 km from Mafinga on Madibira road, Mar. 1988, *Bidgood, Mwasumbi & Vollesen* 610!; Songea District: ± 1.5 km W of Ruanda turnoff on Mbamba Bay road and just E of R. Likuyu crossing, Apr. 1956, *Milne-Redhead & Taylor* 9569!

DISTR. **T** 1, 4, 7, 8; SE Congo-Kinshasa, Angola, Zambia, Malawi, N Mozambique

HAB. *Brachystegia* and *Brachystegia-Uapaca* woodland and associated grassland on rocky hillsides and sandy soils; 450–1550 m

USES. None recorded on herbarium specimen

CONSERVATION NOTES. This species is locally common in the miombo woodlands of S Tanzania and particularly N Malawi. It is probably under-collected because it is night-flowering; it may therefore be more widespread than currently recorded. Much suitable habitat remains within its range and it is not considered threatened: Least Concern (LC).

NOTE. In some populations from **T** 8 (e.g. *Milne-Redhead & Taylor* 9569, 9715; *Bidgood et al.* 2090), the sterile bracts are narrower than usual and somewhat falcate, sometimes remaining green at maturity and the reticulate marginal venation can be absent. This form is potentially confusable with *L. scabra* but is separable in being night-flowering, the corolla being larger and hairy on the palate of the lower lip, at least some of the inflorescences being terminal and the lowermost fertile bracts not being disproportionately larger than the remaining bracts in each inflorescence.

The type location is highly isolated from the East African populations of this species and no subsequent material from Angola has been seen. The single corolla present on the holotype measures only ± 10 mm long which is considerably smaller than recorded in our region, although it is possibly immature. More material is clearly required from Angola to further investigate this difference.

9. **Lepidagathis nemorosa** *S. Moore* in J.B. 48: 253 (1910). Type: Congo-Kinshasa, Lofoi R., *Kässner* 2655 (BM!, holo.; K!, P!, iso.)

Suffruticose perennial, with few to many erect or decumbent branches from a woody rootstock, 10–60 cm tall; stems antrorse-pubescent to crisped- or spreading-pilose, or glabrous except for a line of hairs at nodes. Leaves subsessile or petiole to 18 mm long; blade glossy above, elliptic, narrowly oblong-elliptic to -lanceolate or oblanceolate, 3.3–9 cm long, 0.9–3 cm wide, base cuneate or attenuate, apex acute to rounded, surfaces glabrous or pubescent mainly on principal veins; lateral veins 4–6 pairs. Inflorescences axillary, fasciculate, usually opposite, subsessile; sterile bracts green or brown with age, oblong, lanceolate or spatulate, 4–9.5 mm long, 1–3 mm wide, crisped-pubescent at least on margin, with prominent parallel venation; fertile bracts and bracteoles equal, together enveloping lower portion of calyx, green in lower half and (black-) brown in upper half or turning brown-scarious throughout with age, ovate-elliptic or rarely lanceolate, 5–10 mm long, 2.5–4 mm wide, apex rounded to attenuate, mucronulate, ciliate, sometimes also shortly hairy on surface. Calyx coloration as fertile bracts, lobes glabrous or often with short cilia, surface sometimes puberulent in upper half; anticous pair of lobes free, oblong-lanceolate, 8–13.5 mm long, 2.5–3.7 mm wide, margin somewhat inrolled, apex acute or attenuate; posticous lobe oblong-elliptic to -lanceolate, 3.5–5 mm wide, margin strongly inrolled and partially enveloping other lobes. Corolla 20–28 mm long, white, pale blue or mauve, with pink to purple markings on lower lip; throat and lower lip pubescent outside; tube 10–17 mm long; upper lip 6–8 mm long; lower lip 9.5–12.5 mm long including lobes 6–8.5 mm long. Anticous (longer) stamens with filaments 7.5–10.5 mm long, thecae 2–2.8 mm long, parallel; posticous stamens with vestigial second theca. Ovary glabrous or with few short hairs at apex; style glabrous; stigma with spoon-shaped lobe. Capsule 10–11 mm long, glabrous or with few hairs at apex; seeds ± 2.7 mm long, 2 mm wide.

TANZANIA. Mpanda District: 7 km on Inyonga–Ilunda road, Feb. 2009, *Bidgood, Leliyo & Vollesen* 7953!; Ufipa District: road to timber camp, Mar. 1959, *Richards* 11003!; Mbeya District: 56 km on Tunduma–Sumbawanga road, Apr. 2006, *Bidgood et al.* 5310!

DISTR. **T** 4, 7; S Congo-Kinshasa, Zambia, N Malawi, N Zimbabwe

HAB. *Brachystegia-Isoberlinia* woodland and road verges on sandy soil; 1100–1850 m

USES. None recorded on herbarium specimens

CONSERVATION NOTES. Although scarce in our region (only known from four specimens), this species is common and widespread in the miombo woodlands of Zambia and is not considered threatened: Least Concern (LC).

SYN. *L. lindaviana* Buscal. & Muschl. in E.J. 49: 494 (1913); Piscicelli, Nella Regione Laghi Equatoriali: 116, 117 incl. fig. (1913). Type: ? Zambia, Lake Bangweulu [Banguelo], *von Aosta* 1073 (B†, holo.), **syn. nov.**

L. ringoetii De Wild. in F.R. 13: 146 (1914). Type: Congo-Kinshasa, Katanga, Shinsenda, *Ringoet* 511 (BR!, holo.), **syn. nov.**

L. persimilis S. Moore in J.B. 67: 51 (1929). Type: Zimbabwe, Mwami [Miami], *Rand* 77 (BM!, holo.), **syn. nov.**

Group C "Lepidagathis" -species 10–18

Inflorescences basal, axillary and/or terminal, fasciculate to spiciform unilateral cymes, solitary or often compounded into dense synflorescences; one of each pair of bracts sterile. Corolla with upper lip straight; lower lip with a membranous portion on either side of the centre of the palate, (usually) with an adjacent line of silky hairs (fig. 47/9, p. 319). Stamens all bithecous; filaments with subsessile or short-stalked glands mainly in upper half or glabrous; anther thecae somewhat offset and oblique, with short hairs and subsessile glands dorsally and with a line of cilia along the suture (sometimes sparse). Stigma capitate, bilobed. Capsule 2-seeded, face ovate; seeds with hygroscopic hairs cream- or pale buff-coloured.

NOTE. Species delimitation is highly problematic within this group. J.K. Morton, in studying the taxa with basal inflorescences in the African savannas, noted that "though clearly recognisable variational peaks can be defined…they are connected by a series of intermediate plants which make it very difficult to produce a satisfactory taxonomic treatment" (J.L.S. 96: 333 (1988)). Nowhere is this more so than in East Africa where a range of seemingly distinct taxa occur which in some regions grow in close proximity to one another without intergrading, whilst in other regions appear to hybridise freely (though with little or no within-population variation). The most significant variation between many of these taxa is in growth habit and the position and form of the inflorescences, and it is quite possible that these are influenced primarily by ecological factors, most notably local variation in fire regime. By applying a broad species concept, this group could be reduced to just three taxa within our region: *L. collina* (including *L. diversa*) and *L. peniculifera* from the West African savannas that extend into Uganda and W Kenya, and *L. andersoniana* (to include species 13–17) from the miombo woodlands and montane grasslands of Tanzania south to Zimbabwe, with most taxa reduced to varieties or ecological subspecies. However, the influence of ecology is very difficult to infer from the study of herbarium material alone. A full revision of the group, including the species from West African and the Indian Subcontinent and involving extensive field studies, is therefore required to gain a full understanding of taxon delimitation and suitable taxonomic ranks within this group. In view of this, I have chosen to maintain the "recognisable variational peaks" as distinct species here. Nine species are currently recognised within this group in our region, but it is conceded that not all specimens will readily key out due to the occasional presence of intermediate specimens between most of the taxa.

10. **Lepidagathis collina** (*Endl.*) *Milne-Redh.* in K.B. 8: 119 (1953); F.P.S. 3: 182 (1956); Heine in F.W.T.A. ed. 2, 2: 416 (1963) pro parte excl. *L. diversa*; U.K.W.F. ed. 1: 595 (fig.), 596 (1974); U.K.W.F. ed. 2: 271, pl. 118 (1994). Type: Sudan, Gebbel Accara, by R. Turnad, *Russegger* s.n. (not traced)

Prostrate or procumbent pyrophyte, stems to 60 cm long radiating from woody base, sometimes rooting at nodes, current year's growth often immature at flowering; stems 4-ridged, shortly pubescent and/or pilose at least on ridges. Leaves subsessile; blade linear, oblanceolate, narrowly oblong or (ovate-) elliptic, 2–8 cm long, 0.4–1.5 cm wide, base cuneate to rounded, margin entire or minutely toothed, apex acute or obtuse, apiculate, glabrous or with sparse short to long pale hairs mainly on margin; lateral veins 4–6 pairs, ascending and anastomosing. Inflorescences basal and/or in the lower axils and at branching nodes, compounded into globose heads or cushions, 2–8 cm in diameter, sessile or peduncle to 10 mm long, this with scale-like sterile bracts; sterile bracts of cyme brown-scarious, oblong-ovate or -lanceolate, 12–18 mm long, 3–7 mm wide, apex attenuate or shortly acuminate, straight or somewhat falcate, margin ciliate with short to long white hairs, surface glabrous or minutely pubescent; fertile bracts and bracteoles subequal, as sterile bracts but often paler and subhyaline, rarely green in upper half, (oblong-) lanceolate, 2–5 mm wide, often more densely hairy including long hairs on midrib. Calyx brown towards base, green above, surface with dense short-spreading and/or long-ascending pale hairs, longest on margins, lobes 10.5–16 mm long, anticous pair oblong-lanceolate or -elliptic, 2–4.5 mm wide, apex attenuate or acuminate, straight or somewhat falcate; posticous lobe oblong-ovate (-lanceolate), 3–5.5 mm wide, apex straight, surface with 3–5(–9) ± prominent subparallel veins. Corolla 14.5–19 mm long, white to yellow with many brown or purple markings in throat and on limb, median lobe of lower lip (?always) bright yellow; throat and limb pubescent outside; tube 8–11.5 mm long; upper lip 4–6 mm long; lower lip 5–9 mm long including lobes ± 2.5 mm long. Anticous (longer) stamens with filaments 3.5–5 mm long, thecae 1.4–1.8 mm long. Ovary pubescent in upper half; style with few hairs in upper half and with subsessile glands towards base. Capsule 6.5–7 mm long, shortly hairy towards apex; seeds ± 3 mm long, 2.3 mm wide.

UGANDA. West Nile District: near Atar, 10 km W of Pakwach, Feb. 1969, *Lye & Lester* 2202!; Busoga District: 1/4 mile E of Bugiri Rest House, Aug. 1952, *G.H.S. Wood* 350!; Mengo District: Mpogo, Jan. 1920, *Dummer* 4382!

KENYA. Uasin Gishu District: Soy, Aug. 1932, *Mainwaring* 2207!; Trans-Nzoia District: Moi's [Hoey's] Bridge, Aug. 1963, *Heriz-Smith & Paulo* 853!; Trans-Nzoia/North Kavirondo District: S Elgon, Sep. 1944, *Tweedie* 642!

DISTR. **U** 1, 3, 4; **K** 3, 5; Senegal to Sudan and Ethiopia

HAB. Open fire-prone grassland, rocky outcrops including ironstone pavement with sparse grass, flowering Aug.–Sept. and Jan.–Mar.; 700–2150 m

USES. None recorded on herbarium specimens

CONSERVATION NOTES. A common and widespread species of the West African savannas: Least Concern (LC).

SYN. *Russegera collina* Endl., Nov. Stirp. Mus. Dec.: 38 (1839) & Ic. Gen. Pl.: t. 94 (1839) & p. xii (1841)

Lepidagathis radicalis Nees in DC., Prodr. 11: 255 (1847). Types: Ethiopia, near Adwa [Adoam], *Schimper* 2: 1072 (BM!, G!, K!; M!, STU!, syn.) & Devrarina, *Quartin Dillon* s.n. (P, syn.)

L. myrtifolia S. Moore in J.B. 18: 38 (1880); C.B. Clarke in F.T.A. 5: 124 (1899). Type: Sudan, "Gir", *Schweinfurth* 2493 (K!, ?holo.; P!, iso.), **syn. nov**.

L. schweinfurthii Lindau in E.J. 20: 16 (1894); C.B. Clarke in F.T.A. 5: 123 (1899); F.P.S. 3: 182 (1956). Type: Sudan, "Djur", *Schweinfurth* 2339 (B†, holo.; K!, iso.)

L. ampliata C.B. Clarke in F.T.A. 5: 123 (1899). Type: Kenya, Kavirondo, *Scott-Elliot* 7072 (K!, holo.; BM!, iso.)

L. hamiltoniana Wall. subsp. *collina* (Endl.) J.K. Morton in J.L.S. 96: 341 (1988); Friis & Vollesen in Biol. Skr. 51(2): 449 (2005); Ensermu in Fl. Eth. 5: 427 (2006), **syn. nov**.

[*L. hamiltoniana* subsp. *hamiltoniana sensu* J.K. Morton in J.L.S. 96: 340 (1988) pro parte quoad spec. ex Africa, *non L. hamiltoniana* Wall.]

NOTE. Morton (l.c.) divided the African *L. collina* into two taxa based upon differences in stem and leaf size, leaf venation and inflorescence size. He considered the West African (Guinea Bissau to Sudan) material, in which the plants have long stems with long, slender, usually 3-veined leaves and large basal inflorescences, to be inseparable from *L. hamiltoniana* Wall. from India. *L. collina sensu* stricto, ranging from Central African Republic to Kenya, was separated by being smaller in all parts and having a pinnate leaf venation. However, occasional intermediates were noted, hence Morton considered the two taxa only subspecifically distinct and reduced *L. collina* to a subspecies of *L. hamiltoniana*.

Following a reassessment of this complex here, I do not feel that the recognition of two taxa in Africa is supportable. The differences in stem, leaf and inflorescence size are all clinal and the variation in leaf venation is largely linked to the length/width ratio of the leaf; the form with strongly anastomosing pinnate veins merges into the 3-veined form as the leaves become more slender. Indeed, Morton conceded that plants from West Africa with the broadest leaves could have a pinnate venation. Whilst the African material is very similar to *L. hamiltoniana*, close examination reveals some differences. Of most note is the consistently hairy ovary in the African material, this being glabrous in the Indian specimens. In addition, the limited flowering material available for *L. hamiltoniana* appears to have smaller corollas than the African plants, often less than 10 mm (–13.5 mm) long. *L. hamiltoniana* is in fact one of several closely related Indian taxa that appear very similar to the African material, including the type of the genus *L. cristata* Willd., of which Morton noted "the East African subsp. *collina* has its counterpart in India in the form of *L. cristata*" (p. 334). Relationships between taxa are clearly complex in this group and not fully understood. It therefore seems premature to include *L. collina* within *L. hamiltoniana* until a full revision of this group has been carried out, and so the name *L. collina* is resurrected here. The description and synonymy presented above cover only the forms relevant to the F.T.E.A. region.

The form previously separated as *L. myrtifolia* (in our region represented by e.g. *Dummer* 4382, **U** 4) has short, broad leaves but intermediates also occur in Uganda.

11. **Lepidagathis diversa** *C.B. Clarke* in F.T.A. 5: 126 (1899); F.P.S. 3: 183 (1956); J.K. Morton in J.L.S. 96: 337, 343, fig. 5 (1988); Friis & Vollesen in Biol. Skr. 51(2): 449 (2005); Ensermu in Fl. Eth. 5: 427, fig. 166.35 (2006). Type: Uganda, West Nile District, Madi, *Speke & Grant* 657 (K!, holo.)

Erect perennial herb, 20–70 cm tall; stems unbranched, 4-ridged when young, spreading- or antrorse-pubescent or -pilose. Leaves subsessile; blade (linear-) lanceolate, 9–15 cm long, 0.7–2 cm wide, base acute to rounded, margin minutely toothed, apex acute-apiculate, indumentum as stems, most dense on veins beneath; lateral and tertiary veins ascending and anastomosing, prominent beneath. Inflorescences in upper axils, sometimes also reduced inflorescences at lowermost axils, of several unilateral cymes compounded into dense globose heads, 2.5–5 cm in diameter; bracts scarious, (pale) brown, sterile bracts ovate or lanceolate, 9–13.5 mm long, 2.5–4.5 mm wide, narrowed into a long falcate acumen, margin white-pilose at least in upper half, surface minutely pubescent at least on acumen; fertile bracts 12–17 mm long, often more densely hairy; bracteoles lanceolate, 11.5–17 mm long, 1.5–3 mm wide. Calyx lobes green with brown apex, margin and often main veins densely white-pilose, surface densely short-pubescent; anticous lobes lanceolate, 13.5–17 mm long, 2–3.5 mm wide, narrowed into a falcate or recurved acumen; posticous lobe ovate (-lanceolate), 15–18 mm long, 3.5–4.5 mm wide, apex less strongly falcate, surface with 3–5(–7) ± prominent subparallel veins. Corolla 14–18 mm long, white or pale yellow, with brown markings in throat and on lower lip, upper lip brown with paler markings, median lobe of lower lip yellow; throat and limb densely pubescent outside; tube 9–12 mm long; upper lip 3.5–5.5 mm long; lower lip 5–6 mm long including lobes 2–3.5 mm long. Anticous (longer) stamens with filaments 4–6 mm long, thecae 1.5–2 mm long. Ovary pubescent in upper half; style with few hairs mainly in upper half, with or without subsessile glands towards base. Capsule 5.5–6.5 mm long, sparsely hairy towards apex; seeds 2.5–3.3 mm long, 2–2.7 mm wide.

UGANDA. West Nile District: W Madi, Mt Otze, Oct. 1959, *Scott* in EA 11783!; Acholi District: Pawe [Palaro], Gulu, Dec. 1931, *Hancock* 2392!; Bunyoro District: Kigulya, Oct. 1933, *Hazel* 343!

DISTR. U 1, 2, 4; Ghana, Benin, Nigeria, Cameroon, E Congo-Kinshasa, Sudan, W Ethiopia

HAB. Open and lightly wooded grassland, roadsides and secondary growth on abandoned farmland, flowering before the main rains, Sept.–Feb.; 900–1100 m

USES. None recorded on herbarium specimens

CONSERVATION NOTES. A widespread species, uncommon in our region but much more frequent in West African savannas: Least Concern (LC).

SYN. [*L. mollis sensu* Oliver in Trans. Linn. Soc., Bot. 29: 128 (1875) & *sensu* Lindau in P.O.A., C: 368 (1895), *non* T. Anderson]
[*L. collina sensu* Heine in F.W.T.A. ed. 2, 2: 416 (1963) pro parte, *non* (Endl.) Milne-Redh.]

NOTE. There is no doubt that *L. collina* and *L. diversa* are closely related; indeed, Heine (l.c.) treated them as synonymous, attributing the differences in habit to ecological factors. In resurrecting *L. diversa*, Morton (l.c.) pointed also to the more falcate bracts in that species. This applies equally to the anticous calyx lobes, these having a pronounced, strongly falcate or recurved acumen in *L. diversa* and an attenuate or shortly acuminate, straight or weakly falcate apex in *L. collina*. Within the Flora region there appears to be no overlap in habit and inflorescence position between the two taxa, but occasional specimens from West Africa appear intermediate; for example, the type of *L. mollis* T. Anderson from Nigeria (*Barter* 955, K!) and from Cameroon (*Talbot* s.n., BM!) have basal inflorescences typical of *L. collina* but the stems are more erect, with the upper inflorescences tending towards those of *L. diversa*. Champluvier (BR) considers plants from N Congo-Kinshasa (Fl. Afr. Centr. regions VI and VII) with wholly axillary synflorescences on trailing stems as *L. diversa* but the bracts and calyces are not strongly falcate here and I would consider these plants better placed in *L. collina*.

12. **Lepidagathis peniculifera** *S. Moore* in J.B. 18: 39 (1880); C.B. Clarke in F.T.A. 5: 127 (1899); F.P.S. 3: 183 (1956). Type: Sudan, Mittu, Reggo, *Schweinfurth* 2794 (K!, holo.)

Erect or trailing suffruticose herb from a woody base, 25–90 cm tall; stems 4-ridged, subterete and stout with age, glabrous except for a line of long white hairs at the nodes. Leaves sometimes immature at flowering, sessile; blade drying blackish, linear, 5–7 cm long, 0.2–0.5 cm wide, base cuneate, apex acute-apiculate, glabrous or margin with few long white hairs, sometimes with more hairs when young; 3-veined from base but often only midrib prominent. Inflorescences terminal and/or in the upper leaf axils, 1–4 per axil, each a dense unilateral spike, 2.5–6 cm long, subsessile, flowers clearly held in two rows; bracts black or brown-scarious, sterile bracts oblong-ovate, 10–12 mm long, 4.5–5.5 mm wide, apex caudate for 3.5–6.5 mm, often falcate, surface with minute white hairs, with dense long white silky hairs immediately below the acumen within, margin white-ciliate; fertile bracts somewhat wider, with a curved line of long white or cream silky hairs along the midrib; bracteoles oblong, 9–11 mm long, 3.5–5 mm wide, long-caudate, those on exterior of inflorescence with hairs as sterile bracts, those on interior with hairs as fertile bracts. Calyx lobes brown-scarious with darker acumen, with silky cream-coloured subappressed hairs most dense and longest in the upper half and along the margin; anticous lobes oblong-elliptic, 9–11 mm long, somewhat unequal in width, outer lobe of a pair 3.5–5 mm wide, inner lobe 2.5–4 mm wide, apices long-caudate, often recurved; posticous lobe elliptic, 5–7 mm wide. Corolla 18–20 mm long, pink or purple with darker purple markings on lower lip; throat and limb densely silky-hairy outside; tube 10–11.5 mm long; upper lip 6.5–7.5 mm long; lower lip ± 8 mm long including lobes ± 4 mm long. Anticous (longer) stamens with filaments 6–7 mm long, thecae 2–2.5 mm long, suture cilia conspicuous. Ovary glabrous; style with sparse long hairs in lower half and minute ring of hairs at base. Capsule ± 8.5 mm long, glabrous; seeds ± 5 mm long, 3.5 mm wide.

UGANDA. Karamoja District: Labwor Hills, Dec. 1963, *J. Wilson* 1552!
DISTR. **U** 1; ?Cameroon, Central African Republic, NE Congo-Kinshasa, S Sudan
HAB. *Loudetia* grassland over laterite, stony hillsides; altitude unknown
USES. None recorded on herbarium specimens
CONSERVATION NOTES. This is a rather scarce species of the eastern Guinean savanna, known from few collections. However, it is probably under-recorded due to limited botanical exploration over much of its range, and threats are unknown: Data Deficient (DD).

SYN. *L. perglabra* C.B. Clarke in F.T.A. 5: 125 (1899); F.P.S. 3: 182 (1956). Type: Sudan, Ngoli, *Schweinfurth* 4076 pro parte (K!, holo.), **syn. nov.**
L. variegata Benoist in Not. Syst., Paris 13: 198 (1948). Type: Central African Republic, near R. Kaba, 75 km N of Bambari, *Tisserant* 1319 (P!, holo.), **syn. nov.**

NOTE. Clarke (l.c.) recorded the "floral leaves" (bracts) as being without tails or very shortly mucronate. However, the caudate apices to the bracts, bracteoles and calyces in this species are brittle and easily broken off in dried material as is clearly evident in the type specimen. However, some specimens from Sudan and Congo-Kinshasa, otherwise assignable to this species, truly lack the caudate apices and should perhaps be treated as a discrete variety.
The type of *L. perglabra* is highly immature but the habit and indumentum of the stems, leaves and young inflorescences closely matches other material of *L. peniculifera.*

13. **Lepidagathis andersoniana** *Lindau* in E.J. 20: 16 (1894); Lindau in P.O.A.: 368 (1895) as *L. andersonii*; C.B. Clarke in F.T.A. 5: 126 (1899) pro parte excl. *Whyte* s.n. ex Fort Hill, Malawi; Champluvier in Fl. Rwanda 3: 471, fig. 144.4 (1985). Types: ?Tanzania, Ussindje (?Usinge), *Stuhlmann* 3525 (B†, syn.); Malawi, Shire Highlands, *Buchanan* 325 (B†, syn.; K!, isosyn.); Malawi, without precise locality, *Buchanan* 774 & 832 (both B†, syn.; BM!, K!, isosyn.)

Erect or straggling suffruticose herb, few- to much-branched from a woody base, (15–)40–130 cm tall; stems weakly 4-ridged when young, subterete and often stout with age, white (-buff) pilose to lanate when young, ± glabrescent. Leaves subsessile; blade often drying black, elliptic, lanceolate or linear, 3.5–17 cm long, 0.3–2.5 cm wide, base cuneate to obtuse, margin entire or minutely toothed, apex acute-apiculate, indumentum as stems when young, hairs persisting on margin and veins beneath or glabrescent; prominently 3(–5)-veined from base. Inflorescences axillary in upper half of stems, 1(–2) per axil, often in opposite pairs, each a

FIG. 47. *LEPIDAGATHIS ANDERSONIANA* — **1**, habit; **2**, sterile bract from towards base of inflorescence. *L. ERIOCEPHALA* — **3**, habit; **4**, sterile bract from towards base of inflorescence. *L. FISCHERI* — **5**, habit. *L. LANATOGLABRA* — **6**, habit; **7**, abaxial leaf surface, showing primary venation; **8**, dissected calyx, outer surface, posticous lobe outermost left; **9**, dissected corolla with stamens; **10**, detail of one of the abaxial pair of stamens; **11**, capsule valve, lateral and face views; **12**, detail of mature seed. 1 & 2 from *Milne-Redhead & Taylor* 9598; 3 & 4 from *Richards* 13798; 5 from *Bidgood et al.* 7058; 6 from *Kayombo* 1017; 7–12 from *Richards* 13537. Drawn by Andrew Brown.

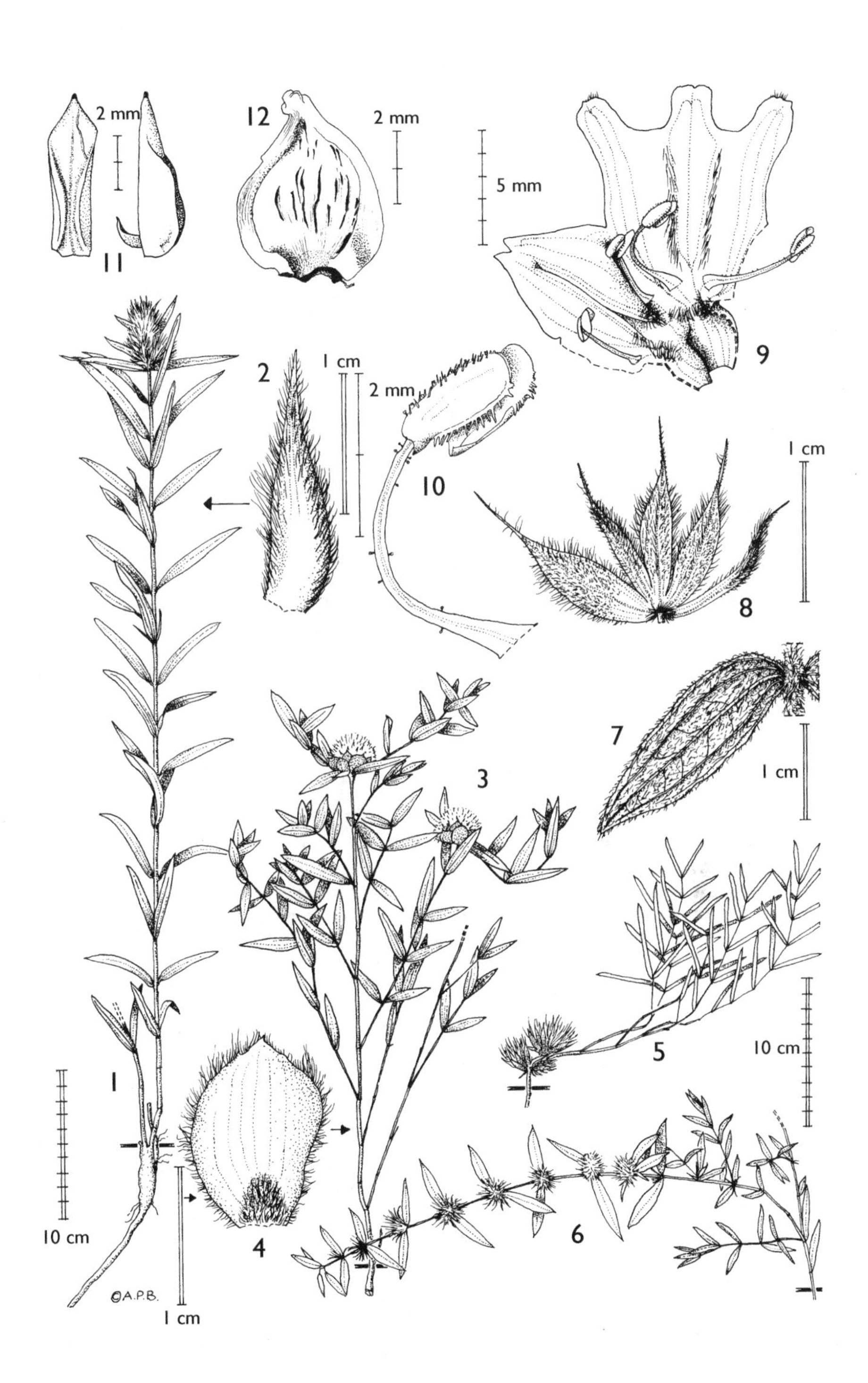
2 mm
11
12
2 mm
5 mm
9
2
1 cm
2 mm
10
1 cm
8
7
1 cm
3
5
10 cm
1
4
6
10 cm
1 cm
©A.P.B.

congested unilateral spike or fascicle, 1.5–5.5 cm long; peduncle 0–7 mm long; sterile bracts black- or brown-scarious, (oblong-) ovate to lanceolate, 11–25 mm long, 2.5–7(–11) mm wide, apex long-attenuate or acuminate, straight or somewhat falcate, white- or buff-plumose at least on margin; fertile bracts similar but often somewhat smaller, surface more densely hairy; bracteoles brown-scarious or one of a pair green with dark brown apex, narrowly elliptic or lanceolate, 11–21 mm long, 1.5–6 mm wide, with linear acumen. Calyx lobes green with dark brown (-purple) acumen, indumentum as fertile bracts, dense; anticous lobes oblong-elliptic or lanceolate, 9–16 mm long, 2–4.5 mm wide, linear acumen sometimes recurved; posticous lobe ovate or lanceolate, 10.5–17 mm long, 2.5–7.5 mm wide, venation inconspicuous. Corolla 11–16 mm long, white, grey or pale purple, lobes usually purple, limb with purple markings towards margin and with a yellow (-orange) median patch on lower lip; throat and limb silky-hairy outside; tube 6–9 mm long; upper lip 4–6 mm long; lower lip 4.5–7.5 mm long including lobes 1.5–3 mm long. Anticous (longer) stamens with filaments 3–4.5 mm long, thecae 1.4–1.8 mm long, suture cilia short and often sparse. Ovary glabrous; style with long hairs and subsessile glands in lower half. Capsule 7–8 mm long, glabrous; seeds 4–4.5 mm long, 3–3.5 mm wide. Fig. 47/1, 2. p. 319.

TANZANIA. Ngara District: Kirushya, Bugufi, Feb. 1960, *Tanner* 4725!; Kigoma District: Kasye Forest, Mar. 1994, *Bidgood, Mbago & Vollesen* 2994!; Songea District: ± 3 km NE of Kigonsera, Apr. 1956, *Milne-Redhead & Taylor* 9598!

DISTR. **T** 1, 4, 8; SE Congo-Kinshasa, Rwanda, Burundi, S Malawi, N Mozambique

HAB. *Brachystegia* and other miombo woodland, woodland clearings and fire-prone grassland, typically on sandy soils or rocky hillsides; usually flowering during and immediately after the main rains, Feb.–June (–July); 200–1800 m

USES. None recorded on herbarium specimens

CONSERVATION NOTES. Widespread and locally common. Its favoured habitat remains widespread and it appears unthreatened so long as regular burning regimes are maintained: Least Concern (LC).

NOTE. *Richards* 20440 from the Itigi–Chunya road agrees with *L. andersoniana* in bract and bracteole shape but tends towards *L. eriocephala* in the more branched, less robust stems and the more congested synflorescences.

Detailed field observation may reveal additional characters of diagnostic value within the *L. andersoniana* complex. For example, it appears that the patterning on the corolla is stronger and more complex in *L. andersoniana* and *L. eriocephala* than in other members of the group. However, notes on flower colour are often very limited on herbarium vouchers; e.g. the flowers of *L. fischeri* and *L. lanatoglabra* are often recorded merely as "white", but even on the dried material, both yellow/orange and purple/brown markings are clearly discernable on the lower lip.

14. **Lepidagathis eriocephala** *Lindau* in E.J. 30: 409 (1901). Type: Tanzania, Mbeya District, Usafwa [Usafua], Pungulumo [Bunguluma] Mt, *Goetze* 1083 (B†, holo.; BM!, BR!, iso.)

Erect or decumbent, rarely prostrate, suffruticose herb, few- or often much-branched from a woody base, 20–80 cm tall; stems 4-ridged, white-pilose when young at least at nodes, glabrescent. Leaves subsessile; blade elliptic to narrowly lanceolate, 2.5–7.5 cm long, 0.4–1.3 cm wide, base cuneate to rounded, margin entire or minutely toothed, apex acute-apiculate, surface glabrous or sparsely white-pilose mainly on margin and midrib beneath; prominently 3-veined from base. Inflorescences axillary in upper half of stems, or rarely basal (see note), dense globose heads 3–6.5 cm in diameter; sterile bracts dark brown-scarious, broadly oblong-elliptic to rounded or ± obovate, 11–17 mm long, 5–13.5 mm wide, outer surface convex, apex abruptly narrowed into an apiculum or short acumen, margin white-plumose; fertile bracts similar but oblong to spatulate, 3.5–7 mm wide, apex abruptly acuminate, margin and midrib densely white-plumose;

bracteoles as fertile bracts but 1.3–4.5 mm wide. Calyx lobes green with dark brown (-purple) acumen, surface and margin densely white-pilose in upper half; anticous lobes narrowly oblong-elliptic, lanceolate or oblanceolate, 9.5–11.5 mm long, 1.5–2.5 mm wide; posticous lobe oblong-elliptic or -lanceolate, 10–13 mm long, 2–4 mm wide, parallel-veined. Corolla 11.5–15 mm long, white or grey with purple markings on the limb and with a yellow median patch on lower lip; limb silky-hairy outside; tube 7.5–9.5 mm long; upper lip 4–5.5 mm long; lower lip 4–6 mm long including lobes 1.5–3.5 mm long. Anticous stamens with filaments 3–4.5 mm long, thecae 1.2–1.5 mm long. Ovary glabrous; style with long hairs and subsessile glands in lower half. Capsule ± 6.5 mm long, glabrous; only immature seeds seen. Fig. 47/ 3, 4, p. 319.

TANZANIA. Ufipa District: Tatanda, Mbaa Hill, Apr. 1997, *Bidgood et al.* 3463!; Mbeya District: Mbozi, Sept. 1936, *Burtt* 6128! & Pungulumo [Pungaluma] Hills, May 1990, *Kayombo* 992!

DISTR. **T** 4, 7; N Malawi, E Zambia (Nyika Plateau: see note), W Mozambique

HAB. *Brachystegia* and *Uapaca* woodland, fire-prone grassland and abandoned cultivation, typically on sandy or gravelly soils or on rocky hillsides; usually flowering during and immediately after the main rains, Mar.–May, occasionally later, July, Sept.; 1300–1850 m

USES. None recorded on herbarium specimens

CONSERVATION NOTES. Although certainly localised, this species is known historically from over 10 sites. Suitable habitat remains widespread within its range but growing human population pressure in S Tanzania and particularly N Malawi are likely to have led to the decline of some sites for this species. It is therefore provisionally assessed as Near Threatened (NT). However, this species may well prove to be more common in Mozambique than is currently known.

SYN. [*L. andersoniana sensu* C.B. Clarke in F.T.A. 5: 126 (1899) pro parte quoad *Whyte* s.n. ex Fort Hill, Malawi, *non* Lindau]

NOTE. This species is very close to *L. andersoniana*, which it replaces in southernmost Tanzania and northern Malawi, but it is clearly recognisable by the more dense, globose axillary heads, the individual unilateral cymes not being clearly discernable as in *L. andersoniana*, and by the broader, convex and more abruptly narrowed sterile bracts.

In our region the synflorescences are always held away from the ground but in N and W Mozambique the plants are sometimes prostrate, the flowering heads still restricted to the upper half of the stems but appressed to the ground. Several collections from the Nyika Plateau of Malawi and Zambia closely resemble *L. eriocephala* in the shape of the sterile bracts but the flowering heads are held at or towards the base of the stems, thus tending towards *L. fischeri*; these populations may be of hybrid origin or may be an ecological variant of *L. eriocephala* in response to differences in fire regime.

15. **Lepidagathis lanatoglabra** *C.B. Clarke* in F.T.A. 5: 124 (1899). Types: Malawi, Tanganyika Plateau, *Whyte* s.n. & Khondowe to Karonga, *Whyte* s.n. & North Nyassa, Songwe and Karongas, *Whyte* s.n. (all K!, syn.)

Prostrate, procumbent or weakly decumbent suffrutex, branching from a woody base, branches to 30–60 cm long; stems slender, 4-ridged, sparsely to densely pale pilose when young or glabrous. Leaves (sub)sessile; blade linear, lanceolate, ovate or elliptic, 2.7–7 cm long, 0.2–2 cm wide, base cuneate to rounded, margin entire, apex acute-apiculate, glabrous or pale-pilose, often lanate when young; prominently 3-veined from base. Inflorescences axillary either throughout the stem or mainly at the lower nodes, held at ground level, cymes fasciculate, 1–3 per axil forming heads 1.5–3.7 cm in diameter, subsessile; bracts brown-scarious with darker, sometimes purple-tinged acumen, rarely green when young, sterile bracts lanceolate, 6–15 mm long, 1.5–5 mm wide, apex attenuate or with a linear acumen, straight or somewhat curved, white-pilose mainly or only on margin; fertile bracts similar but often hyaline when young, 11.5–15 mm long, linear acumen more pronounced, often more densely hairy on the prominent midrib; bracteoles as fertile bracts but 1.5–3 mm wide, those on the exterior of the inflorescence sometimes green-tinged and with fewer hairs on the midrib. Calyx green, paler towards base, with a dark purple (-brown) acumen; anticous lobes elliptic-lanceolate, 9–15 mm long, 1.5–3 mm wide, acumen linear, ± straight, surface with mixed short and long ascending white hairs, with dense long ± spreading hairs on margin except on acumen; posticous lobe ovate or lanceolate, 2–4 mm wide, with 5 or 7 parallel veins ± prominent. Corolla 11–14.5 mm long, white or more rarely blue or violet, palate of lower lip with yellow to orange markings centrally and purple markings towards margin and/or in the throat; limb silky-hairy outside; tube 7–8.5 mm long; upper lip 3–4.5 mm long; lower lip 3.5–6 mm long including lobes 1.5–3 mm long. Anticous stamens with filaments 3.7–4.7 mm long, thecae 1.2–1.7 mm long. Ovary glabrous; style with long hairs and subsessile glands in lower half. Capsule 6–7.5 mm long, glabrous; seeds 3.3–4 mm long, 2.5–3 mm wide. Fig. 47/6–12, p. 319.

TANZANIA. Mbeya District: 51 km W of Tunduma, June 1960, *Leach & Brunton* 10095! & Pungulumo [Pungaluma] Hills, May 1990, *Kayombo* 1017!; Rungwe District: Matema, Ipinda, Unyakyusa, June 1972, *Leedal* 1168!

DISTR. **T** 4, 7; SE Congo-Kinshasa, Zambia, Malawi, N Mozambique

HAB. *Brachystegia* and other miombo woodland, degraded woodland and grassland, typically on rocky hillsides and/or sandy soils, also on roadsides and as a weed of cultivation, preferring areas of short and/or sparse vegetation; usually flowering towards the end of or after the main rains, (Apr.–) May–July (–Sept.); 600–1600 m

USES. None recorded on herbarium specimens

CONSERVATION NOTES. A widespread and locally common species in the miombo woodlands of the northern Zambesian regional centre of endemism, this species appears to tolerate or even prefer some disturbance and is not considered threatened: Least Concern (LC).

SYN. *L. lanatoglabra* C.B. Clarke var. *latifolia* C.B. Clarke in F.T.A. 5: 125 (1899). Type: Malawi, Tanganyika Plateau, Chitipa [Fort Hill], *Whyte* s.n. (K!, holo.), **syn. nov.**

L. sparsiceps C.B. Clarke in F.T.A. 5: 124 (1899). Types: Malawi, Mpata and commencement of Tanganyika Plateau, *Whyte* s.n. & Nyika Plateau, *Whyte* s.n. & Manganja Hills, *Kirk* s.n. (all K!, syn.), **syn. nov.**

L. nematocephala Lindau in E.J. 30: 409 (1901). Type: Tanzania, Mbeya District, Usafwa [Usafua], Pungulumo [Bunguluma] Mt, *Goetze* 1084 (B†, holo.; BM!, iso.), **syn. nov.**

NOTE. This species is very close to *L. fischeri* and the variation may prove to be continuous between the two; they are currently separated mainly on the position of the synflorescences.

In the syntypes of *L. sparsiceps* the bracts, bracteoles and calyces are more densely long white-ciliate, particularly on the acuminate apices, than in other material of *L. lanatoglabra*. This character tends towards both *L. fischeri* and *L. eriocephala* but the many small axillary inflorescences and small, lanceolate sterile bracts are as those of *L. lanatoglabra*, hence it is synonymised here. Similar plants but with broader sterile bracts from central Malawi (e.g. *Brummitt* 9588, K!) however appear largely intermediate with *L. eriocephala* and further studies are required from that region to quantify the extent of intergrading.

L. randii S. Moore (including *L. dicomoides* Hutch.), from Zambia and Zimbabwe, differs in the plants usually being erect or decumbent with the inflorescences in the upper axils of the stem, not held at ground level. In addition, it has more dense, globose or hemispheric synflorescences with the bracteoles and calyx lobes having a more pronounced linear acumen with many long ciliate hairs; in *L. lanatoglabra* the long ciliate hairs are dense on the widened basal portion of the calyx lobes but are often sparse or largely absent on the acumen, which is mainly minutely hairy (but see note on *L. sparsiceps* above). Occasional intermediate specimens are encountered in central Zambia.

16. **Lepidagathis fischeri** *C.B. Clarke* in F.T.A. 5: 123 (1899). Types: Tanzania, Singida District, Unyamwezi, [Usuri] Ussure, *Fischer* 490 (B†, holo.); Tabora District, 23 km on Ipole to Inyanga road, *Bidgood, Leliyo & Vollesen* 7079 (K!, neo., chosen here; CAS!, DSM!, NHT!, isoneo.)

Suffrutex with prostrate, decumbent or erect leafy stems branching from a woody rhizomatous base and/or rootstock, leafy stems 5–40 cm long, rarely undeveloped during flowering; stems 4-ridged, sparsely or rarely densely pale pilose when young or glabrous throughout. Leaves (sub)sessile; blade linear to broadly (ovate-) elliptic, 1.5–13 cm long, 0.1–3 cm wide, base cuneate to rounded or subcordate, margin entire, apex acute-apiculate, glabrous or sparsely pale-pilose, rarely densely so when young; prominently 3(–5)-veined from base, then often with scalariform tertiary veins visible, sometimes only midrib prominent in linear-leaved variants. Inflorescences mainly basal and/or at the lower, usually leafless nodes, often compounded into dense cushion-like synflorescences, 3–11.5 cm in diameter, subsessile or on short peduncles with reduced sterile bracts; sterile floral bracts brown-scarious or green to purple-brown when young, lanceolate, 11–26 mm long, 2.8–6.5 mm wide, apex attenuate or acuminate, margin white-ciliate to -plumose, surface glabrous or often minutely pubescent; fertile bracts similar but often thinner, 12–23 mm long, 2.5–4 mm wide, acumen often more pronounced, long pilose hairs often also on midrib; bracteoles as fertile bracts but 2–3.3 mm wide, those on exterior of inflorescence often green with a darker acumen and with fewer hairs on the midrib. Calyx lobes green, usually with a dark purple or brown acumen, surface with short or mixed short and long ascending white hairs, margin with dense long ± spreading hairs, often shorter on acumen; anticous lobes (elliptic-) lanceolate, 10.5–18 mm long, 2–3.7 mm wide, acumen linear, ± straight; posticous lobe 12–19 mm long, 2.5–5 mm wide, with 5 or 7 parallel veins ± prominent. Corolla 11.5–18 mm long, white or pale purple, palate of lower lip with yellow or orange markings centrally and purple markings towards margin and/or in the throat; limb silky-hairy outside; tube 5.5–11 mm long; upper lip 3.5–5 mm long; lower lip 4.5–7 mm long including lobes 2.5–3.7 mm long. Anticous stamens with filaments 3.5–4.5 mm long, thecae 1.3–1.8 mm long. Ovary glabrous; style with few hairs and subsessile glands in lower half. Capsule 8–9 mm long, glabrous; seeds ± 5 mm long, 3.5 mm wide. Fig. 47/5, p. 319.

TANZANIA. Mpanda District: Katavi National Park–Sumbawanga, Nov. 2006, *Luke & Luke* 11629!; Ufipa District: 21 km on Kipili–Namanyere road, May 1997, *Bidgood et al.* 3817!; Iringa District: Mufindi, Irunda Hill, Aug. 1984, *Bridson & Congdon* 551!

DISTR. **T** 1, 4, 5, 7; SE Congo-Kinshasa, Zambia, N Malawi

HAB. *Brachystegia* and other miombo woodland and grassland including recently burnt areas, montane grassland and open forest margins, often on rocky hillsides or over laterite, also commonly on roadside banks and bare or disturbed ground with short grassland; flowering after the main rains or immediately following burning, Apr.–Aug.(–Nov.); 850–2300 m

USES. None recorded on herbarium specimens

CONSERVATION NOTES. A widespread and locally common species in the miombo woodlands and montane grasslands of the northern Zambesian regional centre of endemism, this species appears to tolerate or even prefer some disturbance and is not considered threatened: Least Concern (LC).

SYN. *L. lindauiana* De Wild. in Ann. Mus. Congo-Kinshasa, Bot., sér. 4: 145 (1903). Type: Congo-Kinshasa, Lukafu, *Verdick* 520 (BR!, holo.), **syn. nov.**

L. rogersii Turrill in K.B.: 360 (1912); J.K. Morton in J.L.S. 96: 334, 342, fig. 3 (1988) pro parte. Type: Congo-Kinshasa, Sakania, *Rogers* 10032 (K!, holo.; BM!, GRA!, NBG!, iso.), **syn. nov.**

NOTE. *L. fischeri* as circumscribed here is a rather variable species. The most widespread form has slender, linear or lanceolate leaves, less than 2 cm wide and the synflorescences are basal or crowded in the lower, leafless axils. Plants from the NW part of its range (W Tanzania, e.g. *Bidgood et al.* 6743, **T** 4) have broader, ovate to elliptic leaves over 2 cm wide and can have the inflorescences less crowded or rarely regularly spaced in the leafy axils (*Bidgood et al.* 3676). A particularly striking form from the Mufindi highlands, **T** 7 (e.g. *Bridson & Congdon* 551) also has ovate or elliptic leaves but here they are only 1.5–2.5 cm long (elsewhere usually over 4 cm long) on prostrate stems, and the large synflorescences are densely white-plumose. This form is geographically isolated from other populations of *L. fischeri* and perhaps warrants subspecific status.

The original type specimen of *L. fischeri*, destroyed in Berlin during World War II, was collected from **T** 5. No material of this taxon (nor any material from group C of *Lepidagathis*) has since been collected from that region. However, the description agrees with this taxon as currently circumscribed and a neotype is therefore selected from SW Tanzania.

17. **Lepidagathis sp. A** (= *Richards* 15115)

Erect, decumbent or spreading suffruticose herb; stems somewhat 4-ridged when young, soon terete, glabrous or pale yellow-pilose. Leaves sessile; blade ovate, lanceolate or oblong-elliptic, 5.5–7 cm long, 1.3–2.5 cm wide, base obtuse to subcordate, margin entire, apex acute-apiculate, glabrous or pilose on margin when young; prominently 3-veined from base and with scalariform tertiary venation. Inflorescences axillary, restricted to upper parts of stems or more widespread, cymes fasciculate, clustered into globose heads 3–5 cm in diameter when mature, sessile; sterile bracts at first green but turning brown-scarious, lanceolate, 18–24 mm long, 3–5 mm wide, apex attenuate, densely yellow-pilose at least on margin; fertile bracts and bracteoles similar but green with prominent pale midrib, narrowly lanceolate, 18–22 mm long, 1.5–3 mm wide, attenuate into a linear acumen, yellow-pilose throughout. Calyx lobes green throughout, ± densely long yellow-pilose; anticous lobes linear-lanceolate, 15–19 mm long, ± 2.5 mm wide, acumen linear; posticous lobe lanceolate, 3.5–4.5 mm wide, with 3 or 5 prominent parallel veins. Corolla 13.5–15 mm long, white or cream, sometimes green-veined, palate of lower lip with orange markings centrally and purple markings towards margin; limb silky-hairy outside; tube 8–10 mm long; upper lip 3.5–5.5 mm long; lower lip 4.5–5.5 mm long including lobes 2.5–3 mm long. Anticous (longer) stamens with filaments 4–4.5 mm long, thecae ± 1.5 mm long. Ovary with few sessile glands towards apex; style with long hairs in lower half. Capsule and seeds not seen.

TANZANIA. Ufipa District: path [from] Chaputuka village, Kalambo Falls to Timber Camp, Feb. 1965, *Richards* 19694!

DISTR. **T** 4; NE Zambia

HAB. Amongst grass in miombo woodland over gritty soils; flowering towards /at the end of the main rains, Apr.–May; ± 1300 m

USES. None recorded on herbarium specimens

CONSERVATION NOTES. This taxon is currently known from only three collections within a very narrow range along the Tanzania-Zambia border; it may prove to be threatened but is currently assessed as Data Deficient (DD).

NOTE. The single specimen of this taxon from our region is immature; however mature flowering material is available from just over the border in Zambia (*Richards* 15115, 24484, both K!). It appears quite distinct from all other taxa within the *L. andersoniana* complex due to its yellow inflorescence indumentum and long, slender calyx lobes which are green throughout. However, *Richards* 6346 from the nearby Kawa River, Ufipa District, appears intermediate between this and the broad-leaved variant of *L. fischeri* found in adjacent areas of SW Tanzania. This taxon may therefore prove to be an extreme variant of that species, but further studies of the populations from the Tanzania/Zambia border region are required to reach any further conclusions.

18. **Lepidagathis sp. B** (= *Richards & Arasululu* 26234)

Pyrophytic herb, leafy stems usually developing during or after flowering, erect or decumbent from a woody rootstock, this sometimes creeping; stems 4-ridged, white-pilose or largely glabrous. Leaves subsessile, often immature or absent at flowering; blade linear to elliptic, 3–10 cm long, 0.3–1.5 cm wide, base cuneate, margin entire, apex acute-apiculate, glabrous or white-pilose; 3-veined from base. Inflorescences basal, cymes fasciculate or shortly spiciform, to 4 cm long, usually clustered into cushions 2.5–8 cm in diameter, subsessile or shortly pedunculate, peduncle with triangular scale-like sterile bracts; sterile floral bracts often inconspicuous, brown-scarious, triangular to lanceolate, 4.5–20 mm long, 1.5–4.5 mm wide, apex acute or attenuate, white-pilose at least on margin; fertile bracts similar but pale-hyaline or green when young, sometimes scarious with age, densely pilose; bracteoles green throughout or one of a pair hyaline and with brown apex, (linear-) lanceolate, elliptic or oblanceolate, 5.5–20 mm long, 1–4 mm wide, apex acute or shortly acuminate, surface densely white-pilose. Calyx lobes green, with or often without a short brown acumen, surface densely white-pilose and with minute subsessile glands; anticous lobes lanceolate, elliptic or oblanceolate, 8.5–21 mm long, 1.8–4 mm wide; posticous lobe ovate (-elliptic) to lanceolate, 2.3–4.5 mm wide, with 5 or 7 parallel veins ± prominent. Corolla 13–16 mm long, white, palate of lower lip and/or throat with yellow markings centrally and brown or purple markings towards margin, limb silky-hairy outside; tube 8–9.5 mm long; upper lip 3.5–5 mm long; lower lip 4.5–7 mm long including lobes 2.5–3.5 mm long. Anticous (longer) stamens with filaments 3–5 mm long, thecae 1.4–2.1 mm long. Ovary glabrous; style with long hairs and subsessile glands in lower half. Capsule ± 7.5 mm long, glabrous; seeds not seen.

TANZANIA. Buha District: 25 km on Kasulu–Kibondo road, June 2000, *Bidgood, Leliyo & Vollesen* 4746!; Mpanda District, near turning off main road to Lake Katabe, 13 km from Sitalike, *Richards & Arasululu* 26234!; Iringa District: 85 km from Iringa on Mbeya road, Aug. 1966, *Gillett* 17386!

DISTR. **T** 4, 7; SE Congo-Kinshasa, NW Zambia

HAB. Fire-prone *Brachystegia-Isoberlinia* woodland and grassland, usually flowering soon after burning, (June–) Aug–Oct (–Nov); 1000–2000 m

USES. None recorded on herbarium specimens

CONSERVATION NOTES. This taxon occurs in two disjunct areas: SW Tanzania and the border region of S Congo and NW Zambia. It is perhaps under-recorded due to its close association with burning events. Threats to its habitat seem limited, although cessation of regular burning would almost certainly have a negative impact. In view of its uncertain taxonomic status it is currently assessed as Data Deficient (DD).

SYN. [*L. rogersii sensu* J.K. Morton in J.L.S. 96: 334, 342 (1988) pro parte, *non* Turrill]

NOTE. Species B differs from *L. fischeri* in the bracteoles and calyces largely lacking a linear acumen and being green throughout and usually having smaller, inconspicuous sterile bracts which leave the calyces clearly exposed. The calyx lobes are also more evenly long-pilose. The resultant inflorescence appears very distinctive, with a grey-green appearance; whilst in *L. fischeri* the inflorescence typically appears brown or purplish due to the more conspicuous, soon-scarious bracts and the darker calyx apices. In addition, this taxon is often more strictly pyrophytic than *L. fischeri*, although *Bidgood et al.* 7376 from the Tabora–Bukoma road, **T** 4 was collected from rocky hillside with no evidence of recent burning (K. Vollesen *pers. comm.*).

Within our region, this taxon appears not to intergrade with *L. fischeri*. Indeed, in the Mufindi Highlands there are clearly two taxa involved which are very easily separable (see also note to *L. fischeri*). However, in the many collections of this complex from S Congo held at BR, *sp. B* appears to intergrade with typical *L. fischeri*.

33. BARLERIA

L., Sp. Pl.: 636 (1753); Nees in DC., Prodr. 11: 223 (1847); C.B. Clarke in F.T.A. 5: 140 (1899); Obermeijer in Ann. Transv. Mus. 15: 127 (1933); M.-J. Balkwill & K. Balkwill in Proc. XIVth AETFAT congress, Wageningen: 393–408 (1996) & in K.B. 52: 551 (1997) & in J. Biogeogr. 25: 95–110 (1998); Darbyshire in K.B. 63: 601–611 (2009)

Perennial herbs, subshrubs or climbers, evergreen or deciduous. Stems 4-angular or subterete; cystoliths ± conspicuous, many, often occuring in 2 adjacent cells and sometimes appearing to cross. Leaves opposite-decussate, pairs equal or somewhat anisophyllous. Inflorescences simple and axillary or compounded into a terminal synflorescence; cymose with monochasial and/or dichasial branching or reduced to a single flower; bract pairs (sub-)equal, foliaceous or reduced and/or highly modified; bracteole pairs equal or dimorphic, often highly modified from the bracts. Calyx divided almost to the base, 4-lobed, anticous and posticous (outer) lobes usually considerably larger and proportionally broader than the lateral (inner) lobes, anticous lobe often emarginate or bifid. Corolla usually showy; tube cylindrical throughout or campanulate to funnel-shaped above attachment point of stamens; limb 5-lobed or rarely the adaxial pair fully fused, subregular to zygomorphic (see note). Fertile stamens (in our region) 2, abaxial; filaments arising from various levels within the corolla tube, twisted through 180° and crossing near the base; anthers bithecous, thecae parallel, muticous, exserted or rarely included within the corolla tube; staminodes 2 or 3, adaxial staminode often less well developed than the lateral pair or absent, with or without antherodes, these often producing a few pollen grains. Disk cupular. Ovary 2-locular, 2 ovules per locule or one aborting very early; style filiform; stigma subcapitate to linear, either 2-lobed, lobes then sometimes subconfluent, or with 1 lobe much-reduced or absent. Capsule ± compressed laterally, fertile portion rounded to fusiform, with or without a prominent sterile apical beak, placental base inelastic but lateral walls sometimes thin and tearing from the thickened flanks at dehiscence. Seeds 2 or 4 per capsule, held on retinacula, usually (ovate-) discoid and with surfaces covered in hygroscopic hairs, rarely only subflattened and with unevenly distributed minute hairs.

A genus of ± 300 species, mainly in the palaeotropics and subtropics with one species in the neotropics; most diverse in eastern and southern Africa. Many of the species are very striking in flower, yet surprisingly few are widely cultivated; Jex-Blake (Gardening in E. Afr., 4th ed., 1957) noted "the genus is sadly liable to disease" (p. 104).

The infrageneric classification proposed by M.-J. & K. Balkwill (in K.B. 52: 551 (1997)) is followed here with minor modifications; the division of the seven sections into two subgenera is not, however, deemed worthwhile. Species are keyed out separately within each section. Sections in *Barleria* are easily separated when mature fruits are available and the key to sections for fruiting material should be used where possible. However, the genus is more often observed and collected when in flower due to its showy corollas and to the persistence of the calyx lobes which often hide the fruit. A key to sections for flowering material is therefore also presented. As flowering material is often required for identification to species, ideal collections would comprise both flowers and fruits.

NOTES ON THE DESCRIPTIONS

- Calyces – the shape and indumentum of the posticous lobe is only described if differing from that of the anticous lobe.
- Corollas – the corolla limb is rather variable in *Barleria* and can be divided into several arrangements:
  - Subregular – the lobes of the corolla are divided from the tube at ± the same point and are subequal in size, although the adaxial pair are often slightly narrower and the abaxial lobe is often slightly broader than the lateral pair (e.g. fig. 57/1, 6).
  - "2+3" arrangement – the adaxial pair of lobes are partially fused, forming a 2-lobed upper lip; the abaxial and lateral lobes form a 3-lobed lower lip (e.g. fig. 54/8). In some cases in sect. *Barleria* there is an arrangement inbetween this and "subregular" in which the adaxial pair of lobes are not partially fused but differ sufficiently from the lateral and abaxial lobes to form a weakly defined 2-lobed upper lip (e.g. fig. 48/1, p. 338)
  - "1+3" arrangement – an extreme form of the "2+3" arrangement in which the adaxial pair of lobes fully fuse
  - "4+1" arrangement – the abaxial corolla lobe splits away from the tube considerably earlier than the lateral and adaxial lobes, the resultant limb being highly zygomorphic with a 1-lobed lower lip and 4-lobed upper lip (e.g. fig. 55/4 & 10, p. 410; fig. 56/1 & 3, p. 428). This is the most common arrangement encountered in the Flora region.
  - "*Cavirostrata*-type" – in species of sect. *Cavirostrata* from our region, the lobes split from the tube at ± the same point but the two adaxial lobes are widely divergent to form a highly zygomorphic limb (fig. 53/9, p. 388)
- The length of the corolla tube is measured from the base to the point at which the first corolla lobe splits; in species with the "4+1" arrangement, the length by

which the abaxial lobe is offset from the remaining lobes (this being an "open tube") is also recorded. The length of the corolla is measured from the base of the tube to the apex of the lateral lobes when flattened.

- Androecium – in our region, the lateral staminodes are much reduced in comparison to the fertile pair of stamens, but they often bear antherodes which produce some pollen grains. This has led some authors to treat them as stamens. Therefore, although they are always referred to as staminodes in the descriptions and section/species keys here, the key to genera (Vollesen, F.T.E.A. Acanthaceae part 1: 2–11) includes *Barleria* in both leads to couplet 8 which separates genera with 2 and 4 fertile stamens. In very rare cases, abnormal flowers occur in which all 5 stamens are developed, for example in *Verdcourt* 487 (*Barleria ventricosa* Nees aggregate). In some species from outside our region, four fertile stamens are present (*fide* M.-J. & K. Balkwill in K.B. 52: 539 (1997)).

KEY TO SECTIONS OF *BARLERIA*: FRUITS PRESENT

1. Seeds not or only partially flattened, surfaces black with unevenly distributed minute pale grey hairs, glabrescent; capsule potentially 4-seeded*, prominently beaked .... sect. IV. **Cavirostrata** p.385
   Seeds discoid (surface often ovate), clothed in cream, golden, bronze or purple-brown hygroscopic hairs; capsule 2-seeded or if 4-seeded then lacking a prominent beak 2
2. Capsule potentially 4-seeded, lacking a prominent beak, septum woody except for a shallow membranous portion above the upper retinacula .... sect. I. **Barleria** p.329
   Capsule 2-seeded, septum largely membranous or if largely or wholly woody then capsule with a prominent beak 3
3. Capsule with septum largely membranous, apex lacking a prominent beak or if rarely beaked then stem and leaf indumentum stellate or dendritic; walls of capsule often tearing from thickened flanks at dehiscence 4
   Capsule with septum woody throughout or with a shallow membranous portion above the retinacula, apex prominently beak, stem and leaf indumentum never stellate or dendritic; walls of capsule not tearing from flanks 5
4. Indumentum of simple hairs; stigma subcapitate or clavate .... sect. II. **Fissimura** p. 355
   Indumentum of predominantly stellate or dendritic hairs, often with a long central "arm" (fig. 52/4); stigma linear .... sect. III. **Stellatohirta** p.379
5. Capsule with a shallow membranous portion above the retinacula; plants unarmed; leaves usually lacking sunken glands beneath .... sect. V. **Somalia** p.387
   Capsule septum woody throughout; plants usually spiny, more rarely unarmed; leaves with sunken glands visible beneath .... sect. VI. **Prionitis** p.415

* One or more seeds often abort in all sections but it is easy to determine the potential number of seeds by checking for reduced retinacula and for comparing the size of the mature seeds to the size of the locule.

KEY TO SECTIONS OF *BARLERIA*: FLOWERS ONLY PRESENT

1. Corolla limb with the sinus between the two adaxial lobes at a markedly wider angle than the other sinuses (fig. 53/9); outer calyx lobes with only the palmate or parallel principal veins conspicuous, margin entire . . . . sect. IV. **Cavirostrata** p.385
   Corolla limb with the sinus between the two adaxial lobes not at a notably wider angle than the other sinuses, or if somewhat so then outer calyx lobes with conspicuous palmate-reticulate venation and/or margin toothed or spinose 2
2. Indumentum stellate or dendritic, hairs often with a long central "arm" (fig. 52/4); outer calyx lobes with only the palmate or subparallel principal veins conspicuous; flowers held in well-defined globose to cylindrical synflorescences with each axil single-flowered . . . . sect. III. **Stellatohirta** p.379
   Indumentum of simple and/or medifixed (biramous) hairs, or if stellate then outer calyx lobes with conspicuous palmate-reticulate venation and flowers held in unilateral (rarely single-flowered) axillary cymes sometimes clustered towards stem apices 3
3. Plants armed with 2–4(–8)-rayed axillary spines, prominently stalked or subsessile (fig. 57/2, p. 437), clearly differentiated from the bracteoles, or if these absent then *either* corolla yellow, orange or apricot and outer calyx lobes lanceolate, ovate or oblong-elliptic, up to 8 mm wide, base never cordate, *or* if corolla white or cream then salver-shaped and calyx pale grey-green or whitish (76. *B. marginata*); leaves with minute sunken glands visible beneath . . . . sect. VI. **Prionitis** p.415
   Plants unarmed or if axillary spines present then these paired, being the persistent spinose bracteoles of aborted or old inflorescences, simple or with pinnate lateral spines, not rayed; corolla white, pink, blue, purple or red, or if yellow (43. *B. calophylloides*) then outer calyx lobes broadly cordiform, 14.5–30 mm wide; if corolla white and salver-shaped then calyx not as above; leaves usually lacking sunken glands beneath 4
4. Plants with bracteoles spinose or spine-tipped, those of old or aborted inflorescences often persisting as paired axillary spines, or if unarmed then *either* plants of coastal bushland with corolla bright red to rose-pink (18. *B. repens*) *or* corolla white, salver-shaped, held in axillary unilateral (rarely single-flowered) cymes *or* if characters otherwise as sect. *Fissimura* (see below) then stigma linear (1. *B. holstii*) . . . . sect. I. **Barleria** p.329
   Plants unarmed; combinations of characters not as above 5

5. Corolla usually drying blue with darker venation or blue-black throughout; tube abruptly, rarely gradually, funnel-shaped or campanulate above the attachment point of the stamens, cylindrical below; limb in "4+1" arrangement; stigma subcapitate or clavate; lateral staminodes usually with antherodes .... sect. II. **Fissimura** p.355
   Corolla not drying as above; tube cylindrical, sometimes narrowly widened towards the mouth but never funnel-shaped or campanulate above the attachment point of the stamens; limb arrangement variable; stigma linear; all staminodes lacking antherodes .... sect. V. **Somalia** p.387

SECT. I. BARLERIA

M.-J. Balkwill & K. Balkwill in K.B. 52: 555 (1997)

Wahabia Fenzl in Flora 27: 312 (1844)
Barleriacanthus Oerst. in Vidensk. Medd. Dansk. Naturhist. Foren. Kjobenh.: 136 (1854)
Barleriosiphon Oerst. in Vidensk. Medd. Dansk. Naturhist. Foren. Kjobenh.: 136 (1854)
Dicranacanthus Oerst. in Vidensk. Medd. Dansk. Naturhist. Foren. Kjobenh.: 136 (1854)
Barleria subgen. *Acanthoidea sensu* C.B. Clarke in F.T.A. 5: 141 (1899) pro parte excl. *B. flava*
Barleria sect. *Eubarleria* subsects. *Pungentes, Aculeatae, Innocuae* and *Heterotrichae sensu* Obermeijer in Ann. Transv. Mus. 15: 133–138 (1933)
Barleria sect. *Chrysothrix* M. Balkwill in J. Biogeogr. 25: 110 (1998) pro parte (see note); M. & K. Balkwill in K.B. 52: 558 (1997), *idem*

Plants unarmed or often with the spinose bracteoles of aborted or old inflorescences persisting as paired axillary spines. Indumentum simple or stellate. Leaves often with broad sessile glands towards base beneath, sometimes also on bracteoles and calyx. Inflorescences axillary, unilateral (scorpioid) or single-flowered cymes; bracts foliaceous. Calyx often scarious, outer lobes with conspicuous palmate-reticulate venation. Corolla white, blue, purple or red, often drying blue with darker venation or blue-black throughout; limb in "4+1", "2+3" or subregular arrangement. Staminodes 3, lateral pair with antherodes well-developed or absent, adaxial staminode with antherode often absent. Stigma linear or subcapitate, apex ± bilobed. Capsule usually drying black, laterally flattened, oblong-fusiform, without a prominent beak; lateral walls usually remaining attached to flanks at dehiscence; septum with a shallow membranous portion above upper retinacula, elsewhere woody. Seeds 4 (or fewer by abortion), discoid with dense matted hygroscopic hairs, often purplish, brown or bronze.

A section of ± 100 species, most diverse in S and SE Africa. The taxonomy in our region is complicated by the presence of several species complexes (notably *B. grandicalyx-B. ramulosa, B. crassa-B. nyasensis, B. paolii* complex and *B. acanthoides-B. homoiotricha*).

B. holstii and *B. whytei* have previously been placed within sect. *Chrysothrix* which M. Balkwill separates from sect. *Barleria* by the presence of appressed yellow hairs on the stem and a compound terminal synflorescence (not an axillary scorpioid or 1-flowered cyme). However, many species within sect. *Barleria* have a similar stem indumentum, though often interspersed within a shorter, white indumentum, and this character appears to hold little taxonomic value. With regard to inflorescence form, the two East African taxa fall within a group of species, including the Indian *B. strigosa* Willd. and *B. nitida* Nees, in which the inflorescences are clearly scorpioid, and, though sometimes clustered towards the stem apex, are axillary and not compounded into a terminal synflorescence. They are therefore inseparable from the inflorescence form of sect. *Barleria* and are included within this section here.

1. Indumentum of simple hairs only 2
   Indumentum including at least some long-armed stellate hairs on the leaves, stems and/or calyces (check young parts as stellate base sometimes caducous) 19
2. Bracteoles ovate, elliptic or lanceolate, pairs often unequal in size, the larger bracteole 3–16 mm wide at the widest point excluding, if present, lateral spines or teeth 3
   Bracteoles linear (-lanceolate), oblanceolate or reduced to spines, less than or up to 3 mm wide at the widest point excluding, if present, lateral spines or teeth 7
3. Leaves elliptic or subpandurate, 6.5–18 cm long, apex unarmed; stems with mainly coarse yellowish or buff-coloured ascending or appressed hairs; shorter spreading or curved hairs sparse or absent 4
   Leaves obovate to (oblong-) elliptic, 0.9–5.5 cm long, apex usually mucronate or spinose; stems with dense short white retrorse, antrorse or spreading hairs, with or without interspersed coarse yellowish hairs 5
4. Leaf base attenuate or cuneate; corolla tube infundibuliform with stamens attached in the lower half . . . . . . 1. *B. holstii* p.332
   Leaf base abruptly cordate, amplexicaul when young; corolla tube cylindrical, barely expanded towards the mouth, with stamens attached in the upper half . . . . . . 2. *B. whytei* p.333
5. Corolla white or cream (drying blue-black), tube over 40 mm long, narrowly cylindrical throughout . . . . . 16. *B. inclusa* p.352
   Corolla pale blue or mauve, rarely whitish, tube up to 40 mm long, somewhat widened towards the mouth 6
6. Outer calyx lobes with margin subentire or with short teeth, not coarsely ciliate, submarginal spines usually present; corolla lobes 7–11 mm long, very short in comparison to the 33–40 mm long tube . . 14. *B. mucronifolia* p.350
   Outer calyx lobes with margin laciniate and coarsely ciliate, submarginal spines absent; corolla lobes 13–19 mm long, not so disproportionately shorter than the 25–34 mm long tube . . . . . . 15. *B. steudneri* p.351
7. Bracteoles and (if present) axillary spines simple, margin entire 8
   Bracteoles and axillary spines with 1–several lateral spines or teeth, pinnately arranged or on one (inner) margin only 14
8. Corolla bright red to rose-pink; bracteoles unarmed 18. *B. repens* p.354
   Corolla variously white, blue, mauve or purple; bracteoles spine-tipped 9
9. Corolla in "4+1" arrangement with the abaxial lobe offset by 4–8.5 mm; leaf apex at most minutely apiculate, not spinose or mucronate; naturalised ornamental . . . . . . 17. *B. cristata* p.353
   Corolla in "2+3" arrangement or (?)subregular*; leaf apex spinose or mucronate; native species 10

10. Flowers held in several-flowered unilateral cymes; anthers ± 7 mm long .... 7. *B. blepharoides* p.342
    Flowers solitary or more rarely in 2–3-flowered unilateral cymes (then usually only 1 flower mature at a given time); anthers 2.5–5.5 mm long 11
11. Stamens attached in upper half of corolla tube, this cylindrical throughout or narrowly expanded towards mouth; stems with mixed longer coarse ascending or spreading hairs and shorter finer spreading to retrorse hairs of variable density 12
    Stamens attached in lower half of or midway along corolla tube, this funnel-shaped above the attachment point; stems with many appressed or strongly ascending buff-coloured or yellowish strigose hairs 13
12. Suffruticose perennial of fire-prone habitats, producing several annual branches from a woody rootstock, usually few-branched above, not very spiny; leaves 1.7–6.7 cm long; calyx somewhat to strongly accrescent in fruit, outer calyx lobes then to 34 mm long .... 3. *B. grandicalyx* p.334
    Much-branched (sub)shrub, very spiny between short internodes; leaves 0.6–2.7 cm long; calyx only weakly accrescent in fruit, outer lobes to 19 mm long .... 4. *B. ramulosa* p.336
13. Leaf apex obtuse or rounded, shortly apiculate; corolla 30–36 mm long including tube up to 20 mm long; lateral calyx lobes 3–5.5 mm long .... 5. *B. crassa* p.339
    Most leaves with apex acute and conspicuously mucronate; corolla 40–50 mm long, including tube over 25 mm long; lateral calyx lobes 7.5–13 mm long .... 6. *B. nyasensis* p.340
14. Bracteoles and axillary spines highly asymmetric, with 1–3(–4) lateral spines on one (inner) margin only, the outer margin entire or with minute teeth (fig. 49/2, 6, p. 346) 15
    Bracteoles and axillary spines with lateral spines or teeth more evenly distributed, 1–5 per side, opposite or alternate 16
15. Outer calyx lobes broadly ovate or suborbicular, 6.5–16 mm wide; corolla tube 27–45 mm long, narrowly cylindrical throughout, limb subregular, abaxial lobe not offset .... 10. *B. gracilispina* p.345
    Outer calyx lobes lanceolate to narrowly oblong-elliptic, 0.8–3(–4.5) mm wide; corolla tube 12–21 mm long, narrowly widened towards the mouth; limb in "4+1" arrangement with abaxial lobe offset by up to 2.5 mm .... 11. *B. paolioides* p.347
16. Leaf apex at most minutely apiculate, not spinose; axillary spines absent; corolla limb in "4+1" arrangement, tube campanulate above attachment point of stamens; naturalised ornamental .... 17. *B. cristata* p.353
    Leaf apex mucronate or spinose; axillary spines present; corolla limb subregular or if in "4+1" arrangement then corolla tube cylindrical and only narrowly widened towards the mouth; native species 17

17. Corolla tube 18–27 mm long, narrowly campanulate in upper half; flowers always solitary at each axil; outer calyx lobes 2–6 mm wide excluding marginal spines . . . . 12. *B. delamerei* p.348
    Corolla tube 33–95 mm long, narrowly cylindrical throughout or somewhat widened towards mouth; flowers solitary or in 3–15-flowered unilateral cymes; outer calyx lobes (3–)5–17 mm wide excluding marginal teeth or spines 18
18. Corolla white or cream, tube over 50 mm long, narrowly cylindrical throughout; outer calyx lobes lacking submarginal spines (but margin sometimes spinose); stems usually with sparse to dense glandular hairs, rarely absent . . . . 13. *B. acanthoides* p.349
    Corolla usually pale blue or mauve, tube up to 40 mm long, cylindrical but somewhat widened towards mouth; outer calyx lobes often with submarginal spines; stem indumentum eglandular only . . . . 14. *B. mucronifolia* p.350
19. Outer calyx lobes with margin entire or shortly toothed, not spinulose, sometimes involute, outer surfaces lacking crisped eglandular hairs towards base; stellate hairs on leaves buff-golden to golden-brown; old stems never floccose . . . . . . 8. *B. fulvostellata* p.342
    Outer calyx lobes with coarse spinulose teeth along margin, never involute, outer surfaces with crisped eglandular hairs ± many towards base; stellate hairs on leaves white-grey; old stems sometimes floccose 9. *B. spinulosa* p.343

* Limb arrangement unknown in 7. *B. blepharioides* but the protologue makes no mention of the abaxial lobe being offset from the lateral and adaxial lobes

1. **Barleria holstii** *Lindau* in E.J. 20: 19 (1894); C.B. Clarke in F.T.A. 5: 166 (1899); Vollesen in Opera Bot. 59: 79 (1980); Makholela in Figueiredo & Smith, Pl. Angola, Strelitzia 22: 21 (2008). Type: Tanzania, Lushoto District, Mashewa [Mascheua], *Holst* 3516 (B†, holo.; K!, M!, W!, iso.)

Perennial herb or subshrub, 50–200 cm tall; stems sparsely to densely buff- or yellow-strigose, hair bases becoming somewhat swollen with age and persisting on the mature woody stems. Axillary spines absent. Leaves subsessile, elliptic, 6.5–16 cm long, 2.5–5.5 cm wide, base attenuate or cuneate, apex attenuate or acuminate, surfaces strigose, especially on margin and veins beneath; lateral veins 5–6 pairs. Inflorescences unilateral, (1–)2–8-flowered, congested, to 2 cm long; bracteole pairs unequal and offset, the larger ovate to elliptic, often asymmetric, 12–22 mm long, (3–)6–12 mm wide, the smaller proportionally narrower, more strongly asymmetric and declinate, each with margin denticulate, teeth narrow and with an apical bristle, surfaces pale green or tinged purple. Anticous calyx lobe ovate or elliptic, 12–20 mm long, 7–12.5 mm wide, base obtuse or rounded, margin as bracteoles, apex emarginate or truncate, surface strigose, venation usually darker than surface; posticuous lobe broadly ovate, 17–30 mm long, 10.5–19 mm wide, base rounded, apex acute or subattenuate; lateral lobes lanceolate, 3.5–6 mm long. Corolla white, pale blue or pale purple, 30–43 mm long, pubescent outside with spreading eglandular and glandular hairs; tube 19–23 mm long, funnel-shaped above attachment point of stamens; limb in "4+1" arrangement; abaxial lobe offset by 3–5 mm, 12–22 mm long, 8.5–18 mm wide; lateral and adaxial lobes 9.5–18 mm long, 6.5–12.5 mm wide. Stamens attached 7–9.5 mm from base of corolla tube; filaments 18–29 mm long; anthers 2.5–3.2 mm long; lateral staminodes 1–3 mm long, pubescent, antherodes 0.5–0.9 mm long. Pistil glabrous; stigma linear, 1–2 mm long. Capsule 13.5–15 mm long, glabrous; seeds 5.5–6 mm long and wide.

TANZANIA. Lushoto District: lower slopes of Mavumbi peak, NE of Mashewa, July 1953, *Drummond & Hemsley* 3204!; Morogoro District: ± 10 km S of Matambwe, Apr. 1979, *Vollesen* in MRC 3723!; Lindi District: Rondo Plateau, June 1995, *Clarke* 3!

DISTR. **T** 3, 6–8; Angola, W Zambia, Malawi

HAB. Lowland riverine and coastal forest and thicket; 100–900(–1300) m

USES. "Root used as purgative" (**T** 3; *Koritschoner* 951)

CONSERVATION NOTES. This species is locally common in lowland forest and along seasonal rivers in east Tanzania, with isolated sites in Zambia, Malawi and Angola (the latter *fide* Makholela, l.c.). Although this habitat is often threatened by human encroachment, several sites for this species are within protected areas, e.g. Mikumi and Selous National Parks, and are therefore afforded some protection. However, populations in the more heavily populated regions at the northern end of this species' range, particularly in the foothills of the Usambara Mts, may be under threat or lost. It is therefore provisionally assessed as Near Threatened (NT).

2. **Barleria whytei** *S. Moore* in J.B. 41: 138 (1903); M. & K. Balkwill in K.B. 52: 558, fig. 10F (1997). Type: Kenya, Kwale District, S of Mombasa, *Whyte* s.n. (BM!, holo.; K!, ?iso.)

Perennial herb or subshrub, 40–75 cm tall; stems buff-yellow strigose when young, hair bases becoming somewhat swollen with age and persisting on mature woody stems. Axillary spines absent. Leaves subsessile, blade elliptic or subpandurate, 7.5–18 cm long, 2.5–7 cm wide, base abruptly cordate, somewhat amplexicaul at least when young, apex acuminate, surfaces sparsely strigose at least on margin and veins beneath; lateral veins 4–6 pairs. Inflorescences often clustered towards stem apices, unilateral, 1–8-flowered, congested, to 4.5 cm long; bracteoles paler than leaves at least towards base, ovate, pairs unequal and offset, the larger 10–17 mm long, 4–11.5 mm wide, margin entire. Anticous calyx lobe broadly ovate, 14–20 mm long, 8–15 mm wide, base rounded or subcordate, margin entire or obscurely denticulate, apex acute, obtuse or subattenuate, surface paler towards base, strigose particularly on margin and principal veins, hairs on margin sometimes with a swollen base; posticuous lobe 16–29 mm long, 10–20 mm wide; lateral lobes lanceolate, 4.5–8 mm long. Corolla white, salver-shaped, 32–41 mm long, pubescent outside with spreading eglandular and glandular hairs and occasional strigose hairs; tube narrowly cylindrical, 22–30 mm long, barely expanded towards mouth; limb subregular, widely spreading; abaxial lobe 8.5–10.5 mm long, 8–11 mm wide; lateral pair as abaxial lobe but 5–7.5 mm wide; adaxial pair 5–6 mm wide. Stamens attached in upper half of corolla tube; filaments 11.5–17 mm long, glabrous; anthers 2–2.5 mm long; lateral staminodes 1–2.5 mm long, pubescent towards base, antherodes ± 0.8 mm long. Pistil glabrous; stigma linear, 0.6–1 mm long. Capsule 12–16 mm long, glabrous; seeds 4–6 mm long and wide.

KENYA. Kwale District: Shimoni, Aug. 1953, *Drummond & Hemsley* 3923! & Kaya Bogowa, Oct. 1991, *Luke* 2907! & Shimoni, May 1999, *Luke & Saidi* 5829!

TANZANIA. Rufiji District: Mafia Island, S coast, Aug. 1932, *Schlieben* 2676! & E seaboard of Mafia Island, Oct. 1990, *Frontier Tanzania* 1425!

DISTR. **K** 7; **T** 6; not known elsewhere

HAB. Coastal forest on coral rag; 0–10 m

USES. None recorded on herbarium specimens

CONSERVATION NOTES. This species is recorded from two small and disjunct areas: the SE Kenyan coast and Mafia Island off Tanzania; in both areas it is confined to the highly specialised low-canopy coastal forests on raised coral deposits. It was recorded as locally common at Shimoni (the probable type location) by Quentin Luke as recently as the late 1990s and was also noted as plentiful on Mafia Island by Schlieben in the 1930s. However, its environment is threatened at least in Kenya through tourism-related construction, hence this species is considered Endangered (EN B2ab(iii)). Interestingly, this species has been cultivated in Malindi, **K** 7 by Ann Robertson with much success.

NOTE. *Schlieben* 2676 (B!, BM!, BR!) is annotated "*Barleria mafiensis* Mildbr." but this name appears never to have been published and would, in any case, be a later synonym of *B. whytei*.

3. **Barleria grandicalyx** *Lindau* in E.J. 20: 25 (1894); C.B. Clarke in F.T.A. 5: 149 (1899); F.P.S. 3: 169 (1956); U.K.W.F. ed. 1: 592 (fig.), 593 (1974); Champluvier in Fl. Rwanda, Sperm. 3: 438, fig. 137.3 (1985); U.K.W.F. ed. 2: 272, t. 118 (1994); Ensermu in Fl. Eth. 5: 408 (2006). Type: Sudan, Gir [Gin], *Schweinfurth* 2512 (K!, lecto., chosen here; P!, W!, isolecto.)

Erect to trailing suffruticose perennial, several-branched from a woody rootstock, 10–100 cm tall; stems with coarse spreading or ascending hairs and short paler spreading or retrorse hairs, rarely also glandular-pubescent. Axillary spines simple, paired. Leaves (ovate-) elliptic or obovate-elliptic, 1.7–6.7 cm long, 0.6–3 cm wide, base cuneate to obtuse, margin sometimes revolute when young, apex acute to rounded or shortly attenuate, spinulose, surfaces with sparse to dense spreading to subappressed hairs, veins beneath often strigose, usually also with short crisped or patent hairs along margin, midrib and margin rarely glandular-pubescent; lateral veins 3–6 pairs; petiole to 9 mm long or absent. Flowers solitary or rarely in 2–3-flowered unilateral cymes; bracteoles spinose, 4–20 mm long, to 1(–2) mm wide, margin entire; flowers sessile or pedicels to 3.5 mm long. Calyx ± strongly accrescent; outer lobes subequal, (oblong-) ovate or elliptic to somewhat obovate, 10–21 mm long, 4–13 mm wide in flower, up to 34 mm long and 23 mm wide in fruit, base rounded or cordate, more rarely obtuse or cuneate, margin subentire or with spinulose teeth, apex spinose, anticous lobe sometimes bifidly so, surfaces strigulose, usually with short pale crisped or patent hairs on margin and scattered glandular hairs; lateral lobes linear-lanceolate, 6–10.5 mm long. Corolla white, blue or purple, 23–43 mm long, pubescent or pilose outside, hairs mixed eglandular and glandular or mainly the latter; tube 11.5–29 mm long, barely expanded above attachment point of stamens, declinate below; limb in weak "2+3" arrangement; abaxial lobe 7–21 mm long, 4.5–12.5 mm wide; lateral lobes as abaxial but 6–14 mm wide; adaxial lobes 7–18 mm long, 4.5–10 mm wide. Stamens attached 7.5–19 mm from base of corolla tube; filaments 9–18 mm long; anthers exserted, 2.5–4.5(–5.5) mm long; lateral staminodes 1.5–4 mm long, pilose, antherodes barely developed or to 1.2(–2) mm long. Ovary glabrous or with ascending hairs towards apex; style largely glabrous; stigma linear or clavate, often curved, 0.7–2 mm long. Capsule 13.5–24 mm long, indumentum as ovary; seeds 5–8 mm long and wide.

DISTR. (for species) **U** 1, 3; **K** 2–6; **T** 1–3, 5; Central African Republic, Congo-Kinshasa, Rwanda, Burundi, Sudan, Ethiopia

CONSERVATION NOTES. This species is locally common and widespread in eastern and central Africa over a large altitudinal range. It is tolerant of habitat disturbance, favouring regularly burnt grasslands which remains a common habitat in traditionally farmed regions: Least Concern (LC).

SYN. *B. mucronata* Lindau in E.J. 20: 24 (1894); C.B. Clarke in F.T.A. 5: 149 (1899). Type: Tanzania, Lushoto District, Usambara Mts, Mashewa [Mascheua], *Holst* 8853 (K!, lecto., chosen here; M!, W!, isolecto.)

B. kilimandscharica Lindau in E.J. 20: 26 (1894). Type: Tanzania, Moshi District, Kilimanjaro, *Meyer* 57 (B†, holo.)

B. grandicalyx Lindau var. *vix-dentata* C.B. Clarke in F.T.A. 5: 149 (1899); F.P.S. 3: 169 (1956). Type: Sudan, Upper Nile, *Freeman & Lucas* 88 (K!, lecto., chosen here – see note), **syn. nov.**

NOTE. This widespread and polymorphic species is broadly divisible into three allopatric variants in our region which are treated separately below. However, a more complete analysis across this species' full range, particularly in eastern Congo-Kinshasa, is required before formal recognition of subspecies.

Of Clarke's 6 syntypes of var. *vix-dentata*, *Schweinfurth* 2846 and *Freeman & Lucas* 88 from Sudan (form *a*) and *Smith* s.n. from Kilimanjaro (form *b*) are referable here. The remaining three syntypes are from Angola, of which *Johnston* s.n. (K) and *Welwitsch* 5048 (BM) have been seen. These gatherings clearly differs from *B. grandicalyx*, being more closely related to *B. crassa* C.B. Clarke. In light of this, the Angolan material is rejected from *B. grandicalyx* and var. *vix-dentata* is lectotypified appropriately. The degree of toothing of the calyx margin is variable within *B. grandicalyx* and it is not possible to sustain varieties based upon this character, hence var. *vix-dentata* is synonymised.

KEY TO INFRASPECIFIC VARIANTS

1. Outer calyx lobes oblong-elliptic or slightly -obovate or -ovate, base cuneate or acute, margin subentire or with small narrow teeth towards apex only ..... subsp. *c "mikumiensis"*
   Outer calyx lobes (oblong-) ovate with base rounded or cordate or if ovate-elliptic with base acute then margin with spinulose teeth 2
2. Stem and lower leaf surface with mainly long ascending or spreading coarse hairs; short spreading or retrorse hairs often sparse; corolla tube barely longer than lobes, the latter 11.5–21 mm long ..... subsp. *a "grandicalyx"*
   Stem and lower leaf surface with dense short spreading or retrorse hairs in addition to sparse to dense long coarse hairs; corolla tube clearly longer than the lobes, the latter 7–13 mm long ..... subsp. *b "mucronata"*

a. subsp. *a "grandicalyx"*

Stem and lower leaf surface with mainly long, ascending or spreading, coarse hairs; short spreading or retrorse hairs often sparse. Outer calyx lobes (oblong-) ovate with base rounded or cordate, margin subentire or with ± conspicuous spinulose teeth, strongly accrescent in fruit, anticous lobe 14.5–21 mm long and 8.5–13 mm wide in flower, 20–34 mm long and 18–23 mm wide in fruit. Corolla blue or purple, rarely white; tube 13.5–23 mm long, barely longer than the 11.5–21 mm long lobes. Capsule 16–24 mm long.

UGANDA. West Nile District: Aringa County, near Baringa, Feb. 1969, *Lye* 2153!; Acholi District: Kitgum, Sept. 1940, *Purseglove* 1011!; Teso District: Serere, Nov. 1931, *Chandler* 155!
KENYA. West Suk District: Kapenguria, May 1932, *Napier* 1949!; Trans-Nzoia District: Kitale, Mar. 1953, *Bogdan* 3671!; North Kavirondo District: SW slopes of Mt Elgon, 1934, *Tweedie* 121!
DISTR. **U** 1, 3, **K** 2, 3, 5; Central African Republic, Sudan, Ethiopia
HAB. Short grassland and associated open *Acacia-Combretum* woodland, usually with a regular burning regime, grass-dominated rocky hillsides; 700–2300 m

b. subsp. *b "mucronata"*

Stem and lower leaf surface with dense short spreading or retrorse hairs in addition to sparse to dense long, ascending or spreading, coarse hairs. Outer calyx lobes (oblong-) ovate with base rounded or cordate, or rarely ovate-elliptic with base acute or obtuse, margin usually with ± conspicuous spinulose teeth, weakly to strongly accrescent in fruit, anticous lobe 11–19 mm long and 7–11 mm wide in flower, 16–33 mm long and 9.5–20 mm wide in fruit. Corolla white or more rarely pale blue; tube 18–29 mm long, clearly longer than the 7–13 mm long lobes. Capsule 16–23 mm long.

KENYA. Embu District: E of Kanyuambora, Jan. 1966, *Gillett* 17036!; Kisumu-Londiani District: Kisumu, Sept. 1958, *Tweedie* 1691!; Masai District: Elgese Gonyek, near Mara R., Apr. 1961, *Glover, Gwynne & Samuel* 532!
TANZANIA. Mwanza District: Solima, Nassa, Oct. 1952, *Tanner* 1097!; Moshi District: Engare Nairobi, Lake Magadini, Feb. 1969, *Richards* 23879!; Pare District: Mkomazi Game Reserve, Ibaya Hill, May 1995, *Abdallah & Vollesen* 95/71!
DISTR. **K** 4–6; **T** 1–3, 5; (?Burundi, Rwanda, Congo-Kinshasa)

HAB. As form *a*; 1000–1950 m
USES. "Browsed by donkeys; spiny branches used to strain off leeches from water" (**K** 6; *Glover et al.* 532); "seeds mixed with grain seeds before cultivation to encourage growth; analogy from the clumps into which it grows" (**T** 1; *Tanner* 1097)

NOTE. Corolla size is very variable within subsp. *b*. Plants with strikingly small corollas are recorded from SC Kenya mainly in the Masai Mara (e.g. *Glover et al.* 298); a gradual increase in corolla size is recorded radiating out from this area. The long corolla tube relative to the limb is however constant in this form.

Although the calyx is usually strongly accrescent in this variant, some collections from the eastern part of its range (e.g. *Verdcourt* 519, **K** 4 and *Abdallah & Vollesen* 95/71, **T** 3) have only a weakly accrescent calyx; such plants sometimes have more elliptic outer calyx lobes than typical, with an obtuse or acute base. They therefore tend towards subsp. *c* but are readily separated by having a conspicuously toothed margin to the calyx lobes and a longer corolla tube in relation to the limb.

c. subsp. *c "mikumiensis"*

Stem and lower leaf surface with mainly long, ascending or spreading coarse hairs; short spreading or retrorse hairs often sparse. Outer calyx lobes oblong-elliptic to slightly -obovate or -ovate, base cuneate or acute, margin subentire or with small narrow teeth only towards apex, rather weakly accrescent, anticous lobe 10–15 mm long and 4–7.5 mm wide in flower, 11.5–22 mm long and 5–10 mm wide in fruit. Corolla blue or purple; tube 12–17 mm long, subequal in length to the lobes 11.5–16 mm long. Capsule 13.5–19 mm long.

TANZANIA. Kilosa District: Mikumi National Park, 5 km W of Park H.Q., June 1980, *Vollesen* 80/11!; Morogoro District: W of Ngerengere R., Mar. 1926, *Peter* 39240! & Kingolwira, July 1935, *Burtt* 5167!
DISTR. **T** 6; not known elsewhere
HAB. Open grassland with scattered trees or scrub, sometimes on seasonally inundated soils, *Acacia-Combretum* woodland; 450–600 m

SYN. [*B.* sp. 1 *sensu* Thomson in Speke, Journ. Disc. source Nile, Append.: 643 (1863)]
[*B. acanthoides sensu* Oliver in Trans. Linn. Soc., Bot. 29: 127 (1875), *non* Vahl]
[*B. buxifolia sensu* C.B. Clarke in F.T.A. 5: 150 (1899), *non* L.]

NOTE. *B. grandicalyx* and *B. ramulosa* (sp. 4 below) are undoubtably closely related and separation by dichotomous key is difficult due in part to the variability within each species and the resultant inconsistency of diagnostic characters. The principal difference is in the habit: *B. grandicalyx* is a suffrutex of fire-prone habitats with annual branching from a woody base, whilst *B. ramulosa* is shrubby and much-branched throughout. However, some specimens of the former, presumably from areas where the regular fire regime has been suppressed, develop more branched, presumably perennial stems. *B. ramulosa* also tends to have shorter internodes and so is often much more spiny. Subsp. *a* and *b* of *B. grandicalyx* have more ovate and strongly accrescent calyx lobes than in *B. ramulosa* and tend to have larger, more oblong-elliptic (not obovate) leaves. However, subsp. *c* is somewhat intermediate in these characters and it is possible that this form is a result of intergrading between the two species. On close inspection, *B. grandicalyx* usually has short, crisped or spreading hairs along the margins of the leaves and outer calyx lobes, different to the surface indumentum; this character is always absent in *B. ramulosa*. However, it can be very inconspicuous in the former and only detectable by microscopic study.

4. **Barleria ramulosa** *C.B. Clarke* in F.T.A. 5: 150 (1899); Blundell, Wild Fl. E. Afr.: 388, t. 798 (1987); K.T.S.L.: 599 (1994); U.K.W.F. ed. 2: 272 (1994); Ensermu in Fl. Eth. 5: 408 (2006). Types: Kenya, Machakos/Kitui District, Ukambani, *Scott-Elliot* 6312 (BM!, K!, syn.) & Kwale District, Duruma, *Hildebrandt* 2339 (B†, syn.)

Spiny (sub)shrub, 30–250 cm tall; stems with sparse to dense short whitish ± retrorse hairs and interspersed coarse buff ascending or spreading hairs, sometimes also glandular-pubescent; internodes short. Axillary spines simple, paired, many. Leaves obovate (-elliptic) to orbicular, 0.6–2.7 cm long, 0.3–1.7 cm wide, base cuneate, attenuate or rounded, margin somewhat revolute, apex rounded, obtuse or shallowly emarginate, spinose, upper surface ± sparsely pubescent with short coarse spreading hairs, veins beneath strigose, with many pale (sub)spreading hairs between the veins particularly when young, sometimes also glandular-pubescent along the margin; lateral veins 3–4(–5) pairs; petiole short or absent. Flowers solitary, subsessile or peduncle to 3.5 mm long; bracteoles spinose, 7–20 mm long, to 1 mm wide, margin entire. Calyx scarious with age and somewhat accrescent; outer lobes subequal, (oblong-) elliptic to somewhat ovate or obovate, 8–14 mm long, 2.5–8 mm wide in flower, up to 19 mm long and 13.5 mm wide in fruit, base cuneate to obtuse or rounded, margin spinosely toothed or subentire, apex spinose, anticous lobe sometimes bifidly so, surfaces strigose, often with interspersed shorter spreading hairs and sometimes glandular hairs; lateral lobes linear-lanceolate, 5–9.5 mm long. Corolla white, pale blue or lilac, 21–41 mm long, pilose outside, hairs mainly glandular; tube 11.5–25 mm long, barely widened above attachment point of stamens; limb in weak "2+3" arrangement; abaxial and lateral lobes 7.5–16 mm long, 5–10.5 mm wide; adaxial lobes 7–14.5 mm long, 4.5–10 mm wide. Stamens attached 7.5–18 mm from base of corolla tube; filaments 11–18 mm long; anthers exserted, 3–5.3 mm long; lateral staminodes 1–3.5 mm long, pilose, antherodes 0.3–1 mm long or undeveloped. Ovary glabrous or with few hairs at apex; style glabrous; stigma linear, 0.7–2.7 mm long. Capsule 11–20 mm long, glabrous or largely so; seeds 3.5–7 mm long and wide. Fig. 48, p. 338.

a. var. **ramulosa**

Stems lacking glandular hairs. Leaf blade obovate (-elliptic), base cuneate to attenuate. Flowers subsessile. Outer calyx lobes (oblong-) elliptic or somewhat ovate, margin spinose or denticulate, glandular hairs absent or few along margin. Lateral staminodes 1–1.7 mm long, pilose; antherodes 0.3–1 mm long. Capsule 11–15(–17) mm long. Fig. 48/1–8, p. 338.

KENYA. Machakos District: near Chiemu, Ngulia, Tsavo National Park (West), Aug. 1963, *Verdcourt* 3699!; Kwale District: between Samburu and Mackinnon Road, Sept. 1953, *Drummond & Hemsley* 4082!; Kilifi District: Dakabuka, Dec. 1990, *Luke & Robertson* 2567!

TANZANIA. Masai District: Engare Nanyuki, Feb. 1969, *Richards* 23920!; Lushoto District: Mazinde, May 1953, *Drummond & Hemsley* 2336! & 16 km on Mkomazi to Mombo road, Apr. 1988, *Bidgood & Vollesen* 1263!

DISTR. **K** 1, 4, 6, 7; **T** 2, 3; E Ethiopia

HAB. Dry open bushland and wooded grassland with *Acacia-Commiphora*, *Boscia-Dobera* etc., often on rocky or sandy soils, sometimes along roadsides or in abandoned cultivation; 0–1750 m

USES. "Roots give a stomach medicine" (**K** 7; *Graham* V326!)

CONSERVATION NOTES. *B. ramulosa* appears locally common, particularly in the border area of E Kenya and Tanzania. It favours dry, open habitats and, due to its very spiny nature, is likely to be tolerant of some grazing pressure: Least Concern (LC).

NOTE. Plants from the coastal region of Kenya and Tanzania usually have conspicuous marginal spines on the outer calyx lobes and the corollas are usually white. Those from interior N Tanzania have much smaller calyx spines sometimes reduced to minute teeth and the flowers are usually smaller and pale blue. The distinction however appears insufficiently consistent for formal separation.

b. var. **dispersa** *I. Darbysh.* **var. nov.** a varietate typica floribus pedicellis 1–4.5 mm longis instructis (non subsessilibus), lobis calycis exterioribus multis pilis glandulosis vestitis (non pilis carentibus neque ad marginem tantum hirsutis), staminodiis lateralibus 2.3–3.5 mm (non 1–1.7 mm) longis et antherodiis vix effectis differt. Type: Kenya, Northern Frontier District, 27 km from Marsabit on Isiolo road, *Verdcourt & Dale* 2219 (K!, holo.; EA!, iso.)

Stems often with many glandular hairs. Leaf blade orbicular or obovate-elliptic, base rounded to cuneate. Flowers on pedicels 1–4.5 mm long. Outer calyx lobes narrowly oblong or somewhat obovate, margin entire or minutely denticulate towards apex, glandular hairs many on surface and margin. Lateral staminodes 2.3–3.5 mm long, with minute glands and scattered hairs; antherodes barely developed. Capsule 14–20 mm long. Fig. 48/9.

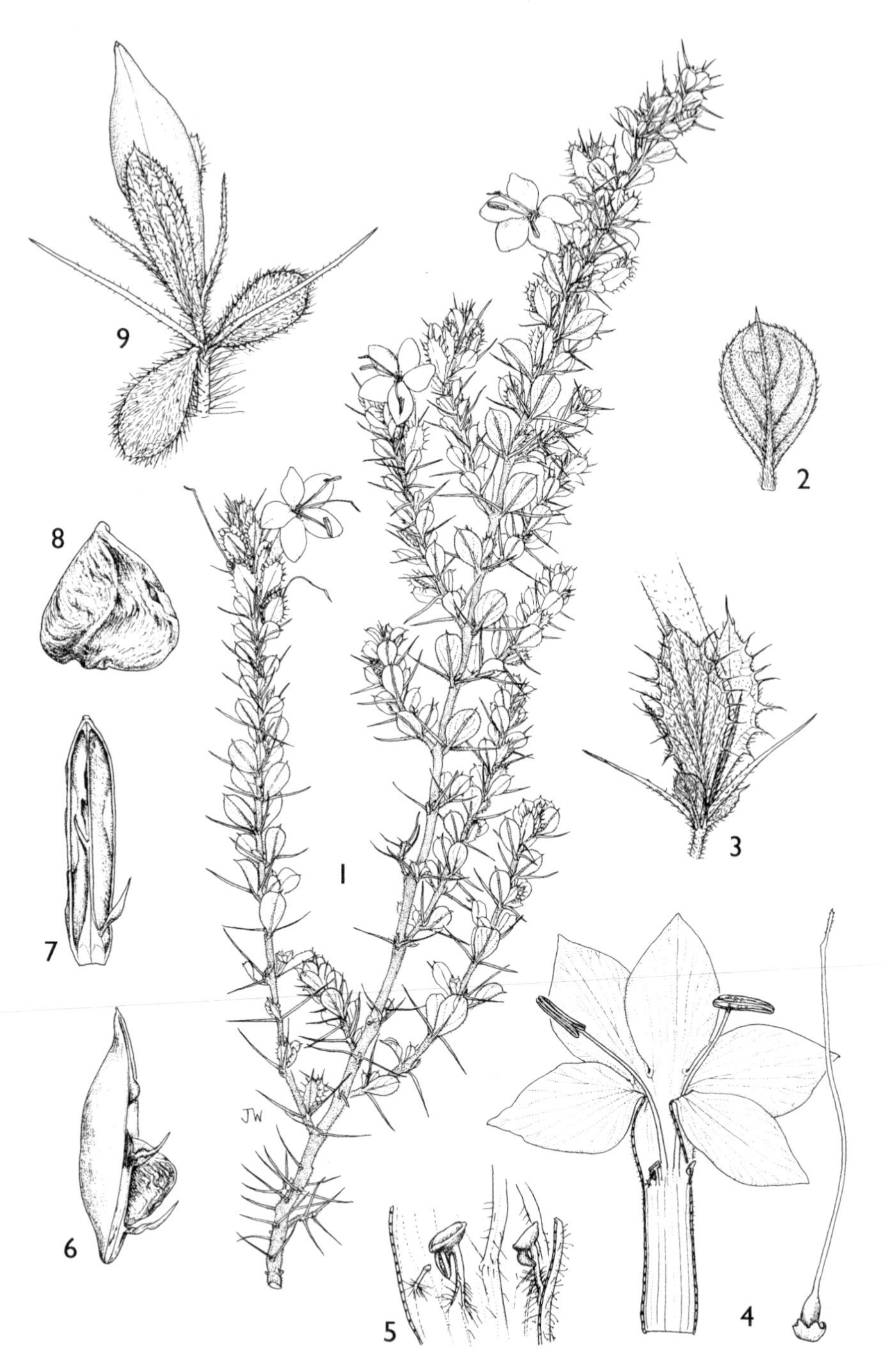

FIG. 48. *BARLERIA RAMULOSA* var. *RAMULOSA* — **1**, habit, × ²/₃; **2**, leaf, lower surface, × 2; **3**, bracts, bracteoles and calyx with base of corolla tube, × 2; **4**, dissected corolla with androecium and pistil, abaxial lobe central, × 2; **5**, detail of staminodes, × 6; **6**, capsule valve, side view with one seed in place, × 3; **7**, capsule valve, internal face, × 3; **8**, seed, × 5. *B. RAMULOSA* var. *DISPERSA* — **9**, bracts, bracteoles, calyx and capsule, × 2. 1 & 2 from *Hooper & Townsend* 1280; 3–5 from *Verdcourt* 3192; 6–8 from *Tweedie* 4040; 9 from *Verdcourt & Dale* 2216. Drawn by Juliet Williamson.

KENYA. Northern Frontier District: Marsabit, May 1958, *T. Adamson* 19! & 27 km from Marsabit on Isiolo road, July 1958, *Verdcourt & Dale* 2219! (type); Masai District: Nguruman Escarpment, Feb. 1976, *Hutson* 17!

TANZANIA. Masai District: Kitumbeine, Jan. 1936, *Greenway* 4271! & Ol Doinyo Sambu, Digodigo, Nguruman, Sept. 1944, *Bally* 3853! & Ndaleta, Kitumbeine, Monduli, Nov. 1966, *Carmichael* 1301!

DISTR. **K** 1, 6; **T** 2; not known elsewhere

HAB. *Acacia-Commiphora* wooded grassland, dry scrub; 1350–1750 m

USES. "Used to cover water when cattle are drinking to prevent water insects entering their mouths and lodging in their throats" (**T** 2; *Greenway* 4271)

CONSERVATION NOTES. The distribution of var. *dispersa* is unusual, being apparently absent from much suitable dry bushland in the central Kenyan Rift Valley. It is currently known only from seven collections within two isolated population centres. It favours dry, open bushland, a habitat common within its range, and its spiny habit is a good defense against overgrazing by cattle or wild herbivors. It is therefore provisionally assessed as of Least Concern (LC).

SYN. [*B.* sp. (= *Ensermu K. & Zerihun W.* 612) *sensu* Ensermu in Fl. Eth. 5: 409 (2006) pro parte quoad spec. ex Kenya]

NOTE. It is tempting to treat this distinctive variety as a separate species. However, *Gillett* 13961 from Furroli on the Kenya/Ethiopia border appears intermediate between the two taxa; it has many glandular hairs on the calyx and long staminodes which largely lack hairs, typical of var. *dispersa*, but has subsessile flowers, developed antherodes and smaller fruits more typical of var. *ramulosa*.

Ensermu provisionally included the N Kenyan specimens of this variety within his *B.* sp. (= *Ensermu K. & Zerihun W.* 612). However, the Ethiopian material differs in having longer axillary spines and bracteoles (the latter 18–25 mm long), longer lateral calyx lobes (10.5–14.5 mm long), larger corollas (42–54 mm long) with more oblong lobes and a strikingly broader tube, longer staminal filaments (23–31 mm long) which are rather densely hairy at the base, longer staminodes which are densely hairy and larger capsules (22–26 mm long). The Ethiopian taxon is most closely related to *B. longissima* Lindau, also endemic to southern Ethiopia and differing principally in having a much longer corolla tube, similar to that of *B. acanthoides*.

5. **Barleria crassa** *C.B. Clarke* in F.T.A. 5: 151 (1899). Type: Zimbabwe, South African Goldfields, *Baines* s.n. (K!, holo.)

Perennial herb or subshrub, 45–200 cm tall; stems buff- or yellowish-strigose with interspersed shorter spreading hairs when young. Axillary spines simple, paired, sometimes absent. Leaves elliptic, obovate or suborbicular, 0.7–4.5 cm long, 0.4–1.6 cm wide, base obtuse to cuneate or subattenuate, margin revolute when young, apex obtuse, rounded or rarely acute, apiculate, upper surface with coarse ± spreading hairs, lower surface strigulose on the veins and with many spreading whitish hairs elsewhere at least when young; lateral veins 3–6 pairs, these and reticulate tertiary venation ± prominent beneath; petiole short or absent. Flowers solitary or in 2–3-flowered unilateral cymes; bracteoles linear-lanceolate, 5–18 mm long, 1–2.5 mm wide, margin entire, apex spinose; flowers subsessile. Calyx often purplish towards apex and on veins when young, scarious with age; outer lobes subequal, broadly ovate, 9.5–23 mm long, 7–16 mm wide, base truncate or cordate, margin denticulate to spinulose, rarely subentire, apex obtuse to attenuate in outline, ± mucronate, anticous lobe rarely emarginate, surfaces strigulose and with scattered minute stalked glands; lateral lobes lanceolate, 3–8.5 mm long. Corolla white, blue or mauve, (22–)26–39 mm long, glandular-pilose outside or sparsely so; tube (14–)16–24 mm long, funnel-shaped above attachment point of stamens; limb in weak "2+3" arrangement; abaxial lobe 9–18 mm long, 10–17 mm wide; lateral lobes as abaxial but 9–16 mm wide; adaxial lobes (6–)7.5–15 mm long, (4–)6–10 mm wide. Stamens attached (6–)7.5–12 mm from base of corolla tube; filaments (12–)16–24 mm long; anthers exserted, 3–4.8 mm long; lateral staminodes 0.7–3.7 mm long, with minute stalked glands and ± pilose, antherodes 0.4–1.4 mm long or barely developed. Ovary glabrous; style declinate-strigose at base; stigma linear, 1–2 mm long. Capsule (11.5–)14.5–17 mm long, glabrous; only immature seeds seen.

DISTR. (for species) **T** 4; Angola, Zambia, Malawi, Zimbabwe

SYN. *B. venosa* Oberm. in Ann. Transv. Mus. 15: 168 (1933). Type: Zimbabwe, *van Son* in *Vernay-Lang Kalahari Expedition* s.n. (PRE, holo.), **syn. nov.**

subsp. **mbalensis** *I. Darbysh.* **subsp. nov.** a subsp. *crassa* lobis calycis exterioribus magis conspicue atque spinulose dentatis, lobis calycis lateralibus brevioribus, corolla pallidiore, tubo corollae quam limbo vix longiore et staminodiis longioribus differt. Type: Zambia, Kalambo Falls, *Bullock* 3000 (K!, holo. & iso.)

Leaf apex obtuse or rounded. Outer calyx lobes with margin spinulose; lateral lobes 3–5.5 mm long. Corolla white, pale blue or pinkish; sparsely pilose or largely glabrous outside; tube 16–20 mm long, lobes of lower lip 14–18 mm long. Lateral staminodes 3–3.7 mm long, largely lacking pilose hairs; antherodes barely developed or to 0.5 mm long.

TANZANIA. Ufipa District: Kasanga road, June 1957, *Richards* 10132! (pro parte, mixed coll. with *B. spinulosa* Klotzsch)

DISTR. **T** 4; NE Zambia

HAB. *Brachystegia* and *Pterocarpus* woodland, dry rocky hillsides; 900–1250 m

USES. None recorded on herbarium specimens

CONSERVATION NOTES. *B. crassa* is a widespread and locally common species, most often recorded on dry rocky slopes amongst boulders and so unlikely to be threatened by agricultural activity: Least Concern (LC). Subsp. *mbalensis* is much more localised but appears not uncommon in NE Zambia and is probably under-recorded in SW Tanzania. It favours similar habitats to the typical subspecies and may not be threatened but is currently assessed as Data Deficient (DD) as more information on distribution and threats are required.

NOTE. Subsp. *crassa*, widespread in Zambia and Zimbabwe, differs from subsp. *mbalensis* in having denticulate or subentire outer calyx lobes, lateral calyx lobes 4.5–8.5 mm long, a darker blue or mauve corolla with the tube (17–)19–24 mm long and clearly longer than the lower lip, and pilose staminodes 1–2 mm long with well-developed antherodes 0.7–1.4 mm long. Plants from the Kawambwa region of N Zambia, however, superficially resemble subsp. *mbalensis* in sharing the spinulose calyx margin, but have longer lateral calyx lobes, a somewhat longer corolla tube, shorter, pilose staminodes (but with poorly developed antherodes) and a more hairy exterior to the corolla. These plants therefore appear somewhat intermediate between subsp. *mbalensis* and subsp. *crassa*. Subsp. *mbalensis* is one of several variants occurring in the NE part of the species' range and further subspecies may need to be recognised from outside our region.

6. **Barleria nyasensis** *C.B. Clarke* in F.T.A. 5: 150 (1899). Type: Mozambique, mountains E of Lake Nyasa, *Johnson* s.n. (K!, holo.)

Perennial herb or subshrub, 30–170 cm tall; stems buff- or yellowish-strigose with interspersed shorter spreading hairs when young. Axillary spines largely absent; if present then simple. Leaves oblong-elliptic, 2–5 cm long, 0.7–1.5 cm wide, base acute or subattenuate, margin revolute when young, apex ± acute, spinulose, upper surface with coarse spreading hairs, lower surface strigose on the veins and with many spreading whitish hairs when young; lateral veins 4–6 pairs, these and reticulate tertiary venation ± prominent beneath; petiole short or absent. Flowers solitary or in 2–3-flowered unilateral cymes; bracteoles linear-lanceolate, 10–23 mm long, 1–2.7 mm wide, margin entire, apex spinose; flowers subsessile. Calyx scarious with age; outer lobes subequal, broadly ovate (-elliptic), 14.5–26 mm long, 10.5–18 mm wide, base truncate-attenuate or cordate, margin spinulose, apex obtuse to attenuate in outline, surface strigose, often with scattered minute stalked glands; lateral lobes lanceolate-attenuate, 7.5–13 mm long. Corolla white or pale blue (–mauve), 40–50 mm long, glandular-pilose outside; tube (25–)28–34 mm long, funnel-shaped above attachment point of stamens; limb in weak "2+3" arrangement; abaxial and lateral lobes 12–16 mm long, 11.5–15 mm wide; adaxial lobes 11.5–14.5 mm long, 7–9.5 mm wide. Stamens attached 10–16 mm from base of corolla tube; filaments 19–22 mm long; anthers exserted, 4–5 mm long; lateral staminodes 2–2.8 mm long, pilose and with minute stalked glands, antherodes barely developed. Ovary glabrous; style declinate-strigose at base or glabrous; stigma linear, 1–1.5 mm long. Capsule and seeds not seen.

TANZANIA. Mbeya/Njombe District: Kimani R., Nyengenge waterfall, June 1991, *P. Lovett & Kayombo* 463!; Njombe District: Chimala–Matamba road, Ndumbi Gorge, May 1986, *Congdon* 83! & 4 km up track to Matamba, into Poroto Mts S of Chimala, June 1990, *Carter, Abdallah & Newton* 2486!

DISTR. **T** 7; NW Mozambique

HAB. *Brachystegia* and *Uapaca* woodland, fringes of riverine woodland, often on dry, rocky slopes; 1000–2150 m

USES. None recorded on herbarium specimens

CONSERVATION NOTES. This is a local species, all recent records being confined to a very small area of southern Tanzania where it is, however, locally common. For example, it is common and conspicuous in mid-altitude woodland between Chimala and the Kipengere Mts (*pers. obs.*). It has not been recollected in Mozambique since the type collection (for which no precise locality data were recorded), but may still be extant in the under-collected north of the country. As it favours dry, rocky hillsides of low agricultural potential, its habitat may not have diminished greatly, although over-grazing is a potential threat. It is assessed as of Least Concern (LC).

NOTE. This species is close to *B. crassa*, particularly subsp. *mbalensis* with which it shares the spinulose calyx margin and pale corollas. It could be treated as a further subspecies within *B. crassa* but is maintained here on the basis of the conspicuously larger flowers and the more acute, conspicuously spinulose leaf apices.

6a. **Barleria nyasensis** × **B. spinulosa**

Perennial herb, ± 100 cm tall; stems with many coarse buff-coloured ascending hairs and interspersed short patent glandular hairs when young. Axillary spines absent. Leaves subsessile, elliptic-lanceolate, 1.5–2.5 cm long, 0.7–1 cm wide, base acute, margin somewhat revolute, apex acute-apiculate, upper surface with coarse ascending hairs, lower surface strigose on the veins and with many spreading whitish hairs and occasional long-armed stellate hairs when young; lateral veins 4–5 pairs. Flowers in 2–4-flowered unilateral cymes; bracteoles linear-lanceolate, 11–15 mm long, ± 1.5 mm wide, margin entire, apex spinose; flowers subsessile. Calyx soon scarious; outer lobes subequal, broadly ovate, 13.5–16 mm long, 9.5–11 mm wide, base truncate or subcordate, margin and apex spinose, surface with coarse ascending hairs mainly on veins, occasionally with a stellate base, with scattered short patent glandular hairs; lateral lobes lanceolate, 5.5 mm long. Corolla pale blue, 32–36 mm long, glandular-pilose outside mainly on lateral lobes; tube 20–24 mm long, narrowly funnel-shaped above attachment point of stamens; limb in weak "2+3" arrangement; abaxial lobe 11.5 mm long, 11 mm wide; lateral lobes 10.5 mm long, 9 mm wide; adaxial lobes 10 mm long, 8 mm wide. Stamens attached 12 mm from base of corolla tube; filaments 19 mm long; anthers 4 mm long; lateral staminodes 3–4 mm long, pilose, antherodes barely developed. Ovary glabrous; style densely declinate-strigose at base; stigma linear, ± 1.7 mm long.

TANZANIA. Mbeya District: Makete, Usalimwani, May 2008, *Suleiman et al.* 3570!

DISTR. **T** 7; not known elsewhere

HAB. Dry open *Brachystegia* and *Uapaca* woodland; 1400 m

NOTE. The cited specimen was collected by the author with H.O. Suleiman (DSM). A single plant was found in an area with abundant *B. nyasensis* and occasional *B. spinulosa*. It is clearly intermediate between the two, having the habit, leaves and bracteoles of *B. nyasensis* but having a partially stellate indumentum, more flowers on some of the inflorescences, smaller corollas and a more coarsely spinose calyx margin, all tending towards *B. spinulosa*.

7. **Barleria blepharoides** *Lindau* in E.J. 20: 24 (1894); C.B. Clarke in F.T.A. 5: 151 (1899). Type: Tanzania, Tabora District, Unyamwezi, Rubuga, *Stuhlmann* 496 (B†, holo.)

Shrub; stems rounded, somewhat furrowed, long-pubescent. Leaves ovate, to 2.5 cm long, 1.2 cm wide, base usually rounded, apex spinulose, both surfaces densely hairy; petiole short. Inflorescences several-flowered unilateral cymes; bracteoles reduced to stout, simple spines, 10–20 mm long, straight. Calyx blue, becoming pale with age; anticous lobe ovate, ± 22 mm long, 16 mm wide, margin spinose, surface hairy, 4-veined from base and with reticulate secondary venation; posticous lobe ± 25 mm long; lateral lobes lanceolate, ± 15 mm long. Corolla blue, ± 36 mm long, somewhat hairy; tube ± 18 mm long, 4 mm in diameter towards base, 5 mm in diameter above; limb 5-lobed, rounded, 13–18 mm long, 12–15 mm wide. Stamens with filaments ± 20 mm long, broad; anthers exserted, ± 7 mm long; lateral staminodes ± 4 mm long, antherodes small. Ovary ± 3 mm long; style ± 40 mm long; stigma not recorded. Capsule and seeds unknown.

TANZANIA. Tabora District: Unyamwezi, Rubuga, Mar. 1890, *Stuhlmann* 496
DISTR. **T** 4; not known elsewhere
HAB. Not recorded
USES. None recorded on herbarium specimen
CONSERVATION NOTES. With no extant collections and no information on its ecology, this species must currently be assessed as Data Deficient (DD), but it will probably prove to be threatened when and if it is rediscovered.

NOTE. Although there is no extant material of this taxon, the protologue description does not match any material seen in *Barleria* and so it is maintained as a good species here, the description being adapted from those of Lindau and Clarke (l.c.). Lindau originally placed it in sect. *Prionitis* but considered it close to *B. acanthoides* and allies; from the presence of unilateral cymes, spinose bracteoles and broadly ovate, reticulate-veined, spinose calyces it clearly falls within sect. *Barleria*. It is unique within this section in East Africa in having the combination of simple-spinose bracteoles and well-developed unilateral cymes.

8. **Barleria fulvostellata** *C.B. Clarke* in F.T.A. 5: 163 (1899); Vollesen in Opera Bot. 59: 79 (1980); M. & K. Balkwill in K.B. 52: 555, figs. 6K & 9D$_4$ (1997); Darbyshire in K.B. 64: 675 (2010). Type: Malawi, Nyika Plateau, *Whyte* 181 (K!, lecto., chosen by Darbyshire, l.c.)

Perennial suffruticose herb or subshrub, 30–150 cm tall; stems with dense golden (-brown) long-armed stellate or dendritic hairs, stellate base rarely caducous, often also with patent glandular hairs. Axillary spines absent. Leaves elliptic, ovate or suborbicular, 2–13.5 cm long, 1.3–6 cm wide, base truncate, subcordate, cuneate or shortly attenuate, apex rounded to subattenuate, surfaces densely stellate-pubescent, hairs long-armed, stellate base rarely caducous, margin sometimes also with patent glandular hairs; lateral veins 3–6 pairs, these and the reticulate tertiary venation ± prominent beneath; petiole to 17 mm long or absent. Flowers solitary or in 2–12-flowered congested unilateral cymes, sometimes becoming branched at maturity, often clustered towards stem apices; bracteoles linear (-lanceolate), narrowly elliptic or oblanceolate, 5–23 mm long, 0.5–5.5 mm wide, margin entire or minutely toothed, apex ± mucronate; flowers subsessile or pedicels to 4.5 mm long. Calyx scarious with age; anticous lobe (oblong-) elliptic, obovate or ovate-orbicular, 8–22 mm long, 4.5–17 mm wide, base cuneate to rounded, margin subentire or denticulate, often involute, apex rounded to attenuate or shallowly emarginate, indumentum as leaves; posticuous lobe slightly larger, apex rounded to attenuate, ± mucronulate; lateral lobes lanceolate, ovate or oblong-elliptic, 6–14.5 mm long. Corolla pale blue to purple or rarely white, 26–50 mm long, pilose outside, hairs mixed glandular and eglandular hairs or mainly the latter; tube cylindrical, 14–29 mm long, ± declinate below attachment point of stamens; limb in weak "2+3" arrangement; abaxial and lateral lobes 12.5–22 mm long, 9–14.5 mm wide; adaxial lobes 8–20 mm long, 6.5–11.5 mm wide. Stamens attached 8.5–16 mm from base of corolla tube; filaments 12–28 mm long; anthers 3–6 mm long; lateral staminodes 1–3.5 mm long, pubescent or pilose and with minute stalked glands, antherodes barely developed or rarely to 1 mm long. Ovary pubescent towards apex; style pubescent towards base; stigma 0.5–1.5 mm long. Capsule 13–17 mm long, sparsely pilose towards apex or with an apical tuft of hairs only; seeds 4–5.5 mm long, 3.8–5 mm wide.

DISTR. (for species) **T** 6–8; E Zambia, Malawi, N Mozambique

NOTE. *B. fulvostellata* is a variable species with four subspecies currently recognised (see Darbyshire, l.c.). Within our region, two subspecies are readily separable:

a. subsp. **fulvostellata**, Darbyshire in K.B. 64: 676 (2010)

Stellate hairs with a conspicuous fasciculate-stellate base, visible to the naked eye, and a short arm 0.5–0.8 mm long. Mature leaves broadly ovate to suborbicular, length/width ratio 1–1.7/1, base rounded, truncate or subcordate. Inflorescences 1–4-flowered. Outer calyx lobes tardily scarious, oblong-elliptic, margin subentire or minutely toothed, often strongly involute. Anthers 5–6 mm long.

TANZANIA. Mbeya District: 122 km on Mbeya–Iringa road, May 1963, *Boaler* 904! & Igawa, Chimala, Apr. 2001, *Congdon* 600!; Mbeya/Njombe District: 113 km on Chimala–Iringa road, Mar. 1988, *Bidgood, Mwasumbi & Vollesen* 717!
DISTR. **T** 7; N Malawi
HAB. *Acacia-Commiphora* and *Brachystegia* woodland; 1150–1200 m
USES. None recorded on herbarium specimens
CONSERVATION NOTES. Assessed as Endangered (EN B2ab(iii)) by Darbyshire (l.c.).

b. subsp. **scariosa** *I. Darbysh.* in K.B. 64: 678 (2010). Type: Tanzania, Rufiji District, Beho Beho, *Vollesen* in MRC 4617 (C!, holo.; DSM!, EA, K!, iso.)

Stellate hairs with a fewer branched and less conspicuous stellate base often not visible to the naked eye, sometimes caducous, and with a longer arm, at least some over 1 mm long. Mature leaves ovate-elliptic to broadly elliptic, length/width ratio 1.6–2.7/1, base attenuate or cuneate to obtuse or rounded. Inflorescences (2–)3–12-flowered, often branched at maturity. Outer calyx lobes thin and soon scarious, variously obovate to ovate-orbicular, margin toothed or scarcely so, less strongly or not involute. Anthers 3–5.2 mm long.

TANZANIA. Rufiji District: Kibesa, Aug. 1976, *Vollesen* in MRC 3888! & near Beho Beho camp, Aug. 1993, *Luke* 3742!; Lindi District: Nachingwea, Aug. 1952, *Anderson* 790!
DISTR. **T** 6, 8; N Mozambique
HAB. Woodland and wooded grassland; 100–900 m
USES. None recorded on herbarium specimens
CONSERVATION NOTES. Assessed as of Least Concern (LC) by Darbyshire (l.c.).

NOTE. In subsp. *scariosa*, northern populations (**T** 6) have narrower, ± obovate outer calyx lobes and a somewhat lax, branched inflorescence at maturity whilst southern plants (**T** 8 and Mozambique) have broadly ovate or suborbicular outer calyx lobes and a more congested inflorescence. Intermediate specimens, however, occur around the **T** 6/8 border and the variation appears to be clinal.

9. **Barleria spinulosa** *Klotzsch* in Peters, Reise Mossamb., Bot. 1: 208 (1861); C.B. Clarke in F.T.A. 5: 152 (1899); Obermeijer in Ann. Transv. Mus. 15: 174 (1933); Vollesen in Opera Bot. 59: 79 (1980). Type: Mozambique, Quirimba [Kerimba] Island, *Peters* s.n. (B†, holo.; PRE!, iso.)

Erect or scrambling perennial herb, 30–150 cm tall; stems with dense short patent glandular and eglandular hairs and interspersed buff (-yellow) coarse hairs with or without a stellate base; older stems sometimes floccose. Axillary spines absent. Leaves sometimes immature at flowering, ovate or elliptic, (4–)6–9(–13) cm long, (1.8–)3–6 cm wide, base attenuate, apex subattenuate to acuminate, rarely obtuse, young leaves densely pale grey stellate-pubescent, stellate base often caducous particularly above, ± glabrescent, margin sometimes also with patent glandular hairs; lateral veins 4–6 pairs; petiole 4–32 mm long. Inflorescences 2–7-flowered congested unilateral cymes; bracteoles linear-lanceolate, pairs unequal, the larger 10–23 mm long, 1–2.5 mm wide, margin with widely-spaced teeth, apex spinose; flowers subsessile. Calyx often purplish towards apex and on veins, soon scarious; anticous lobe ovate (-orbicular) or somewhat oblong, 10–16 mm long, (4.5–)6–14 mm wide, base truncate or shortly attenuate, margin with coarse spinulose teeth, apex acute to rounded or emarginate in outline, surfaces with sparse simple and long-armed stellate hairs, basal portion with ± dense short crisped eglandular hairs, glandular hairs often restricted to margin; posticuous lobe 13–19 mm long, apex attenuate, spinose; lateral lobes lanceolate-attenuate, 5–6.5 mm long. Corolla 27–37 mm long, limb blue-mauve, tube or rarely whole flower white, glandular- and eglandular-pilose outside; tube cylindrical, 16–22 mm long, declinate below attachment point of stamens; limb in weak "2+3" arrangement, lobes each 7.5–13.5 mm long, 4.5–8 mm wide, adaxial pair rather widely divergent. Stamens attached 11–16 mm from base of corolla tube; filaments 11–12 mm long; anthers 3.3–4.5 mm long; lateral staminodes 1.3–2.5 mm long, pilose, antherodes 0.7–0.9 mm long. Ovary glabrous except apical ring of minute crisped hairs; style glabrous or sparsely pubescent near base; stigma 0.5–1 mm long. Capsule 11–13.5 mm long, glabrous or largely so; seeds 3–4.5 mm long and wide.

TANZANIA. Ufipa District: Milepa, May 1951, *Bullock* 3909!; Mbeya/Njombe District: 42 km on Chimala–Iringa road, Apr. 2006, *Bidgood et al.* 5308!; Kilwa District: Selous Game Reserve, Nakilala, June 1971, *Ludanga* 1304!

DISTR. **T** 4, 6–8; Zambia, Malawi, Mozambique, Zimbabwe, Botswana, South Africa

HAB. Dry woodland and bushland, often of *Brachystegia*, *Combretum* and/or *Terminalia*, particularly in riverine habitats, rarely in moist lowland forest; 150–1350 m

USES. None recorded on herbarium specimens

CONSERVATION NOTES. A widespread and locally abundant species, being particularly well represented in Malawi. Although loss of low- and mid-altitude woodland will inevitably have caused local population reductions, it is not considered threatened: Least Concern (LC).

SYN. *B. squarrosa* Klotzsch in Peters, Reise Mossamb., Bot. 1: 207 (1861). Type: Mozambique, Rios de Sena, Tete, *Peters* s.n. (B†, holo.)

B. consanguinea Klotzsch in Peters, Reise Mossamb., Bot. 1: 206 (1861); C.B. Clarke in F.T.A. 5: 153 (1899). Type: Mozambique, Rios de Sena, *Peters* s.n. (B†, holo.), **syn. nov.**

B. clivorum C.B. Clarke in F.T.A. 5: 153 (1899). Types: Mozambique, Lower Zambesi, Chupanga, *Kirk* s.n. & Malawi, between Mpata and commencement of Tanganyika Plateau, *Whyte* s.n. (both K!, syn.), **syn. nov.**

NOTE. Variation in stem indumentum, leaf size and calyx lobe shape have led to the past recognition of several taxa here. Klotzsch recognised three species: *B. squarrosa*, *B. consanguinea* and *B. spinulosa* but provided little in the way of diagnoses. The former name was applied to a specimen with narrow, oblong outer calyx lobes; Clarke (l.c.) however considered this to be at the narrowest extreme of a clinal variation in calyx lobe shape in *B. spinulosa*. He maintained *B. consanguinea* based upon the presence of long spreading stem hairs. This indumentum type is however widely recorded in *B. spinulosa* particularly from the northern part of its range and some specimens display an intermediate indumentum type, hence it is considered to hold little taxonomic significance. Clarke's *B. clivorum* was separated from *B. spinulosa* based entirely on its large leaf size, but the leaves of the syntypes fall within the full range now recorded for *B. spinulosa*.

B. kirkii T. Anderson, described from the Zambesi Valley in Mozambique, is closely allied to *B. spinulosa* and is doubtfuly distinct (see discussion by Brummitt in Bot. Mag. 181, t. 707 (1971)). It typically has smaller and proportionally narrower outer calyx lobes, oblong or somewhat obovate in shape, with smaller teeth often restricted to the upper half of each lobe. It tends to have a somewhat larger corolla, with the lobes of the upper lip being distinctly smaller than those of the lower lip, and the anthers are marginally larger. These differences however appear inconsistent, with intermediate specimens recorded particularly from the Zambesi basin. It is therefore likely that *B. kirkii* warrants at most varietal status within *B. spinulosa*.

10. **Barleria gracilispina** (*Fiori*) *I. Darbysh.*, **comb. et stat. nov**. Types: Somalia, near Goriei, *Paoli* 617 (FT!, holo.; K!, iso.); Uganda, Karamoja District, Moroto, *Verdcourt* 814 (K!, epi.; EA!, isoepi.), see note

Spiny subshrub, 20–60(–100) cm tall; stems with dense short white appressed, retrorse or subspreading hairs. Axillary spines paired, asymmetric with 1–2 lateral spines towards base of inner margin. Leaves obovate or oblanceolate, 1–3.3 cm long, 0.4–1.6 cm wide, base attenuate or cuneate, apex rounded to acute or emarginate, spinose, surfaces with sparse to many coarse appressed or ascending hairs and minute appressed or crisped white hairs, these sometimes restricted to the margin and midrib when mature; lateral veins 3–4 pairs; petiole short or absent. Flowers solitary, sessile; bracteoles spinose, asymmetric with 1–3(–4) lateral spines on inner margin towards base, entire or with minute teeth on outer margin, 5–17(–26) mm long, 1–3 mm wide towards base excluding lateral spines. Calyx purplish and with dark venation when young, soon scarious; anticous lobe broadly ovate or suborbicular, 8.5–16 mm long, 6.5–16 mm wide, base obtuse, rounded or shallowly cordate, apex acute to rounded, ± mucronulate, margin subentire or with ± conspicuous bristle-like teeth and submarginal spines, indumentum as leaves, crisped white hairs mainly towards the base; posticuous lobe with a ± more conspicuous mucro; lateral lobes (ovate-) lanceolate, 4.5–8.5 mm long. Night-flowering; corolla white or cream, drying blue-black, 35–52 mm long, glandular- and eglandular-pubescent outside; tube 27–45 mm long, narrowly cylindrical; limb subregular; abaxial and lateral lobes 5.5–7.5 mm long, 4.5–6.5 mm wide, adaxial lobes 5–7 mm long, 3.5–5 mm wide. Stamens attached 17–22 mm from base of corolla tube; filaments (11.5–)15–19 mm long; anthers held at corolla mouth or shortly exserted, 1.7–2.3 mm long; lateral staminodes 1.7–4.2 mm long, pilose, antherodes 0.5–0.7 mm long. Ovary with few appressed hairs towards apex and an apical ring of minute crisped hairs; style largely glabrous; stigma subcapitate, 0.2–0.5 mm long. Capsule (7.5–)9–12.5 mm long, with few appressed hairs towards apex or glabrous; seeds 3–4.2 mm long, 2.7–3.5 mm wide. Fig. 49/1–5, p. 346.

UGANDA. Karamoja District: near Amudat, June 1939, *A.S. Thomas* 2986! & Kangole, Dec. 1957, *Wilson* 411! & Amudat, Karasuk, June 1960, *Tweedie* 2022!

KENYA. Northern Frontier District: Dandu, June 1952, *Gillett* 13427!; Turkana District: 3 km from Kacheliba, Oct. 1964, *Leippert* 5053!; Tana River District: Nairobi–Garissa road, 13 km E of Hatama Corner, May 1974, *Gillett & Gachathi* 20522!

TANZANIA. Masai District: track to Moshi and Oldonyo Ngailoni hill, Jan. 1969, *Richards* 23669!; Lushoto District: Kilimele Lake–Mashewa, Aug. 1915, *Peter* 60270! & Mkomazi Game Reserve, 3 km N of Umba R. Camp, June 1996, *Abdallah, Mboya & Vollesen* 96/172!

DISTR. **U** 1; **K** 1–4, 6, 7; **T** 2, 3; S Somalia

HAB. Dense or more commonly open *Acacia-Commiphora*, *Commiphora-Boswellia* and *Euphorbia* woodland, grassland and degraded areas including roadsides and overgrazed land, sometimes on black clays; 30–1650 m

USES. "Eaten by sheep and goats" (**K** 2; *McDonald* s.n.); "Boran: use it as fever remedy - the root is boiled. Kamba: [for] stomach trouble - the root is boiled and rubbed into cuts on the tummy" (**K** 4; *J. Adamson* 586)

CONSERVATION NOTES. This species is widespread and fairly common within its range and although overgrazing may have led to some decline in habitat quality, it can persist in degraded areas. It is not considered threatened: Least Concern (LC).

SYN. *B. acanthoides* Vahl var. *gracilispina* Fiori in Result. Sci. Miss. Stef.-Paoli, Coll. Bot.: 139 (1916)
[*B. acanthoides sensu* U.K.W.F. ed. 2: 272, pl. 118 (1994) pro parte incl. plate]
[*B. paolii sensu* Hedrén in Fl. Somalia 3: 431 (2006) pro parte quoad *Thulin et al.* 7641, *non* Fiori – see note]

NOTE. Hedrén (l.c.) treated this and the following species as conspecific under a broadly circumscribed *B. paolii*, remarking only upon the unusual variation in the length of the corolla tube. However, within our region, two partially sympatric and easily separable taxa are recorded within the *B. paolii* complex based upon variation in calyx shape, corolla tube length and shape and arrangement of the corolla limb (see key), and I feel that they must be separated at the

FIG. 49. *BARLERIA GRACILISPINA* — **1**, habit, × $^2/_3$; **2**, bracteoles and calyx, × 2; **3**, dissected corolla with androecium, abaxial lobe held to the left, × 1.5; **4**, capsule valve with seed, × 3; **5**, mature seed, × 4. *B. PAOLIOIDES* — **6**, flower and flower bud in situ with bracteoles, × 1.5. *B. ACANTHOIDES* — **7**, inflorescence with flower, × 1. 1 & 2 from *Faden* 74/758; 3 from *Verdcourt* 814; 4 & 5 from *Abdallah et al.* 96/172; 6 from *Gillett* 13379; 7 from *Beentje* 2160. Drawn by Juliet Williamson.

species rank. However, the type gathering of *B. paolii* from S Somalia (*Paoli* 320, FT!) is somewhat intermediate in character, having the long corolla tube (25–30 mm) of *B. gracilispina*, but having a narrowly widened upper portion to the tube and slightly offset abaxial lobe as in *B. paolioides*, and having intermediate outer calyx lobes: elliptic, ± 5 mm wide (see fig. 3, Bull. Soc. Bot. Ital.: 54 (1915)). It is therefore quite possible that *B. paolii sensu stricto* is of hybrid origin or alternatively a third, closely allied species. *Gillett* 13376 from NE Kenya appears to match the corolla of the type of *B. paolii*, the single (dissected) corolla on the K sheet having a broadened tube upwards and offset abaxial lobe, but it has broader calyx lobes typical of *B. gracilispina*. This latter gathering was made from near to the type locality of *B. paolioides* and also within the range of *B. gracilispina* and so could, again, be considered of hybrid origin. Good material of this group from Somalia is scarce, but it is hypothesised that, whilst the two species are quite distinct and do not hybridise over most of their range, a small zone of hybridisation occurs at the northern limit of their range in S Somalia and just extends into NE Kenya.

The original type of *B. acanthoides* var. *gracilispina* is in fruit only. Whilst it clearly matches the taxon circumscribed here, having the broadly ovate calyx lobes, corollas provide the most important diagnostic character within the *B. paolii* complex; *Verdcourt* 814 is therefore chosen as an epitype of this species as it has excellently preserved flowers.

11. **Barleria paolioides** *I. Darbysh.* **sp. nov.** *B. gracilispinae* (Fiori) I. Darbysh. similis sed tubo corollae 11.5–21 mm (non 27–45 mm) longo sursum anguste dilatato (non ubique anguste cylindrico), limbo asymmetrico (non subregulari), lobo abaxiali corollae a lobis ceteris usque 2.5 mm distanti (non subregulari) et lobis calycis exterioribus lanceolatis vel anguste oblongo-ellipticis (non late ovatis neque suborbicularibus) 0.8–4.5 mm (non 6.5–16 mm) latis differt. Type: Kenya, Northern Frontier District, 30 km SSW of El Wak on Wajir road, *Gillett* 13379 (K!, holo.; B!, BR!, EA!, K!, iso.)

Spiny subshrub, 20–60 cm tall; stems with minute white retrorse or appressed hairs. Axillary spines paired, with 1–2 lateral spines towards base of inner margin. Leaves obovate, 0.6–1.8 cm long, 0.3–0.8 cm wide, base attenuate or cuneate, apex shallowly emarginate or rounded to acute, spinose, surfaces with coarse ascending hairs and minute appressed or crisped white hairs, the latter often restricted to margin and midrib when mature; lateral veins 3–4 pairs; petiole short or absent. Flowers solitary, sessile; bracteoles as axillary spines, 2.5–11 mm long, to 1 mm wide excluding lateral spines. Calyx drying purplish when young, soon scarious; anticous lobe lanceolate or narrowly oblong-elliptic, 4.5–9.5 mm long, 0.8–3(–4.5) mm wide, margin subentire or usually with bristle-like teeth or spines mainly in the upper half, sometimes also with submarginal spines, apex acute or shortly bifid, indumentum as leaves; posticuous lobe 5–10.5 mm long, 1–3.5(–4.5) mm wide, apex ± mucronate; lateral lobes linear-lanceolate, 3.5–7 mm long. Night-flowering; corolla white or cream, drying blue-black, 19–29 mm long, pubescent outside with descending eglandular hairs and spreading glandular hairs; tube 12–21 mm long, narrowly widened above attachment point of stamens; limb in weak "4+1" arrangement; abaxial lobe offset by 1.5–2.5 mm, 7.5–8.5 mm long, 4.5–5.5 mm wide; lateral lobes as abaxial but 6.5–8 mm long; adaxial lobes 6–7 mm long, 3.3–4.5 mm wide. Stamens attached 6–9 mm from base of corolla tube; filaments 11–14 mm long; anthers exserted, 1.5–1.8 mm long; lateral staminodes 1.8–2.3 mm long, pilose, antherodes 0.5–0.8 mm long. Ovary with few to many appressed hairs towards apex and an apical ring of minute crisped hairs; style largely glabrous; stigma subcapitate, 0.3–0.4 mm long. Capsule 8–11.5 mm long, indumentum as ovary or glabrescent; seeds 3–4 mm long and wide. Fig. 49/6, p. 346.

KENYA. Northern Frontier District: NE of Dadaab, Nov. 1978, *Brenan, Gillett & Kanuri* 14794!; Tana River District: Tana River National Primate Reserve, middle road 2.5 km, March 1990, *Luke et al.* in TPR 379!; Kwale District: Mackinnon Road, Buchuma area, Aug. 1965, *Williams Sangai* 832!

TANZANIA. Pare District: Mkomazi, Sept. 1987, *Ruffo* 2495! & Mkomazi Game Reserve, Ngurunga Dam to Marimbosho Hill, June 1996, *Abdallah, Mboya & Vollesen* 96/114!; Rufiji District: ± 10 km SE of Beho Beho, June 1977, *Vollesen* in MRC 4642!

DISTR. **K** 1, 7; **T** 3, 6; E Ethiopia, S Somalia

HAB. Open *Acacia-Commiphora*, *Commiphora-Boswellia-Lannea* and *Terminalia* woodland; grassland including seasonally inundated areas; 30–850 m

USES. None recorded on herbarium specimens

CONSERVATION NOTES. This species is rather widespread but apparently scarce, being known from ± 20 localities. However, it can be locally frequent to abundant in favourable habitats. It is not recorded from degraded habitats so may not tolerate disturbance, but large areas of suitable habitat remain within its core range of dry eastern Kenya. It is therefore not considered threatened: Least Concern (LC).

SYN. [*B.* sp. nov. aff. *spinisepala sensu* Vollesen in Opera Bot. 59: 79 (1980)]
[*B. lanceata sensu* Ensermu in Fl. Eth. 5: 408 (2006) pro parte, *non* (Forssk.) C. Chr.]
[*B. paolii sensu* Hedrén in Fl. Somalia 3: 431 (2006) pro parte, *non* Fiori]
[*B. spinisepala sensu* Hedrén in Fl. Somalia 3: 431 (2006) pro parte quoad *Thulin & Bashir Mohamed* 7108, *non* E.A. Bruce - see note]

NOTE. Ensermu (l.c.) treated the Ethiopian plants (Bale region) of this taxon under *B. lanceata* (Forssk.) C. Chr. Although similar, *B. lanceata* differs in having a dense, fine white indumentum which gives the mature plants a greyish appearance and in the spines and bracteoles of that species being considerably longer and with more evenly distributed lateral spines on both the outer and inner margins towards the base. *B. lanceata* is restricted to the Red Sea coast and adjacent lowlands in S Arabia and NE Africa.

Thulin & Bashir Mohamed 7108 from S Somalia, wrongly identified as *B. spinisepala* (= *B. delamerei*; sp. 12 below) by Hedrén (l.c.), has long outer calyx lobes (anticous to 13 mm) with long marginal spines (to 6 mm) but is otherwise a good match for *B. paolioides*; it should perhaps be considered as a distinct variety.

12. **Barleria delamerei** *S. Moore* in J.B. 38: 206 (1900); Baker & C.B. Clarke in F.T.A. 5: 513 (1900). Type: Kenya, Northern Frontier District, near Lake Marsabit, *Delamere* s.n. (BM!, holo.)

Spiny subshrub, branches prostrate or ascending to 60 cm tall; stems shortly white-pubescent, hairs spreading or retrorse, with interspersed coarse yellowish ascending hairs, later glabrescent. Axillary spines paired, ± curved, margin pinnately spinose. Leaves obovate, 0.7–3.2 cm long, 0.4–1.7 cm wide, base attenuate or cuneate, margin entire or irregularly toothed due to swollen hair bases, apex rounded to attenuate, spinose, surfaces strigose and with short white ± crisped hairs restricted to margin or more widespread; lateral veins 3–5 pairs; petiole short or absent. Flowers solitary, sessile; bracteoles spinose, linear or lanceolate, 7.5–24 mm long, 0.7–2(–3) mm wide excluding lateral spines 2–6 mm long, 1–5 on each margin. Calyx soon scarious; anticous lobe lanceolate, oblong or elliptic, 8–21 mm long, 2–6 mm wide, margin spinose, apex narrowed into 1–2(–3) spines, indumentum as leaves, often with scattered stalked glands; posticuous lobe 12–24 mm long, apex long-spinose; lateral lobes lanceolate, 4.5–10(–13) mm long. Corolla limb pale blue, mauve or purple, throat and tube paler, 26–40 mm long, glandular- and eglandular-pubescent outside; tube 18–27 mm long, narrowly campanulate above attachment point of stamens; limb subregular, lobes 8.5–13.5 mm long, abaxial lobe 8–12 mm wide, lateral lobes 8–10 mm wide, adaxial lobes 6.5–9 mm wide. Stamens attached ± midway along corolla tube; filaments 10–22 mm long; anthers exserted, 2–2.5 mm long; lateral staminodes 2–3.5(–6) mm long, pilose, antherodes 0.7–1.2 mm long. Ovary with few appressed hairs towards apex and with apical ring of minute crisped hairs; style sparsely pubescent towards base; stigma subcapitate, 0.5–0.8 mm long. Capsule 11–15 mm long, glabrous or with few appressed hairs towards apex; seeds 4–5.5 mm long and wide.

KENYA. Northern Frontier District: Mt Marsabit, Gof Redo, June 1960, *Oteke* 8!; Baringo District: Lake Bogoria [Hannington]–Mogotia km 17, Jan. 1969, *Napper & Faden* 1813!; Masai District: 72 km from Namanga on Kajiado road, Nov. 1960, *Verdcourt* 3012!

TANZANIA. Masai District: track to Kitumbeine, Jan. 1969, *Richards* 23727! & road from Longido to Engare Naibor, Mar. 1970, *Richards* 25664!; Masai/Mbulu District: Lake Manyara, June 1961, *Rodrigues* C.60!

DISTR. **K** 1, 3, 4, 6; **T** 2; S Ethiopia
HAB. *Acacia-Commiphora*, *-Terminalia* and *-Balanites* bushland including heavily grazed areas, grazed grassland, dry rocky hillsides; 1000–2150 m
USES. None recorded on herbarium specimens
CONSERVATION NOTES. This species is locally common within its range and although it can be found in undisturbed dry woodland it appears to thrive in areas of disturbance, probably being impalatable to grazers due to its extreme spininess and so able to dominant heavily grazed areas: Least Concern (LC).

SYN. *B. spinisepala* E.A. Bruce in K.B.: 98 (1932); U.K.W.F. ed. 1: 593 (1974); Blundell in Wild Fl. E. Afr.: 388 (1987); U.K.W.F. ed. 2: 272 (1994); Ensermu in Fl. Eth. 5: 409 (2006). Type: Kenya, Masai District, Kajiado, *Napier* 755 (K!, holo.; EA!, iso.), **syn. nov.**
B. mucronifolia Lindau var. *spinulifolia* Fiori in Miss. Biol. Borana, Racc. Bot.: 217 (1939). Type: Ethiopia, Sidamo, Yabello, *Cufodontis* 449 (FT, holo.; W!, iso.), **syn. nov.**

NOTE. Plants from Ethiopia and N Kenya (including the type) have broader, oblong to elliptic calyx lobes (3–)3.5–6 mm wide with up to 9 marginal spines per side and with several parallel to palmate principal veins prominent. Further south, the calyx lobes are more lanceolate, 2–3.5 mm wide, with only 2–4 marginal spines per side and usually only 1–2 principal veins prominent; the type of *B. spinisepala* is representative of this form. Two subspecies could be recognised, though there is some overlap in N Kenya and calyx shape is known to vary considerably within other species in this section.

13. **Barleria acanthoides** *Vahl* in Symb. Bot. 1: 47 (1790); Nees in DC., Prodr. 11: 240 (1847); C.B. Clarke in F.T.A. 5: 152 (1899); F.P.S. 3: 169 (1956); U.K.W.F. ed. 1: 593 (1974) pro parte, excl. fig.; U.K.W.F. ed. 2: 272 (1994) pro parte; Wood, Handb. Yemen Fl.: 272 (1997); M. & K. Balkwill in K.B. 52: 555, fig. $9B_3$ (1997); Friis & Vollesen in Biol. Skr. 51 (2): 437 (2005); Hedrén in Fl. Somalia 3: 428 (2006); Ensermu in Fl. Eth. 5: 411 (2006). Type: Yemen, *Forsskål* s.n. (C, microfiche 116: II. 7–8 pro parte, right-hand specimen!, lecto., chosen here, K!, photo.)

Spiny subshrub, 10–60(–100) cm tall; stems with dense short white retrorse to spreading eglandular hairs and sparse to dense patent glandular hairs, the latter rarely absent. Axillary spines paired, margin pinnately spinose. Leaves (oblong-) obovate, 1.3–4.7(–7.5) cm long, 0.6–2 cm wide, base attenuate or cuneate, apex rounded, obtuse or shallowly emarginate, spinose, surfaces with many short white crisped or spreading hairs particularly when young, with coarse ascending hairs above, veins strigose beneath, often with patent glandular hairs at least along margin; lateral veins 4–5 pairs; petiole to 9 mm long or absent. Flowers solitary or often in 3–11-flowered unilateral cymes 1–4.5 cm long; bracteoles spinose, linear-lanceolate, pairs ± unequal, the larger 6.5–33 mm long, 0.5–2(–3) mm wide, the smaller often declinate, both with margin denticulate to pinnately spinose, 1–4 teeth or spines per side; flowers subsessile. Calyx initially drying green or purple-tinged with darker venation, soon scarious; anticous lobe ± broadly ovate or elliptic or rarely oblong-lanceolate, 11–20 mm long, (3–)5–13.5 mm wide, base obtuse to subcordate or rarely acute, margin subentire or with short to long spinose teeth, sometimes involute, apex acute to rounded or emarginate, apiculate or mucronate, surface with short spreading or crisped white hairs and sometimes with scattered patent glandular hairs, veins sparsely strigose; posticous lobe with apex acute to attenuate, mucronate; lateral lobes lanceolate, 5–9(–11) mm long. Night flowering; corolla white or cream, drying blue-black, (60–)70–108 mm long, eglandular- and glandular-pilose outside; tube cylindrical, (54–)60–95 mm long; limb subregular; abaxial lobe 4.5–9.5 mm long, 6–11 mm wide, lateral lobes 5–11 mm long, 5–9.5 mm wide; adaxial pair as abaxial lobe but 5–8 mm wide. Stamens attached 27–47 mm from base of corolla tube; filaments 31–46 mm long; anthers held immediately above or shortly below the mouth, 3–4.3 mm long; lateral staminodes 8.5–23 mm long, pilose, antherodes 0.5–1(–1.6) mm long. Ovary with few short ascending hairs towards apex; style with minute crisped hairs towards base; stigma subcapitate to clavate, 0.5–0.8 mm long. Capsule 9.5–14 mm long, indumentum as ovary; seeds 3.3–4 mm long and wide. Fig. 49/7, p. 346.

UGANDA. Acholi District: Agoro Rest Camp, Chua, June 1942, *Eggeling* 5097!; Karamoja District: Kangole, Dec. 1957, *Wilson* 412! & Nakiloro, Moroto, June 1963, *Kertland* s.n.!

KENYA. Northern Frontier District: Samburu National Park, Dec. 1969, *Gillett* 18957!; Embu District: Ishiara, June 1985, *Beentje* 2160!; Masai District: 16 km Olorgesailie– Nairobi [road], Aug. 1957, *Verdcourt* 1832!

TANZANIA. Masai District: gorge and stream near road to Engaruka, Feb. 1970, *Richards* 25553!; Pare Distirct: Hedaru–Mheza, Feb. 1930, *Greenway* 2103! & Lake Kalimawe, Jan. 1967, *Richards* 21920!

DISTR. **U** 1; **K** 1–4, 6, 7; **T** 2, 3; Sudan, Ethiopia, Eritrea, Djibouti, Somalia; Egypt, Arabia, India & Pakistan

HAB. Open or degraded *Acacia-Commiphora* and *Combretum* bushland often on dry rocky slopes, edges of seasonal wetlands; 200–1400(–2250) m

USES. "Ritual use" (**K** 1; *Sato* 417); "roots used for cure of stomach ache" (**K** 2; *Mwangangi & Gwynne* 1019); "use for closing large holes in wood in bee hives [to] keep out animals" (**K** 3; *Ott* 10); browsed by livestock/animals (many)

CONSERVATION NOTES. Widespread and often abundant, tolerant of high grazing pressure: Least Concern (LC).

SYN. [*B. lanceata sensu* Hepper & Friis, Pl. Forsskål's Fl. Aegypt.-Arab.: 64 (1994) pro parte, *non* (Forssk.) C. Chr.]

NOTE. A variable species with several regional forms recognisable. Typical plants of *B. acanthoides* have several-flowered unilateral cymes, each with broad elliptic or ovate outer calyx lobes and narrow bracteoles with short lateral teeth/spines. However, collections with single-flowered inflorescences are not uncommon; these plants tend to be more spiny, both the bracteoles and the outer calyx lobes often having well-developed marginal spines. A particularly distinctive form occurs in the Turkana region (e.g. *Modha* 17; *Hemming* 3000). Here the plants often appear pale grey due to the dense crisped-white indumentum, and the inflorescences are often single-flowered with long bracteoles (usually over 20 mm) and relatively narrow outer calyx lobes (3–7 mm wide) which have a more involute margin than typical. Similar plants are recorded from N Ethiopia (e.g. *Gilbert* 2354, Harar region). Intermediates between this and typical *B. acanthoides* are however not uncommon, and in *Gilbert et al.* 5649 the two forms occur within the same population.

Forsskål's original material seen by Vahl is a mixed collection, with two pieces referable to *B. mucronifolia*; the error is likely to have resulted from the absence of flowers on any of the specimens. Although the protologue of *B. acanthoides* is not diagnostic, the name *B. acanthoides* has subsequently been applied consistently to the taxon matching the right hand specimen on sheet 116: II. 7–8, which is therefore selected as the lectotype here.

14. **Barleria mucronifolia** *Lindau* in Ann. Ist. Bot. Roma 6: 71 (1896); Wood, Handb. Yemen Fl.: 272 (1997); Hedrén in Fl. Somalia 3: 429 (2006). Type: Somalia/Ethiopia, Bore, Ganane near Hamari, *Riva* 1098 (FT!, holo.)

Spiny subshrub, 30–60(–150) cm tall; stems with ± dense short white hairs variously retrorse, antrorse and/or spreading. Axillary spines paired, margin pinnately spinose. Leaves obovate to oblong-elliptic, 0.9–1.8 cm long, 0.4–1.3 cm wide, base attenuate or acute, margin entire or with irregular teeth formed from swollen hair bases, apex rounded, obtuse or shallowly emarginate, spinose, strigose mainly on margin and veins beneath, with short white crisped hairs widespread or restricted to the margin, sometimes also with short spreading hairs between the veins; lateral veins 3–4 pairs; petiole to 5 mm long. Inflorescences unilateral, (1–)3–15-flowered, 1–5 cm long, congested; bracteoles (ovate-) lanceolate, pairs unequal, the larger held ± upright, 11–26 mm long, 1.5–8.5 mm wide, the smaller adpressed to the inflorescence axis or declinate, each with margin spinosely toothed, 1–5 spines per side, apex spinose, the basal pair of bracteoles of each inflorescence proportionally narrower with more conspicuous marginal spines; flowers subsessile. Calyx initially drying dark purplish but soon scarious; anticous lobe ovate, 14–20 mm long, 9–13 mm wide, base cordate, margin ± involute, subentire or with short teeth and short submarginal spines, apex rounded, obtuse or emarginate, veins sparsely strigulose, with short hairs elsewhere; posticuous lobe 15–25 mm long, 9–17 mm wide, submarginal spines more conspicuous, apex attenuate into a ± long spine; lateral lobes lanceolate, 6–8.5 mm long. Corolla pale blue or mauve, rarely cream, 44–53 mm long, eglandular-pubescent outside; tube 33–40 mm long, narrowly funnel-shaped towards mouth; limb in weak "4+1" arrangement; abaxial lobe offset by 3–5 mm, 8.5–11 mm long, 7–8.5 mm wide; lateral and adaxial lobes 7–10 mm long, 5.5–8 mm wide. Stamens attached 17–24 mm from base of corolla tube; filaments 15–21 mm long; anthers shortly exserted, 2–2.8 mm long; lateral staminodes 2.5–3.3 mm long, indumentum as stamens, antherodes 0.6–1.1 mm long. Ovary with short hairs in upper half; style with minute crisped hairs towards base; stigma subcapitate, 0.35–0.8 mm long. Capsule 12.5–19 mm long, shortly pubescent towards apex; seeds 5–6 mm long and wide.

Kenya. Northern Frontier District: NW flank of Danissa Hill, Dec. 1971, *Bally & Smith* 14610! & 30 km on the Ramu–Malka Mari road, May 1978, *Gilbert & Thulin* 1551!

Distr. **K** 1; Ethiopia, Somalia; Yemen

Hab. Dense or open *Acacia-Commiphora* bushland, dry rocky slopes and plains including degraded areas; 400–650 m

Uses. None recorded on herbarium specimens

Conservation notes. This species appears locally common, particularly in Somalia. It is probably under-recorded in NE Kenya, perhaps being overlooked due to its similarity to the more widespread *B. acanthoides*. As it favours dry bushland and rocky ground of low agricultural potential, and is tolerant of some disturbance, it is currently considered unthreatened: Least Concern (LC).

Syn. *B. homoiotricha* C.B. Clarke in F.T.A. 5: 154 (1899); Ensermu in Fl. Eth. 5: 411 (2006). Types: Somalia, Djedaynio, *Cole* s.n. (K!, syn.) & Somalia, *Gillet & Aylmer* s.n. (syn., not traced)
B. iodocephala Chiov. in Fl. Somala: 261 (1929). Type: Somalia, Migiurtini coast, between Uarsimòghe and Tudi, *Puccioni & Stefanini* 682 (FT!, holo.)

Note. The type of *B. mucronifolia* has more long hairs on the leaves and stems than the other material seen (Clarke l.c. describing it as villous) but otherwise looks a good match; I have therefore followed Hedrén in synonymising *B. homoiotricha* within *B. mucronifolia*.

From the many collections of open, non-withered flowers this species would appear to be day-flowering. Indeed, *Ash* 1237 (Harar, Ethiopia), notes "flowers opening a.m. with sunrise; by evening buds fully developed but do not open until a.m. next day". However, *Ironside-Wood* 5/73/158 (Sheikh, Somalia) noted "possibly opening in the evening, but certainly falling in the early morning".

15. **Barleria steudneri** *C.B. Clarke* in F.T.A. 5: 153 (1899); El Ghazali in Sudan Journ. Sci. 3: 22, fig. 3 (1988); Ensermu in Fl. Eth. 5: 413 (2006). Type: Eritrea, Keren, *Steudner* 1508 (B†, holo.)

Trailing subshrub with decumbent flowering branches, 15–75 cm tall; stems with dense short white ± retrorse hairs and interspersed coarse yellowish hairs. Axillary "spines" ovate or lanceolate, paired, often declinate, margin spinose. Leaves (elliptic-) obovate, 2–5.5 cm long, 1–2.2 cm wide, base cuneate or attenuate, margin entire or with irregular teeth formed from swollen hair bases, apex rounded or obtuse, mucronate, strigose particularly on margin and veins beneath, often also with short white ± crisped hairs at least on the margin and midrib; lateral veins 4–5 pairs; petiole to 5 mm long or absent. Inflorescences unilateral, several- to many-flowered, 2–6 cm long, congested; bracteoles scarious with age, ovate or oblong-lanceolate, pairs somewhat unequal, the larger held ± erect, 21–36 mm long, 7.5–16 mm wide, the smaller usually adpressed to inflorescence axis, each with margin and apex spinose; flowers subsessile. Calyx scarious with age; anticous lobe (ovate-) elliptic, 18–23(–26) mm long, 9.5–16 mm wide, base rounded or obtuse, margin laciniate, apex rounded to acute or rarely attenuate, indumentum as leaves but densely and coarsely ciliate; posticous lobe 22–33(–36) mm long, 12–17 mm wide, apex attenuate-spinose; lateral lobes linear-lanceolate, 8–12.5 mm long. Corolla pale blue or mauve, rarely whitish, 40–52 mm long, pubescent outside, hairs mainly eglandular; tube 25–34 mm long, narrowly funnel-shaped above attachment point of stamens; limb in weak "2+3" arrangement; abaxial and lateral lobes 14.5–19 mm long, 8.5–16 mm wide; adaxial lobes 13–19 mm long, 7–12 mm wide. Stamens attached ± midway along corolla tube; filaments 12–18 mm long; anthers shortly exserted, 3–3.8 mm long; lateral staminodes 5–8 mm long, pilose, antherodes 1–1.8 mm long. Ovary largely glabrous except for apical ring of minute hairs; style glabrous or sparsely pubescent towards base; stigma subcapitate, 0.5–1 mm long. Capsule 17–20 mm long, with few appressed hairs at apex; only immature seeds seen.

UGANDA. Acholi District: Rom, Chua, Dec. 1935, *Eggeling* 2361!; Karamoja District: Lodoketemit, Pian County, July 1959, *Kerfoot* 1298! & Amudat, Nov. 1964, *Tweedie* 2934!
KENYA. Kenya/Ethiopia border: Somatal Pass, Dec. 1982, *Powys* G39!; Turkana District: Kenailmet, Karasuk, June 1959, *Symes* 587! & Karasuk-Turkana, Sept. 1968, *Ossent* in EA 14062!
DISTR. **U** 1; **K** 1, 2; Sudan, S Ethiopia, Eritrea (see note)
HAB. Dry grassland and open *Acacia-Commiphora* or *Euphorbia-Acacia* woodland, *Sansevieria* thickets, sometimes on eroded soils; 1100–1800 m
USES. No specifics recorded on herbarium specimens
CONSERVATION NOTES. Although highly restricted in range, this species can be common in southern Ethiopia (*Bidgood et al.* 4966) and northeast Uganda (*Kerfoot* 1298, 3606, 4470; *Liebenberg* 182). It is also probably under-recorded due to the limited botanical exploration within its range and due to its tendency to flower during the dry season (Vollesen, *pers. comm.*). Suitable habitat is still widespread within its range: Least Concern (LC).

SYN. [*B. spinulosa sensu* Broun & Massey, Fl. Sudan: 342 (1929), *non* Klotzsch]
[*B. capitata sensu* Cufodontis in B.J.B.B. 34, Suppl.: 942 (1964) & ? *sensu* C.B. Clarke in F.T.A. 5: 153 (1899) pro parte quoad *Schweinfurth* 1071 (B†) ex Eritrea (based upon geography), *non* Klotzsch]

NOTE. The type of *B. steudneri* was presumably destroyed in WWII and no further material has been seen from Eritrea. Specimens referred to *B. steudneri* by E. Milne-Redhead at K, and later by Ensermu (l.c.) are highly isolated geographically from the type locality. Although the protologue description of *B. steudneri* largely agrees with this material, particularly with respect to the "shaggy" calyx lobes (presumed to refer to the dense coarse marginal hairs) there are differences, notably that the bracteole pairs are recorded as more highly unequal in size (the smaller recorded as only $^1/_3$ inch long, this much smaller than in our material) and in the corolla being recorded as "more than 1 in. long" (closer to 2 inches in our material). Further collections from Eritrea may therefore reveal that there are two taxa involved.

16. **Barleria inclusa** *I. Darbysh.* **sp. nov.** *B. capitatae* Klotzsch similis sed tubo corollae breviore 50–78 mm (non 85–125 mm) longo, staminibus ad tubum corollae medium vel sub medium (non ad partem superam tertiam) affixis, antheris in tubo corollae inclusis vel ad orem retentis (non breviter exsertis), calycibus minoribus et lobo postico calycis 19–33 mm (non 30–43 mm) longo differt. Type: Tanzania, Tabora District, 29 km on Tabora–Nzega road, *Bidgood, Hoenselaar, Leliyo & Vollesen* 5928 (K !, holo.; BR!, CAS!, DSM!, EA!, K !, MO!, NHT!, iso.)

Trailing or straggling perennial herb or subshrub with ± decumbent flowering branches to 30–60 cm tall; stems with dense short white retrorse or spreading hairs and interspersed coarse yellowish hairs, the latter sometimes restricted to the nodes. Axillary "spines", if present, (ovate-) lanceolate, paired, often declinate, margin spinose. Leaves elliptic or obovate, 2–5 cm long, 0.5–1.5 cm wide, base cuneate or attenuate, apex acute to rounded or shortly attenuate, ± mucronate, surfaces strigose mainly on the margin and veins beneath, usually also with short white crisped hairs at least on the margin and midrib; lateral veins 4–5 pairs; petiole to 6 mm long or absent. Inflorescences congested unilateral cymes, 1.5–8 cm long, 3–16-flowered, axis often inrolled at fruiting; bracteoles oblong-lanceolate or ovate, ± asymmetric, pairs highly unequal, the larger held ± erect, 15–30 mm long, 4.5–15 mm wide, the smaller declinate or adpressed to inflorescence axis, each with margin coarsely hairy to spinosely dentate, apex apiculate to spinose. Anticous calyx lobe broadly ovate (-elliptic), 16–26 mm long, 6–18 mm wide, base obtuse to cordate, margin subentire to laciniate, apex acute or attenuate, ± apiculate, indumentum as leaves but with many crisped white hairs; posticuous lobe 19–33 mm long, 8–22 mm wide, apex ± mucronate, surface with submarginal spines conspicuous at maturity; lateral lobes linear-lanceolate, 9–13.5 mm long. Night-flowering; corolla white or cream, drying blue-black, 59–95 mm long, glandular- and eglandular-pubescent outside; tube cylindrical, (43–)50–78 mm long; limb subregular, lobes 9–22 mm long, 8–18 mm wide. Stamens attached midway or in the lower half of the corolla tube; filaments 15–45 mm long; anthers included within corolla tube or held at mouth, 3.5–4.5 mm long; lateral staminodes 2.5–6.5 mm long, pilose, antherodes 0.5–1 mm long. Ovary glabrous or sparsely hairy towards apex, with apical ring of minute hairs; style puberulous at base; stigma subcapitate, 0.5–0.7 mm long. Capsule 14–19 mm long, glabrous or with few appressed hairs towards apex; seeds 7–8 mm long, ± 6.5 mm wide.

KENYA. Masai District: Chyulu Plains, Kwadisha Hill, Mar. 1993, *Luke* 2537!; Tana River District: Kora Game Reserve, May 1977, *Gillett* 21110!; Kwale District: near Taru, between Samburu and Mackinnon road, Sept. 1953, *Drummond & Hemsley* 4138!

TANZANIA. Arusha District: Ngare Nanyuki, 15 km from Momela, Dec. 1970, *Greenway & Kanuri* 14823!; Dodoma District, Kazikazi, May 1932, *Burtt* 3586!; Chunya District, Tsetse Research Camp, 8 km NW of Saza, May 1990, *Carter, Abdallah & Newton* 2438!

DISTR. **K** 4, 6, 7; **T** 1, 2, 4, 5, 7; not known elsewhere

HAB. Open dry *Acacia-Commiphora* and *Combretum* woodland and grassland, dry rocky hillsides, termite mounds, sometimes on seasonally wet ground, also a weed in disturbed grassland and rangeland; 1000–1550 m

USES. None recorded on herbarium specimens

CONSERVATION NOTES. This species is restricted to the Flora region and appears rather uncommon through much of its range, being known to the author from less than 30 collections. However, it appears tolerant of, or even to favour, some habitat distubance and can cope with high grazing pressure, sometimes becoming weedy in rangeland (*Pase & Mumiukha* 1982). It is therefore not considered threatened: Least Concern (LC).

SYN. [*B. capitata sensu* Agnew, U.K.W.F. ed. 2: 272 (1994), *non* Klotzsch]

NOTE. This species is closely related to *B. capitata* Klotzsch from the Zambesi basin and has previously been included within that taxon. However, *B. inclusa* has a consistently shorter corolla tube (50–78 mm long versus 85–125 mm), the stamens are attached at or below midway along the corolla tube (not in the upper third) and the anthers remain included within the corolla tube or at most partially exserted at the mouth (versus shortly exserted) at anthesis. In addition, the calyces are smaller (posticous lobe 19–33 mm long versus 30–43 mm) and the white stem indumentum is somewhat longer with the hairs retrorse or spreading (versus hairs minute and either all antrorse or mixed antrorse and retrorse).

This species appears largely divisible into two forms. Plants from the northern part of the range (**K**, **T** 2) have only 3–7-flowered inflorescences, stamens with filaments 17–45 mm long, anthers held near the apex of the corolla tube or at the mouth, and small corolla lobes 9–16 mm long. Southern populations (**T** 1, 4, 5, 7) have 5–16-flowered inflorescences, shorter stamens 15–17 mm long with the anthers included within the tube 4.5–9 mm below the corolla mouth, and usually larger corolla lobes, (11.5–)15–22 mm long. Within the latter form, plants from the eastern and southern extremes of the range (e.g. *Kuchar* 23597; *Carter et al.* 2438) are largely spineless, the calyx and bracteole margins subentire; elsewhere this species is usually spiny. However, much of the material seen to date is rather scant and better quality collections from across its range are required before any infraspecific taxa are recognised.

17. **Barleria cristata** *L.*, Sp. Pl., ed. 1, 2: 636 (1753); Sims in Bot. Mag. 39, t. 1615 (1814); C.B. Clarke in Fl. Brit. Ind. 4: 488 (1884). Type: India, Herb. Linnaeus No. 805.12 (LINN!, lecto., selected by Brummitt & Vollesen in Taxon 41: 558 (1992))

Subshrub, 50–200 cm tall; stems with coarse buff-yellow ascending hairs and shorter, finer, ± retrorse hairs. Axillary spines absent. Leaves elliptic, ovate or lanceolate, 3.8–12.5 cm long, 1.2–4 cm wide, base and apex attenuate or acute, the latter minutely apiculate, surfaces strigose, hairs many on margin and veins beneath; lateral veins 4–6 pairs; petiole 2–12 mm long. Inflorescences 1(–2) per axil, each with flowers solitary or in 2–4-flowered congested unilateral cymes; bracteoles linear-lanceolate, 8.5–16 mm long, 1–1.8 mm wide, margin with short slender teeth developing at maturity, apex spinose; flowers sessile. Calyx initially green or tinged purplish, paler towards base, soon scarious; anticous lobe ovate-trullate, 14.5–19 mm long, 6.5–8.5 mm wide, base and apex acute or attenuate or the latter rarely emarginate, apiculate, margin spinulose, surface strigulose particularly on the veins, with patent glandular hairs many along the margin; posticuous lobe 17–23 mm long; lateral lobes lanceolate, 6–8 mm long. Corolla blue, mauve or white, 46–75 mm long, glandular-pilose outside; tube 31–43 mm long, campanulate above attachment point of stamens; limb in "4+1" arrangement; abaxial lobe offset by 4–8.5 mm, 14–25 mm long, 11.5–20 mm wide; lateral and adaxial lobes 11–22 mm long, 8–15 mm wide. Stamens attached ± midway along corolla tube; filaments 20–25 mm long; anthers 1.8–3 mm long; lateral staminodes ± 3 mm long, pilose, antherodes ± 0.5 mm long. Pistil glabrous or style with few hairs towards base; stigma subcapitate, 0.4–0.7 mm long. Capsule 13.5–17 mm long, glabrous; seeds ± 5 mm long, 4.5 mm wide.

TANZANIA. Zanzibar, Mjini District, Nov. 1999, *Fakih & Abdulla* 488!

DISTR. **Z**; India and Himalaya to S China and Thailand, widely cultivated and sometimes naturalised elsewhere in the tropics

HAB. "In wet area"; ± sea level

USES. Widely cultivated as an ornamental

CONSERVATION NOTES. This species is common within its native range in tropical and subtropical Asia and is also widely naturalised elsewhere: Least Concern (LC).

NOTE. The cited specimen is almost certainly derived from cultivation but it appears to be naturalised on Zanzibar. Elsewhere in our region, this species is recorded from gardens and nurseries in SE Kenya and E Tanzania. Within its native range, it is very variable in terms of leaf shape and size, indumentum, inflorescence density and calyx morphology. However, the East African material is rather uniform and matches cultivated plants from elsewhere in the tropics; the description is based upon this form.

18. **Barleria repens** *Nees* in DC., Prodr. 11: 230 (1847) pro parte (see note); J.D. Hooker in Bot. Mag. 113, t. 6954 (1887) pro parte excl. *Wakefield* s.n.; C.B. Clarke in F.T.A. 5: 166 (1899) pro parte excl. *Bojer* s.n. ex Pemba Island; Obermeijer in Ann. Transv. Mus. 15: 168 (1933); U.O.P.Z.: 139 (1949); M. & K. Balkwill in K.B. 52: 557, figs. 2C & $9A_7$ (1997). Type: Mozambique, Raza Island, *Forbes* s.n. (K !, lecto., chosen here; K !, isolecto.)

Trailing or scandent subshrub, branches 15–350 cm long; stems with buff or yellowish, appressed or ascending hairs, usually also with shorter ± retrorse hairs when young. Axillary spines absent. Leaves often somewhat anisophyllous, (ovate-) elliptic or somewhat obovate, 1.8–7.3 cm long, 0.7–3 cm wide, base attenuate, apex subattenuate to obtuse or rounded, ± apiculate, surfaces strigose, hairs many on margin and veins beneath; lateral veins 4–6 pairs; petiole 4–11 mm long. Flowers solitary or rarely in 2–3-flowered congested unilateral cymes; bracteoles linear or oblanceolate, 1.5–9.5(–17) mm long, 0.2–1.5(–3) mm wide, unarmed, entire; flowers subsessile. Calyx initially green (-brown), soon scarious, accrescent; anticous lobe broadly ovate, (12–)15–23 mm long, (7.5–)9.5–14 mm wide in flower, up to 29.5 mm long and 23 mm wide in fruit, base rounded or cordate, apex acute- or obtuse-apiculate, rarely emarginate, margin entire or with minute teeth formed by swollen hair bases, veins and margin sparsely strigulose particularly towards base; posticuous lobe (13.5–)17–29 mm long in flower, to 35 mm in fruit, apex acute-apiculate; lateral lobes lanceolate, 4–7 mm long. Corolla bright red to rose-pink, (38–)43–61 mm long, glandular- and eglandular-pilose outside; tube

27–38 mm long, campanulate above attachment point of stamens; limb in "4+1" arrangement; abaxial lobe offset by (3–)7–9.5 mm, (11–)15–21 mm long, 10–17 mm wide; lateral lobes (10–)11.5–17 mm long, 8.5–14 mm wide; adaxial lobes (8.5–)11–16 mm long, 5.5–10.5 mm wide. Stamens attached ± midway along corolla tube; filaments (14–)16–22 mm long; anthers 2.5–3.5 mm long; lateral staminodes 1–3 mm long, pilose, antherodes 0.5–1.2 mm long. Pistil glabrous; stigma subcapitate, 0.4–0.7 mm long. Capsule 15–19 mm long, glabrous; seeds 3.8–5 mm long and wide.

KENYA. Kilifi District: Mida, Sept. 1929, *Graham* 2102! & Mida to Arabuko area, Dec. 1990, *Luke & Robertson* 2598!

TANZANIA. Uzaramo District: Kisiju, 65 km S of Dar es Salaam, Sept. 1977, *Wingfield* 4191!; Rufiji District: Kirongwe, Aug. 1937, *Greenway* 5159!; Kilwa District: Kilwa Island, Mar. 1885, *Kirk* s.n.!; Zanzibar, Mbweni, June 1960, *Faulkner* 2615!

DISTR. **K** 7; **T** 6, 8; **Z**; coastal Mozambique & Kwa-Zulu Natal

HAB. Understorey of dry coastal forest and plantations, margins of coastal scrub and lowland woodland, on sandy soils including stabilised sand dunes, occasionally naturalised from cultivation near the coast; 0–100 m

USES. Cultivated as an ornamental

CONSERVATION NOTES. This species is widespread along the Indian Ocean coast of Africa south of the Equator and, although uncommon in our region, it is locally abundant in Mozambique and South Africa where it favours scrubland on stabilised dunes. It is not considered threatened: Least Concern (LC).

SYN. *B. querimbensis* Klotzsch in Peters, Reise Mossamb., Bot. 1: 205 (1861). Type: Mozambique, Quirimba [Kerimba] Island, *Peters* s.n. (B†, holo.), **syn. nov.**

B. swynnertonii S. Moore in J.L.S. 40: 160 (1911). Type: Mozambique, Beira, *Swynnerton* 1958 (BM!, holo.), **syn. nov.**

NOTE. The protologue of *Barleria repens* was based upon two specimens: *Bojer* s.n. from Pemba Island and *Forbes* s.n. from Raza Island, incorrectly cited by Nees as "In insulæ Pembae (Raza insulæ ad oras orientales Africæ australioris) locis humidis (Forbes! in h. Hook.)". The Bojer collection is sterile but clearly differs from the Forbes material, most notably in stem and leaf indumentum. The more ample fruiting Forbes collection, held on two sheets at K (one shared with *Bojer* s.n.), clearly forms the basis for the original description and for the future application of the name *Barleria repens* and is therefore chosen as the lectotype here. The identity of the sterile Bojer specimen remains uncertain; it is clearly an *Acanthaceae*, having opposite leaves with cystoliths, but does not match any *Barleria* species.

SECT. II. FISSIMURA

M. Balkwill in J. Biogeogr. 25: 110 (1998); M. & K. Balkwill in K.B. 52: 569 (1997)

Barlerites Oerst. in Vidensk. Medd. Dansk. Naturhist. Foren. Kjobenh.: 25 (1854) pro parte

Barleria subgen. *Eu-Barleria* '*Villosae*' *sensu* C.B. Clarke in F.T.A. 5: 144 (1899) pro parte excl. *B. antunesi*, *B. holstii*, *B. limnogeton*, *B. querimbensis*, *B. repens* and *B. rotundisepala*

Barleria sect. *Eu-Barleria* subsect. *Dispermae sensu* Obermeijer in Ann. Transv. Mus. 15: 137 (1933)

Axillary spines absent. Indumentum simple. Inflorescences unilateral, dichasial or single-flowered cymes, axillary or compounded into a terminal synflorescence; bracts foliaceous, reduced in taxa with a terminal synflorescence. Outer calyx lobes with palmate-reticulate venation; lateral lobes linear-lanceolate. Corolla white, blue, purple or rarely red-purple, usually drying blue with darker venation or blue-black throughout; tube cylindrical below attachment point of stamens, with a funnel-shaped or campanulate throat above; limb in "4+1" arrangement. Staminodes 3, lateral pair with antherodes well developed or rarely absent; adaxial staminode with antherode usually absent. Stigma subcapitate or clavate, of 2 subconfluent lobes. Capsule drying black, laterally flattened, fusiform or somewhat obovate, without a prominent beak; lateral walls thin, usually partially tearing from the thickened flanks at dehiscence; septum largely membranous. Seeds 2, discoid with dense (purplish-) brown, bronze or golden hygroscopic hairs.

A section of 20–25 species, confined to Africa and Arabia and most diverse in our region. The taxonomy of this section is complex with several of the species polymorphic and the majority closely related to one another, rendering formal delimitation of taxa extremely difficult. Central to the problem is how to treat the *B. ventricosa* aggregate, which contains a number of variants that, in parts of their range, appear quite distinct and grow in close proximity without hybridisation but, in other areas, appear to merge entirely. The resultant variation in *B. ventricosa sensu lato* is so great that it is difficult to know where to draw the line and I accept that several of the species maintained as distinct from *B. ventricosa* here could be considered further extreme variants within that aggregate; I have, however, tried to avoid lumping taxa wherever possible. A similar situation is true for *B. submollis* which is, in itself, closely allied to *B. ventricosa*.

Due to the nature of the variation in sect. *Fissimura*, identification by dichotomous key can be difficult. I have therefore added notes on potential identification pitfalls to several of the species. Geography can help and so is included in the key.

1. Anthers and stigma extremely long-exserted, held well beyond the corolla lobes, filaments 35–65 mm long, longer than the entire corolla; **U** 2, 4, **T** 1, 4 . . . . 19. *B. brownii* p.358
   Anthers and stigma exserted from the corolla tube but either not extended beyond the lobes or only shortly so, filaments 8–41 mm long, shorter than the entire corolla 2
2. Suffruticose perennial of fire-prone habitats, producing 1 to several annual shoots from a woody base and rootstock; **T** 1, 4, 5, 7 . . . . 24. *B. boehmii* p.368
   Perennial herb or subshrub, variously erect, trailing, scrambling or scandent, not suffruticose 3
3. Leaves ± broadly ovate, base rounded, cordate or obtuse; calyx indumentum spreading; stems with a mixture of sparse to many long coarse spreading hairs, many short retrorse or spreading hairs and few to many (rarely absent) short patent glandular hairs, or if dominated by spreading long golden-yellow hairs (**T** 3 variant of *B. submollis*) then leaf length/width ratio 1–1.5/1, corolla with basal cylindrical portion of tube 4–7 mm long and capsule 9–12.5 mm long 4
   Leaves ovate to elliptic (-obovate), base acute, cuneate or attenuate, or if obtuse to subcordate then longer hairs on stems and calyces appressed or strongly ascending, or if spreading and golden-yellow then leaf length/width ratio over 1.5 (–2.5)/1 and/or corolla with basal cylindrical portion 9–12.5 mm long and capsule 13–17 mm long 6
4. Corolla 16–21 mm long, basal cylindrical portion of tube ± equal in length to expanded throat; **K** 7, **T** 3, 6, 8 . . . . 27. *B. usambarica* p.371
   Corolla (25–)35–75 mm long, basal cylindrical portion of tube clearly shorter than expanded throat 5

5. Stem indumentum dominated by dense short white retrorse hairs, often also with many patent glandular hairs, long spreading hairs sparse; corolla 56–75 mm long; anthers (3.5–)4–5 mm long; calyx lobes with many patent glandular hairs throughout; **K** 1, 3, 4 . . 26. *B. robertsoniae* p.370
   Stem indumentum variable but not dominated by white retrorse hairs: short retrorse hairs colourless and less conspicuous, glandular hairs usually many, long spreading hairs often many or dominant; corolla (25–)35–61 mm long; anthers 2–3.7 mm long; calyx lobes usually with glandular hairs only many along margin, sometimes absent; **U** 1, **K** 1–4, 6, 7, **T** 2, 3 . . . . 25. *B. submollis* p.369
6. Corolla red-purple or rose-red; tube somewhat declinate, gradually and narrowly expanded upwards, 2–3 times longer than the abaxial lobe; lobes held forwards, barely spreading at anthesis (fig. 51/5); **T** 6–8 . . . . 31. *B. lukwangulensis* p.376
   Corolla white, blue, mauve or purple; tube straight, divided into a basal cylindrical portion and funnel-shaped, campanulate or broadly cylindrical throat, less than or up to 2 times longer than the abaxial lobe, lobes ± widely spreading at anthesis 7
7. Outer calyx lobes glabrous except along margin, or if with sparse hairs on the principal veins then pedicels 5–11 mm long above bracteoles, flowers always solitary and plants soon woody with pale greyish or sandy-coloured bark 8
   Outer calyx lobes variously sparsely to densely hairy, never glabrous; if hairs restricted to principal veins then not with the combination of conspicuously pedicellate, solitary flowers and pale woody stems 9
8. Scandent to 2–3 m in coastal forest, leafy stems herbaceous; corolla white; posticous calyx lobe with apex acute; flowers solitary or in 2–3-flowered cymes, often clustered towards stem apices; **K** 7, **T** 3 . . . . 29. *B. lukei* p.373
   Subshrub to 50 cm tall or prostrate perennial in dry *Brachystegia* woodland; leafy stems soon woody with pale greyish or sandy-coloured bark; corolla blue with white throat; posticous calyx lobe with apex attenuate; flowers solitary, axillary; **T** 5, 7 32. *B. pseudosomalia* p.377
9. Corolla white, basal cylindrical portion of tube 9–12.5 mm long, ± as long as the broadly funnel-shaped throat; stems, leaves and calyces with long, spreading or ascending coarse golden-yellow or golden-brown hairs; **T** 3, 6, 8 . . . . 28. *B. amanensis* p.372
   Corolla blue, mauve or purple or if white then basal cylindrical portion of tube clearly shorter than the expanded throat, less than 9 mm long or if longer then expanded throat broadly cylindrical below the mouth and indumentum paler, buff or yellowish, not golden-yellow or golden-brown 10

10. Plants drying green-black or brown-black; corolla drying blue-black; calyces with rather stiff, ± appressed pale buff (-yellow) strigose hairs largely restricted to the principal veins and margins 11
 Plants drying green to dark green or if drying blackish then corolla drying pinkish or blue with conspicuous darker venation and/or plants conspicuously hairy, particularly on the calyces, the hairs less stiff, ascending or subspreading 12
11. Scandent perennial herb or subshrub to 8 m tall in moist montane forest; outer calyx lobes with base obtuse, rounded or shortly attenuate, margin entire or shortly and irregularly toothed; leaf length/width ratio over 2/1; basal cylindrical portion of corolla tube 6.5–11.5 mm long; **T** 2, 5, 6 . . . . . 30. *B. scandens* p.374
 Much-branched shrub of dry thicket; outer calyx lobes with base subcordate, margin usually consicuously toothed; leaf length/width ratio under 2/1; basal cylindrical portion of corolla tube ± 5.5 mm long; **K** 1 . . . . . 33. *B. sp. A* p.378
12. Outer calyx lobes with a conspicuous pale patch towards the base, contrasting with the dark reticulate venation; ovary and immature capsule with short-stalked glands towards apex; **T** 4 . . . . . 20. *B. neurophylla* p.360
 Outer calyx lobes not conspicuously paler towards base; ovary and capsule glabrous 13
13. Corolla 38–63 mm (usually over 45 mm) long including tube 25–37 mm, tube rapidly expanded into a broadly cylindrical throat above attachment of stamens; **K** 1, 3, 4, 7, **T** 2 . . . . . 23. *B. volkensii* p.367
 Corolla in our region 15–40 mm long including tube 9.5–22 mm, throat funnel-shaped to convexly so 14
14. Outer calyx lobes with base attenuate, cuneate or rounded, margin entire or if toothed then 2–7 teeth per side, teeth longest at the widest point of the lobes; rarely with up to 12 regular teeth per side but then corolla larger (28 mm*; corolla range 15–40 mm elsewhere in this species); widespread (see separate key for variants) . . . . . 21. *B. ventricosa* aggregate p.360
 Outer calyx lobes with base (sub)cordate, margin regularly and conspicuously denticulate, 9–15 teeth per side; corolla 17–24 mm long; **K** 1 . . . 22. *B. boranensis* p.366

* see note to *B. boranensis*, p.366

19. **Barleria brownii** *S. Moore* in J.B. 46: 73 (1908); Heine in F.W.T.A. ed. 2, 2: 420 (1963) & in Fl. Gabon 13: 166, fig. 33 (1966); Friis & Vollesen in Biol. Skr. 51 (2): 437 (2005); Ensermu in Fl. Eth. 5: 404, fig. 166.26 (2006). Type: Uganda, Mengo District, Entebbe, *Brown* 313 (BM!, holo.; K!, iso.)

Straggling or scandent perennial herb or shrub, 40–600 cm tall. Young stems with short retrorse hairs in two opposite lines, nodes and uppermost internodes sparsely yellow-strigose, soon glabrescent. Leaves ovate to oblong-elliptic, 6–13 cm long, 3–6.5 cm wide, base cuneate, shortly attenuate or rounded, apex acuminate or attenuate, glabrous except for occasional appressed hairs on principal veins; lateral veins 4–6 pairs; petiole 10–37 mm long. Inflorescences of 2–9-flowered, unilateral or partially dichasial cymes compounded into a paniculate or spiciform terminal synflorescence 2–12 cm long; bracts broadly ovate to somewhat obovate, 6–25 mm long, 3.5–23 mm wide; bracteoles linear, oblanceolate or elliptic, 2–18 mm long, 0.3–7 mm wide. Anticous calyx lobe oblong-elliptic to -obovate, rarely ovate-orbicular, 7.5–11(–20) mm long, 3–6(–18) mm wide, base acute to rounded, margin entire, apex attenuate to rounded or shallowly emarginate, surface sparsely appressed-pubescent, often with few long glandular hairs particularly towards apex; posticous lobe 8–14(–23) mm long, 4–8.5(–21) mm wide, margin often involute, apex usually with a short acumen, rarely rounded; lateral lobes 5–7.5(–12.5) mm long. Corolla white or blue, throat often tinged pink or purple, 30–50 mm long, glandular-pilose and with fewer eglandular hairs outside; tube 13.5–21 mm long, campanulate above attachment point of stamens; abaxial lobe offset by 2.5–6.5 mm, 16–27 mm long, 11.5–15 mm wide; lateral lobes 14.5–26 mm long, 7.5–12.5 mm wide; adaxial lobes as lateral pair but 4.5–8 mm wide. Stamens attached 5.5–9 mm from base of corolla tube, long exserted; filaments 35–65 mm long; anthers 2.5–4 mm long; lateral staminodes 1.5–9.5 mm long, with minute stalked glands and occasional hairs towards base or throughout, antherodes 1.25–2 mm long. Pistil glabrous; stigma long exserted, clavate, 0.4–0.9 mm long. Capsule 12.5–14.5 mm long, glabrous; seeds ± 6 mm long, 5 mm wide.

UGANDA. Bunyoro District: Budongo Forest, Nov. 1938, *Loveridge* 116!; Busoga District: Kiryamuli near Mutai, N of Jinja, Nov. 1952, *Wood* 500!; Masaka District: Malabigambo Forest, 6 km SSW of Katera, Oct. 1953, *Drummond & Hemsley* 4548!

TANZANIA. Bukoba District: Minziro Forest Reserve, June 1958, *Procter* 963! & Minziro, July 1958, *Makwilo Semkiwa* 52!; Kigoma District: Lubalisi village, Ikubulu subvillage, Ntakata Forest, Aug. 2005, *Abeid & Pondamali* 2000! (see note)

DISTR. **U** 2–4; **T** 1, 4; Ghana to SW Ethiopia, Congo-Kinshasa and Angola

HAB. Undergrowth of moist forest including swamp and riverine forest, often along forest margins or in secondary growth; 900–1250 m

USES. None recorded on herbarium specimens

CONSERVATION NOTES. Widespread in the Guineo-Congolian forests and well-represented in herbarium collections, this species is probably not uncommon. It also appears tolerant of moderate disturbance, apparently becoming numerous in disturbed, regenerating forest: Least Concern (LC).

SYN. *B. longistyla* Lindau in Z.A.E. 2: 298 (1911). Type: Congo-Kinshasa, Kapangapanga, between Beni and Muera, *Mildbraed* 2303 (B†, holo.; BR!, iso.), **syn. nov.**

B. talbotii S. Moore in Cat. Talbot's Nigerian Pl.: 86 (1913). Type: Nigeria, Oban, *Talbot* 1396 (BM!, holo.; K!, iso.) – see note

NOTE. *Abeid & Pondamali* 2000 (**T** 4) differs from all other material in our region in having large, ovate-orbicular outer calyx lobes measuring 15–23 × 13–21 mm, length/width ratio 1–1.2/1 (1.4–2.5/1 in other material from our region). Similar specimens are recorded from West Africa (notably Nigeria, Bioko and W Congo-Kinshasa) and were previously separated as *B. talbotii* S. Moore. In Congo-Kinshasa, this form has been provisionally named as *B. brownii* subsp. *latisepala* by Champluvier on herbarium sheets at BR. However, intermediate specimens are rather widely recorded in West and Central Africa. *B. brownii* is therefore treated as a single entity here, though the two forms are highly distinct within our region.

20. **Barleria neurophylla** *C.B. Clarke* in F.T.A. 5: 166 (1899). Type: Congo-Kinshasa, Lake Tanganyika, Kavala Island, *Carson* s.n. (K!, holo.)

Erect, scrambling or procumbent perennial herb, 10–180 cm long; stems with rather sparse long-ascending or spreading coarse pale buff or yellowish hairs and with a line of shorter crisped hairs in opposite furrows, later glabrescent. Leaves elliptic or somewhat obovate, 6–13 cm long, 2.3–5.3 cm wide, base cuneate or attenuate, apex (sub)attenuate, surfaces with sparse ascending or spreading long hairs; lateral veins 4–6 pairs, prominent beneath; petiole 0–5 mm long. Inflorescences axillary, 1–2-flowered, sometimes clustered towards branch apices, subsessile; bracts foliaceous; bracteoles linear (-lanceolate), 2–5.5 mm long, to 0.4 mm wide. Outer calyx lobes conspicuously pale at base and with darker reticulate venation, scarious in fruit; anticous lobe broadly ovate or triangular, 9.5–17 mm long, 9–18 mm wide, base shortly attenuate to rounded, margin entire or with minute teeth formed by swollen hair bases, apex shallowly emarginate, indumentum as leaves but more dense on principal veins and margin; posticous lobe 10–19 mm long, apex acute or obtuse, ± apiculate; lateral lobes 6–8 mm long. Corolla pale blue to purple, throat with purple or reddish lines or spots, 32–36 mm long, sparsely glandular-pilose outside; tube 19–21 mm long, funnel-shaped above attachment point of stamens; abaxial lobe offset by 5–6 mm, 12–14.5 mm long, 11–13 mm wide; lateral lobes 9.5–11.5 mm long, 6–7 mm wide; adaxial lobes 8–9 mm long, 3.5–4.5 mm wide. Stamens attached ± 7.5 mm from base of corolla tube; filaments ± 21 mm long; anthers 2.7–3.2 mm long; lateral staminodes ± 2.5 mm long, pubescent, antherodes ± 1 mm long. Ovary with minute short-stalked glands in upper half; style glabrous; stigma subcapitate, 0.5–0.7 mm long. Capsule 10.5–11.5 mm long, with sparse stalked glands towards apex or glabrescent; seeds ± 6 mm long, 5 mm wide.

TANZANIA. Kigoma District: Kabogo Mts, May 1963, *Azuma* in *Kyoto Univ. Expedition* 542! & Ntakata area, Aug. 2005, *Mwangoka & Ramadhani* 4052!; Mpanda District: Mahali Mts, Kasiha Valley, July 1959, *Newbould & Harley* 4503!

DISTR. **T** 4; E Congo-Kinshasa

HAB. By streams and pools in riverine forest, lakeshore rock crevices; 750–1250 m

USES. None recorded on herbarium specimens

CONSERVATION NOTES. *B. neurophylla* appears restricted to the shores of central Lake Tanganyika and adjacent river valleys. This area remains under-studied botanically, particularly on the western (Congo) shore, hence the abundance of this species is difficult to judge. Expeditions into the Mahali Mts during the 1950s produced several records for this species and it may be locally frequent there; this area is now a National Park and is therefore afforded some protection. It was recorded as abundant at Ntakata, Kigoma in 2005 (*Mwangoka & Ramadhani* 4052). However, significant losses of riverine forest have occurred in this region outside of protected areas, which may well have resulted in losses of some populations. With fewer than 10 localities currently known, this species is therefore assessed as Vulnerable (VU B2ab(iii)).

NOTE. The pale basal portion to the outer calyx lobes, in which the darker, green-black palmate principal veins and scalariform tertiary veins are conspicuous, helps to distinguish this species from others in this section, though this character is best observed in flowering material before the calyx turns scarious. This patterning is rather similar to that found in some species of sect. *Barleria*; indeed *B. neurophylla* has previously been placed close to *B. holstii* Lindau, but is readily separable from that species even in the absence of fruits by the short, linear (-lanceolate), not large ovate, bracteoles. Within sect. *Fissimura*, *B. neurophylla* appears most closely allied to *B. ruellioides* T. Anderson from West Africa.

21. **Barleria ventricosa** *Nees* aggregate, Nees in DC., Prodr. 11: 230 (1847); C.B. Clarke in F.T.A. 5: 164 (1899) pro parte excl. *Fischer* 135; F.P.S. 3: 170 (1956); F.P.U.: 139 (1971); U.K.W.F. ed. 1: 594, fig. p. 592 (1974); Champluvier in Fl. Rwanda 3: 438, fig. 137.2 (1985); Blundell, Wild Fl. E. Afr.: 388, pl. 846 (1987); U.K.W.F. ed. 2: 273, pl. 119 (1994); K.T.S.L.: 600 (1994); Friis & Vollesen in Biol. Skr. 51 (2): 437 (2005);

Hedrén in Fl. Somalia 3: 434 (2006); Ensermu in Fl. Eth. 5: 404, fig. 166.27 (2006). Types: Ethiopia, Tigray, Mt [Scholoda] Soloda, *Schimper* I.42 (?GZU, syn.; BR!, HBG!, K!, M!, isosyn.) & sine loc., *Schimper* III.1903 (?GZU, syn.; BR!, K!, M!, MPU!) & Mt Kubbi, *Schimper* II.797 (?GZU, syn.; BR!, isosyn.); Ethiopia, *Quartin Dillon* s.n. (P, syn.)

Trailing, scrambling, scandent or erect perennial herb or shrub, 10–500 cm tall; stems with sparse to many appressed or ascending, more rarely subspreading, coarse buff or yellowish hairs and with finer, shorter retrorse or spreading hairs restricted to opposite furrows or more widespread, sometimes with sparse to many patent glandular hairs when young. Leaves ovate or elliptic, 1.5–13 cm long, 1–7.5 cm wide, base attenuate or cuneate, upper leaves sometimes rounded or subcordate, apex attenuate, acuminate or acute, surfaces with sparse to dense coarse pale buff (-yellow) or rich-brown ascending hairs at least on the veins, sometimes with finer, more spreading hairs between the veins beneath; lateral veins 3–6 pairs; petiole 5–40 mm long or absent. Inflorescences axillary or congested into a verticillate or dense spiciform terminal synflorescence 2–8.5 cm long, each axil 1–7-flowered, unilateral or partially dichasial; bracts foliaceous or, in the synflorescence, much-reduced, then typically 9–35 mm long, 4–14 mm wide; bracteoles linear, spatulate or narrowly elliptic, 0.7–16 mm long, 0.2–4 mm wide; flowers sessile or pedicels to 4.5 mm long. Anticous calyx lobe elliptic, rhombic, ovate or rounded, 4.5–17 mm long, 2.5–14.5 mm wide, base attenuate, cuneate or rounded, margin entire or toothed, teeth with an apical bristle, apex bifid for up to 6.5 mm or acute to attenuate, indumentum as leaves beneath, with or without few to many patent glandular hairs; posticous lobe 6–24 mm long, 3–17 mm wide, apex acute, attenuate or obtuse, apiculate; lateral lobes 4.5–12 mm long. Corolla pale to bright blue, blue-purple or rarely white, 15–40 mm long, upper portion of tube and limb glandular-pilose and with retrorse eglandular hairs outside; tube 9.5–22 mm long, funnel-shaped above attachment point of stamens; abaxial lobe offset by 1.5–7 mm, 6–18 mm long, 5.5–14.5 mm wide; lateral lobes 4.5–14 mm long, 3.5–11.5 mm wide; adaxial lobes as lateral pair but 2.5–8.5 mm wide. Stamens attached 3–10 mm from base of corolla tube; filaments 8–24 mm long; anthers 1.7–4 mm long; lateral staminodes 1–2.5 mm long, pilose and with or without short-stalked glands, antherodes 0.5–1.4 mm long. Pistil glabrous; stigma subcapitate or clavate, 0.3–0.9 mm long. Capsule 7–14.5 mm long, glabrous; seeds 4.3–6.7 mm long, 3.8–5.5 mm wide. Fig. 50, p. 362.

DISTR. (for aggregate species) **U** 1–4; **K** 1–7; **T** 1–3, 5–7; Sudan, Ethiopia, Eritrea, Somalia, Rwanda, Congo-Kinshasa, Zimbabwe, South Africa; Yemen

CONSERVATION NOTES. *B. ventricosa sensu lato* is both widespread and locally abundant in a range of habitats: Least Concern (LC).

SYN. [*B. mollis* R. Br. in Salt, Voy. Abyss. App. 4: 64 (1814), *nom. nud.*]
[*Ruellia* no. 2 *sensu* Thomson in Speke, Journ. Disc. source Nile. Append.: 643 (1863)]
B. grantii Oliv. in Trans. Linn. Soc., London 29: 127, pl. 127 (1875); C.B. Clarke in F.T.A. 5: 164 (1899). Type: Uganda, West Nile District, Madi, banks of Nile, *Grant* 665 (K!, holo., number not listed in protologue), **syn. nov.**
[*B. submollis* Lindau in E.J. 20: 21 (1894) pro parte quoad *Holst* 8935, *non* lectotype chosen below]
B. stuhlmannii Lindau in E.J. 20: 20 (1894); C.B. Clarke in F.T.A. 5: 167 (1899); Champluvier in Fl. Rwanda 3: 438, fig. 137.4 (1985). Types: Tanzania, Mpwapwa, *Stuhlmann* 289 (B†, holo.); Mpwapwa, *Hornby* 109 (K!, neo., chosen here), **syn. nov.**
B. vix-dentata C.B. Clarke in F.T.A. 5: 165 (1899); Friis & Vollesen in Biol. Skr. 51 (2): 438 (2005); Ensermu in Fl. Eth. 5: 406 (2006). Type: Uganda, without precise locality, *C.T. Wilson* 37 (K!, lecto., chosen here), **syn. nov.**
B. micrantha C.B. Clarke in F.T.A. 5: 168 (1899); Blundell, Wild Fl. E. Afr.: 387 (1987); U.K.W.F. ed. 2: 273 (1994). Type: Uganda, E side of Lake Albert & Edward, *Scott-Elliot* s.n. (K!, holo.), **syn. nov.**
B. scindens Oberm. in Ann. Transv. Mus. 15: 171, pl. I.7 (1907); Mapaura & Timberlake, Checklist Zimb. Vasc. Pl.: 13 (2004). Type: Zimbabwe, Bulawayo, *Rogers* 13592 (PRE holotype; BOL! isotype); **syn. nov.**

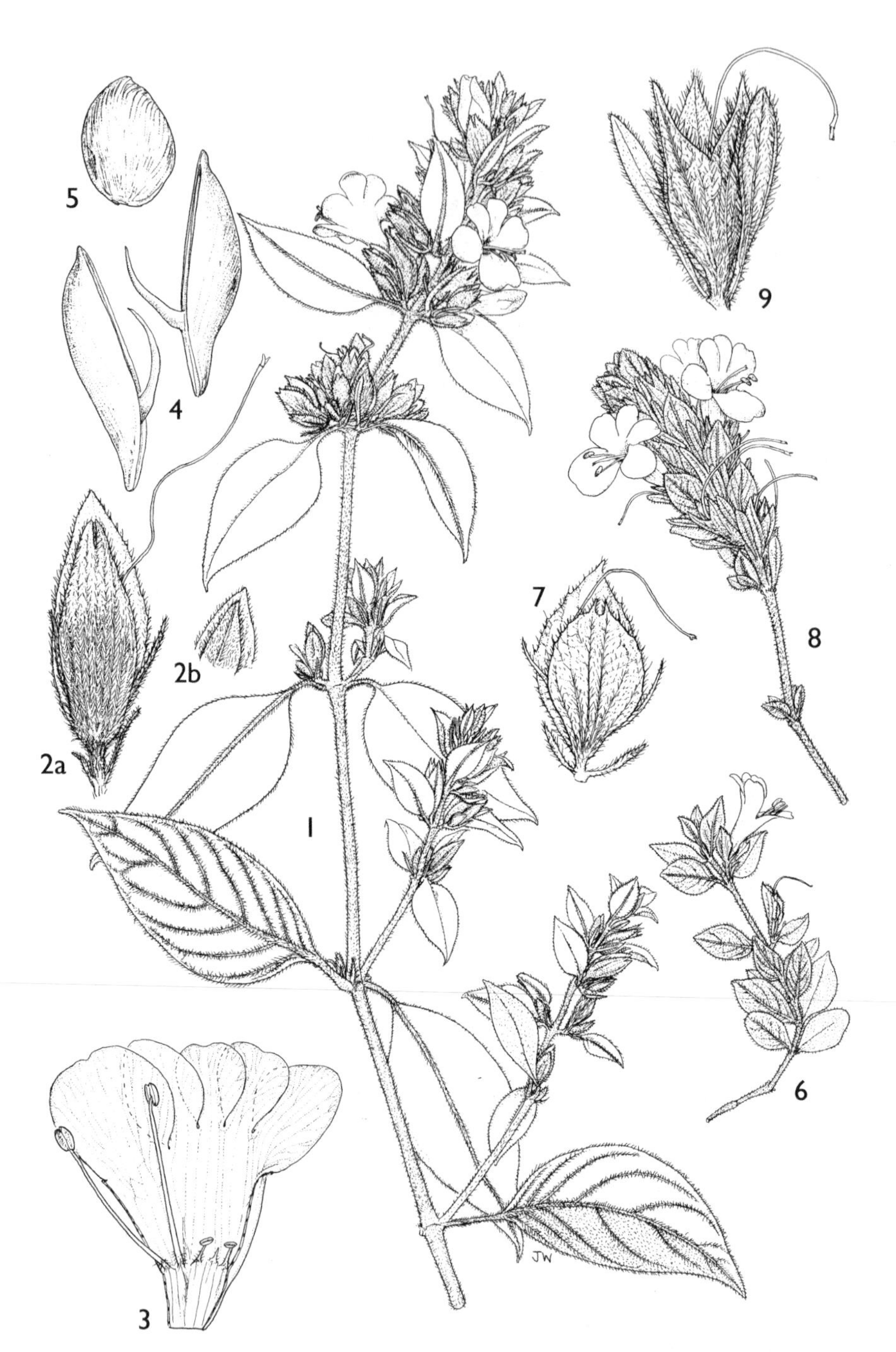

FIG. 50. *BARLERIA VENTRICOSA* aggregate: "*B. KENIENSIS*" — **1**, habit, × $^2/_3$; **2a**, calyx and bracteoles with pistil, × 2; **2b**, apex of calyx, variant with entire anticous lobe, × 2; **3**, dissected corolla with androecium, abaxial lobe to the left, × $1^1/_2$; **4**, capsule, × 3; **5**, seed, × 3. "*B. STUHLMANNII*" — **6**, habit, short lateral flowering branch, × $^2/_3$; **7**, calyx and bracteoles with pistil, × 2. "*B. VIX-DENTATA*" — **8**, inflorescence, × $^2/_3$; **9**, calyx and bracteoles with pistil, × 2. 1 from *Verdcourt* 3719; 2 from *Verdcourt* 2149; 3 from *Congdon* 375; 4 & 5 from *Tweedie* 3789; 6 from *Kerfoot* 3479; 7 from *Luke* 7556; 8 & 9 from *Lind* 2112. Drawn by Juliet Williamson.

B. bagshawei S. Moore in J.B. 48: 252 (1910). Types: Uganda, shore of "Albert Lake Edward", E side, *Bagshawe* 1412; Bunyoro District, bank of Nile at Foweira, *Bagshawe* 1577 (both BM!, syn.), **syn. nov.**

B. keniensis Mildbr. in N.B.G.B. 9: 500 (1926). Type: Kenya, Kiambu District, Kikuyu, *F. Thomas* III.30 (B†, holo.; K!, iso.), **syn. nov.**

B. stuhlmannii Lindau var. *obtusata* Mildbr. in N.B.G.B. 9: 499 (1926). Types: Kenya, North Nyeri District, Cole's Mill, *R. & T. Fries* 1045 & near Forest Station, *R. & T. Fries* 534 pro parte (both B†, syn.), **syn. nov.**

NOTE. On a regional scale in our area, *B. ventricosa* can often be divided into discrete "taxa", yet these merge entirely in other areas. These variants are therefore informally recognized and keyed out below, though it must be accepted that throughout the F.T.E.A. range not all specimens will key out readily and in some regions the majority of populations are intermediate in character and must be treated as *B. ventricosa sensu lato*. Of particular note, in the highlands flanking the central Kenyan Rift "*B. keniensis*" (*B. ventricosa* in U.K.W.F.) and "*B. stuhlmannii*" (*B. micrantha* in U.K.W.F.) are two discrete and easily separable taxa, apparently not hybridising. The former is often, though not exclusively, in wetter, more luxuriant habitats such as montane forest margins, as opposed to, for example, drier bushland. However, in the Kenya/Tanzania border region a whole swathe of intermediate populations occur in abundance (including the "large-flowered form" of *B. micrantha* recorded in U.K.W.F.) and the distinguishing characters break down entirely. In the highlands of Ethiopia, many forms of *B. ventricosa* are recorded with the variation appearing clinal.

KEY TO INFRASPECIFIC VARIANTS/SPECIES WITHIN THE AGGREGATE (OMITTING INTERMEDIATE FORMS)

1. Outer calyx lobes usually with a ± prominently toothed margin, more rarely entire; inflorescences axillary, rarely clustered towards stem apices; corolla 15–27 mm long; leaf apex acute, obtuse or subattenuate .... d. "*B. stuhlmannii*"
   Outer calyx lobes with margin entire or minutely toothed; inflorescences axillary or congested into verticillate or spiciform synflorescences; corolla 25–40 mm; leaf apex usually attenuate or acuminate 2
2. Young shoots and inflorescences often with a silky or velutinous buff, yellowish or yellow-brown indumentum; anticous calyx lobe with apex acute, attenuate or shortly notched for up to 2.5 mm, length/width ratio 1.8–3.8/1; cymes 1–7-flowered, usually congested into a terminal synflorescence with flowers often held on one side .... c. "*B. keniensis*"
   Young shoots and inflorescence without a silky or velutinous indumentum, variously sparsely to more densely hairy; anticous calyx lobe with apex bifid for 3–6 mm or if more shallowly so then length/width ratio 1–1.9/1; cymes 1–3(–4)-flowered, axillary or congested into a verticillate or spiciform synflorescence 3
3. Synflorescence slender, densely spiciform, not verticillate, each node 1(–2)-flowered; anticous calyx lobe with length/width ratio 1.8–2.8/1, apex bifid for 3.5–6.5 mm . b. "*B. vix-dentata*"
   Cymes axillary or if in a synflorescence then usually verticillate, each node 1–3(–4)-flowered; anticous calyx lobe with length/width ratio 1–1.9/1, apex bifid for 1–4 mm or shallowly emarginate .... a. "*B. ventricosa*"

a. "**B. ventricosa** Nees" (including *B. grantii*, *B. bagshawei*)

Leaf apex usually attenuate or acuminate. Cymes axillary or congested into a loose (rarely dense) verticillate synflorescence, each axil 1–3(–4)-flowered; bracteoles linear, spatulate or lanceolate, 3.5–18 mm long, 0.5–4.5 mm wide. Anticous calyx lobe elliptic, broadly ovate or rounded, length/width ratio 1–1.9/1, margin entire or minutely toothed, apex bifid for 1–4 mm or shallowly emarginate. Corolla 26–40 mm long. Capsule 10–12 mm long.

UGANDA. West Nile District: E Madi, Zoka Forest, Jan. 1952, *Leggat* 41!; Busoga District: 5 km from Namasagali on Jinja road, Dec. 1951, *Norman* 75!; Mengo District: km 68, Kampala–Masindi road, Jan. 1963, *Tallantire* 651!
KENYA. Northern Frontier District: Mt Nyiru, July 1960, *Kerfoot* 1997!; Turkana District: Murua Nysigar peak, Sept. 1963, *Paulo* 1039!; Uasin Gishu District: Kipkarren, Jan. 1932, *Brodhurst-Hill* 709!
DISTR. **U** 1–4; **K** 1–3, 5; Congo-Kinshasa, Rwanda, S Sudan, Ethiopia, Somalia; Yemen
HAB. Grassland, wooded grassland and thicket, *Juniperus* forest and particularly forest margins; 650–2450 m
USES. "Vermifuge – leaves eaten" (**U** 2; *Jarrett* 219)

NOTE. The description here covers only the variation recorded within the F.T.E.A. region, where the most common form, previously separated as *B. grantii* and *B. bagshawei*, is rather sparsely hairy and has broad outer calyx lobes. This matches much of the material from W Ethiopia and Sudan. In N and C Ethiopia (including the type material) and extending into **K** 1 & 2, more hairy forms occur, often with proportionally narrower calyx lobes. However, a whole range of intermediates are recorded from Ethiopia.

b. "**B. vix-dentata** C.B. Clarke"

Leaf apex usually attenuate or acuminate. Cymes congested into a dense spiciform terminal synflorescence, each axil 1(–2)-flowered; bracteoles spatulate or narrowly elliptic, held erect and adpressed to the calyces, 10–16 mm long, 1.3–4 mm wide. Anticous calyx lobe elliptic or rhombic, length/width ratio 1.8–2.8/1, margin entire, apex bifid for 3.5–6.5 mm. Corolla 28–32 mm long. Capsule 10–11 mm long. Fig. 50/8 & 9, p. 362.

UGANDA. Karamoja District: Kidepo National Park, Dec. 1972, *Synnott* 1381!; Mengo District: 16 km E Kakoge, Dec. 1955, *Langdale-Brown* 1685!; Mubende District: near Kakumiro, Jan. 1957, *Lind* 2112!
DISTR. **U** 1, 4; ?Central African Republic (see note), NE Congo-Kinshasa, S Sudan, ?SW Ethiopia
HAB. *Combretum-Terminalia* and *Acacia* woodland and wooded *Hyparrhenia* grassland, open submontane and gallery forest, roadsides; 1100–1700 m
USES. None recorded on herbarium specimens

NOTE. Whilst "typical" populations of *vix-dentata* are easily recognisable, due mainly to the dense, slender synflorescence, intermediates with typical *ventricosa* are recorded in SW Ethiopia (including the material cited by Ensermu under *B. vix-dentata* which I have seen) and in Uganda (e.g. *Snowden* 156 from Singo, **U** 4).

The West African *B. villosa* S. Moore is close to "*B. vix-dentata*", sharing the dense synflorescence and long bracteoles, but tends to be larger and more hairy in all parts. However, plants from Central African Republic and adjacent areas of Congo-Kinshasa appear somewhat intermediate, being as hairy as *B. villosa* but with the inflorescences and flowers similar in size and shape to *vix-dentata*. Any wider revision of the *B. ventricosa* complex would therefore have to include *B. villosa* and the closely allied West African *B. opaca* Nees.

Of Clarke's four original syntypes, *Scott Elliot* 6928, 7023 & 7940 are referable to "*B. keniensis*", whilst *Wilson* 37 (which is the most informative material) has formed the basis for the subsequent application of the name *B. vix-dentata*; this latter specimen is therefore chosen as the lectotype here.

c. "**B. keniensis** Mildbr."

Young leaves, shoots and calyces usually with a ± dense, silky or velutinous buff, yellowish or yellow-brown indumentum. Leaf apex attenuate or acuminate. Cymes axillary or usually congested into a verticillate terminal synflorescence with flowers often held on one side, each axil 1–7-flowered; bracteoles linear or narrowly spatulate, 3–11.5 mm long, 0.2–1.5 mm wide.

Anticous calyx lobe (oblong-) elliptic or somewhat lanceolate, length/width ratio 1.8–3.8/1, margin entire or minutely toothed, apex acute, attenuate or notched for up to 2.5 mm. Corolla (20–)25–35 mm long. Capsule 10–14.5 mm long. Fig. 50/1–5, p. 362.

UGANDA. Toro District: Kibale Forest, July 1938, *A.S. Thomas* 2281!; Mbale District: Budadiri, [Bugishu] Bugisu, Jan. 1932, *Chandler* 515!; Mubende District: 4–5 km E of Mubende centrum, June 1969, *Lye & Rwaburindore* 3402!

KENYA. Uasin Gishu District: Moiben, Dec. 1931, *Brodhurst-Hill* 670!; Embu District: Embu, Aug. 1932, *Graham* 2140!; North Kavirondo District: Kakamega Forest, Dec. 1956, *Verdcourt* 1654!

TANZANIA. Bukoba District: Ruiga River Forest Reserve, July 1958, *Procter* 972! & Minziro, Kagera, Nov. 1994, *Congdon* 375!; Biharamulo District: 50 km on Biharamulo–Muleba road, July 2000, *Bidgood, Leliyo & Vollesen* 4898!

DISTR. **U** 1–4; **K** 1, 3–6; **T** 1; E Congo-Kinshasa, Rwanda

HAB. Undergrowth of montane forest, often of drier types such as *Juniperus*, *Podocarpus* and *Prunus-Olea-Albizia-Ocotea*, often at forest margins and in clearances, riverine forest and thicket; dense bushland and thicket on rocky slopes and roadsides, thickets under exotic plantations, secondary growth on fallow land, more rarely in open woodland and grassland; 1050–2750 m

USES. "Used for making green colour for raffia" (**U** 2: *Thomas* 2281); "grazed by all domestic stock" (**K** 6: *Glover et al.* 2540, 2607)

NOTE. This form is usually quite distinct in our region. The dense silky indumentum on the young shoots and inflorescences is, however, shared by some variants of *B. ventricosa* from Ethiopia, including some of the type material, but there the cymes are always 1–3-flowered, and the plants tend to dry a paler green, "*B. keniensis*" often having up to 7 flowers and tending to dry dark green or green-black. However, there are occasional intermediates between *B. ventricosa sensu stricto* and "*B. keniensis*" in NE Uganda and N Kenya and the variation is almost certainly clinal.

Most of the type material of *B. keniensis* was destroyed in the Berlin herbarium fire but a small scrap is extant at Kew. The calyces clearly match the taxon described above, although the immature leaves are unusually broad and with a truncate base.

d. "**B. stuhlmannii** Lindau" (including *B. micrantha*)

Leaf apex acute, obtuse or subattenuate. Cymes axillary, rarely clustered towards stem apices when young, each axil 1–3-flowered; bracteoles linear or oblanceolate, 0.5–9.5 mm long, 0.2–1 mm wide. Anticous calyx lobe ovate, elliptic, rhombic or subrounded, length/width ratio 0.9–2.5/1, margin usually conspicuously toothed, 2–7 teeth per side, rarely entire, apex acute, emarginate or notched for up to 2.5 mm. Corolla 15–27 mm long. Capsule 8.5–12 mm long. Fig. 50/6 & 7.

UGANDA. Busoga District: Igwe mutalla, 16 km SSE of Bugiri, 3 km W of the Bugiri–Buswale road, Dec. 1952, *Wood* 523!

KENYA. Naivasha District: [Ol] Longonot Estate, Jan. 1962, *Kerfoot* 3479!; North Nyeri District: Zawadi Estate, ± 7 km on Nyeri–Kiganjo road, June 1974, *Faden & Faden* 74/842!; Masai District: Olarro Camp, Feb. 2001, *Luke & Luke* 7321!

TANZANIA. Masai District: Lolkisale, June 1965, *Leippert* 5879!; Mpwapwa District: 3–5 km on Kibatwe–Motta track, Apr. 1988, *Bidgood, Mwasumbi & Vollesen* 1008!; Iringa District: Mafinga–Madibira road between Sadani and Igoma, near Ndembera R., July 1990, *Congdon* 285!

DISTR. **U** 2, 3, **K** 1, 3–6, 7, **T** 1–3, 5–7; E Congo-Kinshasa, Rwanda, Burundi; N Somalia (*Gillett* 4574)

HAB. A wide variety of open or dense dry woodland and thicket; drier types of montane and submontane forest particularly along margins and in secondary growth; dry rocky hillsides and eroded slopes, termite mounds, roadsides; 750–2500 m

USES. "Grazed by domestic stock" (**K** 6; *Glover et al.* 452, 1589, 1626, 1848, 1918; *Glover & Samuel* 3102)

NOTE. Central Kenyan material (often named *B. micrantha*) differs somewhat from C & S Tanzania material in the plants tending to be more slender and often trailing, in being less hairy including fewer glandular hairs and in having longer bracteoles (up to 9.5 mm versus up to 4 mm long). However, there is considerable overlap in all these characters. The form from Rwanda, Burundi and Congo exactly matches the Kenyan populations. Very similar plants, but with entire calyx lobes, are recorded from N Somalia (*Gillett* 4574). Plants from the Itigi Thicket (Singida/Dodoma District, e.g. *Polhill & Paulo* 2189) have the most prominent marginal teeth on the outer calyx lobes but otherwise agree closely with the Tanzanian material of "*B. stuhlmannii*". *Schlieben* 817 (BR) and 1057A (B, BR, EA, K) from Ulanga District, **T** 6 have very broad ovate outer calyx lobes and closely resemble "*B. scindens*" from Zimbabwe and Transvaal, which is otherwise a good match for "*B. stuhlmannii*".

Hornby 109 is chosen as the neotype of *B. stuhlmannii* as it is from the same location as the original type and bears the note from E. Milne-Redhead "*Barleria stuhlmannii* Lindau (compared with the type and agrees well)".

22. **Barleria boranensis** *Fiori* in Miss. Biol. Borana, Racc. Bot.: 217, fig. 66 (1939); Ensermu in Fl. Eth. 5: 407 (2006). Type: Ethiopia, Sidamo, Moyale, *Cufodontis* 666 (FT, holo.; W!, iso.)

Erect or scrambling perennial herb, 30–100 cm tall; stems with appressed coarse buff or yellowish hairs and with a line of shorter crisped hairs in opposite furrows. Leaves sometimes immature at flowering, ovate (-elliptic), 2–5.5 cm long, 1.3–2.8 cm wide, base attenuate to rounded, apex acute or subattenuate, surfaces with coarse ascending hairs rather many when young, sparse at maturity; lateral veins 4–5 pairs; petiole 2–9 mm long. Inflorescences axillary, flowers solitary or in 2–3-flowered unilateral cymes, subsessile; bracts foliaceous; bracteoles linear or oblanceolate, 2.5–6.5 mm long, 0.3–2 mm wide; pedicels to 2 mm long. Calyx accrescent; anticous lobe broadly ovate or orbicular, 8–12.5 mm long and wide in flower, to 17 mm long and 19 mm wide in fruit, base (sub)cordate, margin conspicuously dentate, 9–15 apiculate teeth per side, each with an apical bristle, apex truncate, emarginate or obtuse, indumentum as leaves but shorter, palmate veins prominent in fruit; posticous lobe 9–14.5 mm long in flower, to 18 mm long in fruit, apex acute or obtuse, apiculate; lateral lobes 3–5 mm long in flower, to 8.5 mm long in fruit. Corolla white or pale purple, sometimes with purple streaking in throat, 17–24 mm long, limb with sparse coarse eglandular hairs and finer glandular hairs outside; tube 10–14 mm long, broadly funnel-shaped above attachment point of stamens; abaxial lobe offset by 2.5–3.5 mm, 6.5–8 mm long, 5.5–7.5 mm wide; lateral lobes 4–6 mm long, 3.5–5.5 mm wide; adaxial lobes as lateral pair but 3–4 mm wide. Stamens attached 3.5–5.5 mm from base of corolla tube; filaments 10–15 mm long; anthers 1.5–2.3 mm long; lateral staminodes 1–1.5 mm long, pilose, antherodes 0.5–0.8 mm long. Pistil glabrous; stigma subcapitate, 0.3–0.5 mm long. Capsule 10–11.5 mm long, glabrous; seeds 6–6.5 mm long, 4.7–5.5 mm wide.

KENYA. Northern Frontier District: Moyale, July 1952, *Gillett* 13470! & *idem*, July 1952, *Gillett* 13634! & 32 km N of Wajir, Sept. 1953, *Bally* 9089!

DISTR. **K** 1; S Ethiopia

HAB. Dry woodland and bushland of *Acacia, Commiphora, Combretum, Ficus, Cussonia, Dichrostachys* and/or *Terminalia*, on granitic rocky hillsides or on black clay soils; 1050–1550 m

USES. None recorded on herbarium specimens

CONSERVATION NOTES. This species appears rather scarce within a limited range in the Ethiopia-Kenya border region. It may however be under-recorded due to limited collection in this region. Human populations are generally low within its range, although over-grazing may be a threat in the Moyale region on both sides of the border. It is therefore provisionally assessed as Data Deficient (DD), with more data on current distribution and abundance desirable.

SYN. *B. boranensis* Fiori forma *leucosepala* Fiori in Miss. Biol. Borana, Racc. Bot.: 217, fig. 67 (1939). Type: Ethiopia, Sidamo, Negelle, *Cufodontis* 217 (FT, holo.; W!, iso.)

NOTE. This species is close to the *B. ventricosa* aggregate, particularly "*B. stuhlmannii*", but is readily separated by the differences listed in the key; in addition, the calyces are more noticeably accrescent in fruit. *Purseglove* 3518 from Lake Edward Plains, **U** 2, has up to 12 short teeth on each side of the calyx margin similar to *B. boranensis* but the corolla is larger (28 mm) and the calyx lobes are broadly obtuse at the base; this specimen is therefore placed within *B. ventricosa*.

23. **Barleria volkensii** *Lindau* in E.J. 20: 22 (1894); C.B. Clarke in F.T.A. 5: 167 (1899) pro parte quoad *Volkens* 380; U.K.W.F. ed. 1: 594 (1974); K.T.S.L.: 600 (1994); U.K.W.F. ed. 2: 273 (1994). Type: Tanzania, Kilimanjaro, Kwa Kinabo, *Volkens* 380 (B†, holo.; BM!, K!, iso.)

Straggling or scandent perennial herb or subshrub, 40–250 cm tall; stems with sparse to many ± appressed coarse pale buff or yellowish hairs, with a line of shorter crisped hairs in opposite furrows, sometimes with scattered patent glandular hairs when young. Leaves ovate or elliptic, 3.5–11.5 cm long, 1.5–6.2 cm wide, base attenuate or cuneate, apex (sub)attenuate or acuminate, margin and veins beneath with ± appressed hairs, often with shorter and more spreading hairs between the veins beneath, more sparse or glabrous above; lateral veins (3–)4–6(–8) pairs; petiole 3–35 mm long. Inflorescences axillary, flowers (1–)2–5 in unilateral or (partially) dichasial cymes, often clustered towards branch apices; bracts foliaceous; bracteoles linear, oblanceolate or elliptic-lanceolate, 1.5–18 mm long, 0.2–4.5 mm wide. Outer calyx lobes often paler towards base, somewhat accrescent; anticous lobe broadly ovate, elliptic or suborbicular, 13.5–24 mm long, 8.5–18 mm wide, base rounded, obtuse or subattenuate, margin with ± well-developed slender teeth, these apiculate and with an apical bristle, rarely subentire, apex emarginate or bifid for up to 3.5 mm, rarely entire, indumentum as leaves beneath or hairs restricted to veins, patent glandular hairs sometimes present; posticous lobe ovate, 15–26 mm long, 10–17 mm wide, apex acute or obtuse, conspicuously apiculate; lateral lobes 5.5–12.5(–17) mm long. Corolla pale blue to mauve or white, 38–63 mm long, glandular-pilose outside; tube 25–37 mm long, abruptly expanded above attachment point of stamens, throat broadly cylindrical; abaxial lobe offset by 4.5–8.5 mm, 13–20 mm long, 11.5–16 mm wide; lateral lobes 10–18 mm long, 7–13 mm wide; adaxial lobes as lateral pair but 6–9 mm wide. Stamens attached 7–11 mm from base of corolla tube; filaments 27–41 mm long; anthers 3–4.7 mm long; lateral staminodes 1.5–6 mm long, pilose, antherodes 0.7–1.3 mm long. Pistil glabrous; stigma clavate, 0.5–0.9 mm long. Capsule 12–13 mm long, glabrous; seeds 6–7 mm long, 5.5–6 mm wide.

KENYA. Laikipia District: ± 30 km N of Rumuruti, Nov. 1978, *Hepper & Jaeger* 6641!; Kiambu District: Kukui, June 1902, *Kässner* 990!; Teita District: Taita Hills, Wusi, May 1931, *Napier* 1080!

TANZANIA. Masai District: Kitumbeine Mt, Mar. 1969, *Richards* 24240!; Mbulu District: Lake Manyara National Park, May 1965, *Greenway & Kanuri* 11801!; Moshi District: E Kilimanjaro, Mrao Rombo, June 1927, *Haarer* 516!

DISTR. **K** 1, 3, 4, 7; **T** 2; not known elsewhere

HAB. In shade in *Acacia* bushland, riverine thicket, drier types of montane forest, often on rocky slopes; 900–2150 m

USES. None recorded on herbarium specimens

CONSERVATION NOTES. This taxon is rather widespread in East Africa and can be locally common in a variety of habitats. It is not considered threatened: Least Concern (LC).

NOTE. *B. volkensii* is usually easily identifiable in our region through the combination of large corollas (usually 45 mm or more) with a long, broadly cylindrical throat and large, broad, usually conspicuously toothed calyces. However, it may ultimately have be considered a further variant within the *B. ventricosa* aggregate. Large-flowered forms of *B. ventricosa* recorded from Ethiopia (e.g. *Friis et al.* 8752, Sidamo) are similar to *B. volkensii*, sharing the cylindrical expanded throat and differing only in the proportionally narrower and more elliptic, typically hairier calyces with the margin always entire. *Gilbert et al.* 5575 from the Ndoto Mts, **K** 1 tends towards this form. In addition, occasional smaller-flowered plants of *B. volkensii* are recorded in our region (e.g. *Bally* 5674, **K** 1; *Karani* 117, **T** 2), though in some cases these are recorded from plants with stunted growth, perhaps flowering under stress.

Plants from Mt Marsabit, **K** 1 (e.g. *Williams* in EAH 11023) are quite distinct from "typical" *B. volkensii* in having sparse to many glandular hairs on the calyx (versus glandular hairs absent), the anticous calyx lobe being narrower than the posticous lobe (versus broader and often partially folded around posticous lobe), and have somewhat larger anthers (3.5–4.7 mm versus 3–3.8 mm) and longer lateral staminodes (3–6 mm versus 1.5–2.5 mm). The Marsabit plants could be considered a separate subspecies if *B. volkensii* is maintained as distinct from *B. ventricosa* following a full revision of this group.

Kerfoot 2559 from the Mathews Range, **K** 1, and *Mwachala* in EW 431 from the Taita Hills, **K** 7, appear close to *B. volkensii* but differs in having proportionately broader leaves with a rounded base which are reminiscent of *B. submollis* and *B. robertsoniae.* It is possible that these specimens are of hybrid origin.

24. **Barleria boehmii** *Lindau* in E.J. 20: 19 (1894); C.B. Clarke in F.T.A. 5: 167 (1899). Types: Tanzania, Tabora District, Kakoma, *Böhm* 25 (B†, holo.) & *idem, Lloyd* 10 (K!, neo., selected here – see note)

Suffruticose herb, 1–many ± erect stems 25–60(–150) cm tall from a woody rootstock; stems with sparse to dense long-appressed to spreading coarse buff or golden hairs, with a line of inconspicuous short crisped hairs in opposite furrows. Leaves narrowly elliptic (-obovate) to broadly ovate, 2–6 cm long, 1–3.8 cm wide, base sometimes asymmetric, cuneate to rounded, apex acute, obtuse or rarely rounded, surfaces with rather sparse ascending to spreading hairs, many on margin and veins beneath; lateral veins 3–4 pairs; petiole to 4 mm long. Inflorescences of 1–3(–several)-flowered axillary cymes, sometimes congested into a loose thyrsoid or spiciform terminal synflorescence 2–9.5 cm long; bracts foliaceous but much reduced at the upper axils where 9–22 mm long, 2.5–8.5 mm wide; bracteoles linear, oblanceolate or narrowly elliptic, 3–21 mm long, 0.2–4.5 mm wide. Anticous calyx lobe ovate, elliptic or rarely ovate-orbicular, 8.5–21 mm long, 3–10.5(–14.5) mm wide, base attenuate to rounded or rarely subcordate, margin entire, apex emarginate or rarely acute, indumentum as leaves, sometimes also glandular-pilose; posticous lobe 9.5–24 mm long, 4–16 mm wide, apex acute to rounded; lateral lobes 5–13 mm long. Corolla pale blue to mauve or white, 21–45 mm long, glandular-pubescent outside; tube 11.5–21 mm long, broadly campanulate above attachment point of stamens; abaxial lobe offset by 5–12 mm, 8–19 mm long, 6–13.5 mm wide; lateral lobes 6–13.5 mm long, 5–11.5 mm wide; adaxial lobes 5.5–12 mm long, 3–8.5 mm wide. Stamens attached 6–12 mm from base of corolla tube; filaments 12–23 mm long; anthers 2–3.3 mm long; lateral staminodes 1–7 mm long, pubescent towards base, antherodes barely developed or to 1 mm long. Pistil glabrous; stigma subcapitate, 0.3–0.7 mm long. Capsule 9–13.5 mm long, glabrous. ?Immature seeds ± 5 mm long and wide.

TANZANIA. Mwanza District: Geita, near Bukoba, Nov. 1944, *Bally* 4078!; Mpanda District: 8 km on Mpanda–Uvinza road, Mar. 1994, *Bidgood, Mbago & Vollesen* 2679!; Mbeya District: Tunduma road, 13 km from Tunduma, *Richards* 17068!

DISTR. **T** 1, 4, 5, 7; Burundi, NE Zambia, N Malawi

HAB. *Brachystegia* woodland, rough grassland and roadsides; 750–1550 m

USES. None recorded on herbarium specimens

CONSERVATION NOTES. This species is widespread and locally common in the miombo woodland and associated grasslands of W Tanzania, and appears tolerant of some disturbance, having been recorded from road verges (*Richards* 17068) and from degraded miombo (*Verdcourt* 3305). It is not considered threatened: Least Concern (LC).

NOTE. *B. boehmii* is the miombo woodland equivalent of *B. ventricosa.* It varies considerably in leaf shape and flower and fruit size in our region but there are no consistent trends identifiable that would allow recognition of infraspecific taxa. Small-flowered and -fruited specimens, more common in the north of the range, have poorly developed staminodes whereas large-flowered specimens, more common in the south and west, have long staminodes with developed antherodes. Intermediates do, however, occur (e.g. *Verdcourt* 3305, **T** 1) and corolla size can vary considerably even within a single population (e.g. *Richards* 7107, **T** 4). *Bidgood et al.* 2679 and 5433, both from the Mpanda–Uvinza road (**T** 4), differ from other material in having glandular hairs on the bracteoles and outer calyx lobes, very many in the latter specimen. A striking form with broadly ovate leaves, rounded at the base, is recorded from the Kigoma area (e.g. *Pirozynski* 455), but plants with both this leaf shape and the more typical narrower, elliptic leaves are recorded from N Malawi and are clearly conspecific.

Lloyd 10 is here chosen as the neotype of *B. boehmii* to aid future circumscription of this species. The original type, destroyed in the WW II fire at Berlin, was collected from the same locality. Edgar Milne-Redhead who saw both the type and the *Lloyd* gathering recorded the latter as "*B. boehmii* Lindau *forma*", but the variation in stem indumentum and leaf shape he noted fall well within the full range of variation now known in this taxon.

25. **Barleria submollis** *Lindau* in E. J. 20: 21 (1894) pro parte quoad *Holst* 3515, ? 3577A (not traced) & *Volkens* 314; U.K.W.F. ed. 1: 594 (1974); Blundell, Wild Fl. E. Afr.: 388, pl.799 (1987); U.K.W.F. ed. 2: 273, pl.118 (1994); Friis & Vollesen in Biol. Skr. 51 (2): 437 (2005); Ensermu in Fl. Eth. 5: 407 (2006). Type: Tanzania, Moshi District, Lake Chala, *Volkens* 314 (K!, lecto., chosen here; BM!, isolecto.)

Trailing, scrambling or scandent perennial herb or subshrub, 20–180 cm tall; stems with ± many short retrorse or spreading eglandular hairs usually intermixed with many patent glandular hairs and few to many long pale-buff to golden-yellow spreading eglandular hairs. Leaves ± broadly ovate, 1.2–6 cm long, 0.8–4.5 cm wide, base rounded, shallowly cordate or broadly obtuse, apex rounded to shortly attenuate, surfaces spreading-pubescent, often dense beneath, usually also with glandular hairs along the margin or more widespread when young; lateral veins 3–5 pairs; petiole 3–30 mm long. Inflorescences axillary, flowers solitary or in 2–3-flowered unilateral cymes; bracts foliaceous; bracteoles linear or oblanceolate, 1.5–11 mm long, 0.2–4 mm wide; flowers subsessile or pedicels to 5 mm long. Anticous calyx lobe ± broadly ovate (-elliptic), 7–26 mm long, 4.5–22 mm wide, base rounded, cordate or shortly attenuate, margin entire or dentate, teeth with an apical bristle, apex emarginate, more rarely acute or obtuse, surface sparsely or often densely spreading-pubescent, with patent glandular hairs most dense at or restricted to the margin, rarely absent; posticous lobe 7.5–27 mm long, apex acute to rounded or subattenuate; lateral lobes 6–13.5 mm long. Corolla white to pale blue or mauve, often with red or purple parallel lines in the throat, (25–)35–61 mm long, rather sparsely eglandular- and glandular-pilose outside; tube (12.5–)16–38 mm long, funnel-shaped or rather narrowly so above attachment point of stamens; abaxial lobe offset by 2–7.5 mm, 9–21 mm long, 7–17 mm wide; lateral lobes as abaxial lobe but 6–14.5 mm wide; adaxial pair 8–19 mm long, 6–11 mm wide. Stamens attached 4–13(–16) mm from base of corolla tube; filaments (13–)20–33 mm long; anthers (2–)2.5–3.7 mm long; lateral staminodes 1.5–4 mm long, pilose, antherodes 0.5–1.5 mm long. Pistil glabrous; stigma subcapitate to shortly clavate, 0.3–1 mm long. Capsule 9–12.5 mm long, glabrous; seeds 5–6.5 mm long, 4.5–5.5 mm wide.

Uganda. Karamoja District: foothills of Mt Moroto, Oct. 1952, *Verdcourt* 811! & Lodoketemit, Nov. 1962, *Kerfoot* 4471!

Kenya. Northern Frontier District: Dandu, June 1952, *Gillett* 13418!; Kitui District: Kibwezi–Kitui road, Yatta Gap, Apr. 1969, *Napper* 2024!; Masai District: Chyulu Hills, Ol Donyo Wuas Camp, July 1990, *Luke* 2443!

Tanzania. Mbulu District: Mto wa Mchanga, Lake Manyara National Park, Mar. 1964, *Greenway & Kanuri* 11375!; Pare District: Mkomazi Game Reserve, Viteweni Hill, Apr. 1995, *Abdallah & Vollesen* 95/29!; Lushoto District: W Usambara Mts, W of Mgwashi village, Mar. 1984, *Borhidi et al.* 84481!

Distr. **U** 1; **K** 1–4, 6, 7; **T** 2, 3; S Sudan (fide Friis & Vollesen, l.c.), S Ethiopia and S Somalia

Hab. Dry bushland, particularly *Acacia-Commiphora*, lightly wooded grassland, riverine thicket, coastal forest margins, roadsides; 50–1700(–2300) m

Uses. None recorded on herbarium specimens

Conservation notes. A widespread and fairly common species in a variety of habitats: Least Concern (LC).

Syn. [*B. repens sensu* Hook. f. in Bot. Mag. 113, t. 6954 (1887) pro parte quoad *Wakefield* s.n. in text, *non* Nees]

[*B. boivinii sensu* Lindau in P.O.A. C: 369 (1895), *non* T. Anderson]

[*B. volkensii sensu* C.B. Clarke in F.T.A. 5: 167 (1899) pro parte excl. *Volkens* 380, *non* Lindau]

[*B. stuhlmannii sensu* C.B. Clarke in F.T.A. 5: 167 (1899) pro parte quoad *Holst* 3515, *non* Lindau]

B. umbrosa Lindau in E.J. 33: 188 (1902). Type: Somalia, Salakle, *Ellenbeck* 2248 (B†, holo.), **syn. nov.**

B. scassellatii Fiori in Bull. Soc. Bot. Ital.: 56, fig. 5 (1915). Types: Somalia, Arrar, *Scassellati* 14 & Urufle, *Scassellati* 83 (both FT, syn.) & Zingibar, *Paoli* 373 (FT, syn.; K!, isosyn.), **syn. nov.**

[*B. micrantha sensu* Hedrén in Fl. Somalia 3: 434, fig. 291 (2006) pro parte quoad spec. ex Somalia, *non* C.B. Clarke]

NOTE. Three species are recognised here within the complex of taxa with a dense spreading indumentum and broadly ovate leaves with an abruptly narrowed base: *B. submollis*, *B. robertsoniae* and *B. usambarica*. However, *B. submollis* as circumscribed here is polymorphic, with several characters varying independently of one another to produce a range of regional forms. Whilst formal recognition of infraspecific taxa is difficult, some of the more striking forms are worthy of note:

Corollas: usually over 35 mm long in this species but smaller-flowered populations (25–33 mm) are not infrequent in **K** 6 (e.g. *Gillett* 18691) and **T** 2 (e.g. *Bigger* 1984) and also in Ethiopia and S Somalia. Of the larger-flowered forms, most populations from **T** 2 & 3 have a shorter cylindrical base to the corolla tube than elsewhere (4–7 mm versus 7.5–13(–16) mm).

Calyces: coastal forest populations generally have larger outer calyx lobes than inland forms, the extreme case being populations from NE Kenya (*Gillespie* 313) and SE Somalia previously separated as *B. scassellatii*, where the outer lobes are 18–27 × 15–22 mm, deeply cordate at the base and with a conspicuously dentate margin. These plants are also less hairy than most populations of *B. submollis* and are perhaps worthy of distinction, but intermediate specimens are noted from coastal Kenya, e.g. *Robertson* 6985. Large, deeply cordate calyces are also found in several populations from **T** 2 (e.g. *Richards* 24985) but here they are often entire and the plants are more densely hairy throughout.

Stem indumentum: in most populations, the long coarse spreading hairs on the stems are rather sparse and pale buff (-yellow) in colour, but those from the foothills of the Usambara Mts have dense golden-yellow spreading hairs particularly on the young stems, rather similar to *B. amanensis* (see also note to that species).

26. **Barleria robertsoniae** *I. Darbysh.* **sp. nov.** *B. submolli* Lindau similis sed indumento caulis e pilis densis brevibus albis retrorsis praecipue constanti, caulibus maturis mox lignosis, corolla majore 56–75 mm (non 25–61 mm) longa, antheris majoribus (3.5–)4–5 mm (non 2–3.7 mm) longis, lobis calycis pilis multis patentibus glandulosis ubique vestitis (non ad marginem tantum pilos multos glandulosos ferentibus neque pilis carentibus) differt. Type: Kenya, North Nyeri District, Ngare Ndare Farm, *Robertson, A. Dyer & R. Dyer* 7417 (K!, holo.)

Erect or scrambling perennial herb or subshrub, 60–100 cm tall; stems soon woody, with dense white retrorse hairs and occasional coarse pale buff spreading or ascending hairs; young stems usually also with many short patent glandular hairs, rarely absent. Leaves broadly ovate, 2–4.5 cm long, 1.5–4 cm wide, base rounded or shallowly cordate, apex obtuse to subattenuate, surfaces densely pubescent, usually also with patent glandular hairs at least along the margin; lateral veins 4–5 pairs; petiole 3–14 mm long. Inflorescences axillary, flowers solitary or in 2–3-flowered unilateral cymes, subsessile; bracts foliaceous but often reduced and sessile, usually with more glandular hairs than the leaves; bracteoles oblanceolate or spatulate, 7–14.5 mm long, 1.5–4 mm wide; pedicels to 1.5 mm long. Anticous calyx lobe broadly ovate or elliptic, 11.5–17 mm long, 8.5–14.5 mm wide, base shortly attenuate to rounded, margin entire, apex emarginate or truncate, surface densely spreading eglandular- and glandular-pubescent; posticous lobe 12.5–20 mm long, apex acute or obtuse, minutely apiculate; lateral lobes 10–15 mm long. Corolla white, cream or rarely mauve, with parallel dark red lines in the throat, 55–75 mm long, glandular- and eglandular-pilose outside; tube 32–42 mm long, gradually and rather narrowly campanulate above attachment point of stamens; abaxial lobe offset by 5.5–10.5 mm, 19–27 mm long, 12–18 mm wide; lateral lobes as abaxial lobe but 12.5–20 mm wide;

adaxial lobes 18–25 mm long, 11–15 mm wide. Stamens attached 11.5–17 mm from base of corolla tube; filaments 35–41 mm long; anthers (3.5–)4–5 mm long; lateral staminodes 3.5–5 mm long, pilose, antherodes 0.7–1.3 mm long. Ovary glabrous or with few minute stalked glands towards apex; style glabrous; stigma shortly clavate or subcapitate, 0.4–1 mm long. Capsule ± 12.5 mm long, indumentum as ovary; only immature seeds seen.

KENYA. Northern Frontier District: Samburu Country, June 1937, *Jex-Blake* in CM 6920! & Wamba, Dec. 1971, *Bally & Smith* 14739!; Laikipia District: Rumuruti, ± 50 km N near Colchecio Lodge, Nov. 1978, *Hepper & Jaeger* 6617!

DISTR. **K** 1, 3, 4; not known elsewhere

HAB. *Acacia-Commiphora* bushland, open wooded grassland, secondary thicket; 900–2450 m

USES. None recorded on herbarium specimens

CONSERVATION NOTES. This species appears scarce within a highly limited range in northern Kenya, being known to the author from 12 collections from 11 localities. Parts of its range are, however, under-collected and its favoured habitat remains widespread within its range; it also appears tolerant of some disturbance, having been recorded from secondary thicket at Wamba (*Bally & Smith* 14739). Based on its scarcity, it is provisionally assessed as Near Threatened (NT) but this may be downgraded to Least Concern if more sites for this species are uncovered.

NOTE. This species is named in honour of Ann Robertson, one of the most prominent contemporary field botanists and conservationists in Kenya and co-collector of the type specimen. It is clearly closely related to *B. submollis* and could be considered a subspecies of that polymorphic taxon, but it is readily separated by the combination of characters given in the key and is immediately recognisable by its striking large flowers and white indumentum. Ann Robertson (*pers. comm.*) also points towards the more robust, woody stems found in this species, which give it quite a different appearance in the field, a character not readily observable in most herbarium material.

27. **Barleria usambarica** *Lindau* in E.J. 20: 21 (1894); C.B. Clarke in F.T.A. 5: 168 (1899); Vollesen in Opera Bot. 59: 79 (1980). Type: Tanzania, Tanga District, Duga, *Holst* 3165 (B†, holo.; K!, M!, W!, iso.)

Perennial herb, prostrate or scrambling to 100 cm tall; stems densely hairy with mixed long spreading coarse hairs, shorter retrorse or spreading hairs and patent glandular hairs, the latter sometimes sparse. Leaves broadly ovate, 1.3–3.8 cm long, 0.8–2.5 cm wide, base rounded or shallowly cordate, apex obtuse to subattenuate, surfaces with coarse ± spreading hairs, dense at least when young, longest on margin and veins beneath, usually with few glandular hairs mainly along the margin; lateral veins 3–4 pairs; petiole to 15 mm long. Inflorescences axillary, flowers solitary or in 2–3-flowered unilateral cymes, subsessile; bracts foliaceous; bracteoles linear, 0.7–3.3 mm long, ± 0.2 mm wide; pedicels to 1.5(–3) mm long. Anticous calyx lobe broadly ovate (-orbicular), 6–11 mm long, 4–9 mm wide, shortly attenuate into a rounded base, margin subentire or minutely toothed, apex emarginate or rarely obtuse to subattenuate, indumentum as leaves; posticous lobe to 12 mm long, apex obtuse to subattenuate, ± apiculate; lateral lobes 3.5–7 mm long. Corolla pale blue, mauve or rarely white, 16–21 mm long, sparsely glandular-pilose and with occasional eglandular hairs outside; tube 8.5–12.5 mm long, broadly funnel-shaped above attachment point of stamens; abaxial lobe offset by 2.5–3.5 mm, 7–9.5 mm long, 5.5–9 mm wide; lateral lobes 5–8 mm long, 3.5–5.5 mm wide; adaxial lobes 4–7 mm long, 3–4 mm wide. Stamens attached 5–7 mm from base of corolla tube; filaments 8–12 mm long; anthers 1.8–2.5 mm long; lateral staminodes 1–2.2 mm long, with mixed eglandular and shorter glandular hairs, antherodes 0.6–0.9 mm long. Ovary glabrous or with few short glandular hairs towards apex; style glabrous or with few hairs towards base; stigma shortly clavate, 0.3–0.6 mm long. Capsule 6.5–10 mm long, indumentum as ovary; seeds 4–5 mm long, 3.5–4.5 mm wide.

KENYA. Kilifi District: Mangea Hill, Jan. 1992, *Robertson* 6558!; Tana River District: Garsen–Witu road, near Nyangore Bridge, July 1974, *R.B. & A.J. Faden* 74/1157!

TANZANIA. Pangani District: Segera Forest, Aug. 1968, *Faulkner* 4130!; Morogoro District: 15 km NE of Kingolwira Station, Apr. 1954, *Welch* 227!; Kilwa District: Kingupira, June 1975, *Vollesen* in MRC 2395!

DISTR. **K** 7, **T** 3, 6, 8; not known elsewhere

HAB. Lowland forest margins and associated bushland and grassland, riverine thicket, rocky hillsides; 0–600(–?1400) m

USES. None recorded on herbarium specimens

CONSERVATION NOTES. *B. usambarica* appears to be locally frequent within its restricted range. Although lowland forest and bushland clearance has been widespread in the coastal regions of Kenya and Tanzania in recent decades, this species appears tolerant of some disturbance, often being recorded from margins or clearances. It is therefore assessed as of Least Concern (LC).

NOTE. *B. usambarica* resembles a miniature form of *B. submollis*, to which it is undoubtably closely related. However, in addition to the smaller flowers, the corolla tube is rather different, the basal cylindrical portion being longer than or equal to the expanded throat, whilst in *B. submollis* this is considerably shorter.

28. **Barleria amanensis** *Lindau* in E.J. 43: 352 (1909). Types: Tanzania, Amani, *Braun* 754 & 1145 (B†, syn.) & Sigi valley, Amani, *Braun* 1871 (B†, syn.; EA 4 sheets!, K 2 sheets!, isosyn.)

Scrambling or scandent perennial herb or subshrub, to 200–400 cm tall; stems with ± many ascending or spreading coarse golden-yellow or golden-brown hairs and shorter crisped, often retrorse pale hairs, sometimes with scattered glandular hairs when young. Leaves ovate (-elliptic), 3–8 cm long, 1.8–4.5 cm wide, base attenuate to subcordate, apex attenuate or acuminate, rarely obtuse, surfaces with few to many ascending or spreading golden-yellow hairs, sometimes with few glandular and crisped-eglandular hairs towards base beneath and along margin; lateral veins 3–6 pairs; petiole 5–20 mm long. Inflorescences axillary, flowers solitary or often in 2–3-flowered unilateral cymes, usually clustered into congested terminal synflorescences 2–14 cm long with flowers mainly held on one side; bracts foliaceous but often reduced; bracteoles linear (-lanceolate), 0.8–5 mm long, 0.2–0.6 mm wide, rarely oblanceolate and up to 9.5 mm long and 1.7 mm wide on inflorescences at lower axils; pedicels 1–4.5 mm long. Calyx somewhat scarious in fruit, anticous lobe ovate or elliptic to broadly so, 10–21 mm long, 7–16 mm wide, base attenuate to rounded or subcordate, margin entire or with minute teeth formed by swollen hair bases, apex usually broadly bifid for up to 4 mm, indumentum as leaves, hairs restricted to principal veins and margin or more widespread, usually with short crisped hairs and/or occasional glandular hairs at base and margin or more widespread; posticous lobe ovate to broadly so, 15–30 mm long, 9–22 mm wide, apex ± acute, surface often paler towards base; lateral lobes 4–13.5 mm long. Corolla white, (33–)38–54 mm long, glandular-pilose outside; tube (18–)21–26 mm long, broadly funnel-shaped above attachment point of stamens; abaxial lobe offset by 4.5–10 mm, 13.5–22 mm long, 15–23 mm wide; lateral lobes 13–18 mm long, 10.5–17 mm wide; adaxial lobes as lateral pair but to 14.5 mm wide. Stamens attached 9–12.5 mm from base of corolla tube; filaments (16–)21–30 mm long; anthers 2.5–3.5 mm long; lateral staminodes 1.5–2.5 mm long, pilose, antherodes absent or to 1 mm long. Pistil glabrous or style hairy towards base; stigma subcapitate, 0.4–0.9 mm long. Capsule 13–17 mm long, glabrous; seeds ± 7 mm long, 6 mm wide.

TANZANIA. Lushoto District: Usambara Mts, Kwamkuyu R., Jan. 1931, *Greenway* 2798! & lower slopes of Mavumbi peak, NE of Mashewa, July 1953, *Drummond & Hemsley* 3203!; Ulanga District: Mahenge Plateau, Feb. 1932, *Schlieben* 1746!

DISTR. **T** 3, 6, 8; not known elsewhere

HAB. Moist or drier forest, often growing along streams, rarely in more open rocky woodland; 400–1200 m

USES. "Root [used] for gonorrhea" (**T** 3; *Koritschoner* 656); "root accelerates birth" (sine loc., *Koritschoner* 757)

CONSERVATION NOTES. *B. amanensis* is currently known from five sites in low- to mid-altitude forest in the Usambara Mts and from five scattered localities further south in Tanzania. It has been most often collected along the Sigi and Kwamkuyo Rivers in the vicinity of Amani. This area has been heavily denuded through timber extraction and conversion to agriculture and it is likely that most of the sites for *B. amanensis* are threatened or lost. Further south, two of the sites, Shikurufumi and Rondo Plateau, are designated as Forest Reserves but past disturbance at the latter site has been severe and agricultural encroachment is a future threat. It is therefore considered Vulnerable (VU B2ab(iii)).

SYN. [*B.* sp. ?nov. (= *Clarke* 11) *sensu* G.P. Clarke in Syst. Geogr. Pl. 71: 1069 (2001)]

NOTE. Care should be taken to separate this species from the **T** 3 form of *B. submollis* with a dense yellow indumentum (see note to that species), which however has usually smaller corollas (29–40 mm versus (33–)38–54 mm) with a shorter cylindrical base to the tube (4–7 mm versus 9–12.5) and less broadly funnel-shaped throat, smaller fruits (9–12.5 mm versus 13–17 mm) and axillary inflorescences (versus usually held in a terminal synfloresence). *B. amanensis* is usually found in forest whilst *B. submollis* is recorded from more open and drier habitats; however, one collection of the former (*Schlieben* 1746) is recorded from open woodland.

One of the two B sheets of *Schlieben* 1746 includes a branch on which the inflorescences are more uniformly axillary, not forming a terminal synflorescence, and the outer calyx lobes and bracteoles are larger than typical, but in the remaining material of this collection the more typical inflorescence type is evident. Similar variation in inflorescence arrangement has been recorded in *B. ventricosa.*

29. **Barleria lukei** *I. Darbysh.* **sp. nov.** *B. amanensi* Lindau similis sed calycibus glabris (non conspicue aureis hirsutis) praeter ad marginem sparse hirsutis et lobis calycis anticis latioribus ad basin magis manifeste cordatis differt. Type: Tanzania, Pangani District: Bushiri, *Faulkner* K672 (K!, holo.)

Scandent perennial herb or shrub, to 200–300 cm tall; stems usually with two opposite lines of short retrorse hairs and occasional glandular hairs when young, soon glabrescent except for a line of longer, coarse ascending yellow hairs at the nodes. Leaves ovate (-lanceolate), 5.5–9 cm long, 2–4 cm wide, base cuneate, shortly attenuate or subrounded, apex attenuate or acuminate, surfaces glabrous except for occasional hairs along midrib and margin; lateral veins 4–5 pairs; petiole 7–30 mm long. Inflorescences axillary but often clustered towards apex of short lateral branches, flowers solitary or in 2–3-flowered unilateral cymes; bracts foliaceous but often reduced; bracteoles linear, then 0.8–4 mm long, 0.2–0.4 mm wide, or those of lower inflorescences sometimes narrowly elliptic or ovate, then to 7.5 mm long and 3.5 mm wide; pedicels 3–8.5 mm long, shortly retrorse-pubescent or largely glabrous. Calyx somewhat accrescent and scarious in fruit; anticous lobe broadly ovate, 15–22 mm long, 15–21 mm wide, base (sub)cordate, margin entire or irregularly toothed, teeth with an apical bristle, apex emarginate, surface glabrous except for few inconspicuous hairs along margin at base and apex, principal veins prominent with age; posticous lobe 20–33 mm long, 16–22 mm wide, apex ± acute, apiculate; lateral lobes linear-lanceolate, 6–10 mm long. Corolla white, 43–62 mm long, glandular-pilose outside; tube 22–32 mm long, broadly funnel-shaped above attachment point of stamens; abaxial lobe offset by 7–11 mm, this and the lateral lobes each 16–22 mm long, 14.5–18 mm wide; adaxial lobes 15–18 mm long, 11–13.5 mm wide. Stamens attached 10–15 mm from base of corolla tube; filaments 23–30 mm long; anthers 3–3.5 mm long; lateral staminodes ± 1.5 mm long, pilose, antherodes 0.8–1.1 mm long. Pistil glabrous; stigma shortly clavate, 0.5–0.9 mm long. Capsule 12.5–14 mm long, glabrous; seeds 6.3–6.8 mm long, 6–6.3 mm wide.

KENYA. Kwale District: Shimba Hills, Mwele Mdogo Forest, 19 km SW of Kwale, Aug. 1953, *Drummond & Hemsley* 3976! & Shimba Hills, Mkongani Forest, May 1968, *Magogo & Glover* 1031!; Kilifi District: Kaya Jibana, SW slope, & fl. in hort. Robertson, Dec. 1990, *Robertson & Luke* 2636!

TANZANIA. Pangani District: Bushiri, July 1950, *Faulkner* K672! & Msaraza, Madanga, July 1955, *Tanner* 1997! & Mwera Estate, Langoni Forest, Aug. 1982, *Hawthorne* 1387!

DISTR. **K** 7; **T** 3; not known elsewhere

HAB. Lowland forest including secondary growth and margins; 50–450 m

USES. None recorded on herbarium specimens

CONSERVATION NOTES. This species is known from eleven localities (eight from herbarium material, plus three further sight records, Q. Luke, *pers. comm.*), most of the records being from the Shimba Hills and the Kayas of Kilifi District. The former site is now protected as a National Reserve whilst the Kayas are protected as National Monuments, but both areas have suffered from past deforestation and the latter remain threatened by encroachment. The Gongoni site is very close to a titanium mining site, but is protected as a Forest Reserve. The Tanzanian coastal forests have been heavily denuded and it is likely that this species is threatened there. This species is therefore provisionally assessed as Near Threatened (NT) but it may well prove to be Vulnerable under criterion A if accurate data on forest loss become available, or under criterion B if a survey of this species' current distribution finds some of the historical populations no longer extant.

SYN. [*B.* sp. nr. *amaniensis sensu* Luke in Journ. E. Afr. Nat. Hist. Soc. 94: 91 (2005)]

NOTE. This species is the coastal forest counterpart of *B. amanensis*. They share a scandent habit, flowers typically clustered towards the apex of the branches and large, broadly infundibuliform white corollas. *B. lukei* is, however, easily separated by lacking the coarse golden indumentum of *B. amanensis* on the stem internodes, pedicels and calyces. In addition *B. lukei* usually has broader outer calyx lobes (15–21 mm wide versus 7–16 mm) with a more clearly cordate base and longer pedicels (3–8.5 mm versus 1–4.5 mm). The outer calyx lobes of *B. lukei* are glabrous except for occasional hairs on the margin, a character otherwise shared in this section only by some specimens of *B. pseudosomalia*.

The species epithet "*lukei*" honours Quentin Luke, one of the foremost field botanists in East Africa and co-collector of several specimens of this species.

30. **Barleria scandens** *I. Darbysh.* **sp. nov.** *B. amanensi* Lindau similis sed caulibus foliisque calycibusque pilis sparsioribus pallidioribus appressioribus vestitis, inflorescentiis semper axillaribus manifeste dispersis (non in synflorescentiis congestis terminalibus dispositis) et corollis caesiis vel malvinis (non albis) differt. Type: Tanzania, Morogoro District, W slopes of Nguru Mts above Maskati, *Bidgood, Mwasumbi, Pócs & Vollesen* 446 (K!, holo.; DSM!, EA!, WAG, iso.)

Scandent perennial herb or shrub, (60–)180–800 cm tall, drying green-black or brown-black; stems with two opposite lines of short crisped, often retrorse hairs and occasional glandular hairs, young stems and nodes with additional appressed coarse buff-yellow hairs, glabrescent. Leaves elliptic to somewhat ovate or obovate, 3.5–9 cm long, 1.4–3.3 cm wide, base ± long-attenuate, apex attenuate or acuminate, surfaces with sparse appressed hairs mainly on margin and principal veins beneath and with short crisped hairs along midrib above; lateral veins 4–5 pairs; petiole to 10 mm long or blade sessile. Inflorescences axillary, flowers solitary or in 2–4-flowered unilateral cymes; bracts foliaceous; bracteoles (linear-) oblanceolate or narrowly elliptic, 7–18 mm long, 0.7–2.5 mm wide; pedicels 1–3 mm long. Calyx glossy in fruit; anticous lobe broadly ovate or elliptic, 13.5–20 mm long, 11–15 mm wide, base obtuse, rounded or shortly attenuate, margin entire or with few irregular teeth, apex obtuse, rounded or bifid for up to 3.5 mm, surface with appressed hairs along the margin and principal veins particularly towards the base; posticous lobe ovate, 18–24 mm long, apex acute or attenuate, apiculate; lateral lobes linear-lanceolate, 6–13.5 mm long. Corolla blue or mauve, drying blue-black, 30–50 mm long, glandular-pilose and with coarse eglandular hairs outside; tube 17–24 mm long, constricted immediately below attachment point of the stamens then funnel-shaped above; abaxial lobe offset by 3–8 mm, 13.5–22 mm long, 11.5–17 mm wide; lateral lobes 11–20 mm long, 10–17 mm wide; adaxial lobes 11–17 mm long, 8.5–11 mm wide. Stamens attached 6.5–11.5 mm from base of corolla tube; filaments 20–26 mm long; anthers 2.5–3.5 mm long; lateral staminodes 3–4.5 mm long, with sparse eglandular and glandular hairs, antherodes 0.8–1 mm long. Pistil glabrous; stigma subcapitate, 0.5–0.7 mm long. Capsule 13 mm long, glabrous; only immature seeds seen. Fig. 51/1–4, p. 375.

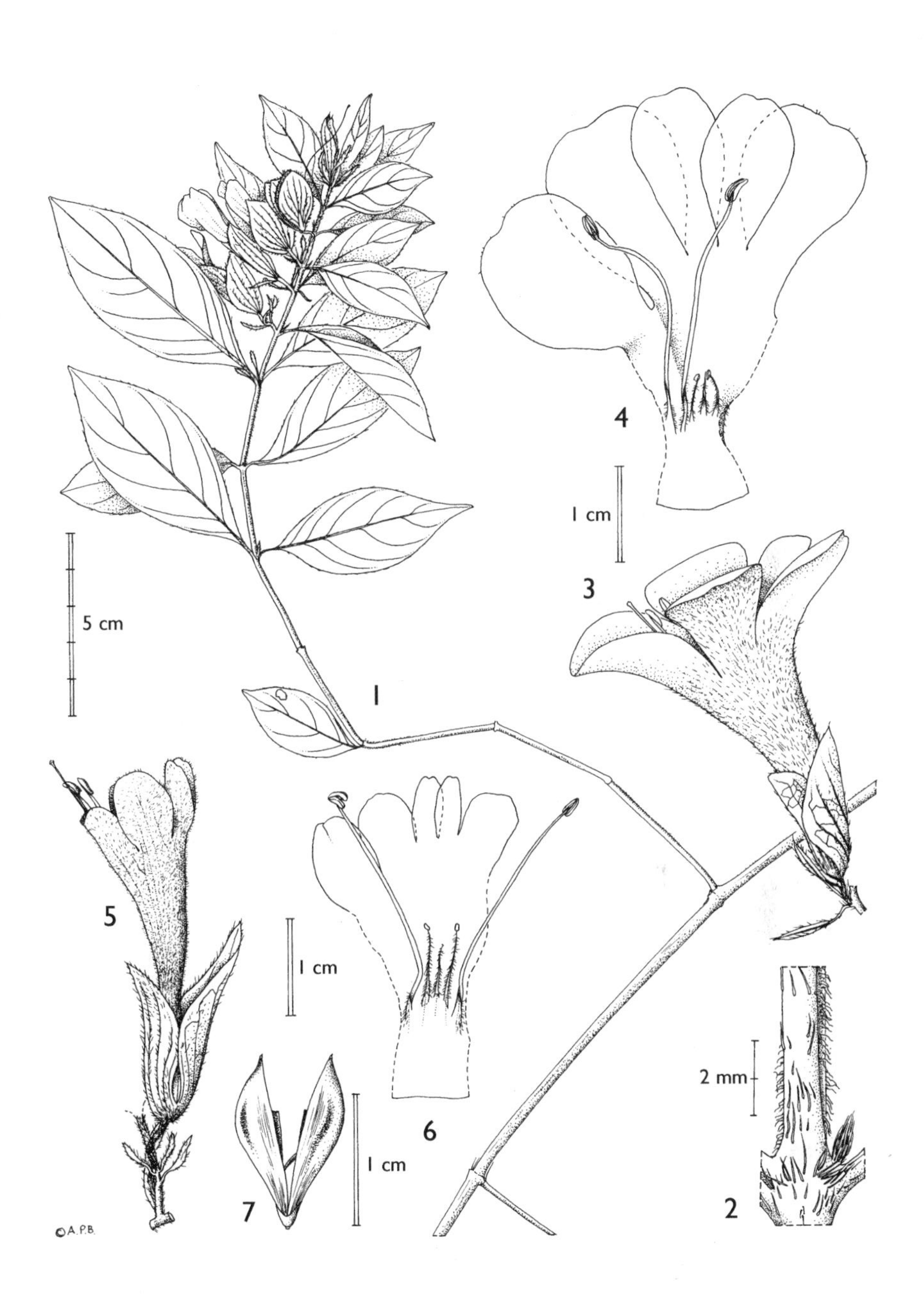

FIG. 51. *BARLERIA SCANDENS* — **1**, habit; **2**, indumentum on young stem; **3**, flower in situ, lateral view; **4**, dissected corolla, abaxial lobe held to the left. *BARLERIA LUKWANGULENSIS* — **5**, flower in situ, lateral view (arranged vertically for convenience of position – natural position would be horizontal); **6**, dissected corolla, abaxial lobe held to the left; **7**, capsule. 1–4 from *Bidgood et al.* 446; 5 & 6 from *Congdon* 209; 7 from *Archbold* 2548. Drawn by Andrew Brown.

TANZANIA. Mpwapwa District: Rubeho Mts, between Mafwemera and Dibulilo subvillages, May 2005, *Kindeketa, Kayombo & Laizer* 2535!; Morogoro District: Nguru Mts, Mkombola–Maskati path, Mar. 1999, *Congdon* 543!; Kilosa District: Ukaguru Mts, Mamiwa Forest Reserve, Mnyera Peak, July 1972, *Mabberley, Pócs & Salehe* 1308!

DISTR. **T** 2, 5, 6; not known elsewhere

HAB. Moist montane forest including *Albizia-Podocarpus* forest, forest margins including edges of bamboo forest; 1400–2150 m

USES. None recorded on herbarium specimens

CONSERVATION NOTES. *B. scandens* is found mainly in the central Eastern Arc mountain chain. It has been most frequently collected in the Nguru Mts but this is most likely due to this range having been more thoroughly botanised than the adjacent Ukaguru and Rubeho Mts. Indeed, in the latter range Kindeketa *et al.* (no. 2535) noted this species as locally abundant at one site. Sizable areas of montane forest remain intact in the vicinity of Maskati in the Ngurus (K. Vollesen, *pers. comm.*). However, continued forest clearance on the lower slopes of this range may have reduced populations of this species towards the lower end of its altitudinal range. A similar situation is likely to apply to populations in the Rubeho and Ukaguru Mts. With less than ten localities known in total, it is therefore assessed as Vulnerable (VU B2ab(iii)).

NOTE. This species is the montane counterpart of *B. amanensis*. They share a similar calyx morphology and corolla size but are readily separable by a number of characters. In *B. scandens* the coarse eglandular stem hairs are pale buff (-yellow), shorter and appressed and are sparse or absent on the mature stems, whilst in *B. amanensis* they are golden-yellow or golden-brown, longer, ascending or spreading and many throughout; the leaves and outer calyx lobes are similarly more hairy. In *B. amanensis* the flowers are often held in a congested terminal synflorescence with reduced bracts whilst in B. *scandens* the cymes remain more clearly spaced and axillary. In addition, *B. scandens* has blue or mauve (not white) corollas with a less broadly funnel-shaped throat, longer bracteoles (7–18 mm long, not 0.8–5(–9.5) mm) and longer staminodes (3–4.5 mm long versus 1.5–2.5 mm) which are sparsely hairy, not rather densely pilose.

Care should also be taken to separate this species from montane forest forms of *B. ventricosa* "*B. stuhlmannii*" from **T** 2 (e.g. *Simon et al.* 471) which are sparsely hairy and dry a blackish-green colour. Amongst several differences, these plants have more ovate leaves, at most shortly attenuate at the base and acute to subattenuate at the apex, the bracteoles are shorter (to 3–6 mm long), at least some of the calyces are toothed and the corollas are smaller (to ± 25 mm long). Other scandent, montane forest forms of *B. ventricosa* from our region fall within "*B. keniensis*" and are easily separated from *B. scandens* by, for example, the much more hairy young stems and calyces.

The isolated populations in **T** 2 appear to have smaller corollas (± 30 mm long) than those of the Eastern Arc Mts (40–50 mm long), although more flowering material from across its range is required to confirm this trend.

31. **Barleria lukwangulensis** *I. Darbysh.* **sp. nov.** *B. scandenti* I. Darbysh. corolla rubro-purpurea (non caesia neque malvina), tubo declinato gradatim atque anguste expanso (non recto abrupte infundibuliformi), limbo vix (non late) patenti, lobo abaxiali 6–10 mm (non 13.5–22 mm) longo ab lobis ceteris 1.5–3.5 mm tantum (non 3–8 mm) distanti differt; M. & K. Balkwill in K.B. 52: 569, figs. 2M & 3E (1997). Type: Tanzania, Morogoro District, Uluguru Mts, Lukwangule Plateau, *Schlieben* 3478 (B!, holo.; B!, BM, BR, iso.)

Scandent perennial herb or shrub, 100–600 cm tall; stems with appressed coarse golden-brown or buff-coloured hairs and with inconspicuous short crisped hairs concentrated in opposite furrows. Leaves ovate or elliptic, 3–8 cm long, 1.5–3 cm wide, base attenuate or with a narrowly cuneate extension, apex subattenuate to acuminate, surfaces with ± sparse coarse appressed or ascending hairs; lateral veins (3–)4(–5) pairs; petiole 3–15 mm long. Inflorescences axillary, flowers solitary or in 2–3-flowered unilateral cymes; bracts foliaceous; bracteoles narrowly elliptic, narrowly oblanceolate or linear, 3–16 mm long, 0.2–5 mm wide; pedicels 2.5–7 mm long. Anticous calyx lobe ovate-elliptic to broadly elliptic, 12–20 mm long, 6–13.5 mm wide, base obtuse or acute, margin entire, apex emarginate or rarely truncate to acute, indumentum as leaves but shorter, venation inconspicuous; posticous lobe ovate (-elliptic) to broadly so, 14.5–23 mm long, 7.5–14.5 mm wide, apex acute or subattenuate, minutely apiculate; lateral lobes lanceolate, 5–7.5 mm long. Corolla red-purple, or rose-red, base of tube white, 23–34 mm long, glandular-pilose and with coarse eglandular hairs outside; tube 16–24 mm long, somewhat declinate, gradually and rather narrowly expanded upwards; limb with lobes barely spreading; abaxial lobe offset by 1.5–3.5 mm, 6–10 mm long and wide; lateral lobes 5–9.5 mm long, 4.5–9 mm wide; adaxial lobes as lateral pair but 2.5–6.5 mm wide. Stamens attached 6.5–9 mm from base of corolla tube; filaments 16–23 mm long; anthers 2–3.3 mm long; lateral staminodes 3–7 mm long, pilose, antherodes barely developed or to 0.9 mm long. Pistil glabrous; stigma subcapitate, 0.2–0.4 mm long. Capsule 13–14 mm long, glabrous; only immature seeds seen. Fig. 51/5–7, p. 375.

TANZANIA. Morogoro District: Uluguru Mts, Lukwangule Plateau above Chenzema Mission, Mar. 1953, *Drummond & Hemsley* 1528!; Iringa District: Mufindi, Kigogo Forest Reserve, June 1988, *Congdon* 209!; Songea District: Matengo Hills, Lupembe Hill, May 1956, *Milne-Redhead & Taylor* 10519!

DISTR. **T** 6–8; not known elsewhere

HAB. Montane moist forest, often in areas with dense undergrowth including disturbed patches; 1500–2400 m

USES. None recorded on herbarium specimens

CONSERVATION NOTES. This species has been most often collected from the Lukwangule Plateau of the Uluguru Mts and the Mufindi Escarpment, though these are well-botanised sites and *B. lukwangulensis* is probably under-recorded in, for example, the less well known Udzungwa range. Although suitable habitat at its lower altitudinal limit is likely to have retracted due to forest clearance, much montane forest remains in its core range of over 2000 m altitude. It also appears tolerant of minor habitat disturbance, perhaps because its scandent habit allows it to compete well for light availability amongst the dense understorey of disturbed forest. It is not considered threatened at present: Least Concern (LC).

SYN. [*B. lukafuensis sensu* M. & K. Balkwill in J. Biogeogr. 25: fig. 12a (1998), *non* De Wild.]

NOTE. In the absence of flowers, *B. lukwangulensis* is essentially inseparable from *B. scandens*. However, the red-purple or rose-red corolla with a proportionally long, slender, slightly declinate tube and small, barely spreading lobes readily separate it from that species and indeed from all other *Barleria* taxa. It is one of several *Acanthaceae* taxa, believed to be bird-pollinated, centred on the Uluguru Mts that display convergent evolution (see Darbyshire, K.B. 63: 373–374 (2009)). Indeed, without close inspection, flowering material of *B. lukwangulensis* could be confused with the distantly related *Dicliptera grandiflora* Gilli which shares the red corolla with a proportionally long, gradually expanded tube and has paired cymule bracts superficially similar to the outer calyx lobes of *B. lukwangulensis*. It is possible that *B. lukwangulensis* evolved from *B. scandens* in the southern portion of the Eastern Arc chain as a specialisation to sunbird pollination; the two species are currently allopatric.

Populations from the Mufindi escarpment and the Matengo Hills differ somewhat from those of the Uluguru Mts in having narrower outer calyx lobes (length/width ratio typically (1.7–)2/1, versus up to 1.5/1) and in the adaxial pair of corolla lobes being clearly narrower than the lateral pair, these being subequal in the Uluguru plants.

The name *Barleria lukwangulensis* was first applied to this taxon in the 1930s by Mildbraed in an annotation on *Schlieben* 3478 but was never published; it is therefore validated here.

32. **Barleria pseudosomalia** *I. Darbysh.* **sp. nov.** *B. scandenti* I. Darbysh. similis sed habitu breviore subfruticoso (non scandenti), caulibus mox lignosis cortice etiam ramorum foliosorum pallide cinerea vel arenaceo-colorati (non ramis foliosis herbaceis atque in siccitate nigricantibus), inflorescentiis semper unifloris (non floribus in cymis ad maturitatem unilateralibus 2–4-floris dispositis), calyce pallide brunneo in fructu scarioso, longitudine lobi calycis antici plus minusve latitudinem aequanti (non quam latitudine majore) ad basin (sub)cordato (non rotundato neque obtuso neque breviter attenuato), pedicellis longioribus 5–11 mm (non 1–3 mm) et antheris longioribus 4–4.5 mm (non 2.5–3.5 mm) differt. Type: Tanzania, Mpwapwa District, 4 km on Kibakwe–Motta track, *Bidgood, Mwasumbi & Vollesen* 1001 (K!, holo.; C, CAS, DSM!, EA!, K!, iso.)

Subshrub or prostrate perennial herb, to 50 cm tall; stems soon woody with pale grey or sandy-coloured bark, young stems with coarse appressed buff-yellow hairs and sometimes with scattered short retrorse hairs, glabrescent. Leaves (ovate-) elliptic to narrowly so, 5.5–10.5 cm long, 2–3.8 cm wide, base long-attenuate, apex acuminate, surfaces with coarse appressed or ascending hairs along principal veins and margin and scattered above, sparse on mature leaves; lateral veins 4–5 pairs; petiole to 6 mm long or blade sessile. Inflorescences axillary, flowers solitary; bracts foliaceous; bracteoles often caducous, linear or narrowly elliptic-lanceolate, 2.5–10 mm long, 0.2–1.5 mm wide; pedicels 5–11 mm long. Calyx scarious in fruit; anticous lobe broadly ovate, 16–23 mm long and wide, base (sub)cordate, margin entire, apex truncate or emarginate, margin ciliate with appressed or ascending hairs, elsewhere glabrous or principal veins with few hairs; posticous lobe 20–29 mm long, base rounded or shallowly cordate, apex attenuate; lateral lobes linear-lanceolate, 5.5–10 mm long. Corolla (pale) blue, throat white, 42–48 mm long, sparsely glandular-pilose outside; tube 20–25 mm long, broadly funnel-shaped above attachment point of stamens; abaxial lobe offset by ± 9–10 mm, 20–23 mm long, 12–14 mm wide; lateral lobes 14–15 mm long, 11.5–13 mm wide; adaxial lobes as lateral pair but ± 9 mm wide. Stamens attached 7.5–8 mm from base of corolla tube; filaments ± 24 mm long; anthers 3.5–4.5 mm long; lateral staminodes 2–2.5 mm long, pilose, antherodes 1–1.2 mm long. Pistil glabrous; stigma subcapitate, 0.5–0.7 mm long. Capsule 13–14.5 mm long, glabrous; seeds 7.5–8.5 mm long, 7–8 mm wide.

TANZANIA. Mpwapwa District: 4 km on Kibakwe–Motta track, Apr. 1988, *Bidgood, Mwasumbi & Vollesen* 1001! (type) & between Ikuyu and Mangalisa, Apr. 1988, *J. Lovett & Congdon* 3192!; Iringa District: Iringa to Dodoma road N of Isimani, Jan. 1988, *J. Lovett & Congdon* 2976!

DISTR. **T** 5, 7; not known elsewhere

HAB. Dry woodland and scrub with *Brachystegia microphylla*; 1200–1250 m

USES. None recorded on herbarium specimens

CONSERVATION NOTES. This species is currently known from only three locations on the dry lower slopes of the Rubeho Mts in the rainshadow of the Eastern Arc chain. It appears to have been discovered as late as 1988 (remarkably the only year in which the species has been collected, with two locations being found by separate collecting teams on the same date!). Suitable habitat is, however, widespread in the drier parts of these hills and the adjacent Ruaha River valley. The paucity of known collections is perhaps, therefore, more a reflection of under-collection in this isolated region. With so little information currently available on its abundance and ecological requirements, particularly its tolerance of grazing pressure, it must be assessed as Data Deficient (DD) but may well prove to be threatened by habitat loss.

NOTE. The combination of soon-woody stems with pale bark, conspicuously pedicellate, single-flowered cymes, and very broad outer calyx lobes together easily separate this species from all others in sect. *Fissimura*. In fact, in the absence of fruits it would be easy to place this species within sect. *Somalia* as it is superficially similar to, for example, *B. mackenii* Hook.f., *B. galpinii* C.B. Clarke and *B. granarii* I. Darbysh. (see p. 401). However, the 2-seeded, unbeaked capsules clearly place this species in sect. *Fissimura* and this is confirmed by the presence of lateral staminodes with developed antherodes and a subcapitate stigma.

33. **Barleria sp. A** (= *Magogo* 1352)

Shrub, ± 180 cm tall, much-branched; stems with appressed coarse pale buff-coloured hairs, sparse at maturity. Leaves ? immature at flowering, subsessile, elliptic, 1.5–2 cm long, ± 1 cm wide, base acute or attenuate, apex subattenuate, surfaces strigose, mainly on the principal veins or more scattered above; lateral veins 4 pairs. Inflorescences axillary towards branch apices, each 1–2-flowered, subsessile; bracts foliaceous; bracteoles (linear-) oblanceolate, 4–8.5 mm long, 0.5–2 mm wide. Anticous calyx lobe broadly ovate, 11.5–13.5 mm long, 10–13 mm wide, base subcordate, margin usually conspicuously 6–9-toothed, teeth apiculate and with an apical bristle, apex broadly bifid for up to 2 mm, each segment mucronulate, shortly and sparsely strigose on the palmate principal veins and margin; posticous lobe 12.5–16 mm long, 9–12 mm wide, apex acute-mucronulate; lateral lobes 6.5–9 mm long. Corolla pale blue, drying blue-black, ± 32 mm long, sparsely glandular-pilose and with appressed eglandular hairs mainly on lateral lobes outside; tube ± 19 mm long, broadly funnel-shaped above attachment point of stamens; abaxial lobe offset by ± 5 mm, 11.5–13 mm long and wide; lateral lobes 8–10 mm long and wide; adaxial lobes as lateral pair but 7–7.5 mm wide. Stamens attached ± 5.5 mm from base of corolla tube; filaments ± 24 mm long; anthers ± 3.5 mm long; lateral staminodes ± 3 mm long, pilose, antherodes 0.65–1 mm long. Pistil glabrous; stigma subcapitate, 0.4–0.5 mm long. Capsule and seeds not seen.

KENYA. Northern Frontier District: Marsabit–Isiolo road, May 1970, *Magogo* 1352!
DISTR. **K** 1; not known elsewhere
HAB. Thicket edge; 1050 m
USES. None recorded on herbarium specimen
CONSERVATION NOTES. Currently known only from the above collection, Magogo recorded it as common but it has never been recollected despite this being an accessible and well-botanised locality. More data on its distribution and abundance are required: Data Deficient (DD).

NOTE. Although close to several other species in sect. *Fissimura*, this specimen is not an exact match for any of the currently recognised taxa. The indumentum, and the fact that the plants dry blackish, place it close to *B. scandens* but it clearly differs in the characters listed in the key. The broad, subcordate, usually toothed calyx lobes are reminiscent of *B. boranensis* but the teeth are less regular, the hairs on the calyx are sparser and more strictly appressed and the corollas are much larger than in that species.

SECT. III. STELLATOHIRTA

M. Balkwill in J. Biogeogr. 25: 110 (1998); M. & K. Balkwill in K.B. 52: 569 (1997); Darbyshire in K.B. 63: 261–268 (2008)

Barleria subgen. *Eu-Barleria* "*Stellato-hirtae*" *sensu* C.B. Clarke in F.T.A. 5: 143 (1899) pro parte excl. *B. fulvostellata*
Barleria sect. *Eubarleria* subsect. *Thamnotrichae sensu* Obermeijer in Ann. Transv. Mus. 15: 139 (1933)

Axillary spines absent. Indumentum stellate or dendritic. Inflorescences single-flowered subsessile cymes compounded into a dense terminal synflorescence; bracts highly modified. Calyx lobes with margins entire or somewhat irregular, not toothed. Corolla white, blue or purple; limb in "4+1" arrangement. Stamens attached ± midway along corolla tube; staminodes 2(–3), lateral pair with antherodes well developed or rarely absent; if adaxial staminode present then minute, antherode absent. Stigma linear, single-lobed. Capsule laterally flattened, fusiform, unbeaked or shortly beaked; lateral walls often thin, remaining attached to or tearing slightly from the thickened flanks at dehiscence; septum largely membranous. Seeds 2, discoid with dense buff-coloured hygroscopic hairs.

± 15 species confined to Africa and Arabia, with centres of diversity in our region and in Angola. M. & K. Balkwill (in K.B. 52, 1997) recorded the capsules of this section as differing from those of sect. *Fissimura* in that the lateral walls do not split away from the thickened raphes at dehiscence. However, tearing is clearly visible in the capsules of *B. splendens* within sect. *Stellatohirta*, therefore this character is not used in the key to sections.

1. Bracteoles and outer calyx lobes with spinose apices 2
   Bracteoles and outer calyx lobes lacking spinose apices 3
2. Bracts narrowly elliptic, bracteoles lanceolate, both lacking an arista; posticous calyx lobe gradually acuminate; upper leaf surface with whitish long-armed stellate hairs, stellate base persistent . . . . 34. *B. taitensis*
   Bracts and bracteoles conspicuously aristate above an obovate or oblanceolate basal portion; posticous calyx lobe abruptly long-caudate; upper leaf surface with golden long-armed stellate hairs, stellate base ± caducous . . . . 35. *B. aristata*
3. Synflorescences elongate, (2.5–)5–11.5 cm long; anticous calyx lobe deeply bifid with segments 5.5–15 mm long; ovary and capsule hairy towards apex . . . . 38. *B. splendens*
   Synflorescences subglobose, 1.5–4.5 cm in diameter; anticous calyx lobe more shallowly bifid with segments 1.5–4.5 mm long; ovary and (where known*) capsule glabrous 4
4. Corolla 40–45 mm long with lobes ± 12 mm wide, tube cylindrical, barely expanded towards mouth; bracteoles and calyces turning scarious at maturity . . . . 36. *B. sp. B*
   Corolla 55–75 mm long with lobes 19–30 mm wide, tube funnel-shaped above attachment point of stamens; bracteoles and calyx not turning scarious . . . . 37. *B. aenea*

* the capsule of *B.* sp. B is unknown but its ovary is glabrous

34. **Barleria taitensis** *S. Moore* in J.B. 40: 343 (1902); Blundell, Wild Fl. E. Afr.: 388, pl. 121 (1987); U.K.W.F. ed. 2: 272 (1994); K.T.S.L.: 599 (1994). Type: Kenya, Machakos District, Kiumbi R. [Makindo], *Kässner* 600 (BM!, holo.; K!, MO!, iso.)

Erect or scrambling subshrub, 30–250 cm tall. Young stems with dense golden antrorsely long-armed dendritic hairs. Leaves (ovate-) elliptic to lanceolate, 2.5–13 cm long, 0.8–3.3 cm wide, base cuneate, apex acute or obtuse, ± apiculate, surfaces with whitish long-armed stellate hairs, shorter and more dense beneath, margin and principal veins beneath with longer golden dendritic hairs; lateral veins 4–6 pairs, impressed above, prominent beneath; petiole to 4 mm long or absent. Synflorescences conical, subcapitate or rarely more elongate, 1.5–17 cm long; bracts narrowly (oblong-) elliptic, 11–21 mm long, 2.5–6.5 mm wide, apex spinose, ± outcurved, indumentum as lower leaf, dense; bracteoles lanceolate, 8.5–16 mm long, 1.7–2.5(–4) mm wide. Anticous calyx lobe subrhombic, 14–22 mm long, 7.5–16 mm wide, base and apex attenuate, the latter bifid with segments 1–4.5 mm long with divergent spinose tips, venation palmate, prominent, indumentum as bracts; posticous lobe ovate, 15–21 mm long, 4.5–7 mm wide, apex ± long-acuminate with a spinose tip; lateral lobes linear-lanceolate, 6.5–10 mm long. Corolla white or flushed pink or blue, more rarely bright blue, 27–59 mm long, pubescent towards apex of tube and on lateral lobes outside; tube narrowly cylindrical, 10–30 mm long; abaxial lobe offset by 2–4.5 mm; each lobe 13–25 mm long, abaxial lobe 7.5–12 mm wide, lateral lobes 6–11.5 mm wide, adaxial lobes 5.5–10 mm wide. Staminal filaments 15–33 mm long, shortly pubescent in lower half and with minute glandular hairs at base; anthers 3–4 mm long; lateral staminodes 1.3–2.5 mm long, antherodes 0.7–1.3 mm long. Ovary glabrous; style puberulous towards base; stigma 1–2.2 mm long. Capsule 10–16 mm long including 1.5–2 mm beak, glabrous; seeds 6.5–9.5 mm long, 4–6.5 mm wide.

KENYA. Machakos District: Bushwhackers Camp, June 1958, *Irwin* 409!; Kitui District: Mutomo–Kitui km 16, Mar. 1969, *Napper* 1948!; Teita District: Voi, Dec. 1961, *Polhill & Paulo* 935!

TANZANIA. Lushoto District: Umba Valley, Dec. 1892, *Smith* s.n.! & Mkomazi Game Reserve, 3 km N of Umba River Camp, June 1996, *Abdallah, Mboya & Vollesen* 96/171!; Dodoma District: 84 km on Kondoa–Dodoma road, June 2006, *Bidgood et al.* 6341!

DISTR. **K** 4, 7; **T** 3, 5, 7; Angola (see note)

HAB. Dry *Acacia-Commiphora* bushland and associated grassland, margins of riverine bushland, often on rocky hillsides on sandy soils; 350–1200 m

USES. None recorded on herbarium specimens

CONSERVATION NOTES. A locally abundant species, in northern Tanzania *Abdallah et al.* (96/171) noted it as "common ... from Kisima eastwards, often in great numbers where it occurs". In Kenya, it is particularly common in the Tsavo area, where its bushland habitat is well-protected: Least Concern (LC).

SYN. *B. stellato-tomentosa* S. Moore var. *ukambensis* Lindau in E.J. 20: 23 (1894). Syntypes: Kenya, Machakos/Kitui District, Ukambani, *Hildebrandt* 2722 & 2722a; Teita District, Ndara, *Hildebrandt* 2457 (all B†)

[*B. salicifolia sensu* C.B. Clarke in F.T.A. 5: 162 (1899) pro parte quoad *Hildebrandt* 2722, 2722a & 2457 ex Kenya & *Smith* s.n. ex Tanzania, *non* S. Moore]

[*B. argentea sensu* Blundell, Wild Fl. E. Afr.: pl. 119 (1987), *non* Balf.f]

NOTE. Specimens from Dodoma District, **T** 5 (*Bidgood et al.* 6341; *Hansen* 634; *Thulin & Mhoro* 557) differ from typical *B. taitensis* in having more lanceolate leaves, corollas with a short tube (10.5–16 mm long) and somewhat larger capsules and seeds. They are possibly matched by material from the Ruaha National Park, **T** 7, but the two specimens seen from there (*Greenway & Kanuri* in EAH 14389; *Bjørnstad* 1861) are either sterile or in bud. It is possible that two subspecies are recognisable in our area but more material is needed to determine the consistency of these differences.

Variation in flower colour is significant in our region and even within populations; Diana Napper noted that a colony from Kitui (*Napper* 1948) had flowers "± 65% white, 35% blue, the tone paling with degree of exposure to direct sun". However, Spencer Moore's note (in J.B. 40: 342 (1902)) that the flowers can be yellow is erroneous.

Var. *occidentalis* S. Moore, restricted to Angola and Namibia, is similar to the **T** 5 material in having small, blue corollas with a short tube but differs in the calyx being smaller, the posticous lobe being less strongly acuminate, and in the spinose apices of the bracteoles and calyces being less conspicuous. As it is geographically allopatric from the East African plants, subspecies rank would seem more appropriate. Plants previously named as *B. taitensis* from southern Zambia and Zimbabwe (e.g. in Mapaura & Timberlake (eds.), Checkl. Zimb. Vasc. Pl.: 13 (2004)) are close to var. *occidentalis*, but are distinct in the anticous calyx lobe being only minutely divided with the segments parallel rather than divergent; such plants are referable to *B. rogersii* S. Moore.

35. **Barleria aristata** *I. Darbysh.* in K.B. 63: 261, figs. 1 & 2 (2008). Type: Tanzania, Iringa District, 17 km on Kitonga–Morogoro road, *Bidgood, Darbyshire, Hoenselaar, Leliyo, Sanchez-Ken & Vollesen* 5027 (K!, holo.; BR!, CAS!, DSM!, EA!, MO!, NHT!, iso.)

Erect or scrambling subshrub, 30–200 cm tall. Young stems with dense buff to golden stellate or dendritic hairs, some with a golden ascending arm. Leaves narrowly ovate-elliptic or lanceolate, 3–8.5 cm long, 1–2 cm wide, base cuneate or acute, apex acute or subattenuate, upper surface with golden long-armed stellate hairs, stellate base ± caducous, lower surface whitish stellate-pubescent and with golden long-armed dendritic hairs on principal veins and margin; lateral veins (3–)4–5 pairs, ± impressed above, prominent beneath; petiole to 5 mm long or absent. Synflorescences conical or subglobose, 1.5–4 cm long; bracts broadly obovate-aristate, 16–21 mm long including arista, 5.5–8 mm wide, apex spinose, indumentum as lower leaf with additional scattered subsessile glands, long-armed hairs on margin and veins with a bulbous dendritic base; bracteoles asymmetrically obovate- to oblanceolate-aristate, 12–17 mm long, 2–4.5 mm wide. Anticous calyx lobe subrhombic, 15–22 mm long, 8.5–11(–15) mm wide, base and apex attenuate, the latter bifid with segments aristate, 4–7.5 mm long with ± divergent spinose tips, venation palmate, prominent, indumentum as bracts in upper half, with sparse, predominantly simple hairs below; posticous lobe obovate at base, caudate above, 16–23 mm long, 4–6.5 mm wide, apex spinose, venation parallel, long-armed stellate hairs restricted to upper half of caudate portion, largely glabrous towards base except for simple hairs along margin; lateral lobes linear-lanceolate, 7–11 mm long. Corolla blue (-purple) with a paler throat, (28–)35–42 mm long, pubescent towards apex of tube and on lateral lobes outside; tube narrowly cylindrical, (12–)14.5–17 mm long; abaxial lobe offset by 2–4 mm; each lobe 15–22 mm long, abaxial lobe 9–16 mm wide, lateral lobes 8.5–13 mm wide, adaxial lobes 7–12.5 mm wide. Staminal filaments 18–24 mm long, shortly pubescent in lower half and with minute glandular hairs at base; anthers 3–4 mm long; lateral staminodes 1.2–1.6 mm long, antherodes 0.2–0.8 mm long. Ovary glabrous; style puberulous towards base; stigma 1.3–1.8 mm long. Capsule 13.5–16 mm long including 1.7–2.7 mm beak, glabrous; seeds 6.5–7 mm long, ± 6.5 mm wide. Fig. 52, p. 383.

TANZANIA. Mpwapwa District: 9 km S of Mpwapwa on Gulwe track, Apr. 1988, *Bidgood, Mwasumbi & Vollesen* 954!; Kilosa District: Great Ruaha River Gorge, 48 km W of Mikumi on Tanzania–Zambia Highway, May 1990, *Carter, Abdallah & Newton* 2245!; Iringa District: 82 km on Iringa to Dar es Salaam road, near Mahenge village, Mar. 1988, *Bidgood, Mwasumbi & Vollesen* 573!

DISTR. **T** 5–7; not known elsewhere

HAB. Dry bushland, most often of *Acacia-Commiphora*, including roadsides and degraded areas; 500–1200 m

USES. None recorded on herbarium specimens

CONSERVATION NOTES. Assessed as of Least Concern (LC) by Darbyshire (l.c.).

36. **Barleria** sp. **B** (= *Mwangulango* 925)

Perennial to 90 cm tall, much-branched. Young stems with dense golden dendritic hairs, few with an antrorse arm. Leaves subcoriaceous, ovate or elliptic, 4.8–9 cm long, 2.2–4.5 cm wide, base obtuse to cuneate, apex acute, upper surface with buff long-armed dendritic hairs, lower surface with pale stellate or dendritic hairs and with scattered golden dendritic hairs on principal veins and margin; lateral veins 4–6 pairs, prominent beneath; petiole to 3.5 mm long or absent. Synflorescences subglobose, 1.5–2.5 cm in diameter; bracts broadly ovate, 14–19 mm long, 7–14 mm wide, with dense pale dendritic hairs, often long-armed particularly on principal veins and margin, reticulate venation prominent; bracteoles turning scarious at maturity, narrowly ovate or lanceolate, (10.5–)15–19 mm long, (1–)5–6.5 mm wide, indumentum as bracts but hairs whitish and with many long-armed hairs. Calyx turning scarious at maturity; anticous lobe elliptic, 16–21 mm long, 7–11.5 mm wide, apex bifid with segments rounded or obtusely deltate, 2–4.5 mm long, venation subparallel, prominent, indumentum as bracteoles, long-armed hairs dense on margin and principal veins; posticous lobe marginally longer, apex acute to obtuse; lateral lobes subulate, 11–13.5 mm long. Corolla purple, 40–45 mm long, glabrous outside; tube narrowly cylindrical, 19–23 mm long, lobes each ± 20–25 mm long, 12 mm wide. Staminal filaments 18–23 mm long, shortly pubescent in lower half and with minute glandular hairs towards base; anthers ± 4.5 mm long; lateral staminodes 3.5–5 mm long, antherodes barely developed. Ovary glabrous; style puberulous at base; stigma ± 3 mm long. Capsule and seeds not seen.

TANZANIA. Ufipa District: Katavi-Rukwa protected areas, Ufipa Escarpment, May 2002, *Mwangulango* 925!

DISTR. **T** 4; not known elsewhere

HAB. Unrecorded; 1000 m

USES. None recorded on herbarium specimens

CONSERVATION NOTES. Although SW Tanzania remains under-collected, this species is clearly rare, having been found at only one locality where it was recorded as frequent. No information on habitat preferences or threats to this site were recorded on the specimen. It is therefore currently assessed as Data Deficient (DD) but may well prove to be threatened.

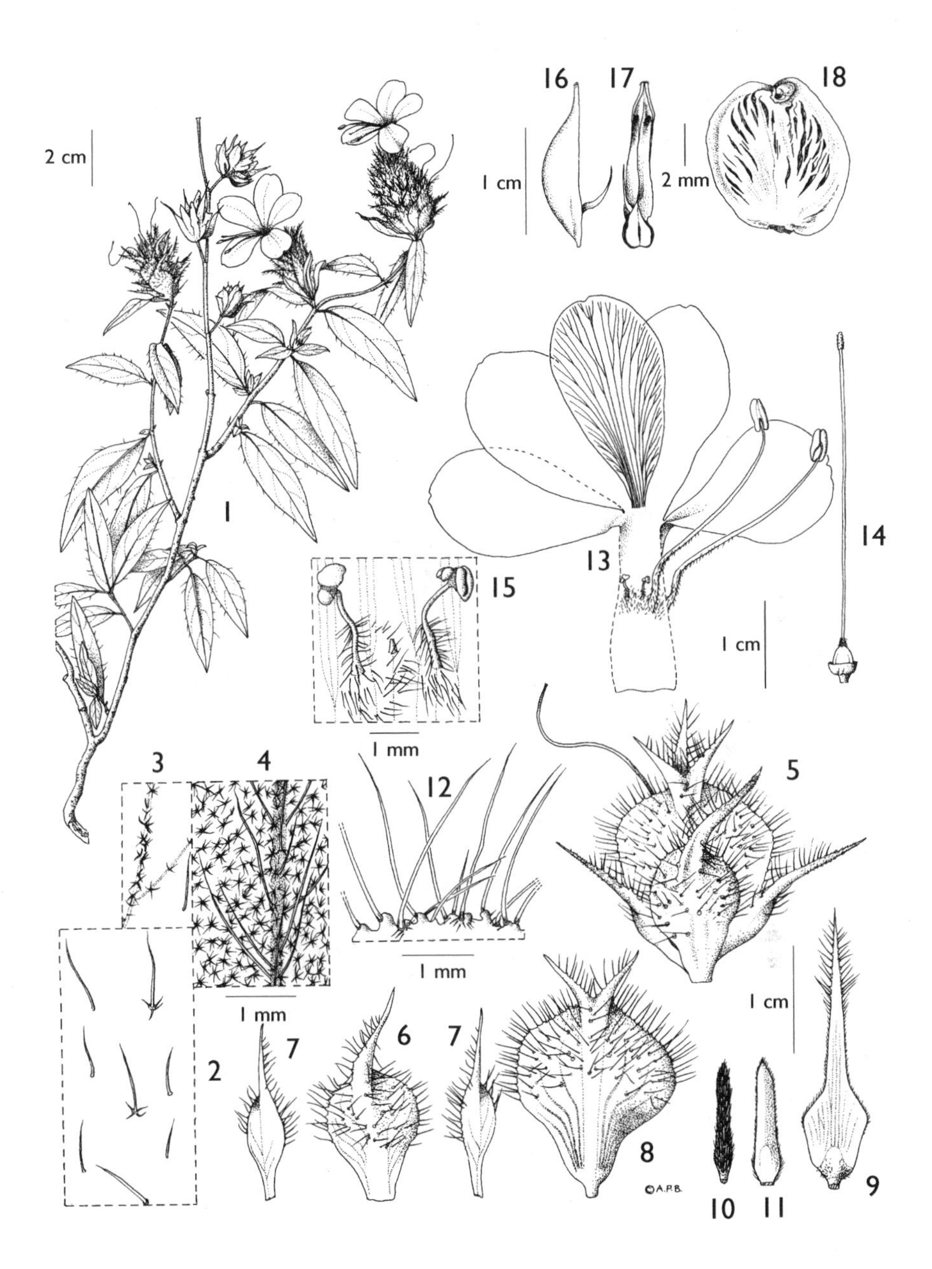

FIG. 52. *BARLERIA ARISTATA* — **1**, habit; **2**, leaf indumentum, upper surface; **3**, leaf indumentum, upper surface at midrib; **4**, leaf indumentum, lower surface at midrib; **5**, single cyme, outer view; **6**, bract; **7**, bracteoles; **8**, anticous calyx lobe; **9**, posticous calyx lobe; **10**, lateral calyx lobe, unflattened, inner surface; **11**, lateral calyx lobe, flattened, outer surface; **12**, detail of indumentum on upper margin of anticous calyx lobe; **13**, dissected corolla with stamens and staminodes, abaxial lobe to the right; **14**, pistil; **15**, detail of staminodes; **16**, valve of capsule, lateral view; **17**, valve of capsule, interior view; **18**, seed. 1 from *Bidgood et al.* 954 and from photographs; 2–15 from *Bidgood et al.* 5027; 16–18 from *Hornby* 604A. Drawn by Andrew Brown. Reproduced from Kew Bulletin 63: 263 (2008) with permission of the Trustees of the Royal Botanic Gardens, Kew.

NOTE. The single specimen seen has only one mature flower each on the K and MO sheets and the lobes are somewhat damaged in both so precise measurements were not possible. Although almost certainly new to science, further material including fruits and flowers are required for its description.

37. **Barleria aenea** *I. Darbysh.* in K.B. 63: 264, figs. 3 A–H & 4 (2008). Type: Tanzania, Ufipa District, 45 km on Namanyere–Karonga road, *Bidgood, Mbago & Vollesen* 2655 (K!, holo.; BR!, C, CAS, DSM!, EA!, K!, NHT, iso.)

Suffruticose perennial with many erect stems from a woody rootstock, 45–75 cm tall; stems with dense golden-bronze dendritic hairs, many with a long antrorse arm, ± persistent. Leaves (ovate-) elliptic or somewhat obovate, 6.5–12.5 cm long, 2–5.5 cm wide, base cuneate or attenuate, apex acute or subattenuate, upper surface with pale long-armed stellate hairs, stellate base ± caducous, lower surface pale stellate-pubescent and with buff to golden long-armed dendritic hairs on principal veins and margin; lateral veins 5–7 pairs, these and the reticulate tertiary venation prominent beneath; petiole 2–10 mm long. Synflorescences subglobose, 2.5–4.5 cm in diameter; bracts (ovate-) elliptic, 19–34 mm long, 5.5–18 mm wide, with dense golden-bronze long-armed dendritic hairs particularly on margin and principal veins, interspersed with paler stellate hairs; bracteoles ± narrowly (ovate-) elliptic, 21–28 mm long, 6–10 mm wide. Anticous calyx lobe ovate-elliptic, 19–22 mm long, 7.5–10.5 mm wide, apex bifid with segments obtusely deltate, 1.5–3 mm long, venation subparallel, inconspicuous, indumentum as bracts but sparse towards base; posticous lobe elliptic, 20–23 mm long, 7.5–9.5 mm wide, apex acute or subattenuate, long-armed stellate hairs restricted to the upper third, largely glabrous towards base except for simple hairs along margin; lateral lobes subulate, 9.5–11.5 mm long. Corolla blue or purple with a whitish throat and tube, whole flower rarely white, 55–75 mm long, glabrous outside or sparsely pubescent towards apex of lateral lobes; tube 21–25 mm long, cylindrical in lower half, funnel-shaped in upper half; abaxial lobe offset by 7–12 mm, lobes each 23–36 mm long, 19–30 mm wide. Staminal filaments (19–)26–32 mm long, with many subsessile glands and scattered short hairs in the lower two thirds; anthers (4–)4.5–5.5 mm long; lateral staminodes 5.5–7 mm long, antherodes to 0.5 mm long. Ovary glabrous; style puberulous towards base; stigma 3.2–4 mm long. Capsule ± 16 mm long, apex attenuate, glabrous; seeds 10–10.5 mm long, ± 7 mm wide.

TANZANIA. Ufipa District: 45 km on Namanyere–Karonga road, Mar. 1994, *Bidgood, Mbago & Vollesen* 2655! (type)

DISTR. **T** 4; NE Zambia

HAB. *Brachystegia* woodland; 950 m

USES. None recorded on herbarium specimens

CONSERVATION NOTES. Assessed as Data Deficient (DD) by Darbyshire (l.c.), this species is nevertheless likely to be threatened due to its highly limited range.

38. **Barleria splendens** *E.A. Bruce* in K.B.: 305 (1934); M. & K. Balkwill in K.B. 52: 570, figs. 5L & 14P (1997) & in J. Biogeogr. 25: fig. 12d (1998). Type: Tanzania, Shinyanga District, Tinde Hills, *Burtt* 2409 (K!, holo.)

Subshrub, 60–150(–210) cm tall, much-branched. Young stems with dense dendritic hairs, many with buff to golden appressed-ascending arms. Leaves ovate, 4–9 cm long, 1.5–4.5 cm wide, base attenuate, apex acute or obtuse, upper surfaces pale stellate-pubescent, lower surface with dense whitish dendritic hairs, principal veins and margin with additional golden long-armed dendritic hairs; lateral veins (4–)5–6(–7) pairs, prominent beneath; petiole 5–15 mm long. Synflorescences cylindrical to conical, (2.5–)5–11.5 cm long, 4-faceted; bracts broadly ovate or elliptic, 14–20 mm long, 7.5–15 mm wide, indumentum as lower leaf surface with additional sparse to dense glandular hairs, venation pinnate on the bracts of the lower cymes, 3-veined upwards, prominent; bracteoles narrowly elliptic, lanceolate or oblanceolate, 14.5–21 mm long, 2–5(–6.5) mm wide, more densely ciliate than bracts. Anticous calyx lobe elliptic, 15–24 mm long, 6–12 mm wide, apex deeply bifid, segments lanceolate, 5.5–15 mm long, venation palmate, prominent, indumentum as bracteoles; posticous lobe (ovate-) elliptic, 16–24 mm long, 6–13 mm wide, apex acute or subattenuate; lateral lobes lanceolate, 10–16 mm long. Corolla (pale) blue, (25–)32–39 mm long, eglandular- and ± glandular-pubescent towards apex of tube and on lateral lobes; tube cylindrical, 12–17 mm long; abaxial lobe offset by 2–2.5 mm; each lobe (11–)16–20 mm long, abaxial lobe 8–13.5 mm wide, lateral and adaxial lobes 6.5–12 mm wide. Staminal filaments (16–)19–25 mm long, shortly pubescent in lower half and with minute glandular hairs at base; anthers 3.6–4.3 mm long; lateral staminodes 0.6–3.2 mm long, antherodes undeveloped or 0.5–1 mm long. Ovary pilose towards apex; style glabrous or pubescent at base; stigma 0.7–1.3 mm long. Capsule (14.5–)17–20 mm long including 3.5–5 mm long pubescent beak; seeds 6.5–9 mm long, 6.5–8 mm wide.

TANZANIA. Shinyanga District: Mwantine Hills, May 1931, *Burtt* 2457!; Maswa District: Komali Hill, Shanwa, June 1935, *Burtt* 5137!; Dodoma District: 3 km on Manyoni–Kilimatinde road, Apr. 1988, *Bidgood, Mwasumbi & Vollesen* 1096!

DISTR. **T** 1, 5; not known elsewhere

HAB. Rocky hillsides with thickets of variously *Combretum, Commiphora, Grewia, Acacia, Brachystegia microphylla* and/or *Pseudoprosopsis*, often on granite; 1150–1250 m

USES. None recorded on herbarium specimens

CONSERVATION NOTES. Currently known from only seven locations within three broad subpopulations in North/central Tanzania. It was recorded as locally common in the Shinyanga and Kilimatinde areas by Burtt in the 1930s, but there are no recent abundance data. The bushlands of this region are rather poorly studied botanically and with suitable habitat being widespread there, it is likely that *B. splendens* is more frequent than currently known. Threats to the habitat are unknown. It is provisionally assessed as Data Deficient (DD) but may well prove to be unthreatened.

NOTE. Specimens from Shinyanga District (**T** 1) tend to have much fewer glandular hairs on the bracts, bracteoles and outer calyx lobes than those from Dodoma and Mpwapwa Districts (**T** 5), and also have well-developed antherodes on the lateral staminode pair (0.5–1 mm long), these being largely undeveloped in the **T** 5 material. However, with so few collections currently available (only nine seen in total), it is unclear as to how consistent these differences are.

SECT. IV. CAVIROSTRATA

M. Balkwill in Journ. Biogeogr. 25: 110 (1998); M. & K. Balkwill in K.B. 52: 566 (1997); Darbyshire in K.B. 63: 601–611 (2009)

Barlerianthus Oerst. in Vidensk. Medd. Dansk. Naturhist. Foren. Kjobenh.: 136 (1854)

Axillary spines absent. Indumentum simple or stellate. Inflorescences of 1–2-flowered subsessile cymes compounded into a verticillate or strobilate terminal synflorescence; bracts foliaceous or reduced. Calyx lobe margins entire. Corolla white, blue or mauve; tube subcylindrical, somewhat widened towards base and apex; limb 5-lobed, sinus between adaxial pair of lobes markedly wider than other sinuses. Staminodes 3, antherodes absent. Stigma linear, single-lobed. Capsule barely compressed laterally, narrowed into a prominent apical beak, this hollow towards base; lateral walls remaining attached to flanks at dehiscence; septum with a shallow membranous portion above upper retinacula, elsewhere woody. Seeds 4 (or 2 by abortion), subglobose or subellipsoid, partially or barely flattened, with minute blunt hairs unevely distributed on the surface, glabrescent.

A section of 9–10 species, 5 restricted to the Indian subcontinent and 4–5 to tropical Africa. In addition to the two species recorded from our region, *B. descampsii* Lindau has been recorded from NE Zambia close to the Tanzania border and may well occur in the miombo woodlands of **T** 4 (see note under *B. richardsiae*). These 3 species form a discrete group, to which the above description applies (see Darbyshire, l.c.).

Indumentum predominantly stellate; synflorescence congested throughout; bracts broadly obovate to elliptic, largely enclosing the calyces .... 39. *B. grandipetala*
Indumentum simple; synflorescence (at least on the principal branches) with cymes more distantly spaced at least in the lower portion; bracts in the upper portion much reduced and narrowly elliptic to oblanceolate, not enclosing the calyces .... 40. *B. richardsiae*

39. **Barleria grandipetala** *De Wild.* in B.J.B.B. 5: 10 (1915); Darbyshire in K.B. 63: 611 (2009). Type: Congo-Kinshasa, Katanga, *Descamps* s.n. (BR!, holo., K!, photo.)

Suffruticose perennial, 1–few erect branches from a woody rootstock, 30–75 cm tall; stems with pale golden spreading or ascending long-armed stellate hairs. Leaves opposite; ovate, elliptic or lowermost pairs somewhat obovate, 5.5–15 cm long, 2–5.7 cm wide, base obtuse to cuneate, apex acute or obtuse, upper surface with long-armed stellate hairs, stellate base ± caducous, lower surface whitish to pale golden stellate-pubescent, many hairs long-armed, most dense on principal veins; lateral veins (3–)4–5 pairs; petiole to 8 mm long. Synflorescences congested, cylindrical or on lateral branches subglobose, 3–11.5 cm long, flowers solitary at each axil; bracts towards base of synflorescence foliaceous, those in the upper portion ± broadly obovate to elliptic, 15–34 mm long, 8.5–25 mm wide; bracteoles oblanceolate, narrowly elliptic or linear, 13–34 mm long, 2–8 mm wide. Anticous calyx lobe obovate, 16–23 mm long, 6–10(–15) mm wide, base attenuate, apex bifid for (1–)2–4(–5.5) mm, indumentum as lower leaf surface, venation palmate, prominent; posticous lobe 17–24 mm long, apex broadly obtuse to acute; lateral lobes lanceolate, 7.5–12.5 mm long. Corolla pale blue or mauve with a white tube, 37–62 mm long, shortly glandular-pubescent outside; tube 13–27 mm long; limb with lobes each 19–33 mm long, abaxial lobe 7.5–11 mm wide, lateral and adaxial lobes 10–14.5 mm wide. Stamens attached 5–8.5 mm from base of corolla tube; filaments 19–25 mm long, with few to many short-stalked glands throughout; anthers (4.5–)5.5–7 mm long; lateral staminodes 3–7 mm long. Ovary appressed-pubescent towards apex; style pubescent in lower half or throughout; stigma 1.5–2 mm long. Capsule 14–16 mm long, including beak 3.5–4.5 mm long, appressed-pubescent in upper half; seeds black, subellipsoid, 3.3–4 mm long, 3–3.5 mm wide.

TANZANIA. Biharamulo District: Nyantakara, Feb. 1959, *Procter* 1149!; Kigoma District: Kigoma, Kibirizi, Apr. 1994, *Bidgood & Vollesen* 3042!; Kahama District: Mkweni, Feb. 1937, *Burtt* 5553!
DISTR. **T** 1, 4; SE Congo-Kinshasa, Burundi
HAB. *Brachystegia* woodland, open *Julbernardia-Parinari* bushland and associated grassland; 800–1500 m
USES. None recorded on herbarium specimens
CONSERVATION NOTES. Widespread in westernmost Tanzania and well-collected there, despite only limited botanical exploration. It has been recorded as locally common at both Katonga (*Gobbo & Zacharia* 152) and Mkweni (*Burtt* 5553). Its abundance in Congo and Burundi is unknown, though it is probably under-recorded in the former, and may also be present in NE Zambia. Extensive areas of suitable habitat remain intact within its range and this species appears tolerant of some disturbance, having been recorded from degraded bushland (*Bidgood & Vollesen* 3042). It is not considered threatened at present: Least Concern (LC).

40. **Barleria richardsiae** *I. Darbysh.* in K.B. 63: 608, fig. 2 (2009). Type: Tanzania, Ufipa District: 33 km on Namanyere–Karonga road, *Bidgood, Mbago & Vollesen* 2649 (K!, holo.; BR!, C!, CAS, DSM!, EA!, K!, NHT, P, iso.)

Suffruticose perennial, with several erect branches from a woody rootstock, 50–100 cm tall; stems densely hairy with shorter, whitish or buff-coloured, spreading or retrorse hairs and longer, buff or yellow, ascending or spreading hairs. Leaves often in whorls of 3–4, ovate, narrowly elliptic or lanceolate, 6–10.5 cm long, 1–3.5 cm wide, base cuneate to obtuse, apex acute or subattenuate, surfaces with coarse ascending hairs, most dense on margin and veins beneath, midrib with additional shorter spreading hairs; lateral veins 3–5 pairs; petiole to 5 mm long or absent. Synflorescences thyrsoid, (3–)10–28 cm long, cymes congested towards apex, more widely spaced below, each with flowers solitary or paired; bracts foliaceous, becoming smaller upwards, those of uppermost axils 14–23 mm long; bracteoles oblanceolate, narrowly obovate-elliptic or linear, 12–31 mm long, 1.5–7 mm wide. Anticous calyx lobe oblong or obovate, 13–23 mm long, 5–12 mm wide, base cuneate or acute, apex shallowly notched for up to 3 mm, surface with many coarse ascending hairs particularly on veins and margin, hair bases ± swollen, and with few interspersed short spreading hairs, venation parallel or narrowly palmate, prominent but sometimes obscured by the indumentum; posticous lobe 16–25 mm long, apex acute to rounded; lateral lobes (ovate-) lanceolate, 6–14 mm long. Corolla mauve or lilac with whitish tube, 40–54 mm long, shortly glandular-pubescent outside; tube 18–24 mm long; limb with lobes each 20–29 mm long, abaxial lobe 11.5–18 mm wide, lateral lobes 12.5–20 mm wide, adaxial lobes 9–16 mm wide. Stamens attached 6–8.5 mm from base of corolla tube; filaments 21–25 mm long, with many short-stalked glands in upper half, sparse below; anthers 5.5–7.5 mm long; lateral staminodes 1.5–3.7 mm long. Ovary appressed-pubescent towards apex; style pubescent in lower half or throughout; stigma 1–1.6 mm long. Capsule 16–19 mm long including beak 4.5–6 mm long, appressed- or ascending-pubescent in upper half; seeds black, subellipsoid, 3.8–4.3 mm long, 3–3.8 mm wide. Fig. 53, p. 388.

TANZANIA. Mpanda District: Kampisa R., Feb. 1996, *Congdon* 456!; Ufipa District: 16 km on Namanyere–Kipili road, May 1997, *Bidgood et al.* 3678! & 8 km on Tatanda–Sumbawanga road, Apr. 2006, *Bidgood et al.* 5602!

DISTR. **T** 4; NE Zambia

HAB. *Brachystegia*, *Isoberlinia*, *Pterocarpus* and/or *Uapaca* woodland on sandy soils or rocky slopes; 900–1750 m

USES. None recorded on herbarium specimens

CONSERVATION NOTES. Assessed as Near Threatened (NT) by Darbyshire (l.c.).

NOTE. *B. descampsii* of northern Zambia and southern Congo-Kinshasa is similar to this species but is readily separated by its stellate indumentum and a more dense indumentum to the more congested synflorescence.

SECT. V. SOMALIA

(Oliv.) Lindau in E. & P.Pf. 4 (IIIB): 315 (1895); C.B. Clarke in F.T.A. 5: 142 (1899); Obermeijer in Ann. Transv. Mus. 15: 130 (1933); M. & K. Balkwill in K.B. 52: 563 (1997); Darbyshire & Ndangalasi in Journ. E. Afr. Nat. Hist. 97: 123–134 (2009)

Somalia Oliv. in Hook. Ic. Pl. 16, t. 1528 (1886)
Barleria subgen. *Eu-Barleria* "*Glabratae*" *sensu* C.B. Clarke in F.T.A. 5: 143 (1899) pro parte, excl. *B. grandis* and *B. marginata*

Axillary spines absent. Axillary buds often densely pale-hairy. Indumentum simple and/or of medifixed or anvil-shaped (biramous with one long arm and one short arm or the latter reduced to a swelling) hairs, or glabrous. Inflorescences 1–several-flowered dichasial cymes, either simple and axillary, then bracts

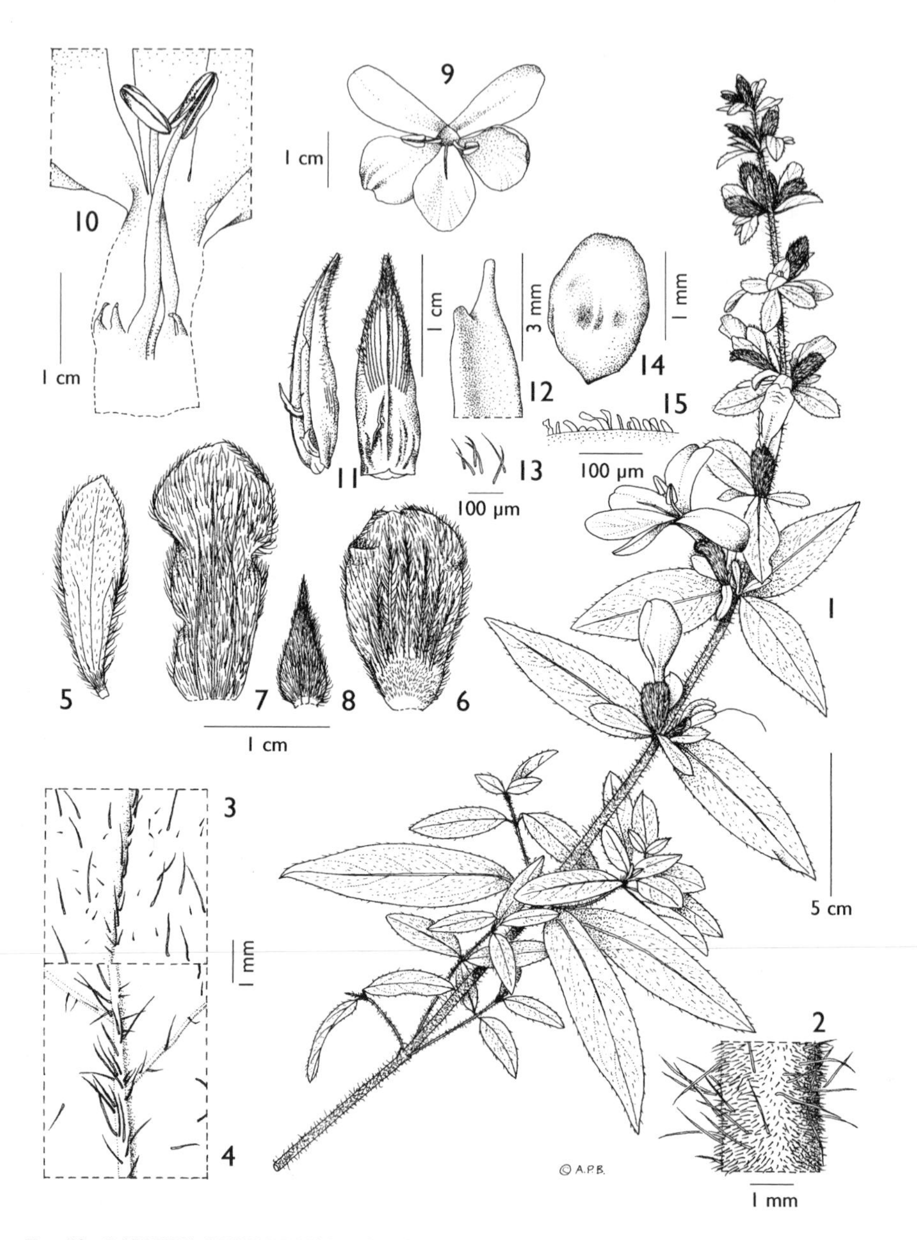

FIG. 53. *BARLERIA RICHARDSIAE* — **1**, habit; **2**, stem indumentum; **3**, adaxial leaf indumentum at midrib; **4**, abaxial leaf indumentum at midrib; **5**, bracteole, inner surface; **6**, anticous calyx lobe, outer surface; **7**, lateral calyx lobe, outer surface; **8**, posticous calyx lobe, outer surface; **9**, face view of corolla; **10**, dissected corolla tube with attached stamens and staminodes; **11**, fruit, lateral and exterior view of single valve; **12**, distal section of upper retinaculum; **13**, capsule indumentum; **14**, seed; **15**, seed indumentum in profile. 1–8 from *Richards* 10804, 9 from photograph of *Bidgood et al.* 2649, 10 from *Richards* 19017, 11–15 from *Bidgood et al.* 3678. Drawn by Andrew Brown. Reproduced from Kew Bulletin 63: 607 (2009) with permission of the Trustees of the Royal Botanic Gardens, Kew.

foliaceous, or compounded into terminal synflorescences, then bracts ± modified. Calyx lobe margins entire. Corolla white, blue, mauve or purple, rarely yellow, sometimes with darker guidelines; limb variously in "4+1", "2+3", "1+3" or subregular arrangement. Staminodes 2–3, antherodes absent. Stigma linear, apex entire or minutely bilobed. Capsule drying (pale) brown, laterally compressed, fertile portion orbicular, ± abruptly narrowed into a prominent solid apical beak; lateral walls remaining attached to flanks at dehiscence, septum with a membranous portion above the retinacula, elsewhere woody. Seeds 2, discoid with dense cream or buff-coloured wavy hygroscopic hairs.

A section of ± 50 species, most diverse in continental Africa, also recorded in Madagascar, Arabia and the Indian subcontinent. Of the seven groups of species recognised in our region by Darbyshire & Ndangalasi (l.c.), spp. 55–62 fall within group A; spp. 41–44 within group B; sp. 45 within group C; sp. 46–49 within group D; spp. 50 & 51 within group E; sp. 52 in group F and spp. 53 & 54 in group G. Of these, group A, in which the species all have medifixed (biramous) hairs, forms a particularly discrete group.

1. Medifixed hairs absent; indumentum of simple hairs only or plants largely glabrous 2
   Medifixed hairs present at least on the leaf margin and midrib beneath and outer calyx lobes, often more widespread 16
2. Corolla yellow . . . . 43. *B. calophylloides* p.395
   Corolla white, pink, mauve or blue 3
3. Stems with dense short white or whitish-buff patent or retrorse hairs up to 0.5(–0.8) mm long 15
   Stem indumentum variable but not as above: hairs longer and more sparse (or soon glabrescent) or if dense then ascending to appressed and/or crisped or woolly 4
4. Corolla 4-lobed ("1+3" arrangement), adaxial lobe emarginate . . . . 51. *B. lugardii* p.402
   Corolla 5-lobed, adaxial pair of lobes not or only partially fused, free for 5 mm or more 5
5. Anticous calyx lobe over 10 mm wide or if narrower (to (4–)7.5 mm in *B. lactiflora*) then mature leaves glabrous; inflorescence lacking glandular hairs or if these present then anticous calyx lobe ovate-orbicular; plants never white-woolly 6
   Anticous calyx lobe 2–6.5 mm wide; leaves hairy at least on the margin and midrib; glandular hairs present at least on the bracteoles and calyces, or if absent then plants with woolly white indumentum; anticous calyx lobe never ovate-orbicular 10
6. Corolla 48–67 mm long including tube 25–33 mm long; ovary and capsule (where known*) glabrous 7
   Corolla 23–42 mm long including tube 11–24 mm long; ovary and capsule pubescent 8
7. Leaves, bracteoles and outer calyx lobes largely glabrous except sometimes for scattered hairs along the margins (lateral calyx lobes always ciliate), base of outer calyx lobes deeply cordate . . . . 43. *B. calophylloides* p.395
   Leaves, bracteoles and calyces coarsely hairy, base of outer calyx lobes rounded or shallowly cordate . . . 44. *B. vollesenii* p.397

* Capsule not seen in *B. vollesenii*, but the ovary is glabrous and as ovary and capsule indumentum are consistent in *Barleria*, it can be inferred that the capsule will be glabrous.

8. Subshrub; anticous calyx lobe ovate-orbicular, ± as long as wide; corolla limb in "2+3" arrangement, adaxial lobes partially fused .... 50. *B. granarii* p.401
   Suffruticose perennials with annual branches from a woody base; anticous calyx lobe ovate, elliptic or subrhombic, longer than wide; corolla subregular, adaxial lobes not partially fused 9
9. Mature leaves glabrous, largest leaves oblong-elliptic to ovate, 8.5–16 cm long; outer calyx lobes long-ciliate, hairs straight with ± conspicuously swollen bases .... 41. *B. lactiflora* p.392
   Mature leaves hairy at least on the midrib beneath, often more widespread, largest leaves obovate to elliptic, 4.5–10 cm long; outer calyx lobes with at least some crisped hairs on the margin, often more widespread, lacking swollen bases .... 42. *B. venenata* p.393
10. Stems and leaves white-woolly; calyces lacking glandular hairs, anticous lobe ± 20 mm long; anthers ± 6.5 mm long .... 49. *B. sp. C* p.401
   Stems and leaves with indumentum variable but never white-woolly; calyces with patent glandular hairs present (sometimes sparse), anticous lobe 9.5–18 mm long; anthers 2.3–5 mm long 11
11. Corolla limb in "2+3" arrangement, adaxial pair of lobes partially fused, the free portions linear-oblong, 5–7.5 mm long, 1–2 mm wide, corolla throat with mauve guidelines .... 52. *B. griseoviridis* p.403
   Corolla limb subregular or in "4+1" arrangement, adaxial pair of lobes not or barely fused, elliptic or obovate, 12.5–23 mm long, 4.5–7.5 mm wide, throat lacking guidelines 12
12. Ovary and capsule glabrous or with few apical hairs; corolla limb in "4+1" arrangement, abaxial lobe offset by 2.5–5 mm; anticous calyx lobe with apex entire or notched for up to 2.5 mm ...... 45. *B. hirta* p.397
   Ovary and capsule densely pubescent distally; corolla limb subregular; anticous calyx lobe bifid for 2–10 mm 13
13. Flowers axillary, solitary; leaves to 4 cm long, length less than twice the width .... 48. *B. diplotricha* p.400
   Flowers compounded into a well-defined terminal synflorescence; leaves 4.5–13 cm long, length over twice (–6 times) the width 14
14. Terminal synflorescence strobilate, with broadly ovate to obovate bracts highly modified from the narrowly oblong-elliptic to lanceolate leaves; indumentum of stem and synflorescence predominantly silvery; anticous calyx lobe 7–10 mm long, apex bifid for 2–4.5 mm .... 46. *B. limnogeton* p.398
   Terminal synflorescence subcapitate, conical, subcylindrical or thyrsoid, not strobilate, bracts much-reduced in relation to leaves but not so highly modified in shape; indumentum of stem and synflorescence predominantly buff-golden; anticous calyx lobe 9.5–17 mm long, apex bifid for 4–9 mm .... 47. *B. mpandensis* p.399

15. Anticous calyx lobe linear-lanceolate, up to 2 mm wide, with 2 prominent parallel veins; calyx and bracteoles with mixed eglandular and glandular hairs . . . . 53. *B. angustiloba* p.404
    Anticous calyx lobes (obovate-) elliptic, 3–7(–10) mm wide, with several palmate veins; calyx and bracteoles usually with eglandular hairs only . . . 54. *B. maculata* p.405
16. Stems largely glabrous except for occasional appressed medifixed and simple hairs around the nodes and scattered glandular hairs when young; plants of coastal bushland (**T** 8) . . . . 55. *B. laceratiflora* p.406
    Stems more densely hairy throughout; plants of dry bushland and grassland, not coastal (widespread but only extending south to **T** 2 & 3) 17
17. Outer calyx lobes oblong (-elliptic) to somewhat obovate, widest at or above the midpoint, anticous lobe as long as or slightly shorter than the lateral lobes; capsule with hairs of variable length including markedly longer hairs on the fertile portion . . . . 60. *B. hirtifructa* p.411
    Outer calyx lobes variously linear or lanceolate to ovate or suborbicular, widest at the base or below the midpoint, or if narrowly oblanceolate then anticous lobe clearly longer than the lateral lobes; capsule puberulous to pubescent 18
18. Outer calyx lobes linear, lanceolate, triangular or narrowly oblanceolate, posticous lobe 1–2.5 mm wide 19
    Outer calyx lobes ovate to suborbicular, posticous lobe 3–18 mm wide 21
19. Leaf indumentum of mixed subappressed medifixed hairs and rather long spreading eglandular hairs, often dense (patent glandular hairs often additionally present in both leads); outer calyx lobes linear or narrowly oblanceolate 59. *B. sp. E* p.411
    Leaf indumentum mainly of (sub-)appressed medifixed hairs, evenly distributed or restricted to the margin and midrib; spreading eglandular hairs, if present, very short; outer calyx lobes linear-lanceolate to triangular 20
20. Leaves linear, oblanceolate, obovate or oblong-elliptic, apiculum inconspicuous; corolla 11.5–23 mm long; anthers less than 2 mm long; capsule puberulous . . . . 56. *B. argentea* p.406
    Leaves (except sometimes the uppermost pairs) ovate or oblong-ovate, apiculum conspicuous; corolla 19–28 mm long; anthers 2–3 mm long; capsule pubescent . . . . 58. *B. masaiensis* p.408
21. Posticous calyx lobe with apex attenuate, base obtuse or rounded; inflorescences congested, with peduncle to 5 mm long and flowers subsessile . . . 57. *B. sp. D* p.408
    Posticous calyx lobe with apex acute to obtuse, apiculate, base cordate or if rounded then inflorescence more lax: peduncle 8 mm long or more, pedicels 5 mm long or more 22

22. Outer calyx lobes with base rounded, truncate or at most shallowly cordate; cymes lax, peduncle 8 mm long or more, pedicels 5 mm long or more; corolla (17–)21–40 mm long, lateral and adaxial lobes elliptic to obovate, over (7–)9 mm long .. 61. *B. hochstetteri* p.412
   Outer calyx lobes with base cordate; cymes congested, peduncle up to 8(–13) mm long, pedicels up to 5(–10) mm long; corolla 15–21 mm long, lateral and adaxial lobes suborbicular, up to 6.5 mm long .... 62. *B. orbicularis* p.414

41. **Barleria lactiflora** *Brummitt & Seyani* in K.B. 32: 723 (1978). Type: Malawi, Mzimba District, 8 km past Lunyangwa R., 9 km SW of M14, *Pawek* 10899 (K!, holo.; MAL, MO!, PRE!, WAG!, iso.)

Suffruticose herb, several to many ± erect stems from a woody rootstock, 30–60(–100) cm tall; stems with 2 opposite lines of ascending hairs when young, ± glabrescent. Leaves subsessile, somewhat coriaceous, narrowly oblong-elliptic to more broadly ovate, 8.5–16 cm long, 2–6 cm wide, base cuneate to obtuse or rounded, apex acute to rounded, lower pairs often smaller and somewhat obovate, surfaces glabrous when mature; lateral veins 5–7(–8) pairs. Flowers compounded into a dense terminal capitate or cylindrical synflorescence, 3.5–9.5 cm long, sometimes with additional solitary flowers in the uppermost leafy axils; bracts ovate, elliptic or somewhat obovate, 15–33 mm long, 3–14 mm wide, margin long-ciliate at least in the lower half, sometimes also on the veins, hairs usually with swollen bases; bracteoles as bracts but ovate-lanceolate to oblanceolate, 1.5–7.5 mm wide, hairs on margin many; flowers subsessile. Anticous calyx lobe elliptic or subrhombic, 16–29 mm long, (4–)7.5–12(–17) mm wide, base obtuse or acute, apex bifid with deltate segments 4.5–8.5(–11) mm long, rarely obtuse or shallowly emarginate, indumentum as bracteoles, venation palmate, ± prominent; posticous lobe ovate, 16–31 mm long, (3.5–)5.5–11(–14.5) mm wide, apex acute or obtuse; lateral lobes lanceolate, 7–13.5 mm long. Corolla white, blue or mauve, 27–42 mm long, pubescent on the limb outside, hairs mainly eglandular; tube cylindrical, 11.5–17 mm long, somewhat expanded towards mouth; limb subregular, lobes 13.5–25 mm long, abaxial lobe 9–14 mm wide, lateral lobes 7.5–13.5 mm wide, adaxial lobes 5.5–10.5 mm wide. Stamens attached ± midway along corolla tube; filaments 13–23 mm long; anthers 3–5 mm long; lateral staminodes 1–3(–5.5) mm long. Ovary densely pubescent distally; style pubescent at base; stigma curved, 2–3.3 mm long. Capsule 17–22 mm long including 8.5–11 mm beak, pubescent particularly on beak; seeds ± 8 mm long, 6–7.5 mm wide.

TANZANIA. Ufipa District: Ilemba–Kasamvu, Mar. 1950, *Bullock* 2694!; Dodoma District: Itigi road 105 km from Itigi, Mar. 1965, *Richards* 19850!; Chunya District: 8 km on Makongolosi–Mbeya road, Feb. 1994, *Bidgood, Mbago & Vollesen* 2317!

DISTR. **T** 4, 5, 7; N Malawi

HAB. *Brachystegia*, *Combretum-Terminalia* and *Acacia* woodland and wooded grassland on sandy soils and rocky hillsides, sometimes on roadsides; 900–1700 m (extending up 2400 m on the Nyika Plateau of N Malawi)

USES. None recorded on herbarium specimens

CONSERVATION NOTES. Widespread in W Tanzania and N Malawi and represented by many herbarium collections. It has been recorded as locally common by some collectors and appears adaptable to both wooded and more open, disturbed environments. It is not considered threatened: Least Concern (LC).

NOTE. Populations from the Songwe valley (e.g. *Bidgood et al.* 683; *Mhoro* 2256) have narrower bracts, bracteoles and calyces than typical, these being densely hairy, the swollen hair bases less conspicuous than usual. Plants from the Ruaha National Park (*Richards* 21253, *Bjørnstad* 1617) have proportionately broader leaves than usual; the former gathering is additionately unusual in having a shallowly emarginate or obtuse, not clearly bifid, apex to the anticous calyx lobe.

On *Mpemba* in C.A.W.M. 4631 (EA!) the flower colour is recorded as yellow but this must be in error, possibly due to confusion with *B. calophylloides*.

42. **Barleria venenata** *I. Darbysh.* **sp. nov.** *B. lactiflorae* Brummitt & Seyani similis sed foliis obovatis vel obovato-ellipticis (non oblongo-ellipticis neque ovatis) subtus saltem secus costam sed plerumque etiam alibi pilosis (non glabris), lobis calycis exterioribus pilos ad basin inflatos non ferentibus atque ad marginem venasque saltem aliquos pilos crispatos ferentibus (non pilis omnibus rectis ad basin inflatis) differt. Type: Tanzania, Musoma District, Lobo Hill Trace, *Greenway, Turner & Owen* 10308 (K!, holo.; EA!, iso.)

Suffruticose herb, several to many erect or decumbent stems from a woody rootstock, 10–45 cm tall; stems sparsely to densely pale-pilose to -tomentellous when young, glabrescent. Leaves obovate or elliptic, 4.5–10 cm long, 1.5–4.5 cm wide, base cuneate-attenuate, apex obtuse, rounded or rarely acute, surfaces whitish-pilose, most dense or sometimes restricted to the midrib and lateral veins beneath; lateral veins 4–5(–6) pairs; petiole to 6(–10) mm long or absent. Inflorescences single-flowered, axillary but often compounded into a loose terminal synflorescence to 5 cm long; bracts of the lower cymes foliaceous, those above often spatulate or oblong, 14–30 mm long, 3.5–12 mm wide; bracteoles lanceolate, linear or oblanceolate, 12–30 mm long, 1.5–5.5 mm wide, whitish-pilose to -tomentellous throughout or restricted to the margin and midrib; flowers subsessile. Anticous calyx lobe ovate or elliptic, 17–27 mm long, 10.5–17 mm wide, base obtuse or rounded, apex bifid with deltate segments 0.5–4 mm long, more rarely emarginate or rounded, indumentum as bracteoles, with at least some crisped hairs on the margin and principal veins, venation palmate, prominent; posticous lobe 18–29 mm long, 9.5–15 mm wide, apex acute or obtuse; lateral lobes lanceolate, 8–13 mm long. Corolla 26–42 mm long, white or rarely pale mauve; glabrous or limb sparsely pubescent outside with or without scattered short glandular hairs; tube cylindrical, 15–24 mm long, barely expanded towards mouth; limb subregular; abaxial lobe 12–18 mm long, 7–10.5 mm wide, lateral lobes as abaxial but 6–10 mm wide, adaxial lobes 10–16 mm long, 5–7 mm wide. Stamens attached at or slightly above midway along corolla tube; filaments 12–19 mm long; anthers 3.3–4.5 mm long; lateral staminodes 1–2.5 mm long. Ovary densely pubescent in upper half; style pubescent towards base; stigma 2–3.5 mm long. Capsule 16–21 mm long including 7–10.5 mm beak, pubescent particularly on beak; seeds 7–8 mm long, 6–7 mm wide. Fig. 54/11–20, p. 394.

Kenya. Masai District: Masai Mara Game Reserve, Mara Wildlife Research Station, Mar. 1979, *Kuchar* 10959! & *idem*, Aug. 1979, *Kuchar* 12148! & Sand R. S–SW of Keekorok, Apr. 1980, *Kuchar* 13331!

Tanzania. Mbulu District: Tarangire National Park, top of Boundary Hill, Feb. 1970, *Richards* 25395!; Handeni District: Kwamkono, June 1966, *Archbold* 810!; Tabora District: 33 km on Tabora–Itigi road, May 2006, *Bidgood et al.* 5995!

Distr. **K** 6; **T** 1, 2–6, ?7 (see note), 8; not known elsewhere

Hab. Grassland, often in open woodland and thickets of *Acacia, Commiphora, Combretum* or *Brachystegia-Pterocarpus*, or in dry forest, also roadsides and shambas; 450–1500(–1950) m

Uses. "Sukuma: for poisoning" (**T** 1; *Tanner* 1495)

Conservation notes. This species is locally common in the dry interior of N Tanzania and S Kenya, and well represented in herbarium collections from the region. Populations in the coastal lowlands of Tanzania appear less frequent, though it may be under-recorded in the southeastern part of its range due to limited collecting in that area. It is found in a variety of habitats and appears tolerant of some disturbance and is not therefore considered threatened: Least Concern (LC).

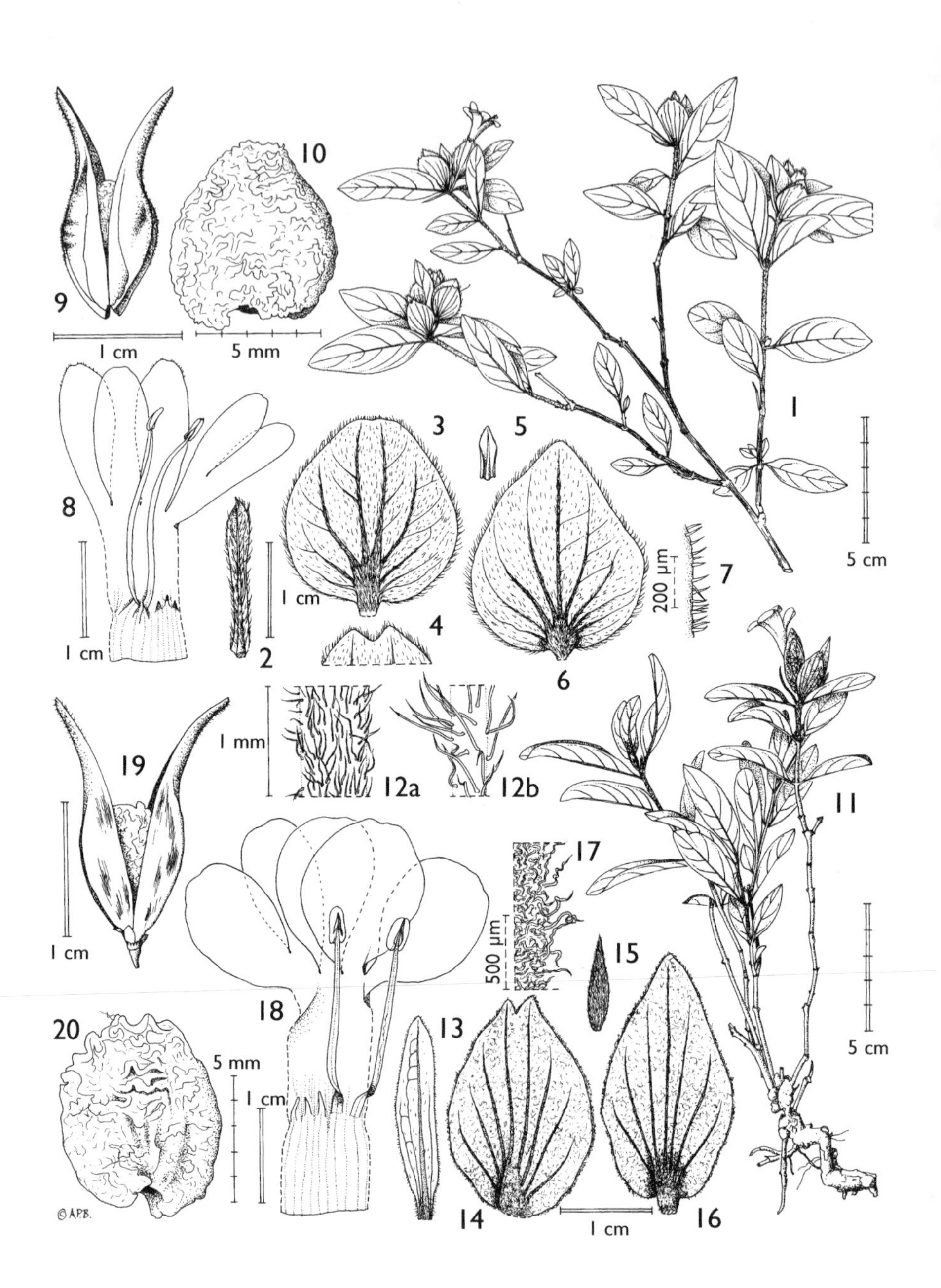

FIG. 54. *BARLERIA GRANARII* — **1**, habit; **2**, bracteole; **3**, anticous calyx lobe; **4**, anticous calyx lobe, variation in apex; **5**, lateral calyx lobe; **6**, posticous calyx lobe; **7**, profile of indumentum of lateral calyx lobe; **8**, dissected corolla with androecium; **9**, capsule; **10**, seed. *BARLERIA VENENATA* — **11**, habit; **12a** & **12b**, indumentum of midrib on abaxial leaf surface; **13**, bracteole; **14**, anticous calyx lobe; **15**, lateral calyx lobe; **16**, posticous calyx lobe; **17**, indumentum of margin of posticous calyx lobe; **18**, dissected corolla with androecium; **19**, capsule; **20**, seed. 1 from *Bidgood et al.* 654, 2–7 from *Polhill & Paulo* 2046; 8 from *Bidgood et al.* 1171; 9 & 10 from *Lovett et al.* 2167; 11 & 12a from *Greenway et al.* 10308; 12b from *Welch* 233; 13–18 from *Richards* 25395; 19 & 20 from *Wallace* 631. Drawn by Andrew Brown.

SYN. *B.* sp. nov. aff. *calophylloides* (= Welch 233) *sensu* Vollesen in Opera Bot. 59: 79 (1980)]
[*B.* sp. M *sensu* U.K.W.F. ed. 2: 272 (1994)]

NOTE. *B. venenata* is related to *B. lactiflora* but has thinner, generally smaller leaves, at least some of the upper cauline leaves are obovate (versus oblong-elliptic to broadly ovate) and pilose at least on the midrib beneath and usually more widespread (versus glabrous). *B. lactiflora* has a more dense terminal synflorescence; in *B. venenata* the lower fertile axes are more clearly spaced and/or the terminal cluster is fewer-flowered. The inflorescence indumentum also differs: in *B. lactiflora* the bracts, bracteoles and outer calyx lobes have long, straight hairs usually with a swollen, bulbous base on the margin and principal veins. In *B. venenata* at least some of the hairs on the calyx margin, and often on the bracts and bracteoles, are crisped. In more densely hairy specimens, the inflorescence can be tomentellous. If straight hairs are present, they lack the bulbous bases seen in *B. lactiflora*.

This species has a rather unusual distribution with two disjunct populations, one in the interior of Tanzania and southern Kenya and the second restricted to lowland woodland and forest of the Tanzanian coastal belt. Interior plants are often more densely hairy than coastal ones, which can be vegetatively largely glabrous. In addition, coastal plants have many glandular hairs on the lateral calyx lobes, these being absent or sparse on interior plants. However, some intermediate populations occur, particularly in the Mbulu area, **T** 2 (e.g. *Richards* 24380).

A single fruiting specimen from Njombe District (**T** 7, *Bidgood & Vollesen* 2202) appears close to this taxon but differs in having more broadly elliptic leaves (to 5 cm wide) and very broad outer calyx lobes (anticous lobe 21–27 × 17–21 mm) with a subcordate base. Flowering material is required to confirm the placement of this population.

43. **Barleria calophylloides** *Lindau* in E.J. 20: 17 (1894); C.B. Clarke in F.T.A. 5: 159 (1899); M. & K. Balkwill in K.B. 52: 565, fig. 12H (1997). Types: Tanzania, Biharamulo District, Bukome, *Stuhlmann* 3431 & Tabora District, Igonda [Gonda], *Böhm* 164 (both B†, syn.); Kahama District, 22 km E of Ushirombo, Kahama–Biharamulo road, *Boaler* 456 (K!, neo., chosen here; EA!, isoneo.)

Suffruticose herb, several to many prostrate or decumbent branches from a woody rootstock, to 50 cm long; stems with 2 opposite lines of short pubescence when young and/or crisped-pilose, glabrescent. Leaves subcoriaceous, ovate or (oblong-) elliptic, 4–12.5 cm long, 2–6.5 cm wide, base rounded or shallowly cordate to cuneate or attenuate, apex acute, obtuse or rounded, apiculate, lower leaves often smaller and somewhat obovate, surfaces glabrous or sparsely pilose on the margin towards the base; lateral veins 4–7 pairs; petiole to 12 mm long or absent. Inflorescences single-flowered, axillary but often crowded towards stem apex; bracts foliaceous; bracteoles often held patent to the calyx, obovate, oblanceolate or lanceolate, 4–25 mm long, 0.5–10.5 mm wide, glabrous or margin pilose; pedicels 2–4.5 mm long. Anticous calyx lobe broadly ovate, 19–35 mm long, 14.5–29 mm wide, base cordate, apex emarginate or rounded, surfaces glabrous or with few appressed marginal hairs at apex, rarely with scattered crisped pilose hairs on margin, venation palmate, prominent; posticous lobe 22–41 mm long, 15–30 mm wide, apex obtuse or rounded; lateral lobes lanceolate, 7–13.5(–15) mm long, expanded and hyaline towards base, ciliate. Corolla yellow or white, 48–62 mm long, glabrous outside; tube cylindrical, 25–33 mm long, barely expanded towards mouth; limb subregular; abaxial lobe offset by ± 2 mm, 18–27 mm long, 9.5–18 mm wide; lateral lobes as abaxial but 22–29 mm long; adaxial lobes as abaxial but 7–14 mm wide. Stamens attached ± 10 mm from base of corolla tube; filaments 20–28 mm long; anthers 3–5 mm long; lateral staminodes 0.3–2 mm long. Pistil glabrous; stigma 2–3.3 mm long. Capsule 16–23 mm long including 6.5–9 mm beak, glabrous; seeds 9–9.5 mm long, 8–8.5 mm wide.

a. subsp. **calophylloides**

Stems with two lines of short pubescence; leaves and bracteoles glabrous. Upper cauline leaves ovate or (oblong-) elliptic, largest leaf (5–)6.5–12.5 cm long. Corolla pale yellow to sulphur yellow, rarely ?white (see note).

TANZANIA. Ufipa District: 7 km on Namanyere–Karonga road, Mar. 1994, *Bidgood, Mbago & Vollesen* 2617!; Dodoma District: Kazikazi, Mar. 1933, *Burtt* 4619!; Iringa District: Great North Road, 11 km N of Iringa, Feb. 1962, *Polhill & Paulo* 1365!

DISTR. **T** 1, 2, 4, 5, 7; NE Zambia

HAB. Open and dense woodland, particularly of *Brachystegia, Combretum, Terminalia, Isoberlinia* and *Julbernardia*, usually on sandy soils; (850–)1000–1650(–1850) m

USES. None recorded on herbarium specimens

CONSERVATION NOTES. This subspecies is widespread in the miombo woodlands of W Tanzania and adjacent NE Zambia, and is noted by several collectors to be locally common. It appears tolerant of some disturbance, being recorded from both pristine and degraded woodland. Large areas of suitable habitat remain intact within its range, particularly in the more remote areas of W Tanzania. It is therefore not considered threatened: Least Concern (LC).

NOTE. *Boaler* 456 is here selected as the neotype following the loss of the original syntypes at the Berlin herbarium during World War II. The Boaler collection is good material, typical of this taxon, and its location lies between that of the two original syntypes.

Peter 45972 (B, K!) from Unyamwezi in Tanzania, otherwise agreeing closely with subsp. *calophylloides*, is recorded as having white flowers; this may be in error as no other white-flowered specimens have been recorded.

b. subsp. **pilosa** *I. Darbysh.* **subsp. nov.** a subspecie typica corolla alba (non flava), caulibus et marginibus folii bracteolique pilos longos multicellulares ferentibus et foliis minoribus in proportione latioribus differt. Type: Tanzania, Njombe/Mbeya District, 54 km on Chimala–Iringa road, *Bidgood, Darbyshire, Hoenselaar, Leliyo, Sanchez-Ken & Vollesen* 5292 (K!, holo.; BR!, CAS!, DSM!, NHT!, iso.)

Stems with long crisped pilose hairs, multicellular with conspicuous cell-walls, these also present towards base of leaf and bracteole margins, rarely also on outer calyx lobe margins. Upper cauline leaves broadly ovate or elliptic, largest leaf 4–6.7 cm long. Corolla white.

TANZANIA. Mbeya District: Ihango, Mswiswi, lower Usafwa, Feb. 1980, *Leedal* 5849! & Igawa, near Chimala, Mar. 1992, *Congdon* 325!; Njombe/Mbeya District: 54 km on Chimala–Iringa road, Apr. 2006, *Bidgood et al.* 5292! (type)

DISTR. **T** 7; Malawi

HAB. *Brachystegia* and *Parinari* woodland on sandy or stony soils; 1100–1400 m

USES. None recorded on herbarium specimens

CONSERVATION NOTES. This subspecies has a highly restricted distribution and is currently known from only seven localities. It is clearly scarce in Tanzania, having only been discovered there recently. Much of the miombo woodland in this region has been replaced by agriculture in recent times, particularly along the Tanzania–Zambia highway, and extant populations are considered highly threatened there. High human population pressure in Malawi poses similar threats to the woodland in that country. It is therefore assessed as Vulnerable (VU B2ab(iii)).

SYN. [*B. polyneura sensu* Brummitt & Seyani in K.B. 32: 726 (1978) pro parte quoad spec. ex Malawi, *non* S. Moore]

NOTE. Although the two subspecies are usually easily separable, a population from Mgori, Singida District, appears intermediate (*Kuchar* 22108, fr. & 23418, fl., both K!, MO!). These plants have sulphur-yellow flowers as in subsp. *calophylloides* but tend towards subsp. *pilosa* in having small leaves (5–6 cm long) and in the fruiting specimen having occasional long-pilose hairs on the young stems and bracteole margins.

Salubeni 1009 from Lilongwe District, Malawi is recorded as having orange flowers, though the Kew sheet of this specimen is in fruit and it is probable that either the label data are erroneous or that they refer to wilted flowers which often turn an orange-brown colour.

Brummitt & Seyani (l.c.) assigned the Malawi material to *B. polyneura* S. Moore, a species otherwise known only from Angola. They stated that the only difference between the highly disjunct populations was the cuneate leaf base in the Malawi material, this being rounded in Angolan specimens. However, several additional differences to *B. polyneura* are notable on closer inspection, including having a differing stem indumentum (*B. polyneura* having short straight hairs on the young stems only), the lateral calyx lobes being ciliate, not glabrous as in *B. polyneura*, and in particular, the few well-preserved flowers seen on the type of *B. polyneura* (*Welwitsch* 5029, BM! holotype, K! isotype) being much smaller, 35 mm long with the tube 13.5 mm long, versus 50–60 mm long with tube 25–33 mm long in subsp. *pilosa*.

44. **Barleria vollesenii** *I. Darbysh.* **sp. nov.** *B. calophylloidi* Lindau similis sed planta omnino grosse pilosa (non admodum glabra), lobis exterioribus calycis ad basin rotundatis vel nonnihil cordatis (non profunde cordatis), corolla longiore 62–67 mm (non 48–62 mm) longa, lobo abaxiali corollae ab lobis ceteris 10–11.5 mm (non ± 2 mm) distanti differt. Type: Tanzania, Masasi District, Chivirikiti village to Mbangala Forest Reserve, *Bidgood, Abdallah & Vollesen* 1988 (K!, holo.; BR!, C, CAS, EA!, K!, LISC, MO, NHT, P, UPS, WAG, iso.)

Suffruticose herb, several ascending or erect unbranched stems from a woody rootstock, 20–60 cm tall; stems with long pale coarse hairs, both spreading and subappressed, the latter ± concentrated in two opposite lines. Leaves oblong-elliptic or somewhat lanceolate, 7–11 cm long, 1.8–2.8 cm wide, base cuneate, apex acute or obtuse, apiculate, indumentum as stem, hairs most dense on margin and veins beneath; lateral veins 5–6 pairs; petiole to 4 mm long or absent. Inflorescences single-flowered, axillary but crowded towards stem apices; bracts foliaceous; bracteoles narrowly oblong-elliptic or somewhat obovate, 14–40 mm long, 2–10 mm wide, indumentum as leaves; pedicels 1–3 mm long. Anticous calyx lobe broadly ovate, 25–35 mm long, 11–15 mm wide, base rounded or shallowly cordate, apex notched with deltate-apiculate segments 1.5–4 mm long, indumentum as leaves, hairs rather dense on veins and margin, hair bases often swollen; venation palmate, ± prominent; posticous lobe 31–41 mm long, 14–17 mm wide, apex acute-apiculate; lateral lobes linear-lanceolate, 14–15 mm long. Corolla white, 62–67 mm long, limb with inconspicuous short glandular and eglandular hairs outside; tube cylindrical, 26–29 mm long, somewhat expanded towards mouth; limb in "4+1" arrangement; abaxial lobe offset by 10–11.5 mm, ± 30 mm long, 17 mm wide; lateral lobes ± 28 mm long, 13.5 mm wide; adaxial lobes ± 24 mm long, 13 mm wide. Stamens attached ± 9 mm from base of corolla tube; filaments 27–31 mm long; anthers 5–5.5 mm long; lateral staminodes 2–3 mm long. Pistil glabrous; stigma curved, ± 2.5 mm long. Capsule and seeds not seen.

TANZANIA. Masasi District: Chivirikiti village to Mbangala Forest Reserve, Mar. 1991, *Bidgood, Abdallah & Vollesen* 1988! (type)
DISTR. **T** 8; N Mozambique
HAB. Rocky outcrops in *Brachystegia* woodland; 300 m
USES. None recorded on herbarium specimens
CONSERVATION NOTES. Although currently known from only two collections, this may be more a reflection of limited botanical exploration in the Tanzania-Mozambique border area rather than of this species' true abundance and distribution. Its range is, however, clearly highly limited as it does not extend into either the better-known central coastal region of Tanzania or to the extensive higher altitude *Brachystegia* woodlands further west. No information is currently available as to whether its habitat is threatened. In light of such uncertainties, it is provisionally assessed as Data Deficient (DD) but may well prove to be threatened.

NOTE. With its very large, pure white corolla and its coarse indumentum, *B. vollesenii* is a highly distinctive species that is unlikely to be confused with any others.

45. **Barleria hirta** *Oberm.* in Ann. Transv. Mus. 15: 148 (1933); Plowes & Drummond, Wild. Fl. Zimbabwe, revised ed.: pl. 166 (1990); M. & K. Balkwill in K.B. 52: 565, fig. 12E (1997). Type: Zimbabwe, Umtali, Odzani River Valley, *Teague* 8 (BOL!, holo.; GRA, K!, iso.)

Suffruticose herb, several erect or ascending stems from a woody rootstock, 20–70 cm tall; stems wiry, whitish-pilose, hairs spreading and/or subappressed, young stems also glandular-pubescent. Axillary buds rather densely white-pilose. Leaves lanceolate, 3.5–12 cm long, 0.5–2.8 cm wide, base shortly attenuate, apex acute, apiculate, lower surface subappressed-pilose particularly on principal veins and margin, hairs more scattered or absent above, some hairs anvil-shaped; lateral veins 4–5 pairs; petiole 2–7 mm long. Inflorescences single-flowered, axillary but often crowded towards stem apex; bracts foliaceous, much reduced upwards, with scattered glandular hairs; bracteoles linear or oblanceolate, 6–21 mm long, 0.7–2.5 mm wide, with mixed ascending or appressed hairs and spreading glandular hairs; flowers sessile. Anticous calyx lobe oblong-ovate to linear-lanceolate, 8.5–18 mm long, 2.5–5 mm wide, apex entire or notched for 0.5–2.5 mm, indumentum as bracteoles, venation parallel; posticous lobe 9.5–19 mm long, 2–4.5 mm wide, apex acute or attenuate; lateral lobes lanceolate, 7.5–11 mm long, widened and with a hyaline margin towards base. Corolla pale to rich blue-purple or rarely white, 25–42 mm long, pubescent principally on lateral lobes outside, hairs mainly glandular; tube cylindrical, 12.5–22 mm long, somewhat expanded towards mouth; limb in "4+1" arrangement; abaxial lobe offset by 2.5–5 mm; each lobe 12.5–20 mm long, abaxial lobe 8–10 mm wide, lateral lobes 6.5–9.5 mm wide, adaxial lobes 4.5–7 mm wide. Stamens attached 5–6 mm from base of corolla tube; filaments 16–25 mm long; anthers 2.3–3.7 mm long; lateral staminodes 0.5–1 mm long. Ovary glabrous or with few apical hairs; style glabrous; stigma 1.2–2.7 mm long. Only immature capsule seen, ± 16 mm long including 6 mm beak, glabrous; only immature seeds seen.

TANZANIA. Lushoto District: Monga Forest, June 1970, *Faulkner* 4375!; Songea District: near R. Lipalangilo ± 8 km N of Lumecha bridge, May 1956, *Milne-Redhead & Taylor* 10000!
DISTR. **T** 3, 8; S Malawi, E & S Zimbabwe
HAB. *Brachystegia-Uapaca* woodland and open forest with a grassy understorey; 350–1000 m
USES. None recorded on herbarium specimens
CONSERVATION NOTES. Although widely distributed, this species is absent from many areas within its range, the populations often being highly isolated. It is common only in Zimbabwe where it is well-represented in herbarium material. In Tanzania and Malawi it is apparently scarce, with only three populations known. It is provisionally assessed as Least Concern (LC), but more information is desirable on the factors preventing this species from colonising many seemingly suitable sites.

NOTE. Significant regional variation is recorded within *B. hirta*. The single collection from Lushoto District is closest to plants from Zimbabwe, sharing the linear-lanceolate outer calyx lobes (length/width ratio usually over 4/1), large flowers (over 35 mm long), long anthers (2.7–3.7 mm long) and a short stigma (1.2–1.6 mm long). They differ only in the Lushoto plant having somewhat larger leaves which are glabrous, not sparsely pilose, above. The Songea specimen is closer to specimens from Mt Michiru in Malawi, having shorter, more ovate outer calyx lobes (length/width ratio 2.5–4/1), smaller flowers (28–33 mm long) and a longer stigma (1.8–2.7 mm long). The Songea plants however have small anthers (± 2.3 mm long) and a glabrous upper leaf surface whilst those of Michiru match the Zimbabwe populations in these characters. Regional subspecies may well be discernable but more material from Tanzania and Malawi is desirable.

46. **Barleria limnogeton** *S. Moore* in J.L.S. 15: 95 (1876); C.B. Clarke in F.T.A. 5: 164 (1899). Type: Tanzania, Kigoma District: S of Kawele, *Cameron* s.n. (K!, holo.)

Suffruticose subshrub, several-branched from a woody rootstock, 40–80 cm tall; stems densely silvery-sericeous, or hairs buff-coloured when young. Leaves pale grey-green beneath, darker above, narrowly oblong-elliptic or lanceolate, 7–13 cm long, 1.5–3 cm wide, base and apex acute or obtuse, apex apiculate, surfaces silvery-sericeous, additionally tomentellous beneath; lateral veins 5–7 pairs; petiole to 7.5 mm long or absent. Inflorescences single-flowered, compounded into a strobilate terminal synflorescence, 3.5–8.5 cm long, often 4-faceted; bracts broadly ovate to somewhat obovate, 14–20 mm long, 8.5–20 mm wide, indumentum as leaves, with or without scattered glandular hairs, midrib and often lateral veins prominent; bracteoles linear or lanceolate, 11.5–18 mm long, 1–4.5 mm wide; flowers sessile. Anticous calyx lobe ovate, 7–10 mm long, 2–3 mm wide, apex bifid with parallel linear segments 2–4.5 mm long, indumentum as bracts, venation inconspicuous; posticous lobe 8.5–11 mm long, ± 3 mm wide, apex acuminate; lateral lobes linear-lanceolate, 6–8.5 mm long. Corolla pale mauve or white, 27–37 mm long, tube apex and limb pubescent outside with mixed glandular and eglandular hairs; tube cylindrical, 9–14 mm long, barely expanded towards mouth; limb subregular; abaxial lobe 13.5–19 mm long, 9.5–12.5 mm wide, lateral lobes as abaxial but 8.5–11 mm wide, adaxial lobes shortly fused at base, 16–20 mm long, 5.5–7.5 mm wide. Stamens attached ± midway along corolla tube; filaments 12–17 mm long; anthers 3.5–5 mm long; lateral staminodes 1–3 mm long. Ovary densely pubescent; style pubescent, particularly towards base; stigma ± 1 mm long. Capsule ± 17 mm long including 7 mm beak, pubescent particularly on beak; only immature seeds seen.

TANZANIA. Kigoma District: Busondo, Tongwe, July 1926, *Grant* s.n.!; Mpanda District: ± 195 km from Mpanda on Uvinza road, June 1980, *Hooper, Townsend & Mwasumbi* 1965! & Mpanda–Uvinza road, Uzondo Plateau, May 2000, *Bidgood, Leliyo & Vollesen* 4564!

DISTR. **T** 4; not known elsewhere

HAB. *Brachystegia* and *Parinari* woodland on sandy or gravelly soils; 1450–1650 m

USES. None recorded on herbarium specimens

CONSERVATION NOTES. Currently known from a very limited range on the E side of Lake Tanganyika. Three of the five Tanzanian collections seen are recent, and extensive stands of miombo woodland remain along the Mpanda–Uvinza road. Shamba agriculture is however beginning to encroach into this habitat, posing a potential future threat there. It is therefore assessed as Vulnerable (VU D2).

NOTE. The type collection has minor differences to the other material, most notably in having more clearly petiolate leaves, somewhat smaller bracts which become slightly obovate upwards and which lack glandular hairs (though these are only sparsely present on *Bidgood et al.* 4139), and more linear bracteoles 1–2 mm wide, not 3–4.5 mm.

B. limnogeton is superficially similar to *B. splendens* E.A. Bruce (Sect. *Stellatohirta*), with which it shares an elongate strobilate terminal synflorescence with a dense pale indumentum to the highly modified bracts. It is however readily separable from that species in, amongst several differences, having a simple (not stellate) indumentum, proportionately narrower leaves and a much smaller calyx.

47. **Barleria mpandensis** *I. Darbysh.* in Journ. E. Afr. Nat. Hist. 97: 129 (2009). Type: Tanzania, Mpanda District, Kabungu, *Semsei "per" Herring* 151 (K!, holo.; EA!, K!, iso.)

Suffruticose herb, with several erect or ascending stems from a woody rootstock, 15–75 cm tall; stems with dense buff-golden appressed or ascending hairs. Leaves ovate, lanceolate or elliptic, 4.5–12.5 cm long, 1–4.5 cm wide, base obtuse to cuneate, apex acute or obtuse, indumentum as stem but less dense, hairs many on midrib, margin and lateral veins beneath, additionally pale-tomentellous beneath at least on the uppermost leaves; lateral veins 4–8 pairs; petiole to 10 mm long or absent. Inflorescences 1–several-flowered, compounded into a subcapitate, conical, cylindrical or thyrsoid terminal synflorescence, 1.5–28 cm long; bracts ovate to obovate, 9–33 mm long, 3.5–16 mm wide, indumentum as uppermost leaves, additionaly patent glandular-pubescent particularly towards apex; bracteoles as bracts but linear, oblanceolate or narrowly elliptic, 9.5–18 mm long, 0.7–4.5 mm wide; flowers sessile. Anticous calyx lobe oblong-ovate or oblong-elliptic, 9.5–17 mm long, 2.5–6.5 mm wide, apex deeply bifid with lanceolate segments, 4–9 mm long, indumentum as bracts, venation subparallel, inconspicuous; posticous lobe ovate (-elliptic), 9.5–19 mm long, apex acute; lateral lobes linear-lanceolate, 9.5–16 mm long. Corolla white, pale pink or lilac, 23–32 mm long, pubescent primarily on the lateral lobes outside with mixed glandular and eglandular hairs; tube cylindrical, 9–12 mm long, barely expanded towards mouth; limb subregular, lobes 13–21 mm long, abaxial lobe 8–11.5 mm wide, lateral lobes 7.5–9.5 mm wide, adaxial lobes 5–6.5 mm wide. Stamens attached 4–6.5 mm from base of corolla tube; filaments 12–20 mm long; anthers 3.3–4.7 mm long; lateral staminodes 1–2.5 mm long. Ovary densely pubescent distally; style pubescent in lower half, densely so towards base; stigma 1–2 mm long. Capsule ± 16-16.5 mm long including 8 mm beak, densely pubescent; only immature seeds seen.

a. subsp. **mpandensis**; Darbyshire in Journ. E. Afr. Nat. Hist. 97: 130, fig. 2 J, K & N–T (2009)

Plants 15–35 cm tall. Leaf surface pale tomentellous beneath only on the uppermost leaves, most leaves only with appressed buff-golden hairs; lateral veins 4–5 pairs. Synflorescences subcapitate to subcylindrical, 1.5–7 cm long, 1 flower per axil. Corolla 28–32 mm long.

TANZANIA. Mpanda District: Mpanda–Uvinza road, May 2000, *Bidgood, Leliyo & Vollesen* 4579! & 75 km on Mpanda–Uvinza road, June 2000, *Bidgood, Leliyo & Vollesen* 4706! & Mbala, 6 km NE of Sambala village, Aug. 2005, *Mwangoka & Anderson* 4171!
DISTR. T 4; not known elsewhere
HAB. *Brachystegia* and *Acacia-Combretum* woodland on sandy or clayey soil, riverine forest with *Parinari* and *Anthocleista*; 1200–1750 m
USES. None recorded on herbarium specimens
CONSERVATION NOTES. Assessed as Data Deficient (DD) by Darbyshire (l.c.).

b. subsp. **tomentella** *I. Darbysh.* in Journ. E. Afr. Nat. Hist. 97: 130, fig. 2 L & M (2009). Type: Tanzania, Kigoma District: Kigoma to Kasulu, *Verdcourt* 2808 (K!, holo.; BR!, EA!, K!, iso.)

Plants 60–75 cm tall. Leaf surface pale tomentellous beneath throughout in addition to the appressed buff-golden hairs; lateral veins 5–8 pairs. Synflorescences conical to thyrsoid, 6–28 cm long, more than 1 flower per axil. Corolla 23–25 mm long.

TANZANIA. Kigoma District: between Kalinzi and Kigoma on road to Usumbura, July 1960, *Verdcourt* 2798! & Kigoma to Kasulu, km 37, July 1960, *Verdcourt* 2808! (type)
DISTR. T 4; not known elsewhere
HAB. *Protea, Brachystegia, Combretum, Strychnos* & *Diplorhynchus* woodland with understorey of *Hyparrhenia*; ± 1200 m
USES. None recorded on herbarium specimens
CONSERVATION NOTES. Assessed as Data Deficient (DD) by Darbyshire (l.c.).

48. **Barleria diplotricha** *I. Darbysh. & Ndang.* in Journ. E. Afr. Nat. Hist. 97: 128, fig. 1 (2009). Type: Tanzania, Buha District, Kibondo, Malagarasi-Moyowosi [Muyovosi] Ramsar Site, *Ndangalasi & Suleiman* 1112 (K!, holo.; DSM, iso.)

Suffruticose herb, several unbranched erect or ascending stems from a woody rootstock, to 30–50 cm tall; stems with dense longer pale buff ascending hairs and shorter pale crisped hairs. Leaves subsessile, discolorous, dark green above, drying blackish, white-green beneath, elliptic, 3.3–4 cm long, 1.9–2.5 cm wide, base obtuse or rounded, apex obtuse, with rather sparse long ascending hairs above, these more dense beneath particularly on main veins and margin, densely pale-tomentellous between the veins beneath; lateral veins 4–5 pairs. Inflorescences axillary, single-flowered, sessile; bracts foliaceous; bracteoles linear to narrowly oblong-elliptic, 10–14 mm long, 1.5–3 mm wide, ascending-pubescent on margin and midrib with shorter, crisped hairs and patent glandular hairs elsewhere. Anticous calyx lobe narrowly ovate-elliptic, 11.5–14 mm long, 3–4 mm wide, apex deeply bifid with linear-lanceolate segments 5–6.5 mm long, indumentum as bracteoles, venation inconspicuous; posticous lobe narrowly ovate, 13–14.5 mm long, 4–4.5 mm wide, apex (sub)attenuate; lateral lobes linear-lanceolate, 9.5–11 mm long. Corolla lilac with white throat and pale yellow-green base to tube, ± 33 mm long, limb pubescent outside with mixed glandular and eglandular hairs; tube cylindrical, ± 10.5 mm long; limb subregular; abaxial lobe ± 21 mm long, 9 mm wide; lateral lobes ± 23 mm long, 7 mm wide; adaxial lobes as lateral pair but 4.5 mm wide, fused for ± 2 mm. Stamens attached ± 3 mm from base of corolla tube; filaments ± 19 mm long; anthers 4.3–5 mm long; lateral staminodes 4–5 mm long. Ovary densely pubescent distally; style pubescent, densely so in lower half; stigma ± 1 mm long. Capsule and seeds not seen.

TANZANIA. Buha District: Kibondo, Malagarasi-Moyowosi [Muyovosi] Ramsar Site, ± 10 km NNW of Migungani area, northern edge of Kabera swamp, along seasonal track to Kazenga Rangers Post, Oct. 2007, *Ndangalasi & Suleiman* 1112! (type)
DISTR. T 4; known only from the type gathering

HAB. Edges of recently burned *Brachystegia* woodland in transition to seasonal wetlands on black cotton soils; ± 1100 m
USES. None recorded on herbarium specimen
CONSERVATION NOTES. Assessed as Vulnerable (VU D2) by Darbyshire & Ndangalasi (l.c.).

49. **Barleria** sp. **C** (= *Harger* s.n.)

Herb; stems with dense woolly white indumentum. Leaves ovate, ± 2.5 cm long, 1.7 cm wide, base obtuse, apex subattenuate, indumentum as stem, most dense on principal veins and margin; lateral veins 4–5 pairs; petiole to 2 mm long. Inflorescences axillary and subterminal, single-flowered, sessile; bracts foliaceous, uppermost bracts somewhat obovate; bracteoles linear-lanceolate, ± 18 mm long, 2 mm wide. Anticous calyx lobe oblong-lanceolate, ± 20 mm long, 4.5 mm wide, base acute, apex deeply bifid with narrowly lanceolate segments ± 10 mm long, indumentum as stem but less dense, venation parallel; posticous lobe marginally longer, apex acute; lateral lobes linear-lanceolate, ± 12 mm long. Corolla mauve, white in throat; limb ?subregular, lobes ± 25 mm long, lateral pair ± 10 mm wide, adaxial pair ± 7 mm wide, abaxial lobe apparently broadest but damaged in material seen. Stamens with anthers ± 6.5 mm long. Immature capsule 14.5 mm long, tapered into prominent beak, densely pubescent.

KENYA. Elgeyo District: Elgeyo Escarpment, without date, *Harger* s.n.!
DISTR. **K** 3; not known elsewhere
HAB. Not recorded; 2450 m
USES. None recorded on herbarium specimen
CONSERVATION NOTES. Known only from the single specimen cited, this species is clearly rare but no information on its habitat requirements or current status are available: Data Deficient (DD).

NOTE. This taxon is known only from a single scrap and associated illustration. Only the lobes of the single corolla are preserved and the one fruit is immature and damaged. It is a highly distinctive taxon, the woolly indumentum easily separating it from all other *Barleria* species in the region; I have been unable to match it with any known species. Its placement in sect. *Somalia* seems appropriate in view of the immature capsule morphology (prominently beaked and densely hairy) and the similarity of the calyx and illustrated flower to those of *B. mpandensis* and *B. diplotricha*; more ample material is however needed for confirmation.

50. **Barleria granarii** *I. Darbysh.* **sp. nov.** *B. mackenii* Hook.f. similis sed caulibus maturis glabrescentibus (non dense albo-puberulis), calyce exteriori omnino pubescenti atque ad venas marginemque strigoso, corolla minore alba, fauce corollae signis roseis vel purpureis notata (non omnino malvina neque purpurea) et tubo corollae cylindrico atque multo angustiore (non late campanulato) differt. Type: Tanzania, Dodoma District, 8 km on Kilimatinde–Dodoma road, *Bidgood, Mwasumbi & Vollesen* 1171 (K!, holo.; C!, CAS, DSM!, EA!, NHT, iso.)

Subshrub, 20–100 cm tall. Leafy stems reddish-brown, ascending-pubescent, with some anvil-shaped, subappressed hairs, young stems often also patent glandular-pubescent; mature stems with pale grey-brown bark, glabrescent. Axillary buds densely buff-pubescent. Leaves ovate (-elliptic), 4.5–11 cm long, 2–5.7 cm wide, base attenuate, apex acute or obtuse, apiculate, surfaces coarsely pubescent, most dense on margin and veins beneath, sometimes with short glandular hairs towards base; lateral veins 4–6 pairs; petiole to 10(–20) mm long or absent. Inflorescences single-flowered, axillary but often crowded in upper axils; bracts foliaceous, reduced and suborbicular to obovate upwards; bracteoles linear or oblanceolate, 7–19 mm long, 1–3(–6.5) mm wide, indumentum as leaves but with hairs longer on midrib and margin, glandular hairs often many; flowers sessile. Anticous calyx lobe broadly ovate-orbicular, 13–21 mm long and wide in flower, base rounded or shallowly cordate, apex shallowly notched, truncate or rounded, surface pubescent, principal veins and margin with longer strigose hairs, hairs on margin usually with a swollen base, often also with scattered glandular hairs particularly towards the apex, venation palmate, prominent; posticous lobe broadly ovate, 15–25 mm long, 12–20 mm wide, apex attenuate to obtuse, apiculate; lateral lobes linear-lanceolate, 4.5–9 mm long, puberulent. Corolla white with pink or purple guidelines, 23–33 mm long, lateral lobes pubescent outside, hairs mainly eglandular; tube cylindrical, 11–15 mm long, expanded somewhat towards mouth; limb in "2+3" arrangement; abaxial lobe offset by 1.5–4 mm, this and the lateral lobes each 10–18 mm long, 6–10.5 mm wide; adaxial lobes fused for 4–6 mm, free portions spatulate, 7.5–13.5 mm long, 3.5–6 mm wide. Stamens attached 3–5 mm from base of corolla tube; filaments 16–24 mm long; anthers 2–3 mm long; lateral staminodes ± 0.5 mm long. Ovary puberulous distally; style pubescent towards base; stigma 1.7–3 mm long. Capsule 15–20 mm long including 6.5–8 mm beak, shortly pubescent; seeds 6–8 mm long, 5.5–7.5 mm wide. Fig. 54/1–10, p. 394.

TANZANIA. Mpwapwa District: 1–2 km NE of Gulwe village, Apr. 1988, *Pócs* 88062/H!; Kilosa District: Ruaha valley, 400 km from Dar es Salaam, Mar. 1986, *Bidgood & J. Lovett* 261!; Iringa District: Great N road, Nyangolo Scarp, 50 km N of Iringa, Apr. 1962, *Polhill & Paulo* 2046!

DISTR. **T** (?)2, 5–7; not known elsewhere

HAB. Dry bushland and riverine woodland, most commonly of *Acacia*, *Commiphora* and/or *Combretum*, more rarely of *Brachystegia-Isoberlinia*, often on rocky outcrops or sandy/gravelly soils, sometimes on roadsides; 500–1500 m

USES. None recorded on herbarium specimens

CONSERVATION NOTES. Although endemic to Tanzania, this species is widespread and locally common in the dry centre of the country. Its favoured dry, rocky woodland habitats have limited agricultural potential and degradation through human activity is likely to be limited. It also appears adaptable to some disturbance, having been recorded from roadsides. Least Concern (LC).

NOTE. *B. granarii* is easily separated from all other species in our region. From the presence of anvil-shaped hairs on the stems, densely hairy axillary buds and partially fused adaxial corolla lobes, it is almost certainly most closely related to *B. lugardii* but, amongst several differences from that species, the outer calyx lobes are much broader and the limb is 5-lobed not 4-lobed. On cursory inspection, it most closely resembles *B. mackenii* Hook.f., recorded from Zambia and Zimbabwe to South Africa; that species however clearly differs in having a dense white-puberulent indumentum to the mature stems, the calyx being hairy only on the principal veins and margin and the corolla being mauve to purple and larger with a much broader, campanulate upper portion to the tube.

The provenance of the single record from Mt Kilimanjaro (*Chambers* s.n.) must be in doubt due both to its isolation from the remainder of the range and to the fact that it has never been recollected at this well-botanised site.

The species epithet "*granarii*" ("grainstore": seedbank) honours the conservation work of the Millennium Seedbank Partnership (MSBP) in Tanzania. The seeds of this and a range of other endemic and/or endangered species were banked during the May 2008 MSBP Expedition in which the author participated.

51. **Barleria lugardii** *C.B. Clarke* in F.T.A. 5: 161 (1899); Vollesen in Opera Bot. 59: 79 (1980); M. & K. Balkwill in K.B. 52: 539, figs. 2J, 6U & 12L (1997). Type: Botswana, Khwebe, *Lugard* 128 (K!, holo.)

Subshrub, 10–100 cm tall; stems with white anvil-shaped ascending hairs, young stems often also patent glandular-pubescent, mature stems with pale greyish bark, glabrescent. Axillary buds densely white-strigose. Leaves ovate-elliptic, 3–9 cm long, 1–4 cm wide, base attenuate, apex acute, obtuse or somewhat attenuate, apiculate, margin and midrib beneath sparsely strigose, sometimes with short glandular hairs towards the base; lateral veins 3–5 pairs; petiole to 7 mm long or absent. Inflorescences 1–3(–several)-flowered, axillary but in densely-flowered plants forming a thyrsoid terminal synflorescence; bracts foliaceous, often narrowing upwards; bracteoles linear(–oblanceolate) to narrowly elliptic, 6–22 mm long, 0.5–3.5 mm wide, indumentum as leaves but with glandular hairs often more widespread; flowers sessile. Anticous calyx lobe ovate, 5.5–14 mm long, 2–6.5 mm wide, apex attenuate, entire or bifid with linear segments 0.5–4.5 mm long, strigose on margin and principal veins, often with scattered glandular hairs particularly towards apex, venation palmate to subparallel; posticous lobe as anticous but 6.5–17 mm long, 2.5–7.5 mm wide, apex attenuate; lateral lobes lanceolate, 3.5–6.5 mm long. Corolla white, sometimes with pink to purple guidelines, 16–29 mm long, pubescent on the lateral lobes outside with mixed glandular and eglandular hairs; tube cylindric, 7–12.5 mm long, expanded somewhat towards the mouth; limb in "1+3" arrangement; abaxial lobe offset by 0.5–2.5 mm, this and the lateral pair 9–18 mm long, 4.5–9 mm wide, adaxial lobe oblong-elliptic, 7.5–16 mm long, 3.5–6.5 mm wide, apex emarginate. Stamens attached 2–3.5 mm from base of corolla tube; filaments 13–23 mm long; anthers 2–3 mm long; lateral staminodes 0.5–1 mm long. Ovary densely pubescent in upper half; style pubescent at base; stigma 1–1.5 mm long. Capsule 12–16 mm long including 5.5–8 mm beak, shortly pubescent particularly on beak; seeds 5.5–7 mm long, 4.5–6 mm wide.

TANZANIA. Singida District: E of Issuna on the Singida–Manyoni road, Apr. 1964, *Greenway & Polhill* 11729!; Kilosa District: 6 km on Malolo–Kisanga track, Apr. 1998, *Bidgood, Mwasumbi & Vollesen* 889!; Iringa District: Great N Road between Kisinge and Nyangolo, 38 km N of Iringa, Apr. 1962, *Polhill & Paulo* 2036!

DISTR. **T** 1, 5–8; Zimbabwe, Botswana, Namibia

HAB. Dry bushland and riverine woodland of variously *Acacia*, *Commiphora*, *Combretum*, *Terminalia* and/or *Lannea*, more rarely of *Brachystegia*; (250–)650–1600 m

USES. None recorded on herbarium specimens

CONSERVATION NOTES. Although absent from large areas of its overall range, this species appears locally common in Tanzania and the Botswana-Zimbabwe borderlands, where it is found in a variety of open to dense bushland types. It is not considered threatened: Least Concern (LC).

SYN. *B. breyeri* Oberm. in Ann. Transv. Mus. 15: 151 (1933). Type: Namibia, Klein Namutoni, *Breyer* in Transvaal Mus. 20642 (PRE-TRV, holo.)

NOTE. In this species the two adaxial lobes of the corolla fuse to form a single, emarginate lobe; a similar corolla form is recorded in *B. rehmannii* C.B. Clarke from South Africa and sometimes, though not always, in *B. angustiloba* (see sp. 53).

Two forms of this species are widely recorded in Tanzania. Plants with small (16–20 mm long) pure white corollas and proportionally broader calyx lobes, the anticous lobe often only obscurely divided or undivided at the apex, are recorded from **T** 7–8. These closely match plants from Botswana, Zimbabwe and Namibia, including the type. Plants with large (24–29 mm long) corollas often spotted pink or purple in the throat and with proportionally narrower calyx lobes, the anticous lobe usually clearly bifid, are recorded from **T** 1, 5, 6 and 7, generally at lower altitudes but with some overlap. However, some specimens are intermediate (e.g. *Greenway & Polhill* 11729 and *Kuchar* 24522, both from Singida District, **T** 5), with corollas in the range 21–23 mm long. The variation therefore appears to be clinal rather than divisable into two discrete taxa.

52. **Barleria griseoviridis** *I. Darbysh.* in Journ. E. Afr. Nat. Hist. 97: 132, fig. 2 A–I (2009). Type: Tanzania, Ufipa District, 21 km on Kipili–Namanyere road, *Bidgood, Sitoni, Vollesen & Whitehouse* 3813 (K!, holo.; BR!, C, CAS, DSM!, EA!, K!, MO, NHT, P, US, WAG, iso.)

Suffruticose herb, several ascending stems from a woody rootstock, 20–35 cm tall; stems hispid and with patent glandular hairs when young. Leaves subsessile, oblanceolate (-elliptic), 7.5–11.5 cm long, 1.5–2.8 cm wide, base cuneate, apex obtuse or acute, minutely apiculate, midrib beneath and margin with subspreading to appressed somewhat hispid hairs, young leaves also glandular-pubescent; lateral veins 4–5 pairs. Inflorescences single-flowered, compounded into a conical terminal synflorescence 2.5–5 cm long; bracts obovate 16–23 mm long, 5–9 mm wide, densely glandular-pubescent outside and with subappressed eglandular hairs on margin and midrib; bracteoles linear to oblanceolate, 13–18 mm long, 1.5–2.5 mm wide; flowers sessile. Anticous calyx lobe ovate-lanceolate, 12.5–14.5 mm long, 2.5–3.5 mm wide, apex acute, entire or shortly notched, indumentum as bracts but with eglandular hairs many, venation parallel; posticous lobe 14–15 mm long, apex acute; lateral lobes linear-lanceolate, 10–12 mm long. Corolla white with mauve guidelines, 20–23 mm long, lateral lobes pubescent outside with mainly eglandular hairs; tube rather broadly cylindrical, 11–11.5 mm long; limb strongly zygomorphic, in "2+3" arrangement; abaxial lobe 9–10 mm long, ± 6 mm wide; lateral lobes 9.5–11 mm long, 2.8–4 mm wide; adaxial lobes partially fused, free portions linear-oblong, 5–7.5 mm long, 1–2 mm wide. Stamens attached 3–3.5 mm from base of corolla tube; filaments 16–18 mm long; anthers ± 2.5 mm long; staminodes barely developed. Ovary densely pubescent; style glabrous; stigma ± 1 mm long. Immature capsule only seen, ± 15 mm long including 6.5 mm beak, shortly pubescent; seeds not seen.

TANZANIA. Ufipa District: 21 km on Kipili–Namanyere road, May 1997, *Bidgood et al.* 3813! (type)
DISTR. **T** 4; not known elsewhere
HAB. *Brachystegia* woodland on sandy soil; 850 m
USES. None recorded on herbarium specimen
CONSERVATION NOTES. Assessed as Vulnerable (VU D2) by Darbyshire (l.c.).

53. **Barleria angustiloba** *Lindau* in E.J. 20: 20 (1894). Type: "East Africa", *Fischer* 135 (B†, holo.; HBG!, iso.)

Erect or scrambling subshrub, 30–120 cm tall; stems densely short white-pubescent, hairs spreading or retrorse, to 0.5 mm long, often with scattered glandular hairs when young. Leaves ovate or suborbicular, 2–6 cm long, 1–2.7 cm wide, base shortly attenuate, apex acute or obtuse to rounded or shallowly emarginate, minutely apiculate, indumentum as stem, hairs somewhat longer and coarser on veins beneath; lateral veins 3–5 pairs; petiole 2–15 mm long. Inflorescences axillary, 1–3-flowered, sessile; bracts foliaceous; bracteoles (oblong-) obovate, 7.5–20 mm long, 1.5–6 mm wide, indumentum as leaves with additional scattered glandular hairs outside; flowers sessile. Anticous calyx lobe linear-lanceolate, 7–13 mm long in flower, up to 17 mm in fruit, 0.7–2 mm wide, apex acute or bifid with linear segments 0.5–5.5 mm long, indumentum as bracteoles, hairs longest on margin, venation parallel with 2 prominent veins outside; posticous lobe as anticous but apex acute, only the midrib prominent outside; lateral lobes linear-lanceolate, 6.5–12 mm long in flower, up to 15 mm in fruit. Corolla white, pale blue or mauve, with red to purple guidelines, 23–35 mm long, upper half of tube and limb pubescent outside, hairs mainly eglandular; tube cylindrical, 8.5–12 mm long, somewhat expanded towards mouth; limb in "2+3" or "1+3" arrangement; abaxial lobe offset by 1–2.5 mm, 12.5–20 mm long, 5.5–8.5 mm wide; lateral lobes as abaxial but 4.5–7 mm wide; adaxial lobes either completely fused or more commonly partially so, free portions linear-spatulate, 9–17 mm long, 2–4 mm wide. Stamens attached 2.5–4 mm from base of corolla tube; filaments 13.5–22 mm long; anthers 2.3–3.3 mm long; lateral staminodes 0.3–1.5 mm long. Ovary densely pubescent; style pubescent at base; stigma 1–2 mm long. Capsule 11–14.5 mm long including 3.5–6 mm beak, pubescent; seeds 4–5 mm long and wide.

KENYA. Laikipia District: Rumuruti, ± 50 km N near Colchecio Lodge edge of scarp, Nov. 1978, *Hepper & Jaeger* 6616!; Masai District: 54 km on Nairobi–Magadi road, May 1988, *Bidgood & Vollesen* 1272! & Chyulu Hills, Ol Donyo Wuas Camp, July 1990, *Luke* 2444!
TANZANIA. Shinyanga District: Mwantine Hills, May 1931, *Burtt* 2456!; Musoma District: 8 km from Seronera on the Lake Magadi road, May 1961, *Greenway* 10175!; Masai District: Olbalbal Escarpment, Mar. 1961, *Newbould* 5734!
DISTR. **K** 3, 4, 6; **T** 1, 2, 5; not known elsewhere
HAB. Dry, rocky slopes with open wooded grassland, typically with *Acacia* and/or *Commiphora*; 900–1800 m
USES. "Browsed by sheep and goats" (**K** 6; *Glover & Samuel* 2911)
CONSERVATION NOTES. This species can be locally common within its limited range, e.g. *Greenway et al.* 10585 recorded it as a local dominant at the Moru Kopjes of Musoma District, **T** 1. Due to the low agricultural potential of its dry, rocky habitat, it is unlikely to be threatened by human activity: Least Concern (LC).

SYN. [*B. ventricosa sensu* C.B. Clarke in F.T.A. 5: 164 (1899) pro parte quoad *Fischer* 135, *non* Nees]
[*B.* sp. K *sensu* U.K.W.F. ed. 1: 594 (1974) & ed. 2: 272 (1994)]

NOTE. In the majority of specimens of this species, the adaxial corolla lobes are only partially fused. However, in those from the Mwantine Hills, Shinyanga District (*Burtt* 2456 & 3474!) and from Ngaserai Hill, Masai District (*Mbano & Willy* in CAWM 5707!) they are fully fused to form a single obovate lobe. A specimen from Mt Meru, Arusha District (*Richards* 23525!) however has corollas with both bilobed and entire upper "lips".

Plants from north of Lake Magadi in southern Kenya (e.g. *Bidgood & Vollesen* 1272) have smaller, more rounded leaves than typical (similar to those of *B. maculata* S. Moore; see below) but are otherwise inseparable from the remaining material.

54. **Barleria maculata** *S. Moore* in J.B. 54: 289 (1916). Type: Kenya, Teita District, Maktau, *Buchanan* s.n. (BM!, holo.)

Erect or scrambling perennial herb or subshrub, 30–90 cm tall; stems densely short-pubescent, hairs white or sometimes pale buff when young, spreading to retrorse, to 0.5(–0.8) mm long. Leaves broadly ovate-elliptic, suborbicular or somewhat obovate, 1.2–3.8 cm long, 0.6–2.2 cm wide, base shortly attenuate to obtuse, apex obtuse or rounded, often minutely apiculate, indumentum as stem but less dense, margin sometimes with clearly longer hairs with a swollen base; lateral veins 3–4 pairs; petiole 2–9 mm long. Inflorescences axillary, 1(–2)-flowered, sessile; bracts foliaceous; bracteoles broadly obovate or rarely oblanceolate, 5.5–21 mm long, 2–7.5 mm wide, indumentum as leaves. Anticous calyx lobe (obovate-)elliptic to narrowly so, 5.5–16 mm long, 3–11 mm wide in flower, apex bifid with deltate segments 1.5–3.5 mm long, indumentum as leaves, hairs longest on the margin where sometimes with a swollen base, surface palmately 4–6-veined; posticous lobe obovate or oblong-elliptic, 6.5–18 mm long, 3–8 mm wide, apex obtuse or acute, surface palmately 3–7-veined; lateral lobes linear-lanceolate, 2.5–7.5 mm long. Corolla white, pale blue or mauve, often with purple guidelines, 21–33 mm long, limb pubescent outside, most dense on the lateral lobes; tube cylindrical, 8.5–12.5 mm long; limb in "2+3" arrangement; abaxial lobe offset by 3–4.5 mm, 10–20 mm long, 5–8.5 mm wide; lateral lobes as abaxial but 3.5–8 mm wide; adaxial lobes fused for up to 1.5 mm, spatulate to obovate, 9–17 mm long, 3–6 mm wide. Stamens attached 4–6 mm from base of corolla tube; filaments 11.5–20 mm long; anthers 2–3.5 mm long; lateral staminodes 0.5–2.8 mm long. Ovary pubescent distally or largely glabrous; style glabrous; stigma 1–2 mm long. Capsule 12.5–16 mm long including 5-6.5 mm beak, pubescent mainly on beak, sometimes sparse; seeds 5–6.5 mm long, 4.5–5 mm wide.

KENYA. Teita District: Mzinga Hill, Voi, Jan. 1964, *Verdcourt* 3894! & Tsavo National Park East, Mudanda Rock, S end, Jan. 1967, *Greenway & Kanuri* 12980!; Kwale District: Mackinnon Road, Sept. 1953, *Drummond & Hemsley* 1953!

TANZANIA. Pare District: Koko Hill, Same, Feb. 1957, *Wilson* 28! & near Same Secondary School, 1965, *Wingfield* 716!

DISTR. **K** 4, 7; **T** 3; not known elsewhere

HAB. Dry, rocky or sandy slopes with open wooded grassland, typically with *Acacia-Commiphora*; 80–900 m

USES. None recorded on herbarium specimens

CONSERVATION NOTES. A local species known from rather few gatherings. Its stronghold, however, is in the well protected bushlands of Tsavo National Park where its only likely threat is over-grazing by wild game. It is not currently considered threatened: Least Concern (LC).

NOTE. *B. maculata* is the eastern, lowland vicariant of *B. angustiloba* but is readily separable by the broader outer calyx lobes with palmate venation. It also usually lacks glandular hairs on the inflorescence, though these are present in *Bally* 13486 from Lugard Falls and *Luke* 2529 from Dakawachu, Kilifi. In addition, the corolla of *B. maculata* is less hairy outside, particularly on the tube, the abaxial corolla lobe is more strongly offset from the lateral pair and the adaxial lobes are fused for a shorter distance and are broader, the filaments are attached higher up the corolla tube and the ovary and capsule are less hairy.

Luke 3492 from Mbunguni, Kwale District, Kenya has oblanceolate bracteoles and elongate, proportionally narrow outer calyx lobes with a more acute apex than typical. It is also unusual in having long marginal hairs on the leaves, bracteoles and calyx, with swollen bases similar to those recorded in *B. lactiflora* and *B. granarii*. The plants in this specimen branch only from the base whilst *B. maculata* is usually more branched throughout. Further material is desirable from this location.

55. **Barleria laceratiflora** *Lindau* in E.J. 38: 68 (1905). Type: Tanzania, Lindi District, Ras Rungi, *Busse* 2367 (B†, holo.; EA!, iso.)

Suffruticose perennial, 15–30 cm tall; stems with few appressed medifixed and simple hairs at and below the nodes and scattered short patent glandular hairs when young, elsewhere glabrous. Leaves elliptic, 4.3–7 cm long, 1.4–2 cm wide, base cuneate-attenuate, apex acute, apiculate, midrib and lateral veins beneath and margin with few stiff appressed medifixed hairs, elsewhere glabrous; lateral veins 4 pairs; petiole to 6 mm long. Inflorescences 1–3-flowered, axillary; cymes subsessile or peduncle to 2.5 mm long; bracts foliaceous but often reduced and proportionally narrower; bracteoles oblanceolate, 6–17 mm long, 0.5–3.5 mm wide; flowers subsessile. Anticous calyx lobe elliptic, 8.5–10 mm long, 4–5.5 mm wide, apex bifid for 2.5–5 mm, surface with appressed medifixed hairs mainly on margin and veins, usually with few short glandular hairs towards base and on margin, 2 principal veins prominent; posticous lobe 12–14.5 mm long, 5.5–6.5 mm wide, apex rounded or obtuse, apiculate, 3(–5)-veined from base; lateral lobes lanceolate, 5–6 mm long. Corolla pale blue, 19–24 mm long, glandular-pubescent outside mainly on limb; tube cylindrical, 6.5–7 mm long; limb in "4+1" arrangement; abaxial lobe offset by 6–7 mm, 10–13 mm long, 5.5 mm wide; lateral lobes 8–10 mm long, 4–4.5 mm wide; adaxial lobes 5.5–8 mm long, 2.5–3.5 mm wide. Stamens attached 2.5–3.5 mm from base of corolla tube; filaments 12.5–14 mm long; anthers 3.2–3.6 mm long; lateral staminodes 0.5–0.8 mm long. Ovary puberulous; style glabrous; stigma 1–1.6 mm long. Capsule and seeds not seen.

TANZANIA. Lindi District: Ras Rungi, May 1903, *Busse* 2367! (type)
DISTR. **T** 8; N Mozambique
HAB. Coastal bushland on sandy and coral-derived soils; sea-level
USES. None recorded on herbarium specimens
CONSERVATION NOTES. Only two specimens have been seen, the second being from Goa Island, Mozambique, collected in 1947 (*Gomes e Sousa* 3503) where it was recorded as common. Extensive dune systems with bushland are still found along the coast near Lindi, but the Mozambique site may have been degraded by the growth in population and tourism on this part of the coast. It is therefore quite possibly threatened but more data on current distribution and threats are required: Data Deficient (DD).

56. **Barleria argentea** *Balf. f.* in Proc. Roy. Soc. Edin. 12: 86 (1883) & in Trans. Roy. Soc. Edin. 33: 214 (1888); C.B. Clarke in F.T.A. 5: 155 (1899); U.K.W.F., ed. 2: 273, t. 119 (1994); M. & K. Balkwill in K.B. 52: 565, figs. 2R, 12G & 12P (1997); Wood, Handb. Yemen flora: 271 (1997); Hédren in Fl. Somalia 3: 435 (2006); Ensermu in F.E.E. 5: 422, fig. 166.34 (2006). Type: Yemen, Socotra, *Balfour* 544 (E, holo.; K!, iso.)

Compact perennial herb or subshrub, 5–30(–60) cm tall, often much-branched from the base; stems with ± dense short white retrorse or patent hairs and/or appressed medifixed hairs, with few to many patent glandular hairs. Leaves linear, oblanceolate, obovate or oblong-elliptic, 0.7–6.5 cm long, 0.2–1.5 cm wide, base cuneate or attenuate, apex acute to rounded, indumentum of stiff appressed medifixed hairs 0.4–1.6 mm long, evenly distributed or restricted to margin and midrib beneath, often also with short glandular hairs along margin and/or scattered, rarely dense, short patent eglandular hairs; lateral veins inconspicuous; petiole to 15 mm long or absent. Inflorescences axillary, 1–3(–7)-flowered, 1–2 per axil; sessile or peduncle to 20 mm long; bracts foliaceous; bracteoles linear, lanceolate, oblanceolate or obovate,

2–15(–33) mm long, 0.2–2.5(–4.5) mm wide; flowers sessile or pedicels to 2 mm long. Anticous calyx lobe triangular or lanceolate, 2.5–7(–8.5) mm long, 1–2.5 mm wide, apex acute or attenuate, often shortly notched, indumentum as leaves, glandular hairs sparse to rather dense, stiff appressed hairs often simple along margin, 2 parallel principal veins prominent or veins inconspicuous; posticous lobe 3–10(–13) mm long, apex acute or attenuate, often only midrib prominent; lateral lobes linear-lanceolate, 2–5(–6.5) mm long. Corolla pale blue to purple or white, with red to purple guidelines, 11.5–23 mm long, pubescent outside on upper tube and limb, hairs mainly eglandular; tube 5–11.5 mm long, ± expanded towards mouth; limb in "4+1" arrangement, abaxial lobe offset by 3–6 mm, 6–11 mm long, 4–7.5 mm wide; lateral lobes 3.5–8 mm long, 2.5–5.5 mm wide; adaxial lobes 2.5–7 mm long, 1.5–4.5 mm wide. Stamens attached 1.5–4 mm from base of corolla tube; filaments 8.5–13.5 mm long; anthers 0.9–1.9 mm long; lateral staminodes 0.25–1 mm long. Ovary puberulous; style largely glabrous; stigma 0.3–0.9 mm long. Capsule 7.5–12.5(–14) mm long including 3–6.5(–7.5) mm beak, puberulous; seeds 3–5.5 mm long, 2.5–4.5 mm wide.

UGANDA. Karamoja District: Kosike, near Nabilatuk, July 1957, *Dyson-Hudson* 285! & Chakwi R., Karamoja road, June 1959, *Symes* 594! & Lodoketemit, near Moroto, Jan. 1964, *Napper* 1708!
KENYA. Northern Frontier District: 54 km NE of El Wak, Dec. 1971, *Bally & Smith* 14573!; Turkana District: Oropoi, Feb. 1965, *Newbould* 6956!; Tana River District: Garissa–Galole road, 37 km S of fork for Nairobi, Jan. 1972, *Gillett* 19514!
TANZANIA. Masai District: Engaruka, Feb. 1970, *Richards* 25500!; Lushoto District: 8 km SE of Mkomazi, May 1953, *Drummond & Hemsley* 2377! & 8 km on Mkomazi–Mombo road, Apr. 1988, *Bidgood & Vollesen* 1266!
DISTR. **U** 1; **K** 1–4, 6, 7; **T** 2, 3; S Sudan, Ethiopia, Somalia; Socotra, Yemen & Saudi Arabia
HAB. Open *Acacia-Commiphora* woodland, semi-desert bushland and grassland, including disturbed and eroded areas; 100–1700 m
USES. "Whole plant used for fodder" (**K** 1; *Mwangangi* 5069); "animal fodder" (**K** 3; *Timberlake* 442, 815, 857)
CONSERVATION NOTES. Widespread and locally common in a variety of dry habitats and appearing tolerant of disturbance; it is not considered threatened: Least Concern (LC).

SYN. *Somalia diffusa* Oliv. in Hook. Ic. Pl. 16, t. 1528 (1886). Type: Somalia, *James & Thrupp* s.n. (K!, holo.)
Barleria diffusa (Oliv.) Lindau in E.J. 20: 27 (1894), in obs. & in E. & P.Pf. 4, 3B: 315 (1895); U.K.W.F. ed. 1: 593 (1974)
B. schweinfurthiana Lindau in E.J. 20: 26 (1894). Type: Tanzania, Pare District, W of Pare Mts, *Höhnel* 6 (B†, holo.)
[*B. yemensis* Lindau in E. & P.Pf. 4, 3B: 315 (1895) as *B. yemense*, nom. nud.]

NOTE. F.T.E.A. species 56–62 form part of a complex which is most diverse in NE Africa and Arabia and in much need of revisional treatment. Many taxa have been described and species delimitation is complicated by the regular occurrence of intermediate specimens between the recognised taxa, suggesting that hybridisation is not uncommon (see for example Hédren in Fl. Somalia 3: 435–438). It it is quite possible that at least some of the currently recognised species are better considered distinct at only an infraspecific rank. However, pending further study I have maintained them as separate species. Four taxa in the Flora area do not fit within any of the species as currently delimited - two are (in my opinion) clearly distinct and so described as new species here, the other two require further collections and a more detailed study of the whole group before a firm decision on their status can be made.

B. argentea itself is a remarkably variable species, with a wide range of regional and population-level forms. The formal recognition of infraspecific taxa would seem premature in the absence of a full revision, but the more significant variants in our region are worthy of note. Most populations have subsessile inflorescences, but those from NE Kenya are often pedunculate; they also tend to have the most sparse indumentum with the leaves having medifixed hairs only on the margin and midrib. These populations fall into two groups: those with a stem indumentum dominated by short spreading simple eglandular hairs and those dominated by appressed medifixed hairs. The latter is close to that of *B. parviflora* T. Anderson, described from Ethiopia. Indeed, Hedrén (in Fl. Somalia 3: 435) suggested that the two taxa may prove conspecific. However, Ensermu (in Fl. Eth. 5: 423) restricts *B. parviflora* to specimens with a rounded to cordate leaf base, ovate bracteoles and a glabrous pistil, none of which are recorded in the Kenyan material.

Of the forms with subsessile inflorescences, there is a general NW-SE trend of increasing leaf width, increasing hairiness and decreasing size of the corollas (11.5–16 mm long in **K** 7 and **T** 3 versus 14.5–23 mm in **U** 1 and **K** 2), capsules (7.5–9.5 mm long versus 10–12.5 mm) and seeds (3–4 mm long versus 4.5–5 mm). However, intermediate specimens are fairly frequent in central Kenya.

A particularly distinctive form from SE Kenya and NE Tanzania has glandular hairs to 1.5 mm long on the calyx margin and often also the bracteoles, stem and leaf margin, e.g. *Harvey & Vollesen* 66 (**K** 4) and *Polhill & Paulo* 929 (**K** 7). This "glandular-pilose" form also has narrower, more oblanceolate leaves and larger flowers and fruits than the more widespread "southeastern" form (see above). It is therefore a good candidate for varietal status, though intermediates again occur (e.g. *Friis & Hansen* 2629, **K** 7; *Geilinger* 4245, **T** 2).

57. **Barleria** sp. **D** (= *Tweedie* 2573)

Erect or straggling perennial herb, 10–90 cm tall; stems with short retrorse or patent hairs and patent glandular hairs. Leaves ovate or oblong-elliptic, 1.5–6 cm long, 0.6–2.5 cm wide, base shortly attenuate, apex acute to rounded, apiculate, indumentum of stiff appressed medifixed hairs 1–1.7 mm long, often sparse when mature, with patent glandular hairs along margin, rarely with scattered short patent eglandular hairs when young; lateral veins 2–4 pairs; petiole 2–16 mm long. Inflorescences axillary, 1–3-flowered, 1–2 per axil; sessile or peduncle to 5 mm long; bracts foliaceous; bracteoles linear or oblanceolate, 3–14 mm long, 0.2–1.8 mm wide; flowers subsessile. Anticous calyx lobe broadly ovate or suborbicular, 4.5–8.5 mm long, 2.5–4.5 mm wide, base obtuse or rounded, apex attenuate, usually notched for up to 1.5 mm, indumentum as leaves but glandular hairs many, appressed hairs often simple along margin, venation subparallel or palmate, 2 or 4 veins prominent; posticous lobe 6–11 mm long, 3–5.5 mm wide, apex attenuate, 3 or 5 veins prominent; lateral lobes lanceolate, 2.5–4 mm long. Corolla white or blue (–mauve), with red to purple guidelines, 12.5–20 mm long, pubescent outside on upper tube and limb, hairs mainly eglandular; tube 5–7.5 mm long, expanded towards mouth; limb in "4+1" arrangement; abaxial lobe offset by 3.5–6.5 mm, 6.5–12 mm long, 4.5–7 mm wide; lateral lobes 4–7.5 mm long, 2.5–4 mm wide; adaxial lobes, 3.5–7 mm long, 2.5–3 mm wide. Stamens attached ± 1.5 mm from base of corolla tube; filaments 9.5–15 mm long; anthers 1.8–2.7 mm long; lateral staminodes 0.5–1 mm long. Ovary puberulous; style largely glabrous; stigma 1–1.5 mm long. Capsule 9.5–10 mm long including ± 4.5 mm beak, puberulous; seeds ± 4 mm long, 3.5 mm wide.

KENYA. Meru District: Meru Game Reserve, Jan. 1966, *J. Adamson* 5!; Teita District: Voi, May 1930, *Napier* 1007! & *idem*, Feb. 1963, *Tweedie* 2573!

DISTR. **K** 4, 7; not known elsewhere

HAB. Open *Acacia-Commiphora* bushland and associated disturbed grasslands, edges of cultivation; 400–750 m

USES. None recorded on herbarium specimens

CONSERVATION NOTES. Data Deficient (DD).

NOTE. This taxon is close to *B. argentea* of which it is perhaps only a variety, but is provisionally separated here pending further analysis of this group. It is separated primarily by its broad, ovate to suborbicular outer calyx lobes. In addition, the anthers and stigma are often larger in this taxon, and in several specimens the leaves are larger and ovate and the plants are taller and more straggling (e.g. *Tweedie* 2573).

The type of *B. benadirensis* (Fiori) Chiov. from S Somalia (*Paoli* 819, FT!) has similar calyces and inflorescences to this taxon but differs in being a compact dwarf shrub, more hairy throughout.

58. **Barleria masaiensis** *I. Darbysh.* **sp. nov.** *B. argenteae* Balf. f. similis sed habitu robustiore, foliis caulinis inferis (oblongo-)ovatis (non linearis neque oblanceolatis neque obovatis neque anguste oblongis) 1–3 cm (non 0.2–1.5 cm) latis, apiculo magis conspicuo, inflorescentiis 3–9 (non plerumque 1–3) floris, corolla majore

19–28 mm (non 11.5–23 mm) longa, antheris majoribus 2–3 mm (non 1–2 mm) longis et capsula pubescenti (non puberula) differt. Typus: Kenya, Masai District, ± 16 km, Olorgesailie [Orgasalik]–Nairobi, *Verdcourt* 1829 (K!, holo.; EA!, iso.)

Erect perennial herb or subshrub, 20–90 cm tall; stems with dense short white patent or retrorse hairs, patent glandular hairs and occasional subappressed medifixed hairs. Leaves (oblong-) ovate or those towards the stem apex elliptic, 2.2–6.5 cm long, 1–3 cm wide, base attenuate or acute, apex acute or obtuse, apiculate, surfaces with ± dense, (sub-)appressed medifixed hairs 1–2 mm long, sometimes also with short patent eglandular and/or glandular hairs, the latter particularly along the margin; lateral veins 3–4 pairs; petiole 6–23 mm long. Inflorescences axillary, 3–9-flowered, 1–2 per axil, congested; peduncle 2–9(–18) mm long; bracts foliaceous; bracteoles linear or oblanceolate, 3–12.5 mm long, 0.3–1.5 mm wide, the lowermost pair of the dichasium larger, 10.5–30 mm long, 0.5–8.5 mm wide; pedicels 0.5–2.5 mm long. Anticous calyx lobe (linear-) lanceolate, (3.5–)5.5–8.5 mm long, 1–2 mm wide, apex acute or more rarely notched, densely pubescent with short spreading eglandular hairs and longer spreading glandular hairs, often also with subappressed simple or medifixed hairs, 2 parallel principal veins, often inconspicuous; posticous lobe (4.5–)6–10 mm long, apex acute, only the midrib visible; lateral lobes linear-lanceolate, (4–)5–8 mm long. Corolla white or pale blue to mauve, with purple guidelines, 19–28 mm long, upper tube and limb pubescent outside, hairs mainly eglandular; tube 7–12 mm long, somewhat expanded towards mouth; limb in "4+1" arrangement; abaxial lobe offset by 5–7.5 mm, 10.5–13 mm long, 6–8.5 mm wide; lateral lobes 7.5–10 mm long, 4.5–7.5 mm wide; adaxial lobes 8–9.5 mm long, 3.5–6 mm wide. Stamens attached 2.5–3.5 mm from base of corolla tube; filaments 17–21 mm long; anthers 2–3 mm long; lateral staminodes 0.5–2.5 mm long. Ovary densely pubescent; style pubescent towards base; stigma 0.7–1 mm long. Capsule 11.5–15 mm long including 5.5–8 mm beak, pubescent; seeds 4–5 mm long, 3.8–4.7 mm wide. Fig. 55/1–6, p. 410.

KENYA. Northern Frontier District: 27 km S of Mado Gashi, Jan. 1972, *Bally & Smith* 14968!; Masai District: Ngong Hills on new Magadi road, June 1939, *Bally* 21! & Olorgesailie [Ol Orgasailie], Apr. 1969, *Napper, Greenway & Kanuri* 1991!

TANZANIA. Masai District: Engaruka [Enganika], Feb. 1932, *St. Clair-Thompson* 337! & Mozinik [Monik] Plateau, rift wall W of Lake Natron, July 1962, *Newbould* 6210! & *idem*, Nov. 1962, *Newbould* 6327!

DISTR. **K** 1, 6; **T** 2; not known elsewhere

HAB. Open *Acacia-Commiphora*, *Combretum* and *Cordia* bushland or dry grassland and scrub, sometimes on eroded soils such as along dry river-beds or on gulley slopes; 300–1450 m

USES. None recorded on herbarium specimens

CONSERVATION NOTES. This species has a very limited distribution, largely centred upon the Rift Valley around Lakes Natron and Magadi, with an isolated population in northern Kenya. It is probably under-collected in the Lake Naton region, being most often recorded from the well-botanised Nairobi–Magadi road where the habitat is similar. As this species is tolerant of some erosion, the most likely serious threat to its favoured habitat, it is unlikely to have experienced significant population declines due to human pressure: Least Concern (LC).

SYN. [*B.* sp. G *sensu* U.K.W.F. ed. 1: 593 (1974) & ed. 2: 273 (1994)]
[*B.* sp. aff. *glandulifera sensu* M. & K. Balkwill in J. Biogeogr. 25: 105, fig. 10a (1998)]

NOTE. Although closely related to *B. argentea*, this species is easily separated by comparison to herbarium material, being more robust with larger, more ovate leaves with a conspicuous apiculus, having more dense inflorescences, larger corollas and anthers and more conspicuously hairy capsules. The two species are sympatric in the Kenya-Tanzania border region around Lakes Magadi and Natron; the "southeastern" form of *B. argentea* found there has subsessile inflorescences with very small calyces, corollas and fruits (see note to that species) and so is easily separable from *B. masaiensis*.

The single isolated specimen from **K** 1 (*Bally & Smith* 14968) differs from the S Kenyan and N Tanzanian material in being less hairy, particularly on the leaves and in having the shortest staminodes. It otherwise closely matches the remaining material.

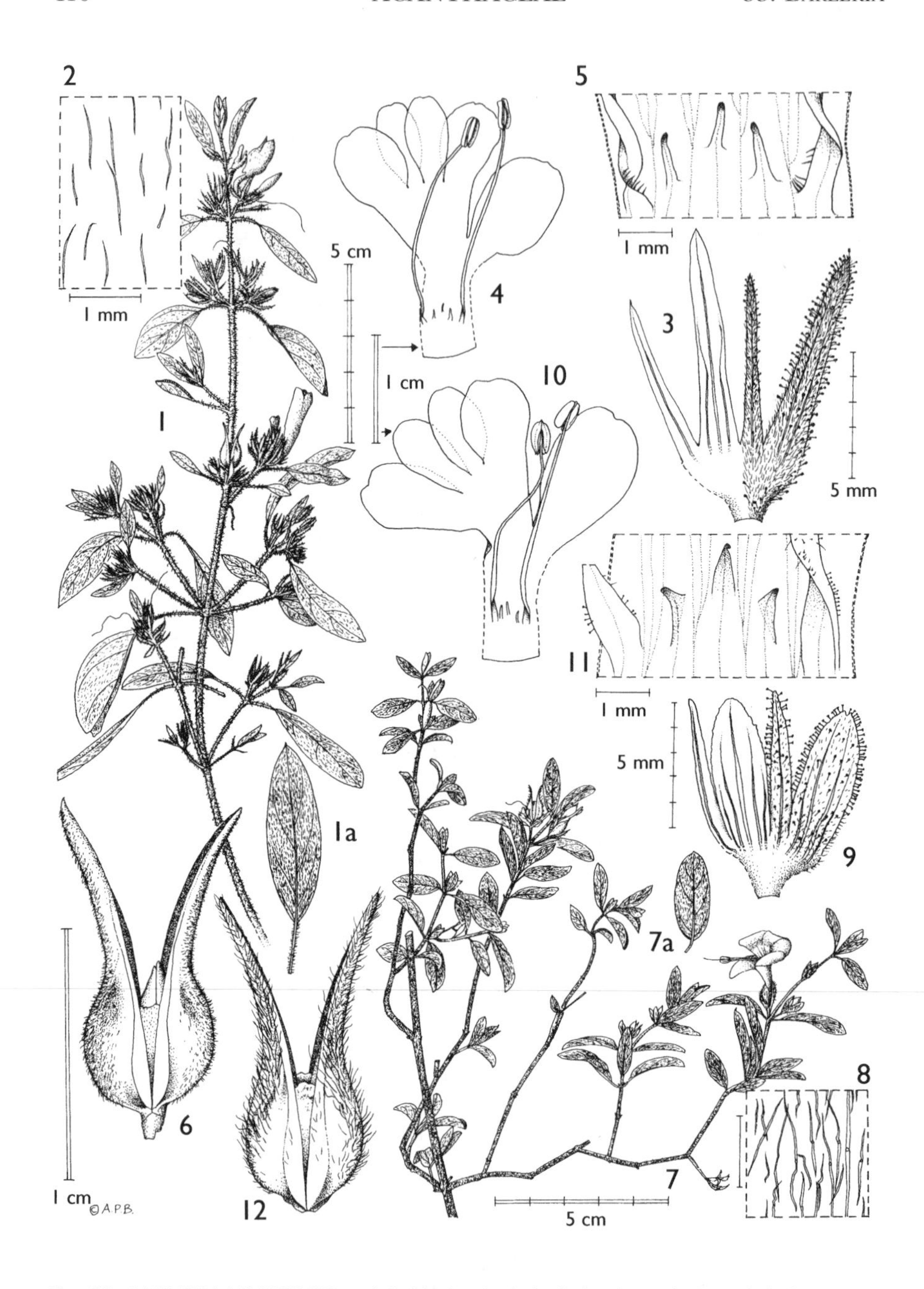

FIG. 55. *BARLERIA MASAIENSIS* — **1**, habit; **1a**, single leaf, showing apiculum; **2**, indumentum, adaxial surface of leaf; **3**, dissected calyx, outer surface, anticous lobe to the right, hairs omitted from the left half to show venation; **4**, dissected corolla with androecium, abaxial lobe to the right; **5**, detail of staminodes and staminal filament bases; **6**, capsule. *B. HIRTIFRUCTA* — **7**, habit; **7a**, single leaf; **8**, indumentum, adaxial surface of leaf; **9**, dissected calyx, outer surface, anticous lobe to the right, hairs omitted from the left half to show venation; **10**, dissected corolla with androecium, abaxial lobe to the right; **11**, detail of staminodes and staminal filament bases; **12**, capsule. 1 & 2 from *Verdcourt* 1464; 3 from *Greenway* 13534; 4 & 5 from *Newbould* 6210; 6 from *Greenway* 11762; 7 & 8 from *Hepper & Jaeger* 6955; 9–11 from *Mathew* 6683; 12 from *Gilbert et al.* 5663. Drawn by Andrew Brown.

59. **Barleria** sp. **E** (= *Rawlins* 771)

Spreading perennial herb or subshrub to 35 cm tall; stems with dense short white patent or retrorse hairs and few to many patent glandular hairs. Leaves (oblong-) elliptic or ovate, 1.2–3.5 cm long, 0.6–2.1 cm wide, base abruptly attenuate or obtuse, apex rounded or obtuse, apiculate, surfaces densely hairy with subappressed medifixed hairs 1–1.7 mm long, short to long patent hairs and short glandular hairs, the latter particularly along margin; lateral veins 3–4 pairs; petiole 2–12 mm long. Inflorescences axillary, 1–3-flowered, 1–2 per axil; peduncle 1.5–5 mm long; bracts foliaceous; bracteoles obovate to oblanceolate, 5.5–14 mm long, 0.8–5 mm wide; pedicels to 2.5 mm long. Anticous calyx lobe linear, 6.5–9 mm long, 1–1.5 mm wide, apex acute or notched, with dense short patent eglandular and glandular hairs and many appressed stiff medifixed or simple hairs, venation inconspicuous; posticous lobe marginally longer, sometimes narrowly oblanceolate and to 2.2 mm wide, apex acute; lateral lobes linear-lanceolate, 5.5–8.5 mm long. Corolla white, 22–24 mm long, upper tube and limb densely pubescent and with few glandular hairs; tube 9.5–10 mm long, barely expanded towards mouth; limb in "4+1" arrangement; abaxial lobe offset by 4–5 mm, 10.5–12 mm long, ± 6 mm wide, lateral and adaxial lobes ± 8.5–9 mm long, 3.5–4 mm wide. Stamens attached 2.5–3 mm from base of corolla tube; filaments 16–19 mm long; anthers 1.5–2 mm long; lateral staminodes ± 0.5 mm long. Ovary densely appressed-pubescent; style pubescent towards base; stigma 0.8–1.2 mm long. Capsule 12.5–13 mm long including 6.5 mm beak, pubescent with hairs of variable length; seeds ± 4 mm long and wide.

KENYA. Kilifi District: Galana Ranch S, June 1975, *Bally* 16859! & Galana Ranch, S of Hoshingo Hill, June 1975, *Bally* 16864!; Tana River District: Karawa, S of Garsen, June 1959, *Rawlins* 771!
DISTR. **K** 7; S Somalia
HAB. *Acacia-Commiphora* bushland and open *Euphorbia-Terminalia-Commiphora* bushland; ± 100 m
USES. None recorded on herbarium specimens
CONSERVATION NOTES. Currently known from the three collections cited plus a single specimen (*Gillett et al.* 25239) from S Somalia. It was recorded as locally common in *Acacia-Commiphora* bushland at the Somalia site, but data on abundance are lacking for the Kenyan populations. Data Deficient (DD).

SYN. [*B. clinopodium sensu* Hedrén in Fl. Somalia 3: 438 (2006) pro parte quoad *Gillett et al.* 25239 [cited as 25236] ex Somalia, *non* Fiori]

NOTE. This taxon more closely resembles plants from central and S Somalia assigned to *B. glandulifera* Lindau by Hedrén (Fl. Somalia 3: 436). However, that taxon has oblanceolate to obovate outer calyx lobes 2–5 mm wide, larger flowers (corolla 28–43 mm long, anthers 2.4–3 mm) and lacks simple spreading hairs on the leaves (the medifixed hairs often spreading). The type of *B. glandulifera* (*Hildebrandt* 860e) is from N Somalia and, although most of the collection was destroyed in WWII, the extant fragment at K is clearly different to the taxon from central and S Somalia, having a completely different indumentum. Nor is the C/S Somalian species an exact match for the type of *B. insericata* Chiov. (*Puccioni & Stefanini* 373, FT!), included in synonymy within *B. glandulifera* by Hedrén, which has dense appressed medifixed hairs on the stems, lacking the short spreading glandular and eglandular hairs of the former. Clearly this group requires further investigation before names can be meaningfully applied.

Within our region, this taxon is most likely to be confused with *B. masaiensis* but has smaller leaves with a more rounded apex, fewer-flowered inflorescences, more linear outer calyx lobes with more dense appressed coarse hairs, a more densely hairy exterior to the corolla and smaller anthers.

60. **Barleria hirtifructa** *I. Darbysh.* **sp. nov.** *B. argenteae* Balf. f. atque *B. masaiensi* I. Darbysh. similis sed lobis exterioribus calycis oblongis vel oblongo-ellipticis vel subovatis (non triangularis neque lineari-lanceolatis) ad medium vel super (non ad basin) latissimis 1.7–3.5 mm (non 1–2.5 mm) latis, lobo postico ad apicem obtuso vel subacuto (non anguste acuto) differt. A *B. argentea* etiam antheris majoribus 2–3.3 mm (non 1–2 mm) longis et capsula pubescenti pilis ad longitudinem variabilibus vestita (non puberula), parte fecunda capsulae pilis longioribus ascendentibus etiam

instructa differt. A *B. masaiensi* lobo antico calycis lobos laterales aequanti vel quam eis paulo breviore (non quam eis valde longiore) etiam differt. Type: Kenya, Northern Frontier District, Mt Kulal, near lower Gatab airstrip, *Hepper & Jaeger* 6955 (K!, holo.)

Subshrub, 20–60 cm tall; stems soon woody, with dense short white patent or retrorse hairs and few to many patent glandular hairs, often also with scattered appressed medifixed hairs. Leaves oblong-elliptic or somewhat obovate, 1.5–5 cm long, 0.5–2 cm wide, base acute or attenuate, apex rounded, obtuse or rarely acute, apiculate, surfaces with evenly distributed appressed medifixed hairs 0.7–2 mm long, often also with short patent eglandular and/or glandular hairs, the latter particularly along the margin; lateral veins 3–4 pairs; petiole 1.5–11 mm long. Inflorescences axillary, 1–3(–10)-flowered, solitary; peduncle 0.5–13(–27) mm long; bracts foliaceous; bracteoles obovate, 3.5–11(–22) mm long, 0.8–5(–7) mm wide; flowers subsessile or pedicels to 4 mm long. Anticous calyx lobe oblong, oblong-elliptic or somewhat obovate, 3.5–7 mm long, 1.7–3.5 mm wide, apex obtuse, acute or shallowly notched, densely glandular-pubescent particularly towards margin, also with short patent hairs and appressed coarse medifixed or simple hairs, venation parallel, often 2 principal veins visible; posticous lobe marginally longer, apex acute to obtuse, 1 or 3 principal veins visible; lateral lobes lanceolate, 3.5–7 mm long. Corolla white, blue or mauve, with red or purple guidelines, 17–28 mm long, upper tube and limb pubescent outside, hairs mainly eglandular; tube 7–12.5 mm long, barely expanded towards mouth; limb in "4+1" arrangement, abaxial lobe offset by 4–7 mm, 7.5–16 mm long, 4–8.5 mm wide; lateral lobes 5.5–9.5 mm long, 3–7 mm wide; adaxial lobes as lateral pair but 2.5–5 mm wide. Stamens attached 2–3.5 mm from base of corolla tube; filaments 11–20 mm long; anthers 2–3.3 mm long; lateral staminodes 0.4–0.8 mm long. Ovary densely pubescent; style with few hairs at base; stigma 1–1.6 mm long. Capsule 11–13.5 mm long including 4.5–7 mm beak, pubescent including long subappressed hairs on fertile portion; seeds 3.5–4 mm long, 3–3.7 mm wide. Fig. 55/7–12, p. 410.

KENYA. Northern Frontier District: near Mugurr, SW [Rudolph] Turkana, June 1970, *Mathew* 6683! & Ngurunit, Sept. 1976, *Herlocker* 158!; Meru District: Isiolo–Wajir road just S of [Shaba Dogo] Shaptiga hill, July 1974, *Faden* 74/962!

DISTR. **K** 1, 4; not known elsewhere

HAB. Dry *Acacia-Commiphora* bushland, rocky outcrops and rocky semi-desert; 750–1250 m

USES. None recorded on herbarium specimens

CONSERVATION NOTES. This taxon is apparently restricted to the area between Mt Kulal and Isiolo in the dry bushlands of northern Kenya, where it is known from less than 10 localities. However, this area is largely under-collected and *B. hirtifructa* may well prove to be more common than the herbarium data currently suggest. Much of its range is rather remote with only minimal human impact, although the populations arounds Isiolo may have declined due to expansion of that town and recent road construction in the surrounding area. It is currently assessed as Data Deficient (DD) but is likely to eventually prove to be unthreatened.

NOTE. The combination of small, proportionately broad outer calyx lobes with an obtuse or broadly acute apex, subequal in length to the lateral lobes, and the presence of long hairs on the fertile portion of the capsule separate this taxon from all others in this complex. It is otherwise close to *B. masaiensis* of which it appears to be a northern vicariant.

Hepper & Jaeger 6734 from South Horr, **K** 1 has very short, conspicuously glandular outer calyx lobes and a shrubby habit as in *B. hirtifructa* but tends towards *B. argentea* in the calyx lobes being narrowly triangular, widest at the base and in having small anthers; this could be of hybrid origin.

Plants from the vicinity of Isiolo in the southern part of its range differ in having the largest leaves (those from northern populations max. 3 cm long), inflorescences with more flowers (northern populations 1–3-flowered) and corollas towards the smaller end of the size range with small anthers (2–2.5 mm long, northern populations 2.5–3.3 mm long).

61. **Barleria hochstetteri** *Nees* in DC., Prodr. 11: 231 (1847); C.B. Clarke in Fl. Brit. Ind. 4: 483 (1884) & in F.T.A. 5: 157 (1899); F.P.S. 3: 169 (1956); Wood, Handb. Yemen Fl.: 272 (1997); Hedrén in Fl. Somalia 3: 435 (2006); Ensermu in Fl. Eth. 5: 423 (2006). Type: Sudan, Kordofan, *Kotschy* 159 (K! lecto., chosen here; BM!, M!, P!, W!, WAG!, isolecto.)

Subshrub, 10–80 cm tall; stems with dense short white patent or retrorse hairs, patent glandular hairs and appressed medifixed hairs. Leaves oblong-elliptic to -ovate, rarely lanceolate, 2–5.5(–7) cm long, 0.8–2(–3.5) cm wide, base attenuate, apex obtuse, acute or rounded, apiculate, surfaces with evenly distributed appressed medifixed hairs 1–2 mm long, usually also with short patent eglandular and/or glandular hairs, the latter particularly along margin; lateral veins (2–)3 pairs; petiole 2–10(–17) mm long. Inflorescences axillary, 1–3(–7)-flowered, solitary, ± lax; peduncle 8–25(–38) mm long; bracts foliaceous; bracteoles oblanceolate or obovate, 3.5–25 mm long, 0.5–10 mm wide; pedicels 5–15(–27) mm long. Anticous calyx lobe ovate, (4.5–)8–13.5 mm long, (3–)4.5–9.5 mm wide, base rounded to shallowly cordate, apex acute or notched, appressed medifixed hairs often confined to principal veins outside, with short patent eglandular and glandular hairs often many, the latter most dense on margin, venation palmate; posticous lobe (5–)8–17 mm long, (3.5–)6–12 mm wide, apex acute or obtuse, apiculate; lateral lobes lanceolate, (2.5–)3.5–5 mm long. Corolla white, pale blue or mauve, often with purple guidelines, (17–)21–40 mm long, eglandular- and/or glandular-pubescent on lateral lobes outside; tube (8–)11–16 mm long, somewhat expanded towards mouth; limb in "4+1" arrangement; abaxial lobe offset by 4.5–7 mm, (8.5–)10.5–22 mm long, 6.5–14 mm wide; lateral lobes (7–)9–17 mm long, 6–12.5 mm wide; adaxial lobes as lateral pair but 4.5–9.5 mm wide. Stamens attached 1.5–3 mm from base of corolla tube; filaments (13–)18–25 mm long; anthers 2–3(–3.8) mm long; lateral staminodes 0.4–1.5 mm long. Ovary puberulous towards apex; style with few hairs at base; stigma (0.7–)1–1.5 mm long. Capsule 10–13.5 mm long including 4.5–8 mm beak, shortly pubescent; seeds 3–4 mm long and wide.

Kenya. Northern Frontier District: Dela, June 1951, *Kirrika* 76!

Distr. **K** 1; Sudan, Eritrea, Ethiopia, Djibouti, Somalia; Egypt, Yemen, Saudi Arabia & India

Hab. *Acacia-Commiphora* bushland and open dry grassland and scrub, often on red sandy and alluvial soils; ± 400 m

Uses. None recorded on herbarium specimens

Conservation notes. Widespread and locally common in NE Africa and Arabia. It is apparently scarce in N Kenya but may be under-recorded due to limited botanical exploration there. Its favoured habitats remain widespread and often undisturbed: Least Concern (LC).

Syn. [*B. diandra* Nees in DC., Prodr. 11: 231 (1847), *nom. nud.* in syn.]

B. rivae Lindau in Ann. Ist. Bot. Roma 6: 74 (1896), as "*rivaei*". Type: Ethiopia-Somalia border, Daua R., *Riva* 1082 (FT!, holo.)

B. pirottae Lindau in Ann. Ist. Bot. Roma 6: 75 (1896) as "*pirottaei*". Type: Ethiopia, Danna Torrent, *Riva* 819 (FT!, holo.)

Note. This species is closely related to *B. orbicularis* from which separation can be difficult. *B. hochstetteri* usually has a more lax inflorescence with larger flowers and with the outer calyx lobes somewhat narrower, less rounded in the upper half and with a rounded, truncate or at most shallowly cordate base. These calyx characters have most often been used in past keys (e.g. C.B. Clarke in F.T.A.; Hédren in Fl. Somalia) but there is certainly some overlap in the length/width ratio of the lobes. *B. hochstetteri* usually has many short patent eglandular and glandular hairs on the outer calyx lobes whilst these are either absent or confined to the margin in *B. orbicularis*, giving the calyx of *B. hochstetteri* a more velutinous appearance. However, *Bally* 9088a from N of Wajir (**K** 1), which otherwise matches *B. orbicularis*, has rather many short patent hairs on the calyx. Some specimens of *B. hochstetteri* from Sudan (e.g. *Kotschy* 119) have more contracted cymes with shorter pedicels than typical. Such specimens tend also to have corollas towards the lower end of the size range for this species. Even smaller corollas (down to 17.5 mm long but not as small as the 12 mm recorded by Hedrén in Fl. Somalia 3: 435) are recorded in gatherings from NW Somalia and Djibouti; these plants however have very small calyx lobes with a rounded or somewhat obtuse base, very different from *B. orbicularis*. Plants with a similar calyx from **K** 1 are represented by a single specimen of young material (*Gilbert & Thulin* 1590) which is notably very hairy throughout; mature flowering material is desirable from that site to confirm its placement here but the long pedicels are typical of this species. It is notable that in small-flowered plants, the proportions of the corolla are different to *B. orbicularis* with the latter species having shorter, more rounded lobes and a more campanulate tube.

62. **Barleria orbicularis** *T. Anderson* in J.L.S. 7: 29 (1864); C.B. Clarke in F.T.A. 5: 157 (1899); Hedrén in Fl. Somalia 3: 436 (2006); Ensermu in Fl. Eth. 5: 425 (2006). Type: Ethiopia, near Gageros, *Schimper* 2189 (BM!, holo.; BR!, K!, iso.)

Compact subshrub, 20–50 cm tall; stems with dense short white patent hairs and often also short patent glandular hairs, and with many appressed medifixed hairs on the upper internodes. Leaves oblong-elliptic to -obovate, 1.5–5.3 cm long, 0.7–2 cm wide, base attenuate or cuneate, apex rounded, apiculate, surfaces with evenly distributed appressed medifixed hairs 1.3–2.2 mm long, with or without short patent eglandular and/or glandular hairs; lateral veins 3(–4) pairs; petiole 3–10 mm long. Inflorescences axillary, 1–3(–several)-flowered, solitary, congested; primary peduncle 1–8(–13) mm long; bracts foliaceous; bracteoles elliptic or obovate, 3–15(–20) mm long, 0.6–6.5(–8) mm wide, apex often declinate; pedicels 1–5(–8.5) mm long. Anticous calyx lobe broadly ovate (–orbicular), 6.5–18 mm long, 4.5–18 mm wide, base cordate, apex acute, obtuse or shallowly notched, apiculate, with appressed medifixed hairs outside, and appressed simple hairs on the margin, rarely with additional short patent eglandular hairs, short glandular hairs if present largely restricted to the margin, venation palmate; posticous lobe 7–21 mm long, 4.5–18 mm wide, apex acute or obtuse, apiculate; lateral lobes linear-lanceolate, 2.5–5 mm long. Corolla white, pale blue or pink, often with purple guidelines, 15–21 mm long, lateral lobes pubescent outside, adaxial lobes with occasional hairs; tube 9–12 mm long, gradually expanded towards mouth; limb in "4+1" arrangement; abaxial lobe offset by 1.5–4 mm, 5.5–9 mm long, 4–7 mm wide; lateral lobes 4–6.5 mm long, 3.5–5.5 mm wide; adaxial lobes 4–5.5 mm long, 2.5–4.5 mm wide. Stamens attached 2.5–4.5 mm from base of corolla tube; filaments 10–13 mm long; anthers 1.6–2.3 mm long; lateral staminodes 0.3–1.3 mm long. Ovary puberulous distally; style with few hairs at base; stigma 0.7–1.2 mm long. Capsule 9.5–12.5 mm long including 4–6 mm beak, shortly pubescent; seeds 3.5–5 mm long, 3–4.5 mm wide.

KENYA. Northern Frontier District: Dandu, May 1952, *Gillett* 13066! & 17 km SW of Damasa Pan, May 1952, *Gillett* 13329! & 5 km E of Banisa and Derkali roads, May 1978, *Gilbert & Thulin* 1461!
DISTR. **K** 1; Ethiopia, Somalia; Yemen (see note)
HAB. *Acacia-Commiphora* bushland, often on red sandy and alluvial soils; 300–850 m
USES. "Whole plant smoked for healing epilepsy" (**K** 1; *Mwangangi* 5072)
CONSERVATION NOTES. Widespread in lowland eastern Ethiopia and northern and southern Somalia where it is represented by many herbarium collections. Although appearing scarce in northern Kenya it may be under-recorded due to only limited botanical exploration in that region. Its favoured dry *Acacia-Commiphora* bushland habitat remains widespread. Least Concern (LC).

SYN. *B. cardiocalyx* Solms-Laub. in Beitr. Fl. Aeth.: 105 (1867). Type ? as for *B. orbicularis*
B. chlamydocalyx Lindau in Ann. Ist. Bot. Roma 6: 73 (1896). Type: Ethiopia, Bela near Daua R., *Riva* 1464 (FT!, holo.)
B. rotundisepala Rendle in J.B. 35: 376 (1897); C.B. Clarke in F.T.A. 5: 168 (1899). Type: Somalia, Wagga Mt, *Lort Phillips* s.n. (BM!, holo.), but see note

NOTE. Calyx size and indumentum are variable in this species; plants with calyces at the largest end of the range are recorded from Ethiopia (including the type); these tend to have the most deeply cordate base and often lack glandular hairs on the calyx lobes and stems. Plants from Kenya have calyces at the smaller end of the range, with a more shallowly cordate base and with at least some glandular hairs present. Intermediate specimens are common in Ethiopia and Somalia.

Plants recorded as *B. orbicularis* from Yemen (Wood, Handb. Yemen flora: 272 (1997)) have larger corollas than the African material, 23–24 mm long, with the lobes proportionately larger. These plants also have shorter medifixed hairs on the leaves (± 1 mm long) and have fewer hairs on the lateral corolla lobes. This material seems intermediate between *B. orbicularis* and *B. hochstetteri*. Hedrén (in Fl. Somalia 3: 130 (2006)) records corollas to 25 mm long for *B. orbicularis* but it is unclear as to whether or not this includes the Yemen material.

B. rotundisepala is only tentatively placed in synonymy here. The type has small corollas with rounded lobes and outer calyx lobes with a rounded apex as in typical *B. orbicularis* but the calyces are more densely hairy and more shallowly cordate than typical, tending towards *B. hochstetteri*.

SECT. VI. PRIONITIS

(Oerst.) Nees in DC., Prodr. 11: 237 (1847) pro parte; Obermeijer in Ann. Transv. Mus. 15: 129 (1933); M. & K. Balkwill in K.B. 52: 561 (1997)

Prionitis Oerst. in Vidensk. Medd. Dansk. Naturhist. Foren. Kjobenh.: 24 (1854)
Barleria subgen. *Prionitis sensu* C.B. Clarke in F.T.A. 5: 141 (1899) pro parte, excl. *B. hereroensis* and *B. triacantha*

Plants armed with 2–4(–8)-rayed axillary spines, or unarmed; indumentum simple or with some anvil-shaped hairs, or absent. Leaves often yellowish-green or glaucous, lower surface with minute sunken glands throughout, differentiated from the broad sessile glands usually present towards the base beneath. Inflorescences 1–several-flowered dichasial cymes, axillary, then bracts foliaceous, or compounded into terminal synflorescences, then bracts ± modified. Calyx lobes often spine-tipped, margin ± entire. Corolla yellow, orange, scarlet, white or more rarely blue to purple; limb subregular or in "4+1" arrangement. Staminodes 2–3, lateral pair with antherodes present, adaxial staminode, if present, much reduced and antherode absent. Stigma linear or clavate, apex entire or minutely bilobed. Capsule drying (pale) brown, laterally compressed, with a pronounced solid apical beak; lateral walls remaining attached to flanks at dehiscence, septum woody throughout. Seeds 2, ovate-discoid, with dense buff to golden hygroscopic hairs.

A section of ± 45 species, most diverse in NE and East Africa and on Madagascar. The East African species have received very little attention since C.B. Clarke's Flora of Tropical Africa account and as a result many distinct taxa have remained undescribed and several taxonomic errors have persisted.

Although the taxa are more discrete and generally less variable in this section than in, for example, sect. *Fissimura* it is not always clear as to at what rank the different entities should be recognised. Past precedent was set by Brummitt & J.R.I. Wood (in K.B. 38: 436 (1983)) who recognised a very broadly circumscribed *Barleria prionitis* L., with eight subspecies from across tropical Asia and Africa and, in doing so, reduced the rank of several taxa previously recognised as good species. They also suggested that several more subspecies of *B. prionitis* may be recognised within the complex group of taxa in NE Africa following further study. Whilst the eight "subspecies" are clearly closely related, they are easily separable by a range of characters including significant floral differences, and are in the main highly isolated from one another both geographically and ecologically. Furthermore, they seem to be no more similar to one another than they are to many other entities within sect. *Prionitis*. I am therefore of the opinion that at least the five African subspecies of *B. prionitis* are best elevated to species rank, a view first suggested, though not executed, by Ensermu (in Fl. Eth. 5: 414 (2006)). Each already has the species combination formally published with the exception of the East African subsp. *tanzaniana* Brummitt & J.R.I. Wood; the new status for that taxon is therefore published below. It should be noted that if the broader circumscription of *B. prionitis* was followed, at least *B. faulknerae*, *B. maritima*, *B. tanzaniana* and *B. subregularis* would have to be reduced to subspecies status; this would, however, seem unsatisfactory given the significant differences displayed between these taxa.

1. Corolla limb strongly zygomorphic, limb in "4+1" arrangement with abaxial lobe offset by 6–20 mm or if rarely by less (down to 4.5 mm) then this offset at least half as long as the corolla tube* 2
   Corolla limb subregular, lobes splitting from the tube at ± the same point or if the abaxial lobe somewhat offset (for up to 4 mm) then this considerably less than half as long as the corolla tube 13
2. Corolla puberulous to pilose outside at least on the upper portion of the tube and base of the limb 3
   Corolla glabrous outside except occasionally for few hairs on the lobe margins 10

* Corolla not known in sp. F (=*Luke* 4375) but it is presumed to key out in this half of the lead as it is close to *B. rhynchocarpa* which has a strongly zygomorphic corolla.

3. Axillary spines absent or rarely few vestigial spines to 2 mm long present 4
   Axillary spines present, 2–4(–5)-rayed, considerably longer than 2 mm 7
4. Bracts with a dense indumentum of short spreading eglandular hairs and short to long spreading glandular hairs 5
   Bract indumentum less dense: margin and veins strigose, surface with or without subsessile glands and/or short spreading glandular hairs but always lacking short spreading eglandular hairs 6
5. Leaves petiolate, ovate or elliptic, length/width ratio 1.8–2.7/1 .... 63. *B. rhynchocarpa* p.418
   Leaves subsessile, narrowly oblong-elliptic, length/width ratio over 3/1 .... 64. *B. sp. F* p.419
6. Corolla glandular- and eglandular-pilose outside, tube 19–22 mm long; outer calyx lobes with many glandular hairs in the upper half, parallel venation visible .... 65. *B. faulknerae* p.420
   Corolla minutely eglandular- and glandular-pubescent outside, tube 12–17 mm long; outer calyx lobes with few glandular hairs restricted to apical portion or absent, venation inconspicuous 66. *B. maritima* p.421
7. Capsule puberulous; calyx lobes strigose outside; corolla (in our area) with a darker purple-brown to blackish patch at the base of the upper 4 lobes 70. *B. proxima* p.425
   Capsule glabrous; calyx lobes with surface glabrous or with few sessile or stalked glands, eglandular hairs absent or restricted to margin, acumen and/or midrib; corolla lacking dark patch at the base of the upper lobes 8
8. Axillary spines orange or red-brown, stalk inconspicuous, up to 1 mm long, rays 2(–4), often curved; flowers held in a well-defined terminal synflorescence .... 66. *B. maritima* p.421
   Axillary spines grey or whitish, conspicuously stalked for 2–18 mm, rays 4(–5), straight; flowers axillary, clustered towards the stem apices but not forming a well-defined synflorescence 9
9. Bracts and bracteoles with broad sessile glands absent or few present at base; leaves variously linear, oblanceolate or oblong-elliptic, length/width ratio 2.7–17/1 excluding mucro; staminal filaments to 22 mm long .... 74. *B. linearifolia* p.433
   Bracts and particularly bracteoles with conspicuous scattered broad sessile glands; leaves oblong-elliptic to somewhat obovate, length/width ratio 1.6–3.5/1 excluding mucro; staminal filaments ± 30–32 mm long .... 75. *B. sp. G* p.434
10. Abaxial corolla lobe longer than lateral lobes 11
    Abaxial corolla lobe shorter than lateral lobes, usually conspicuously reduced 12

11. Axillary spines harsh and usually many, longest rays 16–38 mm; stems and young leaves with many short spreading hairs; outer calyx lobes to 6.5 mm long, with many short ascending hairs and scattered broad sessile glands outside; corolla limb white . . . . 69. *B. trispinosa* p.424
    Axillary spines smaller and more sparse, longest rays 3.5–17 mm; stems and young leaves lacking short spreading hairs, the former glabrous or upper internodes with crisped hairs on two opposite sides; outer calyx lobes 8–12.5 mm long, glabrous or with few minute subsessile glands towards apex; corolla limb (pinkish-) yellow or orange . . . . 67. *B. tanzaniana* p.422
12. Leaves obovate (-rounded), oblanceolate or (oblong-) elliptic, length/width ratio less than 4/1 excluding mucro; abaxial corolla lobe elliptic or obovate, 2.5–5.5 mm wide; flowers held in a ± well-defined verticillate, spiciform or subcapitate synflorescence, each node with flowers solitary or more often in 3–7-flowered cymes . . . . 71. *B. polhillii* p.426
    Leaves linear-lanceolate to narrowly oblong, length/width ratio 6–16/1 excluding mucro; abaxial corolla lobe lanceolate or subulate, 1–2 mm wide; flowers axillary, solitary . . . . 72. *B. brevispina* p.431
13. Corolla tube glabrous outside, if less than 17 mm long then corolla white, blue or purple; flowers compounded into a well-defined, spiciform or subcapitate terminal synflorescence 14
    Corolla tube hairy outside, usually rather densely and conspicuously so or if sparse and minute (*B. athiensis*) then tube up to 16 mm long and yellow (-brown) or yellow-orange; flowers axillary, often clustered towards stem apices but not forming a well-defined terminal synflorescence 15
14. Corolla white, blue or purple, tube 11.5–20 mm long; synflorescence with a 3–7-flowered dichasium at each axil when mature, the short lateral peduncles of each cyme fused to the subtending bracteole . . . . 73. *B. quadrispina* p.432
    Corolla yellow, orange, salmon-coloured or very rarely white, tube (17–)20–30 mm long; synflorescence with a solitary flower at each axil 77. *B. eranthemoides* p.436
15. Corolla tube 11–18 mm long 16
    Corolla tube 19–33 mm long 17
16. Erect or straggling perennial, 50–150 cm tall; axillary spines sparse or absent, minute, rays to 7 mm long; leaves ovate or elliptic, the largest 5.5–15 cm long, apex subattenuate to acuminate with a minute mucro . . . . 68. *B. subregularis* p.423
    Dwarf subshrub, 10–50 cm tall; axillary spines many and conspicuous, longest rays 15–35 mm long; leaves oblanceolate or obovate, the largest less than 5 cm long, apex acute or obtuse and with a conspicuous mucro . . . . 78. *B. athiensis* p.439

17. Axillary spines either absent or, if present, then rays on a conspicuous stalk (1.5–)4.5–17 mm long; corolla white or cream (rarely ?yellow); largest leaves 4.5–10.5 cm long excluding apical spine, length/width ratio 4–12/1; calyx lobes largely glabrous outside excluding margin ..... 76. *B. marginata* p.435

Axillary spines always present, rays subsessile; corolla yellow or pale orange; largest leaves less than 4.5 cm long excluding apical spine, length/width ratio less than 3.5/1; calyx lobes shortly eglandular-pubescent outside, often also with glandular hairs towards apex 18

18. Much-branched shrub; leaves broadly obovate, length/width ratio less than or up to 2/1; mature capsule 11–14.5 mm long ..... 79. *B. holubii* p.440

Suffruticose perennial, branching mainly at base; leaves oblong-elliptic or oblong-obovate, length/width ratio over 2/1; mature capsule (14-)17–22 mm long ..... 80. *B. penelopeana* p.441

63. **Barleria rhynchocarpa** *Klotzsch* in Peters, Reise Mossamb., Bot. 1: 204 (1861); Vollesen in F.T.E.A. Acanthaceae 1: 139 (2008). Types: Mozambique, Quirimba Is., *Peters* s.n. (B†, holo.); Mozambique, Ibo Is., *Pedro & Pedrogão* 5046 (EA!, neo., chosen here)

Erect or decumbent perennial herb or subshrub, 15–120 cm tall; stems minutely pubescent on two opposite sides or more widespread when young, soon glabrescent. Axillary spines absent. Leaves ovate or elliptic, 6–11.5 cm long, 3–5.8 cm wide, base attenuate, apex acute or obtuse, apiculate, glabrous or margin and veins beneath sparsely strigose, sometimes shortly spreading-pubescent when young; lateral veins prominent beneath; petiole 5–30 mm long. Flowers held in a ± cylindrical terminal synflorescence, 3–10(-15) cm long, each axil 1(–2)-flowered, subsessile; bracts imbricate, (oblong-) elliptic to obovate or lower pairs ovate, 12.5–25 mm long, 5–18 mm wide, apex obtuse or rounded, mucronulate, surface densely pubescent with mixed shorter eglandular and longer glandular hairs, margin often glandular-pilose; bracteoles as bracts but narrowly (elliptic-) lanceolate, 8.5–20 mm long, 1–3.5 mm wide. Anticous calyx lobe oblong-elliptic (–ovate), 12.5–21 mm long, 4.5–8 mm wide, apex acute or subattenuate, often notched, indumentum as bracts, venation subparallel; posticous lobe more ovate, apex often attenuate; lateral lobes linear-lanceolate, 10.5–16 mm long. Corolla yellow, orange or apricot, 45–62 mm long, pilose outside, some hairs minutely gland-tipped; tube 20–28 mm long; limb in "4+1" arrangement; abaxial lobe offset by 9–14 mm, 17–23 mm long, 10–13.5 mm wide; lateral lobes 13.5–18 mm long, 11.5–15 mm wide; adaxial lobes as lateral pair but 8.5–12.5 mm wide. Stamens attached 12–13 mm from base of corolla tube; filaments 22–27 mm long, pubescent in lower two thirds; anthers 3.2–4.3 mm long; lateral staminodes 1.5–3.5 mm long, pilose, antherodes 1–1.3 mm long. Pistil glabrous; stigma 0.7–1 mm long. Capsule 16–20 mm long including 4.5–7 mm beak, glabrous; seeds 9–11 mm long, 6.5–8.5 mm wide.

TANZANIA. Pangani District: Langoni, Mwera, June 1956, *Tanner* 2916!; Uzaramo District: Dar es Salaam, Oyster Bay, June 1968, *Batty* 171!; Kilwa District: Kingupira, Sept. 1976, *Vollesen* in MRC 4000!; Zanzibar, Chukwani, May 1959, *Faulkner* 2268!

DISTR. **T** 3, 6, 8; **Z**; NE Mozambique

HAB. Coastal bushland and grassland, foreshores, lowland riverine forest and thicket; 0–150 m

USES. Cultivated as an ornamental (Zanzibar, *R.O. Williams* 197)

CONSERVATION NOTES. This species is locally frequent and may form large stands on Zanzibar and the adjacent mainland coast in a variety of habitats. Inland, it is found only in dense riparian vegetation. Coastal sites are threatened by habitat destruction through expansion of tourism and growth of coastal towns including Dar es Salaam. Much lowland riverine forest has been cleared in Tanzania but some of the known inland sites for this species are in Forest Reserves and the Selous National Park and so are afforded some protection. It is provisionally assessed as Near Threatened (NT) with more data on distribution and current population trends desirable.

SYN. *Crossandra nilotica* var. *acuminata sensu* C.B. Clarke in F.T.A. 5: 115 (1899) pro parte quoad *Peters* s.n., *non C. nilotica* Oliv. var. *acuminata* Lindau]
B. sacleuxii Benoist in Not. Syst., Paris 2: 17 (1911); U.O.P.Z.: 139 (1949); Brummitt in Bot. Mag. 182, t. 773 (1979); Vollesen in Opera Bot. 59: 79 (1980); M. & K. Balkwill in K.B. 52: 561 (1997) & in Journ. Biogeogr. 25: fig. 5d (1998). Type: Tanzania, Zanzibar, *Sacleux* 545 (P!, holo., BR, EA & K photo.!), **syn. nov.**
Crossandra rhynchocarpa (Klotzsch) Cuf., E.P.A.: 955 (1964) pro parte quoad typus
[*B. ukamensis sensu* da Silva *et al.*, Prelim. Checkl. Vasc. Pl. Moz.: 18 (2004), *non* Lindau]

NOTE. The name *Barleria rhynchocarpa* was synonymised under *Crossandra nilotica* Oliv. by C.B. Clarke but in view of the description of the 4-partite calyx in the protologue this is clearly erroneous (see Vollesen, l.c.). The original Peters specimen was destroyed during World War II. However, the protologue description matches the taxon as circumscribed here and it is notable that the inflorescences of this taxon and *Crossandra nilotica* look very similar in the absence of flowers. The name *Barleria rhynchocarpa* is therefore applied here and is neotypified using the only known extant specimen known from the islands of NE Mozambique.

Brummitt (l.c.) suggested that *B.* (*sacleuxii*) *rhynchocarpa* may prove to be only subspecifically distinct from *B. crossandriformis* C.B. Clarke of S Zimbabwe and South Africa. The two species are vegetatively largely inseparable, although *B. crossandriformis* often has small 2–4-rayed axillary spines, always absent in *B. rhynchocarpa*. However, *B. crossandriformis* has smaller corollas 25–35 mm long (Brummitt recorded the corolla of *B. rhynchocarpa* as (35–)40–56 mm but in the wild-collected specimens seen the mature corollas are not less than 45 mm long) and shortly pubescent, not glabrous, capsules. The two taxa are therefore readily separable and are maintained as specifically distinct here.

64. **Barleria** sp. **F** (= *Luke* 4375)

Herb resembling *B. rhynchocarpa* but differing in: leaves subsessile, narrowly oblong-elliptic, to 7.5 cm long and 2 cm wide, base acute; bracts, bracteoles and calyces with somewhat fewer and uniformly short glandular hairs, the eglandular hairs more ascending particularly on the margins; outer calyx lobes with apex acuminate, anticous lobe bifid for over a quarter of its length. Young corolla bud only seen, eglandular- and glandular-pubescent. Fruits not seen.

KENYA. Teita District: Tsavo Park (East), Sala–Galdessa, June 1995, *Luke & Luke* 4375!

DISTR. **K** 7; not known elsewhere

HAB. Not recorded

USES. None recorded on herbarium specimen

CONSERVATION NOTES. This taxon is clearly scarce, having been recorded only once despite much past collection in the Tsavo area. Further data on the single known population, together with a more complete knowledge of the taxonomy of this species, are required prior to a full conservation assessment though it is likely to prove threatened: Data Deficient (DD).

NOTE. Although only a single immature synflorescence is present on the unicate specimen, the material is sufficient to indicate that it is distinct from *B. rhynchocarpa*. Mature fertile material is required.

65. **Barleria faulknerae** *I. Darbysh.* **sp. nov.** *B. rhynchocarpae* Klotzsch similis sed inflorescentia pilos patentes glandulosos tantum ferenti (nec indumentum densum brevem eglandulatum etiam ferenti), bracteis pro rata angustioribus atque ad apicem acutis vel attenuatis (nec obtusis neque rotundatis), mucrone bracteae magis manifesto, lobo abaxiali corollae 18–19 mm (nec 9–14 mm) magis asymmetrice disposito, staminibus 8.5–9.5 mm (nec 12–13 mm) a basi tubi corollae dispositis differt. Type: Tanzania, Zanzibar, Fumba, *Faulkner* 2977 (K!, holo.)

Erect or decumbent perennial herb; stems largely glabrous or uppermost internodes sparsely and minutely pubescent on two opposite sides, nodes strigose. Axillary spines absent or few nodes with vestigial 2-rayed spine to 2 mm long. Leaves ovate or elliptic, 6–10 cm long, 2.5–4.2 cm wide, base attenuate, apex acute or rarely obtuse, mucronulate, glabrous except margin and veins beneath sparsely strigose; petiole to 10 mm long. Flowers held in a congested, ± conical terminal synflorescence, 2.5–4.5 cm long, each axil single-flowered, subsessile; bracts oblong-obovate (-elliptic), 18–25 mm long, 3.5–9 mm wide, apex acute or shortly attenuate, mucronate, surface strigose on veins and margin and with ± sparse sessile to short-stalked capitate glands, usually stalked on margin; bracteoles as bracts but (elliptic-) lanceolate, 11.5–19 mm long, 1–2.5 mm wide, glandular hairs many particularly towards apex, with occasional short spreading eglandular hairs on margin. Anticous calyx lobe ovate to lanceolate, 12–16 mm long, 4.5–6 mm wide, apex attenuate into mucro, indumentum as bracteoles but lacking appressed hairs, venation parallel; posticous lobe as anticous but 15–17 mm long, mucro longer; lateral lobes lanceolate-attenuate, 12–14.5 mm long. Corolla yellow-orange, 57–61 mm long, pilose outside with many glandular hairs; tube 19–22 mm long; limb in "4+1" arrangement; abaxial lobe offset by 18–19 mm, 22–26 mm long, 9–11.5 mm wide; lateral lobes 19–22 mm long, 12.5–14.5 mm wide; adaxial lobes as lateral pair but 11.5–12.5 mm wide. Stamens attached 8.5–9.5 mm from base of corolla tube; filaments 27.5–30 mm long, pubescent in lower two thirds; anthers 3.5–4.2 mm long; lateral staminodes 2–2.5 mm long, pilose, antherodes 1–1.2 mm long. Pistil glabrous; stigma 0.8–1.2 mm long. Immature capsule only seen.

TANZANIA. Uzaramo District: Dar es Salaam, Oyster Bay, Oct. 1975, *Wingfield* 3167!; Zanzibar, Chukwani, June 1959, *Faulkner* 2291! & Fumba, July 1961, *Faulkner* 2876!

DISTR. **T** 6; **Z**; not known elsewhere

HAB. Foreshore and adjacent coastal bushland on sand; sea level

USES. None recorded on herbarium specimens

CONSERVATION NOTES. This species is apparently restricted to two sites on the coast of Zanzibar and one in Dar es Salaam, where the populations are almost certainly threatened by the expansion of tourism and urbanisation. No recent collections have been made. It is currently considered Endangered (EN B1ab(iii)+B2ab(iii)) with data on current status and threats urgently required.

NOTE. This species is sympatric with both *B. rhynchocarpa* and *B. maritima* in coastal scrub and grows in mixed colonies with the former. Although readily separable from both, it is somewhat intermediate between the two in several characters and may therefore be of hybrid origin, but is treated as a full species here pending further analysis in the field. It differs from *B. rhynchocarpa* in the inflorescence lacking a dense short eglandular pubescence and having fewer and shorter-stalked or sessile glands, in the bracts being proportionally narrower with a more acute or attenuate apex and more pronounced mucro, in the abaxial lobe of the corolla being more strongly offset (18–19 mm, not 9–14 mm) and in the stamens being attached lower down the corolla tube (8.5–9.5 mm from the base, not 12–13 mm). From spineless plants of *B. maritima* it is most readily separated by having many and longer glandular hairs on the calyx and corolla, more oblong bracts which are less strongly narrowed towards the base, more conspicuous venation on the outer calyx lobes, and in the corolla tube being longer (19–22 mm, not 12–17 mm) and the lobes broader (lateral lobes 12.5–14.5 mm wide, not 7–10.5 mm).

66. **Barleria maritima** *I. Darbysh.* **sp. nov.** *B. tanzanianae* (Brummitt & J.R.I. Wood) I. Darbysh. similis sed bracteis atque bracteolis erectis vel ascendentibus (nec patentis), corolla minute hirsuta (nec glabra), lobo abaxiali corollae 14–20 mm (nec 5.5–13.5 mm) magis asymmetrice disposito, antheris 3–4 mm (nec 2.3–3 mm) longis, caulibus superioribus glabris vel pilos minutos patentos (nec pilos crispos plerumque retrorsos) ferentibus differt. Type: Kenya, Kwale District, Twiga, 19 km S of Mombasa, *Verdcourt* 3907 (K!, holo.; EA!, iso.)

Straggling, trailing or erect shrub, (25–)50–250 cm tall; stems largely glabrous or upper internodes with short spreading hairs on two opposite sides, nodes strigose. Axillary spines (rarely absent) orange to red-brown, stalk to 1 mm long, rays 2(–4), often curved, each 1.5–12.5 mm long. Leaves oblong-elliptic to somewhat ovate or obovate, 3–8 cm long, 1.5–3.5 cm wide, base attenuate to obtuse, apex acute to rounded or shallowly emarginate, mucronulate, margin, midrib and veins beneath sparsely strigose; petiole to 12 mm long. Flowers held in a spiciform or subcapitate terminal synflorescence, congested or more lax with age, 1.5–8.5 cm long, each axil 1–3-flowered, subsessile or peduncle to 4.5 mm long; bracts held upright or ascending, oblanceolate, (13.5–)17–35 mm long, 2–12 mm wide, apex mucronate, indumentum as leaves beneath; bracteoles linear or oblanceolate, 8.5–19 mm long, 1–4.5 mm wide. Anticous calyx lobe (ovate-) lanceolate, 8.5–14 mm long, 2.5–6 mm wide, apex (sub)attenuate into mucro, this sometimes shortly bifid, largely glabrous except for short hairs along margin and often few short-stalked glands on acumen, venation inconspicuous; posticous lobe as anticous but mucro longer; lateral lobes lanceolate-attenuate, 7.5–12.5 mm long, with few to many short-stalked glands towards apex. Corolla yellow or orange, 42–59 mm long, minutely eglandular- and glandular-pubescent outside; tube 12–17 mm long; limb in "4+1" arrangement; abaxial lobe offset by 14–20 mm, 18–25 mm long, 4.5–9 mm wide; lateral lobes 13–23 mm long, 7–10.5 mm wide; adaxial lobes as lateral pair but 5.5–10 mm wide. Stamens attached 6.5–9 mm from base of corolla tube; filaments 23–30 mm long, minutely pubescent towards base; anthers 3–4 mm long; lateral staminodes 0.8–1.5(–4) mm long, pilose, antherodes 0.7–1.1(–1.7) mm long. Pistil glabrous; stigma 0.8–1.3 mm long. Capsule 16–20 mm long including 5–7 mm beak, glabrous; seeds 8–8.5 mm long, 5.5–6.5 mm wide.

KENYA. Kwale District: Ukunda, Aug. 1957, *Symes* 189!; Mombasa District: Likoni, Oct. 1933, *Mainwaring* 3045!; Kilifi District: Kilifi, Sept. 1932, *Napier* 2301!

TANZANIA. Uzaramo District: Dar es Salaam, Feb. 1874, *Hildebrandt* 1223! & Msasani, May 1968, *Batty* 63! & Bongoyo Island, Dar es Salaam, July 1969, *Mwasumbi* 10537!; Zanzibar, Prison Island, 1931, *Vaughan* 1403!

DISTR. **K** 7; **T** 6; **Z**; not known elsewhere

HAB. Low bushland and scrub on coral rag and on calcareous sand along the shoreline; ± sea level

USES. "Leaves used as medicine for wasting sick in goats" (**K** 7; *Moggridge* 197)

CONSERVATION NOTES. This species is highly localised in coastal East Africa, but is known historically from many sites. Several locations are likely to be threatened or have been lost due to the expansion of Mombasa, Malindi and Dar es Salaam and the growth of tourism e.g. on Zanzibar. It is currently assessed as Near Threatened (NT).

SYN. *B. setigera* Rendle var. *brevispina* C.B. Clarke in F.T.A. 5: 148 (1899) excl. *Salt* s.n. ex Ethiopia & *Volkens* 482 ex Pangani. Type: Kenya, Mombasa, *Hildebrandt* 2002b (K!, lecto., chosen here; B!, BM!, isolecto.), **syn. nov.**

[*B. prionitis* L. subsp. *tanzaniana* Brummitt & J.R.I. Wood in K.B. 38: 439 (1983) pro parte quoad *Harris* 285]

[*B. setigera sensu* K.T.S.L.: 599 (1994), *non* Rendle]

NOTE. *B. maritima* is closely related to *B. tanzaniana* which it replaces along parts of the East African coast. It differs in the bracts and bracteoles being held upright or ascending and therefore partially enclosing the calyces (not ± spreading and with the calyx clearly visible), the corolla being minutely hairy (not glabrous) with a more strongly offset abaxial lobe (14–20 mm versus 5.5–13.5 mm) and proportionally narrower lobes, the anthers being larger (3–4 mm, not 2.3–3 mm), the hairs of the upper internodes, if present, being minute and spreading (not crisped and typically retrorse), at least the lateral lobes of the calyx bearing short-stalked glands (versus with subsessile glands or glabrous), and the axillary spines often being recurved (versus straight).

Clarke (l.c.) based his *B. setigera* var. *brevispina* upon six syntypes of which four are referable here (*Hildebrandt* 1223 and 2002b; *Stuhlmann* 471 & *Kirk* s.n.). Of the remaining two, *Volkens* 482 from Pangani is referable to *B. polhillii* subsp. *nidus-avis* (described below) whilst *Salt* s.n. from Ethiopia is a variant of *B.* (*prionitis* subsp.) *induta* C.B. Clarke or closely allied taxon. *B. setigera* Rendle *sensu stricto* is synonymous with *B. quadrispina* Lindau and not closely related to this species.

The provenance of *Grote* 4077, labelled as from Kilimanjaro at 2000 m, must be in doubt as this specimen is clearly referable to this otherwise coastal species.

67. **Barleria tanzaniana** (*Brummitt & J.R.I. Wood*) *I. Darbysh.*, **comb. et stat. nov.** Type: Tanzania, Pangani District, 2 km NE of Pangani, *Drummond & Hemsley* 3320 (K!, holo.; B!, BR!, EA!, iso.)

Scandent or spreading subshrub, 40–500 cm tall; stems glabrous or upper internodes with crisped, often retrorse hairs on two opposite sides, nodes often strigose. Axillary spines ± sparse, stalk to 1 mm long, rays 2(–4), straight, longest ray 3.5–17 mm long. Leaves elliptic (-obovate), 4.8–13.5 cm long, 1.7–5.2 cm wide, base cuneate-attenuate, apex acute or attenuate, mucronate, glabrous or margin, midrib and veins beneath sparsely strigose; lateral veins prominent beneath; petiole to 12 mm long. Flowers held in a spiciform terminal synflorescence, initially congested but soon lax, 2–16 cm long, each axil single-flowered, sessile or peduncle to 3 mm long and pedicels to 1.5 mm long; bracts ± spreading, lower pairs foliaceous, soon becoming smaller upwards where oblanceolate, 12.5–32(–42) mm long, 2–11(–14) mm wide, apex acute or shortly attenuate, mucronate, indumentum as leaves beneath, ± few broad sessile glands towards base; bracteoles spinose, 4–14 mm long, 0.4–1.8 mm wide. Outer calyx lobes subequal, ovate or lanceolate, 8–12.5 mm long, 3.5–6.5 mm wide, apex acute or attenuate, mucronulate, anticous lobe often bifidly so, glabrous or with few minute subsessile glands towards apex, venation parallel, often inconspicuous; lateral lobes lanceolate-attenuate, 6.5–10.5 mm long, usually with minute subsessile glands and/or eglandular hairs towards apex. Corolla yellow, orange or apricot, (28–)33–44 mm long, glabrous outside; tube 9–12 mm long; limb in "4+1" arrangement; abaxial lobe offset by (5.5–)7.5–13.5 mm, (13–)17–23 mm long, 7–12 mm wide; lateral lobes as abaxial but (11.5–)15–21 mm long; adaxial lobes as lateral pair but 4.5–8.5 mm wide. Stamens attached 4.5–7 mm from base of corolla tube; filaments 21–29.5 mm long, minutely pubescent in lower half; anthers 2.3–3 mm long; lateral staminodes 1.5–3.5 mm long, pubescent or pilose, antherodes 0.9–1.2 mm long. Pistil glabrous; stigma 0.6–1 mm long. Capsule 14–19 mm long including 4–7 mm beak, glabrous; seeds 6.5–8 mm long, 5–6.7 mm wide.

TANZANIA. Lushoto District: forest bordering Kumba swamp ± 5 km S of Mashewa, June 1953, *Drummond & Hemsley* 3076!; Morogoro District: 32 km on Morogoro–Dar es Salaam road, July 1958, *Tweedie* 1602!; Kilwa District: Kingupira, June 1975, *Vollesen* in MRC 2498!; Pemba, Wete road, km 8, Aug. 1929, *Vaughan* 592!

DISTR. **T** 3, 6, 8; **P** (?cultivated); N Mozambique

HAB. Riverine forest and thicket, open lowland forest and margins, mixed woodland with e.g. *Combretum* or rarely *Brachystegia* woodland, roadsides; 0–700(–1350) m

USES. "For disease which blisters hands and feet: thorns push blisters and leaves wilted on them" (**T** 3; *Tanner* 2000); "used as medicine" (**T** 6; *Kibure et al.* 1006); "leaves crushed and liquid drunk for swollen testicles. Drops put in sore eyes" (**T** 6; *Greenway* 4978); "for cough: leaves cooked and mixed with maize flour and eaten. For syphilis: leaves pounded up, mixed with water and blown into penis" (**T** 3; *Tanner* 3012); "medicinal plant" (**T** 6; *Haerdi* 304/0)

CONSERVATION NOTES. Rather common in a variety of habitats in the lowlands of E Tanzania and tolerant of some disturbance: Least Concern (LC).

SYN. [*B. diacantha sensu* C.B. Clarke in F.T.A. 5: 145 (1899) pro parte quoad *Holst* 3213 & *Stuhlmann* 8530, *non* Nees]
[*B. prionitis sensu* Vollesen in Opera Bot. 59: 79 (1980), *non* L.]
B. prionitis L. subsp. *tanzaniana* Brummitt & J.R.I. Wood in K.B. 38: 439 (1983), excl. *Harris* 285

NOTE. Of the African taxa previously treated as subspecies of *B. prionitis* (see introductory notes to this section), *B. tanzaniana* appears taxonomically closest to the true *B. prionitis* of India but is easily separated by having a glabrous (not pubescent) corolla, an acute or attenuate (not conspicuously acuminate) apex to the outer calyx lobes, finer, often 2-rayed (not coarse, typically 4-rayed) spines, a somewhat more lax synflorescence (although with much overlap between the early stages of flowering in *B. tanzaniana* and mature synflorescences in *B. prionitis*) and longer, proportionally narrower bracts. *B. prionitis* usually dries green whereas *B. tanzaniana* typically dries a darker and colder (blackish-) green especially on the synflorescence.

68. **Barleria subregularis** *I. Darbysh.* **sp. nov.** *B. ameliae* A. Meeuse similis sed lobo abaxiali corollae 0.5–2 mm (nec 4.5–8 mm) vix asymmetrice disposito, calyce apicem versus atque ad marginem tantum hirsuto (nec ubique strigoso), capsula 17–21 mm (nec 14–16 mm) longa differt. Type: Tanzania, Mpwapwa District, 13 km S of Mpwapwa on Gulwe track, *Bidgood, Mwasumbi & Vollesen* 958 (K!, holo.; C!, DSM!, EA!, iso.)

Erect or straggling perennial herb or subshrub, 50–150 cm tall; stems glabrous or upper internodes with few to many minute spreading eglandular and/or glandular hairs, nodes often strigose; mature stems with whitish flaking epidermis. Axillary spines usually restricted to lower stems, stalk 0.8–3.5 mm long, rays (2–)4(–6), shallowly curved, longest ray 1–7 mm long. Leaves ovate or elliptic, 5.5–15 cm long, 2.5–8 cm wide, base attenuate, apex subattenuate to acuminate, mucronulate, glabrous or margin, midrib and veins beneath sparsely strigose; lateral veins prominent beneath; petiole to 13 mm long. Inflorescences axillary on upper stems, each 1–7-flowered; peduncle 1.5–6(–11.5) mm long, partially fused to bract stalk; bracts of lower cymes foliaceous, reduced and narrowing upwards where typically 15–50 mm long, 2.5–12.5 mm wide, apex mucronate, surface sometimes with scattered short-stalked glands; bracteoles linear (-lanceolate) or those of lowermost cymes oblanceolate, 4–19 mm long, 0.3–1.5(–3.5) mm wide, apex mucronate; pedicels 0.5–4 mm long. Anticous calyx lobe (narrowly) lanceolate, 9–16 mm long, 2.5–5.5 mm wide at base, apex acute or attenuate, apiculate, appressed- to ascending-pubescent in upper half and along midrib, ± scattered subsessile or short-stalked glands, venation inconspicuous; posticous lobe 9.5–18 mm long, apex mucronate; lateral lobes somewhat narrower. Corolla yellow or pale orange, 28–31 mm long, pubescent outside, limb usually also with scattered glandular hairs; tube 15–18 mm long; limb weakly in "4+1" arrangement, appearing subregular; abaxial lobe offset by 0.5–2 mm, 10–12 mm long, 8–9 mm wide; lateral lobes 10–13.5 mm long, 7–9.5 mm wide; adaxial lobes as lateral pair but 6–7.5 mm wide. Stamens attached 8–9 mm from base of corolla tube; filaments 13.5–15 mm long, glabrous except for few hairs at base; anthers 2.8–3.8 mm long; lateral staminodes 1–2 mm long, pilose, antherodes ± 1 mm long. Pistil glabrous; stigma 1–1.8 mm long. Capsule 17–21 mm long including 5–7.5 mm beak, glabrous; seeds 8–10 mm long, 5–7 mm wide.

TANZANIA. Mpwapwa District: 1–2 km NE of Gulwe village, Apr. 1988, *Pócs & Pócs* 88062/R!; Kilosa District: Ruaha valley, 15 km N of Mbuyuni on Malolo track, May 1990, *Carter, Abdallah & Newton* 2266!; Iringa District: Udzungwa Mts National Park, May 2002, *Luke et al.* 8376!
DISTR. **T** 5–7; not known elsewhere
HAB. Dry bushland, woodland and wooded grassland with *Acacia, Commiphora, Adansonia, Euphorbia* and/or *Delonix*, on sandy soils and rocky slopes; 600–1100 m
USES. None recorded on herbarium specimens

CONSERVATION NOTES. This is a highly localised species but has been recorded as locally common on Gulwe Mt (*Hornby* 782). Dry woodland remains widespread in the Ruaha Gorge and Gulwe area though is potentially threatened by overgrazing and disturbance, particularly at the former site which is bisected by the main Tanzania–Zambia highway. It is therefore considered Near Threatened (NT).

NOTE. This distinctive species is close to *B.* (*prionitis* subsp.) *ameliae* A. Meeuse, sharing the minute axillary spines, loosely arranged inflorescences on the upper stems and similar bracteole and calyx morphology. It is most readily separated by corolla form: in *B. ameliae* the abaxial lobe is offset from the remaining lobes by 4.5–8 mm and the corolla is therefore strongly zygomorphic whereas in *B. subregularis* the offset is only 0.5–2 mm and the corolla limb is subregular. In addition, *B. subregularis* usually has shortly pedunculate (not sessile) inflorescences, the calyx lobes are hairy only along the midrib and towards the apex (not evenly strigulose throughout) and the capsule is larger (17–21 mm, not 14–16 mm). *B. ameliae* is a species of riverine woodland, usually dominated by *Colophospermum mopane*, recorded from S Malawi to E Namibia and S Zimbabwe.

B. subregularis has a largely identical distribution to *B. aristata* (sect. *Stellatohirta*) in the southernmost extension of Frank White's Somalia-Masai regional centre of endemism (Vegetation of Africa, 1983). The two have been noted growing together (K. Vollesen, *pers. comm.*), although *B. subregularis* is notably less common than *B. aristata*.

69. **Barleria trispinosa** (*Forssk.*) *Vahl*, Symb. Bot. 1: 46 (1790); C.B. Clarke in F.T.A. 5: 146 (1899); F.P.S. 3: 168 (1956); Brummitt & J.R.I. Wood in K.B. 38: 433 (1983); Hepper & Friis, Pl. Forsskål's Fl. Aegypt.-Arab.: 64 (1994); Wood, Handb. Yemen Fl.: 272 (1997); Hedrén in Fl. Somalia 3: 427 (2006), in notes; Ensermu in Fl. Eth. 5: 415 (2006). Type: Yemen, *Forsskål* 376 (C microfiche 59: II. 1–2, lecto., chosen by Brummitt & J.R.I. Wood (l.c.), K!, photo.)

Spiny subshrub, 25–200 cm tall; stems with dense short spreading or retrorse hairs throughout or soon glabrescent, strigose hairs scattered or restricted to the nodes. Axillary spines on stalk 0.7–4 mm long; rays (2–)3–4, longest ray 16–38 mm long. Leaves (broadly) ovate to elliptic, 1.5–9.5 cm long, 0.8–4.5 cm wide, base cuneate to rounded, apex rounded to acuminate, mucronate, surfaces shortly pubescent, hairs often many beneath when young, margin and veins beneath with coarser appressed or ascending hairs; petiole to 10 mm long. Flowers held in a congested spiciform terminal synflorescence, 1.5–4 cm long, each axil single-flowered, subsessile; bracts often imbricate, variously ovate, elliptic, obovate or oblanceolate, 8–23 mm long, 3–16 mm wide, surface sometimes whitish-green towards base, apex mucronate, indumentum as leaves beneath but often with many broad sessile glands towards base; bracteoles triangular to linear-lanceolate, 1–21 mm long, 0.5–2 mm wide, ± rapidly becoming shorter upwards. Calyx largely concealed by bracts; outer lobes ovate (-elliptic), 4–7(–11) mm long, 3–4 mm wide, apex acute or gradually to abruptly narrowed into a mucro, anticous lobe rarely bifidly so, surfaces with many short ascending or spreading hairs and with scattered sessile glands; lateral lobes oblong to lanceolate, 3–6 mm long. Corolla white, yellow, orange or apricot, 40–60 mm long, glabrous or ± sparsely pubescent outside; tube 13.5–18 mm long; limb in "4+1" arrangement; abaxial lobe offset by 11.5–19 mm, 18–25 mm long, 8–10.5 mm wide; lateral lobes as abaxial but 16–24 mm long; adaxial lobes as lateral pair but 6.5–9 mm wide. Stamens attached 7–8.5 mm from base of corolla tube; filaments 22–29 mm long, pubescent towards base; anthers 3.2–4.2 mm long; lateral staminodes 1.7–3 mm long, pubescent, antherodes 0.7–1.1 mm long. Pistil glabrous or largely so; stigma 0.7–1.4 mm long. Capsule 14.5–20 mm long including 5–6.5 mm beak, glabrous or with scattered ascending hairs; seeds 7–8 mm long, 4–5.5 mm wide.

DISTR. (for species) **U** 1; **K** 1–3; Sudan, Eritrea, Ethiopia; Yemen, Saudi Arabia

SYN. *Justicia trispinosa* Forssk., Pl. Aegypt.-Arab.: 6 (1775)

Barleria diacantha Nees in DC., Prodr. 11: 238 (1847); C.B. Clarke in F.T.A. 5: 145 (1899) pro parte excl. *Schimper* 682 & spec. ex Tanzania; F.P.S. 3: 168, fig. 46 (1956). Type: Ethiopia, *Schimper* 1922 (lecto., chosen by Brummitt & J.R.I. Wood in K.B. 38: 433 (1983), location not specified; B!, BR!, K!, M!, MPU! isolecto.)

B. marghilomaniae Volkens & Schweinf. in Ghika, Cinq Mois Pays Somalis: 216 (1898). Type: Ethiopia (fide Hedrén, l.c.), Salul [Sulul], *Ghika* s.n. (B!, holo., K!, photo.)

subsp. **glandulosissima** *I. Darbysh.* **subsp. nov.** a subspecie typica bracteis glandulas sessiles plures basim versus ferenti, folio ad apicem rotundato vel obtuso vel subacuto (nec acuto neque acuminato), indumento caulis magis persistenti, pagina externa corollae semper glabra (nec plerumque sparsim hirsuta), lobis corollae albis (nec flavis neque aurantiis) differt. Type: Uganda, Karamoja District, 26 km W of Moroto, *Langdale-Brown* 1582 (K!, holo.; EA!, iso.)

Stem indumentum persistent. Leaf blade to 4.5 cm long and 2 cm wide, apex rounded to subacute; bracts with many broad sessile glands towards base, obovate (-elliptic) or oblanceolate, the lowermost pairs sometimes ovate. Bracteoles to 14.5 mm long. Outer calyx lobes 4–6.5 mm long, apex ± abruptly mucronate, margin narrowly whitish-green, surface with short ascending hairs. Corolla pure white or with greenish-yellow throat and tube, glabrous outside. Capsule 14.5–16 mm long.

UGANDA. Karamoja District: Suk Hills, Sep. 1955, *J. Wilson* 188! & Nakiloro, Apr. 1960, *J. Wilson* 937! & Moroto R., without date, *Eggeling* 2962!

KENYA. Northern Frontier District: ± 5 km S of Tuum on W side of Mt Nyiru, Oct. 1978, *Gilbert, Gachathi & Gatheri* 5185!; Turkana District: Nyao-Lodwar area, Sept. 1960, *Paulo* 960!; Laikipia District: Colcheccio, Pelagalagi R. valley, Nov. 1993, *Luke* 3381!

DISTR. **U** 1; **K** 1–3; not known elsewhere

HAB. Dry *Acacia-Commiphora* thicket, open rocky slopes and plains with scattered bushes, grazed bushland and grassland; 800–1800 m

USES. No specific uses recorded on herbarium specimens

CONSERVATION NOTES. This subspecies is locally abundant within its restricted range and probably unthreatened as its favoured habitats are sparsely populated by man, with overgrazing being the only potential threat: Least Concern (LC).

SYN. [*B.* sp. Z *sensu* U.K.W.F. ed. 2: 272 (1994)]
[*B.* sp. D *sensu* K.T.S.L.: 599 (1994)]

NOTE. Subsp. *trispinosa* of southern Arabia and NE Africa is a rather variable taxon. It usually has yellow or orange flowers which are sparsely pubescent outside, although occasional glabrous-flowered plants are recorded and a single collection with white flowers has been seen from Saudi Arabia (*Collenette* 365; K). Subsp. *trispinosa* otherwise differs from our plants in having fewer sessile glands on the bracts, the stem often being glabrescent and the outer calyx lobes usually having an acute to more gradually attenuate apex. Leaf and bract shape are variable in subsp. *trispinosa*: in specimens where the leaves are broadly ovate (-elliptic), similar to subsp. *glandulosissima*, the bracts are typically ovate and so readily differentiated from the (elliptic-) obovate or oblanceolate bracts typically present in at least the upper portion of the synflorescence in subsp. *glandulosissima*. However, in the form of subsp. *trispinosa* previously separated as *B. diacantha*, where the leaves are proportionally narrower and attenuate to acuminate at the apex, the bracts are typically more elliptic or somewhat obovate; in such specimens it is the leaves rather than the bracts that differ from subsp. *glandulosissima*. Based on this evidence, it is possible that "*diacantha*" should also be recognised at an infraspecific rank.

70. **Barleria proxima** *Lindau* in Ann. Ist. Bot. Roma 6: 72 (1896); C.B. Clarke in F.T.A. 5: 145 (1899); Milne-Redhead in Hook., Ic. Pl. 33, t. 3292 (1935); Hepper & Friis, Pl. Pehr Forsskål's Fl. Aegypt.-Arab.: 64 (1994); Wood, Handb. Yemen Fl.: 272 (1997); Hedrén in Fl. Somalia 3: 439, fig. 293 (2006) pro parte; Ensermu in Fl. Eth. 5: 418, fig. 166.31 (2006); Pickering & Patzelt, Field Guide Wild Pl. Oman: 89 (2008). Type: Somalia/Ethiopia, Ogaden, *Riva* 366 (B!, lecto., chosen here, K!, photo.; FT, isolecto.)

Spiny subshrub or perennial herb, 30–75 cm tall; stems with many short spreading hairs or sometimes glabrous except for strigose hairs at and beneath the nodes. Axillary spines subsessile or stalk to 2.5 mm long, rays 4, longest ray 7–20 mm long.

Leaves (oblong-) obovate, 2.3–6.2 cm long, 1.2–3 cm wide, base cuneate (-attenuate), apex obtuse, rounded or shallowly emarginate, mucronate, surfaces strigose at least on margin and midrib beneath, often more widespread and many; petiole to 5 mm long. Flowers held in a spiciform terminal synflorescence, 1.5–5.5(–9) cm long, congested or extending with age; each axil 1(–2)-flowered, sessile; bracts falcate, (narrowly) oblong-obovate to -elliptic, 12–26 mm long, 2–12 mm wide, narrowing and more acute into apical spine upwards, strigose mainly on veins and margin and with many broad sessile glands towards base; bracteoles white, linear-lanceolate, 9.5–18 mm long, 0.5–2.5 mm wide, spinose. Calyx whitish towards apex, pale (yellow-) green below; anticous lobe ovate-lanceolate, 8–14 mm long, 3.5–5.5 mm wide at base, apex (sub)attenuate or more abruptly narrowed into a linear, spinose acumen, surface strigose, venation parallel; posticous lobe 10–15 mm long, acumen often more pronounced; lateral lobes (linear-) lanceolate, 8–11.5 mm long. Corolla orange, scarlet or yellow, with a purple-brown to blackish patch at base of upper four lobes, 31–38 mm long, upper tube and limb puberulous outside; tube 11.5–13.5 mm long; limb in "4+1" arrangement; abaxial lobe offset by 8–10 mm, 7.5–12.5 mm long, 3–4.5 mm wide, lateral and adaxial lobes 10.5–16 mm long, 4.5–8.5 mm wide. Stamens attached 8–9.5 mm from base of corolla tube; filaments 16–17 mm long, pubescent in lower half; anthers 2.8–3.5 mm long; lateral staminodes 0.7–1 mm long, pilose, antherodes 0.4–0.7 mm long. Ovary puberulous towards apex; style glabrous; stigma 0.7–1.2 mm long. Capsule 10.5–14.5 mm long including 4–5.5 mm beak, puberulous; seeds 5–6.5 mm long, 3–5.5 mm wide.

KENYA. Northern Frontier District: 24 km SW of Mandera, May 1952, *Gillett* 13403! & Awara Plain, 30 km SSW of Ramu, Dec. 1971, *Bally & Smith* 14595! & Ramu–Banisa road, 28 km from turning to Banisa, May 1978, *Gilbert & Thulin* 1416!

DISTR. **K** 1; Eritrea, Djibouti, Ethiopia, Somalia; Arabia

HAB. Open *Acacia-Commiphora* bushland on sandy or gravelly soils, sometimes over limestone; 350–600 m

USES. None recorded on herbarium specimens

CONSERVATION NOTES. Although scarce in our region, this species is widespread and locally frequent in the Horn of Africa and in Arabia, where it is represented by many herbarium collections. It is tolerant of some disturbance and grazing pressure, often being recorded in open or degraded bushland: Least Concern (LC).

SYN. [*B. smithii sensu* C.B. Clarke in F.T.A. 5: 147 (1899) pro parte quoad *Hildebrandt* 857, *non* Rendle]

[*B. eranthemoides* C.B. Clarke in F.T.A. 5: 147 (1899) pro parte quoad *Deflers* 1179]

B. bicolor Chiov. in Fl. Somala 2: 352, fig. 201 (1932). Type: Somalia, Uanle Uen, *Guidotti* 62 (MOD, holo.)

NOTE. Significant regional variation is recorded in *B. proxima*. In the southern form (which just extends into NE Kenya and to which the above description refers), the leaves and flowers are relatively large, the bracts have many sessile glands outside and the upper four corolla lobes have a dark patch at their base. In N Somalia and S Arabia, the plants are typically more compact with smaller leaves, bracts often lacking sessile glands and small, unicolorous flowers. However, intermediate forms are recorded; for example in populations from Oman the leaves and flowers are similar in size to the southern form but the bracts lack sessile glands and the corolla is uniformly orange.

Lindau listed five *Riva* specimens in the protologue of *B. proxima*. However, Milne-Redhead (l.c.) informally lectotypified this species by referring to the B sheet of *Riva* 366 as the type; this is formalised here.

71. **Barleria polhillii** *I. Darbysh.* **sp. nov.** *B. proximae* Lindau similis sed capsula atque pagina externa corollae glabris (nec puberulis), lobis calycis externis pro rata angustioribus in loco latissimo 2–3.5 mm (nec 3.5–5.5 mm) latis, acumine spinoso loborum magnis manifesto, pagina externa loborum plerumque glabra vel pilos glandulos paucos sparsos ferentibus (nec strigosa) differt. Type: Kenya, Kilifi District, 40 km N of Malindi on the Garsen road, *Polhill & Paulo* 728 (K!, holo.; B!, BR!, EA!, iso.)

Spiny subshrub or perennial herb, 15–180 cm tall; stems glabrous or with short spreading or retrorse white hairs on two opposite sides or more widespread, nodes often strigose. Axillary spines sessile or stalk to 1.5 mm long, rays (2–)4, straight, longest ray 3.5–20 mm long. Leaves obovate (-rounded), (oblong-) elliptic or oblanceolate, 1–8 cm long, 0.5–2.5 cm wide, base cuneate or attenuate, apex acute to rounded, mucronate, sparsely strigose mainly on margin and veins beneath, rarely with short spreading hairs beneath; petiole to 6 mm long or absent. Flowers held in a ± well-defined subcapitate, spiciform or thyrsoid synflorescence, each axil 1–7-flowered, sessile or peduncle to 4 mm long; bracts of lower nodes often foliaceous, gradually to rapidly reduced upwards where (linear-) oblanceolate to spatulate, 8–36 mm long, 1.5–11 mm wide, apex often falcate, spinose, indumentum as leaves or with many broad sessile glands towards base and/or scattered glandular hairs; bracteoles linear-lanceolate or the foremost pair of the dichasium oblanceolate, 5.5–25 mm long, 0.4–3.5 mm wide, short peduncles of lateral cyme branches (if present) fused to subtending bracteoles. Anticous calyx lobe lanceolate, 6–18 mm long, 2–6 mm wide towards base then narrowed into apical spine, this sometimes bifid, glabrous or with minute hairs on margin and acumen, sometimes with scattered glandular hairs, venation inconspicuous; posticous lobe 8–22 mm long; lateral lobes somewhat narrower. Corolla yellow, orange or salmon-pink, 24–44 mm long, glabrous outside or with few hairs towards base of lobe margins; tube 9–18 mm long; limb in "4+1" arrangement; abaxial lobe offset by 4.5–12.5 mm, elliptic or obovate, 6.5–14.5 mm long, 2.5–6.5 mm wide; lateral lobes 10–20 mm long, 5.5–9.5(–12) mm wide; adaxial lobes as lateral pair but 3.5–9(–10.5) mm wide. Stamens attached 5.5–9 mm from base of corolla tube; filaments 13–26 mm long, glabrous or minutely pubescent in lower half; anthers 2.2–3.5 mm long; lateral staminodes 0.8–2 mm long, pilose or sparsely so, antherodes 0.5–1 mm long. Pistil glabrous; stigma 0.5–0.9(–1.3) mm long. Capsule 12–14.5 mm long including 4.5–6 mm beak, glabrous; seeds ± 6.5 mm long, 4.5 mm wide. Fig. 56, p. 428.

NOTE. In both *B. polhillii* and *B. quadrispina*, the lateral peduncles of each dichasium (if developed) fuse with the subtending bracteoles. The bracteoles of the lateral flowers therefore appear to branch directly from those of the central flower; as further lateral branching develops the bracteoles of the central flower can appear to have two orders of branching (fig. 56/2, p. 428).

This species is named in honour of Roger Polhill, eminent field botanist and plant taxonomist in East Africa and former editor of F.T.E.A. It displays significant regional variation and is here divided into four subspecies:

1. Bracts with rather many broad sessile glands towards base; anticous calyx lobe somewhat shorter than the lateral lobes .... c. subsp. **turkanae**
   Bracts with few or no broad sessile glands towards base; anticous calyx lobe subequal to or marginally longer than the lateral lobes 2
2. Synflorescence dense, shortly spiciform to subcapitate, each axil with a 3–7-flowered dichasium developing; bracteoles ascending, 17–25 mm long .... b. subsp. **nidus-avis**
   Synflorescence laxly thyrsiform or if more dense and spiciform or subcapitate then each axil 1–3-flowered; bracteoles straight, 5.5–18 mm long 3
3. Outer calyx lobes 2.5–3.5 mm wide at the widest point; corolla 24–37 mm long .... a. subsp. **polhillii**
   Outer calyx lobes 4–6 mm wide at the widest point; corolla ± 39–44 mm long .... d. subsp. **latiloba**

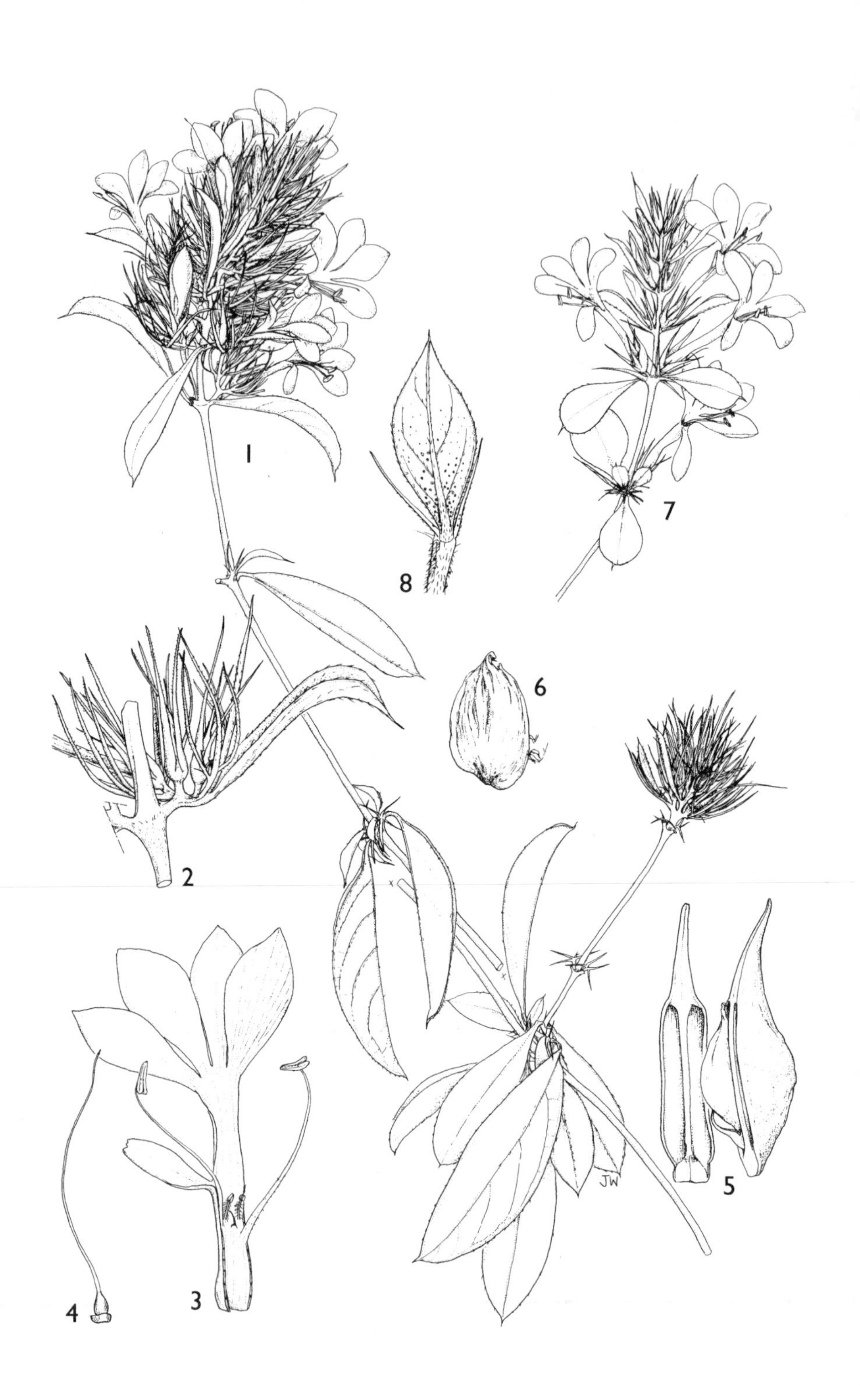
1
2
3
4
5
6
7
8
JW

a. subsp. **polhillii**

Leaves obovate (-rounded) or rarely oblong-obovate, 1.2–4.5 cm long, length/width ratio 1.4–2.6/1. Synflorescence rather lax and thyrsiform or more rarely dense and subcapitate; each axil 1–3-flowered; bracts with no or few broad sessile glands towards base; bracteoles ± straight, 5.5–18 mm long. Anticous calyx lobe 7–13 mm long, 2.5–3.5 mm wide, subequal to lateral lobes; posticous lobe 8–17 mm long. Corolla 24–37 mm long. Fig. 56/7, p. 428.

KENYA. Kilifi District: Garsen road, 43 km N of Galana [Sabaki] R., Sept. 1958, *Moomaw* 905!; Tana River District: Garsen–Malindi road, 1.5 km towards Malindi from the turnoff to Oda, July 1974, *Faden & Faden* 74/1208!; Lamu District: Kui Island, June 1956, *Rawlins* s.n.!

DISTR. **K** 1, 7; E Ethiopia, S Somalia

HAB. Dry woodland and bushland, typically *Acacia-Commiphora*, and associated grassland including grazed areas, usually on sandy soils, also on coral deposits; 0–200(–900) m

USES. None recorded on herbarium specimens

CONSERVATION NOTES. This subspecies is local along the coastal strip of Kenya and S Somalia, extending inland only along the Tana River and in E Ethiopia. It may be under-recorded due to limited collecting in NE Kenya and S Somalia. Suitable habitat remains widespread within its range and it appears tolerant of some grazing disturbance. It is therefore not considered threatened: Least Concern (LC).

SYN. [*B.* sp. A *sensu* Brummitt & J.R.I. Wood in K.B. 38: 440 (1983), in notes]
[*B. prionitis sensu* K.T.S.L.: 599 (1994) pro parte, *non* L.]
[*B. proxima sensu* Hedrén in Fl. Somalia 3: 439 (2006) pro parte including *Kilian & Lobin* 7053, *non* Lindau]
[*B.* sp. (= *Burger* 3373) *sensu* Ensermu in Fl. Eth. 5: 417 (2006) pro parte quoad *Burger* 3373]

NOTE. This taxon has previously been considered close to *B. tanzaniana* (Brummitt & Wood, l.c.) but is readily separated from that species by the smaller, more obovate leaves usually with an obtuse or rounded (not acute or attenuate) apex, the majority of axillary spines being 4-rayed (not 2-rayed), the cymes commonly being 3-flowered at maturity (not always 1-flowered), the outer calyx lobes being narrower and more clearly narrowed upwards, and the abaxial corolla lobe being smaller than (not larger than) the lateral pair. Hedrén (l.c.) placed this taxon within a broadly circumscribed *B. proxima* but it clearly differs from that species in having glabrous capsules and corollas (except occasional hairs on the lobe margins), these both being distinctly hairy in *B. proxima*, and in having narrower outer calyx lobes which lack the strigose indumentum characteristic of *B. proxima*.

Several populations of a taxon similar to this subspecies from Sidamo, Ethiopia (e.g. *Friis et al.* 10926) differ in having white flowers and somewhat longer and narrower leaves and outer calyx lobes. This may be a further subspecies within *B. polhillii* or a closely allied species but requires further investigation.

b. subsp. **nidus-avis** *I. Darbysh.* **subsp. nov**. a subspecie typica synflorescentia magis densa, axilla synflorescentiae 3–7 flores ferenti, bracteolis 17–25 mm (nec 5.5–18 mm tantum) longis et sursum curvis (nec rectis) differt. Type: Kenya, Teita District, Tsavo National Park East, Lugard Falls, Galana R., *Greenway & Kanuri* 12922 (K!, holo.; EA!, iso.)

Leaves oblong-elliptic or somewhat oblanceolate, 3.5–8 cm long, length/width ratio 2.3–3.8/1. Synflorescence dense, shortly spiciform to subcapitate, each axil with a 3–7-flowered cyme (but often only 1 or 3 flowers developed at any given time); bracteoles upcurved, 17–25 mm long. Anticous calyx lobe 11–17 mm long, 2–3.5 mm wide, subequal to or slightly longer than the lateral lobes; posticous lobe 14–22 mm long. Corolla 31–41 mm long. Fig. 56/1–6, p.428.

FIG. 56. *BARLERIA POLHILLII* subsp. *NIDUS-AVIS* — **1**, habit, × $^{2}/_{3}$; **2**, single cyme with subtending bract, showing how lateral peduncles fuse to the subtending bracteoles, × 1.5; **3**, dissected corolla with androecium, abaxial lobe to the right, × 1.5; **4**, pistil, × 1.5; **5**, capsule, × 3; **6** seed, × 3. *B. POLHILLII* subsp. *POLHILLII* — **7**, synflorescence, × $^{2}/_{3}$. *B. POLHILLII* subsp. *TURKANAE* — **8**, bract and bracteoles, with details of the sessile glands on the former, × 3. 1 from *Hucks* 36; 2 from *Dale* 3890; 3 & 4 from *Greenway & Kanuri* 12922; 5 & 6 from *Polhill* 330; 7 from *Polhill & Paulo* 728; 8 from *Popov* 1564. Drawn by Juliet Williamson.

KENYA. Machakos District: Kitani Hill, Mtito Andei, Voi km 17, Mar. 1969, *Napper & Jones* 1975!; Teita District: Mazinga Hill, near Voi, Jan. 1953, *Bally* 8612!; Tana River District: Kora Reserve, Dec. 1982, *van Someren* 882!

TANZANIA. Masai District: Engaruka road, Feb. 1970, *Richards* 25558!; Lushoto District: Kivingo, Jan. 1930, *Greenway* 2043!; Pangani District: near Pangani, July 1906, *Jaeger* 117!

DISTR. **K** 1, 4, 6, 7; **T** 2, 3; not known elsewhere

HAB. Dry woodland and bushland, typically of *Acacia-Commiphora*, and associated grassland, on sandy soils or granitic outcrops; 200–1050 m

USES. None recorded on herbarium specimens

CONSERVATION NOTES. This subspecies is locally common within its restricted range, particularly within the Tsavo National Park where its favoured open bushland habitat is widespread and afforded some protection. It is not considered threatened: Least Concern (LC).

SYN. [*B. setigera* var. *brevispina sensu* C.B. Clarke in F.T.A. 5: 148 (1899) pro parte quoad *Volkens* 482, *non* lectotype (see *B. maritima*)]

[*B.* sp. A *sensu* U.K.W.F. ed. 1: 593, fig. p. 592, without label (1974) & *sensu* Blundell, Wild Fl. E. Afr.: 387 (1987), excl. pl. 239; & *sensu* U.K.W.F. ed. 2: 272 (1994); & *sensu* K.T.S.L.: 600 (1994)]

[*B. eranthemoides sensu* Blundell, Wild Fl. E. Afr.: pl. 238 (1987), excl. description p. 387, *non* C.B. Clarke]

NOTE. Where the ranges of subsp. *polhillii* and subsp. *nidus-avis* meet along the Tana River in Kenya, they appear to intergrade with several intermediate specimens seen. Plants of subsp. *polhillii* from southern Somalia also have a more congested synflorescence, thus tending towards subsp. *nidus-avis*. It is therefore considered most appropriate to treat the two taxa as only subspecifically distinct, although over most of their range they are easily separable.

c. subsp. **turkanae** *I. Darbysh.* **subsp. nov.** a subspecie typica bracteis glandulas sessiles plures basin versus ferenti, lobis lateralibus lobum anticum superantibus (nec subaequantibus) differt. Type: Kenya, Turkana District, Lodwar, *Popov* 1564 (K!, holo.; EA!, iso.)

Leaves elliptic (-obovate), 1–4 cm long, length/width ratio 1.5–2.5/1. Synflorescence spiciform; each axil 1–3-flowered; bracts with many broad sessile glands towards base; bracteoles ± straight, 9–15 mm long. Anticous calyx lobe 6–9.5 mm long, 2–3 mm wide, somewhat shorter than the lateral lobes; posticous lobe 9–13.5 mm long. Corolla 32–38 mm long. Fig. 56/8, p. 428.

KENYA. Northern Frontier District: Marsabit Dist., Dec. 1982, *Sato* 727!; Turkana District: Labur hills, Lokitaung, Apr. 1932, *Champion* T.27! & Lokitaung, May 1953, *Padwa* 215!

DISTR. **K** 1, 2, ?3; not known elsewhere

HAB. Rocky hillsides with open *Acacia*, *Commiphora*, *Terminalia* and/or *Combretum* bushland and scrub including heavily grazed areas, sometimes on sand; 400–950 m

USES. "Camel fodder" (**K** 2; *Champion* T15)

CONSERVATION NOTES. This taxon is largely restricted to the dry hills around Lake Turkana, being currently known from only 10 specimens. It may be locally common here as Champion recorded it as "general throughout [the] area" at Lokitaung. However, little is known of its habitat requirements or current abundance and threats: Data Deficient (DD).

NOTE. The combination of many sessile glands towards the base of the bracts and the largely glabrous calyx in this subspecies may lead to confusion with *B. punctata* Milne-Redh. of E Ethiopia and N & C Somalia. That species however differs in having conspicuously stalked axillary spines and in the abaxial lobe of the corolla being considerably broader and offset from the remaining lobes by only 4.5–6 mm. Some specimens of *B. punctata* (including the type) have very short calyx lobes that lack an acumen; however, in other specimens the lobes are acuminate, though the posticous lobe is always shorter than in *B. polhillii*.

d. subsp. **latiloba** *I. Darbysh.* **subsp. nov.** subsp. *polhillii* similis sed lobis calycis exterioribus latioribus 4–6 mm (non 2.5–3.5 mm) latis et corolla majore 39–44 mm (non 24–37 mm) longa differt. Type: Kenya, Embu District: 24 km N of Ishiara on road to Meru, *Gilbert, Kanuri & Mungai* 5715 (K!, holo.; EA!, iso.)

Leaves obovate (-elliptic), 2.5–3.2 cm long, length/width ratio ± 2.4/1. Synflorescence spiciform or becoming more lax and thyrsiform on the main branches; each axil 1(–3)-flowered; bracts with no or few broad sessile glands towards base; bracteoles ± straight, 13–18 mm long. Anticous calyx lobe 14–18 mm long, 4–6 mm wide, subequal to lateral lobes; posticous lobe 15–20 mm long. Corolla ± 39–44 mm long.

KENYA. North Nyeri/Fort Hall District: Naro Moru–Fort Hall road, Mar. 1933, *Jex-Blake* 3262!; Fort Hall District: Thika–Murang'a [Fort Hall] road, Jan. 1931, *Hill* 2530!; Kitui District: Mutomo Hill, May 1965, *Bally* 12800!

DISTR. **K** 4; not known elsewhere

HAB. Open *Harrisonia-Acacia-Grewia* thicket; bushland and grassland over rocky soils; 650–1850 m

USES. None recorded on herbarium specimens

CONSERVATION NOTES. This subspecies is very local in the dry grassland and bushland in the plains and foothills around Mt Kenya, with only the four specimens cited known to the author, despite this area having been well botanised. Bally however recorded it as common at Mutomo Hill. This part of Kenya is rather heavily populated, with extensive cattle ranching in the grasslands. However, this taxon seems tolerant of grazing pressure, being recorded from overgrazed thicket by Gilbert *et al.* (5715). Based upon the evidence currently available, a provisional assessment of Vulnerable (VU D2) is given, largely on account of its scarcity. However, more data on its current distribution and possible threats are desirable.

SYN. [*B. proxima sensu* U.K.W.F. ed. 1: 593 (1974) & ed. 2: 272 (1994), *non* Lindau]

NOTE. Although superficially similar to subsp. *polhillii*, this taxon is easily separated by the larger calyces and by the fact that the lateral flowers of the cyme, if developed, are sessile, not shortly pedunculate with the peduncles fused to the bracteoles. In the absence of flowers, this subspecies is difficult to separate from *B. eranthemoides* var. *agnewii* but the corolla is completely different.

72. **Barleria brevispina** (*Fiori*) *Hedrén* in Willdenowia 36: 758 (2006) pro parte, & in Fl. Somalia 3: 438 (2006) pro parte (see note). Type: Somalia, Bay Region, between Uenèio and Butiài, *Paoli* 793 (FT!, holo.)

Erect or scandent subshrub, 25–100 cm tall; stems glabrous or upper internodes minutely puberulent on two opposite sides. Axillary spines sometimes sparse, stalk (0.5–)1–3 mm long, rays (3–)4, straight, longest ray 4–15 mm long. Leaves subsessile, linear-lanceolate or narrowly oblong, 3.8–10 cm long, 0.3–1.2 cm wide, base cuneate, apex acute, mucronate, lower surface strigose at least on margin and midrib, hairs simple and/or biramous, sometimes also scattered above; midrib pale and prominent beneath. Inflorescences axillary, sometimes restricted to the uppermost axils, flowers solitary, subsessile or peduncle to 2.5 mm long; bracts foliaceous; bracteoles pale grey-green, linear-lanceolate, 9.5–19 mm long, 1–2 mm wide, apex spinose, glabrous or midrib strigose. Calyx pale grey-green, whitish with age; anticous lobe lanceolate, 14–19 mm long, 3–4 mm wide towards base, apex long-acuminate, spinose, surface glabrous except for occasional strigose hairs on acumen, venation inconspicuous except for prominent midrib on acumen; posticous lobe 15–23 mm long; lateral lobes linear-lanceolate, 13–18 mm long, margin often with few minute glandular hairs. Corolla pale to bright yellow, 30–46 mm long, glabrous outside; tube 7.5–13 mm long; limb in "4+1" arrangement; abaxial lobe offset by 10–14.5 mm, lanceolate or subulate, 3–9 mm long, 1–2 mm wide; lateral lobes 10.5–20 mm long, 5.5–7.5 mm wide; adaxial lobes marginally smaller. Stamens attached 5–7 mm from base of corolla tube; filaments 18–26 mm long, with few hairs towards base; anthers 2.2–3.2 mm long; lateral staminodes 1.2–1.7 mm long, pilose towards base, antherodes 0.7–1 mm long. Pistil glabrous; stigma 0.6–1 mm long. Capsule 12–14 mm long including 4.5–6 mm beak, glabrous; only immature seeds seen.

KENYA. Northern Frontier District: Lag Ola, 45 km W of Ramu on Banisa road, June 1952, *Gillett* 13410! & 30 km on the Ramu–Malka Mari road, May 1978, *Gilbert & Thulin* 1592!

DISTR. **K** 1; Ethiopia, Somalia

HAB. *Acacia-Commiphora* woodland on limestone slopes; 400–450 m

USES. None recorded on herbarium specimens

CONSERVATION NOTES. Although apparently scarce in our region, this species may be more widespread in the under-collected limestone areas of NE Kenya than currently known and it appears locally frequent in the limestone region of southern Somalia. Its favoured habitat of dry rocky bushland remains widespread and largely unthreatened except for potential overgrazing: Least Concern (LC).

SYN. *B. linearifolia* Rendle var. *brevispina* Fiori in Result. Sci. Miss. Stef.-Paoli, Coll. Bot.: 139 (1916) [*B.* sp. (= *Burger* 3373) *sensu* Ensermu in Fl. Eth. 5: 417 (2006) pro parte, including *Ellis* 226]

NOTE. Plants from the northern part of the range (C & N Somalia, E Ethiopia) often have proportionally broader, oblong-elliptic leaves and sometimes broader outer calyx lobes with a more conspicuous pale margin. Kenyan plants agree with the form from S Somalia and S Ethiopia with narrow leaves and calyx lobes; only this form is treated in the above description.

In his delimitation of this species, Hedrén (l.c.) included plants from NE Somalia with a stunted habit (e.g. *Thulin & Warfa* 6146) which, however, clearly differ from *B. brevispina* in having a subregular corolla, the abaxial lobe barely offset from and approximating the size of the remaining lobes. In *B. brevispina sensu stricto* the strongly offset and tiny, narrow abaxial lobe is characteristic. The NE Somalian plants are currently being described as *B. compacta* Malombe & I. Darbysh.

73. **Barleria quadrispina** *Lindau* in Ann. Ist. Bot. Roma 4: 72 (1896); C.B. Clarke in F.T.A. 5: 147 (1899) pro parte excl. *Donaldson-Smith* s.n. ex Turfa; Milne-Redhead in Hook., Ic. Pl. 33, t. 3293 (1935); Hedrén in Fl. Somalia 3: 438, fig. 292 (2006); Ensermu in Fl. Ethiopia & Somalia 5: 416, fig. 166.30 (2006). Type: Ethiopia, Harar, *Robbechi* 10 (B!, lecto., chosen here, K!, photo; FT, isolecto.)

Compact spiny subshrub, 8–40 cm tall; stems largely glabrous or with many short spreading white hairs, nodes often strigose. Axillary spines subsessile or stalk to 12 mm long, rays 4(–6), straight or ascending, longest ray 10–29 mm long. Leaves oblanceolate, obovate or narrowly oblong-elliptic, 3.3–8 cm long, 0.8–2.4 cm wide, base cuneate or attenuate, apex acute to rounded, mucronate, strigose at least on the margin, often also on the veins beneath, sometimes with many short ± spreading hairs beneath; petiole to 9 mm long or absent. Flowers held in a ± dense spiciform or subcapitate terminal synflorescence, 2–6.5(–10) cm long, each axil 3–7-flowered, subsessile or peduncle to 2(–4.5) mm long; bracts of lower axils foliaceous, becoming smaller upwards where green to whitish, (linear-) oblanceolate, 12–40 mm long, 1.5–7 mm wide, apex spinose, margin and midrib with coarse appressed to spreading hairs with or without a bulbous base, surface sometimes with patent glandular hairs and/or few broad sessile glands towards base; bracteoles erect or ascending, linear (-lanceolate) or oblanceolate, 9–32 mm long, 0.5–3 mm wide, short peduncle of lateral cyme branches fused to subtending bracteoles; flowers subsessile or pedicel to 3 mm long. Anticous calyx lobe 5.5–14.5 mm long, basally ovate or lanceolate where 1.5–3(–4) mm wide, then abruptly or rarely gradually narrowed into a flexible spine, this sometimes bifid, glabrous to densely eglandular- and/or glandular pubescent, with or without coarse, ± bulbous-based hairs along margin, venation inconspicuous; posticous lobe 6.5–17 mm long; lateral lobes 5–12.5 mm long. Corolla white, blue or purple, 20–36 mm long, glabrous outside; tube 11.5–20 mm long; limb weakly in "4+1" arrangement or subregular; abaxial lobe offset by 0.5–4 mm, 10–17 mm long, 4.5–12 mm wide; lateral lobes as abaxial but 4–10.5 mm wide; adaxial lobes 3–8 mm wide. Stamens attached 4.5–10 mm from base of corolla tube; filaments 16–23 mm long, shortly pubescent at least in lower half; anthers 1.5–2.5(–3) mm long; lateral staminodes 1–2.7 mm long, pubescent, antherodes 0.5–0.9 mm long. Pistil glabrous; stigma 0.4–0.7 mm long. Capsule 9–11 mm long including 3–3.5 mm beak, glabrous; seeds 3.7–5 mm long, 2.5–4 mm wide.

KENYA. Northern Frontier District: 18 km SW of Mandera, Dec. 1971, *Bally & Smith* 14586! & 55 km SSW of Ramu, Dec. 1971, *Bally & Smith* 14596! & 48 km on Ramu–El Wak road, May 1978, *Gilbert & Thulin* 1628!

DISTR. **K** 1; Ethiopia, Somalia

HAB. *Acacia-Commiphora* bushland and dense or open thicket, on sandy, gravelly or stony soils including over limestone; 250–650 m

USES. None recorded on herbarium specimens

CONSERVATION NOTES. Although scarce in our region, this is a widespread and often common species in Ethiopia and Somalia where it is found in both open and dense bushland. It is not considered threatened: Least Concern (LC).

SYN. *B. setigera* Rendle in J.B. 34: 395 (1896); C.B. Clarke in F.T.A. 5: 148 (1899) pro parte excl. var. *brevispina*. Type: ?Somalia, Darar, *Donaldson-Smith* s.n. (BM!, holo.)
B. setigera Rendle var. *pumila* Rendle in J.B. 34: 396 (1896). Type: ?Somalia, Okoto, *Donaldson-Smith* s.n. (BM!, holo.)
B. waggana Rendle in J.B. 35: 377 (1897); C.B. Clarke in F.T.A. 5: 148 (1899); M. & K. Balkwill in Journ. Biogeogr. 25: fig. 5c (1998). Type: Somalia, Wagar [Wagga] Mt, *Lort-Phillips* s.n. (BM!, holo.)
B. glaucobracteata Hedrén in Willdenowia 36: 758 (2006) & in Fl. Somalia 3: 441 (2006). Type: Somalia, Gedo Region, 7–8 km S of Luuq, *Somali Medicinal Plant Project* SMP 210 (UPS, holo.; K!, iso.), **syn. nov.**

NOTE. *B. quadrispina* is a polymorphic species with significant regional variation in indumentum, leaf shape, calyx length and corolla size and colour. However, these characters appear to vary independently of one another, resulting in many inter-related forms. Plants from SW Somalia described by Hedrén as *B. glaucobracteata* have strikingly pale bracts, bracteoles and calyces and have flowers at the largest end of the size range but are clearly allied to other southern forms of this species and are treated as synonymous here; this form is also recorded in NE Kenya (*Bally & Smith* 14586).

The names *B. quadrispina* and *B. setigera* were both published in 1896, the latter in September. The month of publication is not recorded for the former, though the manuscript is marked as having been submitted for publication in December 1895. It is therefore probable that the name *B. quadrispina* has priority.

Milne-Redhead (l.c.) informally lectotypified this species by referring to the B sheet of *Robbechi* 10 as the type; this specimen is formally selected as the lectotype here. *B. quadrispina* was described from sterile material but *Robbechi* 10 clearly matches the species as circumscribed here.

74. **Barleria linearifolia** *Rendle* in J.B. 34: 397 (1896); Hedrén in Fl. Somalia 3: 441 (2006); Ensermu in Fl. Eth. 5: 419 (2006) pro parte excl. *B. tetraglochin* Milne-Redh. (fig. 166.32). Type: Somalia, Turfa, *Donaldson-Smith* s.n. (BM!, holo.)

Spiny subshrub, 15–60 cm tall or straggling to 200 cm tall; stems largely glabrous or upper internodes sometimes minutely pubescent on two opposite sides, nodes strigose. Axillary spines on stalk 2–18 mm long, rays 4, longest ray 6.5–18 mm long. Leaves subsessile, linear, narrowly oblong-elliptic or oblanceolate, 1.2–11 cm long, 0.3–1.2 cm wide, base acute or cuneate, margin thickened and ± paler than blade, apex acute, mucronate, strigose beneath at least on margin and midrib, often more widespread when young, glabrescent; midrib prominent beneath. Inflorescences axillary but mainly clustered at stem apices, flowers solitary, subsessile; bracts foliaceous; bracteoles often clasping calyx, pale grey-green, lanceolate, 12.5–29 mm long, 1.5–5.5 mm wide, apex spinose, midrib and sometimes margin sparsely strigulose. Calyx pale grey-green, whitish with age; anticous lobe ovate to lanceolate, 10.5–24 mm long, 3–6 mm wide, apex (sub)attenuate, spinose or bifidly so, margin strigulose, elsewhere glabrous or midrib with few hairs, venation ± parallel, sometimes only midrib prominent; posticous lobe 12–26 mm long; lateral lobes linear-lanceolate, 7.5–15 mm long, minutely hairy including some glandular hairs. Corolla yellow, orange, apricot or white, (28–)34–59 mm long, upper tube and base of limb shortly pubescent outside; tube 14.5–26 mm long; limb in "4+1" arrangement; abaxial lobe offset by 6–14 mm, 11.5–16 mm long, 2.5–8.5 mm wide; lateral lobes (7.5–)11–19 mm long, 7–13 mm wide; adaxial lobes as lateral pair but 5.5–11 mm wide. Stamens attached 10.5–15 mm from base of corolla tube; filaments 13–22 mm long, shortly pubescent in lower half; anthers 2.5–3.5 mm long; lateral staminodes 1–2.5 mm long, pilose, antherodes 0.4–1.1 mm long. Pistil glabrous; stigma 1–1.8 mm long. Capsule (12.5–)14.5–16 mm long including 4–5.5 mm beak, glabrous; seeds ± 8.5 mm long, 6 mm wide.

UGANDA. Karamoja District: Kanamugit, without date, *Eggeling* 2929! & Kangole, Dec. 1957, *J. Wilson* 413! & near Moroto, June 1962, *Miller* 576!
KENYA. Northern Frontier District: Furroli, Sep. 1952, *Gillett* 13847!; Meru District: Isiolo, Feb. 1953, *Gillett* 15140!; Tana River District: Hola, Dec. 1972, *Robertson* 1816!

DISTR. **U** 1; **K** 1, 2, 4, 7; SE Sudan, S Ethiopia, Somalia

HAB. *Acacia-Commiphora* bushland and lightly wooded grassland on seasonally inundated black clays, drier red sandy soils or amongst rocks; 30–1350 m

USES. "Preferred food item of camels" (**K** 1; *C.R. Field* 111)

CONSERVATION NOTES. This species is rather widespread but scattered in NE Africa, being most common in N and E Kenya. Its dry or seasonally inundated bushland habitats remain widespread and are not considered highly threatened: Least Concern (LC).

SYN. [*B. quadrispina sensu* C.B. Clarke in F.T.A. 5: 147 (1899) pro parte quoad *Donaldson-Smith* s.n., *non* Lindau]

B. quadrispina Lindau var. *linearifolia* (Rendle) Chiov. in Fl. Somala: 259 (1929)

[*B.* sp. B *sensu* U.K.W.F. ed. 2: 272 (1994)]

NOTE. This species is typically a compact subshrub to 60 cm tall with relatively small flowers, the anticous calyx lobe < 20 mm long and corolla 28–45 mm long. Plants from the Tana River valley (e.g. *Lucas* 26; *Robertson* 1816), however, are straggling to 90–200 cm tall and have larger flowers with the anticous calyx lobe ≥ 20 mm long and corolla over 50 mm long; this form also has the longest, most linear leaves. It is perhaps worthy of subspecific status but more material is desirable to assess the consistency of these differences.

Ensermu (l.c.) treated *B. tetraglochin* Milne-Redh. from central Sudan as a synonym of *B. linearifolia*. However, from the very limited material of *B. tetraglochin* available, it appears to be pyrophytic in habit and the corolla has a less strongly offset and broader abaxial lobe. In addition, the bracteoles have bulbous-based marginal hairs in *B. tetraglochin*, but these are absent in *B. linearifolia*. It is maintained as distinct here, though may ultimately warrant only subspecific status.

B. linearifolia is largely inseparable from *B. marginata* in the absence of flowers (see below). The latter tends to branch mainly from the base (*B. linearifolia* is more shrubby), have many short spreading hairs on the stem and more conspicuous, sometimes bulbous-based coarse hairs on the calyx margin. These characters are, however, not wholly diagnostic and flowers should be sought for reliable identification: *B. marginata* has a subregular, salver-shaped corolla very different to the "4+1" arrangement of *B. linearifolia*.

75. **Barleria** sp. **G** (= *Hepper & Jaeger* 7006)

Compact spiny subshrub, 10–80 cm tall; stems largely glabrous except nodes strigose, upper internodes sometimes minutely pubescent on two opposite sides. Axillary spines on stalk 2.5–15 mm long, rays 4(–5), longest ray 9–23 mm long. Leaves subsessile, oblong-elliptic or somewhat -obovate, 2.2–4.5 cm long, 0.9–2 cm wide, base acute or cuneate, margin somewhat thickened and ± paler than blade, apex obtuse, rounded or acute, mucronate, strigose beneath at least on margin and midrib, sparse at maturity; midrib and sometimes lateral veins prominent beneath. Inflorescences axillary but clustered at stem apices, flowers solitary, subsessile; bracts foliaceous, often with scattered broad sessile glands; bracteoles pale grey-green at maturity, lanceolate, 12.5–26 mm long, 1.5–5.5 mm wide, apex spinose, midrib and margin strigulose, surface with conspicuous scattered broad sessile glands. Calyx pale grey-green, whitish with age; anticous lobe (ovate-) lanceolate, 14–23 mm long, 3.5–8.5 mm wide, apex spinose, margin strigulose or glabrous, surface often with scattered broad sessile glands, venation ± parallel; posticous lobe 14–26 mm long; lateral lobes linear-lanceolate, 10–16 mm long. Corolla yellow-brown or pale orange, ± 50 mm long, upper tube and base of limb shortly pubescent outside; tube ± 19 mm long; limb in "4+1" arrangement; abaxial lobe offset by 12.5–16 mm, ± 18–21 mm long, 5 mm wide; lateral lobes ± 17 mm long, 11–13 mm wide; adaxial lobes as lateral pair but ± 7.5 mm wide. Stamens attached ± 12 mm from base of corolla tube; filaments ± 30–32 mm long, sparsely and shortly pubescent in lower third; anthers 3.5–4.2 mm long; lateral staminodes ± 3 mm long, pilose, antherodes ± 0.8 mm long. Pistil glabrous; stigma ± 0.8 mm long. Capsule 15–18 mm long including 4.5–6 mm beak, glabrous; seeds 8–9 mm long, 6–6.5 mm wide.

KENYA. Northern Frontier District: 5 km from Lotaes, Lake Turkana [Rudolf], Sep. 1944, *Adamson* 127 in *Bally* 3972! & Lake Turkana, E side 30 km S of Loyengalani, Nov. 1978, *Hepper & Jaeger* 7006! & Mt Kulal, Dec. 2001, *Congdon* 611!

DISTR. **K** 1; not known elsewhere

HAB. Dry, rocky "lava desert", amongst scrub in dry river-beds; 350–850 m

USES. "The decoction of the roots is used as a remedy for stomach pains. Grazed by donkeys, sheep, goats, camels and grant's [gazelles]" (**K** 1; *Mwangangi* 1509); "Eaten by camels and donkeys" (**K** 1; *Adamson* 568)

CONSERVATION NOTES. This taxon is highly localised in the region around southern Lake Turkana and Mt Kulal. Its dry, rocky habitat is unlikely to be threatened, but more data are required on its abundance together with clarity over its taxonomic status: Data Deficient (DD).

NOTE. This taxon is close to *B. linearifolia*. It differs in the proportionally broader leaves (length/width ratio 1.6–2.6(–3.5)/1, not 3–17: 1) with a broader apex, in having conspicuous broad sessile glands on the bracts, calyces and particularly the bracteoles (these absent or restricted to few at the base of the bracts and bracteoles in *B. linearifolia*) and in having strikingly long stamens. The two taxa grow together near Loyengalani (*Hepper & Jaeger* 7006 and 7006a) and are both recorded from Mt Kulal. More material of this group is needed from north Kenya to fully assess the status of this taxon.

76. **Barleria marginata** *Oliv.* in Trans. Linn. Soc., Bot. 29: 127, t. 128 (1875); C.B. Clarke in F.T.A. 5: 160 (1899); Vollesen in Opera Bot. 59: 79 (1980); M. & K. Balkwill in K.B. 52: 561, figs. 6Q & 11K (1997). Type: Tanzania, Morogoro District, Matamombo, *Speke & Grant* s.n. (K!, holo.)

Perennial herb with few to many erect or decumbent stems from a woody rootstock, 3–30 cm tall; stems minutely pubescent, hairs often concentrated on two opposite sides, sometimes sparse, strigose at and sometimes below the nodes. Axillary spines absent to many, stalk (1.5–)4.5–17 mm long, rays (2–)4–8, longest ray 4–21 mm long. Leaves subsessile, linear, narrowly elliptic or oblanceolate, 4.5–10.5 cm long, 0.4–2 cm wide, base acute or cuneate, margin thickened and paler than blade, apex acute or obtuse, mucronate, sparsely strigose at least on margin and midrib; midrib prominent beneath. Inflorescences axillary but often clustered towards stem apices, each 1(–3)-flowered, subsessile; bracts foliaceous; bracteoles pale grey-green, (linear-) lanceolate, 14–30 mm long, 1.5–5 mm wide, apex spinose, margin and midrib strigose. Calyx pale grey-green, whitish with age; anticous lobe (oblong-) ovate to lanceolate, 12–23 mm long, 3–8 mm wide, apex attenuate, spinose, margin with coarse ascending, often bulbous-based hairs, elsewhere largely glabrous, venation ± parallel; posticous lobe as anticous but apical spine longer; lateral lobes linear-lanceolate, 7.5–17 mm long. Corolla white or cream, rarely yellow(?),31–49 mm long, rather densely pubescent outside on upper tube and base of limb; tube narrowly cylindrical, 24–33 mm long; limb subregular, lobes widely spreading; abaxial lobe 7.5–15 mm long, 4.5–10 mm wide; lateral lobes 8–17 mm long, 5–11.5 mm wide; adaxial lobes 7–12.5 mm long, 4–7.5 mm wide. Stamens attached 13–21 mm from base of corolla tube; filaments 10.5–13.5 mm long, pubescent in lower two thirds; anthers wholly or only partially exserted, 2–2.8 mm long; lateral staminodes 1.5–3.5 mm long, pilose, antherodes 0.6–0.9 mm long. Pistil glabrous; stigma linear, 0.8–0.9 mm long. Capsule 11.5–17 mm long including 4–5.5 mm beak, glabrous; seeds ± 8 mm long, 5.5–6.5 mm wide.

KENYA. Northern Frontier District: 26 km NE of Habaswein, Apr. 1978, *Gilbert & Thulin* 1122!; Kwale District: Taru desert between Voi and Samburu, Aug. 1961, *Tweedie* 2197! & Kaya Puma, July 2000, *Luke, Mbinda & Mududu* 6346!

TANZANIA. Pare District: Same–Kisiwani road, 16 km E of Same, Feb. 1953, *Drummond & Hemsley* 1328!; Morogoro/Bagamayo District: Dakawa–Sadani, Dec. 1915, *Peter* 60362!; Kilwa District: ± 10 km SW of Kingupira, Dec. 1975, *Vollesen* in MRC 3114!

DISTR. **K** 1, 4, 7; **T** 3, 6, 8; S Ethiopia

HAB. Open *Acacia-Commiphora* and *Terminalia-Spirostachys* bushland and lightly wooded grassland, on seasonally inundated black clay or drier sandy or stony soils; 100–650 m

USES. None recorded on herbarium specimens

CONSERVATION NOTES. This is a rather widespread but patchily distributed and uncommon species. In Ethiopia and Tanzania it appears largely restricted to bushlands and grasslands that are seasonally inundated but some of the Kenyan populations may occur in permanently dry areas. Grazing pressure is perhaps the most likely threat, but it is not currently considered threatened: Least Concern (LC).

SYN. [*Barleria* sp. "dry ground, 7°10'S Oct. 1860" *sensu* Thomson in Speke, Journ. Disc. source Nile, Append.: 643 (1863)]

[*B. eranthemoides* C.B. Clarke in F.T.A. 5: 147 (1899) pro parte quoad *Smith* s.n. ex Kilimanjaro]

[*B.* sp. (= *Mooney* 5618) *sensu* Ensermu in Fl. Eth. 5: 420 (2006)]

NOTE. In southern populations (**T** 6, 8) the axillary spines are sparse and small (stalk to 8.5 mm long) or absent whilst in Kenyan and Ethiopian material they are many and generally larger (stalk 4.5–17 mm long). This variation is possibly correlated with grazing pressure and appears insufficiently consistent to be of taxonomic significance. Plants from the Wajir and Garissa areas of Kenya (e.g. *Stannard & Gilbert* 1050) are notable for having somewhat broader and thinner leaves and in the axillary spines often being 5–8-rayed (usually 4-rayed in other populations). However, they are otherwise inseparable from material from further south.

Populations from Sidamo, Ethiopia are recorded from a higher altitude (1450–1600 m), though in similar habitat. Here, the corollas are smaller than in our region, the tube 24–26 mm long (not 29–33 mm) and the lobes 7–8 mm long (not 8.5–17 mm long).

77. **Barleria eranthemoides** *C.B. Clarke* in F.T.A. 5: 147 (1899) pro parte quoad *Schimper* 111, 759, 2194 & 1685, *Salt* s.n., *James & Thrupp* s.n., *Lort-Phillips* s.n., *Cole* s.n. & *Hildebrandt* 2388; F.P.S. 3: 168 (1956); Heine in F.W.T.A. ed. 2, 2: 421 (1963); F.P.U.: 139 (1971); U.K.W.F. ed. 1: 593, fig. p. 592 (1974); Blundell, Wild Fl. E. Afr.: 387 (1987) pro parte excl. pl. 238; U.K.W.F. ed. 2: 272, pl. 118 (1994); M. & K. Balkwill in K.B. 52: 561, figs. 2N, 5B, 11D, 11S & 11U (1997); Hedrén in Fl. Somalia 3: 441 (2006); Ensermu in Fl. Eth. 5: 420, fig. 166.33 (2006). Type: Ethiopia, *Salt* s.n. (BM!, lecto., **chosen here**)

Spiny subshrub, often forming dense clumps of erect, straggling or trailing stems, 15–150 cm tall; stems glabrous except for inconspicuous minute hairs in opposite furrows of upper internodes, nodes strigose. Axillary spines subsessile, rays (3–)4, longest ray 2–25 mm long. Leaves (oblong-) obovate, oblong-elliptic or oblanceolate, 2–11.5 cm long, 0.8–3.8 cm wide, base attenuate or cuneate, apex rounded to acute, mucronate, sparsely strigose at least on margin and midrib; petiole 2–16 mm long. Flowers held in a dense spiciform to subcapitate terminal synflorescence, 1.5–5 cm long, flowers solitary at each axil, sessile; bracts (linear-) lanceolate, oblong or shallowly pandurate, 10–35 mm long, 1.8–9.5 mm wide, apex mucronate, indumentum as leaves but hairs often more, those on margin sometimes ± spreading and bulbous-based, often with scattered broad sessile glands particularly in lower half; bracteoles (linear-) lanceolate, 7.5–24 mm long, 0.5–3 mm wide, attenuate into apical spine, margin or entire surface pale. Outer calyx lobes subequal, posticous lobe partially enveloping anticous lobe, each 10–22 mm long, 3.5–8 mm wide at base where (ovate-) elliptic to suborbicular, narrowed into a spinose acumen, this rarely bifid on anticous lobe, margin sometimes pale and subhyaline, minutely ciliate,

FIG. 57. *BARLERIA ERANTHEMOIDES* var. *ERANTHEMOIDES* — **1**, habit, × $^{2}/_{3}$; **2**, detail of axillary spine, × 2; **3**, bract, outer surface, × 1.5; **4**, bracteole, × 2; **5**, calyx, four lobes separated, outer surfaces, posticous lobe on the left, × 3; **6**, dissected corolla with androecium, × 1.5; **7**, detail of staminodes and staminal filament bases, × 4; **8**, capsule prior to dehiscence, × 2; **9**, capsule valve, front and side views, × 3; **10**, seed, × 3. *B. ERANTHEMOIDES* var. *AGNEWII* — **11**, bract, outer surface, × 2; **12**, calyx, four lobes separated, outer surfaces, posticous lobe on the left, × 2. 1 from *Mlangwa & Bariki* 1347; 2 & 8–10 from *Verdcourt* 1526; 3–5 from *Dawkins* 650; 6 & 7 from *Richards* 25473; 10 & 11 from *Polhill & Paulo* 961. Drawn by Juliet Williamson.

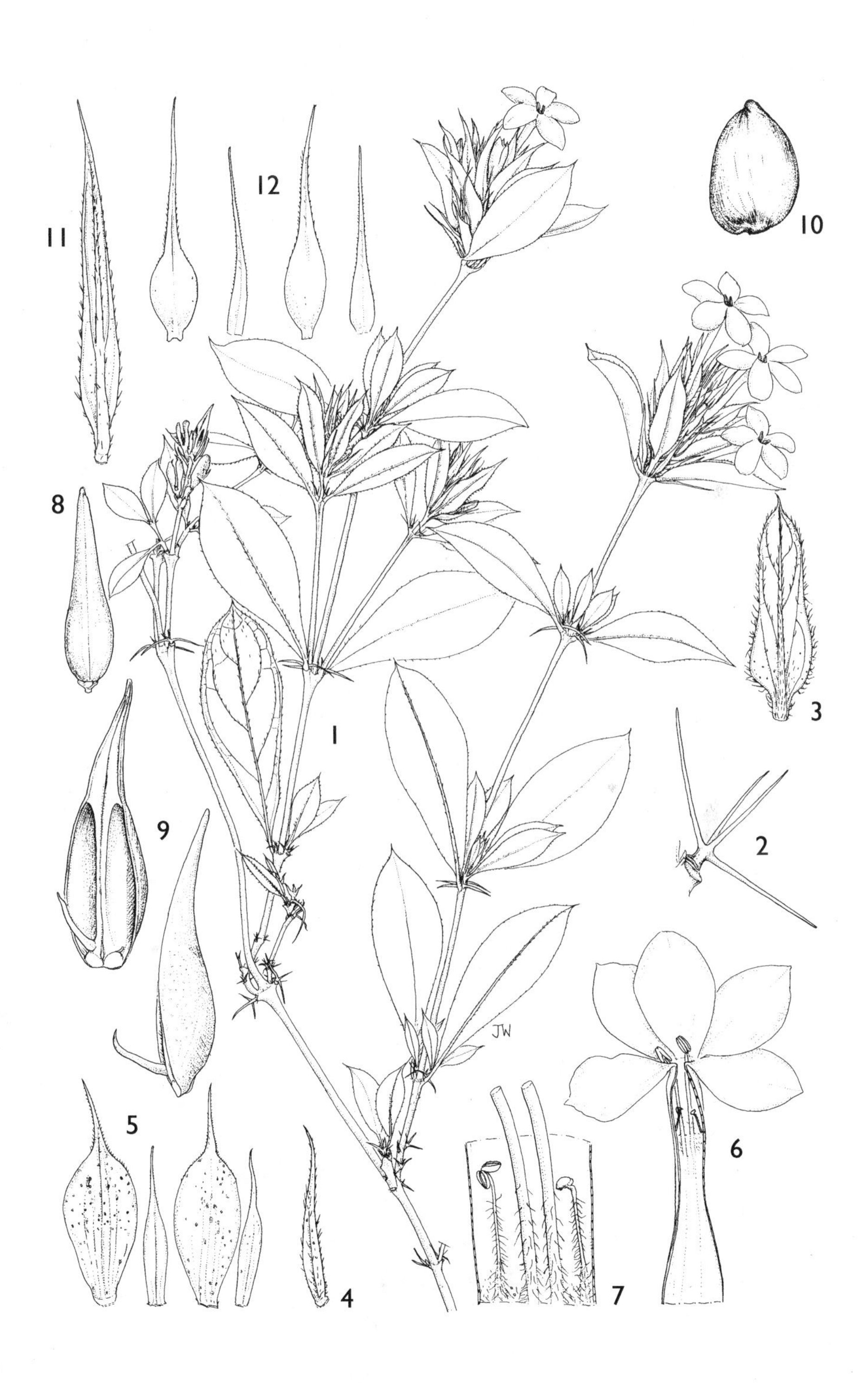
11
12
10
8
1
3
9
2
5
6
4
7
JW

elsewhere glabrous, venation inconspicuous; lateral lobes lanceolate-attenuate to linear-lanceolate, 7–19 mm long. Corolla pale yellow, orange or salmon-pink, rarely white, (27–)31–44 mm long, glabrous outside except sometimes for minute hairs on basal margin of lobes; tube narrowly cylindrical, (17–)20–30 mm long; limb subregular, lobes widely spreading; abaxial and lateral lobes 9–14 mm long, 4.5–8.5 mm wide; adaxial lobes marginally smaller. Stamens attached (10–)12–18 mm from base of corolla tube; filaments (7.5–)9–17 mm long, sparsely pubescent towards base; anthers wholly or only partially exserted, 1.7–2.8 mm long; lateral staminodes 1–5.5 mm long, pilose, antherodes 0.4–1 mm long. Pistil glabrous; stigma 0.4–1.1 mm long. Capsule 11–16 mm long including 4–6.5 mm beak, glabrous; seeds 5.8–6.5 mm long, 3.8–5 mm wide. Fig. 57, p. 437.

a. var. **eranthemoides**

Bracts oblong, subpandurate or lanceolate, length/width ratio 2–5.5(–7)/1, margin with ± spreading or ascending, often bulbous-based hairs. Outer calyx lobes green with paler hyaline margin, this often irregular, apex ± abruptly narrowed into a spinose acumen 1.5–7.5 mm long. Fig. 57/1–10, p. 437.

UGANDA. Karamoja District: Loyoro, Nov. 1939, *A.S. Thomas* 3144! & 1.6 km from Kotido on Loyoro road, Sept. 1950, *Dawkins* 650! & Lodoketemit catchment, July 1958, *Kerfoot* 420!
KENYA. Northern Frontier District: Marsabit, Jan. 1961, *Polhill* 348!; Kitui District: Ndui–Voo, Aug. 1969, *Kimani* 58!; Masai District: Chyulu Plains, below Ol Donyo Wuas camp, June 1991, *Luke* 2842!
TANZANIA. Masai District: Olduwai Gorge, Ngorongoro, Feb. 1966, *Herlocker* 273!; Pare District: Kisiwani, Jan. 1936, *Greenway* 4579!; Lushoto District: 8 km on Mkomazi to Mombo road, Apr. 1988, *Bidgood & Vollesen* 1268!
DISTR. **U** 1; **K** 1–7; **T** 2, 3, 5; Nigeria, Cameroon, Sudan, Ethiopia, Somalia
HAB. Open dry bushland and grassland on sandy soils or on rocky slopes, often becoming abundant in disturbed and overgrazed areas; 400–1850 m
USES. "Eaten by camels" (**K** 1; *Hussein* 21); "food for all animals" (**K** 1; *Itani* 78/44); "eaten by donkeys and rhino" (**K** 6; *Glover & Samuel* 2898); "very important coloniser of bare soil" (**T** 2; *St Clair-Thompson* 207A)
CONSERVATION NOTES. A widespread and often abundant taxon, highly tolerant of disturbance: Least Concern (LC).

SYN. *B. hypocrateriformis* T. Anderson in J.L.S. 7: 28 (1863), *nom. nud.*
[*B. prionitis sensu* T. Anderson in J.L.S. 7: 28 (1863) pro parte quoad *Schimper* 2194, *non* L.]
B. cephalophora Lindau in E.J. 38: 70 (1905) & in E.J. 43: 353 (1909). Type: Tanzania, ?Masai District, Ngirimasi, *Merker* 802 (B!, holo, K!, photo.; cited without number in E.J. 38: 70), **syn. nov**.
[*B.* sp. A *sensu* Blundell, Wild Fl. E. Afr.: pl. 239 (1987), *non* description p. 387]

NOTE. C.B. Clarke (l.c.) informally selected the type as *Salt* s.n. in his notes to the species as he was basing his *B. eranthemoides* upon R. Brown's earlier *nomen nudum*; this specimen is formally chosen as the lectotype here.

Barleria prionitis L. var. *setosa* Klotzsch from Goa Is., Mozambique was reduced to synonymy within *B. eranthemoides* by Clarke (F.T.A. 5: 147), but it is clearly distinct in having, amongst other differences, a shortly hairy corolla tube and lanceolate outer calyx lobes with conspicuous venation and bulbous-based marginal hairs. Specimens from Malawi listed in the protologue of *B. eranthemoides* are referable to *B.* (*prionitis* subsp.) *ameliae* A.Meeuse.

b. var. **agnewii** *I. Darbysh.* **var. nov**. a varietate typica bracteis pro rata angustioribus atque ad marginem pilis basaliter bulbosis carentibus, acumine loborum externorum calycis magis gradatim attenuato 8–15 mm (versus 1.5–7.5 mm) longo differt. Type: Kenya, Kilifi District, Magalini, 5 km E of Marafa, *Luke & Robertson* 2472 (K!, holo.; EA, MO, US, iso.)

Bracts linear-lanceolate, length/width ratio (5.5–)7–12.5/1, margin with appressed or ascending hairs lacking a bulbous base. Outer calyx lobes grey-green, turning whitish with age, margin not paler or hyaline, entire, apex more gradually narrowed into a spinose acumen 8–15 mm long. Fig. 57/11, 12. p. 437.

KENYA. Nairobi District: 24 km E of Nairobi, Jan. 1952, *Bogdan* 3390!; Teita District: Sagala Hill, S of Voi, Dec. 1961, *Polhill & Paulo* 961!; Kilifi District: near Jaribuni Primary School, Kauma, June 1973, *Musyoki & Hansen* 990!

TANZANIA. Pare District: Lembeni, June 1926, *Peter* 41692! & Kiruru, Apr. 1927, *Haarer* 467! & Mkomazi Game Reserve, Kiholo area, Apr. 1995, *Abdallah & Vollesen* 95/56!

DISTR. **K** 4, 7; **T** 2, 3; not known elsewhere

HAB. Open and more dense dry bushland, grassland, often in degraded areas and/or on poor, rocky soils; 25–1800 m

USES. "Medicine" (**K** 7; *Karissa* 49)

CONSERVATION NOTES. This variety is locally common within its restricted range and apparently tolerant of or favouring some disturbance: Least Concern (LC).

SYN. [*B. eranthemoides sensu* C.B. Clarke in F.T.A. 5: 147 (1899) pro parte quoad *Hildebrandt* 2388]
[*B. prionitis sensu* C.B. Clarke in F.T.A. 5: 145 (1899) pro parte quoad *Wakefield* s.n., *non* L.]
[*B.* sp. Q *sensu* U.K.W.F. ed. 1: 593 (1994) & U.K.W.F. ed. 2: 272 (1994) pro parte]

NOTE. This is a very distinct variety, restricted to our region where it is partially sympatric with var. *eranthemoides*. It may warrant full species status, though the leaves, corolla and fruits are largely identical in the two taxa. It is named in honour of Andrew Agnew, esteemed field botanist in Kenya and the first to recognise this as a distinct taxon in his Upland Kenya Wild Flowers (but see note to *B. athiensis*).

78. **Barleria athiensis** *I. Darbysh.* **sp. nov.** *B. eranthemoidi* C.B. Clarke similis sed tubo corollae 11–15 mm (nec (17–)21–30 mm) longo et breviter hirsuto (nec glabro), lobis calycis ad apicem acutis vel attenuatis (nec acumine recto conspicuo ornatis) bracteis non multo foliorum dissimilibus (nec conspicue dissimilibus) differt. Type: Kenya, Machakos District, Athi Plains, *Goyder, Masinde, Meve & Whitehouse* 4003 (K!, holo.; EA, iso.)

Very spiny dwarf subshrub, 10–50 cm tall, much-branched; stems minutely pubescent in two opposite furrows when young, nodes strigose. Axillary spines on stalk 1–4.5 mm long, rays (2–)4(–5), longest ray 15–35 mm long. Leaves oblanceolate to obovate, 1.5–4.7 cm long, 0.5–1.4 cm wide, base cuneate or attenuate, margin somewhat thickened and paler than blade, apex acute or obtuse, mucronate, surfaces glabrous or margin and/or principal veins beneath sparsely strigose; lateral veins ± prominent beneath; petiole to 3.5 mm long or absent. Flowers solitary in the upper axils, together forming a terminal cluster on main and short lateral branches; bracts foliaceous but often reduced, typically 10–20(–26) mm long, 1.5–4(–6.5) mm wide, uppermost pairs sometimes whitsh-green towards base or throughout; bracteoles whitish-green, linear-lanceolate, 10–15(–18) mm long, 0.5–1.5 mm wide, apex spinose; flowers subsessile. Calyx pale grey-green, whitish with age; anticous lobe lanceolate, 9.5–13 mm long, 2–4 mm wide at base, apex acute or attenuate, bifidly spinose, outer surface glabrous or with minute hairs on margin, venation ± parallel, two principal veins often prominent; posticous lobe 10–16 mm long, apex spinose; lateral lobes as anticous but linear-lanceolate. Corolla yellow (-brown) or yellow-orange, 19–26 mm long, upper portion of tube sparsely and minutely pubescent outside; tube narrowly cylindrical, 11–16 mm long; limb subregular; abaxial and lateral lobes 8–12 mm long, 5–7 mm wide; adaxial lobes 7.5–10.5 mm long, 4–5.5 mm wide. Stamens attached 5.5–8.5 mm from base of corolla tube; filaments 9–10 mm long, shortly pubescent; anthers exserted, 2.5–3.5 mm long; staminodes 2(–3), lateral pair 1.2–2 mm long, pilose, antherodes 0.6–1 mm long. Pistil glabrous; stigma 0.9–1.3 mm long. Capsule 13.5–17 mm long including 4–7 mm beak, glabrous; seeds ± 7–8 mm long, 5–6.5 mm wide.

KENYA. Machakos District: SW foot of Mua Hills, Jan. 1965, *Gillett* 16600! & Athi R. on Nairobi–Mombasa road, Feb. 1969, *Napper & Faden* 1840!; Masai District: Kajiado, Feb. 1953, *Drummond & Hemsley* 1245!

DISTR. **K** 3, 4, 6; not known elsewhere

HAB. Open *Acacia-Commiphora* bushland, particularly in areas of stunted growth including *Acacia drepanolobium* bushland, grassland including pasture, on stony hillsides and seasonally inundated sand or black clay; 1500–2000 m

USES. None recorded on herbarium specimens

CONSERVATION NOTES. This species has a highly restricted distribution in central and southern Kenya but has been recorded as locally abundant in Machakos District (*Gillett* 16600). It appears tolerant of some grazing pressure. However, high human population growth and the resultant intensification of agriculture within its range is likely to have led to losses of some populations. This species is therefore currently considered Near Threatened (NT) but may prove to be Vulnerable when updated distribution data become available.

SYN. [*B.* sp. Q *sensu* U.K.W.F. ed. 1: 593 (1994) & U.K.W.F. ed. 2: 272 (1994) pro parte (see note)]

NOTE. Previously considered a form of *B. eranthemoides* but consistently differing in the shorter and minutely hairy corolla tube, the less clearly defined synflorescences with the bracts not clearly modified from the leaves, the lanceolate calyx lobes lacking a prominent acumen, the staminal filaments being hairy for most of their length, not just in the basal portion, the axillary spines being shortly stalked not subsessile and in the more stunted and shrubby habit.

Although the description given for *B.* sp. Q by Agnew in U.K.W.F. (l.c.) clearly refers to *B. eranthemoides* var. *agnewii*, the two specimens cited in ed. 1 are referable here.

79. **Barleria holubii** *C.B. Clarke* in Fl. Cap. 5: 47 (1901); Obermeijer in Ann. Transv. Mus. 15: 146 (1933); M. & K. Balkwill in K.B. 52: 561, figs. 6P, 11E, 11L & 11M (1997). Type: South Africa, Transvaal, Marico District, *Holub* s.n. (K!, holo.)

Very spiny shrub forming dense, often rounded clumps, 15–150 cm tall; stems with short white spreading or retrorse hairs, evenly and densely distributed or concentrated in two opposite lines. Axillary spines subsessile, rays 2–4, longest ray 10–23 mm long. Leaves broadly obovate, 0.8–3 cm long, 0.6–1.7 cm wide, base attenuate or cuneate, apex rounded or abruptly narrowed, mucronate, shortly spreading-pubescent at least beneath when young, with longer ascending or appressed hairs on margin and principal veins beneath; petiole to 3 mm long or absent. Inflorescences axillary towards stem apices, flowers solitary, subsessile; bracts foliaceous; bracteoles spinose, (8.5–)12–21 mm long, bracteoles of old inflorescences persisting. Anticous calyx lobe lanceolate, 8.5–13.5 mm long, 2–3 mm wide at base, apex spinose or bifidly so, surface with white spreading or retrorse hairs, with or without coarser ascending hairs particularly along margin and short patent glandular hairs towards apex, venation parallel; posticous lobe 9–14 mm long; lateral lobes somewhat narrower. Corolla yellow or pale orange, 30–37 mm long, upper portion of tube and limb rather densely pubescent outside, limb with or without occasional glandular hairs; tube narrowly cylindrical, 20–26 mm long; limb subregular, lobes widely spreading; abaxial lobe 6.5–11.5 mm long, 5–6.5 mm wide; lateral lobes 7–12 mm long, 5–6 mm wide, adaxial lobes as abaxial but 3.5–5.5 mm wide. Stamens attached 13–17 mm from base of corolla tube; filaments 8–9.5 mm long, pubescent in lower half; anthers wholly or only partially exserted, 1.9–2.5 mm long; lateral staminodes 0.6–1.5 mm long, pubescent, antherodes 0.6–1 mm long. Pistil glabrous; stigma ± 0.5 mm long. Capsule 11–14.5 mm long including 3–4.5 mm beak, glabrous; seeds ± 7 mm long, 5 mm wide.

DISTR. (for species) **U** 1, **K** 2; W Zimbabwe, E Botswana, N South Africa

subsp. **ugandensis** *I. Darbysh.* **subsp. nov.** a subspecie typica pilis grossis ascendentibus e lobis calycis carentibus, lobis calycis pilos stipitatos glandulos numerosos in dimidio superiore ferentibus differt. Type: Uganda, Acholi District, Ukuti, Chua, *Eggeling* 2418 (K!, holo.; BR!, iso.)

Stems densely hairy throughout. Anticous calyx lobe 8.5–11 mm long, posticous lobe 9–12 mm long, indumentum of short white spreading to retrorse eglandular hairs throughout and many patent glandular hairs towards apex. Corolla with occasional glandular hairs on limb outside.

UGANDA. Karamoja District: Karamoja Plains, without date, *Brasnett* 170! & Loyoro, Aug. 1958, *Wilson* 590! & Loyoro–Kaabong, Aug. 1960, *Wilson* 1049!
KENYA. Turkana District: without precise locality, Oct. 1982, *Ohta* 280!
DISTR. **U** 1; **K** 2; not known elsewhere
HAB. Dry rocky outcrops and plains, dry woodland; 1050–1500 m
USES. Used as a hedge plant (**U** 1; *Liebenberg* 130)
CONSERVATION NOTES. This species is recorded from two small and highly disjunct areas in both of which it is recorded as locally common; but has rarely been collected. Its favoured habitat of rocky hillslopes amongst large boulders has limited agricultural potential and this species is potentially protected from over-grazing by its coarse spines. Therefore, despite its limited range it is probably unthreatened: Least Concern (LC); this assessment applies equally to subsp. *ugandensis*.

NOTE. In subsp. *holubii* from southern Africa the calyx lobes (posticuous 12–14 mm long) have longer, coarser ascending hairs on the margin and have very few or no glandular hairs; it also lacks glandular hairs on the corolla limb.

80. **Barleria penelopeana** *I. Darbysh.* **sp. nov.** *B. casatianae* Buscal. & Muschl. similis sed planta tota atque praecipue lobis externis calycis conspicue hirsutis (nec glabratis), lobis externis calycis brevioribus, lobo postico 10–15 mm (nec 15–22 mm) longo, spinis axillaribus subsessilibus, stipe spinae minus quam 1 mm longo (nec 1–2.5 mm longo) differt. Type: Tanzania, Shinyanga District, Shinyanga, *Koritschoner* 2256 (K!, holo.; EA!, iso.)

Perennial herb with decumbent, few-branched stems from a woody rootstock, 15–45 cm tall; stems with short white spreading or retrorse hairs concentrated in opposite lines or more evenly distributed. Axillary spines subsessile, rays 3–4(–5), longest ray 6.5–21 mm long. Leaves oblong-elliptic to -obovate, 2–4.2 cm long, 0.8–1.7 cm wide, base cuneate or acute, apex acute or obtuse, mucronate, margin and principal veins beneath with coarse spreading or ascending hairs, those on margin often bulbous-based, often also with shorter, finer white spreading hairs beneath; lateral veins prominent beneath; petiole to 3 mm long or absent. Inflorescences axillary towards stem apices, flowers solitary, subsessile; bracts foliaceous, sometimes reduced in upper axils; bracteoles spinose throughout or with a linear-lanceolate blade towards base, (6–)10–18 mm long, to 0.5–2 mm wide. Anticous calyx lobe lanceolate, 9.5–14 mm long, 2.5–4.5 mm wide at base, attenuate into a spinose apex, this sometimes bifid, surface with short white spreading or retrorse hairs throughout and coarser ascending hairs at least on margin and/or midrib, short patent glandular hairs sometimes present in the upper half, parallel venation conspicuous only in fruit; posticous lobe 10–15 mm long; lateral lobes somewhat narrower. Corolla yellow, 31–35 mm long, upper portion of tube pilose, limb with shorter mixed glandular and eglandular hairs; tube narrowly cylindrical, 19–22 mm long; limb subregular, lobes widely spreading, each 10.5–13.5 mm long, 6–7 mm wide. Stamens attached 11.5–14 mm from base of corolla tube; filaments 8–10 mm long, pubescent towards base; anthers exserted, 2–2.5 mm long; lateral staminodes 1–1.5 mm long, pubescent to sparsely so, antherodes 0.6–1.2 mm long. Pistil glabrous; stigma ± 0.5 mm long. Capsule (14-)17–22 mm long including 5–6.5 mm beak, glabrous; seeds ± 9.5 mm long, 7.5 mm wide.

TANZANIA. Mwanza District: Masawe, ? Igshero, Sept. 1952, *Tanner* 967! & Ibondo Forest Reserve, Dec. 1954, *Carmichael* 493!; Shinyanga District: Shinyanga, May–July 1933, *Bax* 303!
DISTR. **T** 1; not known elsewhere
HAB. Grassland, including derived grassy areas following clearance of *Acacia-Commiphora* bushland and heavily grazed areas, *Acacia* woodland; 1100–1350 m
USES. "Sukuma – for pains in stomach; roots dug up, cut into slices, boiled in water with white mtama flour, drink 3 times a day" (**T** 1; *Tanner* 967)
CONSERVATION NOTES. This species is highly localised in N Tanzania and currently known from only five collections. However, it appears to tolerate or even favour disturbance. No information on current abundance and threats are available and it is therefore provisionally assessed as Data Deficient (DD), though may prove to be unthreatened.

NOTE. Closely resembling *B. casatiana* Buscal. & Muschl., endemic to Eritrea, which differs primarily in being much less hairy throughout, the calyx lobes being largely glabrous and the leaves lacking short spreading hairs even when young. It additionally differs in having short-stalked (1–2.5 mm long), not subsessile, axillary spines and in the outer calyx lobes being somewhat longer (anticous 14–19 mm, posticous 15–22 mm long). *B. penelopeana* is also close to *B. holubii* but differs in the characters listed in the key.

This species is named after Mrs Penelope Tanner, wife of Ralph E.S. Tanner; who together collected some 6,000 specimens in Tanzania, and donated these to the Kew herbarium.

TAXON OF DOUBTFUL POSITION

81. **Barleria ukamensis** *Lindau* in P.O.A. C: 368 (1895); C.B. Clarke in F.T.A. 5: 161 (1899). Type: Tanzania, Morogoro District, Ukami, *Stuhlmann* 8154 (B†, holo.)

(?) Subshrub; stems hispidulous. Leaves subsessile, ovate or elliptic, 4–7 cm long, 2.5–3.5 cm wide, base strongly narrowed, apex subobtuse or acute, hispidulous on the nerves beneath. Flowers held in a short terminal synflorescence, rather dense, sparingly hispid; bracts and bracteoles oblong, 18–25 mm long, 4–8 mm wide, apex subobtuse. Anticous and posticous calyx lobes 20–22 mm long, 12–13 mm wide, apex (sub)obtuse, 6–8 principal veins, these hispidulous; lateral lobes 8 mm long. Corolla tube 15 mm long, lobes 16–18 mm long, 7–8 mm wide. Staminal filaments 16 mm long; anthers 4 mm long; staminodes 1.5 mm long. Ovary and base of style glandular-hairy.

NOTE. This name has previously been applied to specimens of *B. rhynchocarpa* from the Tanzanian mainland, but the description clearly does not match that species. Clarke (l.c.), who saw the type, recorded hispidulous stems, subsessile leaves, hispid inflorescences and a glandular hairy pistil, a combination not matched by any *Barleria* treated here. It is most likely to fall within sect. *Somalia*; indeed, Lindau placed it close to *B. calophylloides* in the protologue. However, the type lacked fruits so the sectional placement cannot be confirmed. In view of this uncertainty, this taxon is not included in the keys. The description above is extracted from the protologue and from Clarke (l.c.).

34. ASYSTASIA

Blume, Bijdr.: 796 (1826); Nees in Wall., Pl. As. Rar. 3: 89 (1832) & in DC. Prod. 11: 163 (1847); T. Anderson in J.L.S. Bot. 7: 52 (1863); C.B. Clarke in F.T.A. 5: 130 (1899)

Parasystasia Baillon, Hist. Pl. 10: 460 (1891); Lindau in E. & P. Pf. IV. 3b: 325 (1895)
Asystasiella Lindau in E. & P. Pf. IV, 3b: 326 (1895); S. Moore in J.B. 44: 29 (1906); Iversen in Symb. Bot. Upsal. 29(3): 160 (1991)
Styasasia S. Moore in J.L.S. Bot. 37: 195 (1905)
Salpinctium T. J. Edwards in S. Afr. J. Bot. 55: 7 (1989)

Herbs or shrubs. Leaves opposite, entire (rarely undulate to serrate). Flowers in terminal or axillary spikes, racemes or these merging into panicles; bracts persistent, prominent or not; bracteoles usually minute, more rarely large. Calyx deeply divided (to ± 1 mm from base) into 5 lobes, one lobe longer and slightly wider than the rest. Corolla indistinctly 2-lipped with 5 subequal lobes (lower mid-lobe the largest and usually distinctly rugulate), or moderately to strongly 2-lipped, with 3-lobed lower and 2-lobed upper lip; basal tube cylindric, widening upwards into ± distinct throat. Stamens 4, inserted in throat, subequal to distinctly didynamous included in tube, more rarely exserted; anthers bithecous, thecae slightly superposed, each with 1(–2–3) spurs at the base or apiculate. Style filiform, pubescent in basal part or glabrous; stigma capitate with 2 equal or subequal flat ellipsoid lobes. Capsule 2–4-seeded, clavate with a sterile solid basal stalk which is ± half the total length. Seed compressed (very rarely globose), ± circular in outline, with or without a thickened irregularly crenate margin, surfaces tuberculate, rugose, rugose-tuberculate or smooth.

About 50 species, centred in tropical Africa with the highest diversity in eastern Africa, but also distributed in India, SE Asia, and through Indonesia to New Guinea, Australia and the Pacific. One species introduced to the Americas.

1. Shrubby herbs or shrubs 2
   Perennial (rarely annual) herbs with basal part of stems rooting or with erect or decumbent stems from woody rootstocks 6
2. Corolla indistinctly 2-lipped or with 5 sub-equal lobes, lower lip shorter than tube plus throat; stamens included in throat 3
   Corolla distinctly 2-lipped, the lower lip longer than or as long tube plus throat; stamens exserted 5
3. Flowers sessile or subsessile; bracts ovate, elliptic, obovate or orbicular, 4–9 mm long; corolla 2–3.2 cm long; capsule 1.8–2.7 cm long .... 26. *A. charmian* p.468
   Flowers clearly pedicellate; bracts linear-lanceolate to narrowly triangular, 2–5.5 mm long; corolla 3.5–6.5 cm long; capsule 3–4.8 cm long 4
4. Largest leaf 4–7 cm long; peduncles with non-glandular hairs only; corolla pale blue to lilac, with dark purple or violet lower lip and throat; capsule glabrous, 3.5–4.8 cm long; seed surface smooth ... 1. *A. excellens* p.447
   Largest leaf 11–23 cm long; peduncles with capitate glands and non-glandular hairs; corolla white or very pale pink, with pale purple spots on lower lip; capsule pubescent, 3–3.7 cm long; seed surface reticulate-verrucose .... 2. *A. hedbergii* p.448
5. Corolla 1.3–1.8 cm long; style 7–9 mm long; capsule 2.5–3 cm long, glabrous or with a few short-stalked capitate glands .... 20. *A. africana* p.462
   Corolla 2.5–3.5 cm long; style 2–2.5 cm long; capsule 2.7–3.8 cm long, finely uniformly puberulous with stalked capitate glands .... 21. *A. moorei* p.462
6. Inflorescences with 2–3-flowers only; calyx 1.8–2.1 cm long; capsule glabrous or with a few long hairs near apex .... 7. *A. sp. A* p.451
   Inflorescences usually with more than 3 flowers; calyx 0.2–1.8(–2.5 in fruit) cm long; capsule hairy and/or glandular (glabrous in *A. schliebenii*) 7
7. Bracts narrowly ovate or narrowly elliptic to ovate, elliptic, obovate or orbicular (rarely lanceolate), 4–25 mm long, shorter or longer than calyx 8
   Bracts (apart from lowermost pair) subulate to linear or narrowly triangular, 0.5–4(–6) mm long, shorter than calyx 13
8. Bracts longer than calyx 9
   Bracts shorter than calyx 12
9. Annual herb; corolla 1–1.5 cm long; calyx 4–7 mm long .... 25. *A. mysorensis* p.466
   Perennial herb with several stems from woody rootstock; corolla 1.2–3 cm long; calyx 6–18(–25 in fruit) mm long 10

10. Leaf lamina linear to linear-lanceolate (rarely narrowly ovate); corolla 1.2–1.6(–2) cm long; style 6–9 mm long . . . . 24. *A. lorata* p.465
    Leaf lamina narrowly to broadly ovate or elliptic (rarely lanceolate); corolla 1.8–3 cm long; style 8–12 mm long 11
11. Flowers alternate (only one bract at each node supporting a flower); corolla ± 3 cm long . . . . 22. *A. masaiensis* p.463
    Flowers opposite (both bracts at each node supporting a flower); corolla 1.8–2.7(–3) cm long . . . . 23. *A. guttata* p.464
12. Calyx (6–)8–14 mm long; corolla 2–3.2 cm long; style 1–1.5 cm long; capsule 1.8–2.7 cm long; seed 5–6 × 4.5–5.5 mm . . . . 26. *A. charmian* p.468
    Calyx 5–8(–15 in fruit) mm long; corolla 1–1.9 cm long; style 5–7 mm long; capsule 1.5–1.9 cm long; seed 3.5–5 × 3–4 mm . . . . 27. *A. riparia* p.469
13. Basal cylindric corolla tube 10–32 mm long 14
    Basal cylindric corolla tube 2–7(–8) mm long 17
14. Corolla white with purple veins and markings in throat and on lower lip . . . . 4. *A. malawiana* p.449
    Corolla white to lilac or mauve, without differently coloured markings 15
15. Basal cylindric corolla tube ± 1 cm long; anther thecae ± 2 mm long; style 1.3–2 cm long; capsule 2.2–2.8 cm long; woodland plant . . . . 14. *A. albiflora* p.455
    Basal cylindric corolla tube 2–3.2 cm long; anther thecae 3–3.5 mm long; style 2.5–4.5(–5) cm long; capsule 2.8–3.5 cm long; forest plant 16
16. Flowers alternate (only one bract at each node supporting a flower); corolla 4–5.5 cm long with cylindric tube (2–)2.5–3.2 cm long and throat 1.5–2 cm . . . . 3. *A. vogeliana* p.448
    Flowers opposite (both bracts at each node supporting a flower); corolla ± 4 cm long with cylindric tube ± 2 cm long and throat ± 1 cm . . . . 13. *A. sp. B* p.455
17. Plants from forest habitats (varying from wet evergreen forest and riverine forest to semi-deciduous coastal forest and thicket) 18
    Plants from woodland, bushland, grassland or (rarely) grassy glades in forest, or secondary and weedy habitats including forest margins and clearings 23
18. Corolla limb pure white . . . . 14. *A. albiflora* p.455
    Corolla limb of varying colours but if white then with mauve or purple patch on lower lip 19
19. Calyx, ovary and capsule glabrous . . . . 18. *A. schliebenii* p.460
    Calyx, ovary and capsule hairy and/or glandular 20
20. Corolla 1.4–3.5 cm long with lower cylindric tube and throat clearly differentiated; capsule 2.1–3 cm long 21
    Corolla 0.7–1.4 cm long with lower cylindric tube merging gradually into throat; capsule 1.2–2.3 cm long 22

21. Corolla 1.4–2.6 cm long, cream or yellowish green with brownish purple markings on lower lateral and upper lobes and mauve patch on midlobe; cylindric tube straight; throat 0.6–1 cm long .... 11. *A. tanzaniensis* p.453
    Corolla 2.5–3.5 cm long, white with large purple patch in throat; cylindric tube distinctly curved; throat 1.2–1.5 cm long .... 12. *A. congensis* p.454
22. Corolla 11–14 mm long, yellowish-green or greenish-white, without purple patch on midlobe; flowering stems straggling or scrambling; capsule 1.8–2.3 cm long; seed 4–5 × 3–3.5 mm .... 9. *A. leptostachya* p.452
    Corolla 7–12 mm long, white with purple patch on midlobe; flowering stems erect; capsule 1.2–1.7 cm long; seed 3–4 × 2.5–3 mm .... 10. *A. minutiflora* p.453
23. All racemes axillary 24
    Racemes terminal or terminating axillary branches (rarely also a few axillary) 25
24. Racemes with 1–2(–3) flowers; calyx 8–15 mm long; corolla indumentum of non-glandular hairs .... 19. *A. ansellioides* p.461
    Racemes with 3–5(–7) flowers; calyx 3–5(–7) mm long; corolla indumentum of non-glandular hairs and capitate glands .... 17. *A. calcicola* p.460
25. Calyx, ovary and capsule glabrous .... 18. *A. schliebenii* p.460
    Calyx, ovary and capsule hairy and/or glandular 26
26. Flowers opposite (both bracts at each node supporting a flower) at least in basal part of inflorescence 27
    All flowers alternate (only one bract at each node supporting a flower) 28
27. Corolla 1.3–2 cm long, white to pale mauve, pale pink or purple, with orange to red capitate glands; leaves subsessile, upper with cuneate to attenuate base ... 5. *A. glandulosa* p.450
    Corolla 2.5–4.2 cm long, white, with colourless capitate glands; leaves clearly petiolate, upper with truncate to cordate base .... 14. *A. albiflora* p.455
28. Leaves sessile or subsessile, with linear or linear-lanceolate lamina 29
    Leaves clearly petiolate, with variously shaped lamina 30
29. Calyx 8–14 mm long; capsule indumentum of non-glandular hairs; plant with several erect to decumbent stems; leaves fleshy .... 6. *A. richardsiae* p.450
    Calyx 6–10.5 mm long; capsule indumentum of glandular hairs; plant with 1–2 decumbent or straggling stems; leaves not fleshy .... 8. *A. linearis* p.451
30. Corolla limb pure white 31
    Corolla limb white with violet or purple patch on lower lip or pale mauve to pale blue 32
31. Corolla 2–2.7 cm long; calyx lobes without white margins .... 15. *A. laticapsula* p.456
    Corolla ± 3.5 cm long; calyx lobes with distinct white margins (rare form) .... 14. *A. albiflora* p.455

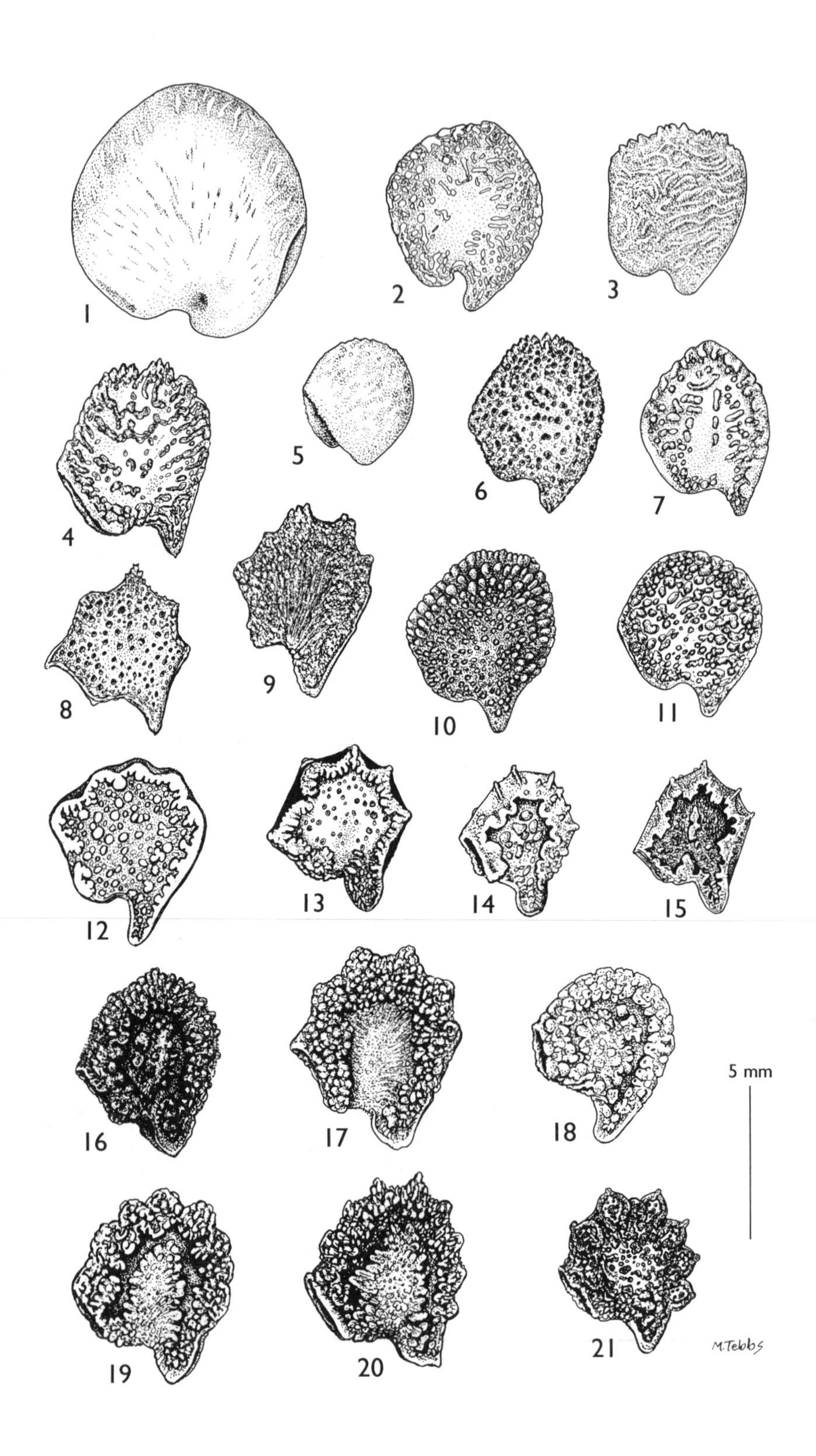
1
2
3
4
5
6
7
8
9
10
11
12
13
14
15
16
17
18
19
20
21
5 mm
M.Tebbs

32. Anther thecae ± 1.5 mm long, apiculate but without spurs; calyx 3–5(–7) mm long; corolla 1–1.5 cm long; style 6–8 mm long; plant on limestone derived soils in NE Kenya .... 17. *A. calcicola* p.460
Anther thecae 2–3 mm long, with small curved basal spurs; calyx 4–12 mm long; corolla 1–2.8 cm long; plant not on limestone derived soils 33
33. Calyx (6–)7–12(–16 in fruit) mm long; anther thecae 2.5–3 mm long; capsule 6–7 mm wide .... 15. *A. laticapsula* p.456
Calyx 4–9 mm long; anther thecae ± 2 mm long; capsule 4–5 mm wide .... 16. *A. gangetica* p.457

1. **Asystasia excellens** *Lindau* in E.J. 33: 190 (1902); Fiori in Chiovenda, Result. Sci. Miss. Stef.-Paoli, Coll. Bot.: 138 (1916); E.P.A.: 956 (1964); Lebrun & Stork, Enum. Pl. Afr. Trop. 4: 468 (1997); Hedrén in Fl. Somalia 3: 398 (2006); Ensermu in Fl. Eth. 5: 398 (2006). Types: Ethiopia, Harerge, Gobelli Valley, *Ellenbeck* 1044, 1047 and 1051 (all B†, syn.) and Modji River, *Ellenbeck* 1099 (B†, syn.). Neotype: Ethiopia, Sidamo, Negelle–Melca Guba road, *Friis et al.* 10948 (K!, neotype, selected here; C, ETH, K!, iso.)

Erect or scrambling shrub to 3 m high; branches grey, with large lenticels, when young sparsely antrorsely sericeous, soon glabrescent. Leaves with petiole to 5 mm long; lamina ovate to broadly elliptic, largest 4–7 × 2–4 cm, base attenuate, decurrent, apex obtuse; subglabrous or sparsely antrorsely sericeous on veins. Racemes terminal, 2–12(–20) cm long, often with 2 branches from base (rarely with several branches forming a loose panicle), antrorsely sericeous or sparsely so; flowers opposite; peduncles 0.5–4 cm long; basal pair(s) of bracts foliaceous, up to 3.5 × 2 cm; middle and upper bracts and bracteoles caducous, linear-lanceolate to narrowly triangular, 2–5.5 mm long, acuminate or acute, subglabrous to sericeous-pubescent; pedicels 2–13 mm long, antrorsely sericeous or sericeous-pubescent. Calyx 4–7 mm long, lobes erect or reflexed, linear-subulate, cuspidate, glabrous or sparsely sericeous, with subsessile capitate glands on inside. Corolla pale blue to lilac with dark purple or violet lower midlobe and throat, with conspicuous purplish veins, 3.5–5.7 cm long, with stalked capitate glands on upper part of cylindric tube and lower part of throat; cylindric tube 5–15 mm long, 2–3 mm wide, throat 1.7–3 cm long, 1.2–2 cm wide at the mouth; limb with 5 spreading to reflexed subequal lobes, lobes 1–1.5 × 1.2–1.7 cm (middle in lower lip to 1.9 cm wide), all with rounded apex. Stamens strongly didynamous, included in throat; filaments 1–1.3 and 1.6–2.1 cm long, glabrous or longer with a row of glandular hairs; thecae ± 4 mm long, ventrally and dorsally with a row of glandular hairs, with (1–)2(–3)-spurs. Style 2–2.5 cm long, pubescent near base, very rarely all along its length. Capsule 3.5–4.8 cm long, glabrous; seed 7–9 × 6–8 mm, margin not thickened, slightly crenulate, surfaces smooth. Fig. 58/1, p. 446.

KENYA. Northern Frontier District: Moyale, 17 Apr. 1952, *Gillett* 12828! & 5 km E of junction of Banissa and Derkali roads, 5 May 1978, *Gilbert & Thulin* 1450!; Lamu District: Kiunga Islands, 1951, *Oxford Univ. Exp.* B10! (see note)

FIG. 58. Seeds of **1**, *ASYSTASIA EXCELLENS*; **2**, *A. HEDBERGII*; **3**, *A. AFRICANA*; **4**, *A. MOOREI*; **5**, *A. LINEARIS*; **6**, *A. LEPTOSTACHYA*; **7**, *A. TANZANIENSIS*; **8**, *A. MINUTIFLORA*; **9**, *A. MALAWIANA*; **10**, *A. GLANDULOSA*; **11**, *A. VOGELIANA*; **12**, *A. CONGENSIS*; **13**, *A. ALBIFLORA*; **14**, *A. ANSELLIOIDES*; **15**, *A. CALCICOLA*; **16**, *A. LATICAPSULA*; **17**, *A. SP. A*; **18**, *A. RIPARIA*; **19**, *A. CHARMIAN*; **20**, *A. GUTTATA*; **21**, *A. LORATA*. All lateral view. 1 from *Gillett* 12828, 2 from *Bullock* 3255, 3 from *Osmaston* 2775, 4 from *Luke* 2703, 5 from *Vollesen* MRC3692, 6 from *Ghogue* 442, 7 from *Mhoro* 419, 8 from *Borhidi* 87235, 9 from *Kayombo* 2346, 10 from *Bidgood* 1301, 11 from *Dawkins* 764, 12 from *Drummond & Hemsley* 4540, 13 from *Bidgood* 4144, 14 from *Luke* 4509, 15 from *Gilbert & Thulin* 1614, 16 from *Abdallah & Vollesen* 95/55, 17 from *Peter* 46123, 18 from *Faden* 68/377, 19 from *Dummer* 5011, 20 from *M.G.Gilbert* 5632, 21 from *Verdcourt* 1457. Drawn by Margaret Tebbs.

DISTR. **K** 1, 7; Ethiopia, Somalia
HAB. *Acacia-Commiphora* woodland and bushland, *Acacia* bushland and scrub on rocky slopes; (25–)850–1300 m

SYN. *A. drake-brockmanii* Turrill in K.B. 1913: 180 (1913); E.P.A.: 956 (1964); K.T.S.L.: 598 (1994); Lebrun & Stork, Enum. Pl. Afr. Trop. 4: 468 (1997). Type: Ethiopia, Bale, Geru Abbas, *Drake-Brockman* 230 (K!, holo.)
A. drake-brockmanii Turrill forma *typica* Chiov., Fl. Somala: 251 (1929) & Fl. Somala 2: 350 (1932). Type: Somalia, Baidoa, *Paoli* 1237 (FT, holo.)
A. drake-brockmanii Turrill forma *lejogyna* Chiov., Fl. Somala: 251 (1929); E.P.A.: 956 (1964). Types: Somalia, Tigieglo, *Stefanini & Puccioni* 231 (FT, syn.) & Baidoa, *Stefanini & Puccioni* 323 (FT, syn.)

NOTE. The exact provenance of the Oxford University Exp. specimen cited above is somewhat doubtful. In Somalia the species is widespread from 25 up to 800 m but it is unlikely that the species occurs on Kiunga Islands itself. It is more likely to have been collected somewhere on the mainland opposite the islands. All other Kenyan material is from the extreme north (Moyale to Mandera).

2. **Asystasia hedbergii** *Ensermu* in Opera Bot. 121: 153, fig. 1 (1993); Lebrun & Stork, Enum. Pl. Afr. Trop. 4: 468 (1997). Type: Tanzania, Mpanda District, 58 km S of Uvinza, *Bullock* 3255 (K!, holo.; K!, iso.)

Erect or straggling shrubby herb or shrub to 2.5 m tall, sparsely branched; young branches glabrous or sparsely puberulous at upper nodes. Leaves with petiole to 3(–4.5) cm long; lamina ovate to elliptic or narrowly so, largest 11–23 × 4–9 cm, base rounded to attenuate, apex acuminate to acute or caudate; glabrous or with a few hairs on midrib. Racemes terminal often with 1–2 branches from base, 5–15 cm long, puberulous with short capitate glands and minute non-glandular hairs; flowers alternate or opposite towards base; peduncles 0.5–3 cm long; bracts and bracteoles subulate to linear, 2–4 mm long, puberulous and with short capitate glands; pedicels 1–6(–10 in fruit) mm long. Calyx 5–10 mm long, lobes subulate, 4–8 mm long, margins hyaline, irregularly lacerate, puberulous and with short capitate glands. Corolla white to pale pink, with pale purple spots on lower lip and pale green tube, 4.5–6.5 cm long, with scattered dark purple capitate glands throughout and non-glandular hairs; cylindric tube 1–1.8 cm long and 2–4 mm wide; throat 2.5–3 cm long and 1–2 cm wide at the mouth; limb with 5 spreading to reflexed subequal lobes, upper and lateral in lower lip 1.2–1.5 × 1.1–1.3 cm, middle in lower lip 1.2–1.8 × 1.3–2.1 cm, all with rounded apex. Stamens didynamous, included in throat, filaments 1–1.2 and 1.4–1.6 cm long, at least one pair with a row of glandular hairs; thecae 4–5 mm long, glabrous, each with 1 spur. Style 2.6–3.2 cm long, glabrous or very rarely with a few hairs at base. Capsule 3–3.7 cm long, puberulous with sub-sessile capitate glands and short non-glandular hairs; seed (immature) 5.5–7 × 4.5–5.5 mm, margin not thickened; surfaces reticulate-verrucose, smooth towards centre. Fig. 58/2, p. 446.

UGANDA. Kigezi District: Kayonza Gorge, March 1945, *Purseglove* 1632! & Ishasha Gorge, March 1946, *Purseglove* 2012! & Feb. 1950, *Purseglove* 3307!
TANZANIA. Kigoma District: Gombe National Park, Mkenke Valley, 17 Sept. 1998, *Gobbo & Daniel* 31!; Mpanda District: Kungwe Mt, Kahoko, 22 July 1959, *Newbould & Harley* 4548! & Ntakatta Forest, 11 June 2000, *Bidgood et al.* 4646!
DISTR. **U** 2; **T** 4; Congo-Kinshasa, Burundi
HAB. Submontane and riverine forest; 750–1550 m

3. **Asystasia vogeliana** *Benth.* in Hooker, Niger Fl.: 479 (1849); T. Anderson in J.L.S. Bot. 7: 53 (1863); C.B. Clarke in F.T.A. 5: 133 (1899); Heine in F.W.T.A., ed. 2, 2: 412 (1963) & in Fl. Gabon 13: 130, pl. 26 (1966); Lebrun & Stork, Enum. Pl. Afr. Trop. 4: 468 (1997); Friis & Vollesen in Biol. Skr. 51: 437 (2005). Type: Equatorial Guinea, Bioko [Fernando Po], *Vogel* 211 (K!, holo.)

Stoloniferous perennial herb, basal part of stems decumbent, rooting at nodes, apical part erect or straggling, to 1(–1.5) m tall, glabrous or sparsely pubescent in two opposite bands. Leaves with petiole to 3.5 cm long; lamina elliptic or narrowly so, largest (4.5–)8–21 × (1.5–)2.5–7 cm, base cuneate (upper) to attenuate and decurrent, apex acute to acuminate or caudate; glabrous. Racemes terminal or also from upper leaf axils and then forming a loose panicle, 1–10 cm long, mostly branched from base, glabrous (downwards) to puberulous with short-stalked capitate glands and minute non-glandular hairs, densest upwards; flowers alternate (often many aborting); peduncle (0.5–)1.5–7 cm long; bracts and bracteoles linear-lanceolate to narrowly triangular, 1–4 mm long, glabrous; pedicels thickened, 5–11 mm long, puberulous or densely so with short-stalked capitate glands. Calyx 5–10 mm long, lobes subulate to linear, puberulous with short-stalked capitate glands. Corolla white to pale lilac, lilac or mauve, tube white, no markings on lower lip, 4–5.5 cm long, sparsely puberulous with short-stalked dark purple capitate glands and scattered non-glandular hairs; cylindric tube (2–)2.5–3.2 cm long and 1–3 mm wide; throat 1.5–2 cm long and 5–8 mm wide at the mouth; limb weakly 2-lipped with subequal lobes 8–11 × 12–15 mm. Stamens didynamous, included in throat, filaments 2.5–5 and 3.5–6 mm long, all glabrous; thecae slightly superposed, 3–3.5 mm long, glabrous, without spur. Style 3.5–4.5(–5) cm long, glabrous. Capsule 2.8–3.5 cm long, puberulous with short-stalked capitate glands and non-glandular hairs; seed 5–6 × 4–5 mm, margin not thickened and without projections, surfaces with tubercles of similar size, central part ± smooth. Fig. 58/11, p. 446.

UGANDA. Bunyoro District: Budongo Forest, Dec. 1933, *Eggeling* 1406!; Busoga District: 20 km N of Jinja, Kagoma, 5 Dec. 1952, *G.H.S. Wood* 550!; Masaka District: 6 km S of Katera, Malabigambo Forest, 2 Oct. 1953, *Drummond & Hemsley* 4559!

TANZANIA. Bukoba District: Minziro Forest Reserve, 4 July 2000, *Bidgood et al.* 4824! & 12 July 2001, *Festo et al.* 1610!

DISTR. **U** 2–4; **T** 1; Guinea, Liberia, Sierra Leone, Ivory Coast, Ghana, Nigeria, Bioko, Cameroon, Gabon, Central African Republic, Congo-Kinshasa, Sudan

HAB. Evergreen forest, often along paths, edges and in clearings, swamp forest, riverine forest; 900–1250 m

4. **Asystasia malawiana** *Brummitt & Chisumpa* in K.B. 32: 703 (1978); Lebrun & Stork, Enum. Pl. Afr. Trop. 4: 468 (1997). Type: Malawi, Ntchisi Forest Reserve, *Brummitt* 9375 (K!, holo.)

Perennial herb with tuberous roots; stems erect or ascending, to 1.5 m high, glabrous to sparsely pubescent, mostly in two bands; lateral branches sometimes bending down, rooting where touching the ground and producing new plants. Leaves with petiole to 8 cm long; lamina ovate to elliptic, largest 8–16 × 4–6.8 cm, base attenuate and decurrent on lower leaves, rounded to subcordate on upper leaves, apex acuminate to cuspidate; glabrous below, with sparse appressed broad many-celled hairs above. Racemes terminal or axillary from upper nodes, 1–4(–10 in fruit) cm long, only rarely branched from base, sparsely puberulous with non-glandular hairs in two bands; flowers alternate, more rarely opposite; peduncles 1–5 cm long; lower bracts sometimes foliaceous, to 1.5 × 1 cm, middle and upper bracts and bracteoles linear-lanceolate to narrowly triangular, 2–4 mm long, glabrous but for the ciliate margins; pedicels 2–8(–10 in fruit) mm long, puberulous or densely so with non-glandular hairs, sometimes with a few dark red to yellow sub-sessile capitate glands. Calyx 4–7(–8.5 in fruit) mm long, lobes linear-lanceolate, puberulous or densely so and with short dark red to yellow capitate glands. Corolla white with purple veins and markings in throat and on lower lip, 2.5–3.7 cm long, sparsely puberulous with orange to red capitate glands and non-glandular hairs; cylindric tube 1–1.4 cm long and 1–2 mm wide, throat 1–1.5 cm long and 0.8–1.1 cm wide at the mouth; limb weakly 2-lipped with subequal sub-orbicular rounded lobes, 6–8 × 6–8 mm. Stamens didynamous, included in throat; filaments 3–4 and 5–6 mm long, glabrous; thecae slightly superposed, ± 2 mm long, dorsally with sub-sessile capitate glands, without spur. Style 16–20 mm long, puberulous near base. Capsule 2.3–2.8 cm long, puberulous with short-stalked capitate glands; seed 5–6 × 4–5 mm, mid-surface nearly smooth, tuberculate towards margins, with 5–6 narrow and smooth channels and 5–6 broad tuberculate ridges which form projections on the margin. Fig. 58/9, p. 446.

TANZANIA. Mbeya District: Mshewe, Ihanga, 8 May 1991, *P. Lovett & Kayombo* 226! & Mbeya Range, E of Mshewe Village, 14 June 1999, *Kayombo* 2346!; Songea District: Mbinga, Kitezi Forest Reserve, 4 March 1987, *Congdon* 162!
DISTR. **T** 7, 8; Malawi, Mozambique
HAB. Montane forest and riverine forest; 1250–1800 m

5. **Asystasia glandulosa** *Lindau* in E.J. 33: 189 (1902); Björnstad in Serengeti Res. Inst. Publ. 215: 25 (1976); Lebrun & Stork, Enum. Pl. Afr. Trop. 4: 468 (1997). Type: Tanzania, Songea District, Milonji River, *Busse* 992 (B†, holo.; EA!, iso.)

Perennial herb with several stems from a creeping woody rootstock; stems to 50(–75) cm long, erect to procumbent, puberulous to pubescent or densely so. Leaves with petiole to 3(–5) mm long; lamina narrowly ovate or narrowly elliptic to ovate or elliptic, largest 5–10 × 1.2–3 cm, base cuneate to attenuate and decurrent, apex rounded to acuminate; sparsely to densely pubescent densest on veins. Racemes terminal, 2–10 cm long, never branched from base, puberulous to pubescent; all flowers opposite; peduncles 1–6 cm long; bracts and bracteoles subulate to narrowly triangular, 1–4 mm long (lowermost pair to 6 mm), puberulous or sparsely so; pedicels 0.5–1.5(–4 in fruit) mm long, puberulous and with stalked capitate glands towards apex. Calyx 4–9 mm long, lobes subulate to linear-lanceolate, puberulous and with capitate glands. Corolla white to pale mauve, pale pink or purple, with purple markings in the throat and on rugula, 1.3–2 cm long, puberulous or sparsely so with capitate glands and non-glandular hairs; cylindric tube 4–6 mm long and 1.5–2 mm wide; throat 7–9 mm long and 5–6 mm wide at the mouth; limb 2-lipped, midlobe in lower lip 6.5–7.5 × 6–7 mm, ovate with rounded apex, lateral lobes 4–5.5 × 4.5–6 mm, upper lip 6–7 mm long with lobes ± 2 × 3 mm. Stamens didynamous, included in throat; filaments 3–4 and 6–7 mm long; thecae slightly superposed, ± 2 mm long, with minute spur or apiculate. Style 10–12 mm long, pubescent. Capsule 2–2.8 cm long, densely puberulous with stalked capitate glands and non-glandular hairs; seed 4–5 × 3–4 mm, with mostly fused tubercles which are evenly distributed apart from concentric band near edge, margin slightly double-ridged, not thickened or toothed, but roughened with the tubercles. Fig. 58/10, p. 446.

TANZANIA. Mpanda District: 10 km on Mpanda–Uruwira road, 9 March 1994, *Bidgood et al.* 2730!; Iringa District: 52 km on Mafinga–Madibira road, 27 Jan. 1991, *Bidgood et al.* 1301!; Songea District: 8 km W of Gumbiro, 26 Jan. 1956, *Milne-Redhead & Taylor* 8436!
DISTR. **T** 4, 5, 7, 8; Congo-Kinshasa, Angola, Zambia, Malawi, Mozambique
HAB. *Brachystegia-Julbernandia* woodland, wooded grassland and bushland, on sandy soils; 750–1400 m

6. **Asystasia richardsiae** *Ensermu* in K.B. 53: 929 (1998). Type: Tanzania, Mpanda District, 5 km on Namanyere–Kipili road, *Bidgood, Mbago & Vollesen* 2602 (K!, holo.; BR!, C!; CAS!, DSM!, EA!, K!, iso.)

Perennial herb with several stems from a woody rootstock with fleshy roots; stems to 30 cm long, erect to decumbent, glabrous to pubescent in two bands, or uniformly hirsute. Leaves slightly fleshy, with petiole 0–2 mm long; lamina lanceolate to narrowly ovate, largest 7–10 × 0.8–1.8 cm, base attenuate to truncate, apex rounded; glabrous to pubescent, densest on veins. Racemes terminal, (0.5–)1–4.5 cm long, not branched or with a single branch from base, glabrous to pubescent in two bands or hirsute; flowers alternate; peduncles 1–6.5 cm long; bracts and bracteoles linear-lanceolate to narrowly triangular, 2–4 mm long, glabrous or pubescent; pedicels 3–5 mm long, subglabrous to pubescent or pilose. Calyx 8–14 mm long, lobes linear, pubescent to hirsute. Corolla cream or white with mauve or red spots on lower midlobe, 1.6–2 cm long, puberulous with non-glandular hairs; cylindric tube 5–6 mm long and ± 2.5 mm wide; throat 5–8 mm long and 5–7 mm wide at the mouth; limb 2-lipped; midlobe in lower lip ovate, ± 6 × 7 mm, rounded, gibbous; lateral lobes obovate to orbicular, ± 5 × 5 mm, rounded at the apex, lobes in upper lip obovate to orbicular, 4–5 × 6–7 mm, rounded. Stamens didynamous, included in throat; filaments ± 3.5 and 5 mm long, glabrous except for few non-glandular hairs at the base; thecae slightly superposed, 2.5–3 mm long, glabrous, with minute spurs. Style 8–10 mm long, pubescent in lower half. Capsule 1.5–2.8 cm long, puberulous or sparsely so with non-glandular hairs; seed not seen.

TANZANIA. Ufipa District, 5 km on Namayere–Kipili road, 3 March 1994, *Bidgood et al.* 2602!; Mpanda District: Kapapa Camp, 28 Oct. 1959, *Richards* 11623!; Tabora District: 15 km on Ipole–Inyonga road, 18 May 2006, *Bidgood et al.* 6070!
DISTR. **T** 4; not known elsewhere
HAB. Short seasonally inundated grassland on greyish sandy or sandy-loamy soil; 750–1400 m

NOTE. Known from relatively few collections from a not very extensive area but, at least in the Ipole-Inyonga area, it is common in suitable habitats (pers. obs.). All the collections, apart from the *Richards* specimen, are recent. The habitat is widespread in the Tabora-Mpanda area and not in any immediate danger.

7. **Asystasia sp. A**

Perennial herb with several stems from a woody rootstock; stems to 45 cm long, creeping, pilose to hirsute. Leaves with petiole 1–3 mm long; lamina ovate to suborbicular, largest ± 3 × 1.5 cm, base cuneate to truncate, apex subacute to rounded; with a few hairs on midrib. Racemes 0.5–2(–4.5) cm long, terminal, but mostly appearing lateral due to displacement by lateral branch, glabrous to pilose mostly in two bands; flowers alternate; peduncles 1–4(–8) cm long; bracts and bracteoles linear to lanceolate, 3–6 mm long, glabrous; pedicels 2–3(–8 in fruit) mm long, glabrous. Calyx 18–21 mm long, lobes filiform, glabrous or with few up to 2 mm long hairs. Corolla not seen. Capsule 2.1–2.3 cm long, glabrous or with a few long non-glandular hairs towards the apex; seed ± 6 × 5 mm, margin with 5–6 projections, surfaces with fused tubercles peripherally and the middle area nearly smooth. Fig. 58/17, p. 446

TANZANIA. Kigoma District: W of Uvinza, 6 Feb. 1926, *Peter* 46123!
DISTR. **T** 4; not known elsewhere
HAB. Not indicated but probably woodland or wooded grassland; 1000 m

NOTE. Known only from this collection. Perhaps related to *A. glandulosa* but differs in the small ovate to sub-orbicular leaves, much longer calyx lobes, glabrous capsules and different seeds. It also resembles *A. richardsiae* in having a woody rootstock and alternately arranged flowers, but differs in its small ovate to sub-orbicular leaves, longer almost glabrous calyx lobes and glabrous capsules. This species has the longest calyx lobes in the genus.

8. **Asystasia linearis** *S. Moore* in J.B. 32: 136 (1894); Lindau in P.O.A. C: 370 (1895); C.B. Clarke in F.T.A. 5: 132 (1899); Vollesen in Opera Bot. 59, fig. 10 (1980). Type: Kenya, Kwale District, Mwache [Mwatchi], Duruma, *Gregory* s.n. (BM!, holo.)

Perennial herb with 1–2 stems from short creeping rootstock with fleshy roots; stems to 75 cm long, decumbent or straggling, not rooting at nodes, glabrous to sparsely puberulous and with tufts of white hairs at the nodes. Leaves with petiole 0–5 mm long; lamina lanceolate to narrowly elliptic, largest 5.5–10.5 × 0.7–1.1(–1.7) cm, base attenuate to cuneate, apex acute to obtuse; glabrous or sparsely puberulous along midrib. Racemes terminal and axillary, (2–)5–13.5 cm long, glabrous or sparsely puberulous with non-glandular hairs, with longer hairs at nodes; flowers alternate; peduncles 1–7 cm long; bracts and bracteoles lanceolate to narrowly triangular, 2–4 mm long, with a few long hairs along the margins; pedicels 2–4 mm long, sparsely puberulous or antrorsely sericeous. Calyx 6–10.5 mm long, lobes linear, sparsely puberulous or antrorsely sericeous. Corolla white with violet markings on the lower midlobe, 1.5–1.8 cm long, retrorsely sericeous-puberulous; cylindric tube 5–6 mm long and ± 2 mm wide; throat 6–8 mm long and ± 5 mm wide at the mouth; limb 2-lipped; midlobe in lower lip ovate, ± 6 × 6 mm, gibbous, lateral lobes oblong, ± 4.5 × 3.5 mm, with rounded apex, lobes in upper lip oblong, ± 6 × 3 mm, with rounded apex. Stamens didynamous, included in throat; filaments ± 3 and 5 mm long; thecae slightly superposed, ± 2 mm long, with spur. Style 9–10 mm long, pubescent in basal half. Capsule 1.8–2.5 cm long, sparsely puberulous with short-stalked capitate glands; seed subglobose, ± 4 × 3 mm, greyish brown mottled with darker grey, smooth, margin not distinct, with small scattered tubercles. Fig. 58/5, p. 446.

KENYA. Mombasa District: Durama, Mwatchi, no date, *Gregory* s.n.!
TANZANIA. Tanga District: Umba Steppe, 19 June 1932, *Geilinger* 500!; Bagamoyo District: Bagamoyo, 31 July 1959, *Mgaza* 288!; Kilwa District: Selous Game Reserve, Kingupira, Lungonya Flood Plain, 13 May 1975, *Vollesen* MRC2294!
DISTR. **K** 7; **T** 3, 6, 8; not known elsewhere
HAB. Tall grassland on black cotton soil; 25–200 m

NOTE. Differs from all other species in the genus by its subglobose seeds but is otherwise morphologically a completely typical *Asystasia*.

9. **Asystasia leptostachya** *Lindau* in E.J. 57: 22 (1920); Lebrun & Stork, Enum. Pl. Afr. Trop. 4: 468 (1997); Vollesen & Darbyshire in Pl. Kupe, Mwanenguba & Bakossi: 222 (2004). Type: Cameroon, Babong, Bajoki, *Ledermann* 1164 (B†, holo.). Neotype: Cameroon, Kupe-Mwanenguba, Ngomboku, *Ghogue* 442 (K!, lecto. selected here, NY!, YA!, iso.)

Stoloniferous perennial herb, basal part of stems decumbent, rooting at nodes, apical part straggling or scrambling, to 75 cm long, subglabrous to pubescent in two opposite bands. Leaves with petiole to 2 cm long; lamina elliptic, largest 8–12 × 2.5–5 cm, base attenuate, decurrent, margin entire, apex acute to acuminate with an obtuse tip; glabrous. Racemes terminal, 1.5–5.5 cm long, simple or with 1 to several branches from base, subglabrous to sparsely puberulous in two bands; flowers alternate; peduncles 0.5–2 cm long; bracts and bracteoles triangular or narrowly so; 0.5–1.5 mm long, pedicels 0.5–1.5 mm long, puberulous occasionally with a few capitate glands. Calyx 2.5–3.5 mm long, lobes lanceolate, minutely puberulous and with short-stalked capitate glands. Corolla yellowish-green or greenish-white, without purple markings on lower lip, 11–14 mm long, with scattered short-stalked capitate glands and non-glandular hairs, rugula glabrous; tube slightly curved, cylindric part 3–5 mm long and 2–3 mm wide, merging almost gradually into throat; throat 4–5 mm long and 3–4 mm wide at the mouth; limb indistinctly 2-lipped, lobes 2–4 × 2–4.5 mm. Stamens didynamous with the anthers of long stamens included in throat or slightly protruding; filaments 2.5–3.5 and 3.5–5 mm long, pubescent with a row of glandular hairs or glabrous; thecae slightly superposed, 1.5–2 mm long, glabrous, with minute spur. Style 4.5–8 mm long, glabrous. Capsule 1.8–2.3 cm long, sparsely puberulous with dark short-stalked capitate glands; seed oblong, 4–5 × 3–3.5 mm, not or slightly toothed at the indistinct margin; surface with ridge like tubercles and very finely puberulous surface. Fig. 58/6, p. 446.

UGANDA. Mengo District: Mawakota, March 1905, *E. Brown* 201!
TANZANIA. Bukoba District: Minziro Forest Reserve, 1 July 2000, *Bidgood et al.* 4772! & 22 Dec. 2000, *Festo et al.* 858! & 12 July 2001, *Festo et al.* 1599!

DISTR. **U** 4; **T** 1; Sierra Leone, Liberia, Ivory Coast, Ghana, Nigeria, Cameroon, Bioko, Gabon, Central African Republic, Congo-Kinshasa
HAB. Lower montane evergreen forest and swamp forest; 1100–1300 m

10. **Asystasia minutiflora** *Ensermu & Vollesen* **sp. nov**. ab *A. leptostachya* caulibus floriferi erectis (nec procumbentibus nec scandentibus), corolla minore 0.7–1.2 cm (nec 1.1–1.4 cm) longa albo sed ad rugulam regione purpurea ornata (nec uniformiter flaveo-viridi nec albo-viride) et semine minore 3–4 × 2.5–3 mm (nec 4–5 × 3–3.5 mm) atque superficie seminis alio modo sculpta differt. Typus: Tanzania, Lindi District, Rondo Plateau, St. Cyprians College, *Bidgood, Abdallah & Vollesen* 1622 (K!, holo.; BR!, C!, CAS!, DSM!, EA!, MO!, NHT!, P!, UPS!, WAG!, iso.)

Stoloniferous perennial herb, basal part of stems decumbent, rooting at nodes, apical part erect, to 30 cm long, puberulous or antrorsely or retrorsely sericeous-puberulous with broad curly many-celled hairs, sometimes in two bands. Leaves with petiole to 1.5 cm long; lamina narrowly ovate to ovate or narrowly elliptic to elliptic, largest 6–8(–11) × 2–3.2 cm, base cuneate to attenuate (and then decurrent), often unequal-sided, margin entire, apex rounded to acuminate; glabrous or with a few hairs (rarely antrorsely sericeous) along midrib. Racemes terminal (sometimes 2 side by side), sometimes laterally displaced, 1–5 cm long, usually unbranched (rarely with a single branch from base), puberulous or sericeous or sparsely so with curly many-celled hairs; flowers alternate; peduncles 0.5–2.5(–3.5) cm long; bracts and bracteoles triangular or narrowly so, 0.5–2 mm long; pedicels 0.5–1.5 mm long, subglabrous to sparsely puberulous and usually with a few capitate glands towards apex. Calyx 2.5–4(–5 in fruit) mm long, lobes linear-lanceolate, puberulous or sparsely so and with short-stalked capitate glands. Corolla white with large purple patch on rugula, 7–12 mm long, with scattered short-stalked capitate glands and hairs, purple patch with long curly hairs; tube slightly curved, cylindric part 2.5–4 mm long and 1–2 mm wide, merging almost gradually into throat; throat 4–5 mm long and 2–3.5 mm wide at the mouth; limb indistinctly 2-lipped, lobes 2–4 × ± 2 mm. Stamens didynamous, included in throat; filaments 1–2 and 2–3 mm long, glabrous; thecae slightly superposed, ± 1 mm long, glabrous, with minute spur. Style 4–6 mm long, glabrous. Capsule 1.2–1.7 cm long, sparsely to densely (rarely) puberulous with stalked capitate glands and hairs (rarely absent); seed 3–4 × 2.5–3 mm, with a large triangular basal "spur", strongly toothed at the distinct margin; surface with evenly distributed spinulose-tuberculate tubercles, surface not puberulous. Fig. 58/8, p. 446.

KENYA. Kwale District: Shimba Hills, Mwele Mdogo Forest, 25 Aug. 1953, *Drummond & Hemsley* 3973! & Shimba Hills, Longo Mwagandi Forest, 22 May 1968, *Magogo & Glover* 1101! & Shimba Hills, Mwele Forest, 4 Jan. 1989, *Luke* 1650!
TANZANIA. Lushoto District: E Usambara Mts, 1.5 km NNE of Amani, 23 July 1953, *Drummond & Hemsley* 3413!; Morogoro District: Nguru Mts, Koruhamba, Nov. 1953, *Paulo* 213!; Lindi District, Rondo Plateau, Rondo Forest Reserve, 10 March 2002, *Congdon* 634!
DISTR. **K** 7; **T** 3, 6, 8; not known elsewhere
HAB. Lowland and intermediate wet evergreen forest, lowland semi-deciduous forest and thicket; 300–1000 m

SYN. *Asystasia sp. H sensu* Iversen in Symb. Bot. Upsal. 29(3): 160 (1991)

11. **Asystasia tanzaniensis** *Ensermu & Vollesen* **sp. nov.** ab *A. minutiflora* atque *A. leptostachya* corolla maiore 1.4–2.6 cm (nec 0.7–1.4 cm) longa, cremea vel flaveo-viridi atque in lobo superiore lobisque lateralibus inferioribus brunneo-purpureo-maculata, sed ad rugulum regione purpurea ornata, calyce maiore 3.5–6 mm (nec 2.5–4 mm) longa, capsula maiore 2.2–2.9 cm (nec 1.2–2.3 cm) longa et semine maiore 5–5.5 × 4–4.5 mm (nec 3–5 × 2.5–3.5 mm) atque superficie seminis alio modo sculpta differt. Typus: Tanzania, Morogoro District, Uluguru Mts, Shikurufumi Forest Reserve, *Mhoro* UMBCP419 (K!, holo.; K!, MO, iso.)

Erect or (more often) scrambling shrubby herb, basal part of stems sometimes stoloniferous and rooting; stems to 2(–3) m long, glabrous or sparsely pubescent in two bands below nodes, sometimes with band of longer hairs at nodes. Leaves with petiole to 3 cm long; lamina ovate to elliptic, largest (6.5–)8–16(–19) × (2–)2.8–7.2(–9) cm, base cuneate to attenuate (and then decurrent), margin subentire to crenate, apex acute to cuspidate with an obtuse tip; glabrous. Racemes terminal (often 2(–3) side by side), sometimes laterally displaced, 1–9(–12) cm long, often with a single branch or bi-furcate from base, glabrous at base, usually sparsely puberulous upwards; flowers alternate, often with many sterile bracts basally in raceme; peduncle 0.5–3 cm long; bracts and bracteoles triangular or narrowly so, 1–2(–3) mm long; pedicels 1.5–4(–6 in fruit) mm long (but see note), puberulous or sparsely so and often with a few capitate glands towards apex. Calyx 3.5–6 mm long (but see note), lobes lanceolate, puberulous or sparsely so and with scattered to dense stalked capitate glands. Corolla cream or yellowish green (? sometimes white) with brownish purple markings on upper and lateral lower lobes and a large mauve patch on rugula, 14–26 mm long, with scattered stalked capitate glands and non-glandular hairs, mauve patch without long curly hairs; tube straight, constricted in apical half, cylindric part 5–6 mm long and 2–3.5 mm wide, clearly distinct from throat; throat 6–10 mm long and 4–8 mm wide at the mouth; limb distinctly 2-lipped, midlobe 5–10 × 4–8 mm, lateral lobes and lobes in upper lip 3–6 × 3–6 mm. Stamens didynamous, included in throat; filaments ± 3 and ± 5 mm long, glabrous; thecae very slightly superposed, 2–2.5 mm long, glabrous, with minute spur or apiculate. Style 6–9 mm long, hairy in basal part. Capsule 2.2–2.9 cm long, sparsely puberulous with stalked capitate glands and non-glandular hairs (rarely absent); seed 5–5.5 × 4–4.5 mm, with a large triangular basal "spur", margin toothed in apical half or only near apex; surface with scattered evenly distributed spinulose-tuberculate tubercles or ridges, surface glabrous. Fig. 58/7, p. 446.

TANZANIA. Morogoro District: Uluguru Mts, above Bunduki, Mgeta River, 1 Jan. 1975, *Polhill & Wingfield* 4640! & Kimboza Forest Reserve, 23 April 1986, *Pócs & Hall* 8653/O!; Iringa District: Udzungwa Mts National Park, "Camp 244" [7°43'S 36°54'E], 6 Oct. 2001, *Luke et al.* 8172!

DISTR. **T** 6, 7; not known elsewhere

HAB. Lowland semideciduous to montane wet evergreen forest, (250–)900–1600(–1850) m

NOTE. In some collections the pedicels become elongated and much thickened and the calyx much enlarged and fruits do not seem to develop. This might well be some sort of gall. The extreme is seen in *Frimodt-Møller* 154 where the pedicels reach a length of 1.5 cm and the calyx reaches 4 cm.

12. **Asystasia congensis** *C.B. Clarke* in F.T.A. 5: 132 (1899). Type: Congo-Kinshasa, Stanley Pool, *Callewaert* s.n. (K!, holo.)

Stoloniferous perennial herb, basal part of stems decumbent, rooting at nodes, apical part erect or ascending, to 50 cm long, sparsely pubescent below nodes and sometimes with band of longer hairs at nodes. Leaves with petiole to 1.5 cm long; lamina ovate to elliptic or narrowly so, largest 9.5–12.5 × 3–4.5 cm, base cuneate to attenuate or rounded to cordate in the upper leaves, margin entire, apex acute to acuminate with an obtuse tip; glabrous or with a few hairs on midrib. Racemes terminal, sometimes 2 side by side (rarely also lateral), 2–10 cm long, simple, glabrous or sparsely puberulous upwards, flowers alternate; peduncles 1–5 cm long; bracts and bracteoles 1–3 mm long, triangular or narrowly so, glabrous to sparsely puberulous, ciliate; pedicels 2–4 mm long, puberulous with stalked capitate glands and non-glandular hairs. Calyx 4–6 mm long, lobes with whitish margin, lanceolate, puberulous with capitate glands and non-glandular hairs. Corolla white with large purple patch in the throat, 2.5–3.5 cm long, sparsely puberulous with non-glandular hairs and capitate glands; cylindric tube 5–7 mm long and ± 2 mm wide, curved, throat 1.2–1.5 cm long and 5–9 mm wide at the mouth; limb indistinctly 2-lipped, lobes in lower lip ± 1 ×

1.2–1.5 cm, lobes in upper lip ± 8 × 7–9 mm. Stamens didynamous, included in throat; filaments 4–6 and 5–7 mm long, glabrous; thecae slightly superposed, ± 2 mm long, glabrous, acute at the base, but not spurred. Style 1.2–1.7 cm long, sparsely pubescent at the base. Capsule 2.1–3 cm long, sparsely puberulous with short capitate glands and a few non-glandular hairs; seed ± 6 × 5 mm, margin thickened with fused tubercles; testa tuberculate with evenly distributed tubercles, smaller and similar in the middle and a mixture of small and large along the margin. Fig. 58/12, p. 446.

Uganda. Masaka District: Masaka Road, km 74, Aug. 1935, *Chandler* 1302! & 6 km SSW of Katera, Malabigambo Forest Reserve, 2 Nov. 1953, *Drummond & Hemsley* 4540! & Sango Bay Forest Reserve, 10 Feb. 2002, *Bayona & Festo* 92/45!
Tanzania. Bukoba District: Minziro Forest Reserve, 6 July 2000, *Bidgood et al.* 4870!
Distr. **U** 4; **T** 1; Cameroon, Congo-Kinshasa
Hab. Evergreen medium altitude forest, swamp forest; 1150–1200 m

Note. *Chandler* 1302 (K) seems to be galled. The calyx is much thickened and the tube much elongated relative to the lobes. See also note after *A. tanzaniensis*.

13. **Asystasia sp. B**

Perennial herb, basal parts absent; apical part of stems ascending, retrorsely sericeous along the four ridges and with bands of longer hairs at nodes. Leaves with petiole to 1.5 cm long; lamina ovate, largest ± 13 × 5.5 cm, base attenuate or rounded to truncate in the upper leaves, margin entire, apex shortly acuminate; with scattered hairs along major veins. Racemes terminal, 2–8.5 cm long, unbranched, retrorsely sericeous along the four ridges; flowers opposite; peduncles 2.5–5.5 cm long; bracts and bracteoles triangular, 1–2 mm long, sparsely sericeous and ciliate; pedicels 2–5(–10 in fruit) mm long, puberulous with short-stalked capitate glands and non-glandular hairs. Calyx 5–7(–9) mm long, lobes lanceolate, acuminate, margin not whitish, puberulous with short-stalked glandular and some non-glandular hairs. Corolla white (?), apparently without markings, ± 4 cm long, sparsely puberulous with short-stalked capitate glands and non-glandular hairs; basal cylindric tube ± 2 cm long and ± 3 mm wide, throat ± 1 cm long and ± 1 cm wide at the mouth; limb with 5 subequal suborbicular lobes 10–12 × 8–14 mm. Stamens slightly didynamous, included in throat, filaments ± 8 and 9 mm long; thecae slightly superposed, ± 3 mm long, glabrous, not spurred. Style ± 2.5 cm long, hairy at base. Capsule (immature) ± 3 cm long, densely puberulous with short-stalked capitate glands and non-glandular hairs; seed not seen.

Kenya. Meru/Tana River District: Tana River, 5 Apr. 1910, *Battiscombe* 250!
Distr. **K** 4/7; not known elsewhere
Hab. Riverine forest; 450 m

Note. This species, which is known only from this collection, is related to *A. congensis*. It differs in having racemes with opposite flowers, larger corolla, larger capsule, different calyx indumentum and calyx lobes without a white margin.

14. **Asystasia albiflora** *Ensermu* in K.B. 53: 930 (1998). Type: Tanzania, Chunya District, near Mbangala Village, *Bidgood, Mbago & Vollesen* 2252 (K!, holo.; BR!, C!, CAS!, DSM!, K!, iso.)

Erect or scrambling perennial herb from irregular woody rootstock, never stoloniferous; stems to 1 m long, subglabrous to densely retrorsely sericeous. Leaves with petioles to 1.5 cm long; lamina narrowly to broadly ovate, largest 6–17 × 2.8–7 cm, base cuneate to attenuate (lower leaves) or cordate to truncate, upper often abruptly narrowed with decurrent wings, margin entire, apex acuminate; glabrous to sparsely puberulous along veins (rarely also on lamina). Racemes terminal, 1.5–15 cm long, simple (rarely branched from base), subglabrous to retrorsely sericeous; flowers opposite or opposite at base and alternate upwards (rarely all alternate), single (rarely in 3-flowered cymules); peduncles 1–9.5 cm long; bracts and bracteoles triangular to linear-lanceolate, 1–6 mm long, glabrous to puberulous and ciliate; pedicels 1–3(–6 in fruit) mm long, glabrous to densely puberulous or retrorsely sericeous. Calyx 5–8.5(–14 in fruit) mm long, lobes with white margins, linear-subulate, subglabrous to densely puberulous and often also with some capitate glands towards base. Corolla very thin and fragile, pure white or with pale yellow throat, 2.5–4.2 cm long, puberulous and with scattered capitate glands; basal cylindric tube 6–10 mm long and 2–3 mm wide at the base, straight or very slightly curved; throat 1–2 cm long and 5–10 mm wide at the mouth; limb indistinctly 2-lipped, lobes in lower lip 0.9–1.5 × 0.9–1.4 cm, lobes in upper lip 0.7–1 × 0.8–1.1 cm. Stamens didynamous, included in throat; filaments 4–6 and 7–10 mm long, glabrous; thecae slightly superposed, ± 2 mm long, acute at base but not spurred. Style 1.3–2 cm long, pubescent at the base and at the apex. Capsule 2.2–2.8 cm long, densely puberulous with capitate glands and non-glandular hairs; seed 5–5.5 × 4–4.5 mm, margin with dense to confluent tubercles forming a thick rim; testa with scattered tubercles near edges, middle part nearly smooth. Fig. 58/13, p. 446.

UGANDA. Bunyoro District: Budongo Forest, Busingiro, Oct. 1933, *Eggeling* 1436!; Mengo District: Mabira Forest, Mulange, Aug. 1922, *Dümmer* 5424! & Mabira Forest, 13 Nov. 1938, *Loveridge* 65!
TANZANIA. Kigoma District: Mt Livandabe [Lubalisi], 28 May 1997, *Bidgood et al.* 4144!; Mpanda District: Katavi National Park, Ilyandi Sand Ridge, 12 Feb. 1962, *Richards* 16098!; Mbeya District: Mshewe Village, Palule River, 21 Dec. 1980, *Lovett et al.* 3799!
DISTR. **U** 2, 4; **T** 4, 7; Central African Republic, Sudan, Congo-Kinshasa, Zambia
HAB. Glades in forest, transition zone between forest (including riverine) and woodland, *Brachystegia* woodland on rocky hills, termite mounds; 750–1700 m

NOTE. *Bidgood et al.* 3013 is somewhat atypical. All the flowers are alternate and the raceme therefore has the scorpioid look characteristic of *A. gangetica*. But the corolla is large and pure white with pale yellow tube. Also the seed has confluent tubercles forming a thick rim as is characteristic of this species. It would be tempting to assume a hybrid origin were it not for the perfectly developed seeds.

15. **Asystasia laticapsula** *Karlström* in Bot. Not. 128: 235, fig. 1 & 2 (1975); [U.K.W.F.: 598 (1974), *nom. nud.*]; Blundell, Wild Fl. E. Afr.: 386 (1987); U.K.W.F., ed. 2: 275, pl. 121 (1994); Lebrun & Stork, Enum. Pl. Afr. Trop. 4: 468 (1997). Type: Kenya, near Nairobi, *Whyte* s.n. (K!, holo.; K!, iso.)

Perennial herb with several decumbent to ascending or erect stems from woody rootstock, never stoloniferous; stems to 50 cm long (rarely scrambling to 2 m), pubescent to pilose or densely so. Leaves with petiole to 1 cm long; lamina narrowly ovate to oblong, largest 2.5–8 × 0.8–3 cm, base attenuate to cuneate, margin entire (rarely crenate), apex obtuse to acute; pubescent or sparsely so. Racemes terminal or terminating short axillary branches, 2–14(–18 in fruit) cm long, simple, puberulous to pubescent or retrorsely sericeous; flowers all alternate; peduncles 1.3–9.5(–13.5) cm long; bracts and bracteoles linear to narrowly triangular, 1–5 mm long; pedicels 1–4(–7 in fruit) mm long, puberulous to pubescent. Calyx (6–)7–12(–16 in fruit) mm long, lobes linear-lanceolate to subulate, without white margin, puberulous to pubescent no capitate glands. Corolla white with violet or purple marks on the gibbous lower midlobe or pure white with pale yellow tube, 2–2.7 cm long, puberulous and with scattered capitate glands; basal cylindric tube 4–6 mm long and 2–3 mm wide, throat 9–11 mm long and 6–8 mm wide at the mouth; limb indistinctly 2-lipped, lobes subequal, 7–10 × 6–9 mm, rounded at the apex. Stamens slightly didynamous, included in throat; filaments 4–5 and 5–6 mm long, glabrous; thecae only very slightly superposed, 2.5–3 mm long, spurred. Style 10–12 mm long, hairy at the base. Capsule 2–2.5 cm long and 6–7 mm wide, puberulous with non-glandular hairs and scattered capitate glands; seed 4–5 × 3.5–4.5 mm, margin with dense to confluent tubercles forming a thick rim and tuberculate edge; testa tuberculate to rugose. Fig. 58/16, p. 446.

KENYA. South Nyeri District: Mwea Rice Irrigation Scheme, 3 June 1976, *Kahurananga & Kibui* 2795!; Nairobi District: Nairobi National Park, near Main Entrance, 25 May 1961, *Verdcourt & Polhill* 3163!; Masai District: Masai Lodge Road, just outside Nairobi National Park on road to Magadi, 4 Dec. 1977, *Gilbert & Stannard* 4944!

TANZANIA. Moshi District: Himo–Moshi road, 13 Apr. 1940, *Vaughan* 2982!; Pare District: Mkomazi Game Reserve, Kisima Hill, 30 Apr. 1995, *Abdallah & Vollesen* 95/66! & Mkomazi Game Reserve, S of Ndea Hill, 5 June 1996, *Vollesen* 96/39!

DISTR. **K** 4, 6; **T** 2, 3; not known elsewhere

HAB. Grassland and wooded grassland, on black cotton soil or on rocky hills; 950–1700 m

NOTE. The Tanzanian material has a white corolla with pale yellow tube but no markings on the palate, while all the Kenyan collections have violet/purple markings on the palate. No other differences have been found.

16. **Asystasia gangetica** (*L.*) *T. Anderson* in Thwaites, Enum. Pl. Zeyl.: 235 (1860) & in J.L.S. Bot. 7: 52 (1867); S. Moore in J.B. 18: 308 (1880); Engler, Hochgebirgsfl. Trop. Afr.: 392 (1892); Lindau in P.O.A. C: 370 (1895) & in Ann. Ist. Bot. Roma 6: 79 (1896); E.P.A.: 957 (1964); U.K.W.F.: 598 (1974); Vollesen in Opera Bot. 59: 78 (1980); Champluvier in Fl. Rwanda 3: 436 (1985); Blundell, Wild Fl. E. Afr.: 386, pl. 797 (1987); Iversen in Symb. Bot. Ups. 29(3): 160 (1991); Ensermu in Proc. XIII Plenary Meeting AETFAT 1: 342 (1994); U.K.W.F., ed. 2: 275 (1994); Lebrun & Stork, Enum. Pl. Afr. Trop. 4: 468 (1997); Friis & Vollesen in Biol. Skr. 51: 436 (2005). Type: India, Ganges, *Herb. Linn.* 28.27 (LINN, lecto., selected by Ensermu, l.c.; K!, photo)

Perennial or suffrutescent herb with irregular rootstock, occasionally rooting from lower nodes; stems creeping-ascending or straggling, to 2(–3) m long, glabrous to densely pubescent or retrorsely sericeous. Leaves with petiole to 3(–5) cm long; lamina ovate or narrowly so (rarely lanceolate), largest (2.5–)3–10.5(–16) × 1–5(–8) cm, base cordate to truncate or lower attenuate to rounded, sometimes abruptly narrowed into a winged petiole, sometimes subhastate, margin entire (rarely crenate or dentate), apex acuminate to obtuse (rarely rounded); subglabrous to pubescent, densest on veins. Racemes terminal or on axillary branches (rarely axillary), 1–20(–25) cm long, simple (rarely branched), subglabrous to pubescent or retrorsely sericeous; all flowers alternate; peduncles 1–8(–9) cm long; bracts and bracteoles triangular to subulate, 0.5–2.5 mm long, sparsely puberulous and distinctly ciliate; pedicels 1–2(–4 in fruit) mm long, subglabrous to pubescent (rarely with capitate glands). Calyx 4–9 mm long, lobes subulate to linear-lanceolate, sparsely puberulous to densely pubescent and sometimes with capitate glands. Corolla white to purple, lower mid-lobe with large violet or purple patch, 1–4 cm long, puberulous or densely so with non-glandular hairs and few to many capitate glands; basal cylindric tube 2–7(–8) mm long and 2–3 mm in diameter; throat 5–11 mm long and 6–10 mm in diameter apically; limb 2-lipped, lobes unequal, 3–10 mm long, midlobe in lower lip the largest. Stamens didynamous, included in tube; filaments 3–5 and 4–6 mm long; thecae slightly offset, ± 2 mm long, spurred. Style 7–15 mm long, puberulous for $^1/_3$ to half of its length. Capsule (1.7–)2–3 × cm long and 4–5 mm wide, puberulous or densely so with capitate glands and non-glandular hairs; seed 4–5.5 × 4–4.5 mm, margin ± distinctly thickened from confluent tubercles, edge straight; surface tuberculate, densest towards margin. Fig. 59, p. 458.

SYN. *Justicia gangetica* L., Amoen. Acad. 4: 299 (1759)

Asystasia coromandeliana Nees in Wall. Pl. As. Rar. 3: 89 (1832) & in DC., Prodr. 11: 165 (1847); Solms-Laubert in Schweinfurth, Beitr. Fl. Aeth.: 104 (1867); C.B. Clarke in Fl. Brit. Ind. 4: 493 (1884) & in F.T.A. 5: 131 (1899) & in Fl. Cap. 5: 42 (1901). Types: India, many syntypes

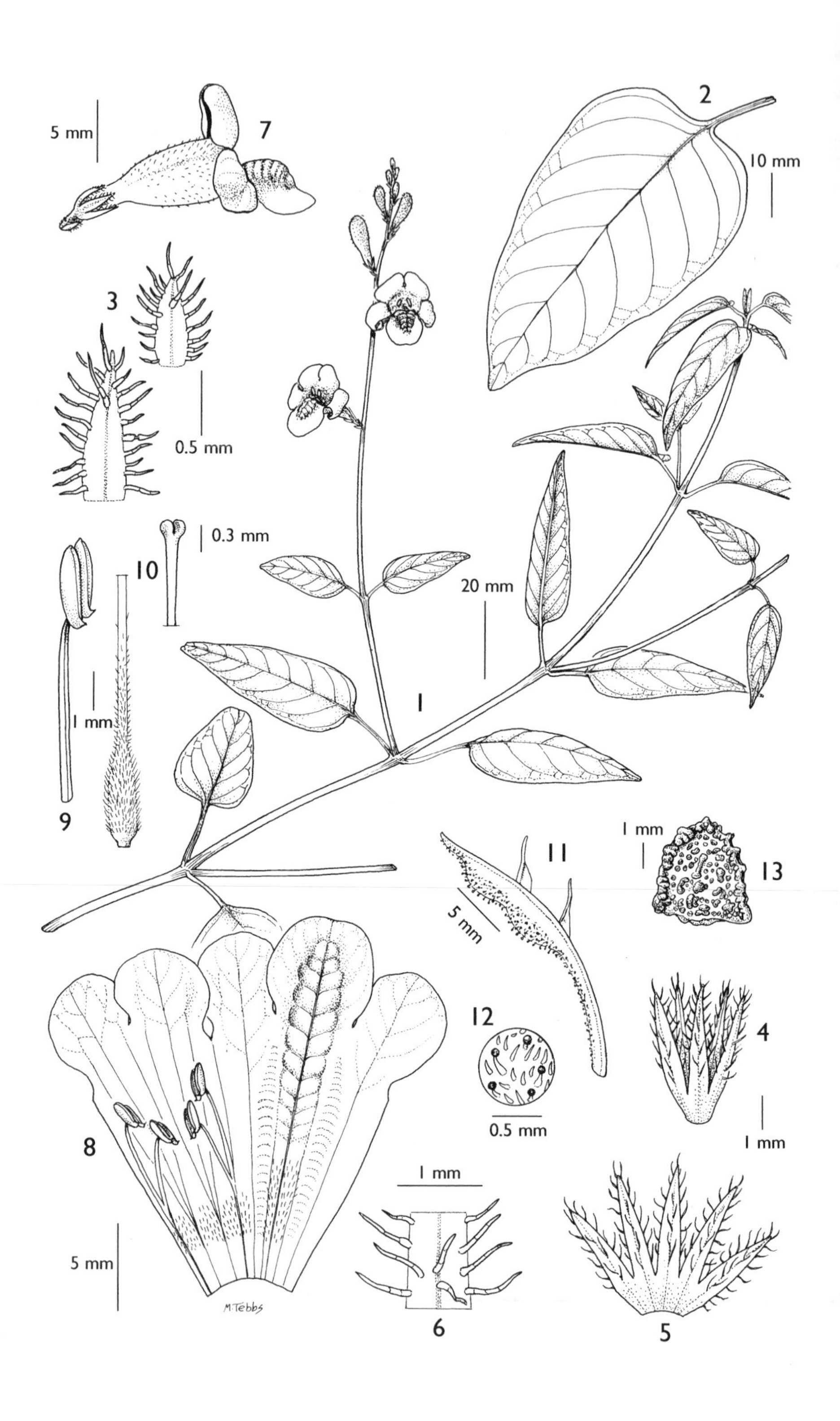
5 mm
7
2
10 mm
3
0.5 mm
0.3 mm
10
20 mm
1 mm
1
9
1 mm
11
13
5 mm
12
4
0.5 mm
1 mm
8
1 mm
5 mm
M. Tebbs
6
5

subsp. **micrantha** (*Nees*) *Ensermu* in Proc. XIII Plenary Meeting AETFAT 1: 343 (1994); Hedrén in Fl. Somalia 3: 399, fig. 271 (2006); Strugnell, Checklist Mt Mulanje: 33 (2006); Ensermu in Fl. Eth. 5: 399 (2006). Type: Sudan, Sennar, *Acerbi* in Herb. DC. 687 (not seen)

Corolla white to pale mauve or pale blue, 1–2.5(–2.8) cm long.

UGANDA. West Nile District: Metu Rest Camp, 20 Sept. 1953, *Chancellor* 299!; Busoga District: Jinja, Aug. 1904, *E. Brown* 74!; Masaka District: Sese Islands, Bugalla Island, 8 Oct. 1958, *Symes* 475!

KENYA. Kisumu District: Kisumu, 9 Aug. 1956, *Napper* 562!; Kilifi District: 8 km S of Malindi, Casuarina Point, 18 Nov. 1992, *Harvey et al.* 54! & Arabuko Sokoke Forest, 25 Nov. 1992, *Robertson & Brummitt* 6753!

TANZANIA. Kigoma District: Bulimba, 27 May 1975, *Kahurananga et al.* 2656!; Uzaramo District: 20 km N of Dar es Salaam, Bahari Beach, 10 Nov. 1986, *Brummitt et al.* 17951!; Kilwa District: Selous Game Reserve, 19 km SSW of Kingupira, 4 April 1976, *Vollesen* MRC3428!; Zanzibar, Massazine, 16 July 1959, *Faulkner* 2303!

DISTR. **U** 1–4; **K** 1–7; **T** 1–8; **Z**; **P**; throughout tropical Africa and South Africa; Madagascar and Mascarene Islands, tropical Arabia, introduced into Malaysia

HAB. Forest margins and clearings and a wide variety of woodland, wooded grassland and bushland, sandy sea shores, roadsides, ditches, plantations, often in disturbed or secondary vegetation; 0–1700 m

SYN. *A. coromandeliana* Nees var. *micrantha* Nees in DC., Prodr. 11: 165 (1847)

A. chelonoides Nees var. *arabica* Nees in DC., Prodr. 11: 164 (1847). Types: Yemen, Djennah & Bahr El Abiad, *Botta* 201 (not seen); Sudan, Fazoghel, *Kotschy* 423 (K!, syn.)

A. podostachys Klotzsch in Peters, Reise Mossamb. Bot.: 199 (1861);Vollesen in Opera Bot. 59: 78 (1980). Type: Tanzania, Zanzibar, *Peters* s.n. (B†, holo.)

A. subhastata Klotzsch in Peters, Reise Mossamb. Bot.: 200 (1861); Vollesen in Opera Bot. 59: 78 (1980). Type: Mozambique, Boror, *Peters* s.n. (B†, holo.)

A. floribunda Klotzsch in Peters, Reise Mossamb. Bot.: 200 (1861). Type: Mozambique, Boror, *Peters* s.n. (B†, holo.)

A. acuminata Klotzsch in Peters, Reise Mossamb. Bot.: 201 (1861). Type: Mozambique, Querimba, *Peters* s.n. (B†, holo.)

A. pubescens Klotzsch in Peters, Reise Mossamb. Bot.: 202 (1861). Type: Mozambique, Anjoana Island and Mossambique Island, *Peters* s.n. (B†, holo.)

A. scabrida Klotzsch in Peters, Reise Mossamb. Bot.: 202 (1861). Type: Mozambique, Mossambique Island, *Peters* s.n. (B†, holo.)

A. multiflora Klotzsch in Peters, Reise Mossamb. Bot.: 203 (1861). Type: Tanzania, Zanzibar, *Peters* s.n. (B†, holo.)

A. querimbensis Klotzsch in Peters, Reise Mossamb. Bot.: 204 (1861). Type: Mozambique, Querimba, *Peters* s.n. (B†, holo.)

A. parvula C.B. Clarke in F.T.A. 5: 132 (1899); Chiovenda, Fl. Somala: 251 (1929) & Fl. Somala 2: 350 (1932); E.P.A.: 957 (1964). Type: Ethiopia, Adda Galla, *James & Thrupp* s.n. (K!, holo.)

A. coromandeliana Nees var. *linearifolia* S. Moore in J.L.S. Bot. 40: 160 (1911). Type: Mozambique, Beira, *Swynnerton* 1949 (BM!, holo.)

A. ansellioides C.B. Clarke var. *lanceolata* Fiori in Result. Sci. Miss. Stef.-Paoli, Coll. Bot.: 138 (1916). Type: Somalia, Bur Eibe, *Stefanini* 1294bis (FT, holo.)

A. pinguifolia T. Edwards in S. Afr. J. Bot. 53 (3): 231 (1987). Type: Mozambique, Inharrime, 15 km south of Inhambane, near Mutamba, *Edwards & Vahrmeyer* 4228 (PRE, holo.)

NOTE. Subsp. *gangetica*, an erect shrubby herb with a purple corolla 2.5–4 cm long, is native to India, Sri Lanka, Thailand, Malaysia, Philippines and the Pacific. It has been introduced (as an ornamental) to Mauritius, Tropical Africa, Florida, central and South America and the West Indies.

FIG. 59. *ASYSTASIA GANGETICA* — **1**, habit; **2**, leaf; **3**, bract (left) and bracteole (right); **4**, calyx; **5**, calyx opened up; **6**, detail of calyx indumentum; **7**, corolla; **8**, corolla, opened up; **9**, apical part of filament and anther; **10**, ovary, basal part of style and stigma; **11**, capsule; **12**, detail of capsule indumentum; **13**, seed. 1 & 3–10 from *Congdon* 380, 2 from *Mwangoka* 4366, 11–12 from *Peter* 60689, 13 from *Abeid* 119. Drawn by Margaret Tebbs.

Considering the large total distribution of this taxon it is relatively uniform. The material from the Flora area is variable but, compared with similarly widespread taxa in other genera, not excessively so. Material from inland areas invariably has a white corolla while material from the coast often (but not always) has a mauve or pale blue corolla. Coastal forms also tend to have smaller flowers.

17. **Asystasia calcicola** *Ensermu & Vollesen* **sp. nov.** ab *A. gangetica* racemis plerumque axillaribus (nec omnibus terminalibus), corolla pallide caerulea vel pallide malvina, filamentis brevioribus antheras minores non calcaratas gerentibus et superficie seminibus alio modo sculpta differt. Crescit in terras calcareas, ubi *A. gangetica* numquam invenitur. Typus: Somalia, 15 km N of El Dhere on road from Aden Yabal, *Gillett & Watson* 24607 (K!, holo.; EA!, iso.)

Annual or perennial herb with up to 10 stems from taproot or woody rootstock; stems erect or semi-erect, to 20 cm long, pubescent or retrorsely sericeous-puberulous or -pubescent, with bands of long hairs at nodes. Leaves with petiole to 7 mm long; lamina ovate or narrowly so, largest 1.5–3.7 × 0.5–1.7 cm, base cuneate (lower) to truncate or subcordate, margin entire to crenate, sometimes subhastate, apex acute to obtuse; pubescent or sparsely so, densest on veins. Racemes axillary or some terminal, 1–5(–8) cm long, simple, puberulous to pubescent or sparsely so; flowers alternate, 3–5(–7) per raceme; peduncles 1.5–3.5 cm long; bracts and bracteoles triangular to subulate, 0.5–1(–2) mm long, glabrous or sparsely puberulous on midrib, distinctly ciliate; pedicels 1–1.5(–3 in fruit) mm long, finely puberulous and apically with capitate glands. Calyx 3–5(–7) mm long, lobes subulate to linear, finely puberulous and with few to many capitate glands, also with scattered to dense long pilose hairs. Corolla pale blue or pale mauve, with darker markings at base of lower lip, 1–1.5 cm long, sparsely puberulous with non-glandular hairs and capitate glands; basal cylindric tube 2–3 mm long and ± 2 mm in diameter; throat 5–8 mm long and 5–7 mm in diameter apically; limb weakly 2-lipped, lobes unequal, 3–5 mm long, midlobe in lower lip the largest. Stamens didynamous, included in tube; filaments ± 3 and 4 mm long; thecae slightly offset, ± 1.5 mm long, not spurred. Style 6–8 mm long, puberulous for ± 1 mm at base. Capsule 1.8–2.3 cm long, finely puberulous with capitate glands and non-glandular hairs; seed 4–5.5 × 3–3.5 mm, margin strongly thickened to form a continuous rim, edge straight or wavy; surface sparsely to densely tuberculate, densest towards centre. Fig. 58/5, p. 446.

KENYA. Northern Frontier District: 48 km on Ramu–El Wak road, 9–10 May 1978, *M.G. Gilbert & Thulin* 1614! & 12 km S of El Wak on Wajir road, 11 May 1978, *M.G. Gilbert & Thulin* 1657!
DISTR. **K** 1; Somalia
HAB. *Acacia-Commiphora-Delonix* bushland on deep sandy limestone derived soil or on rocky limestone hills; 400–650 m

NOTE. This is closely related to the widespread and variable *Asystasia gangetica*. It differs in the mainly axillary racemes, blue or mauve corolla, shorter filaments with smaller anthers and shorter style which is only hairy at the very base as well as in a different seed morphology (Fig. 58, p. 446). Blue or mauve forms of *A. gangetica* occur along the East African coast but always have the characteristics of that species.

Asystasia calcicola also has a different ecology to *A. gangetica*, always growing on limestone rocks or limestone-derived sand, a habitat alien to *A. gangetica*. The old Jurassic limestone plateaus in NE Africa are known to house large numbers of often sparsely collected local endemics.

18. **Asystasia schliebenii** *Mildbr.* in N.B.G.B. 11: 823 (1933); Lebrun & Stork, Enum. Pl. Afr. Trop. 4: 468 (1997). Type: Tanzania, Ulanga District, Mahenge, Liondo, *Schlieben* 2101 (B†, holo.; BM, iso.!)

Perennial herb; stems scrambling or decumbent, to 1 m long, glabrous except for a tuft of white hairs at the nodes, or sparsely puberulous in two bands. Leaves with petiole to 2.5 cm long; lamina ovate, largest 4–9 × 1.5–3 cm, base attenuate to truncate, margin entire, apex acuminate to acute with an obtuse tip; glabrous beneath, with scattered short hairs above. Racemes terminal or terminating short axillary branches, 1–5.5 cm long, 1-sided, glabrous; flowers alternate; peduncles 1–7 cm long; bracts and bracteoles triangular to linear-lanceolate, 1–3 mm long, glabrous except for the sparsely ciliate margins; pedicels 1–3 mm long, glabrous. Calyx 5–9 mm long, lobes linear-subulate, glabrous. Corolla white with purple markings on rugula, 1.8–2.3 cm long sparsely puberulous; basal cylindric tube 4–5 mm long and ± 2.5 mm in diameter; throat 6–9 mm long and ± 9 mm in diameter at mouth, lobes 8–11 mm long, middle in lower lip largest. Stamens didynamous, included or slightly exserted, filaments ± 4 and 5 mm long, glabrous; thecae ± 2.5 mm long, with minute spur. Style ± 12 mm long, hairy in basal 2 mm. Capsule ± 2.2 cm long, glabrous; seed ± 5 × 4 mm, margin slightly lobate, but not distinctly thickened; testa vaguely tuberculate.

TANZANIA. Morogoro District: Nguru Mts, Turiani, Mhonda Mission, 14 Feb. 1988, *J. Lovett & Congdon* 3066!; Ulanga District: Mahenge, Liondo, 18 Apr. 1932, *Schlieben* 2101!

DISTR. **T** 6; not known elsewhere

HAB. Moist evergreen lowland forest, rocky outcrops; 500–900 m

NOTE. *A. schliebenii* is very closely related to *A. gangetica* and may eventually prove to be conspecific with it. The two are very similar in general appearance. They differ in that *A. schliebenii* has glabrous calyx lobes, ovary and capsules. Totally glabrous calyx lobes and a glabrous capsule have never been seen in any of the many collections of *A. gangetica*.

19. **Asystasia ansellioides** *C.B. Clarke* in F.T.A. 5: 136 (1899); Chiovenda, Fl. Somala: 251 (1929) & Fl. Somala 2: 352 (1932); E.P.A.: 956 (1964); Luke & Robertson, Kenya Coastal Forests Checklist Vasc. Pl.: 80 (1993); Lebrun & Stork, Enum. Pl. Afr. Trop. 4: 468 (1997). Type: Kenya, Tana River, *A.S. Thomas* 30 (B†, holo.). Neotype: Kenya, Kwale District, Shimba Hills, Mwalugange Forest Reserve, *Luke et al.* 4509 (K!, neotype, selected here; EA!, iso.)

Much-branched perennial herb; stems to 40 cm long (rarely scrambling to 75 cm), basal part creeping and rooting, apical part erect (rarely scrambling), pubescent to pilose. Leaves with petiole to 2 cm long; lamina broadly to narrowly ovate, largest 2–5.5 × 1–2.5 cm, base attenuate to rounded, margin entire to crenate, apex acute to shortly acuminate; sparsely pubescent to densely pilose. Racemes axillary, 0.5–1(–2) cm long, simple, with 1–2(–3) alternate flowers, sparsely to densely pilose; peduncles 0.5–4 cm long; bracts and bracteoles subulate to narrowly triangular, 1–2 mm long, glabrous but ciliate with long white hairs; pedicels 1–3(–7 in fruit) mm long, pilose. Calyx 8–15 mm long, lobes linear-lanceolate or linear-subulate, sparsely pilose and densely ciliate. Corolla pale pink, pale mauve or white with purple markings in the throat, 1.3–2 cm long, sparsely pubescent or retrorsely sericeous with non-glandular hairs; basal cylindric tube 3–4 mm long and 2–3 mm wide; throat 8–9 mm long and 4–6 mm wide at the mouth; limb indistinctly 2-lipped, lower midlobe the largest, ± 6 × 4.5 mm, lateral lobes ± 6 × 3.5 mm, lobes in upper lip ± 4 × 3 mm. Stamens slightly didynamous; filaments ± 2.5 mm and ± 4 mm long, glabrous; thecae blue, slightly superposed, ± 1.5 mm long, glabrous, not spurred. Style 6–9 mm long, pubescent in the basal 1–2 mm. Capsule 1.5–2.2 cm long, densely puberulous with non-glandular hairs; seed ± 4 × 3 mm, margin very distinctly thickened and with sharp projections, at least 2 apical ones, surfaces rough to distinctly tuberculate. Fig. 58/14, p. 446.

KENYA. Northern Frontier District: Boni Forest, 3 Oct. 1947, *Bally* 6105!; Malindi District: Dagrama, May 1959, *Rawlins* 699!; Kwale District: Shimba Hills, Mwalugange Forest Reserve, 2 June 1996, *Luke et al.* 4509!

TANZANIA. Tanga District: Mtotohovu, near Moa, 10 Sep. 1951, *Greenway* 8705!; Uzaramo District: Dar es Salaam, Kigogo–Ubungo road, 22 July 1971, *Wingfield* 1723! & Dar es Salaam, Makurumula, 24 July. 1971, *Wingfield* 1728!

DISTR. **K** 1, 7; **T** 3, 6; Somalia, Socotra
HAB. Open grassland, *Hyphaene* grassland, open grassy places in lowland *Cynometra* forest and *Brachystegia* woodland, on orange-brown sandy loamy or red sandy soil; 15–200 m

SYN. [*Asystasia coromandeliana sensu* Balfour in Trans. Roy. Soc. Edinb. 31: 217 (1888) pro parte, *non* Nees (1832)]

20. **Asystasia africana** (*S. Moore*) *C.B. Clarke* in F.T.A. 5: 134 (1899) & in Cat. Afr. Pl. Welw. 1: 818 (1900); Heine in Fl. Gabon 13: 141, pl. 28 (1966); Lebrun & Stork, Enum. Pl. Afr. Trop. 4: 468 (1997); Friis & Vollesen in Biol. Skr. 51: 436 (2005). Type: Angola, Pungo Andongo, Quilange, *Welwitsch* 5073 (BM!, lecto., selected here; BM!, K!, iso.)

Erect or scrambling subshrub to 2 m tall, sparsely branched; young branches glabrous to puberulous in two bands and with band of longer brownish hairs at nodes. Leaves with petiole to 3 cm long; lamina ovate or elliptic, largest 10–17 × 3–6 cm, base attenuate, decurrent, margin entire, apex acuminate to caudate with acute to rounded tip; glabrous beneath, with scattered appressed hairs above. Racemes terminal or also axillary from upper axils, 3–5 cm long, simple or branched from lower nodes, merging into panicles up to 7 cm long; sparsely puberulous in two bands; all flowers opposite; peduncles 1–2.7 cm long; bracts and bracteoles linear to narrowly triangular, 1.5–2.5 mm long, glabrous; pedicels 1.5–3.5 mm long, sparsely puberulous and with sub-sessile capitate glands apically. Calyx 3–5 mm long, lobes subulate, sparsely puberulous with short-stalked capitate glands. Corolla white with pink to dark purple markings on the gibbous lower lip, 1.3–1.8 cm long, sparsely puberulous with short-stalked capitate glands; basal cylindric tube 4–6 mm long and 2–2.5 mm wide; throat 3–4 mm long and 3.5–5 mm wide at the mouth; limb strongly 2-lipped, lower lip 6–9 mm long and ± 7 mm wide about the middle, lobes ± 3 × 2 mm, upper lip 7–9 mm long and ± 6 mm wide, lobes ± 1 mm long. Stamens didynamous, exserted, protruding by 1–3 mm from corolla tube; filaments 3–4 and 4–6 mm long, glabrous; anthers purple, thecae 1.5–2 mm long, spurred. Style 7–9 mm long, glabrous. Capsule 2.5–3 cm long, glabrous or with a few short-stalked capitate glands; seed ± 5 × 4 mm; testa uniformly reticulate-verrucose, no distinct rim. Fig. 58/3, p. 446.

UGANDA. Ankole District: Igara, Kalinzu Forest, 6 Jan. 1953, *Osmaston* 2775!; Masaka District: NW side of Lake Nabugabo, 9 Oct. 1953, *Drummond & Hemsley* 4726!
TANZANIA. Bukoba District: Minziro Forest Reserve, 7 Feb. 2002, *Bayona & Festo* 92/10!
DISTR. **U** 2, 4; **T** 1; Cameroon, Gabon, Congo-Kinshasa, Sudan, Angola
HAB. Wet evergreen forest, swamp forest; 1100–1500 m

SYN. *Isochoriste africana* S. Moore in J.B. 18: 309 (1880); Lindau in E. & P. Pf. IV, 3b: 326 (1895)
I. africana S. Moore forma *parvifolia* S. Moore in J.B. 18: 309 (1880). Type: Angola, Pungo Andongo, Quilange, *Welwitsch* 5078 (BM!, holo.)
Styasasia africana (S. Moore) S. Moore in J.L.S. Bot. 37: 195, t.2 (1905)
S. africana (S. Moore) S. Moore var. *parviflora* S. Moore in J.L.S. Bot. 37: 195 (1905). Type: Uganda, Masaka District: Lake Victoria, Misozi [Musozi], *Bagshawe* 54 (BM!, holo.)

NOTE. This species has a large total distribution but is known from rather few collections. Usually only one or two collections are available from each country. It is possible that this species exhibits periodic mass-flowering as seen in a number of other Acanthaceae; e.g. in African species of *Acanthopale*, *Epiclastopelma*, *Isoglossa* and *Mimulopsis*. It may also be noted that both the Ugandan specimens were collected in the same year and it has not been collected since in that country. Kew has two collections from Gabon also both collected in the same year.

21. **Asystasia moorei** *Ensermu* **nom. nov.** Type: Kenya, Mombasa District, Freretown, Mkomani, *W. E. Taylor* s.n. (BM!, holo.)

Erect or scrambling shrubby herb or shrub to 1.5(–2) m tall; young branches glabrous to puberulous in two bands with curly hairs and with bands of longer hairs at nodes. Leaves with petiole to 4 cm long; lamina elliptic to slightly obovate, largest 10–21 × 3.5–8 cm, base cuneate to attenuate and decurrent, margin entire to crenate, apex acuminate to caudate with acute to rounded tip; glabrous. Racemes terminal or also axillary from upper axils and then forming panicles, racemes 1–5 cm long, simple or branched at base, panicles to 8 cm long, puberulous or sparsely so; flowers opposite; peduncles 1–3(–4) cm long, sometimes with a pair of foliaceous bracts; bracts and bracteoles triangular or narrowly so, 1.5–4 mm long, subglabrous; pedicels (2–)3–7 mm long, densely minutely puberulous with short stubby non-glandular hairs. Calyx 3–6 mm long, lobes linear-subulate, sparsely to densely puberulous with stalked capitate glands. Corolla white with large mauve to purple patch on the gibbous lower lip, 2.5–3.5 cm long, puberulous or sparsely so with stalked capitate glands; basal cylindric tube 2–3 mm long and 2–4 mm wide; throat 1–1.2 cm long and 6–8 mm wide at the mouth; limb strongly 2-lipped, lower lip 1.5–2.2 cm long, midlobe gibbous, 7–9 × 7–9 mm, lateral lobes 7–9 × 4–5 mm, folded downwards; upper lip 1.3–1.9 cm long, bifid at the apex. Stamens slightly didynamous, long exserted; filaments 9–10 and 10–11 mm long, all glabrous; thecae hardly superposed, 4–5 mm long, glandular dorsally, spurred. Style 2–2.5 cm long, glabrous. Capsule 2.7–3.8 cm long, finely puberulous with stalked capitate glands or also with some minute non-glandular hairs; seed 6–7 × 4.5–5.5 mm, margin tuberculate, jagged apically; testa with round and ridge-like tubercles. Fig. 58/4, p. 446.

KENYA. Kilifi District: Pangani Forest, 19 Nov. 1978, *Brenan et al.* 14574! & 6 May 1985, *Faden & Beentje* 85/56!; Kwale District: Shimba Hills, Buffalo Ridge, 15 Mar. 1991, *Luke* 2703!

TANZANIA. Pangani District: Bushiri Estate, 24 Nov. 1950, *Faulkner* K731!; Morogoro District: Nguru Mts, Lusunguru Forest Reserve, near Mtibwa Sawmill, 31 Mar. 1953, *Drummond & Hemsley* 1923! & Kanga Mt, 7 Dec. 1987, *Mwasumbi* 13969!

DISTR. **K** 7; **T** 3, 6; not known elsewhere

HAB. Lowland evergreen forest, often along or near streams; 50–900 m

SYN. *Asystasiella africana* S. Moore in J.B. 44: 25 (1906); Iversen in Symb. Bot. Upsal. 29(3): 160 (1991); Luke & Robertson, Kenya Coastal Forests Checklist Vasc. Pl.: 80 (1993); Lebrun & Stork, Enum. Pl. Afr. Trop. 4: 469 (1997), *non Asystasia africana* (S. Moore) C.B. Clarke (1899)

22. **Asystasia masaiensis** *Lindau* in E.J. 43: 354 (1909); Lebrun & Stork, Enum. Pl. Afr. Trop. 4: 468 (1997). Type: Tanzania, Masailand, *Jaeger* 86 (B†, holo.). Neotype: Tanzania, Lushoto District, Mkomazi, *Greenway* 3970 (EA!, neotype, selected here; K!, iso.)

Perennial herb with several ascending to erect stems, probably from rootstock; stems to 50 cm long, retrorsely sericeous in two bands and with bands of longer hairs at nodes, soon glabrescent. Leaves with petiole to 1.3 cm long; lamina ovate, largest 2.5–4.5 × 1–2.5 cm, base attenuate to rounded, margin entire to crenate, apex obtuse to acute; sparsely sericeous along major veins beneath. Racemes terminal, 4–7.5 cm long, simple, 1-sided, retrorsely sericeous or sparsely so; flowers alternate; peduncles 1–3(–7) cm long; bracts ovate-elliptic or narrowly so, 10–15 mm long, acute, cuneate at base, sparsely sericeous and ciliate; bracteoles 5–12 mm long, ovate-elliptic, obtuse to acute at the apex, subglabrous; pedicels 0–1(–3 in fruit) mm long. Calyx 10–15 mm long, lobes linear-lanceolate, densely puberulous with glossy hairs and stalked capitate glands. Corolla yellow or greenish yellow, with red spots in throat, ± 3 cm long, puberulous; basal cylindric tube ± 2 mm long and ± 3 mm wide; throat ± 1.7 cm long and ± 1 cm wide at the mouth; limb with 5 subequal lobes, lower mid-lobe ± 10 × 8 mm; lateral and upper lobes 7–9 × 7–8 mm; all lobes ovate and rounded at the apex. Stamens didynamous, included; filaments ± 4 and 6 mm long, longer pair with a row of glandular hairs; theace slightly superposed, 2.5–3 mm long, spurred. Style 11–12 mm long, glabrous. Capsule (immature) ± 2 cm long; sparsely minutely puberulous with short capitate glands and some non-glandular hairs, seed not seen.

TANZANIA. Lushoto District: Mkomazi, 23 Apr. 1934, *Greenway* 3970! & 5 Apr. 1970, *Archbold* 1222!
DISTR. **T** 3; not known elsewhere
HAB. Under trees and shrubs in *Acacia*-wooded grassland and bushland on sandy soil; 450 m

NOTE. Apart from the destroyed type *Asystasia masaiensis* is known only from these two collections from the Mkomazi area. It is closely related to *A. guttata*, and the two may eventually prove to be conspecific. It differs mainly in having a 1-sided spike (one flower per node) with smaller bracts and bracteoles.

23. **Asystasia guttata** (*Forssk.*) *Brummitt* in K.B. 32: 452 (1978); Champluvier in Fl. Rwanda 3: 436 (1985); Blundell, Wild Fl. E. Afr.: 386, pl. 400 (1987); Luke & Robertson, Kenya Coastal Forests Checklist Vasc. Pl.: 80 (1993); Wood, Handb. Fl. Yemen: 273 (1997); Lebrun & Stork, Enum. Pl. Afr. Trop. 4: 468 (1997); Friis & Vollesen in Biol. Skr. 51: 436 (2005); Hedrén in Fl. Somalia 3: 398 (2006); Ensermu in Fl. Eth. 5: 400 (2006). Type: Yemen, Jabal Khadra, *Forsskål* 369 (C, holo.; K!, photo)

Perennial herb with several stems from woody rootstock (rarely a shrub); stems procumbent, ascending or scrambling (rarely erect), to 1(–2 in scrambling plants) m long, retrorsely sericeous or sericeous-pubescent or sparsely so (rarely subglabrous) all over or in two bands. Leaves with petiole to 3.5 cm long; lamina narrowly to broadly ovate or elliptic (rarely lanceolate), largest 3–8 × (0.8–)1.2–4.2 cm, base attenuate to cuneate, margin entire to crenate, apex acuminate to rounded; sericeous or sparsely so beneath, densest on veins. Spikes terminal, 1.5–8 cm long, simple, subglabrous to densely retrorsely sericeous-pubescent; flowers opposite; peduncles (0–)0.5–6.5(–11) cm long; bracts green or yellowish green, narrowly elliptic or narrowly ovate-elliptic, (8–)13–25 mm long, lowermost often foliaceous, acute to acuminate, sericeous-pubescent or sparsely so, distinctly ciliate; bracteoles (7–)9–15 mm long, lanceolate to ovate-elliptic, acuminate to acute, puberulous (densest towards apex) and ciliate; pedicels 0–1 mm long. Calyx 6–18(–25 in fruit) mm long, lobes subulate to linear-lanceolate, densely puberulous with short capitate glands and non-glandular hairs, ciliate. Corolla in Uganda (and Rwanda) white to pale pink with dark pink markings, in Kenya pale yellow or greenish yellow with purple to dark brown markings in tube and throat, in Tanzania orange reddish brown to bronze red with cream throat, 1.8–2.7(–3) cm long, sparsely to densely puberulous with capitate glands and non-glandular hairs; basal cylindric tube 3–6 mm long and 1.5–3 mm wide; throat 1–1.7 cm long and 6–10 mm wide at the mouth, gibbous below the lower midlobe; limb subactinomorphic with recurved lobes, lower midlobe 6–8 × 7–9 mm, lateral lobes 5–7 × 5–7 mm, lobes in upper lip 4–5 × 5–6 mm. Stamens didynamous, included; filaments 4–8 and 5–10 mm long, glabrous or longer with a row of short-stalked capitate glands; thecae slightly superposed, 2.5–3 mm long, dorsally with sub-sessile capitate glands, spurred. Style 8–12 mm long, hairy at the base or glabrous. Capsule 1.8–2.5(–3 in Uganda) cm long, finely puberulous with short-stalked capitate glands and non-glandular hairs; seed 4.5–6 × 4–5 mm, margin with 6–7 projections, testa rugose-tuberculate along edges and centrally, with a "trench" between. Fig. 58/20, p. 446.

UGANDA. West Nile District: Kobobo, May 1938, *Hazel* 480!; Kigezi District: Kebisoni, March 1950, *Purseglove* 3347!; Kigezi/Ankole District: 15 km beyond Nyaruranje, 29 Mar. 1952, *Norman* 89!
KENYA. Northern Frontier District: Ndoto Mts, Ndigri-Alori, 2 Jan. 1959, *Newbould* 3427!; Turkana District: 32 km on Kacheliba–Karamoja road, 17 June 1953, *Padwa* 288!; Voi District: Tsavo National Park East, Voi Headquarters, 23 Dec. 1966, *Greenway & Kanuri* 12847!

TANZANIA. Kilimanjaro District: Simba Farm, 6 July 1943, *Greenway* 6733! & Engare Nairobi, 19 June 1944, *Greenway* 6847! & Simba Farm, 8 Dec. 1968, *Bigger* 2358!

DISTR. **U** 1–3; **K** 1, 2, 4, 7; **T** 2; Rwanda, Ethiopia, Somalia; Yemen

HAB. *Acacia-Commiphora* woodland, wooded grassland, grassland, limestone hillsides, occasionally becoming a weed; (350–)500–1800(–2100) m

SYN. *Ruellia guttata* Forssk., Fl. Aegypt.-Arab.: 114 (1775); Vahl, Symb. Bot. 1: 72 (1791); Schwartz, Fl. Trop. Arab.: 250 (1939)

Barleria somalensis Franch. in Revoil, Sert. Somal.: 51 (1882). Type: Somalia, *Revoil* s.n. (P, holo.)

Parasystasia somalensis (Franch.) Baill., Hist. Pl. 10: 461 (1891); Lindau in Ann. R. Ist. Bot. Roma 6: 78 (1896)

Asystasia coleae Rolfe in K.B. 1895: 223 (1895); C.B. Clarke in F.T.A. 5: 135 (1899). Type: Somalia, *Cole* s.n. (K!, lecto.; selected by Brummitt in K.B. 32: 452 (1978)

Parasystasia kelleri Lindau in E.J. 24: 321 (1897). Types: Somalia, *Keller* 189 (Z, syn.) & 192 (Z, syn.)

Asystasia somalensis (Franch.) Lebrun & Toussaint, Expl. Parc. Nat. Kagera, Miss. J. Lebrun 1: 129 (1948); E.P.A.: 957 (1964)

NOTE. The differences in corolla colour outlined above are puzzling and it is possible that more than one taxon is involved. But the fact that each colour pattern seems to be confined to a single country makes me wary of making any infraspecific delimitations. The general "gestalt" of the plants do not differ from country to country except that the capsule is larger in Ugandan specimens.

24. **Asystasia lorata** *Ensermu* in Opera Bot. 121: 157 (1993); Lebrun & Stork, Enum. Pl. Afr. Trop. 4: 468 (1997). Type: Kenya, Masai District, W side of Ngong Hills a few miles from the Magadi road, *Verdcourt* 1457 (K!, holo.; BR!, EA!, MO!, PRE!, iso.)

Perennial herb with several stems from a woody rootstock; stems decumbent, ascending or erect, to 75 cm long, sparsely puberulous just below and with band of long hairs at nodes. Leaves subsessile (rarely with petiole to 5 mm long); lamina linear to linear-lanceolate (rarely narrowly ovate), largest 5–10 × 0.2–0.6(–1.4) cm, base decurrent, margin entire or slightly undulate, apex acute; glabrous or sparsely puberulous on veins, fimbriate towards base. Spikes terminal, 2–12(–17) cm long, simple, with opposite (rarely some alternate) flowers, glabrous or sparsely puberulous; peduncles 0.5–8.5 cm long; bracts and bracteoles longer than calyx, lanceolate to narrowly ovate, 9–21 mm long, cuspidate, fimbriate at the margins, otherwise glabrous; pedicels 0–1 mm long. Calyx 7–17 mm long, lobes linear, cuspitate, minutely puberulous with capitate glands and non-glandular hairs. Corolla greyish- or greenish-yellow with brownish or purple veins in throat (also described as orange-brown to brownish pink) and translucent yellow-green patch on lower lip, 1.2–1.6(–2) cm long, sparsely puberulous with capitate glands and non-glandular hairs; basal cylindric tube 3–4 mm long and ± 2 mm wide; throat 5–8(–10) mm long and 4–7 mm wide at the mouth; limb subactinomorphic, lobes reflexed, lower midlobe 4–6 × 4–5 mm, lateral and upper 3–5 × 3–4 mm. Stamens didynamous, included; filaments 3.5–4 and 4.5–5.5 mm long, longer with a row of capitate glands; thecae slightly superposed, ± 2 mm long, dorsally with subsessile capitate glands, spurred. Style 6–9 mm long, puberulous at junction with ovary. Capsule 1.5–2 cm long, finely puberulous with capitate gland and non-glandular hairs; seed 4–4.5 × 3.5–4 mm, margin toothed, testa densely tuberculate, sometimes with "channel" between edge and surface. Fig. 58/21, p. 446.

KENYA. Kiambu District: 27 km on Kikuyu–Narok road, Ndeiya Grazing Scheme, 20 Jan. 1963, *Verdcourt* 3553!; Masai District: Karero, Kenya Marble Quarry, 25 May 1996, *Pearce & Vollesen* 930! & S of Ngong Hills on Magadi Road, 11 Aug. 1951, *Verdcourt* 578!

TANZANIA. Musoma District: Serengeti National Park, 15 km from Seronera on the Nyaraswega-Banagi Hill Circuit, 22 May 1962, *Greenway* 10651!; Masai District: Masandari, 15 July 1962, *Newbould* 6176!

DISTR. **K** 4, 6; **T** 1, 2; not known elsewhere

HAB. *Acacia-Commiphora* woodland, wooded grassland and bushland, grassland, rocky hills, on red sandy clayey loam or on stony sandy soil; (900–)1400–2000 m

SYN. *Asystasia* sp. A of Agnew, U.K.W.F.: 596 (1974); U.K.W.F., ed. 2: 275 (1994)

NOTE. *M.G. Gilbert* 6149 differs from the rest of the material in having a petiolate, narrowly ovate leaf to 1.4 cm wide but is otherwise typical and has the small corolla of this species when compared with *A. guttata.*

25. **Asystasia mysorensis** (*Roth*) *T. Anderson* in J.L.S. Bot. 9: 524 (1867), as "*mysurensis*"; U.K.W.F., ed. 2: 275, pl. 120 (1994); Lebrun & Stork, Enum. Pl. Afr. Trop. 4: 468 (1997); Friis & Vollesen in Biol. Skr. 51: 437 (2005); Ensermu in Fl. Eth. 5: 400 (2006). Type: India, Mysore, *Heyne* s.n. (K!, iso.)

Annual herb; stems erect to decumbent (rarely straggling), to 0.75(–2 in straggling plants) m long, sparsely to densely puberulous to pubescent when young, glabrescent. Leaves with petiole to 2.5(–4) cm long; lamina narrowly to broadly ovate or narrowly to broadly elliptic, largest 2.5–8.5(–11) × 1–3.5(–4.5) cm, base attenuate to cuneate, margin entire, ciliate, apex acuminate to rounded; pubescent or sparsely so along veins beneath, uniformly so above. Spikes terminal, 1–9 cm long, simple, with opposite flowers (rarely some alternate), pubescent or sparsely so; peduncle 0.5–7(–9) cm long; bracts green, lowermost often foliaceous, narrowly ovate or narrowly elliptic to ovate or elliptic, in middle of spike 8–18(–25) mm long, acuminate to acute, sparsely puberulous and fimbriate-ciliate; bracteoles green, lanceolate to narrowly elliptic or narrowly ovate, 5–15 mm long, acuminate, subglabrous to sparsely puberulous and fimbriate-ciliate; pedicels 0–2 mm long. Calyx 4–7 mm long, lobes subulate to linear, puberulous with capitate glands only or with intermixed non-capitate glands. Corolla pure white (rarely pale lemon yellow or pale pink) with greenish-yellow spots on the lower midlobe and palate and sometimes with reddish veins, 1–1.5 cm long, sparsely puberulous with capitate glands; basal cylindric tube ± 2 mm long and 1 mm in diameter; throat 4–6 mm long and 4–5 mm in dimeter at mouth; limb indistinctly 2-lipped, lobes 4–7 × 3–6 mm, subequal, recurved, lower midlobe the largest. Stamens didynamous, included; filaments 3–3.5 and 4.5–5.5 mm long, glabrous (short) or with a row of capitate glands; anthers ± 1.5 mm long, thecae slightly superposed, dorsally with capitate glands, apiculate or minutely spurred. Style 4.5–5.5 mm long, glabrous. Capsule 1.3–2.2 cm long, finely puberulous with capitate and non-capitate glands, with longer hairs at apex; seed 4–5 × 3–4.5 mm, margin toothed (lobate); testa rugose-tuberculate. Fig. 60, p. 467.

UGANDA. West Nile District: 1.5 km E of Omugo Rest Camp, 10 Aug. 1953, *Chancellor* 145!; Teso District: Serere, Nov.-Dec. 1931, *Chandler* 125!; Mengo District: Kampala, Makerere Hill, 13 Oct. 1953, *Drummond & Hemsley* 4746!

KENYA. Nakuru District: Elmenteita, Soysambu Estate, 31 July 1947, *Bogdan* 958!; Nairobi District: Nairobi, 25 May 1951, *Bogdan* 3016!; Kisumu District: Kisumu, May 1963, *Tweedie* 2619!

TANZANIA. Masai District: Endoinyo Emboleh, Osilale, 9 Feb. 1962, *Newbould* 6005!; Mpanda District: 10 km on Karema road from Mpanda–Uvinsa road, 8 March 1994, *Bidgood et al.* 2710!; Mpwapwa District: 6 km N of Kibakwe on Mpwapwa track, 10 Apr. 1988, *Bidgood et al.* 988!

DISTR. **U** 1–4; **K** 3–7; **T** 1–7; Sudan, Ethiopia, Congo-Kinshasa, Rwanda, Burundi, Zambia, Mozambique, Zimbabwe, Botswana, Namibia, South Africa; Yemen and India

HAB. *Acacia-Commiphora, Brachystegia* and *Terminalia-Combretum* woodland, wooded grassland, bushland, grassland, roadsides, widespread weed of cultivation, often in shady places; 500–2100 m

FIG. 60. *ASYSTASIA MYSORENSIS* — **1**, habit; **2**, leaf; **3**, bract; **4**, bracteole; **5**, calyx opened up; **6**, detail of calyx indumentum; **7**, corolla, frontal view; **8**, corolla, opened up; **9**, apical part of filament and anther; **10**, ovary, basal part of style and stigma; **11**, capsule; **12**, seed. 1 & 7 from *Drummond & Hemsley* 4746, 2 & 11 from *Pielou* 52, 3–6 & 8 from *Burtt* 5509, 9 & 10 from *Verdcourt* 3670, 12 from *Bidgood et al* 2205. Drawn by Margaret Tebbs.

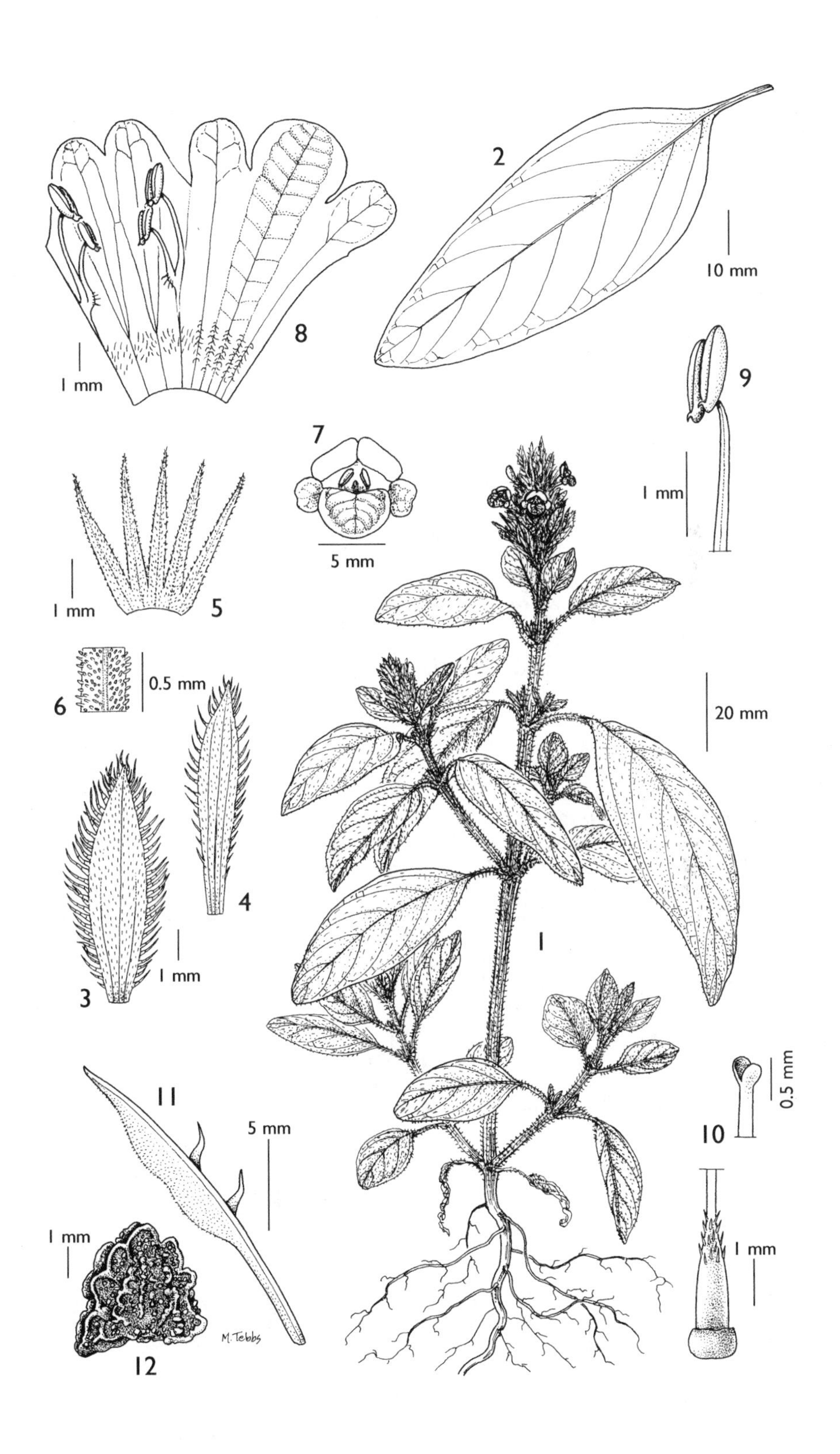
2
10 mm
8
1 mm
9
1 mm
7
5 mm
5
1 mm
6
0.5 mm
20 mm
4
3
1 mm
1
11
5 mm
1 mm
12
M. Tebbs
10
0.5 mm
1 mm

SYN. *Ruellia mysorensis* Roth in Nov. Pl. Sp.: 303 (1821)
Strobilanthes mysorensis (Roth) Nees in Wall. Pl. As. Rar. 3: 86 (1832), as "*mysurensis*" & in DC., Prodr. 11: 192 (1847), as "*mysurensis*"
Adhatoda rostellaria Nees var. *humilis* Nees in DC., Prodr. 11: 397 (1847). Types: Ethiopia, *Schimper* III.1657 (K!, isosyn.), *Schimper* III.1659 (K!, isosyn.), *Schimper* in *Hohenacker* 2220 (K!, isosyn.)
Asystasia schimperi T. Anderson in J.L.S. Bot. 7: 53 (1863); Solms-Laub. in Schweinf., Beitr. Fl. Aeth.: 242 (1867); Oliver in Trans. Linn. Soc. 29: 131 (1875); C.B. Clarke in F.T.A. 5: 135 (1899) & in Fl. Cap. 5: 43 (1901); E.P.A.: 957 (1964); U.K.W.F.: 596 (1974); Champluvier in Fl. Rwanda 3: 436 (1985); Blundell, Wild Fl. E. Afr.: 386, pl. 237 (1987); Wood, Handb. Fl. Yemen: 273 (1997). Type: as for *Adhatoda rostellaria* Nees var. *humilis* Nees
Adhatoda rostrata Solms-Laub. in Schweinf., Beitr. Fl. Aeth.: 104 (1867). Type: as for *Asystasia schimperi*
Asystasia rostrata (Solms-Laub.) Solms-Laub. in Schweinf., Beitr. Fl. Aeth.: 242 (1867); Lindau in P.O.A. C: 370 (1895)
[*A. schimperi* T. Anderson var. *minor* Oliver in Trans. Linn. Soc., ser. 2, Bot. 2: 345 (1887), *nom. nud.*]
A. schimperi T. Anderson var. *grantii* C.B. Clarke in F.T.A. 5: 135 (1899) & in Fl. Cap. 5: 43 (1901); E.P.A.: 957 (1964). Types: Uganda, Bunyoro District, "Unyoro", *Speke & Grant* 603 (K!, syn.); Somalia, *Keller* 189 (not seen). See note

NOTE. C.B. Clarke cites Keller 189 as a syntype of *Asystasia schimperi* var. *grantii*. This collection is also one of the types of *Parasystasia kelleri* Lindau (see p. 465). We have not seen this specimen but according to the description it has considerably larger flowers than *A. mysorensis* which has also never been recorded from Somalia. We consider *P. kelleri* to be a synonym of *A. guttata* and the citation by Clarke to be in error.

Considering its large distribution area this is morphologically a very uniform species.

26. **Asystasia charmian** *S. Moore* in J.B. 18: 308, t. 38 (1880); Lindau in P.O.A. C: 370 (1895); C.B. Clarke in F.T.A. 5: 134 (1899); U.K.W.F.: 596 (1974); Blundell, Wild Fl. E. Afr.: 386, fig. 796 (1987); U.K.W.F., ed. 2: 275 (1994). Type: Kenya, Kitui District, Kitui, *Hildebrandt* 2724 (BM!, holo.; K!, iso.)

Perennial or shrubby herb with several stems from woody rootstock or shrub; stems erect, ascending or straggling to 1(–2 in straggling plants) m long, subglabrous (hairy at nodes) to puberulous or retrorsely sericeous-puberulous. Leaves with petiole to 2.5(–3) cm long; lamina lanceolate to ovate or elliptic (rarely cordiform or orbicular), largest 2.5–13 × 1–4(–5.2) cm, base attenuate to truncate, margin entire or slightly crenate, apex subacuminate to rounded; puberulous or sparsely so, densest along veins. Spikes terminal, 1–16 cm long, simple, with opposite flowers, pubescent to pilose or densely so or sericeous-pubescent; peduncle 1–8(–15) cm long; middle bracts ovate, elliptic, or obovate to orbicular, 4–9 mm long, acute to rounded and apiculate, densely pilose (rarely sparsely pubescent) with long white hairs and short-stalked capitate glands; bracteoles lanceolate to narrowly elliptic, 4–9 mm long, acute to rounded, with similar indumentum; pedicels 0–2(–3 in fruit) mm long. Calyx (6–)8–14 mm long, lobes linear to linear-subulate, pubescent or (usually) pilose with long white hairs and short-stalked capitate glands. Corolla white to pale pink with purple stripes and veins, lower mid-lobe with purplish pink markings, 2–3.2 cm long, with long many-celled pilose hairs and stalked capitate glands; basal cylindric tube 4–9 mm long and 2–3.5 mm wide, throat 8–13 mm long and 6–11 mm wide at the mouth; limb with 5 subequal lobes, lower midlobe the largest, 7–11 × 7–12 mm, lateral lobes 6–9 × 8–10 mm, upper lip 7–8 mm long, lobes 3–4 mm × 4–5 mm, all lobes sub-orbicular with rounded apex. Stamens didynamous, included; filaments 5–8 and 7–11 mm long, glabrous or with capitate glands; thecae slightly superposed, 3–4 mm long, glabrous, spurred. Style 10–15 mm long, pubescent at the base or up to ± half of its length. Capsule 1.8–2.7 cm long, densely puberulous with non-glandular hairs (rarely also with short capitate glands); seed 5–6 × 4.5–5.5 mm, margin with ± 6 broad projections, testa rugose-tuberculate. Fig. 58/19, p. 449.

UGANDA. Karamoja District: Upe, Lokitonyala, Aug. 1958, *J. Wilson* 589!
KENYA. Kitui District: Kibwezi–Kitui road, 11 km after Athi River, 19 March 1969, *Napper & Jones* 1960!; Fort Hall District: Thika, near Ol Doinyo Subuk, Mabuloni Rocks, 12 Dec. 1952, *Verdcourt* 849!; Voi District: Tsavo National Park East, Voi Gate, 5 Jan. 1967, *Greenway & Kanuri* 12967!
TANZANIA. Mbulu District: Lake Manyara National Park, Western Rift Wall, 1 July 1965, *Greenway & Kanuri* 11920!; Same District: Mkomazi Game Reserve, Mbulu area, 5 June 1996, *Vollesen* 99/46!; Kondoa District: Kikori, 27 May 1930, *B.D. Burtt* 2735!
DISTR. **U** 1; **K** 1, 4, 6, 7; **T** 2, 3, 5; Ethiopia
HAB. *Acacia-Commiphora* and *Acacia-Terminalia* wooded grassland and bushland, *Acacia drepanolobium* grassland, on black or dark brown clayey soil; 450–1500 m

SYN. *A. fuchsiifolia* Lindau in E.J. 43: 355 (1909); Lebrun & Stork, Enum. Pl. Afr. Trop. 4: 468 (1997). Type: Tanzania, Pangani, *Jaeger* 113 (B†, holo.)

27. **Asystasia riparia** *Lindau* in E.J. 33: 189 (1902); E.P.A.: 957 (1964); Luke & Robertson, Kenya Coastal Forests Checklist Vasc. Pl.: 80 (1993); Lebrun & Stork, Enum. Pl. Afr. Trop. 4: 468 (1997); Hedrén in Fl. Somalia 3: 399 (2006); Ensermu in Fl. Eth. 5: 401 (2006). Type: Ethiopia, Gamo Gofa, Lake Chamo, *Neumann* 99 (B†, holo.). Neotype: Ethiopia, Sidamo, 20 km NW of Moyale, *Mesfin Tadesse & Vollesen* 4201 (ETH!, neotype, selected here; C!, K!, UPS! iso.)

Annual or perennial herb; stems to 50 cm long, erect or ascending, often much-branched with lower branches spreading or creeping, sparsely to densely retrorsely sericeous, sometimes only in two bands. Leaves with petiole to 2.5 cm long; lamina narrowly to broadly ovate, largest 1.7–6 × 0.8–3.4 cm, base attenuate to cuneate, margin entire or slightly crenate, with scattered long ciliae, apex subacute to shortly acuminate; subglabrous to sericeous, densest on veins. Spikes terminal, 1.5–4 cm long, simple, with opposite flowers, retrorsely sericeous-pubescent; peduncle 0.4–2 cm long; bracts and bracteoles ovate-elliptic, 5–11 mm long, acute to acuminate, subglabrous to sericeous-pubescent, ciliate, lowermost bracts sometimes foliaceous; pedicels 0–1 mm long. Calyx 5–8(–15 in fruit) mm long, lobes subulate to linear, densely puberulous with capitate and non-capitate glandular hairs, ciliate. Corolla dirty white, dull mauve or pale purple, with brownish purple specks in throat and on lower lip, 1–1.9 cm long, with short capitate glands and scattered to dense pilose hairs; basal cylindric tube 3–4 mm long and 2–3 mm wide; throat 4–7 mm long and 3–6 mm wide at mouth, limb with 5 sub-equal lobes 3–8 × 3–6 mm, lower midlobe the largest. Stamens slightly didynamous, included; filaments 3–6 and 4–7 mm long, glabrous or longer with a row of glandular hairs, spurred; thecae 2–3 mm long, slightly superposed, dorsally with glandular hairs. Style 5–7 mm long, pubescent at the base. Capsule 1.5–1.9 cm long, densely puberulous with subsessile capitate glands and longer non-glandular hairs; seed 3.5–5 × 3–4 mm, margin toothed, surface tuberculate, with "channel between edge and face. Fig. 58/18, p. 446.

KENYA. Northern Frontier District: Dandu, 6 May 1952, *Gillett* 13085!; Meru District: Isiolo–Wajir road, 6 km E of junction with Marsabit road, 7 Dec. 1977, *Stannard & Gilbert* 808!; Tana River District: Kora Game Reserve, 27 km from Kampi ya Chui towards Yambandei Hill, 16 May 1983, *Mungai et al.* 215/83!
DISTR. **K** 1, 4, 7; Ethiopia, Somalia
HAB. *Acacia*, *Acacia-Commiphora* and *Acacia-Balanites* woodland, wooded grassland and bushland, *Acacia drepanolobium* grassland, weed of cultivation, on black clay and grey-brown silty soil; 400–1100(–1550) m

SYN. *Asystasia somalica* Gand. in Bull. Soc. Bot. France 69: 349 (1922); E.P.A.: 957 (1964); Lebrun & Stork, Enum. Pl. Afr. Trop. 4: 468 (1997). Type: Somalia, Giumbo, *Fiori* 7 (FT, holo.)
A. ritellii Chiovenda in Fl. Somala 2: 350, fig. 200 (1932); E.P.A.: 957 (1964); Lebrun & Stork, Enum. Pl. Afr. Trop. 4: 468 (1997). Type: Somalia, El Ualud, *Gorini* 491 (FT, holo.)

35. **BRACHYSTEPHANUS**

Nees in DC.,Prodr. 11: 511 (1847); C.B. Clarke in F.T.A. 5: 177 (1899); Figueiredo in K.B. 51: 753–763 (1996); Champluvier & Darbyshire in Syst. Geogr. Pl. 79: 115–192 (2009)

Oreacanthus Benth. in G.P. 2(2): 1104 (1876); C.B. Clarke in F.T.A. 5: 176 (1899); Champluvier & Figueiredo in B.J.B.B. 65: 413–417 (1996)

Perennial herbs or subshrubs, evergreen or rarely deciduous, sometimes monocarpic and cyclically mass-flowering (plietesial); cystoliths many, ± conspicuous. Stems (sub-) 4-angular, often sulcate between the angles when young. Leaves opposite-decussate, pairs equal or rarely anisophyllous, blade margin crenulate or subentire; lower leaves petiolate, uppermost leaves often reduced and subsessile. Inflorescences terminal and sometimes additionally in the upper leaf axils; spiciform or paniculate thyrses, cymule branching dichasial or partially monochasial, or cymules reduced to a single flower; bract pairs equal, free; bracteoles as bracts or much-reduced. Calyx divided almost to the base into 5 linear (-lanceolate) lobes, these subequal to unequal in length. Corolla bilabiate; tube cylindrical or campanulate, long to very short; upper lip either subulate or more broadly ovate to elliptic, apex acute, obtuse, apiculate or emarginate; lower lip oblong to elliptic, apex minutely to conspicuously 3-lobed; rugula absent. Stamens 2; filaments attached at or just below the mouth of the corolla tube, ± long-exserted, glabrous; anthers monothecous, base muticous; staminodes absent. Disk annular or shallowly cupuliform. Ovary oblong-ellipsoid, glabrous or largely so, 2-locular, 2 ovules per locule; style filiform, ± long-exserted, glabrous; stigma either minutely clavate or conspicuously capitate, apex shallowly bilobed. Capsule stipitate, placental base inelastic. Seeds 4 per capsule, or 2 by abortion, held on retinacula, lenticular, tuberculate.

A genus of 22 species distributed in tropical Africa and Madagascar. Two of the three sections recognised by Champluvier are recorded in our region (see key). In addition to the 7 species recorded here, *B. sudanicus* (Friis & Vollesen) Champl. may occur in northernmost Uganda; it is currently known only from the Imatong Mts just over the border in Sudan.

1. Corolla tube shorter than or subequal to limb; upper lip subulate, lower lip conspicuously 3-lobed; stigma minutely clavate, barely wider than style (sect. *Oreacanthus*) 2
   Corolla tube longer than limb; upper lip (ovate-) elliptic, lower lip minutely 3-lobed; stigma capitate, clearly broader than style (sect. *Brachystephanus*) 4
2. Inflorescence ± densely spiciform; bracts conspicuous, variously suborbicular, ovate, elliptic, obovate or obcordiform with attenuate to caudate apex, 4.5–11.5 (–15) mm long . . . . 1. *B. coeruleus*
   Inflorescence laxly spiciform or with the lower cymules shortly pedunculate; bracts inconspicuous, narrowly triangular, less than 2 mm long 3
3. Cymules of inflorescence all (sub)sessile; axes and calyces glandular-puberulous and with scattered long glandular hairs; calyx lobes 5–6.7 mm long; anthers ± 2.3 mm long 2. *B. laxispicatus*
   Cymules of lower portion of inflorescence shortly pedunculate for up to 3 mm; axes and calyces eglandular-puberulous and with scattered short glandular hairs; calyx lobes 2–3.5 mm long; anthers ± 1.5 mm long . . . . 3. *B. schliebenii*

4. Inflorescence a lax panicle, largely glabrous; bracts inconspicuous, narrowly triangular, 1–5 mm long . . . . . . 4. *B. glaberrimus*
   Inflorescence spiciform to subcapitate or if paniculate then highly congested and conspicuously hairy; bracts conspicuous, variously ovate or somewhat obovate to linear-lanceolate, 5–28 mm long 5
5. Calyx, bracts and bracteoles not ciliate, calyx lobes often with scattered glandular hairs; bracts with apex attenuate to caudate; corolla tube 22–38 mm long; staminal filaments 13.5–36 mm long . . . . 5. *B. africanus*
   Calyx lobes and often bracts and bracteoles ciliate with conspicuously multicellular eglandular hairs, glandular hairs absent; bracts with apex acute; corolla tube (8–)14–23 mm long; staminal filaments 6.5–17 mm long 6
6. Corolla bright red (-purple) to rose pink, tube narrowly widened in upper half; cymules of inflorescence each 1(–3)-flowered, inflorescence 0.7–1.5(–2) cm wide; ciliate hairs of bracts, bracteoles and calyces ascending . . . . 6. *B. holstii*
   Corolla pink to purple, tube narrowly cylindrical throughout; cymules of inflorescence each several-flowered, inflorescence 2–3.5 cm wide; ciliate hairs of bracts, bracteoles and calyces ± spreading . . . . 7. *B. roseus*

1. **Brachystephanus coeruleus** *S. Moore* in J.B. 45: 332 (1907); Champluvier & Darbyshire in Syst. Geogr. Pl. 79: 127 (2009). Type: Uganda, Mubende District, Mpamba R., tributary of Ngusi, Lake Albert-Edward, *Bagshawe* 1378 (BM!, holo. & iso.)

Erect, decumbent or scandent perennial herb or subshrub, 35–200 cm tall; stems pubescent in two opposite lines or on paired opposite ridges, hairs short and antrorse or rarely longer and both antrorse and retrorse. Leaves sometimes absent at flowering; blade ovate to elliptic, 8.5–17 cm long, 4–9 cm wide, base (cuneate-) attenuate, apex acuminate, principal veins and margin shortly pubescent, upper surface sparsely pubescent or rarely pilose; lateral veins 7–11 pairs; petiole 1–8 cm long. Inflorescences terminal, spiciform, 2–14.5 cm long, each cymule several-flowered; axis antrorse- to spreading-pubescent, sometimes with additional short glandular hairs and/or long glandular-pilose; bracts pale green with markedly darker acumen and midrib or dark green throughout, often purple-tinged, variously suborbicular, ovate, elliptic, obovate or obcordiform, 4.5–11.5(–15) mm long, 2.3–6.7 mm wide, apex attenuate to curved-caudate, margin and often midrib pubescent, surface sometimes glandular-puberulous and/or long glandular-pilose; bracteoles as bracts but (linear-) lanceolate, 3.5–8(–9.5) mm long, 0.5–2.2 mm wide. Calyx lobes linear, 3.7–7.5 mm long, surface shortly pubescent and ± densely long glandular-pilose. Corolla 10.5–18 mm long, tube white, limb pink, purple, blue or rarely white; limb pilose towards apex; tube subcylindrical, 4.5–7 mm long; upper lip subulate, 6–10.5 mm long, apex obtuse; lower lip 6.5–11 mm long, lobes oblong, 3–5.5 mm long. Staminal filaments (6–)8.5–13 mm long; anthers 1.5–2 mm long. Stigma minute, barely wider than style. Capsule 6–8 mm long, puberulous to sparsely so in upper half; seeds 0.9–1.8 mm wide, rugulose-tuberculate.

a. subsp. **coeruleus**; Champluvier & Darbyshire in Syst. Geogr. Pl. 79: 127, 180, fig. 1 C & F (2009)

Inflorescences typically 8–14.5 cm long at maturity; bracts suborbicular to broadly ovate or elliptic, 4.5–8.5 mm long including attenuate to shortly acuminate apex up to 2 mm long.

UGANDA. Ankole District: Bunyaruguru, Nov. 1938, *Purseglove* 482! & *idem*, July 1939, *Purseglove* 883!; Mubende District: Mpamba R., tributary of Ngusi, Lake Albert-Edward, *Bagshawe* 1378! (type)

KENYA. North Kavirondo District: N Kavirondo, Oct. 1931, *Chater Jack* 142! & Mlaba [Malaba] Forest, Kabaras, Sept. 1964, *Tweedie* 2895!
DISTR. **U** 2, 4; **K** 5; E Congo-Kinshasa
HAB. Forest undergrowth and streamside in gorges; 950–1550 m
USES. None recorded on herbarium specimens
CONSERVATION NOTES. This subspecies appears scarce in the easternmost limits of the Congolian forests, currently known from only six collections, most recently in the 1960s. The Kenyan sites are threatened by the high and increasing human population in the Lake Victoria basin with resultant habitat loss. This taxon is therefore assessed as Endangered (EN B2ab(iii)) but further data on current distribution and threats are desirable.

SYN. *Oreacanthus coeruleus* (S. Moore) Champl. & Figueiredo in B.J.B.B. 65: 416 (1996)

b. subsp. **apiculatus** *Champl.* in Syst. Geogr. Pl. 79: 130, 180, fig. 1 D & G, fig. 31 (2009). Type: Tanzania, Kigoma District, Lugonezi R., Lubalisi, *Kahurananga, Kibuwa & Mungai* 2720 (K!, holo.; BR, EA, K sheets 2–4!, iso.)

Inflorescences to 9.5 cm long at maturity; bracts obovate, obcordiform or broadly elliptic, 6.5–11.5(–15) mm long including prominent curved caudate apex 2–5(–7) mm long.

TANZANIA. Kigoma District: Mt Livandabe [Lubalisi], June 1997, *Bidgood et al.* 4321!; Mpanda District: Ntakatta Forest, June 2000, *Bidgood, Leliyo & Vollesen* 4624!; Mbeya District: E of Itimba village, Kwamwondo forest, June 1991, *Lovett & Kayombo* 445!
DISTR. **T** 4, 7; N Zambia
HAB. Forest and shaded riversides; 1100–2000 m
USES. None recorded on herbarium specimens
CONSERVATION NOTES. This subspecies is currently known from only seven sites. It has been recorded as locally common or even dominant in the herb layer and is almost certainly a plietesial taxon. Populations in the northern part of its range are rather isolated and may remain undisturbed, However, the population from Mbeya Range is threatened by extensive human encroachment there. It is therefore assessed as Vulnerable (VU B2ab(iii)).

NOTE. In subsp. *apiculatus* the long glandular-pilose hairs are often restricted to the calyx lobes. However, *Webb* 66 from the Ufipa Range near Sumbawanga (**T** 4), referable to this subspecies on account of its markedly caudate bracts, has many long glandular hairs throughout the inflorescence as in subsp. *coeruleus*.

The fruits of *Bidgood, Leliyo & Vollesen* 4578 from Mpanda–Uvinza (**T** 4), measuring only 4–5 mm long, appear abnormal and are omitted from the description.

2. **Brachystephanus laxispicatus** *I. Darbysh.* in Syst. Geogr. Pl. 79: 143, fig. 12 (2009). Type: Tanzania, Kigoma District, Kabogo Mts, *Azuma* in *Kyoto University Expedition* 551 (EA!, holo.; KYO, iso.)

Erect perennial herb, 50–150 cm tall; stems shortly antrorse-pubescent in two opposite lines when young, soon glabrescent. Leaves slightly to markedly anisophyllous, (oblong-) elliptic, 12–21 cm long, 4.3–7.5 cm wide, base cuneate (-attenuate), apex acuminate, principal veins shortly and sparsely antrorse-pubescent, elsewhere glabrous; lateral veins 7–11 pairs; petiole 13–60 mm long. Inflorescences (sub)terminal, laxly spiciform, 10.5–18 cm long, each cymule several-flowered, (sub)sessile; axis glandular-puberulous and with interspersed long glandular hairs; bracts inconspicuous, green, narrowly triangular, 1–1.7 mm long, sparsely hairy; bracteoles as bracts but linear and somewhat shorter, more densely hairy. Calyx orange-brown, lobes subequal, linear, 5–6.7 mm long, glandular-puberulous, the gland tips minute, with occasional short to long glandular hairs with more conspicuous gland tips. Corolla 16–18 mm long, tube white, upper lip and median lobe of lower lip purple; limb puberulous outside; tube subcylindrical, 7–7.5 mm long; upper lip subulate, 9.5–10 mm long, apex acute to minutely emarginate; lower lip 9–10.5 mm long, lobes oblong, 2.3–3.5 mm long. Staminal filaments ± 9.5 mm long; anthers ± 2.3 mm long. Stigma minute, barely wider than style. Capsule and seeds not seen.

TANZANIA. Kigoma District: Kabogo Mts, June 1963, *Azuma* in *Kyoto University Expedition* 551!; Mpanda District: Ntakatta Forest, June 2000, *Bidgood, Leliyo & Vollesen* 4655!
DISTR. **T** 4; not known elsewhere
HAB. Riverine forest of *Anthonotha, Newtonia, Pseudospondias, Strombosia* and understorey of *Memecylon jasminoides*; rock exposure by waterfall; ± 1100 m
USES. None recorded on herbarium specimens
CONSERVATION NOTES. Of the two known sites, deforestation has been extensive in the Kabogo area (Pan Africa News 1(2), 1994), threatening this population. However, it is quite possible that *B. laxispicatus* also occurs in the nearby Gombe and Mahale Mts National Parks where suitable habitat remains widespread and protected; it may also occur along the poorly known western flanks of Lake Tanganyika in Congo-Kinshasa. It is assessed as Endangered (EN B1ab(iii), B2ab(iii)) but this may be downgraded if further populations are discovered.

3. **Brachystephanus schliebenii** (*Mildbr.*) *Champl.* in Syst. Geogr. Pl. 79: 147, 184 (2009). Type: Tanzania, Ulanga District, Sali, 35 km S of Mahenge, *Schlieben* 2000 (B†, holo.; HBG, lecto., selected by Champluvier, l.c.; BM!, BR, M!, MO!, P!, isolecto.)

Subshrub; stems minutely antrorse-pubescent on paired opposite ridges, glabrescent. Leaves somewhat anisophyllous, elliptic, 10–23 cm long, 3.5–8 cm wide, base cuneate-attenuate, apex caudate, principal veins puberulous beneath, elsewhere glabrous; lateral veins 8–10 pairs; petiole 8–32 mm long. Inflorescences terminal, laxly spiciform, 14–26 cm long, lower axils with cymes shortly pedunculate for up to 3 mm; axes puberulous and with sparse short glandular hairs; bracts inconspicuous, narrowly triangular, 0.7–1 mm long; bracteoles minute, ± 0.5 mm long; pedicels up to 1.5 mm long. Calyx lobes linear, 2–3.5 mm long, puberulous and with few short glandular hairs. Corolla 12-15 mm long, lobes shortly pubescent outside; tube 5.5–7 mm long; upper lip subulate, ± 6–8 mm long; lower lip ± 6.5–8.5 mm long, lobes to 4 mm long. Staminal filaments ± 6–7 mm long; anthers ± 1.5 mm long. Stigma minute, barely wider than style. Capsule and seeds not seen.

TANZANIA. Ulanga District: Sali, 35 km S of Mahenge, Mar. 1932, *Schlieben* 2000! (type)
DISTR. **T** 6; not known elsewhere
HAB. Wet forest; 1000–1100 m
USES. None recorded on herbarium specimens
CONSERVATION NOTES. No records of this species are known subsequent to the type gathering despite much botanical collecting in east-central Tanzania. Much of the lowland forest has now been lost in this region. This species must therefore be assessed as Critically Endangered (CR B1ab(iii) B2ab(iii)).

SYN. *Oreacanthus schliebenii* Mildbr. in N.B.G.B. 11: 824 (1933); Friis & Vollesen in K.B. 37: 467 (1982)

4. **Brachystephanus glaberrimus** *Champl.* in B.J.B.B. 66: 194, figs. 1–4 (1997); Champluvier & Darbyshire in Syst. Geogr. Pl. 79: 154, 186, fig. 16 (2009). Type: Uganda, Ruwenzori, Msandama, *Maitland* 1015 (K!, holo.)

Erect or scrambling perennial herb or shrub, 150–250 cm tall; stems glabrous or uppermost internodes sparsely puberulent. Leaves (ovate-) elliptic, 12–26 cm long, 5.5–13 cm wide, base narrowly cuneate along upper petiole then abruptly widened, apex acuminate, glabrous or principal veins sparsely puberulous beneath; lateral veins 7–12 pairs; petiole 10–55 mm long. Inflorescences terminal, paniculate, 7–27 cm long, 3–8 cm wide, partial inflorescences dichasial or partially monochasial; axes glabrous or sparsely puberulent, rarely with few scattered glandular hairs; bracts and bracteoles green, narrowly triangular, 1–5 mm long, glabrous. Calyx lobes lanceolate, 4–8.5 mm long, basal portions slightly imbricate, apex acuminate with tip often slightly recurved, glabrous or with few scattered short glandular hairs. Plants often with two distinct flower forms; normal flowers with corolla mauve to violet, 19–25 mm long, glabrous except for minute glands in the throat between the attachment of the stamens; tube cylindrical, shallowly curved, 11–14 mm long; upper lip ovate (-elliptic), 6.5–11 mm long, 3.5–6.5 mm wide, apex involute and apiculate; lower lip ± elliptic, 6–11 mm long, 5–7 mm wide, apex minutely 3-lobed, lobes up to 1(–1.5) mm long. Stamens long-exserted, filaments 14.5–18 mm long; anthers (2.5–)3–3.8 mm long. Abnormal flowers differing in corolla 13.5–19 mm long including tube 7–11.5 mm; either one or both stamens much shorter, filaments of short stamens 1.5–5 mm long. Ovary and style glabrous; stigma capitate, ± 0.3 mm wide. Capsule ± 15 mm long, glabrous; only immature seeds seen, rugulose-tuberculate.

UGANDA. Toro District: Hima Forest, Apr. 1950, *Dale* U799! & Kibale C. Forest Reserve, June 1964, *Ssali* 11!; Kigezi District: Bwindi [Impenetrable] Forest, Mubwindi Swamp, June 1983, *Butynski* s.n.!

DISTR. U 2; E Congo-Kinshasa, Rwanda

HAB. Montane forest undergrowth, swamps; 1500–2250 m

USES. None recorded on herbarium specimens

CONSERVATION NOTES. This species is scarce in the high mountains bordering the Albertine Rift. It is recorded as mass-flowering by Butynski and is almost certainly a plietesial species, so may well be under-recorded due to the significant time periods without mature plants. However, the large fluctuations in numbers of mature individuals associated with a plietesial life cycle, together with the very few known sites (6 to date) render this species Vulnerable (VU B2ac(iv)).

SYN. [*Hypoestes verticillaris sensu* Robyns, F.P.N.A. 2: 298 (1947) pro parte quoad *Humbert* 8306, *non* (L.f.) Roem. & Schult.]

NOTE. On the evidence of the two distinct flower forms recorded, Champluvier (l.c.) concludes that this species can be gynomonoecious. The normal (hermaphrodite) flowers with long-exserted stamens release fertile pollen, whilst in those of the abnormal flowers the anthers remain closed and so do not release the pollen which is collapsed and probably infertile; these latter flowers therefore function as females.

5. **Brachystephanus africanus** *S. Moore* in Trans. Linn. Soc., Bot. 4: 31 (1894); C.B. Clarke in F.T.A. 5: 177 (1899); Champluvier in Fl. Rwanda, Sperm. 3: 444, fig. 140.1 (1985); Figueiredo & Jury in K.B. 51: 753 (1996); White *et al.*, Evergreen Forest Fl. Malawi: 113 (2001); Fischer & Killmann, Pl. Nyungwe N.P. Rwanda: 514 (2008); Champluvier & Darbyshire in Syst. Geogr. Pl. 79: 157, 186, figs. 17 & 35 (2009). Type: Malawi, Mt Mlanje, *Whyte* 56 (BM!, lecto., selected by Figueiredo & Jury, l.c.; K!, isolecto.)

Perennial herb or subshrub, 30–150(–250) cm tall, often decumbent and rooting at lower nodes; stems antrorse-, appressed- or rarely retrorse-pubescent when young or largely glabrous. Leaves ovate, elliptic or oblong, 5.5–24 cm long, 1.8–10 cm wide, base often narrowly cuneate then abruptly attenuate, apex acuminate to caudate, lower surface sparsely to more densely pubescent on veins beneath, upper surface glabrous or with scattered hairs; lateral veins 7–13 pairs; petiole 10–62 mm long. Inflorescences terminal and sometimes also in the upper leaf axils, narrowly spiciform, 5.5–25 cm long, with imbricate bracts at least when young, sometimes more widely spaced with maturity; cymules single-flowered, sessile; axes antrorse-pubescent or glabrous; bracts dark green or tinged pink, purple or red-brown, ovate or lanceolate to oblong-elliptic or somewhat obovate, 5–25 mm long, 1.7–6.5 mm wide, apex attenuate to caudate, straight or incurved, margin narrowly hyaline, glabrous except for minute hairs on acumen, more rarely with scattered glandular hairs and/or eglandular-puberulous; bracteoles lanceolate, 2–6.5(–8) mm long, 0.5–1 mm wide, margin hyaline. Calyx lobes linear, ± unequal, the longest 2.5–17 mm long, usually with scattered glandular hairs in upper half. Corolla (blue-) pink, purple, rarely white, 29–51 mm long; tube and limb shortly pubescent to tomentellous, with scattered longer glandular hairs on limb; tube straight to somewhat curved, narrowly cylindrical, 22–38 mm long; upper lip (ovate-) elliptic, 6.5–13.5 mm long, 5.5–8 mm wide, apex apiculate; lower lip elliptic, 7–14.5 mm long, 6–8.5 mm wide, lobes to 0.5(–1.3) mm long. Staminal filaments 13.5–36 mm long; anthers 1.7–3.5 mm long. Ovary and style glabrous; stigma capitate, 0.2–0.5 mm wide. Capsule 11.5–15 mm long, glabrous; only immature seeds seen, tuberculate, tubercles elongating towards margin and with minute hair-like processes. Fig. 61/1–5, p. 476.

var. **africanus**, *Champl.* in Syst. Geogr. Pl. 79: 158 (2009), modified here.

Bracts ovate, lanceolate, (oblong-) elliptic or at most somewhat oblanceolate, acumen straight or incurved.

UGANDA. Ankole District: Kashoya-Kitomi Forest Reserve, NE of Kyambura R., June 1994, *Poulsen* 561!; Kigezi District: Ishasha Gorge, Feb. 1950, *Purseglove* 3309!; Masaka District: Malabigambo Forest, Sept. 1936, *Eggeling* 3319!

TANZANIA. Bukoba District: Minziro Forest Reserve, footpath to Bulembe village, Sept. 2000, *Festo & Bayona* 769!; Morogoro District: Tegetero, Mar. 1953, *Drummond & Hemsley* 1728!; Iringa District: Udzungwa Mts National Park, June 2002, *Luke & Luke* 8757!

DISTR. **U** 2, 4; **T** 1, 6, 7; Nigeria, Congo-Kinshasa, Rwanda, Burundi, Malawi, Mozambique, Zimbabwe; Madagascar

HAB. Montane and submontane, rarely lowland, forest including streamsides and pathsides, swamp forest, often locally dominant in the understorey; (300–)900–2000 m

USES. None recorded on herbarium specimens

CONSERVATION NOTES. Widespread and often locally abundant: Least Concern (LC).

SYN. *B. bequaertii* De Wild. in Rev. Zool. Bot. Afr. 8, Suppl. Bot.: 36 (1920). Type: Congo-Kinshasa, Masisi, *Bequaert* 6374 (BR!, lecto., chosen by Figueiredo & Jury, l.c.)
B. velutinus De Wild. in Rev. Zool. Bot. Afr. 8, Suppl. Bot.: 36 (1920). Type: Congo-Kinshasa, Masisi–Walikale, *Bequaert* 6570 (BR!, holo.), **syn. nov**.
B. africanus S. Moore var. *velutinus* (De Wild.) Figueiredo in K.B. 51: 756 (1996)
B. africanus S. Moore var. *longibracteatus* Champl. in Syst. Geogr. Pl. 79: 161 (2009). Type: Congo-Kinshasa, Mt Mbese-Mbese, *Léonard* 1964 (BR, holo.), **syn. nov**.

NOTE. Champluvier recognises two varieties within our region based upon the relative length of the calyces and the bracts: the calyces are longer than the bracts in var. *africanus* (fig. 61/2) and vice versa in var. *longibracteatus* (fig. 61/4, 5). However, bract and calyx length appear to vary independently in this species and there is much overlap between the two taxa in terms of their absolute lengths. Further, the amount of variation within each variety is just as great as between them. For example, in Ugandan specimens of var. *longibracteatus* the calyx lobes are very short (2.5–5(–7) mm) whilst in specimens of this variety from the Uluguru Mts they are considerably longer (9–11.5 mm). These latter plants have calyx lobes well within the range for var. *africanus* but have more elongate bracts (often over 20 mm long, these being up to 15 mm long in var. *africanus*). Intermediate material is also rather frequent in our region, for example specimens from Ishasha Gorge, **U** 2 (e.g. *Purseglove* 3309) and from the Nguru Mts, **T** 6 (e.g. *Schlieben* 12268). These varieties are therefore not maintained here. However, var. *recurvatus* Champl. from eastern Congo-Kinshasa and var. *madagascariensis* Figueiredo from Madagascar appear to be good taxa; the former is very likely to occur in SW Uganda and is easily separated by the recurved apices to the bracts (see fig. 61/6).

6. **Brachystephanus holstii** *Lindau* in E.J. 20: 53 (1894); C.B. Clarke in F.T.A. 5: 178 (1899); Figueiredo & Jury in K.B. 51: 760 (1996); Champluvier & Darbyshire in Syst. Geogr. Pl. 79: 174, 186, figs. 28 & 36 (2009). Type: Tanzania, Kilimanjaro, Marungu, *Volkens* 838 (BM!, lecto., selected by Figueiredo & Jury, l.c.)

Perennial herb or subshrub, 10–300 cm tall, often decumbent and rooting at lower nodes, rarely trailing, whole plant often drying blackish; stems retrorse- and/or antrorse- to appressed-pubescent when young, hairs multicellular with conspicuous dark cell walls. Leaves (ovate-) elliptic, 6.5–22 cm long, 2.3–9 cm wide, base cuneate to attenuate, apex acuminate, sparsely pubescent on veins beneath, upper surface

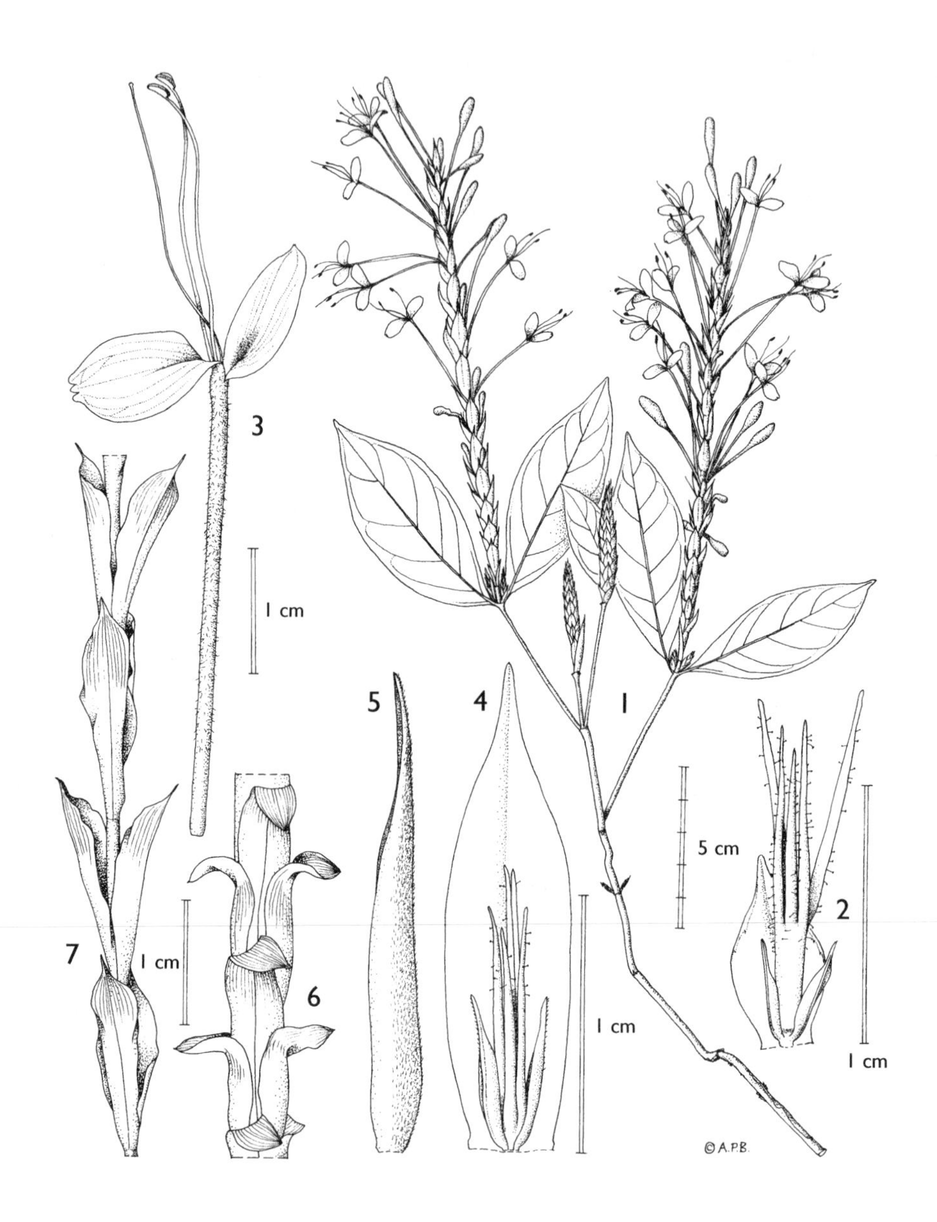

FIG. 61. *BRACHYSTEPHANUS AFRICANUS* — **1**, habit; **2**, bract, bracteoles and calyx; **3**, corolla with stamens and pistil; **4**, bract, bracteoles and calyx, variant with short calyx lobes and long bracts ("var. *longibracteatus*"); **5**, bract, lateral view. *B. AFRICANUS* var. *RECURVATUS* — **6**, partial inflorescence, flowers omitted (variety not yet recorded in our region). *B. AFRICANUS* var. *MADAGASCARIENSIS* — **7**, partial inflorescence, flowers omitted (variety not recorded in our region). 1 & 3 from *Drummond & Hemsley* 1986; 2 from *Bidgood & Vollesen* 2180; 4 & 5 from *Drummond & Hemsley* 1728; 6 from *Troupin* 4757; 7 from *Bernardi* 11994. Drawn by Andrew Brown. Reproduced from Syst. Geogr. Pl. 79: 159 (2009), with permission.

glabrous or sparsely pubescent; lateral veins 9–14 pairs; petiole 10–70 mm long. Inflorescences terminal and sometimes also axillary, densely spiciform or rarely subcapitate, (1.5–)3.5–15 cm long, 0.7–1.5(–2) cm wide, sometimes pendulous when young, cymules single flowered or more rarely 2–3-flowered; axes sparsely to densely antrorse-pubescent, hairs with conspicuous dark brown or purple cell walls; bracts dark green or purple, (linear-) lanceolate, 7.5–28 mm long, 1–4.5 mm wide, margin ciliate to very sparsely so, hairs as on axes, ascending; bracteoles as bracts but to 21 mm long, sometimes oblanceolate. Calyx lobes linear-lanceolate to linear-oblanceolate, subequal to unequal, the longest 6–19 mm long, ciliate and with occasional hairs on surface, hairs as on bracts. Corolla bright red (-purple) to rose pink, (18–)21–32 mm long; puberulous and with scattered longer glandular hairs on upper tube and limb outside; tube curved, (12–)15–23 mm long, narrowly cylindrical towards base, gradually widened in upper half; upper lip ovate to elliptic, 6–11.5 mm long, 4–5.5 mm wide, apex acute or minutely apiculate; lower lip ovate, 6–12 mm long, 4.5–9 mm wide, apex minutely 3-lobed. Staminal filaments 11–17 mm long; anthers 2–3(–3.5) mm long. Stigma capitate, 0.2–0.5 mm wide. Immature capsule only seen, ± 10 mm long, glabrous; seeds not seen.

TANZANIA. Kilimanjaro, S slope between Umbwe and Weru Weru R., Aug. 1932, *Greenway* 3204!; Lushoto District: W Usambara Mts, Shagayu Forest Reserve, NW slope of summit, 2.5 km ENE of Shagayu Sawmill, Mar. 1984, *Borhidi et al.* 84839!; Morogoro District: Uluguru North Forest Reserve, valley of R. Mwere, July 1972, *Mabberley* 1237!

DISTR. **T** 2, 3, 6; E Congo-Kinshasa (see note)

HAB. Wet montane and submontane forest including streamsides, also in bamboo forest; (?400–)950–2400 m

USES. "Root [used] for intestine disorder" (**T** 3; *Koritschoner* 741); "root [used] for gland irritants" (**T** 3; *Koritschoner* 819)

CONSERVATION NOTES. This species is widely recorded in the Eastern Arc Mts of Tanzania, also extending to Kilimanjaro. It can be locally common or even dominant in the forest understorey, particularly in the W Usambara, and appears tolerant of some disturbance (*Linder* 3724). However, many of the lower altitude sites are subject to significant forest loss and it is likely that several historic populations have been lost. It is therefore assessed as Near Threatened (NT).

NOTE. This species is rather variable in the length and indumentum of the bracts and calyx lobes but there are no clear geographic trends. Extreme variants with short, sparsely hairy bracts and calyces (e.g. *Drummond & Hemsley* 1674) appear rather distinct from typical *B. holstii*, but intermediate specimens are also found.

The material from E Congo-Kinshasa currently placed within this species is rather poor, although the few preserved corollas do appear to have the broadened, curved tube characterisitic of *B. holstii*. They differ from most collections from the Eastern Arc in having a broader inflorescence with each bract subtending 2–3 flowers, those from Tanzania usually subtending a single flower. However, several specimens from Tanzania (e.g. *Mabberley* 1237) also have 2–3 flowers per bract. The disjunction in distribution is rather striking and requires further investigation.

Congdon 679 from Kanza, Pare Mts is rather distinctive in having a trailing habit and subcapitate inflorescences (again with up to 3 flowers per cymule); it is possible that this is a result of grazing pressure at this site.

7. **Brachystephanus roseus** *Champl.* in Syst. Geogr. Pl. 79: 178, 190, fig. 30 (2009). Type: Congo-Kinshasa, Kitshanga, *A. Léonard* 2364 (BR!, holo.; K!, iso.)

Perennial herb or subshrub, 15–60(–120) cm tall, whole plant often drying blackish; stems retrorse- to appressed-pubescent when young, hairs multicellular with conspicuous dark cell walls. Leaves elliptic, 7–18 cm long, 3–6.5 cm wide, base (cuneate-) attenuate, apex acuminate, margin and veins beneath pubescent, upper surface glabrous or sparsely pubescent, rarely more densely pubescent throughout; lateral veins 10–11 pairs; petiole 10–32 mm long. Inflorescences terminal, densely spiciform or paniculate with the cymules very shortly pedunculate and congested, 4.5–9 cm long, 2–3.5 cm wide, cymules several-flowered, dichasial or sometimes monochasial; axes densely pubescent, hairs often antrorse, with conspicuous dark brown or purple cell walls; bracts and bracteoles (linear-) lanceolate, 9–24 mm long, 0.8–2.5 mm wide, margin and often midrib pubescent, hairs as on axes but more spreading. Calyx lobes linear, somewhat unequal, the longest 8–12.5 mm long in flower, up to 14 mm long in fruit, ciliate, hairs as on bracts. Corolla pink to purple, 20–24 mm long, puberulous and with scattered short glandular hairs on upper tube and limb outside; tube narrowly cylindrical, curved, 14–16 mm long; upper lip elliptic, 5–7.7 mm long, 2.3–4 mm wide, apex minutely apiculate; lower lip elliptic, 6–8 mm long, 3–5.5 mm wide, apex minutely 3-lobed. Staminal filaments 6.5–14 mm long; anthers 1.5–2 mm long. Stigma capitate, 0.2–0.4 mm wide. Capsule 8.5–12 mm long, glabrous; only immature seeds seen, tuberculate.

UGANDA. Ankole District: Igara, Kalinzu Forest Reserve, Feb. 1953, *Osmaston* 2832!; Kigezi District: Ishasha Gorge, Apr. 1946, *Purseglove* 2037! & 7 km SW of Kirima, along Ishasha R., Sept. 1969, *Lye* 4186!

DISTR. U 2; E Congo-Kinshasa, Rwanda, Burundi

HAB. Montane and submontane forest including riverine and swamp forest, also in drier *Cynometra* forest in E Congo; 1200–1600 m

USES. None recorded on herbarium specimens

CONSERVATION NOTES. This species has a highly restricted range in the mountains and foothills bordering the Albertine Rift. Much of its forest habitat here is threatened by high human populations leading to significant levels of habitat clearance. However, some of its sites are now afforded some protection, such as the Kalinzu and Kashoya-Kitomi Forest Reserves and Bwindi National Park in Uganda. It is therefore assessed as Near Threatened (NT).

SYN. [*B. holstii sensu* Champluvier in Fl. Rwanda, Sperm. 3: 444 (1985) quoad *Reekmans* 4964, *non* Lindau]

36. PSEUDERANTHEMUM

Radlk. in Sitzber. Bayer. Akad. Wiss. 13: 282 (1893); Lindau in E. & P. Pf. IV, 3b: 330 (1895); Milne-Redhead in K.B. 1936: 259 (1936); Champluvier in Syst. Geogr. Pl. 72: 48 (2002)

Pigafetta Adans., Fam. 2: 223 (1763), *nom. rej.*, *non* (Blume) Becc. (1877), *nom. cons.*
Eranthemum sensu C.B. Clarke in F.T.A. 5: 169 (1899) pro parte, *non* L. (1753)

Perennial herbs, subshrubs or shrubs; cystoliths conspicuous, linear; stems subquadrangular or rounded. Leaves opposite, entire to crenate. Flowers single or in 3–7-flowered cymules aggregated into long racemoid cymes. Bracteoles present, narrowly triangular. Calyx divided almost to base into 5 equal linear-lanceolate to narrowly triangular acute to acuminate lobes. Corolla hairy and/or glandular on the outside; basal tube long and linear (rarely short), straight or slightly curved, widened into a short throat; lobes 5, wider and spreading in lower lip, erect in upper lip. Stamens 2, flattened, inserted at base of and held dorsally in throat, included to slightly exserted, glabrous; anthers bithecous, oblong, straight or curved, apiculate or rounded at both ends, glabrous, hairy or glandular on connective, thecae held at different heights. Ovary with 2 ovules per locule; style glabrous; stigma of 2 equal, broadly ellipsoid erect lobes. Capsule (2–)4-seeded, clavate with contracted solid stalk-like basal part, glabrous, seed-bearing part ellipsoid, laterally flattened; retinacula strong, hooked. Seed discoid to ellipsoid, rugose to reticulate on both sides or smooth on outer side.

50–75 species widely distributed in all tropical regions; particularly in SE Asia and in the Pacific Region. The total number of species is very uncertain with several species complexes in SE Asia and the Pacific in severe need of monographic treatment.

P. atropurpureum has been cultivated in Kenya. It differs from the indigenous species in having a white corolla with purple markings combined with a short (± 1 cm long) corolla tube and long (± 5 mm) pedicels. Of the indigenous species *P. ludovicianum* has also been cultivated in Kenya.

1. Cylindric part of corolla tube 3–4 mm long; corolla sharply bent between cylindric part of tube and throat ... 1. *P. campylosiphon*
   Cylindric part of corolla tube (10–)14–35 mm long; corolla straight or very slightly curved ... 2
2. Corolla orange-red to scarlet (rarely yellow); seed deeply reticulate on inner surface, smooth or very slightly reticulate on the outer surface, with a smooth rim ... 5. *P. hildebrandtii*
   Corolla white to mauve or lilac; seed similarly reticulate-rugose on both sides, with a jagged rim ... 3
3. Corolla white with many fine purple dots on lobes (rarely tinged pale mauve); anthers fully exserted ... 2. *P. ludovicianum*
   Corolla white to mauve, without fine purple dots on lobes but sometimes with larger darker patches near centre; anthers included to fully exserted ... 4
4. Plant drying blackish; young stems glabrous; cylindric part of corolla tube 2.5–3 cm long; anthers fully exserted ... 4. *P. usambarensis*
   Plant drying green to greyish green; young stems hairy; cylindric part of corolla tube (1–)1.3–2.1(–2.5) cm long; anthers included to partly exserted ... 3. *P. subviscosum*

1. **Pseuderanthemum campylosiphon** *Mildbr.* in N.B.G.B. 11: 1084 (1934); Milne-Redh. in K.B. 1936: 260 (1936); T.T.C.L.: 14 (1949); Ruffo *et al.*, Cat. Lushoto Herb. Tanzania: 9 (1996); Champluvier in Syst. Geogr. Pl. 72: 48 (2002). Type: Tanzania, Morogoro District: Uluguru Mts, *Schlieben* 2810 (B†, holo.; BM!, iso.)

Erect spindly shrubby herb or shrub to 2 m tall; plant drying black; young stems glabrous (rarely with a thin band of hairs below nodes). Leaves elliptic to obovate, largest 13–19 × 3.5–7.5 cm, base attenuate, decurrent to the 0.5–3 cm long petiole, narrowing gradually to an acuminate apex with an acute to obtuse tip, glabrous. Cymes 5–15 cm long, unbranched, often with additional cymes from upper leaf axils; flowers all solitary (rarely in 3-flowered cymules at lowermost nodes); axis glabrous or with two band of curly brownish hairs; bracts narrowly triangular, curly-ciliate, 1–2 mm long; pedicels 0.5–1(–2.5) mm long, glabrous; bracteoles as bracts, 1–2 mm long. Calyx 3–5 mm long, on the outside sparsely ciliate, inside with dense thick hairs. Corolla white (rarely with brown markings in throat), outside puberulous or densely so with thick hairs; tube strongly bent between basal cylindric part and throat, basal linear tube 3–4 mm long; throat 3–5 mm long, widening slightly upwards, lobes 5–9 × 2–4 mm, oblong-elliptic, rounded to subacute. Stamens included; filaments 1–2 mm long; anthers 1–1.5 mm long, apiculate, with a few hairs on connective. Capsule 2.2–2.5(–3) cm long; seed 5–6 × ± 4 mm, reticulate-verrucose on both side, with a jagged rim.

TANZANIA. Kilosa District: Ukaguru Mts, Mamiwa Forest Reserve, Mandege Forest Station, 29 July 1972, *Mabberley & Pócs* 1266!; Morogoro District: North Uluguru Forest Reserve, Palu, 30 Nov. 1993, *Kisena* 966!; Iringa District: Mwanihana Forest Reserve, above Sanje, 17 June 1986, *J. Lovett et al.* 857!

DISTR. **T** 6, 7; not known elsewhere

HAB. Evergreen montane forest; 1150–1700 m

2. **Pseuderanthemum ludovicianum** (*Büttn.*) *Lindau* in E. & P. Pf. IV, 3b: 330 (1895); Th. Dur. & Schinz, Etud. Fl. Congo: 220 (1896); De Wildeman, Etud. Fl. Katanga 2: 148 (1913) & Contrib. Fl. Katanga: 202 (1921); Milne-Redhead in K.B. 1936: 263, Fig. 5 (1936); F.P.S. 3: 186 (1956); Heine in F.W.T.A., ed. 2, 2: 421 (1963) & in Fl. Gabon 13: 170, pl. 34 (1966); Champluvier in Fl. Rwanda 3: 482 (1985); U.K.W.F., ed. 2: 275 (1994); K.T.S.L.: 606 (1994); Champluvier in Syst. Geogr. Pl. 72: 48 (2002); Friis & Vollesen in Biol. Skr. 51(2): 453 (2005). Type: Congo-Kinshasa, Kasongo Lunda, *Büttner* 460 (B†, holo.; K!, iso.).

Erect shrubby herb or shrub to 2(? –3) m tall; plant drying black; young stems glabrous to sparsely puberulous. Leaves elliptic to obovate, largest (8.5–)12–25(–30) × (3–)5–11(–14) cm, base attenuate to cuneate (rarely rounded), narrowing gradually or abruptly to an acuminate apex with an acuminate to acute tip, glabrous; petiole 0.5–3 cm long. Cymes (2–)5–20(–30) cm long, unbranched (rarely with short branches to 1 cm long from lower nodes), only rarely with additional cymes from upper leaf axils; flowers all in 3-flowered cymules, often with 2 cymules per bract; peduncle 5–20 cm long; axis puberulous or sparsely so with curly brownish hairs; lower pair of bracts sometimes foliaceous, the rest lanceolate to triangular, 3–10 mm long, puberulous-ciliate or sparsely so; pedicels 0.5–2.5 mm long, glabrous; bracteoles as bracts, 1–4 mm long. Calyx 2–4 mm long, on the outside subglabrous to sparsely puberulous, often only sparsely puberulous-ciliate, inside with scattered hairs. Corolla white with many fine purple dots on lobes (rarely tinged pale mauve), outside puberulous or sparsely so with a mixture of hairs and capitate glands, sometimes almost entirely glands; tube straight, basal linear part 1.9–3.5 cm long; throat 1–2 mm long, lobes 6–10 (–12) × 4–7 mm in lower lip, elliptic, rounded. Stamens exserted; filaments 2–3 mm long; anthers black, 1.5–2 mm long, apiculate, glandular on connective. Capsule 2.5–3.2 cm long; seed (pictured by Milne-Redhead, l.c., p. 265, not seen by me) reticulate-verrucose on both sides, with a jagged rim.

UGANDA. Bunyoro District: Budongo Forest, Nov. 1931, *Brasnett* 241!; Mt Elgon, Babungi, Dec. 1939, *Dale* U83!; Masaka District: Malabigambo Forest, 6 km SSW of Katera, 2 Oct. 1953, *Drummond & Hemsley* 4549!

KENYA. North Kavirondo District: Kakamega Forest, 11 Dec. 1956, *Verdcourt* 1681! & 4 Jan. 1968, *Perdue & Kibuwa* 9445! & 24 Jan. 1981, *M.G. Gilbert* 6871!

TANZANIA. Bukoba District: Minziro Forest Reserve, 6 July 2000, *Bidgood et al.* 4872!; Kigoma District: Gombe National Park, Mitumba Valley, 30 Sept. 1999, *Gobbo et al.* 474!; Mpanda District: Kungwe-Mahali Peninsula, Musenabantu Ridge, 17 Aug. 1959, *Harley* 9366!

DISTR. **U** 2–4, **K** 5, **T** 1, 4; Nigeria, Bioko, Cameroon, Central African Republic, Congo-Kinshasa, Burundi, Sudan, Angola

HAB. Undergrowth in wet lowland to montane forest, swamp forest; 750–1600(–1800) m

SYN. *Eranthemum ludovicianum* Büttn. in Verh. Bot. Ver. Brandenb. 32: 41 (1890); C.B. Clarke in F.T.A. 5: 172 (1899); S. Moore in Cat. Pl. Talbot Nigeria: 140 (1913); Hutchinson & Dalziel, F.W.T.A. 2: 262 (1930)

E. ardisioides C.B. Clarke in F.T.A. 5: 173 (1899). Type: Uganda, no locality, *Scott Elliot* 7521 (K!, holo.; BM!, iso.)

E. bilabiale C.B. Clarke in K.B. 1906: 252 (1906). Type: Uganda, Mengo District: Mawokota, *E. Brown* 209 (K!, holo.)

3. **Pseuderanthemum subviscosum** (*C.B. Clarke*) *Stapf* in Bot. Mag. 135: t. 8244 (1909); Lindau in Z.A.E.: 301 (1911); Milne-Redhead in K.B. 1936: 263 (1936); Brenan in Mem. N. Y. Bot. Gard. 9: 23 (1954); Binns, Check List Herb. Fl. Malawi: 15 (1968); Champluvier in Syst. Geogr. Pl. 72: 48 (2002). Type: Mozambique, Makua, Namuli Hills, *Last* s. n. (K!, lecto. selected by Milne-Redhead (l.c.); BM!, iso.)

Erect shrubby herb to 0.5(? –1.5) m tall; plant drying green; young stems puberulous or sparsely so with broad curved hairs with purple walls. Leaves elliptic (rarely linear-lanceolate), largest (4–)8–19 × (1–)2–8.5 cm, base attenuate to cuneate, abruptly (more rarely gradually) narrowing below a subacute to acuminate apex with an acute to obtuse tip, glabrous or with a few hairs on lamina and midrib; petiole to 3 cm long. Cymes 2–17(–25) cm long, unbranched or with lateral cymes to 7 cm long from lower nodes, sometimes also with elongated cymes from upper leaf axils or with sessile cymules here, rarely with the cyme reduced to axillary clusters; flowers solitary or in 3(–7)-flowered cymules towards base of cyme (rarely with 2 cymules per bract); peduncle to 15 cm long, if cyme unbranched with 1(–2) pairs of sterile bracts; axis puberulous or sparsely so with curved (rarely straight) hairs with purple walls, with or without stalked capitate glands; bracts linear to lanceolate, 2–5 mm long, sparsely puberulous-ciliate and sometimes with a few glands; pedicels 1–3 mm long, glabrous to puberulous; bracteoles as bracts, 2–4 mm long. Calyx 3–6 mm long, on the outside sparsely puberulous, with or without stalked capitate glands, inside sparsely puberulous. Corolla white to pale mauve, with or without darker dots on lower lip, outside subglabrous to puberulous and with (rarely without) scattered capitate glands; tube straight or slightly curved, basal linear part (1–) 1.4–2.1(–2.5) cm long; throat 2–4 mm long, lobes 8–13 × 4–7 mm in lower lip, elliptic, rounded or 3-toothed. Stamens included or slightly exserted; filaments ± 2 mm long; anthers pale to purple, 1–1.5 mm long, apiculate or rounded, hairy on connective. Capsule 1.6–2.5(–3) cm long; seed 3–5 × 2–4 mm, densely reticulate-verrucose on both side, with a jagged rim.

UGANDA. Bunyoro District: Budongo Forest, Nov. 1935, *Eggeling* 2273!; Mengo District: Mubango, Sept. 1920, *Dummer* 4465!; Masaka District: Malabigambo Forest, 3 km SSW of Katera, 3 Oct. 1953, *Drummond & Hemsley* 4594!

KENYA. Kilifi District: Gongoni Forest Reserve, 30 Dec. 1993, *Luke* 3947!; Kwale District: Shimba Hills, Longomwagandi Forest, 13 Nov. 1968, *Magogo* 1222 & Shimba Hills, Mwele Mdogo Hill, 28 Dec. 1968, *Glover et al.* 1154!

TANZANIA. Bukoba District: Minziro Forest Reserve, 15 June 1997, *Congdon* 494!; Lushoto District: E Usambara Mts, Amani, 26 Dec. 1956, *Verdcourt* 1735!; Iringa District: Udzungwa Mts, Sanje, 23 July 1983, *Polhill & Lovett* 5115!; Pemba: Ngezi Forest Reserve, Dec. 1983, *Rodgers et al.* 2655!

DISTR. **U** 2, 4; **K** 7; **T** 1–4, 6–8; Cameroon, Gabon, Congo-Brazzaville, Central African Republic, Congo-Kinshasa, Sudan, Ethiopia, Angola, Zambia, Malawi, Mozambique, Zimbabwe, South Africa; Madagascar, Comoro Islands

HAB. Lowland and medium altitude (rarely montane) wet evergreen forest, riverine forest, termite mounds in bushland; near sea level to 1650(–1800) m

SYN. *Eranthemum subviscosum* C.B. Clarke in F.T.A. 5: 173 (1899); S. Moore in J.L.S. Bot. 40: 161 (1911); Eyles in Trans. Roy. Soc. S. Afr. 5: 483 (1916)

E. lindaui C.B. Clarke in F.T.A. 5: 173 (1899). Types: Tanzania, Lushoto District: E Usambara Mts, Mashewa [Maschaua], *Holst* 3494 (K!, syn.) & Nderema, *Holst* 2251 (not seen)

Pseuderanthemum tunicatum sensu Milne-Redhead in K.B. 1936: 264 (1936) pro parte; T.T.C.L.: 14 (1949); Binns, Check List Herb. Fl. Malawi: 15 (1968); Richards & Morony, Check List Fl. Mbala: 233 (1969); Iversen in Symb. Bot. Upsal. 29(3): 162 (1991); Luke & Robertson, Kenya Coast. For. 2. Checklist Vasc. Pl.: 83 (1993); K.T.S.L.: 606 (1994); Ruffo *et al.*, Cat. Lushoto Herb. Tanzania: 9 (1996); White *et al.*, For. Fl. Malawi: 118 (2001); Friis & Vollesen in Biol. Skr. 51(2): 453 (2005); Ensermu in Fl. Eth. 5: 436 (2006), *non* (Afz.) Milne-Redh. (1936)

P. albo-coeruleum Champl. in Syst. Geogr. Pl. 72: 34 (2002). Type: Congo-Brazzaville, Odzala National Park, *Champluvier* 5207 (BR, holo.; K!, iso.)

P. albo-coeruleum Champl. subsp. *robustum* Champl. in Syst. Geogr. Pl. 72: 37 (2002). Type: Tanzania, Morogoro District: Mtibwa Forest Reserve, *Semsei* 1948 (K!, holo.)

NOTE. A very variable species which in the western part of its distribution can be difficult to separate from the West African *P. tunicatum*. The attempt by Champluvier in Syst. Geogr. Pl. 72 (2002) is not entirely convincing and I think a study of the genus throughout Africa is needed before any final conclusions can be drawn.

A form from the Udzungwa Mts (**T** 7) has linear-lanceolate strongly undulate leaves and at first sight looks very distinct. Examples are *Frimodt-Møller* TZ626 and *Luke* 6546. But I have found no separating characters in either flowers or fruits and there are collections intermediate with typical material, e.g. *Luke* 10489.

4. **Pseuderanthemum usambarensis** *Vollesen* **sp. nov.** a *P. subviscoso* planta in sicco atro nec viridi, tubo cylindrico corollae longiore 2.5–3 cm (nec (1–)1.3–2.1(–2.5) cm) et antheris perfecte exsertis (nec in tubo corollae inclusis nec partim tantum exsertis) differt. Type: Tanzania, Lushoto District, East Usambara Mts, Kwamkoro to Kihuhi, *Greenway* 4791 (K!, holo.; EA!, iso.)

Erect shrubby herb to 2 m tall; plant drying black; young stems glabrous. Leaves elliptic to obovate or narrowly so, largest 16–25 × 4–9.5 cm, base attenuate, decurrent, narrowing abruptly below an acuminate apex with an obtuse tip, glabrous or with a few hairs on midrib; petiole ill-defined, up to 2.5 cm long. Cymes 5–14 cm long, unbranched or with short branches to 1 cm long from lower nodes; flowers all in 3–15-flowered cymules or solitary near apex of cyme; peduncle 7.5–25 cm long, with 1(–2) pairs of sterile bracts near or above middle; axis puberulous or sparsely so with capitate glands and with or without non-glandular straight hairs; bracts lanceolate, 2–5 mm long, sparsely puberulous-ciliate and with a few glands; pedicels 0.5–3 mm long, glabrous; bracteoles as bracts, 2–3 mm long. Calyx 4–7 mm long, on the outside sparsely puberulous and with few to many subsessile or stalked capitate glands, inside sparsely puberulous. Corolla white, outside with scattered hairs and capitate glands; tube straight, basal linear part 2.5–3 cm long; throat 3–4 mm long, lobes ± 10 × 6 mm in lower lip, elliptic, subacute. Stamens exserted; filaments ± 2 mm long; anthers pale, ± 1.5 mm long, apiculate, hairy on connective. Capsule 2.2–3 cm long; seed 4–5 × 4–5 mm, densely reticulate-verrucose on both sides, with a jagged rim.

TANZANIA. Lushoto District: E Usambara Mts, Derema to Ngambo, 15 Feb. 1916, *Peter* 60365! & Kwamkoro to Kihuhi, 16 Dec. 1936, *Greenway* 4791! & between Amani and Derema, Hunga Valley, 8 Feb. 1987, *Pócs* 87029/E!

DISTR. **T** 3; not known elsewhere

HAB. Submontane evergreen forest; 850–1100 m

SYN. [*Pseuderanthemum subviscosum sensu* Iversen in Symb. Bot. Upsal. 29(3): 162 (1991), *non* (C.B. Clarke) Stapf (1909)]

5. **Pseuderanthemum hildebrandtii** *Lindau* in E.J. 20: 39 (1894) & in P.O.A. C: 371 (1895) & in E. & P. Pf. IV, 3b: 330 (1895); Milne-Redhead in K.B. 1936: 260 (1936); T.T.C.L.: 14 (1949); Vollesen in Opera Bot. 59: 81 (1980); Iversen in Symb. Bot. Upsal. 29(3): 162 (1991); Luke & Robertson, Kenya Coast. For. 2. Checklist Vasc. Pl.: 83 (1993); K.T.S.L.: 606 (1994); Ruffo *et al.*, Cat. Lushoto Herb. Tanzania: 9 (1996); Champluvier in Syst. Geogr. Pl. 72: 48 (2002). Type: Tanzania, Zanzibar, Kidoti, *Hildebrandt* 981 (K!, lecto., selected by Milne-Redhead (l.c.); BM!, iso.)

Erect shrubby herb or shrub to 2 m tall; plant drying black; young stems pale yellow to orange, smooth and shiny, glabrous to puberulous. Leaves ovate, largest 2.5–16 × 1–7 cm, abruptly narrowed or not below middle, base attenuate and decurrent to cuneate, apex acuminate to obtuse, glabrous to sparsely pubescent, densest along veins; petiole up to 2(–4) cm long. Cymes to 12 cm long, but often much less and sometimes reduced to apparently solitary flowers in upper leaf axils, unbranched or branched, often also with additional cymes from upper leaf axils; flowers solitary or in 3–7-flowered cymules, sometimes with 2 cymules per bract at lower nodes; axis glabrous to sparsely puberulous; lower pair(s) of bracts usually foliaceous (in reduced cymes then looking as if flowers all solitary and axillary), the rest lanceolate to narrowly triangular, 1–3 mm long, glabrous or sparsely puberulous-ciliate; pedicels 1–3(–4) mm long, glabrous; bracteoles as bracts, often with a narrow scarious margin, 1–2 mm long. Calyx 3–5 mm long, on the outside sparsely ciliate, otherwise glabrous, inside puberulous with thick hairs. Corolla orange-red to scarlet (rarely yellow), with darker centre, outside sparsely puberulous (rarely subglabrous) with a mixture of hairs and capitate glands, sometimes almost entirely glands; tube straight,

FIG. 62. *PSEUDERANTHEMUM HILDEBRANDTII* — **1**, habit, × $^2/_3$; **2**, calyx opened up, × 6; **3**, detail of calyx indumentum, × 16; **4**, corolla opened with stamens, × 2; **5**, anther, × 10; **6**, ovary and basal part of style, × 4; **7**, stigma, × 16; **8**, capsule, × 1.5; **9**, seed, × 4. 1–5 from living plant, RBG, Kew, 6–7 from *Batty* 894, 8–9 from *Verdcourt* 1074. Drawn by Margaret Tebbs.

basal linear part 2.2–3.2 cm long; throat 3–4 mm long, lobes elliptic-obovate, 9–16 × 4–8 mm in lower lip, rounded. Stamens exserted; filaments 2–3 mm long; anthers dark purple, 1–1.5 mm long, rounded to apiculate, glabrous on connective. Capsule 1.9–2.7 cm long; seed 4–6 × 3–5 mm, deeply reticulate on the inner side, smooth to slightly reticulate on the outer side, rim smooth. Fig. 62, p. 483.

KENYA. Northern Frontier District: Marsabit, Sokorte Dika, 28 Feb. 1963, *Bally* 12562!; Machakos District: Kibwezi, 30 May 1970, *Faden et al.* 70/146!; Teita District: Taveta, 13 Dec. 1961, *Polhill & Paulo* 983!

TANZANIA. Mbulu District: Lake Manyara National Park, Mto wa Ukindu, 19 Feb. 1964, *Greenway & Kanuri* 11199!; Lushoto District: Korogwe, Karenge Estate, 5 March 1952, *Faulkner* 911!; Morogoro District: Nguru Mts, Mt Kanga, 3 Dec. 1987, *Mwasumbi & Munyenyembe* 13863!; Zanzibar, Kazimbazi, 27 May 1960, *Faulkner* 2572!

DISTR. **K** 1, 4, 6, 7; **T** 2, 3, 6, 8; **Z**; Mozambique

HAB. Coastal, lowland and medium altitude forest, thicket and bushland, riverine forest; near sea level to 1600(–1825) m

SYN. *Eranthemum hildebrandtii* (Lindau) C.B. Clarke in F.T.A. 5: 172 (1899)

NOTE. Very distinct in *Pseuderanthemum* with its red flowers and seeds with two different surfaces. In these characters it seems to tend towards *Ruspolia*. The pale yellow to reddish brown glossy bark and protruding leaf bases are also distinct.

37. RUSPOLIA

Lindau in E. & P. Pf. IV, 3b: 354 (1895) & in Ann. R. Ist. Bot. Roma 6: 79 (1896); C.B. Clarke in F.T.A. 5: 174 (1899); Milne-Redhead in K.B. 1936: 269 (1936)

Eranthemum sensu C.B. Clarke in F.T.A. 5: 169 (1899) pro parte, *non* L. (1753)

Shrubby herbs or shrubs; cystoliths conspicuous, linear; stems subquadrangular or rounded. Leaves opposite, entire or slightly crenate. Flowers single or in 3–7-flowered cymules aggregated into long racemoid cymes. Bracteoles present, narrowly triangular. Calyx divided almost to base into 5 equal subulate to narrowly triangular lobes, often with a long twisted tip. Corolla hairy and/or glandular on the outside; basal tube long, linear, slightly curved, only slightly widened into a short throat, lobes 5, subequal or wider in lower lip, all spreading or deflexed. Stamens 2, inserted at base of and held dorsally in throat, usually slightly exserted, glabrous; anthers monothecous, straight or slightly curved at back, curved ventrally, apiculate or rounded, glabrous. Ovary with 2 ovules per locule; style glabrous; stigma of 2 equal, broadly ellipsoid erect lobes. Capsule (2–)4-seeded, clavate with contracted solid stalk-like basal part, glabrous, seed-bearing part ellipsoid, laterally flattened; retinacula strong, hooked. Seed discoid, inner and outer surfaces different, with a strong raised marginal rim.

5 species, 4 in tropical Africa one of which extends to Madagascar, also 1 endemic in Madagascar.

1. Inflorescence axis, bracts and calyces with stalked capitate glands; corolla outside with long curly hairs, without stalked capitate glands; capsule 2.4–2.6 cm long; seed 5–6 mm long, inner surface with irregular ridges; plant drying green . . . . 4. *R. decurrens*

   Inflorescence axis, bracts and calyces without stalked capitate glands; corolla outside with mixture of long curly hairs and capitate glands or with capitate glands only; capsule 3–4.8 cm long; seed 8–10 mm long 2

2. Corolla outside with mixture of long curly hairs and stalked capitate glands (rarely with stalked glands only); seed on inner surface with irregular ridges; lower nodes in inflorescence 2–5(–8) cm long; plant drying green . . . . 3. *R. seticalyx*
   Corolla outside with stalked capitate glands only; seed on inner surface smooth, with broad marginal rim; lower nodes in inflorescence up to 0.5(–1) cm long; plant drying black 3
3. Pedicels glabrous or sparsely puberulous near base; bracteoles and calyces glabrous or sparsely puberulous-ciliate near apex with straight hairs; capsule 3–4.3 cm long; seed ± 8 × 7 mm . . . . 1. *R. hypocrateriformis*
   Pedicels puberulous throughout; bracteoles and calyces puberulous-ciliate with curly hairs along their whole length; capsule 3.8–4.8 cm long; seed 9–10 × ± 8 mm . . . . 2. *R. australis*

1. **Ruspolia hypocrateriformis** (*Vahl*) *Milne-Redh.* in K.B. 1936: 270 (1936); F.P.S. 3: 187 (1956); Heine in F.W.T.A., ed. 2, 2: 431 (1963); E.P.A.: 958 (1964); Hepper, West Afr. Herb. Isert & Thonning: 19 (1976); Iversen in Symb. Bot. Upsal. 29(3): 162 (1991); Ruffo *et al.*, Cat. Lushoto Herb. Tanzania: 9 (1996); Ensermu in Fl. Eth. 5: 435 (2006). Type: "Guinea", *Thonning* 28 (C, holo. & iso.)

Erect (rarely scrambling) shrubby herb or shrub to 1.25(? –2) m tall; plant drying black; young stems glabrous to sparsely puberulous. Leaves ovate to elliptic, largest 3–9(–11) × 1.5–4(–5.5) cm, gradually narrowed to a long decurrent base, apex acute to acuminate with an obtuse (rarely acute) tip; glabrous or sparsely puberulous along midrib and larger veins, above also with scattered hairs on lamina; petiole to 0.5–2 cm long. Cymes 2–8 cm long, dense, branched with short lateral branches to 5 mm long, sometimes also with lateral cymes from upper leaf axils; lower internodes to 0.5(–1) cm long; flowers all in 3–7-flowered cymules; axis subglabrous to densely sericeous-puberulous, hairs straight, no stalked capitate glands; lower 1–2 pairs of bracts sometimes foliaceous, the rest subulate to narrowly triangular, (6–)8–13 mm long, soon caducous, with straight hairs; pedicels 1–3 mm long, glabrous or sparsely puberulous near base; bracteoles subulate, 6–8 mm long. Calyx (7–)9–13 mm long, glabrous or sparsely puberulous apically on lobes, hairs straight. Corolla crimson-red or flame red (rarely orange red), towards centre with darker spots, outside with scattered stalked capitate glands, no hairs; linear tube 1.8–2.8 cm long; throat 2–3 mm long, lobes elliptic, 7–14 × 4–8 mm, rounded. Filaments 3–4 mm long; anthers purple, 1–1.5 mm long. Capsule 3–4.3 cm long; seed ± 8 × 7 mm, smooth on the outer surface, smooth with a broad raised rim on the inner surface.

UGANDA. West Nile District: Metu, Sept. 1943, *Eggeling* 5312! & 9 Sept. 1953, *Chancellor* 239!

KENYA. Northern Frontier District: Dawa River, Murri, 27 June 1951, *Kirrika* 94!; Lamu District: Kiunga Islands, 1951, *Oxford Univ. Kiunga Exp.* s.n.!; Mombasa District: 16 km S of Mombasa, 10 Dec. 1969, *Bally* 13702! (see note).

DISTR. **U** 1; **K** 1, 7; Senegal, Guinea, Sierra Leone, Mali, Ghana, Togo, Nigeria, Cameroon, Congo-Kinshasa, Ethiopia, Somalia, Angola

HAB. Rocky hillsides (Uganda), riverine thicket, coastal bushland; near sea level to 1000 m

SYN. *Justicia hypocrateriformis* Vahl, Enum. Pl. 1: 165 (1805); Schumach. & Thonn., Beskr. Guin. Pl.: 11 (1827)

Eranthemum hypocrateriforme (Vahl) Roem. & Schult., Syst. 1: 175 (1817); Nees in DC., Prodr. 11: 454 (1847); Benth. in Hook., Niger Fl.: 484 (1849); T. Anderson in J.L.S. Bot. 7: 52 (1863); J. D. Hooker in Bot. Mag. 101: t. 6181 (1875); C.B. Clarke in F.T.A. 5: 171 (1899); Hutchinson & Dalziel, F.W.T.A. 2: 262 (1930)

E. affine Spreng., Syst. Veg., ed. 16, 1: 89 (1824). Type: as for *E. hypocrateriforme*
Pseuderanthemum hypocrateriforme (Vahl) Radlk. in Sitzber. Bayr. Akad. Wiss. 13: 286 (1883); Lindau in E. & P. Pf. IV, 3b: 330 (1895) pro parte; S. Moore in J.B. 68, suppl. 2: 136 (1930)
Ruspolia pseuderanthemoides Lindau in Ann. R. Ist. Bot. Roma 6: 80 (1896); E.P.A.: 958 (1964); K.T.S.L.: 607 (1994); Ruffo *et al.*, Cat. Lushoto Herb. Tanzania: 10 (1996); Hedrén in Fl. Somalia 3: 400 (2006). Type: Ethiopia, Dawa River, Uelbe, *Riva* 1418 (FT, holo.)
[*R. seticalyx sensu* E.P.A.: 958 (1964), *non* (C.B. Clarke) Milne-Redh. (1936)]

NOTE. The material from Uganda and **K** 1 is doubtless all native while the provenance of the material from the Kenya coast is doubtful. It is likely that at least some material from south of Mombasa originated from escaped garden plants, and some collectors have also expressed doubt as to the origin of their material. The species is, however, native in S Somalia. The species has been grown as a garden ornamental in **K** 3, 4, 7; **T** 3, 6 and on Zanzibar.

2. **Ruspolia australis** (*Milne-Redh.*) *Vollesen* **comb. et stat. nov**. Type: South Africa, Wyliespoort, *Schweickerdt & Verdoorn* 441 (PRE, holo.; K!, iso.)

Scrambling shrub to 3(–4) m tall; plant drying black; young stems sparsely puberulous at nodes. Leaves ovate or broadly so, largest 7–12 × 3.2–7 cm, basal part below middle gradually narrowed to a long decurrent base, apex acute to subacuminate; with scattered hairs on midrib and larger veins; petiole ill-defined, to 1 cm long. Cymes 4–8(–12) cm long, dense, branched with short lateral branches to 5 mm long, sometimes also with lateral cymes from upper leaf axils; lower internodes to 1 cm long; flowers all in 3–7-flowered cymules; axis sericeous-puberulous with curly hairs, no stalked capitate glands; lower 1–2 pairs of bracts sometimes foliaceous, the rest narrowly triangular to ovate, 9–12 mm long, persistent, ciliate with curly hairs; pedicels 1–2 mm long, puberulous throughout; bracteoles narrowly triangular, 8–9 mm long, ciliate with curly hairs. Calyx 7–9 mm long, puberulous-ciliate with curly hairs. Corolla crimson-red, towards centre with darker spots, outside with scattered stalked capitate glands, no hairs; linear tube 2.2–2.7 cm long; throat 3–4 mm long, lobes elliptic, 7–12 × 4–7 mm, rounded. Filaments 3–4 mm long; anthers purple, ± 1 mm long. Capsule 3.8–4.8 cm long; seed 9–10 × ± 8 mm, smooth on the outer surface, smooth with a broad raised rim on the inner surface.

TANZANIA. Rufiji District: Matumbi Hills, Kiwengoma Forest, no date, *Frontier* 134!; Masasi District: Masasi, 23 Apr. 1935, *Schlieben* 6383! & near Masasi, Chironga Hill, 8 March 1991, *Bidgood et al.* 1857!

DISTR. **T** 6, 8; Mozambique, Zimbabwe, South Africa

HAB. Dry deciduous forest on rocky hillsides, with *Cussonia, Commiphora* spp., *Milettia*; 450–600 m

SYN. *Ruspolia hypocrateriformis* (Vahl) Milne-Redh. var. *australis* Milne-Redh. in K.B. 1936: 272 (1936)
R. hypocrateriformis sensu Pooley, Wild Fl. Kwazulu-Natal: 78 (1998), *non* (Vahl) Milne-Redh. (1936)

3. **Ruspolia seticalyx** (*C.B. Clarke*) *Milne-Redh.* in K.B. 1936: 270 (1936); T.T.C.L.: 15 (1949); Binns, Check List Herb. Fl. Malawi: 16 (1968); Richards & Morony, Check List Fl. Mbala: 234 (1969); Vollesen in Opera Bot. 59: 81 (1980). Type: Malawi, Kondowa to Karonga, *Whyte* s.n. (K!, lecto.; selected by Milne-Redh., l.c.)

Erect shrubby herb or shrub to 1.25(? –2) m tall; plant drying green; young stems glabrous to puberulous. Leaves ovate to elliptic or broadly so, largest 8–23 × 3–11 cm, abruptly narrowed to a long decurrent base, apex acute to acuminate with an obtuse tip; beneath glabrous to puberulous along midrib and sometimes also larger veins, above glabrous or with uniformly scattered hairs; petiole ill-defined, to 3 cm long. Cymes (2–)5–35 cm long, unbranched, sometimes with lateral cymes from upper leaf axils; lower internodes 2–5(–8) cm long (rarely with some cymes condensed with almost absent internodes); flowers all solitary or in 3(–5)-flowered

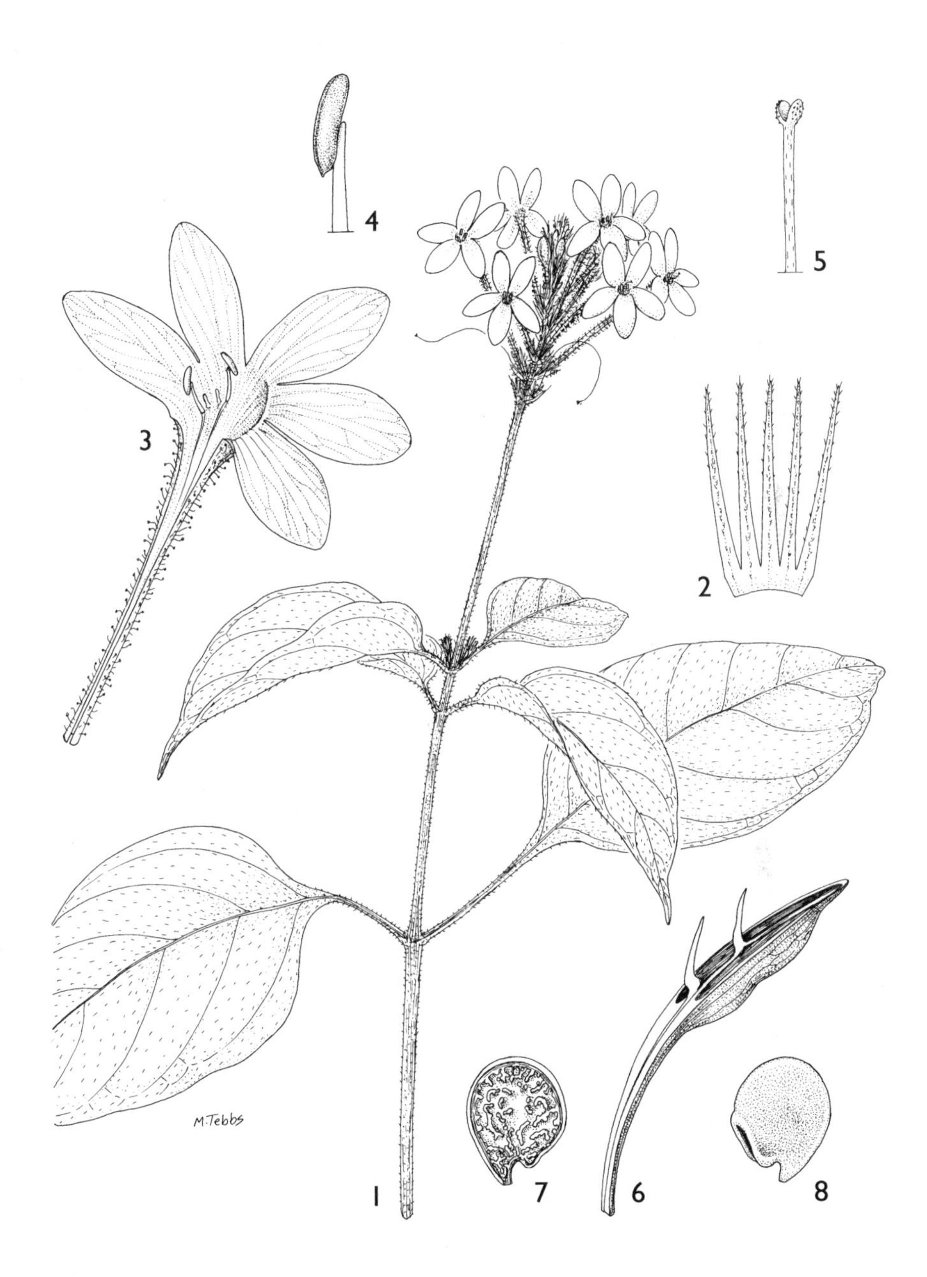

Fig. 63. *RUSPOLIA SETICALYX* — **1**, habit, × $^2/_3$; **2**, calyx, opened up, × 6; **3**, corolla opened up, showing stamens and staminodes, × 2; **4**, anther, × 10; **5**, apical part of style and stigma, × 16; **6**, capsule, × 2; **7**, seed, inner lateral surface, × 2; **8**, seed, outer lateral surface, × 2. 1–5 from *Welch* 576, 6–8 from *Mabberley* 1375. Drawn by Margaret Tebbs.

cymules towards base of cyme (rarely almost to apex); axis subglabrous to puberulous, no stalked capitate glands; lower 1–2 pairs of bracts sometimes foliaceous, the rest subulate to narrowly triangular, 4–10 mm long, indumentum as axis; pedicels 0.5–2(–3 in fruit) mm long; bracteoles subulate to narrowly triangular, 3–5 mm long. Calyx 5–8(–9) mm long, glabrous or sparsely puberulous apically on lobes, no stalked capitate glands. Corolla orange-red to red or scarlet (? rarely orange or yellow), towards centre with many darker spots, outside with a mixture of long curly non-glandular hairs and stalked capitate glands (rarely with glands only); linear tube 2–3 cm long; throat 2–3 mm long, lobes elliptic, 9–14 × 5–9 mm, subacute to rounded. Filaments ± 3 mm long; anthers purple, 1–1.5 mm long. Capsule 3–3.8 cm long; seed 9–10 × 7–8 mm, smooth on the outer surface, with a number of irregular ridges on the inner surface. Fig. 63, p. 487.

TANZANIA. Kigoma District: Lake Tanganyika, Kasye Village, 28 March 1994, *Bidgood et al.* 3007!; Kilosa District: Ukaguru Mts, Jakula River, 4 Aug. 1972, *Mabberley* 1375!; Iringa District: Ruaha National Park, Mt Mpululu, 21 May 1968, *Renvoize & Abdallah* 2303!

DISTR. **T** 1, 2, 4–8; Congo-Kinshasa, Zambia, Malawi, Mozambique, Zimbabwe, Botswana; Madagascar

HAB. Lowland semideciduous *Julbernardia unijugata* forest and thicket, riverine forest and scrub, rocky hills, termite mounds; (150–)500–1250 m

SYN. *Eranthemum seticalyx* C.B. Clarke in F.T.A. 5: 172 (1899)
Pseuderanthemum seticalyx (C.B. Clarke) Stapf in Bot. Mag. 135: t. 8244 (1909)

4. **Ruspolia decurrens** (*Nees*) *Milne-Redh.* in K.B. 1936: 269 (1936); T.T.C.L.: 15 (1949); F.P.S. 3: 187 (1956); E.P.A.: 956 (1964); Vollesen in Opera Bot. 59: 81 (1980); Friis & Vollesen in Biol. Skr. 51(2): 454 (2005); Ensermu in Fl. Eth. 5: 434 (2006). Type: Sudan, Kordofan, Milbes, Choor River, *Kotschy* 276 (K!, lecto.; selected by Milne-Redh., l.c.)

Shrubby herb to 1 m tall, erect or decumbent with ascending flowering branches; plant drying green; young stems sparsely to densely puberulous, with or without stalked capitate glands. Leaves ovate, largest 8–16 × 3–5.5 cm, abruptly narrowed to a long decurrent base, apex acute to acuminate with an obtuse tip; beneath puberulous along midrib and sometimes also larger veins, above glabrous or with uniformly scattered hairs; petiole ill-defined, to 2.5 cm long. Cymes 5–30 cm long, unbranched, lower internodes 2–4(–6) cm long; flowers solitary or in 3-flowered cymules (or 1 flower with 2(–4) undeveloping buds) towards base of cyme; axis puberulous and with stalked capitate glands; lower 1–3 pairs of bracts foliaceous, the rest subulate, 4–6 mm long, indumentum as axis; pedicels 0.5–3 mm long; bracteoles subulate, 4–6 mm long. Calyx 6–9 mm long, puberulous with stalked capitate glands and non-glandular hairs. Corolla pale orange to brick red or pinkish orange, towards centre with many darker spots, outside with long curly non-glandular hairs, no capitate glands; linear tube 2–2.7 cm long; throat 2–3 mm long, lobes elliptic, 8–14 × 5–10 mm, broadly rounded. Filaments ± 3 mm long; anthers purple, 1.5–2 mm long. Capsule 2.4–2.6 cm long; seed 5–6 × 5–6 mm, smooth on the outer surface, with a number of irregular ridges on the inner surface.

TANZANIA. Kilosa District: Mikumi National Park, 3 May 1984, *de Nevers & Charnley* 3410!; Iringa District: between Kilindimo and Izazi, 18 April 1962, *Polhill & Paulo* 2053!; Chunya District: 8 km NW of Saza Tsetse Research Camp, 24 May 1990, *Carter et al.* 2437!

DISTR. **T** 6–8; Central African Republic, Sudan, Ethiopia, Zambia, Malawi, Mozambique, Zimbabwe

HAB. Riverine forest and scrub, shady places in woodland, termite mounds; 100–1000 m

SYN. *Eranthemum decurrens* Nees in DC., Prodr. 11: 453 (1847); A. Rich., Tent. Fl. Abyss. 2: 158 (1851); T. Anderson in J.L.S. Bot. 7: 52 (1863); C.B. Clarke in F.T.A. 5: 170 (1899); Broun & Massey, Fl. Sudan: 343 (1929)

E. senense Klotzsch in Peters, Reise Mossamb. Bot.: 210 (1861); C.B. Clarke in F.T.A. 5: 171 (1899). Type: Mozambique, Lower Zambesi, Sena River, Boror, *Peters* s.n. (B†, holo.)
Pseuderanthemum decurrens (Nees) Radlk. in Sitzber. Bayr. Akad. Wiss. 13: 286 (1883); Lindau in E. & P. Pf. IV, 3b: 330 (1895)
P. senense (Klotzsch) Radlk. in Sitzber. Bayr. Akad. Wiss. 13: 286 (1883); Lindau in E. & P. Pf. IV, 3b: 330 (1895)
P. katangense Champl. in Syst. Geogr. Pl. 72: 47 (2002), quoad *de Nevers & Charnley* 3410, *non* sensu stricto

38. RUTTYA

Harv. in London J. Bot. 1: 27: (1842); Nees in DC., Prodr. 11: 309 (1847); G.P. 2: 1105 (1876); C.B. Clarke in F.T.A. 5: 174 (1899) & in Fl. Cap. 5: 55 (1901)

Haplanthera Hochst. in Flora 26: 71 (1843) & in Bot. Zeit. 2: 853 (1844) (as *Hablanthera*); Nees in DC., Prodr. 11: 308 (1847)

Shrubby herbs or shrubs; cystoliths conspicuous, linear; stems rounded. Leaves opposite, entire. Flowers single, aggregated into short few-flowered to long many-flowered racemoid cymes. Bracteoles present, triangular. Calyx deeply divided into 5 equal subulate to narrowly triangular lobes. Corolla glabrous on the outside; basal tube short and wide, widening slightly into a short throat; lobes 5, lower 3 strongly deflexed against the tube, upper 2 erect, forming a hood. Stamens 2 plus 2 minute to long staminodes, attached near base of and held under upper lip, glabrous, terete to flattened and winged; anthers monothecous, ellipsoid to oblong, straight or curved, with a short spur or mucronate to rounded at base, glabrous. Ovary with 2 ovules per locule; style glabrous; stigma of 2 equal, erect, broadly ellipsoid lobes. Capsule 4-seeded, clavate with contracted solid stalk-like basal part, glabrous, seed-bearing part ellipsoid, laterally flattened; retinacula strong, hooked. Seed discoid, smooth on both sides, with a broad longitudinal ridge on outer side.

Six species. One in N Ethiopia and Eritrea, one widespread in NE Africa and extending to Arabia, one in southern Africa and three on Madagascar.

Ruttya fruticosa *Lindau* in E.J. 20: 45 (1894) & in E. & P. Pf. IV, 3b: 340 (1895) & in P.O.A. C: 372 (1895) & in Ann. Ist. Bot. Roma 6: 81 (1896); Rendle in J.B. 34: 398 (1896); C.B. Clarke in F.T.A. 5: 175 (1899); Glover, Checklist Brit. and Ital. Somal.: 70 (1947); T.T.C.L.: 15 (1949); Turrill in Curtis's Bot. Mag. 172: t. 329 (1958); K.T.S.: 18 (1961); E.P.A.: 965 (1964); Blundell, Wild Fl. E. Afr.: 396, pl. 529 (1987); U.K.W.F. ed. 2: 277, pl. 122 (1994); Audru *et al.*, Pl. Vasc. Djibouti 2, 2: 738 (1994); K.T.S.L.: 607 (1994); Ruffo *et al.*, Cat. Lushoto Herb. Tanzania: 10 (1996); Thulin in Fl. Somalia 3: 400 (2006); Ensermu in Fl. Eth. 5: 452 (2006). Types: Somalia, Ahl Mts, *Hildebrandt* 893 (BM!, isosyn.) & Meid, *Hildebrandt* 1527 (not seen); Kenya, Kitui District: Malemba [Malemboa], *Hildebrandt* 2631 (not seen) & Ukamba, *Fischer* 287 (not seen) & *Fischer* 487 (not seen); Tanzania, Masai District: Lake Chala, *Volkens* 323 (BM!, isosyn.)

Erect or scrambling shrub to 3(–4) m tall; young stems brownish, sparsely puberulous, soon glabrescent; older branches greyish, not or slightly corky. Leaves ovate to elliptic or broadly so, largest 2.5–10.5 × 1.2–3.5(–4.5) cm, suddenly narrowed to a decurrent base, more rarely evenly attenuate, apex acute to rounded; puberulous or sparsely so along midrib and larger veins, otherwise glabrous; petiole 2–5 mm long. Cymes with (2–)4–8 decussate flowers, unbranched or rarely with short 2-flowered branches from basal node; axis to 1.3(–1.7) cm long, subglabrous to sparsely puberulous; lower pair of bracts foliaceous, upper pairs linear to narrowly triangular, 1–3 mm long, puberulous; pedicels 0.5–1.2(–1.5) cm long, glabrous

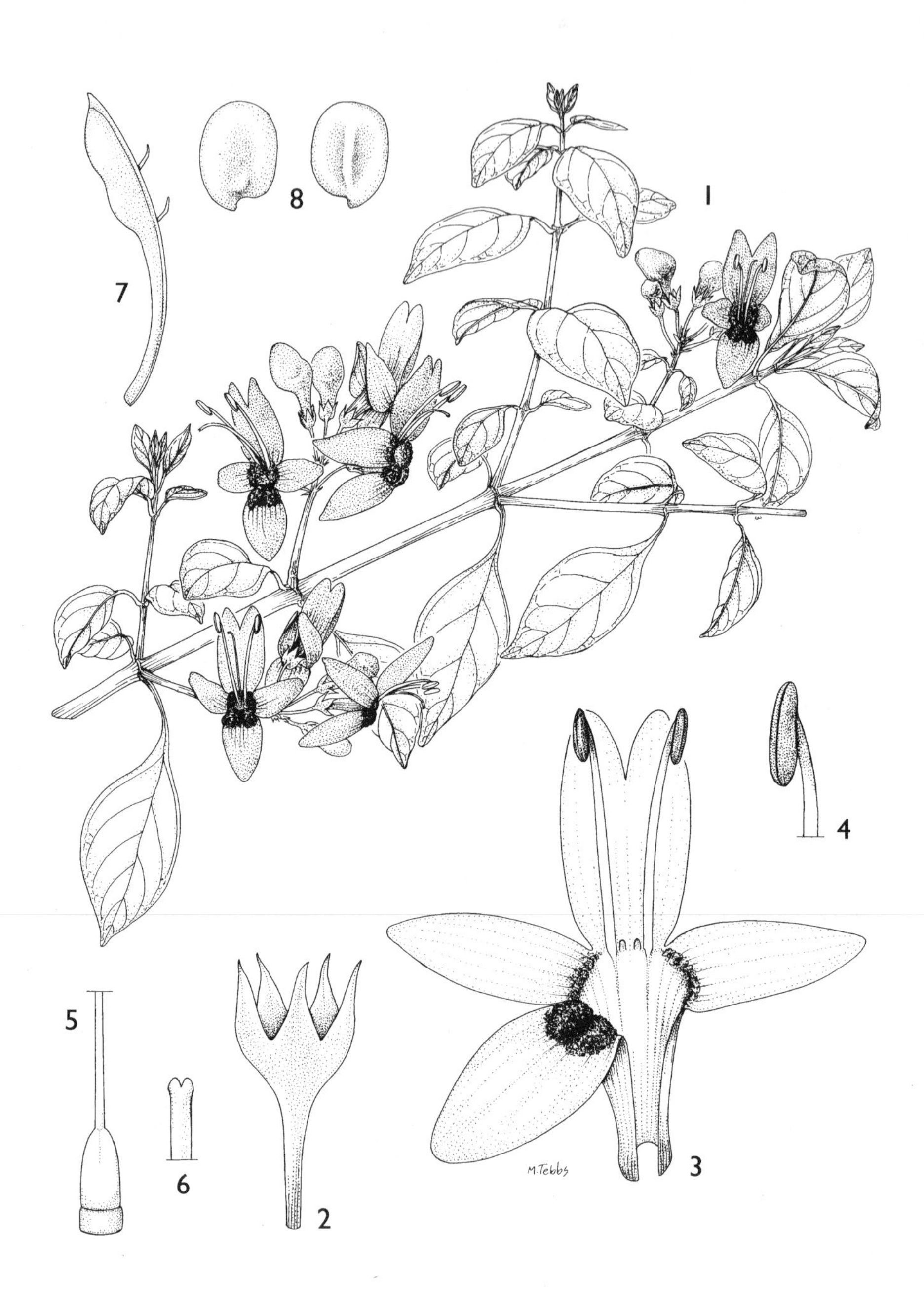

FIG. 64. *RUTTYA FRUTICOSA* — **1**, habit, × $^2/_3$; **2**, calyx, × 4; **3**, corolla opened up, × 2; **4**, anther, × 4; **5**, ovary and basal part of style, × 4; **6**, stigma, × 20; **7**, capsule, × 1; **8**, seed, two different views, × 2. 1 from *Greenway* 11604, 2–6 from *Stewart* 764, 7–8 from *Battiscombe* 102. Drawn by Margaret Tebbs.

(rarely puberulous below bracteoles); bracteoles linear, 1–2 mm long, ciliate. Calyx 3–6 mm long, divided to ± 1 mm from base, glabrous or with sparsely ciliate lobes, lobes narrowly triangular, acuminate. Corolla orange-red to bright red (rarely yellow) with large glossy black nectariferous patch on basal part of lower lip and extending into throat; tube 7–12 mm long, 3–4 mm wide at base, widening gradually from ± 2 mm above base, lobes in lower lip ovate to elliptic, 13–24 × 7–12 mm, subacute to rounded, middle usually wide than lateral, in upper lip 13–22 × 5–8 mm. Filaments yellow, 12–18 mm long, flattened and winged, glabrous; anthers 3–4 mm long, rounded at both ends. Capsule (3.2–)3.8–5 cm long; seed broadly ellipsoid, 7–9 × 5–7 mm. Fig. 64, p. 490.

UGANDA. Karamoja District: E of Kachagalau, Dec. 1957, *Philip* 807!

KENYA. Northern Frontier District: 6 km on Marsabit–Isiolo road, 14 May 1970, *Magogo* 1328!; Machakos District: 3 km N of Kangonde, Nguungi Hill, 18 May 1969, *Kimani* 184!; Masai District: Siyabei River Gorge, Owor Oringenei, 14 July 1962, *Glover & Samuel* 3031!

TANZANIA. Maswa District: Shanwa, Komali Hill, June 1935, *B.D. Burtt* 5200!; Mbulu District: Lake Manyara National Park, Maji Moto Springs, 26 May 1965, *Greenway & Kanuri* 12087!; Dodoma District: 35 km on Itigi–Chunya road, 17 April 1964, *Greenway & Polhill* 11604!

DISTR. **U** 1; **K** 1–4, 6, 7; **T** 1, 2, 5; Ethiopia, Djibouti, Somalia; Yemen, Oman

HAB. Rocky hills and slopes with *Acacia-Commiphora* and *Euphorbia* bushland, rocky streambeds, Itigi thicket, termite mounds in *Brachystegia* woodland, degraded *Juniperus* forest developing into *Euclea* scrub; 600–2000 m

NOTE. This species is commonly cultivated in and around Nairobi. It can be trimmed to form a hedge and will then produce a spectacular display when in flower.

The yellow form is also cultivated. A sheet at Kew (*W.T. Gillis* 11883) bears the name var. *catarina* Gillis and the word Holotype. I have not been able to trace this name and it may never have been published. It is in any case my opinion that this form is not worthy of varietal rank.

39. MONOTHECIUM

Hochst. in Flora 26: 74 (1843); Nees in DC., Prodr. 11: 310 (1847); G.P. 2: 1104 (1876); Lindau in E. & P. Pf. IV, 3b: 340 (1895); C.B. Clarke in F.T.A. 5: 175 (1899)

Anthocometes Nees in DC., Prodr. 11: 311 (1847)

Perennial herbs; cystoliths conspicuous, linear; stems subquadrangular. Leaves opposite, entire. Flowers in elongated terminal racemoid sometimes secund cymes, often with lateral cymes from lower nodes, solitary (often with undeveloping buds) or in 2–3-flowered cymules, subsessile. Bracteoles present, filiform to narrowly ovate. Calyx deeply divided into 5 equal linear acuminate lobes. Corolla puberulous on the outside, strongly 2-lipped with narrow shortly bifid hooded upper lip and spreading shortly 3-lobed oblong-obovate lower lip; basal tube linear, widening gradually into the throat. Stamens 2, attached near base of and held under upper lip, glabrous, terete; anthers monothecous, oblong, slightly curved, rounded or apiculate at base, glabrous. Ovary with 2 ovules per locule; style glabrous; stigma of 2 equal, erect, broadly ellipsoid lobes. Capsule (2–)4-seeded, obovate-clavate with contracted stalk-like basal part, thin-walled, finely puberulous, seed-bearing part ellipsoid; retinacula strong, hooked. Seed discoid, with conical tubercles or ridges.

Four species. Three in Africa, one of which extends to India and Sri Lanka, and one on Madagascar.

It is not absolutely certain that the Madagascan species – of which very little material is available – belongs here. It has the two monothecous anthers characteristic of *Monothecium* but the corolla morphology seems to be quite different.

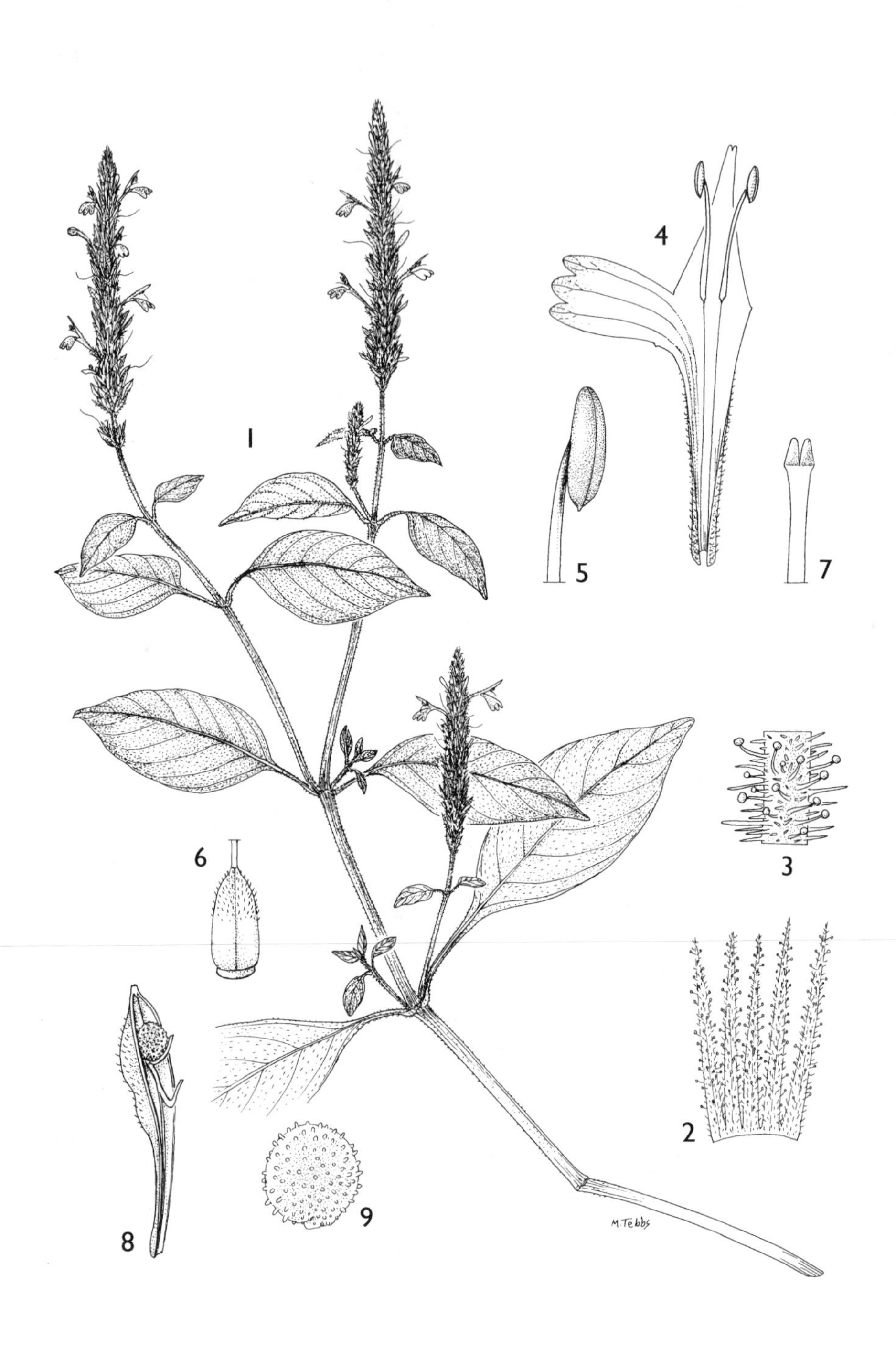

FIG. 65. *MONOTHECIUM GLANDULOSUM* — **1**, habit, × $^2/_3$; **2**, calyx, × 6; **3**, detail of calyx indumentum, × 20; **4**, corolla, × 4; **5**, anther, × 14; **6**, ovary, × 14; **7**, stigma, × 20; **8**, capsule, × 6; **9**, seed, × 14. 1 from *Tweedie* 3385 and *Verdcourt* 1818b, 2–9 from *Bidgood* 4897. Drawn by Margaret Tebbs.

Cymes not secund (flowers in decussate pairs or cymules along the axis); young stems and inflorescences with thin white hairs; inflorescence 2–13 cm long, axis and bracts with stalked capitate glands; corolla tube 8–10 mm long, lips 4–6 mm long; filaments 4–5 mm long; seed with conical tubercles . . . . 1. *M. glandulosum*

Cymes secund (all flowers on one side of the axis); young stems and inflorescences with broad curly purple-walled hairs; inflorescence 1–3(–8) cm long, axis and bracts without stalked capitate glands; corolla tube 4.5–6 mm long, lips 3–4 mm long; filaments 2–3 mm long; seed with concentric ridges of scales . . . . 2. *M. aristatum*

1. **Monothecium glandulosum** *Hochst.* in Flora 26: 74 (1843); Nees in DC., Prodr. 11: 310 (1847); A. Richard, Tent. Fl. Abyss. 2: 152 (1851); T. Anderson in J.L.S. Bot. 7: 45 (1863); Solms Laubert in Schweinfurth, Beitr. Fl. Aeth.: 112 (1867); Engler, Hochgebirgsfl. Trop. Afr.: 392 (1892); C.B. Clarke in F.T.A. 5: 175 (1899); Fries in Wiss. Ergebn. Schwed. Rhodesia-Kongo-Exp. 1911–12 1: 308 (1916); F.P.N.A. 2: 297 (1947); E.P.A.: 965 (1964); Gilbert, Pl. Mt Kilimanjaro: 82 (1970); U.K.W.F., ed. 2: 278 (1994); Champluvier in Fl. Rwanda 3: 477 (1985); Friis & Vollesen in Biol. Skr. 51(2): 451 (2005); Ensermu in Fl. Eth. 5: 438 (2006). Types: Ethiopia, Mt Scholoda, *Schimper* I.346 (BM!, K!, syn.) & Mai Dogale, *Schimper* II.617 (BM!, K!, syn.)

Perennial herb, basal part of stems branched, creeping and rooting, apical flowering part ascending to erect; flowering stems to 50 cm long, sparsely to densely puberulous. Leaves ovate to elliptic or broadly so, largest 2.2–8(–10.5) × 1–4(–5) cm, base attenuate to subcuneate, apex subacute to acuminate; puberulous or sparsely so along midrib and larger veins, above also with scattered hairs on lamina; petiole 0.3–3(–4) cm long. Cymes 2–13 cm long; peduncle 0.5–5(–6) cm long, puberulous and with few to many stalked capitate glands, usually with 1(–2) pairs of sterile bracts near middle and basal bracts also often sterile; axis with similar indumentum; bracts ovate (lower pair rarely foliaceous), acuminate, 4–6.5 mm long, puberulous and with many stalked capitate glands; bracteoles linear to narrowly ovate, 4–6 mm long, indumentum similar. Calyx 5–7 mm long, with similar indumentum, divided to ± 1 mm from base. Corolla white to pale mauve, with two darker patches on lower lip; tube and throat 8–10 mm long, upper lip 4–5 mm long, lower lip 5–6 mm long. Filaments 4–5 mm long, anthers ± 1 mm long, purple. Capsule 6–8 mm long, uniformly puberulous and with stalked glands at apex; seed ± 1 mm in diameter, with conical tubercles. Fig. 65, p. 492.

UGANDA. Ankole District: Ishingiro, June 1939, *Purseglove* 743!; Toro District: Kikorongo, 23 Dec. 1925, *Maitland* 1030!; Busoga District: Jinja, Nov. 1961, *Tweedie* 2259!

KENYA. Northern Frontier District: Marsabit, 14 Feb. 1952, *Gillett* 15116!; Nakuru District: Lake Nakuru National Park, 30 Nov. 1973, *Kutilek* 167!; North Kavirondo District: Bungoma, Dec. 1966, *Tweedie* 3385!

TANZANIA. Biharamulo District: 50 km on Biharamulo–Muleba road, 11 July 2000, *Bidgood et al.* 4897!

DISTR. **U** 2–4; **K** 1, 3, 5, 6; **T** 1, 2; Central African Republic, Congo-Kinshasa, Rwanda, Sudan, Ethiopia

HAB. Montane forest, riverine forest and scrub, termite mounds, moist grassland; 950–1850 m

SYN. *Rostellularia glandulosa* (Hochst.) Nees in DC., Prodr. 11: 373 (1847); A. Rich., Tent. Abyss. 2: 153 (1851)

Hypoestes volkensii Lindau in E.J. 19, Beibl. 47: 47 (1894) & in P.O.A. C: 371 (1895). Types: Tanzania, Kilimanjaro, Kibosho, *Volkens* 1580 (B†, syn..) & Machame [Madschome], Quare River, *Volkens* 1607 (B†, syn.)

Monothecium abbreviatum S. Moore in J.B. 38: 466 (1900). Type: Kenya, Kavirondo, *Scott Elliot* 7106 (BM!, holo.)

2. **Monothecium aristatum** (*Nees*) *T. Anderson* in Thwaites, Enum. Pl. Zeyl.: 234 (1860); S. Moore in J.B. 18: 309 (1880); C.B. Clarke in Fl. Brit. Ind. 4: 524 (1885) & in F.T.A. 5: 176 (1899); Iversen in Symb. Bot. Upsal. 29, 3: 162 (1991); Luke & Robertson, Kenya Coast. For. 2. Checklist Vasc. Pl.: 83 (1993); Cramer in Rev. Handb. Fl. Ceylon 12: 101 (1998); Friis & Vollesen in Biol. Skr. 51(2): 451 (2005). Type: India, *Herb. Wallich* 2481 (K-W!, holo.)

Perennial herb, basal part of stems branched, creeping and rooting, apical flowering part ascending to erect; flowering stems to 50(–70) cm long, sparsely puberulous with broad curly purple-walled hairs. Leaves ovate to elliptic, largest 2.5–8(–11.5) × 1.2–4(–5) cm, base attenuate to cuneate, apex acute to acuminate; with scattered hairs along midrib and larger veins beneath, above subglabrous; petiole 0.3–2.5(–3.5) cm long. Cymes 1–3(–8) cm long, secund, solitary or 2–3 (branched from sterile bracts near middle of peduncle); peduncle of central cyme 0–1.5(–2.5) cm long, puberulous with broad curly purple-walled hairs; axis sparsely puberulous; flowers only on one side of cyme, bracts on other side all sterile (but sometimes with undeveloping buds), fertile bracts ovate, acuminate to cuspidate, 4–6(–8) mm long, subglabrous to sparsely puberulous, with very conspicuous white cystoliths; bracteoles lanceolate, 4–7 mm long, cuspidate, with stalked capitate glands. Calyx 5–7 mm long, subglabrous and with stalked capitate glands, divided to ± 1 mm from base. Corolla white (? rarely pale blue), sometimes with faint mauve patches on lower lip; tube and throat 4.5–6 mm long, upper lip 3–4 mm long, lower lip 3–4 mm long. Filaments 2–3 mm long, anthers ± 1 mm long, purple. Capsule 5–7 mm long, with scattered hairs and capitate glands at apex; seed ± 1 mm in diameter, with concentric ridges of scales.

UGANDA. Bunyoro District: Budongo Forest, 8 March 1973, *Synnott* 1449!; Busoga District: Usoga [Busoga], March 1916, *Dümmer* 2783!; Mengo District: Mabira Forest, Chogwe, Dec. 1922, *Maitland* 518!

KENYA. Lamu District: Mambasasa, Utwani Forest Reserve, 17 Oct. 1957, *Greenway & Rawlins* 9358!; Kwale District: Shimba Hills, Makadara Forest, 9 Jan. 1970, *K. & L. Holm* 8! & Gongoni Forest Reserve, 12 Nov. 1989, *Robertson & Luke* 5970!

TANZANIA. Bukoba District: Minziro Forest Reserve, Mabuye Village, Itara, 29 April 2001, *Festo et al.* 1450!; Lushoto District: Korogwe, Segera Forest, 27 Aug. 1968, *Faulkner* 4129!; Morogoro District: Nguru Mts, Lusunguru Forest Reserve, Mtibwa, 31 March 1953, *Drummond & Hemsley* 1931!

DISTR. **U** 2–4; **K** 7; **T** 1–3, 6; Congo-Kinshasa, Sudan, Angola; India, Sri Lanka

HAB. Lowland and medium altitude wet evergreen forest and swamp forest; 50–1150 m

SYN. *Justicia aristata* Nees in Wallich, Pl. As Rar. 3: 115 (1832), *nom. illeg.*, *non J. aristata* Vahl, Symb. Bot. 2: 20 (1791)

Anthocometes aristata Nees in DC., Prodr. 11: 312 (1847)

40. JUSTICIA

L., Sp. Pl.: 15 (1753) & Gen. Pl., ed. 5: 10 (1754); C.B. Clarke in F.T.A. 5: 179 (1899); V.A.W. Graham in K.B. 43: 581 (1988)

Dianthera L., Sp. Pl. 1: 27 (1753)
Adhatoda Miller, Gard. Dict., Abrid. Ed. 4, 1, alphabetical seq. (1754); C.B. Clarke in F.T.A. 5: 221 (1900)
Gendarussa Nees in Wallich, Pl. As. Rar. 3: 76, 102 (1832)
Rungia Nees in Wallich, Pl. As. Rar. 3: 77, 109 (1832); C.B. Clarke in F.T.A. 5: 252 (1900)
Rhaphidospora Nees in Wallich, Pl. As. Rar. 3: 77, 115 (1832)
Rostellaria Reichenb., Handb.: 190 (1837)
Monechma Hochst. in Flora 24: 374 (1841); C.B. Clarke in F.T.A. 5: 212 (1900)
Tyloglossa Hochst. in Flora 26: 72 (1843)
Duvernoia Nees in DC., Prodr. 11: 322 (1847)
Anisostachya Nees in DC., Prodr. 11: 368 (1847)
Siphonoglossa Oersted in Vid. Medd. Dansk Natur. Foren. 1854: 159 (1855)
Harnieria Solms-Laub. in Sitzber. Ges. Naturf. Fl. Tellur. 4: 62 (1864)
Bentia Rolfe in K.B. 1894: 338 (1894)
Nicoteba Lindau in E. & P. Pf. IV, 3b: 329 (1895)
Calophanoides (C.B. Clarke) Ridley, Fl. Mal. Penins. 2: 592 (1923)
Thamnojusticia Mildbr. in N.B.G.B. 11: 825 (1933)

Herbs, shrubs or (rarely) small trees, erect or scrambling; cystoliths conspicuous or not. Leaves opposite, equal to distinctly anisophyllous (rarely seemingly alternate due to total disappearance of one leaf of each pair), entire to crenate. Flowers in a wide variety of inflorescences from open dichasial cymes (sometimes aggregated into panicles) to subsessile or sessile dichasia aggregated or condensed into various types of racemoid or spiciform sometimes secund cymes with small or large sometimes strobilate bracts or in axillary clusters or solitary; bracts persistent, prominent or not; bracteoles present, large or small. Calyx deeply divided into 5 subequal segments or with 1 segment reduced or almost absent. Corolla glabrous or (usually) hairy outside, often also with capitate glands, inside with band of hairs in upper part of tube; basal tube cylindric, widening upwards into ± distinct throat; distinctly 2-lipped, upper lip shallowly 2-lobed, flat to distinctly hooded, lower lip shallowly to deeply 3-lobed, often with a conspicuous pattern of transverse differently coloured lines ("herringbone" pattern). Stamens 2, no staminodes, inserted in upper part of tube; anthers bithecous, thecae often unequal with upper smaller than lower, usually ± superposed, held parallel or at an angle to each other, lower theca with a white sterile appendage (rarely apiculate). Style filiform, glabrous or hairy; stigma with two ellipsoid erect lobes. Capsule 2–4-seeded, clavate with a solid basal stalk, halves entire during dehiscence or with retinacula splitting from capsule wall or capsule walls splitting from base; retinacula strong, curved. Seed very variable, sphaeroid to discoid, reniform or cordiform, compressed or not, testa smooth or variously ornamented, rugulose, tuberculate, pubescent or echinate.

The largest genus in the *Acanthaceae* and often estimated at around 500 species, but this is almost certainly an underestimate. The present account alone lists 25 new species (or almost 25% of the total in East Africa) and detailed studies in other tropical regions (e.g. Wasshausen & Wood in K.B. 58: 769–831 (2003)) have also resulted in the description of many new species. 600–700 species would seem to be a more realistic guess.

1. Flowers solitary or in 2–4(–6)-flowered dichasia which are sessile or subsessile in axils of vegetative leaves and with clear internodes between whorls, occasionally the internodes contracted towards apex of stem but all "bracts" foliaceous (see Fig. 75, 76, 77) 2
   Inflorescences not as above 39
2. Bracteoles large, 8–16 mm long . . . . . 25. *J. gesneriflora* p.529
   Bracteoles small, 0.5–3(–7) mm long (Sect. *Harnieria*) 3
3. Shrubby herbs or shrubs (at least basal part of stems woody) 4
   Annual or perennial herbs (basal part of stems not woody) 15
4. Corolla lemon yellow to bright yellow . . . . . 66. *J. odora* p.565
   Corolla white, pink, mauve or purple 5
5. Leaves linear to narrowly ovate (rarely ovate); flowers all single; calyx densely whitish sericeous to tomentellous . . . . . 72. *J. elliotii* p.571
   Leaves ovate, elliptic or obovate; some or all axils with more than one flower; calyx puberulous or sericeous-puberulous or sparsely so, sometimes only ciliate 6
6. Anther thecae yellow with distinct brown pigment patches at apex and base of upper theca and at apex of lower theca . . . . . 86. *J. striata* p.584
   Anther thecae yellow to uniformly purple but never with distinct brown pigment patches at apex and base of upper theca and at apex of lower theca 7
7. Corolla ± 7 mm long, tube ± 1 mm longer than upper lip . . . . . 70. *J. sp. F* p.570
   Corolla 8–25 mm long 8
8. Corolla 19–25 mm long, tube 3–4 mm longer than upper lip, lower lip 13–20 × 18–24 mm; anthers dark purple . . . . . 78. *J. ukagurensis* p.576
   Corolla 8–19 mm long, tube or lip same length or tube 1–3 mm longer (rarely upper lip to 1 mm longer), lower lip 4–18 × 5–17 mm; anthers yellow or brown 9
9. Lower corolla lip without distinct white and purple "herring bone" pattern . . . . . 81. *J. diclipteroides* p.578
   Lower corolla lip with distinct white and purple "herring bone" pattern 10
10. Upper internodes usually contracted and forming a racemoid "inflorescence"; capsule 6.5–10 mm long 11
    Upper internodes always well spaced; capsule 9–19 mm long 12
11. Corolla 13–17 mm long; stamens bending out of the corolla and exposing style during female phase of anthesis; leaves less than twice as long as wide; high montane grassland above 2450 m . . . . . 79. *J. afromontana* p.577
    Corolla 9–14 mm long; stamens twisting inwards and pulling anthers tightly around style during female phase of anthesis; leaves 2–3 times as long as wide; plant from below 2450 m . . . . . 87. *J. phyllostachys* p.586

12. Young branches sericeous with appressed hairs; leaves uniformly puberulous on lamina or denser on veins; low altitude dry bushland 13
    Young branches puberulous with spreading or bent hairs; leaves hairy on midrib and veins, glabrous on lamina; lowland or montane forest and forest margins 14
13. Young stems with retrorse hairs; largest leaf 1.5–5.7 × 1–3.5 cm; corolla white to pale mauve, upper lip without dark veins, tube and lip same length; capsule 11–14 mm long .... 69. *J. brevipila* p.569
    Young stems with antrorse hairs; largest leaf 0.7–2.2 × 0.5–1.1(–1.9) cm; corolla bright purple, upper lip with dark veins, tube ± 2 mm longer than lower lip; capsule 9–12 mm long .... 71. *J. phillipsiae* p.570
14. Corolla (8–)10–17 mm long, tube (4–)5–9 mm long and 1.5–4 mm in diameter (if corolla less than 10 mm long then tube ± 1.5 mm in diameter), upper lip without purple streaks ... 67. *J. capensis* p.568
    Corolla 8–11 mm long, tube 4–5 mm long and 2.5–4 mm in diameter, upper lip with purple streaks .... 68. *J. euosmia* p.568
15. Corolla tube 4–15 mm longer than upper lip, expanded only near apex 16
    Corolla tube same length as or up to 3 mm longer than upper lip (rarely upper lip slightly longer than tube), expanded from near base (rarely from near middle) 19
16. Corolla tube ± 4 mm longer than upper lip, upper lip with many purple spots and patches, with teeth 3–4 mm long, tube 2–3 mm diameter in middle, lower lip 13–20 × 18–24 mm; capsule ± 15 mm long .... 78. *J. ukagurensis* p.576
    Corolla tube 4–15 mm longer than upper lip, upper lip with or without darker veins, with teeth 1–2 mm long, tube 1–2 mm diameter in middle, lower lip 4–12 × 4–14 mm; capsule 6.5–11 mm long 17
17. Flowers single (very rarely some nodes with 2 per leaf), bracts absent; bracteoles 1.5–3(–4) mm long; corolla without stalked capitate glands, upper lip without dark veins; capsule 9–11 mm long .... 76. *J. toroensis* p.575
    Flowers single or 2–4 per leaf, bracts present, often conspicuously white ciliate; bracteoles to 1.5 mm long; corolla with stalked capitate glands, upper lip with dark veins; capsule 6.5–10 mm long 18

18. Corolla purple to deep purple (rarely white), tube 6–15 mm longer than upper lip; appendage on lower theca ± 0.5 mm long; leaves puberulous to pubescent or sparsely so; grassland, bushland and woodland . . . . 74. *J. ladanoides* p.572
    Corolla white to pale mauve, tube 4–6 mm longer than upper lip; appendage on lower theca ± 0.3 mm long; leaves subglabrous or with scattered hairs on midrib and larger veins; forest and forest margins . . . . 77. *J. striolata* p.576
19. Corolla with lower lip pale to bright yellow apart from white and purple markings in throat 20
    Corolla with lower lip white to mauve or purple 21
20. Basal part of stems creeping and rooting, apical part scandent or scrambling; largest leaf 4–7(–9.5) cm long, more than twice as long as wide; corolla 9–16 mm long . . . . 80. *J. sulphuriflora* p.577
    Pyrophytic herb with several trailing or ascending unbranched or little-branched stems from woody rootstock; largest leaf 1.7–2.5 cm long, less than twice as long as wide; corolla 9–10 mm long . . . . 88. *J. chalaensis* p.587
21. Only one leaf in a pair supporting a flower; largest leaf 0.15–0.5 cm wide, glabrous (rarely scabrid/puberulous on margins) 22
    Both leaves in a pair supporting flowers; largest leaf (0.2–)0.5–6.8 cm wide (if very rarely a few nodes with only one leaf supporting flowers then leaves wider than 0.5 cm and hairy at least on midrib) 23
22. Largest leaf 1–3 cm long, with strong lateral veins running parallel to midrib; bracts foliaceous, 0.7–2 cm long; bracteoles absent 95. *J. acutifolia* p.591
    Largest leaf 4.5–5 cm long, without lateral veins; bracts absent; bracteoles ± 1 mm long 96. *J. alterniflora* p.592
23. Appendage on lower theca ellipsoid or broadly so in outline, straight or bent 90° relative to theca, broadly rounded 24
    Appendage on lower theca linear in outline, not or very slightly bent relative to theca, acute or often bifid 26
24. Plant annual; flowers single or 2 per axil, bracts present, bracteoles ± 1 mm long; anther appendage straight . . . . 94. *J. mariae* p.591
    Plant perennial; all flowers single, bracts absent, bracteoles 2–7 mm long; anther appendage bent 90° relative to theca 25
25. Plant with several stems from woody rootstock; lamina narrowly elliptic to narrowly ovate (rarely lanceolate); anther appendage broadly ellipsoid in outline, ± 0.5 mm long . 89. *J. lithospermoides* p.588
    Plant with solitary stem from short creeping rhizome; lamina lanceolate or narrowly ovate; anther appendage ellipsoid in outline, ± 0.3 mm long . . . . 90. *J. sp.G* p.588

26. Stamens bending out of the corolla and exposing style during female phase of anthesis 27
Stamens twisting inwards and pulling anthers tightly around style during female phase of anthesis 34
27. Leaves linear to lanceolate, largest 0.2–0.5 cm wide and capsule 7–10 mm long .... 75. *J. leikipiensis* p.575
Leaves lanceolate to ovate, largest 0.4–6.8 cm wide, if less than 0.6 cm wide then capsule 6 mm long or less 28
28. Corolla 3–8.5(–9) mm long, lower lip 2–7 × 3–8 mm; capsule 3.5–9 mm long 29
Corolla 8–22 mm long, lower lip 6–18 × 7–17 mm; capsule 6–14 mm long 31
29. Annual herb with taproot; corolla 3–7 mm long; capsule 3.5–6 mm long .... 84. *J. heterocarpa* p.580
Perennial herb with woody rootstock or creeping rhizome; corolla 6–8.5(–9) mm long; capsule (4.5–)5–9 mm long 30
30. Perennial herb with small woody rootstock; corolla 7–8.5 mm long; capsule 7–9 mm long 83. *J. petterssonii* p.580
Perennial herb with creeping rhizome; corolla 3–8(–9) mm long; capsule 3.5–6 mm long ... 84. *J. heterocarpa* p.580
31. Upper corolla lip hooded, tube widened into throat from ± $^1/_2$ way up; upper internodes usually contracted forming a "pseudo inflorescence" 32
Upper corolla lip flat; tube widened almost from base; upper internodes only very rarely contracted 33
32. Largest leaf 1.2–2.3(–3.5) × 0.8–1.3(–1.8) cm, sparsely uniformly pubescent; corolla 13–17 mm long; capsule 8–10 mm long ... 79. *J. afromontana* p.577
Largest leaf 2.8–8 × 1.2–3 cm, indumentum densest on veins and margin; corolla 15–22 mm long; capsule (9–)10–14 mm long .... 73. *J. pinguior* p.571
33. Leaf base truncate to subcordate; calyx densely pubescent (hairs to 1 mm long); corolla with indumentum of long (to 1 mm) curly hairs 82. *J. kiborianensis* p.579
Leaf base attenuate to truncate (very rarely subcordate); calyx puberulous or sparsely so (hairs to 0.5 mm long); corolla with indumentum of short (to 0.5 mm) straight to slightly curly hairs .... 81. *J. diclipteroides* p.578
34. Anther thecae yellow with distinct brown pigment patches at apex and base of upper theca and at apex of lower theca .... 86. *J. striata* p.584
Anther thecae yellow to uniformly purple but never with distinct brown pigment patches at apex and base of upper theca and at apex of lower theca 35
35. Perennial herbs 36
Annual herbs 37

36. Upper internodes usually contracted, forming a pseudo-racemose "inflorescence"; bracts held erect, covering flowers; corolla 9–14 mm long . . . 87. *J. phyllostachys* p.586
    Upper internodes always well spaced; bracts absent or spreading, not covering flowers; corolla 6–9.5 mm long . . . 85. *J. unyorensis* p.583
37. Bract fused to pedicel making it appear as if flower attached at apex of petiole of bract; peduncles 1–2(–3) mm long. Corolla 3–5 mm long, lower lip 1.5–3 × 2–3.5 mm; capsule 3–4.5(–5) mm long, acute . . . 91. *J. mollugo* p.589
    Bract not fused to pedicel, flower attached at base of petiole of bract; peduncles to 0.5 mm long . . . 38
38. Corolla 5–8.5 mm long, lower lip 4–7 × 5–8 mm; capsule 3.5–6 mm long, acute . . . 92. *J. boaleri* p.589
    Corolla 4–5 mm long, lower lip 2–3 × 2.5–4 mm; capsule 3–4 mm long, obtuse . . . 93. *J. obtusicapsula* p.590
39. Flowers in axillary secund (only 1 bract at each node supporting a flower) spiciform cymes; bracts small (up to 5 mm long), not strobilate; bracteoles small (see Fig. 74) . . . 40
    Inflorescences not as above . . . 55
40. Bracts with white hyaline margin . . . 50. *J. sp. E* p.552
    Bracts without white hyaline margin . . . 41
41. Cymes usually more than one per leaf axil; largest leaf 13.5–18 cm long; bracts and bracteoles obovate-spatulate . . . 52. *J. mkungweensis* p.553
    Cymes always one per leaf axil; largest leaf less than 10(–11) cm long; bracts and bracteoles subulate to linear or narrowly triangular (Sect. *Ansellia*) . . . 42
42. Perennial or shrubby herbs, with woody old stems or with rootstocks or rhizomes . . . 43
    Annual herbs without rootstocks or rhizomes . . . 51
43. Basal old stems distinctly woody with corky bark . . . 44
    Basal old stems herbaceous, not with corky bark . . . 45
44. Corolla 4.5–6.5 mm long; young stems with thin bent hairs with transverse walls not or difficultly visible . . . 57. *J. ornatopila* p.559
    Corolla 8–12 mm long; young stems with long broad glossy curly many-celled hairs with easily visible transverse walls . . . 58. *J. cufodontii* p.560
45. Stems practically always unbranched; capsule 14–16 mm long . . . 59. *J. crassiradix* p.560
    Stems usually branched; capsule 4–12 mm long . . . 46
46. Inflorescences almost invariably 2-flowered; stems hairs with conspicuous purplish transverse walls. Corolla 8–13 mm long and capsule 9–12 mm long . . . 60. *J. nuttii* p.561
    Some or all inflorescences always with 3 or more flowers; stem hairs with colourless transverse walls . . . 47
47. Corolla 4–5.5(–6) mm long and capsule 4–7 mm long . . . 48
    Corolla 5.5–12 mm long; capsule (5–)6–10.5 mm long (if corolla less than 6 mm long then capsule more than 7 mm long) . . . 49

48. Leaves with strongly inrolled margins; capsule glabrous or with a few hairs at the apex; plant with strong woody rootstock .... 56. *J. lorata* p.558
Leaves with ± flat margins; capsule puberulous to below middle; plant without or with weak rootstock .... 61. *J. calyculata* p.561
49. Stems glabrous apart from thin transverse band of hairs at nodes (rarely uniformly puberulous); sepals 2.5–4(–5) and 3.5–6(–7) mm long; capsule (7.5–)8–10.5 mm long; plant with short thin creeping rhizome; grassland or bushland on grey to black clay .... 64. *J. anselliana* p.564
Stems with longitudinal bands of hairs, often also with long curly hairs, or uniformly hairy; sepals 3–7.5 and 5–10(–12) mm long; capsule (5–)6–9 mm long; plant with well developed rootstock; in a variety of woodland, bushland and grassland but usually not on clayey soils 50
50. Stems pubescent to pilose with bands of long broad curly glossy hairs; corolla 8–12 mm long and capsule 8–9 mm long; seed 1.5–2 mm in diameter .... 58. *J. cufodontii* p.560
Stems puberulous or also with scattered long broad curly glossy hairs or uniformly pubescent to tomentose; corolla 6–9(–10) mm long and capsule (5–)6–8 mm long; seed ± 1.5 mm in diameter .... 55. *J. anagalloides* p.556
51. Basal 1.5–2 mm of peduncle fused to petiole of subtending leaf; leaves held like a V (appearing folded in dried specimens); peduncle 2–5(–6) mm long; calyx pilose-ciliate on margins and midribs .... 63. *J. brevipedunculata* p.563
Basal part of peduncle not fused to petiole of subtending leaf; leaves flat; peduncle (3–)5–45(–70) mm long; calyx glabrous or minutely hispid-ciliate on margins 52
52. Stems glabrous apart from thin transverse band of hairs at nodes (very rarely uniformly puberulous); corolla 5.5–8.5 mm long and capsule (7.5–)8–10.5 mm long .... 64. *J. anselliana* p.564
Stems with longitudinal bands of hairs, often also with long curly hairs, never uniformly hairy; corolla 4–5.5 mm long 53
53. Leaves elliptic or narrowly so, widest near middle; capsule (3.5–)4–6.5 mm long ..... 62. *J. exigua* p.562
Leaves lanceolate to ovate, widest below middle; capsule 4.5–10(–11) mm long 54
54. Capsule 4.5–7 mm long; seed 1–1.5 mm in diameter .... 61. *J. calyculata* p.561
Capsule 7–10(–11) mm long; seed ± 2 mm in diameter .... 65. *J. matammensis* p.565
55. Flowers in 1–3(–15)-flowered subsessile dichasia aggregated into clearly defined terminal (or also axillary from upper axils) racemoid cymes; bracts not leaflike nor strobilate (Sect. *Tyloglossa*) (see Fig. 69, 70) 56
Inflorescences not as above 71

56. Plant with all leaves in a basal rosette and ± appressed to the ground 57
    Plant with well-defined leafy stems 58
57. Largest leaf 5–15 cm wide, base cuneate to cordate, central area not conspicuously paler; main axis bracts 6–9 × 3.5–8.5 mm .... 26. *J. fittonioides* p.530
    Largest leaf 1.5–3.5(–5) cm wide, base attenuate and decurrent, central area on both sides of midrib conspicuously paler; main axis bracts 4–5 × 2–3 mm .... 27. *J. drummondii* p.532
58. Corolla bright yellow 59
    Corolla white to pink, mauve or blue 61
59. Young stems whitish tomentose; bracts soon caducous; calyx 7–9 mm long .... 29. *J. gilbertii* p.535
    Young stems subglabrous to densely pubescent or sericeous; bracts persistent; calyx 3–7 (–8) mm long 60
60. Bracts and bracteoles 7–17 mm long; calyx without (rarely with a few) stalked capitate glands; inland plant .... 28. *J. flava* p.532
    Bracts and bracteoles 4–8(–9) mm long; calyx with dense stalked capitate glands; sea shore plant .... 30. *J. baravensis* p.536
61. Annual herb; corolla with conspicuous blue to purple venation on both lips .... 40. *J. kirkiana* p.543
    Perennial herb; corolla not with conspicuous blue to purple venation but often with darker "herring bone" pattern on lower lip 62
62. Lower corolla lip with broad triangular middle lobe and two strap-shaped recoiled and twisted lateral lobes; capsule with longitudinal and transverse ridges .... 21. *J. stachytarphetoides* p.527
    Lower corolla lip with three ± equal short broadly rounded lobes, lateral never strap-shaped, recoiled and twisted; capsule with smooth surface 63
63. Corolla white, usually with pink to mauve "herring bone" pattern on lower lip 64
    Corolla pink to blue or mauve 68
64. Corolla 15–18 mm long 65
    Corolla 7–11 mm long 66
65. Corolla tube 6–9 mm long, slightly shorter than upper lip; bracts, bracteoles and calyx with long-stalked capitate glands with oblong to clavate head .... 38. *J. sp. C* p.542
    Corolla tube 10–13 mm long, more than twice as long as upper lip; bracts, bracteoles and calyx with capitate glands with globose head 39. *J. migeodii* p.542

66. Pyrophytic grassland herb, usually erect; largest leaf 0.3–1.4 cm wide; bracts and bracteoles ovate-triangular; bracts, bracteoles and calyx with capitate glands with globose head; corolla tube (3–5 mm) and upper lip (4–5 mm) ± same length . . . . 34. *J. linearispica* p.539
   Erect or scrambling herbs, usually in forest; largest leaf 2.7–6 cm wide; bracts and bracteoles oblanceolate-obovate; bracts, bracteoles and calyx with long-stalked capitate glands with oblong to clavate head; corolla tube (5–7 mm) conspicuously longer than upper lip (2–4 mm) 67
67. Inflorescence axis and bracts whitish to pale yellowish puberulous; main axis bracts 3–5(–8) mm long; lower corolla lip 4–5 mm long; capsule 6–8 mm long, puberulous apically, glabrous basally . . . . 36. *J. heterotricha* p.540
   Inflorescence axis and bracts densely yellowish puberulous; main axis bracts (5–)6–8(–12) mm long; lower corolla lip 5–7 mm long; capsule 8.5–9.5 mm long, puberulous apically, sericeous basally . . . . 37. *J. bridsoniana* p.541
68. Lower corolla lip with 3 and upper lip with 2 conspicuous dark longitudinal streaks . . . . . . 31. *J. bizuneshiae* p.537
   Corolla lips not with dark longitudinal streaks but with pink to mauve venation in throat or with "herring bone" structure on lower lip 69
69. Corolla pink to mauve with "herring bone" structure on lower lip, 6–13 mm long of which tube 4–8 mm and upper lip 2–5 mm; bracts and bracteoles with long pilose hairs; capsule puberulous . . . . 35. *J. nyassana* p.539
   Corolla pale blue to blue, throat white with pink to mauve venation, 12–17 mm long of which tube 6–10 mm and upper lip 6–8 mm; bracts and bracteoles sericeous or sericeous-puberulous; capsule glabrous 70
70. Bracts, bracteoles and calyx with short straight glandular hairs with globose head; calyx 7–11(–13) mm long; corolla tube (6–9 mm) and upper lip (6–8 mm) ± same length . . . . 32. *J. caerulea* p.537
   Bracts, bracteoles and calyx with long curly glandular hairs with clavate head; calyx (5–)6–8 mm long; corolla tube (9–10) mm distinctly longer than upper lip (6–7 mm) . . . 33. *J. kulalensis* p.538
71. Flowers in dense terminal and axillary spiciform secund (only 1 bract at each node supporting a flower) cymes; calyx of 4 sepals, the 5th reduced to a small tooth (Sect. *Anisostachya*) (see Fig. 73) 72
   Inflorescences various, if dense spiciform cymes then calyx of 5 equal sepals 73

72. Perennial herb; largest leaf 5–12 cm long; fertile bracts 6–7 mm long, with broad white hyaline margin; calyx 4–5 mm long; corolla ± 6 mm long; capsule ± 4.5 mm long; seed echinate ..... 53. *J. roseobracteata* p.554
 Annual herb; largest leaf 2–4.5 cm long; fertile bracts 3–4.5 mm long, without hyaline white margin; calyx 2.5–3 mm long; corolla 2.5–3 mm long; capsule 2.5–3 mm long; seed hairy ..... 54. *J. tenella* p.556
73. Capsule 2-seeded; seed kidney-shaped, testa smooth, glossy, glabrous (rarely hairy); flowers usually in dense terminal and axillary spiciform cymes; bracts usually large, strobilate or not; bracteoles small or large (Sect. *Monechma*) (see Fig. 78) 74
 Capsule 4-seeded; seed usually not kidney-shaped, testa variously sculptured (rarely smooth); inflorescences various; bracteoles small 82
74. Annual herbs; calyx 4–13 mm long; corolla 6–10 mm long; capsule 4–10 mm long, ... 75
 Perennial herbs, shrubby herbs or subshrubs with woody old stems or with rootstocks or rhizomes 77
75. Bracts and calyx conspicuously white-ciliate with hairs (1.5–)2–4 mm long; corolla 7–8 mm long, shorter than the 8–13 mm long calyx; capsule glabrous, 9–10 mm long; bracts foliaceous ..... 97. *J. ciliata* p.592
 Bracts and calyx less conspicuously ciliate with hairs up to 2 mm long; corolla 6–10 mm long, longer than the 4–8 mm long calyx; capsule densely sericeous-puberulous, 4–7.5(–9) mm long; bracts clearly different from leaves 76
76. Young stems finely sericeous with downwardly directed appressed hairs; bracts conspicuously ciliate with ciliae (1–)1.5–2 mm long ..... 98. *J. bracteata* p.593
 Young stems puberulous to pubescent or densely (rarely sparsely) so with spreading or downwardly curved hairs; bracts with ciliae to 1 mm long ..... 100. *J. debilis* p.596
77. Flowers in mostly axillary racemoid cymes; calyx 4–8 mm long; corolla 6–12 mm long; capsule 4–9 mm long 78
 Flowers in terminal racemoid cymes, sometimes also a few lateral from upper axils; calyx 10–19 mm long; corolla 11–20 mm long; capsule 10–17 mm long 79
78. Stem indumentum of upwardly directed hairs; corolla (9–)10–14 mm long; capsule 7–9 mm long ..... 99. *J. eminii* p.596
 Stem indumentum of downwardly directed or spreading hairs; corolla 6–10 mm long; capsule 5–7.5 mm long ..... 100. *J. debilis* p.596
79. Leaves with 5 equally strong longitudinal veins from base to near apex ..... 101. *J. sp. H* p.598
 Leaves with pinnate venation 80

80. Bracts, bracteoles and calyx-lobes with 3 equally strong rib-like longitudinal veins .... 102. *J. tricostata* p.598
Bracts with pinnate venation; bracteoles and calyx-lobes with a single strong central vein 81
81. Leaf-base attenuate, decurrent; corolla 17–20 mm long, upper lip more than twice as long as tube; capsule ± 17 mm long .... 103. *J. attenuifolia* p.599
Leaf-base truncate to subcordate (rarely rounded); corolla 11–16 mm long, tube and upper lip ± same length; capsule 10–14 mm long .... 104. *J. subsessilis* p.599
82. Seed testa smooth and shiny 83
Seed testa variously sculptured 85
83. Flowers in a large open panicle 5–35 cm long; corolla 2.5–3.4 cm long .... 17. *J. salvioides* p.522
Flowers in condensed racemoid cymes 1–4 (–5) cm long 84
84. Corolla 2.5–2.8 cm long; bracteoles ± 5 mm long; shrub from N Kenya .... 18. *J. rendlei* p.524
Corolla less than 2 cm long; bracteoles ± 2 mm long; perennial herb from W Tanzania .... 105. *J. tetrasperma* p.600
85. Flowers in dense terminal (or also from upper leaf axils) spiciform cymes, secund or quadrangular; bracts large, strobilate (Sect. *Betonica* and Sect. *Vascia*) (see Fig. 71, 72) 86
Flowers in various types of open inflorescences (dichasia, racemoid cymes, panicles) with distinct peduncles and non-strobilate bracts (Sect. *Rhaphidospora* and Sect. *Justicia*) (see Fig. 66, 67, 68) 95
86. Both bracts at each node supporting a flower, the spikes quadrangular with flowers pointing in all directions (Sect. *Vascia*) 87
Only 1 bract at each node supporting a flower, the spikes 1-sided with all flowers on one side and only sterile bracts to the other side (Sect. *Betonica*) 91
87. Corolla white, without or with mauve to purple markings or lines on lower lip and in throat; leaves less than three times as long as wide or less than 15 cm long 88
Corolla creamy-green with dark purple streaks; bracts without hyaline white margins; leaves more than three times as long as wide, 20–33 cm long .... 45. *J. lukei* p.547
88. Bracts 6–10(–12) mm long; corolla 8.5–16 mm long 89
Bracts (12–)14–23 mm long; corolla 18–35 mm long; capsule hairy 90
89. Bracts without broad white hyaline margin; calyx 3–5.5(–6.5) mm long; corolla 8.5–11.5 mm long; capsule hairy .... 41. *J. ruwenzoriensis* p.543
Bracts with white hyaline margin; calyx 6–8 mm long, corolla 12–16 mm long; capsule glabrous .... 42. *J. pseudorungia* p.545

90. Young stems retrorsely hairy; calyx with stalked capitate glands; corolla 2.5–3.5 cm long; capsule 17–22 mm long, with retinacula splitting from fruit wall but not elastically rising placentae . . . . 43. *J. schimperiana* p.546
    Young stems spreadingly or antrorsely hairy; calyx without stalked capitate glands; corolla 1.8–2 cm long; capsule 13–18 mm long, with placentae rising elastically from the base . . . 44. *J. grandis* p.547
91. Bracts 2–3.8 cm long; calyx (7–)8–12 mm long; capsule 2–3 cm long; seed 6–7 mm long, slightly sculptured to almost smooth . . . . . . 47. *J. engleriana* p.550
    Bracts 0.6–1.7 cm long; calyx 3–7 mm long; capsule 0.8–1.7(–2.1) cm long; seed 2–3 mm long, densely tuberculate (capsule and seed not known in sp. 48) 92
92. Bracts with hyaline margin; capsule only hairy in apical part, placentae separating from outer capsule wall and rising elastically from the base 93
    Bracts without hyaline margin; capsule uniformly hairy, placentae not separating from outer capsule wall and not rising elastically from the base . . . . 46. *J. betonica* p.549
93. Bracts with conspicuous broad white hyaline margin, not turning brown; calyx sericeous or strigose-pubescent, not glandular; corolla glabrous or hairy on lobes only 94
    Bracts with rather inconspicuous pale green margin, turning brown in fruit; calyx glandular-puberulous; corolla uniformly puberulous . . . . 51. *J. faulknerae* p.553
94. Bracts 8–10 mm long, whitish sericeous to puberulous; calyx sericeous; corolla glabrous 48. *J. paxiana* p.551
    Bracts 6–7 mm long, pubescent with pale yellow hairs; calyx strigose-pubescent; corolla hairy on lobes . . . . 49. *J. sp. D* p.552
95. Corolla over 20 mm long (measured along upper lip) 96
    Corolla 6–19 mm long (measured along upper lip) 101
96. Corolla uniformly red to dark red or purplish red, sometimes with white spots in throat, tube much longer than upper lip 97
    Corolla white to yellow or greenish or pale mauve to pale pink, often with red to purple markings on lower lip, tube and upper lip usually ± the same length 99
97. Erect shrub; young branches antrorsely sericeous or sericeous-puberulous; leaf bases attenuate; peduncles 3–8 cm long; calyx lobes narrowly ovate; corolla glabrous on the outside . . . . . . 14. *J. udzungwaensis* p.520
    Scrambling or scandent shrub; leaf bases subcuneate to truncate; peduncles 0.5–3 cm long; calyx lobes linear-lanceolate 98

98. Young branches puberulous (rarely pubescent); flowers 2(–4) on short lateral branches from upper leaf axils; calyx 7–10(–13) mm long . . . 15. *J. beloperonoides* p.521
   Young branches sparsely antrorsely sericeous; flowers in 1–3-flowered axillary dichasia; calyx 4–5 mm long . . . . 16. *J. sp. B* p.522
99. Flowers in open axillary dichasial cymes; axes without stalked capitate glands; dorsal edge of corolla tube and upper lip forming a straight line; seed verrucose . . . . 5. *J. asystasioides* p.515
   Flowers in racemiform cymes, sometimes merging into large panicles; axes with dense stalked capitate glands; upper corolla lip bent backwards thus forming an angle with tube; seed smooth 100
100. Flowers in short terminal and lateral racemoid cymes 1–3(–5) cm long, not merging into a large panicle; young branches sparsely antrorsely sericeous; leaves subsessile, with broadly rounded apex . . . . 18. *J. rendlei* p.524
   Flowers in elongated racemoid cymes which merges into a large panicle 5–35 cm long; young branches glabrous or with scattered capitate glands; leaves clearly petiolate, with acute to subacuminate apex . . . . 17. *J. salvioides* p.522
101. Perennial almost acaulescent herbs from creeping rootstock with leaves in a rosette; leaf base cordate with one lobe larger than and overlapping the other 102
   Caulescent perennial or shrubby herbs or shrubs; leaves not in a rosette; leaf base usually not cordate, if so then lobes similar 103
102. Leaves crisped-puberulous on midrib and veins, oblong to slightly obovate with broadly rounded to retuse apex . . . . 12. *J. oblongifolia* p.519
   Leaves sericeous on midrib and veins, ovate-cordiform with triangular apex . . . . 13. *J. callopsoidea* p.520
103. Flowers in terminal or axillary panicles 104
   Flowers in axillary dichasia or in terminal and/or axillary racemiform cymes 106
104. Panicles 3–8 cm long, with stalked capitate glands; calyx 3–4 mm long; corolla white . . . 3. *J. microthyrsa* p.512
   Panicles 5–22 cm long, without stalked capitate glands; calyx 5–9 mm long; corolla cream to yellow 105
105. Scandent shrub, not drying black; largest leaf 8–13 cm long, base truncate to subcordate; flowers in terminal panicles; corolla creamy green, 11–14 mm long; capsule 2.4–2.7 cm long . . . . 2. *J. extensa* p.511
   Erect shrub or small tree, drying black; largest leaf 25–33 cm long, base attenuate; flowers in axillary panicles; corolla pale yellow, ± 15 mm long; capsule 4–4.5 cm long . . . . 1. *J. maxima* p.511

106. Flowers in axillary dichasial cymes 107
   Flowers in terminal and/or axillary racemoid cymes 115
107. Leaves glaucous, lamina ovate-oblong to oblong or slightly obovate, base cordate or subcordate with petiole 0.5–1.5 mm long; bracts, bracteoles and calyx lobes with white hyaline margins .... 19. *J. cordata* p.524
   Leaves not glaucous, lamina narrowly to broadly ovate to elliptic, base cuneate to attenuate (rarely truncate); petiole longer; bracts, bracteoles and calyx lobes without white hyaline margins 108
108. Corolla (10–)11–23 mm long 109
   Corolla 6–10 mm long; anther thecae parallel; capsule puberulous 111
109. Anther thecae held at right angles to each other; capsule glabrous .... 5. *J. asystasioides* p.515
   Anther thecae parallel or at a slight angle to each other; capsule puberulous 110
110. Young stems pubescent to tomentose; petiole with stalked capitate glands; calyx without stalked capitate glands; corolla tube same length or shorter than upper lip, lower lip 6–7 mm long; capsule 2.5–3 cm long .... 6. *J. galeata* p.515
   Young stems densely antrorsely sericeous; petiole without stalked capitate glands; calyx with stalked capitate glands; corolla tube longer than upper lip, lower lip 8–13 mm long; capsule 1.8–2.2 cm long .... 7. *J. regis* p.516
111. Corolla basically yellow or yellowish green; seed with dense glochidiate hairs .... 4. *J. scandens* p.513
   Corolla basically white to pale mauve; seed tuberculate 112
112. Stems and inflorescence axes without visible cystoliths; corolla densely puberulous; capsule 13–20 mm long 113
   Stems and inflorescence axes with very distinct cystoliths; corolla glabrous or sparsely puberulous; capsule 9–14 mm long 114
113. Plant drying black; young stems glabrous (rarely hairy just below nodes); peduncle (2–)7–30 mm long; corolla white to creamy green (sometimes tinged purple), tube and upper lip ± same length .... 8. *J. anisophylla* p.517
   Plant not drying black; young stems uniformly sericeous; peduncle ± 1 mm long; corolla pale mauve, tube distinctly longer than upper lip 9. *J. sp. A* p.518
114. Plant not drying black; leaves usually very distinctly anisophyllous with one leaf in a pair often absent; secondary bracts and bracteoles ± 0.5 mm long; calyx 2–4(–5) mm long; corolla 7–8.5 mm long; lowland forest species .... 10. *J. inaequifolia* p.518
   Plant drying black; leaves subequal or somewhat anisophyllous; secondary bracts and bracteoles 2–4 mm long; calyx 6.5–9 mm long; corolla 9–10 mm long; montane forest species .... 11. *J. rodgersii* p.519

115. Corolla upper lip only slightly hooded, tube ± 4 mm longer than upper lip, glabrous on the outside; capsule ± 1 cm long, with placentae separating from outer fruit wall and rising elastically .... 24. *J. gendarussa* p.529
Corolla upper lip strongly hooded, tube and upper lip ± same length (less than 2 mm difference), puberulous on the outside; capsule 1.4–2.7 cm long, placentae not separating from outer fruit wall and not rising elastically 116
116. Flowers in elongated racemiform cymes forming a large loose terminal panicle; corolla held upside down due to bending of pedicel .... 20. *J. interrupta* p.525
Flowers in short axillary and/or terminal racemiform cymes, usually not forming a large panicle; corolla not held upside down 117
117. Cyme rachis and calyx with short-stalked capitate glands; corolla ± 16 mm long; largest leaf 19–26 cm long .... 23. *J. breviracemosa* p.528
Cyme rachis and calyx without capitate glands; corolla 8–12 mm long; largest leaf 4.5–15(–18) cm long 118
118. Young branches and leaves glabrous; bracts and bracteoles 2–3 mm long; corolla 10–12 mm long; capsule 19–27 mm long and seed ± 4 mm in diameter .... 22. *J. francoiseana* p.528
Young branches and leaves subglabrous to densely puberulous (always with some hairs); bracts and bracteoles 4–6(–8) mm long; corolla 8–11 mm long; capsule 14–18 mm long and seed ± 2 mm in diameter .... 21. *J. stachytarphetoides* p.527

The species in this account are mostly grouped in the sections adopted by V.A.W. Graham (in K.B. 43: 551 (1988)). But the sequence of the sections diverges to a large extent and additionally the genera *Monechma* and *Rungia* are now both included in *Justicia*. Also I have doubts about the validity of some of the sections as accepted by Graham; e.g. the distinction between sections *Justicia* and *Rhaphidospora* seems to me somewhat artificial, as does the distinction between sections *Vascia* and *Betonica*. I also consider the *Anisostachya* and *Ansellia* groups to be sufficiently distinct to warrant sectional status.

Sect. **Justicia**; V.A.W. Graham in K.B. 43: 595 (1988) and Sect. **Rhaphidospora** (*Nees*) *T. Anderson* in J.L.S. Bot. 7: 43 (1863); V.A.W. Graham in K.B. 43: 587 (1988)

SYN. Sect. *Gendarussa* C.B. Clarke in F.T.A. 5: 183 (1899)

Flowers in various types of open inflorescences (dichasia, racemoid cymes, sometimes aggregated into panicles), peduncles distinct; rarely seemingly solitary and subsessile but then with large bracts and bracteoles; bracts not imbricate; bracteoles small (rarely large). Calyx with 5 equal lobes. Appendage on lower anther theca short, linear, flat or triangular. Capsule 4-seeded. Seed tuberculate to echinulate (rarely glochidiate). Species 1–25.

Sect. **Tyloglossa** (*Hochst.*) *Lindau* in E. & P. Pf. IV, 3b: 349 (1895); V.A.W. Graham in K.B. 43: 590 (1988)

SYN. Sect. *Rostellularia* C.B. Clarke in F.T.A. 5: 180 (1899)

Flowers in subsessile 1–3(–15)-flowered dichasia aggregated into clearly defined terminal (or also axillary from upper axils) pseudo-racemoid cymes; bracts not imbricate; bracteoles small. Calyx with 5 equal lobes. Appendage on lower anther theca linear, entire or bifurcate. Capsule 4-seeded. Seed verrucose, tuberculate or verrucose-tuberculate. Species 26–40.

Sect. **Vascia** *Lindau* in E. & P. Pf. IV, 3b: 395 (1895); V.A.W. Graham in K.B. 43: 584 (1988)

Flowers single, subsessile, in dense 4-sided racemoid cymes; bracts imbricate, all supporting a flower; bracteoles large, bract-like. Calyx with 5 equal linear to lanceolate acute to cuspidate lobes. Appendage on lower anther theca short and stubby or reduced to a small tooth (rarely elongated). Capsule 4-seeded. Seed reticulately sculptured. Species 41–45.

Sect. **Betonica** (*Nees*) *T. Anderson* in J.L.S. Bot. 7: 38 (1863); C.B. Clarke in F.T.A. 5: 180 (1899); V.A.W. Graham in K.B. 43: 585 (1988)

Flowers single (rarely with 2 additional non-developing lateral buds), subsessile, in dense 1-sided racemoid cymes; bracts imbricate, only one at each node supporting a flower and all flowers turned to the same side; bracteoles large, bract-like. Calyx with 5 equal lobes. Appendage on lower anther theca with elongated acute appendage. Capsule 4-seeded. Seed tuberculate. Species 46–52.

Sect. **Anisostachya** (*Nees*) *Benth.* in G.P. 2: 1110 (1876)

SYN. Sect. *Betonica* (Nees) T. Anderson subsect. *Anisostachya* (Nees) V.A.W. Graham in K.B. 43: 586 (1988)

Flowers single, subsessile, in dense spiciform 1-sided racemoid cymes; bracts imbricate, only one at each node supporting a flower and all flowers turned to the same side; bracteoles small, lanceolate. Calyx with 4 equal linear to lanceolate acute to cuspidate lobes, the 5th lobe reduced to a small tooth; divided to near base. Lower anther theca with flattened, obtuse appendage. Capsule 4-seeded. Seed sparsely hairy or echinate. Species 53–54.

Sect. **Ansellia** *C.B. Clarke* in F.T.A. 5: 183 (1899) & in Fl. Cap. 5: 57 (1901); Ensermu in Symb. Bot. Upsal. 29, 2: 51 (1990)

SYN. Sect. *Rostellaria* T. Anderson subsect. *Ansellia* (C.B. Clarke) V.A.W. Graham in K.B. 43: 598 (1988)

Flowers single, subsessile, in axillary secund racemoid cymes; bracts not imbricate, only one at each node supporting a flower and all flowers turned to the same side; bracteoles minute. Three dorsal calyx lobes longer than two ventral and middle dorsal lobe longer than lateral and sometimes also slightly wider. Lower anther theca with appendage square-cut at apex. Capsule 4-seeded. Seed double reticulate, flattened or not. Species 55–65.

Sect. **Harnieria** (*Solms-Laub.*) *Benth.* in G.P. 2: 1109 (1876); Hedrén in Nordic Journ. Bot. 6: 310 (1986) & in K.B. 43: 349 (1988); V.A.W. Graham in K.B. 43: 591 (1988); Hedrén in Nordic Journ. Bot. 8: 161 (1988) & in B.J.B.B. 58: 129 (1988) & in Bull. Mus. Nat. Hist. Paris, Sect. B Adansonia, ser. 4, 10: 345 (1988) & in J.L.S. 103: 263 (1990) & in Nordic Journ. Bot. 10: 357 (1990)

Syn. Sect. *Calophanoides* C.B. Clarke in F.T.A. 5: 181 (1899)

Flowers single or in 2–4(–6)-flowered cymules in axils of upper leaves; flowers almost always developing on both sides; bracts foliaceous; bracteoles small, subulate to linear or narrowly triangular. Calyx with 5 equal lobes. Lower anther theca with appendage acute and often bifid at apex, more rarely broad and bent (transversally ellipsoid). Capsule 4-seeded. Seed black, tuberculate, with a central ridge. Species 66–96.

Note. The species of the *J. diclipteroides*-group (species 81–88) often develop spiny indehiscent 1-seeded fruits. See Fig. 77, p. 585.

Sect. **Monechma** (*Hochst.*) *T. Anderson* in J.L.S. Bot. 7: 43 (1863); G.P. 2: 1109 (1876); Hedrén in Nordic Journ. Bot. 10: 151 (1990)

Flowers single or in 3(–5)-flowered cymules aggregated into terminal (often also axillary from upper axils) racemoid cymes; bracts foliaceous or differentiated upwards; bracteoles small or similar to bracts. Calyx with 5 equal lobes. Lower anther theca with linear entire or bifurcate appendage. Capsule 2(–4)-seeded. Seed smooth, kidney-shaped, glossy, glabrous (rarely hairy). Species 97–105.

1. **Justicia maxima** (*Lindau*) *S. Moore* in J.B. 54: 289 (1916); V.A.W. Graham in K.B. 43: 589 (1988). Type: Cameroon, Pembo, *Mildbraed* 4226 (B†, holo.)

Erect shrub or small tree to 6 m tall; young stems crisped sericeous-puberulous or densely so, glabrescent; plant drying black. Leaves not anisophyllous; petiole to 5 cm long; lamina ovate or broadly so, largest 25–33 × 9–13.5 cm, base attenuate, decurrent, equal-sided, apex acute with a short blunt tip; sericeous-puberulous or densely so on midrib and veins, without visible cystoliths. Flowers in large axillary panicles composed of dichasia, 15–22 cm long, with foliaceous bracts at lower nodes; peduncles, axes and pedicels crisped-puberulous or densely so; peduncles 5.5–16 cm long; branches to 8 cm long; secondary bracts and bracteoles lanceolate, 5–7 mm long. Calyx 5–7 mm long, sericeous-puberulous, without capitate glands, lobes lanceolate, distinctly 3-veined, with conspicuous white edges, acute. Corolla pale yellow with reddish markings in throat, ± 15 mm long, densely puberulous; tube ± 8 mm long and ± 4 mm diameter; upper lip ± 7 mm long, slightly hooded; lower lip deflexed, ± 8 mm long, deeply 3-lobed, middle lobe much longer and wider. Filaments ± 5 mm long; thecae ± 1.5 mm long, oblong, parallel, ± 75% overlapping, with a few hairs, lower with flat curved appendage ± 0.5 mm long. Capsule 4–4.5 cm long, densely puberulous; seed not seen.

Uganda. Bunyoro District: Budongo Forest, Nov. 1943, *Eggeling* 5490!; Mengo District: 15 km on Kampala–Masaka road, May 1937, *Chandler* 1643!

Distr. **U** 2, 4; Ivory Coast, Ghana, Togo, Benin, Nigeria, Cameroon, Bioko, Gabon, Congo-Kinshasa

Hab. Wet evergreen forest; 1200 m

Syn. *Duvernoia maxima* Lindau in E.J. 49: 405 (1913)
Justicia baronii V.A.W. Graham in K.B. 43: 588 (1988); Vollesen in Cable & Cheek, Pl. Mt Cameroon: 3 (1998). Type: Equatorial Guinea, Bioko, *Mann* 634 (K!, holo.)

Note. The two cited specimens are the most recent collections of this species from Uganda. It needs to be searched for in Budongo Forest and similar habitats in the eastern foothills of the Ruwenzori Mts.

2. **Justicia extensa** *T. Anderson* in J.L.S. Bot. 7: 44 (1862); S. Moore in J.B. 18: 341 (1880); C.B. Clarke in F.T.A. 5: 206 (1900); U.K.W.F.: 604 (1974); U.K.W.F., ed. 2: 280 (1994); Friis & Vollesen in Biol. Skr. 51(2): 446 (2005). Type: Nigeria, Eppah, *Barter* 3301 (K!, holo.)

Erect (rarely) or scandent shrub to 2.5 m tall; young stems finely antrorsely sericeous or sparsely so; plant not drying black. Leaves not or only slightly anisophyllous; petiole 1–5 cm long; lamina ovate or broadly so, largest 8–13 × 4–8.5 cm, base truncate to subcordate, slightly to distinctly unequal-sided, apex acute to subacuminate with obtuse tip; sparsely sericeous on midrib and veins with short almost dot-like cystoliths. Flowers in terminal panicles composed of dichasia, 5–16 cm long, with foliaceous bracts at lowermost nodes; peduncles, axes and pedicels finely sericeous-puberulous or sparsely so; peduncles to 6 cm long; branches to 2.5 cm long; secondary bracts and bracteoles linear-lanceolate to narrowly triangular, 2–4 mm long. Calyx 5.5–9 mm long, finely sericeous and with subsessile capitate glands, lobes lanceolate, distinctly 3-veined, acute. Corolla creamy green with green tube, with large purple patch in throat, 11–14 mm long, densely puberulous; tube 6–8 mm long and 3–4 mm diameter; upper lip 5–6 mm long, slightly hooded; lower lip deflexed, 6–8 mm long, deeply 3-lobed, middle lobe much larger. Filaments 3–4 mm long; thecae ± 1.5 mm long, oblong, parallel, ± 33% overlapping, glabrous, lower with flat curved appendage ± 0.5 mm long. Capsule 2.4–2.7 cm long, densely puberulous; seed tuberculate, ± 6 mm diameter.

UGANDA. Toro District: Fort Portal, 20 Oct. 1906, *Bagshawe* 1265!; Mengo District: Mulange, Oct. 1919, *Dummer* 4325! & Mpigi, Mawokota, Mpanga Forest Reserve, 8 Feb. 1997, *Lye & Katende* 22352!

KENYA. North Kavirondo District: Kakamega Forest, 15 Oct. 1953, *Drummond & Hemsley* 4784! & Yala River Forest Reserve, Quarry Hill, 26 Jan. 1982, *M.G. Gilbert* 6889! & Kisere Forest, 9 Dec. 1984, *Kokwaro* 4338!

DISTR. **U** 2, 4; **K** 5; Guinea, Sierra Leone, Liberia, Ivory Coast, Ghana, Togo, Benin, Nigeria, Cameroon, Bioko, Gabon, Central African Republic, Congo-Kinshasa, Angola

HAB. Wet evergreen intermediate and lower montane forest; 1000–1600 m

SYN. *Duvernoia stuhlmannii* Lindau in E.J. 20: 43 (1894) & in E. & P. Pf. IV, 3b: 339 (1895). Type: Congo-Kinshasa, Semliki, Issange, *Stuhlmann* 2953 (B†, holo.)

D. extensa (T. Anderson) Lindau in P.O.A. C: 372 (1895)

D. dewevrei De Wild. & Th. Dur. in B.S.B.B. 38, Compt. Rend.: 102 (1899). Type: Congo-Kinshasa, Basankussu, Lulonga River, *Dewèvre* s.n. (BR!, holo.)

NOTE. C.B. Clarke in F.T.A. 5: 206 (1900) cites *Stuhlmann* 2953 as originating "between Tanganyika and Victoria Nyanza" which would be in either **T** 1 or 4. But Lindau, when describing *Duvernoia stuhlmannii*, clearly states that it comes from Congo. I have not included Tanzania in the distribution although the species may well occur in the west.

3. **Justicia microthyrsa** *Vollesen* **sp. nov.** a *J. extensa* foliis minoribus anisophyllis ad basin attenuatis (nec truncatis neque subcordatis), panicula minore 3–8 cm (nec 5–16 cm) longa glandulos stipitatos capitatos gerenti (nec glandulis carentibus), calyce minore 3–4 mm (nec 5.5–9 mm) longo et corolla alba (nec cremeo-viride) differt. Typus: Tanzania, Iringa District, Udzungwa Mts National Park, Ndundulu Forest Reserve, *Luke et al.* 10431 (K!, holo.; EA!, MO, NHT, iso.)

Multi-stemmed shrub to 2.5 m tall; young stems sparsely and finely antrorsely sericeous; plant drying black. Leaves slightly to distinctly anisophyllous; petiole to 1.8 cm long; lamina elliptic, largest 8–9 × 3.5–3.8 cm, base attenuate, decurrent, apex acuminate with obtuse tip; sparsely and finely sericeous on midrib and veins. Flowers in terminal panicles composed of dichasia, 3–8 cm long; peduncles and axes sericeous-puberulous and with scattered (peduncle) to dense stalked capitate glands; peduncles 3–6 mm long; branches to 2 cm long; secondary bracts and bracteoles linear-lanceolate, 2–4 mm long. Calyx 3–4 mm long, finely puberulous and with many stalked capitate glands, lobes lanceolate, acute, not 3-veined. Corolla white, ± 15 mm long, puberulous and with dense stalked capitate glands; tube ± 6 mm long, much expanded apically and ± 5 mm diameter; upper lip ± 6 mm long, slightly hooded; lower lip ± 6 mm long. Filaments ± 3 mm long; thecae ± 1.5 mm long, oblong, parallel, ± 75% overlapping, hairy, lower with acute appendage ± 0.5 mm long. Capsule and seed not seen.

TANZANIA. Iringa District: Udzungwa Mountains National Park, Mt Luhomero, 26 Sept. 2000, *Luke et al.* 6676! & Ndundulu Forest Reserve, 10 Sept. 2004, *Luke et al.* 10431!
DISTR. **T** 7; not known elsewhere
HAB. Wet evergreen montane forest; 1250–1500 m

NOTE. Known only from these two collections. The terminal panicle shows its relations with *J. extensa*. It differs from this species in the smaller anisophyllous leaves with attenuate base, in the smaller panicle with stalked capitate glands, in the smaller calyx with dense glands and in the white corolla.

4. **Justicia scandens** *Vahl*, Symb. Bot. 2: 7 (1791); Hansen in Holm-Nielsen *et al.*, Tropical Forest: 204 (1989); Luke & Robertson, Kenya Coast. For. 2. Checklist Vasc. Pl.: 82 (1993); Vollesen *et al.*, Checklist Mkomazi: 83 (1999); Friis & Vollesen in Biol. Skr. 51(2): 448 (2005); Ensermu in Fl. Eth. 5: 461 (2006). Type: India, "Malabar", *Koenig* s.n. (C!, holo.)

Erect to scandent perennial (rarely annual) herb with 1-several stems from creeping rhizome; stems to 1.25 m long, subglabrous to bifariously or uniformly puberulous (rarely pubescent to pilose); plant not drying black. Leaves with petiole to 3.5(–7) cm long; lamina ovate to elliptic or broadly so, largest (5.5–)7–15 × (2–)3.5–8.5 cm, base cuneate to attenuate (rarely truncate), apex acuminate with an acute to obtuse tip; subglabrous to sparsely puberulous (rarely pilose) along midrib and veins. Flowers in lax axillary dichasia, often forming loose panicles towards apex; peduncles, axes and pedicels subglabrous to finely puberulous and with sparse to dense stalked capitate glands (very rarely without); peduncles 1–4(–6) cm long; branches to 2(–3) cm long; main axis bracts foliaceous, smaller, sessile and subcordate to cordate upwards; secondary bracts and bracteoles linear, 0.5–2 mm long. Calyx (2–)3–6 mm long, finely puberulous or sparsely so with short stubby non-capitate glandular hairs and usually ciliate, lobes linear-lanceolate, acute to acuminate. Corolla pale yellow to yellow or yellowish green (? rarely white) with red to purple markings at base of lower lip, 6–10 mm long, puberulous; tube 4–5 mm long and ± 2 mm diameter, slightly curved ventrally; upper lip (2–)3–5 mm long, flat or slightly hooded; lower lip deflexed, 4–6 mm long, deeply 3-lobed. Filaments 2–3 mm long; thecae 0.5–1 mm long, oblong or broadly so, glabrous or hairy, lower with acute curved appendage ± 0.8 mm long. Capsule 12–17 mm long, uniformly puberulous or sparsely so; seed circular in outline, 2–3 mm diameter, with dense glochidiate hairs ± 0.25 mm long. Fig. 66, p. 514.

UGANDA. Karamoja District: Mamalu Rest Camp, Feb. 1963, *J. Wilson* 1361!; Busoga District: Lake Victoria, Lolui Island, 21 May 1964, *Jackson* U139!; Mengo District: Singo, Bukomero, Sept. 1932, *Eggeling* 543!
KENYA. Northern Frontier District: 6 km E of Moyale, 4 July 1952, *Gillett* 13507!; North Kavirondo District: Malaba Forest, March 1963, *Tweedie* 2579!; Kilifi District: Pangani, 27 Feb. 1991, *Luke & Robertson* 2690!
TANZANIA. Musoma District: Orangi River, Retima Pool, 20 May 1962, *Greenway* 10648!; Masai District: Losimingori Hills, Essimongori Forest Reserve, 11 April 2000, *Mollel* 164!; Mpanda District: Mahali Mts, Kasieha Valley, 20 July 1959, *Newbould & Harley* 4506!
DISTR. **U** 1–4; **K** 1, 3–7; **T** 1–8; widespread in tropical Africa, Madagascar; Tropical Asia
HAB. Drier types of lowland to montane forest, riverine forest and scrub, *Acacia* bushland on alluvial plains, lake shores; 50–1850 m

SYN. *Justicia glabra* [Koenig, Hort. Beng.: 4 (1812), *nom. nud.*] ex Roxb., Fl. Ind. (ed. Carey) 1: 132 (1820); C.B. Clarke in F.T.A. 5: 208 (1900); Milne-Redhead in Mem. N. Y. Bot. Gard. 9: 24 (1954); F.P.S. 3: 179 (1956); Binns, Check List Herb. Fl. Malawi: 14 (1968); Richards & Morony, Check List Mbala: 231 (1969); U.K.W.F.: 604 (1974); Vollesen in Opera Bot. 59: 81 (1980); Synnott, Checklist Fl. Budongo For.: 69 (1985); V.A.W. Graham in K.B. 43: 589 (1988); U.K.W.F., ed. 2: 280 (1994); Immelman in Fl. S. Afr. 30 (3, 1): 21 (1995); Ruffo *et al.*, Cat. Lushoto Herb. Tanzania: 6 (1996). Type: as for *J. scandens*
Rhaphidospora glabra (Roxb.) Nees in Wall., Pl. As. Rar. 3: 115 (1832) & in DC., Prodr. 11: 499 (1847); Lindau in P.O.A. C: 370 (1895)

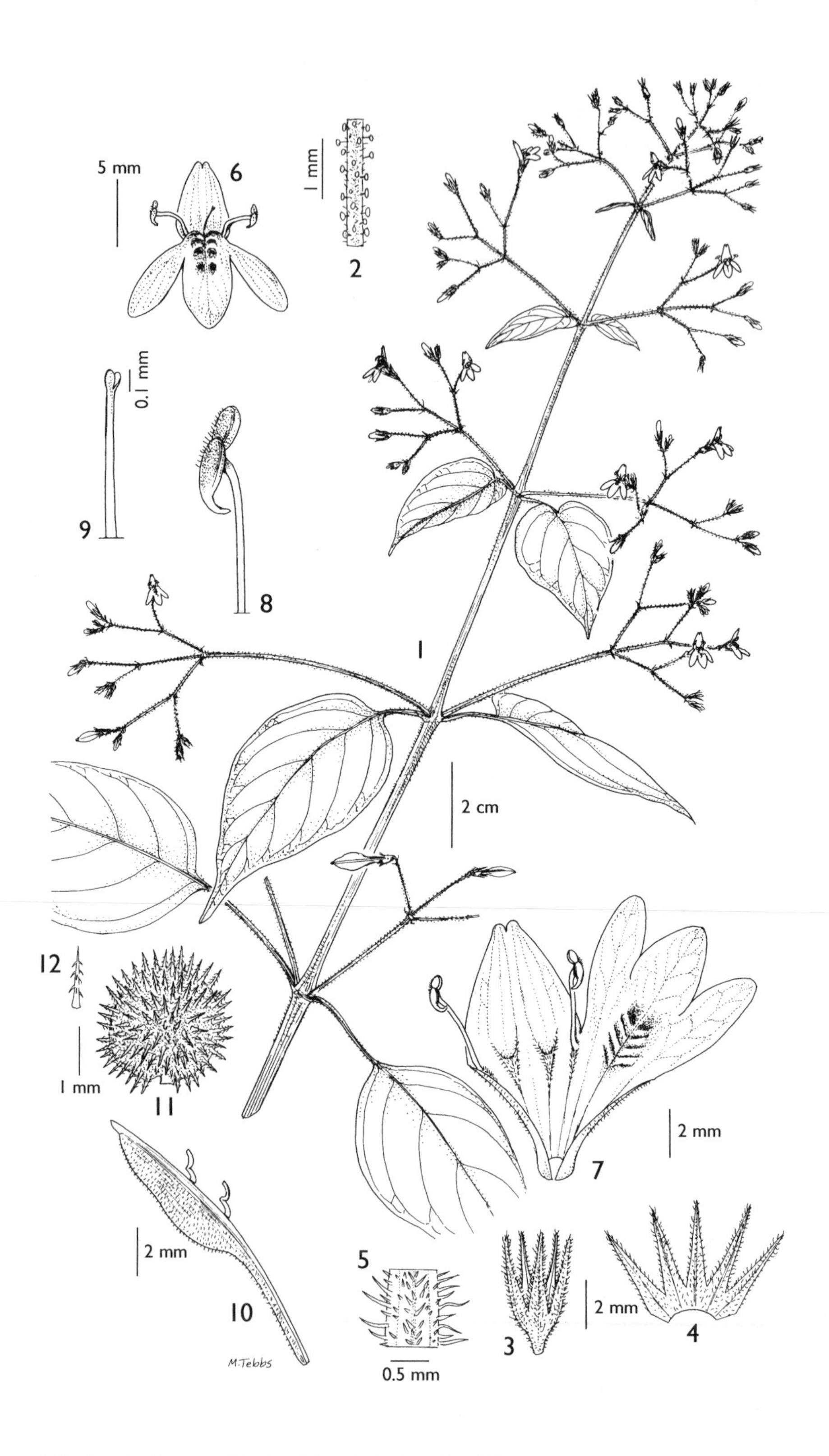
5 mm
6
1 mm
2
0.1 mm
9
8
1
2 cm
12
1 mm
11
2 mm
7
2 mm
10
M.Tebbs
5
0.5 mm
2 mm
3
4

R. abyssinica Nees in DC., Prodr. 11: 500 (1847); A. Richard, Tent. Fl. Abyss. 2: 161 (1850). Type: Ethiopia, Tacazze Valley, *Schimper* II.903 (K!, holo.; BM!, K!, iso.)
Justicia glabra Roxb. var. *pubescens* T. Anderson in J.L.S. Bot. 7: 44 (1863). Type: as for *Rhaphidospora abyssinica* Nees
Rhaphidospora glabra (Roxb.) Nees var. *pubescens* (T. Anderson) Lindau in P.O.A. C: 370 (1895)

5. **Justicia asystasioides** (*Lindau*) *M. E. Steiner* in K.B. 44: 709 (1989); Iversen in Symb. Bot. Upsal. 29(3): 161 (1991); White *et al.*, For. Fl. Malawi: 115 (2001). Types: Tanzania, Lushoto District, W Usambara Mts, Sakare, *Engler* 932 (B†, syn.) & *Engler* 1008 (B†, syn.). Neotype: Tanzania, Lushoto District, W Usambara Mts, Baga Forest Reserve, *Borhidi et al.* 84434 (UPS, neo.; ETH, NHT, iso.)

Erect or straggling shrubby herb or shrub to 3(–4) m tall; young stems sparsely to densely antrorsely sericeous to puberulous or pubescent, hairs broad glossy, with or without purple walls; older branches with straw-coloured bark. Leaves subequal to slightly anisophyllous; petiole 0.5–5.5 cm long; lamina ovate to elliptic, largest 5.5–17 × 2–7 cm, base attenuate, not decurrent, apex acute to cuspidate; sericeous to puberulous or sparsely so on midrib and veins, distinctly ciliate towards base. Flowers in 3–15(–22)-flowered axillary dichasia (rarely 1-flowered with non-developing buds or solitary with no buds); peduncles, axes and pedicels glabrous to puberulous or sericeous-puberulous, no capitate glands; peduncles (0.3–)1–5(–6) cm long; branches 0.3–2 cm long; secondary bracts and bracteoles linear-lanceolate, 1–4 mm long. Calyx 4.5–10 mm long, glabrous to puberulous or sericeous-puberulous, no capitate glands, lobes lanceolate, acute to acuminate. Corolla white with pink to purple markings on lower lip, (10–)12–23 mm long, glabrous (rarely sparsely puberulous on lobes); tube 6–16 mm long and 3–5 mm diameter; upper lip 4–8 mm long, flat to slightly hooded; lower lip deflexed, 5–8 mm long, deeply 3-lobed, middle lobe much longer and wider. Filaments 3–5 mm long, connective enlarged and thecae held at right angle to each other; thecae 1–1.5 mm long, oblong to ellipsoid, glabrous or sparsely hairy, lower apiculate or with short appendage to 0.3 mm long. Capsule 19–27 mm long, glabrous; seed 4.5–5.5 mm diameter, strongly verrucose.

TANZANIA. Lushoto District: E Usambara Mts, Kwamkoro Forest Reserve, 18 Sept. 1986, *Ruffo & Mmari* 1836!; Morogoro District: Nguru Mts, near Turiani, Mkobwe, 29 March 1953, *Drummond & Hemsley* 1874!; Iringa District: Udzungwa Mts, Boma la Mzinga Forest, 19 June 1979, *Mwasumbi* 11862!
DISTR. **T** 3, 6–8; Malawi, Mozambique
HAB. Wet evergreen intermediate and montane forest; (750–)950–2000(?–2900) m

SYN. *Duvernoia asystasioides* Lindau in E.J. 38: 72 (1905); T.T.C.L.: 8 (1949)
Thamnojusticia amabilis Mildbr. in N.B.G.B. 11: 826 (1933). Type: Tanzania, Ulanga District, Mahenge, Sali, *Schlieben* 2202 (B†, holo.; BM!, iso.)
T. grandiflora Mildbr. in N.B.G.B. 12: 100 (1934); Ruffo *et al.*, Cat. Lushoto Herb. Tanzania: 6 (1996). Type: Tanzania, Morogoro District, Nguru Mts, *Schlieben* 4086 (B†, holo.; BM!, iso.)
Justicia amabilis (Mildbr.) V.A.W. Graham in K.B. 43: 595 (1988)
J. mildbraedii V.A.W. Graham in K.B. 43: 596 (1988). Type: as for *Thamnojusticia grandiflora*

6. **Justicia galeata** *Hedrén* in Nordic Journ. Bot. 13: 647 (1993). Type: Kenya, Kilifi District, Dakabuka Hill, *Dale* 1077 (K!, holo.; EA!, iso.)

FIG. 66. *JUSTICIA SCANDENS* — **1**, habit; **2**, detail of pedicel indumentum; **3**, calyx; **4**, calyx opened up; **5**, detail of calyx lobe; **6**, corolla, frontal view; **7**, corolla opened up; **8**, apical part of filament and anther; **9**, apical part of style and stigma; **10**, capsule; **11**, seed; **12**, glochidia from seed. 1, 2 & 8 from *Gillett* 13507, 3–7 & 9 from *Newbould* 3339, 10–12 from *Luke* 2690. Drawn by Margaret Tebbs.

Shrubby herb or shrub to 2 m tall; young stems densely pubescent to tomentose; plant drying black. Leaves slightly to distinctly anisophyllous; petiole with stalked capitate glands, to 2.5(–5.5) cm long; lamina ovate or broadly so to cordiform, largest 2.5–6(–14) × 1.5–4.5(–9) cm, base cuneate to truncate, apex acute to rounded; sericeous-puberulous or densely so on midrib and veins, sparser on lamina. Flowers in 3–7-flowered axillary dichasia (or 1-flowered with 2 non-developing buds), 1 or 2 per node; peduncles, axes and pedicels puberulous to tomentellous; peduncles and branches 1–3 mm long; secondary bracts and bracteoles linear-lanceolate, 1–3 mm long. Calyx 4–7 mm long, puberulous or sparsely so, lobes lanceolate, acuminate. Corolla greenish yellow (? rarely white), when dry with dark bands along edges and centrally on lower lip, 12–15 mm long, puberulous or densely so; tube 5–7 mm long and ± 3 mm diameter, not pouched ventrally; upper lip 7–9 mm long, strongly hooded; lower lip deflexed, 6–7 mm long, deeply 3-lobed, middle lobe much larger. Filaments 4–5 mm long; thecae 1.5–2 mm long, oblong, held at a slight angle to each other, ± 75% overlapping, glabrous or hairy, lower with obtuse appendage ± 0.5 mm long. Capsule 2.5–3 cm long, not breaking irregularly, puberulous; seed circular in outline, ± 4 mm diameter, densely tuberculate to almost echinulate towards apex.

KENYA. Kilifi District: Mangea Hill, 10 April 1988, *Luke* 1077!; Teita District: Taita Hills, Mwandongo Forest, 18 June 1998, *Mwachala* EW984! & Rukinga Ranch, 17 May 2000, *Luke* 6236!

TANZANIA. Mpwapwa District: Kibakwe, 30 June 1938, *Hornby* 866!; Iringa District: Udzungwa Mts National Park [7°38'S 36°33'E], 3 June 2002, *Luke et al.* 8679!

DISTR. **K** 7; **T** 5, 7; not known elsewhere

HAB. Semi-deciduous lowland forest, thickets on rocky hills in *Acacia-Commiphora* bushland; 250–1200 m

SYN. *J. sp. ? nov.* Robertson & Luke, Kenya Coast. For. 2. Checklist Vasc. Pl.: 82 (1993)

NOTE. An unusual distribution but no differences can be found between the two Tanzanian collections and the Kenyan material. It should be noted that although the species occurs in lowland forest in Kenya, it also grows at higher altitudes in the Taita Hills and here in thickets on rocky hills, which is very similar to its habitat in Tanzania.

7. **Justicia regis** *Hedrén* in Nordic Journ. Bot. 10: 265, fig. 1 (1990); U.K.W.F., ed. 2: 280 (1994). Type: Cult. in Hort. Uppsala from seed collected in Kenya, (district not clear), Emening, *King* s.n. (UPS, holo.; K!, iso.)

Perennial or shrubby herb or shrub to 1(–2) m tall; young stems hollow, inflated, with many longitudinal ribs, densely antrorsely whitish sericeous to tomentellous; plant not drying black. Leaves not anisophyllous; petiole to 2(–4.5) cm long; lamina ovate or narrowly so or elliptic, largest 2–6.5(–12) × 1.5–3.5(–7) cm, base cuneate to truncate, apex subacute to rounded or emarginate; subglabrous to sparsely uniformly sericeous. Flowers in contracted axillary dichasia, sometimes ± racemoid due to reductions; peduncles and axes densely antrorsely sericeous; peduncle to 5(–10) mm long; primary bracts foliaceous; branches 1(–3) mm long; secondary bracts and bracteoles linear-lanceolate, 2–4 mm long; pedicels ± 1 mm long, with stalked capitate glands. Calyx 5–8 mm long, sericeous-puberulous and with many stalked capitate glands, lobes linear-lanceolate, cuspidate. Corolla "dirty mauve" or "muddy bluish purple", 11–15 mm long, puberulous; tube 7–8.5 mm long and 4–5 mm diameter, not pouched ventrally; upper lip 5–7 mm long, slightly hooded, tapering upwards; lower lip deflexed, 8–13 mm long, deeply 3-lobed, middle lobe much larger. Filaments 4–6 mm long; thecae 1.5–2 mm long, oblong, parallel, ± 50% overlapping, glabrous, lower with curved appendage ± 1 mm long. Capsule 1.8–2.2 cm long, sometimes breaking irregularly, puberulous; seed (immature) circular in outline, ± 4 mm diameter, tuberculate.

KENYA. Northern Frontier District: Isiolo, Dec. 1956, *J. Adamson* 610!; West Suk District: Kaipapet, Wei Wei, 10 Jan. 1979, *Meyerhoff* 124M!; Ravine District: 24 km E of Eldama Ravine, 8 Feb. 1957, *Bogdan* 4461!

DISTR. **K** 1–3; not known elsewhere

HAB. *Acacia* bushland on rocky hills; 900–1600 m

SYN. *Justicia. sp. D sensu* Agnew, U.K.W.F.: 604 (1974)

8. **Justicia anisophylla** (*Mildbr.*) *Brummitt* in K.B. 45: 281 (1990); Iversen in Symb. Bot. Upsal. 29(3): 161 (1991); Ruffo *et al.*, Cat. Lushoto Herb. Tanzania: 6 (1996). Types: Tanzania, Lushoto District, Amani, *Warnecke* 373 (B†, syn.); Ulanga District, Mahenge, *Schlieben* 2001a (B†, syn.), see note below. Neotype: Tanzania, Lushoto District, E Usambara Mts, Kwamkoro Forest Reserve, *Semsei* 3217 (K!, neo., selected here; EA!, iso.)

Erect (rarely scrambling) perennial or shrubby herb; stems to 1.5 m long, glabrous apart from thin band of hairs at nodes (rarely with two thin lines of hairs just below nodes), cystoliths not visible; plant drying black. Leaves strongly anisophyllous with one of a pair either completely absent or (more commonly) reduced to $^1/_2$ the size of the other or less (rarely both leaves subequal); petiole (of largest leaf) 0.5–2.5(–3.5) cm long; lamina (of largest leaf) ovate to elliptic or broadly (rarely narrowly) so, largest (7–)9.5–17 × 3–7 cm, base attenuate to rounded, usually very unequal-sided, apex acuminate to cuspidate (rarely acute); glabrous (rarely sparsely puberulous on midrib). Flowers in lax few- to many-flowered axillary dichasia towards end of branches, often 2 per node, not forming loose terminal panicles; peduncles, axes and pedicels glabrous or finely puberulous and then also with stalked capitate glands, without visible cystoliths; peduncles (0.2–)0.7–3 cm long or dichasia sessile; branches 0.5–2.5 cm long (usually shorter than peduncle); secondary bracts and bracteoles linear, 0.5–2 mm long. Calyx 3–5 mm long, glabrous to finely sparsely puberulous, lobes linear-lanceolate, acuminate. Corolla tube pale green, lips white to creamy green (sometimes tinged purple) with pink markings on lower lip, 6.5–10 mm long, densely puberulous; tube 3.5–5 mm long and ± 2 mm diameter, slightly pouched ventrally; upper lip 3–5 mm long, slightly hooded; lower lip deflexed, 3–5 mm long, deeply 3-lobed. Filaments 2–4 mm long; thecae 0.5–1 mm long, oblong, parallel, ± 50% overlapping, glabrous or with a few hairs, lower with acute appendage ± 0.75 mm long. Capsule 13–20 mm long, often breaking irregularly, puberulous ; seed circular in outline, ± 2 mm diameter, with elongated to almost hair-like tubercles and with a few glochidiae near apex.

KENYA. Kilifi District: Cha Simba, 16 Feb. 1977, *Faden* 77/421!; Kwale District: Mwadabara Pump House, 17 Oct. 1991, *Luke* 2942!

TANZANIA. Lushoto District: E Usambara Mts, Kiganga-Kwamkoro Forest Reserve, 1 Oct. 1977, *Sigara* 130! & Amani West Forest Reserve, 5 May 1987, *Iversen et al.* 87211!; Rufiji District: Matumbi Hills, Nyamakutwa-Nyamuete Forest Reserve, 11 Nov. 1999, *Kibure* 556!; Zanzibar. Kidoti, Oct. 1873, *Hildebrandt* 1135!

DISTR. **K** 7; **T** 2, 3, 6; **Z**; not known elsewhere

HAB. Lowland and intermediate evergreen and semi-deciduous forest; near sea level to 1150 m

SYN. *Rhaphidospora anisophylla* Mildbr. in N.B.G.B. 12: 520 (1935)

NOTE. *Justicia anisophylla* differs from *J. scandens* in drying black, in having usually strongly anisophyllous leaves, in often (always in isophyllous plants) having more than one cyme per leaf axil.

The identity of the *Schlieben* syntype is not at all certain. Mahenge is a long way outside the rest of the distribution of the species.

9. **Justicia sp. A**

Shrubby herb, 2 m tall; young stems uniformly antrorsely sericeous, cystoliths not visible; plant not drying black. Leaves not or slightly anisophyllous; petiole to 3 cm long; lamina ovate to elliptic, largest ± 9.5 × 5 cm, base cuneate to shortly attenuate, not decurrent, equal sided, apex acute with obtuse tip; sericeous on midrib and veins, sparsely so on lamina. Flowers in 2–7-flowered axillary dichasia, usually 2 per node; peduncles, axes and pedicels sparsely sericeous-puberulous, cystoliths not visible; peduncles and branches ± 1 mm long; secondary bracts and bracteoles linear, ± 1 mm long. Calyx 3.5–4.5 mm long, sparsely puberulous, lobes linear-lanceolate, cuspidate. Corolla "pale mauve", when dry with dark bands along edges and centrally on lower lip, ± 8 mm long, densely puberulous; tube ± 5 mm long and ± 2 mm diameter, not pouched ventrally; upper lip ± 3 mm long, hooded; lower lip deflexed, ± 3 mm long, deeply 3-lobed, middle lobe much larger. Filaments ± 2 mm long; thecae ± 1 mm long, oblong, parallel, ± 33% overlapping, glabrous, lower with appendage ± 0.3 mm long. Capsule 15–17 mm long, breaking irregularly, puberulous; seed circular in outline, ± 3 mm diameter, tuberculate.

TANZANIA. Bukoba District: Minziro Forest Reserve, Kale Hill, 28 April 1994, *Congdon* 365!
DISTR. **T** 1; not known elsewhere
HAB. Tall grassland on low rocky hill surrounded by swamp forest; 1250 m

NOTE. Known only from this collection. Differs from *Justicia anisophylla* in not drying black, in having sericeous (not glabrous) young stems and in having pale mauve (not white to creamy green) corolla with tube longer than the upper lip.

10. **Justicia inaequifolia** *Brummitt* in K.B. 45: 282 (1990); Iversen in Symb. Bot. Upsal. 29(3): 161 (1991); Robertson & Luke, Kenya Coast. For. 2. Checklist Vasc. Pl.: 82 (1993); Ruffo *et al.*, Cat. Lushoto Herb. Tanzania: 6 (1996), as *inaequalis*. Type: Kenya, Kwale District, Jombo Mtn, *Polhill & Robertson* 4833 (K!, holo.; EA!, iso.)

Shrubby herb or shrub to 1.5 m tall; young stems finely antrorsely sericeous-puberulous or sparsely so, with conspicuous cystoliths; plant not drying black. Leaves with conspicuous cystoliths, strongly anisophyllous with one of a pair often completely absent, more rarely up to $^1/_2$ the size of the other (very rarely both leaves subequal); petiole (of largest leaf) 0.3–3 cm long; lamina (of largest leaf) ovate to elliptic or narrowly so, largest (2.5–)3.5–14 × (1.2–)1.5–7.5 cm, base cuneate to rounded, equal to slightly (rarely distinctly) unequal-sided, margin crenate, apex acute to subacuminate with obtuse tip; puberulous or sparsely so on midrib. Flowers in lax few-flowered axillary dichasia towards end of branches, often 2 per node, not forming loose terminal panicles; peduncles, axes and pedicels finely sericeous-puberulous or sparsely so, with conspicuous cystoliths; peduncles (0–)1–4 mm long; branches 0.2–1.5 cm long (longer than peduncle); secondary bracts and bracteoles linear, ± 0.5 mm long. Calyx 2–4(–5) mm long, subglabrous to sparsely sericeous or sericeous-puberulous, lobes linear-lanceolate, acuminate to cuspidate. Corolla white with pink to purple markings on lower lip, tube sometimes tinged green, 7–8.5 mm long, glabrous to sparsely puberulous; tube 3.5–4.5 mm long and 1–2 mm diameter, slightly pouched ventrally; upper lip 3.5–4 mm long, slightly hooded; lower lip deflexed or spreading, 3.5–4 mm long, deeply 3-lobed. Filaments 1–2 mm long; thecae ± 1 mm long, oblong, parallel, ± 50% overlapping, glabrous, lower with minute acute appendage ± 0.3 mm long. Capsule 9–12 mm long, not breaking irregularly, densely puberulous; seed circular in outline, ± 2 mm diameter, densely tuberculate.

KENYA. Kilifi District: Cha Simba, 15 Sept. 1974, *B. R. Adams* 87! & Dzitzoni to Jaribuni, 30 Apr. 1989, *Luke* 1828!; Kwale District: Pengo Hill, 19 Feb. 1968, *Magogo & Glover* 136!
TANZANIA. Tanga District: Amboni Caves, 15 May 1975, *Hepper & Field* 5522!; Lushoto District: 1.5 km on Kisiwani–Muheza road, 29 March 1974, *Faden* 74/338!; Morogoro District: Kanga Mtn, 3 Dec. 1987, *Mwasumbi & Munyenyembe* 13880!

DISTR. **K** 7; **T** 3, 6; not known elsewhere
HAB. Evergreen or semi-deciduous lowland forest, coral rag thicket; near sea level to 450(–850) m

11. **Justicia rodgersii** *Vollesen* **sp. nov.** a *J. inaequifolia* calyce maiore 6.5–9 mm (nec 2–4(–5) mm) longo, corolla maiore 9–10 mm (nec 7–8.5 mm) longa, capsula maiore 13–14 mm (nec 9–12 mm) longa et semine maiore ± 3 mm (nec ± 2 mm) longo differt et in sylvis montium, nec locorum planorum habitat. Typus: Tanzania, Iringa District, Udzungwa Mts National Park, Mt Luhomero, *Luke et al.* 6835 (EA!, holo.; K!, iso.)

Shrubby herb or shrub to 2 m tall; young stems bifariously sericeous-puberulous, cystoliths not visible; older branches with longitudinally ribbed glossy brown bark; plant drying black. Leaves with conspicuous cystoliths, slightly to distinctly anisophyllous but never with one of a pair completely absent; petiole 1–5 cm long; lamina ovate, largest 9–11 × 4–6 cm, base cuneate to attenuate, equal to slightly unequal-sided, margin crenate, apex subacuminate with acute tip; sericeous-puberulous on midrib and veins and distinctly ciliate on margins. Flowers in few-flowered axillary dichasia, 1 per node; peduncles, axes and pedicels puberulous, with conspicuous cystoliths; peduncles 1–4 mm long; branches 3–7 mm long (longer than peduncle); secondary bracts and bracteoles linear, 2–4 mm long. Calyx 6.5–9 mm long, puberulous, lobes linear-lanceolate, cuspidate. Corolla white with reddish streaks on lower lip, 9–10 mm long, glabrous; tube 4.5–5 mm long and ± 2 mm diameter, slightly pouched ventrally; upper lip 4.5–5 mm long, hooded; lower lip deflexed, ± 5 mm long, deeply 3-lobed. Filaments ± 3 mm long; thecae ± 1 mm long, oblong, parallel, ± 50% overlapping, glabrous, lower with minute acute appendage ± 0.3 mm long. Capsule 13–14 mm long, sometimes breaking irregularly, densely puberulous; seed circular in outline, ± 3 mm diameter, tuberculate.

TANZANIA. Iringa District: Udzungwa Mts, West Kilombero Forest Reserve, W of Ruipa River, Oct. 1982, *Rodgers & Hall* 2256! & Mt Luhomero, 2 Oct. 2000, *Luke et al.* 6835! & Ndundulu Forest Reserve, 6 Sept. 2004, *Luke et al.* 10358!
DISTR. **T** 7; Malawi
HAB. Evergreen montane forest; 1450–1800 m

SYN. *Justicia sp. 1 sensu* White *et al.*, For. Fl. Malawi: 116 (2001)

NOTE. Known from four collections from the Udzungwa Mts and one from N Malawi. Differs from *Justicia inaequifolia* in the larger calyx, larger corolla and larger capsule and seed. It is also a montane forest species.

Named in honour of Dr Alan Rodgers, mentor from my formative years in Tanzania in the mid 1970s who sadly died in March 2009, still in his prime. Without his teasing help and encouragement I would probably never have formed such a close relationship with this beautiful country.

12. **Justicia oblongifolia** (*Lindau*) *M.E. Steiner* in K.B. 44: 709 (1989); Iversen in Symb. Bot. Upsal. 29(3): 161 (1991). Type: Tanzania, Lushoto District, E Usambara Mts, Amani, *Warnecke* 226 (B†, holo.; EA!, iso.)

Subacaulescent perennial herb with creeping unbranched rhizome; stems 1–4 cm long, crisped-puberulous; plant not drying black. Leaves in a pseudo-rosette, not anisophyllous; petiole 3–11 cm long; lamina oblong to slightly obovate, largest 6–15 × 2.5–7 cm, base cordate, one lobe often larger than the other and clasping the petiole, apex broadly rounded to retuse; crisped-puberulous or sparsely so, densest on midrib, veins and margins. Flowers in axillary panicles composed of dichasia, 3–6 cm long; peduncles, axes and pedicels puberulous or densely so and with scattered to dense (densest upwards) stalked capitate glands; peduncles 3–16 cm long, longer than panicle; branches to 2 cm long; secondary bracts and bracteoles lanceolate to ovate, 1–2 mm long. Calyx 3–4 mm long, puberulous and with stalked capitate glands, lobes lanceolate, acute. Corolla white, upper lip red towards base,

8–9 mm long, with band of hairs ventrally on tube, otherwise glabrous; tube 4–5 mm long and 1.5–2.5 mm diameter; upper lip 4–5 mm long, slightly hooded; lower lip deflexed, 7–8 mm long, deeply 3-lobed, middle lobe distinctly longer. Filaments ± 1 mm long; thecae ± 1.5 mm long, oblong, parallel, ± 75% overlapping, sparsely hairy, lower with flat triangular appendage ± 0.3 mm long. Capsule ± 14 mm long (fide Lindau), hairy; seed ± 2 mm diameter (fide Lindau), "warted".

TANZANIA. Lushoto District: E Usambara Mts, Amani West Forest Reserve, Feb. 1981, *Beeson* 273! & Mtai Forest Reserve, 4 Aug. 1986, *J. Lovett & Hamilton* 875! & 14 Nov. 1986, *Borhidi et al.* 86741!

DISTR. **T** 3; not known elsewhere

HAB. Wet evergreen intermediate altitude forest with *Cephalosphaera, Allanblackia, Newtonia, Erythrophleum, Synsepalum*, etc.; 800–900 m

SYN. *Rhaphidospora oblongifolia* Lindau in E.J. 38: 71 (1905)

NOTE. Apart from the type known only from a handful of collections. The paucity of material of such a conspicuous plant from a relatively well collected area raises the idea that this might be a species with periodic mass flowering.

13. **Justicia callopsoidea** *Vollesen* **sp. nov.** a *J. oblongifolia* foliis sericeis (nec puberulis), lamina ovato-cordiformis (nec oblonga nec aliquantum obovata) ad apicem triangulari (nec rotundata neque retusa), labio inferiore corollae longiore 7–8 mm (nec ± 4 mm) longo et capsula glabra differt. Typus: Tanzania, Morogoro District, Uluguru Mts, Mkungwe Forest Reserve, E of Kikundu, *Faden* in *Kabuye* 287 (K!, holo.; EA!, iso.)

Subacaulescent perennial herb with creeping unbranched rhizome; stems 3–4 cm long, puberulous; plant drying black. Leaves in a pseudo-rosette, not anisophyllous; petiole 3–8.5 cm long; lamina ovate-cordiform or narrowly so, largest 8–14 × 6–7.5 cm, base cordate, one lobe larger than the other and clasping the petiole, apex triangular; sericeous on midrib and veins beneath. Flowers in axillary panicles composed of dichasia, 2–6 cm long; peduncles, axes and pedicels puberulous and with scattered stalked capitate glands (densest upwards); peduncles 4–9 cm long, longer than panicle; branches to 2 cm long; secondary bracts and bracteoles lanceolate, 1–2 mm long. Calyx 3–4 mm long, puberulous and with stalked capitate glands, lobes lanceolate, acute. Corolla white, ± 8 mm long, with scattered hairs towards apex of lobes; tube ± 4 mm long and ± 2.5 mm diameter apically; upper lip ± 4 mm long, slightly hooded; lower lip flat, ± 4 mm long, deeply 3-lobed, lobes ± same length, middle much wider. Filaments ± 1 mm long; thecae ± 1.5 mm long, parallel, ± 50% overlapping, glabrous, lower with linear appendage ± 0.3 mm long. Capsule ± 13 mm long, glabrous; seed not seen.

TANZANIA. Morogoro District: Uluguru Mts, Mkungwe Forest Reserve, E of Kikundu, 5 July 1970, *Faden* in *Kabuye* 287! & *Pócs et al.* 6218/K!

DISTR. **T** 6; not known elsewhere

HAB. Wet evergreen forest; 800–1100 m

NOTE. Known only from these two collections made on the same day in the same locality. The species is close to *J. oblongifolia*, and further collections may well show the two to be conspecific but at present I prefer to keep them separate. It differs in the indumentum and the differently shaped leaves.

14. **Justicia udzungwaensis** *Vollesen* **sp. nov.** a *J. beloperonoidi* ramulis antrorse sericeis (nec puberulis), foliis ad basin attenuatis (nec subcuneatis neque truncatis), pedunculis longioribus 3–8 cm (nec 0.5–3 cm) longis, lobis calycum anguste ovatis (nec lineari-lanceolatis) et corolla glabra (nec tenuiter puberula) differt. Type: Tanzania, Iringa District, Udzungwa Mts National Park, Mwanihana Forest Reserve, above Sanje Village, *J. Lovett et al.* 854 (K!, holo.; K!, MO, iso.)

Erect shrub to 2 m tall; young stems antrorsely sericeous or sericeous-puberulous with yellowish hairs. Leaves equal or slightly anisophyllous; petiole 0.8–2.5 cm long; lamina elliptic to slightly obovate, largest 8–14 × 2.5–5 cm, base attenuate, not decurrent, apex acute or subacute; sericeous or sericeous-puberulous on midrib. Flowers in 3–7-flowered axillary dichasia (rarely solitary); peduncles and branches antrorsely sericeous or sericeous-puberulous; peduncles 3–8 cm long; branches 1–2(–3.5) cm long; secondary bracts and bracteoles filiform to linear (rarely narrowly obovate), 3–7(–15) mm long. Calyx 6–11 mm long, sparsely and finely sericeous-puberulous on midrib and edges, lobes narrowly ovate (narrowed at base), acuminate. Corolla carmine red with white spots in throat, 2.4–3.1 cm long, glabrous; tube curved dorsally, 1.4–2 cm long and 5–7 mm diameter; upper lip 9–12 mm long, slightly hooded; lower lip spreading or deflexed, 9–15 mm long, deeply 3-lobed, lobes subequal. Filaments 8–10 mm long; thecae parallel, oblong, purple, ± 2 mm long, glabrous, ± 50% overlapping, lower with short appendage to 0.4 mm long, bent at right angle. Capsule and seed not seen.

TANZANIA. Iringa District: Mwanihana Forest Reserve, above Sanje Village, 10 Oct. 1984, *D.W. Thomas* 3824! & 17 June 1986, *J. Lovett et al.* 854! & Udzungwa Mts National Park [7°49'S 36°49'E], 24 Sept. 2001, *Luke et al.* 7772B!

DISTR. **T** 7; not known elsewhere

HAB. Wet evergreen montane forest, including bamboo forest; 1400–1900 m

NOTE. Known only from these three collection. This, the following and probably also *sp. B* form a group of bird-pollinated species restricted to the Uluguru and Udzungwa Mts characterised by dark glossy red flowers. In the same mountains similar bird pollinated species occur also in *Barleria*, *Isoglossa* and *Dicliptera*. It is odd that in the *Acanthaceae* this adaptation to hummingbird pollination has occurred in several genera in this area but not elsewhere in East Africa.

Justicia udzungwaensis differs from *J. beloperonoides* in being an erect shrub with antrorsely sericeous young branches, attenuate leafbase, longer peduncles, ovate calyx lobes and a glabrous corolla.

15. **Justicia beloperonoides** *Lindau* in E.J. 22: 127 (1895) & in E.J. 28: 485 (1900); C.B. Clarke in F.T.A. 5: 205 (1900); T.T.C.L.: 12 (1949). Type: Tanzania, Morogoro District, Uluguru Mts, Lukwangule, *Stuhlmann* 9141 (B†, holo.). Neotype: Tanzania, Morogoro District, Uluguru Mts, E of Mwere River, *Polhill & Wingfield* 4612 (K!, neo., selected here; DSM, iso.)

Scrambling or scandent shrubby herb or shrub to 2.5(–4) m tall; young stems minutely puberulous (rarely pubescent) and with (rarely without) long pilose hairs and/or stalked capitate glands. Leaves equal or slightly anisophyllous; petiole 0.3–1.5(–2.5) cm long; lamina ovate, largest 2–9 × 1.5–4.5 cm, base subcuneate to truncate, apex acute to subacuminate with obtuse tip; minutely puberulous and with or without long pilose hairs on midrib. Flowers usually 2 (rarely 4) on short branches from upper leaf axils; branches with indumentum as stems; peduncles to 1.5 cm long; branches 8 mm long; bracteoles filiform, ± 1 mm long. Calyx 7–10(–13 in fruit) mm long, minutely puberulous to puberulous, densest and longest on margins, occasionally with stalked capitate glands, lobes linear-lanceolate, acuminate. Corolla uniformly crimson red to dark purple (very rarely white with purple markings), (2–)2.5–3 cm long, finely puberulous and with scattered capitate glands; tube slightly curved, (1.3–)1.8–2.2 cm long and 4–6 mm diameter; upper lip 6–10 mm long, slightly hooded; lower lip deflexed, 8–14 mm long, deeply 3-lobed, lobes subequal. Filaments 8–10 mm long; thecae parallel, oblong, ± 1.5 mm long, glabrous or hairy, ± 33% overlapping, lower with short appendage to 0.3 mm long, bent at right angle. Capsule (very immature) ± 15 mm long, glabrous; seed not seen.

TANZANIA. Morogoro District: Uluguru Mts, Bunduki, Salaza Forest, 15 March 1953, *Drummond & Hemsley* 1611! & 15 Nov. 1967, *B. J. Harris* 1133! & Uluguru North Forest Reserve, 5 km NNW of Tegetero Mission, 13 Jan. 2001, *Jannerup & Mhoro* 127!

DISTR. **T** 6; not known elsewhere
HAB. Wet evergreen montane forest; 1500–2000(–2500) m

16. **Justicia sp. B**

Scandent shrub to 2.5 m tall; young stems sparsely and finely antrorsely sericeous, no long hairs, no capitate glands. Leaves slightly anisophyllous; petiole to 1.2 cm long; lamina ovate, largest ± 7 × 3.5 cm, base subcuneate to truncate, often unequal, apex subacuminate with subacute to obtuse tip; sparsely and finely antrorsely sericeous along midrib and larger veins, otherwise glabrous. Flowers in 1–3-flowered axillary dichasia; peduncles 0.5–3 cm long, glabrous; branches to 1.5 cm long, thickened above bracteoles; bracteoles linear, ± 2 mm long. Calyx 4–5 mm long, glabrous or sparsely minutely puberulous, lobes linear-lanceolate, acuminate. Corolla not seen. Capsule ± 2.5 cm long, glabrous; seed (immature) ± 3 × 2 mm, densely tuberculate.

TANZANIA. Iringa District, Udzungwa Mts National Park, Ndundulu Forest Reserve, 14 Sept. 2004, *Luke et al.* 10497!
DISTR. **T** 7; not known elsewhere
HAB. Montane forest; 1550 m

NOTE. Known only from this collection. It would seem to be closest to *J. beloperonoides* differing in stem indumentum, inflorescence and shorter calyx.

17. **Justicia salvioides** *Milne-Redh.* in K.B. 1936: 488 (1936); T.T.C.L.: 12 (1949); Richards & Morony, Check List Mbala: 233 (1969); Bjørnstad in Serengeti Res. Inst. Publ. 215: 26 (1976); Ruffo *et al.*, Cat. Lushoto Herb. Tanzania: 7 (1996). Types: Tanzania, Massaini, *Fischer* 506 (B†, syn.; K!, iso.) & Mpwapwa District, Mlali, *Stuhlmann* 207 (B†, syn.). Lectotype: Tanzania, Massaini (not traced), *Fischer* 506 (K!, lecto., selected here)

Much-branched erect shrubby herb or shrub to 2(?–3) m tall; young stems rounded with many longitudinal shallow pale ribs and green furrows, glabrous or with scattered (rarely dense) stalked capitate glands on uppermost node below inflorescence. Leaves often absent at time of flowering; petiole to 5(–8) mm long; lamina ovate to elliptic or broadly so, largest 3.5–8.5(–13) × 2–4(–6.5) cm, base attenuate, decurrent, apex acute to subacuminate with a blunt to acute tip; glabrous to puberulous with glossy white hairs. Flowers solitary (rarely in 3's) in racemiform cymes congested into large terminal panicles, 5–35 cm long; axes, branches and pedicels densely glandular-puberulous (sticky) and with a layer of fine non-glandular hairs; branches 2–10(–15) cm long; main bracts foliaceous towards base, soon caducous; secondary bracts ovate-elliptic or narrowly so, to 8 mm long; bracteoles linear-lanceolate, 3–6 mm long. Calyx 6–8(–10) mm long, with dense stalked capitate glands and with sparse to dense fine non-glandular hairs, lobes lanceolate, acute. Corolla white to pale mauve with purple markings on lower lip, tube pink on the outside, 2.5–3.4 cm long, puberulous with mixture of glands and hairs; tube 1–1.2 cm long and 4–6 mm diameter, gibbose ventrally; upper lip 1.4–2.3 cm long, bent backwards and strongly curved and hooded; lower lip deflexed, 1.2–2.3 cm long, deeply 3-lobed, middle lobe wider. Filaments held in upper lip; thecae 2–3 mm long,

FIG. 67. *JUSTICIA SALVIOIDES* — **1**, habit, leafless flowering plant; **2**, habit, leafy flowering plant; **3**, large leaf; **4**, detail of inflorescence indumentum; **5**, bract (right) and bracteole (left); **6**, calyx opened up; **7**, detail of calyx indumentum; **8**, corolla opened up; **9**, apical part of filament and anther; **10**, apical part of style and stigma; **11**, capsule; **12**, seed. 1 & 4–10 from *Bidgood et al.* 7481, 2 from *Congdon* 194, 3 from *Burtt* 2505, 11 & 12 from *Congdon* 499. Drawn by Margaret Tebbs.

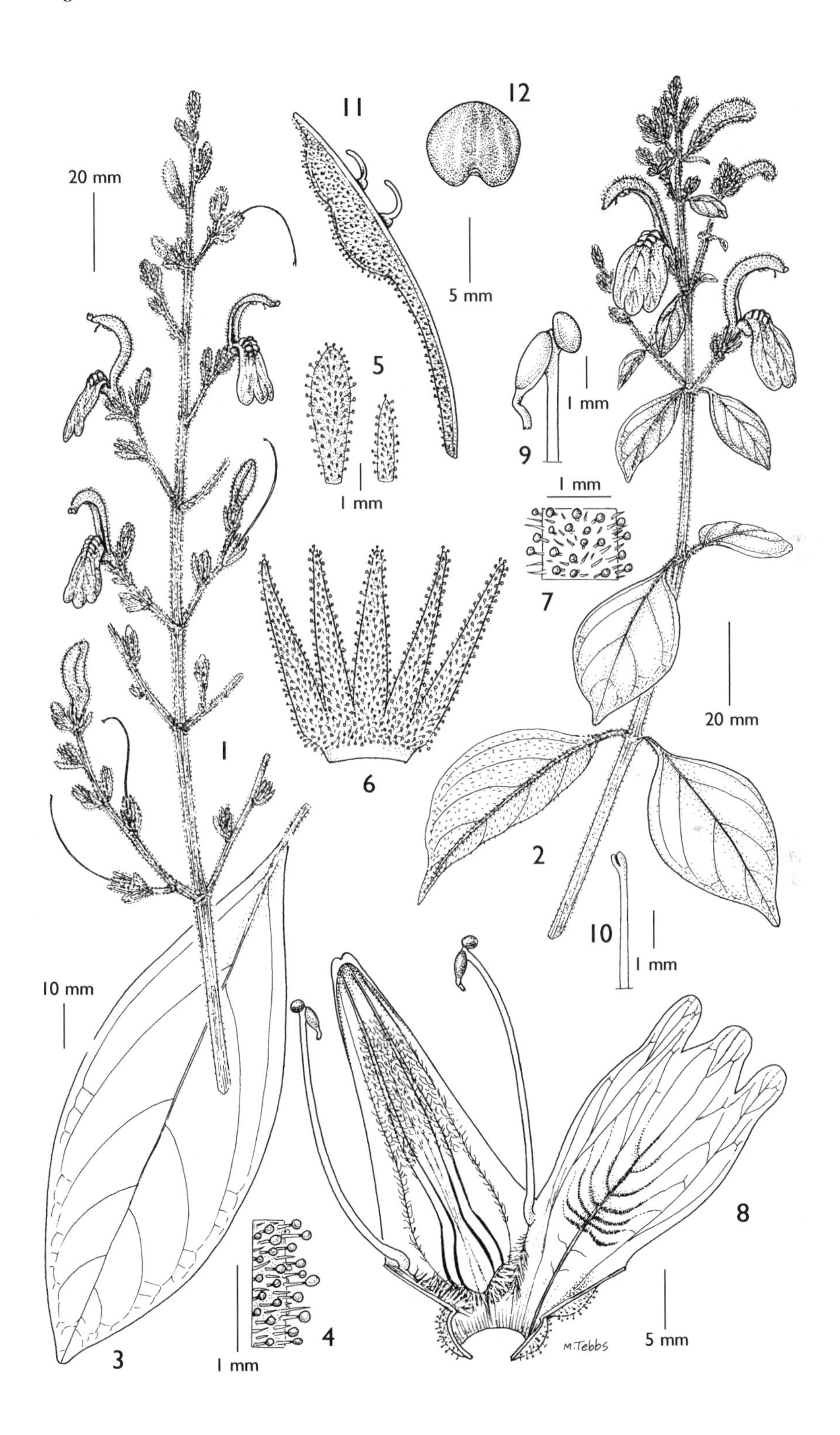
20 mm
11
12
5 mm
5
1 mm
1 mm
9
1 mm
7
6
1
20 mm
2
10
1 mm
10 mm
8
4
3
1 mm
M.Tebbs
5 mm

oblong, parallel or at an angle to each other on a broadened receptacle, with few to many long pilose hairs, lower with bifurcate appendage ± 1.5 mm long. Capsule 2.8–3.4 cm long, glandular-puberulous; seed ("Monechma" type) kidney-shaped, not compressed, ± 5 × 6 mm, smooth. Fig. 67, p. 523.

TANZANIA. Shinyanga District: Shinyanga, Kisumbi Hill, 28 May 1931, *B.D. Burtt* 2505!; Mpwapwa District: Rubeho Mts, Mt Mangaliza, 10 April 1988, *Congdon* 194!; Iringa District: Mbagi, 25 Nov. 1970, *Greenway & Kanuri* 14677!

DISTR. **T** 1, 2, 4, 5, 7; Congo-Kinshasa, Zambia

HAB. Itigi thicket, *Acacia* and *Acacia-Commiphora* bushland, on grey clayey soils; rocky hills with *Grewia* thicket; 750–1500 m

SYN. *Duvernoia salviiflora* Lindau in E.J. 20: 42 (1894) & in P.O.A. C: 372 (1895). Type: as *J. salvioides*

Justicia salviiflora (Lindau) C.B. Clarke in F.T.A. 5: 205 (1900), *non* Kunth (1817)

NOTE. *Justicia salvioides* and *J. rendlei* have typical Sect. *Monechma* seeds, i.e. smooth and kidney-shaped. But they also have a 4-seeded capsule and are obviously not closely related to the species typically associated with this section.

18. **Justicia rendlei** *C.B. Clarke* in F.T.A. 5: 204 (1900); Glover, Check List Brit. and Ital. Somal.: 67 (1947); Ensermu in Fl. Eth. 5: 462 (2006). Type: Ethiopia, Sheik Hussein, *Donaldson Smith* s.n. (BM!, holo.)

Shrub to 2 m tall; young stems sparsely antrorsely sericeous, without stalked capitate glands. Leaves present at time of flowering, subsessile; lamina elliptic or broadly so to broadly obovate, largest ± 9 × 4.5 cm, base attenuate, decurrent, apex broadly rounded; glabrous or sparsely sericeous on midrib. Flowers solitary in short terminal racemiform cymes or also with lateral cymes from upper axils, 1–3(–5) cm long; axes, branches and pedicels with dense subsessile capitate glands and scattered non-glandular hairs; peduncle to 3 mm long; main bracts sometimes foliaceous towards base; secondary bracts ovate-elliptic, to 6 mm long; bracteoles narrowly ovate, ± 5 mm long. Calyx 8–9 mm long, with dense subsessile capitate glands and a few non-glandular hairs, lobes lanceolate, cuspidate. Corolla white with (? always) red to pink markings on lower lip, 2.5–2.8 cm long, puberulous with mixture of short glands and hairs; tube 1.1–1.3 cm long and 2–3 mm diameter, curved dorsally and bent backwards in apical part; upper lip 1.2–1.5 cm long, bent backwards and strongly curved and hooded; lower lip deflexed, 1.1–1.3 cm long, shallowly 3-lobed, middle lobe wider. Filaments held in upper lip; thecae oblong, 2–2.5 mm long, parallel at first, at an angle to each other later, on a broadened receptacle, glabrous, lower with linear appendage ± 3 mm long, upper apiculate. Capsule ± 2 cm long, densely puberulous, no glands; seed ("Monechma" type) kidney-shaped, not compressed, ± 5 × 4 mm, smooth.

KENYA. Northern Frontier District: Dawa River, Murri, 30 June 1951, *Kirrika* 113! & Abakaro, 19 May 1979, *Gulichia* 35!

DISTR. **K** 1; Ethiopia, Somalia

HAB. *Acacia-Commiphora* bushland on rocky limestone hills or limestone pavements; 600 m

SYN. *Duvernoia speciosa* Rendle in J.B. 34: 129, 411 (1896), *non Justicia speciosa* Roxburgh (1820) *nec Justicia speciosa* Lindau (1895). Type: as for *J. rendlei*

[*Justicia grisea sensu* Hedrén in Fl. Somalia 3: 409 (2006) quoad distrib. Kenya, *non* C.B. Clarke (1900)]

19. **Justicia cordata** (*Nees*) *T. Anderson* in J.L.S. Bot. 7: 44 (1863); Engler, Hochgebirgsfl. Trop. Afr.: 393 (1892); C.B. Clarke in F.T.A. 5: 206 (1900); T.T.C.L.: 12 (1949); E.P.A.: 968 (1964); U.K.W.F.: 604 (1974); U.K.W.F., ed. 2: 280, pl. 123 (1994); Vollesen *et al.*, Checklist Mkomazi: 82 (1999); Hedrén in Fl. Somalia 3: 417 (2006); Ensermu in Fl. Eth. 5: 462 (2006). Type: Ethiopia, Tacazze Valley, Selassaquilla, *Schimper* II.1250 (K!, holo.; BM!, iso.)

Much-branched erect or scrambling shrub to 1.5 m tall, often forming untidy bushes; young stems conspicuously longitudinally ridged, glabrous but for a few hairs at nodes (rarely puberulous or pubescent). Leaves glaucous; petiole 0.5–1.5 mm long; lamina ovate-oblong to (normally) oblong or slightly obovate, largest 2–5(–6.5) × 1–2.5(–3) cm, base cordate or subcordate, apex subacute to broadly rounded or retuse; sometimes with a few ciliae near base and on petiole, otherwise glabrous. Flowers in 1–5-flowered axillary dichasia, often only a single flower with non-developing extra buds; peduncle 0.5–5 cm long, glabrous; bracts and bracteoles triangular or narrowly so, 2–3 mm long, with white scarious margins; pedicels strongly ridged, 1–6 mm long, glabrous, sometimes bending backwards to make flower resupinate. Calyx 4–6 mm long, with ciliate lobes, sometimes sparsely hairy on the inside, lobes lanceolate to triangular, with strongly raised midrib and conspicuous scarious white margins, acute. Corolla erect or held upside down, white to pale yellow or yellowish green, with distinct pink to mauve lines and patches on both lips, 13–19 mm long, puberulous or densely so; tube 5–9 mm long and 3–4 mm diameter, slightly gibbose ventrally; upper lip 8–12 mm long, strongly hooded; lower lip deflexed, 7–12 mm long, deeply 3-lobed, lobes elliptic-oblong. Filaments held in upper lip; thecae oblong, parallel, 2–2.5 mm long, hairy on edges, lower with rounded appendage 0.5–1 mm long. Capsule 18–29 mm long, glabrous; seed not compressed, ovoid-cordiform, 3–4 × ± 3 mm, rugose. Fig. 68, p. 526.

UGANDA. Karamoja District: Pirri, July 1930, *Liebenberg* 185! & Loyaro, Sept. 1943, *Dale* 362! & Aug. 1960, *Wilson* 1040!

KENYA. Naivasha District: Gilgil, Sept. 1933, *Napier* 2841!; Machakos District: Kibwezi, 2 June 1969, *Bally* 13313!; Masai District: Chyulu Plains, Oleeilelu Hill, 9 June 1991, *Luke* 2858!

TANZANIA. Musoma District: 5 km on Seronera–Banagi track, 3 April 1961, *Greenway* 9954!; Mbulu District: Lake Manyara National Park, Neale River, 22 Nov. 1966, *Richards* 21615!; Pare District: Mkomazi Game Reserve, Mbulu area, 27 April 1995, *Abdallah & Vollesen* 95/18!

DISTR. **U** 1; **K** 1, 3, 4, 6, 7; **T** 1–3; Eritrea, Ethiopia, Somalia

HAB. *Acacia*, *Acacia-Commiphora*, *Acacia-Combretum* bushland on seasonally waterlogged alkaline clay, on rocky hills and escarpments, lava flows; 750–2100 m

SYN. *J. cynanchifolia* R. Br. in Salt, Voy. Abyss., Append.: 62 (1814), *nom. nud.*

Leptostachya cordata Nees in DC., Prodr. 11: 378 (1847)

Rhaphidospora cordata (Nees) Nees in DC., Prodr. 11: 499 (1847); A. Richard, Tent. Fl. Abyss. 2: 161 (1850); Solms Laubert in Schweinfurth, Beitr. Fl. Aeth.: 113, 244 (1863); Lindau in P.O.A. C: 370 (1895)

Justicia masaiensis C.B. Clarke in F.T.A. 5: 207 (1900). Type: Kenya, *Scott Elliot* 6634 (BM!, holo.)

J. cordata (Nees) T. Anderson var. *pubescens* S. Moore in J.B. 40: 345 (1902). Type: Kenya, Machakos District: Sultan Hamud, *Kässner* 654 (BM!, holo.; K!, iso.)

20. **Justicia interrupta** (*Lindau*) *C.B. Clarke* in F.T.A. 5: 207 (1900); T.T.C.L.: 12 (1949); Vollesen in Opera Bot. 59: 81 (1980); Iversen in Symb. Bot. Upsal. 29(3): 161 (1991); Ruffo *et al.*, Cat. Lushoto Herb. Tanzania: 6 (1996). Type: Tanzania, Morogoro District, Uluguru Mts, Ukami, *Stuhlmann* 8967 (B†, holo.). Neotype: Tanzania, Pangani District, Potwe Forest, *Semsei* 3138 (K!, neo., selected here; EA!, iso.)

Shrub, usually sparsely branched, to 2(–3) m tall; young stems glabrous to sparsely antrorsely sericeous-puberulous. Leaves with petiole to 4.5(–6) cm long; lamina ovate to elliptic or broadly so, largest (5.5–)10–26(–31) × (2.5–)5.5–14(–17) cm, base attenuate, decurrent, apex subacute to acuminate with a rounded tip; glabrous or with a few hairs on midrib. Flowers solitary or in 2–3(–5)-flowered cymules in racemiform cymes forming a large lax terminal panicle to 40 cm long with lateral branches to 35 cm long; axes and branches puberulous or sericeous-puberulous; peduncle not clearly defined; main bracts foliaceous towards base, decreasing upwards; secondary bracts and bracteoles linear-lanceolate to narrowly triangular, 2–5 mm long; pedicels 2–3(–5) mm long, puberulous and with subsessile capitate

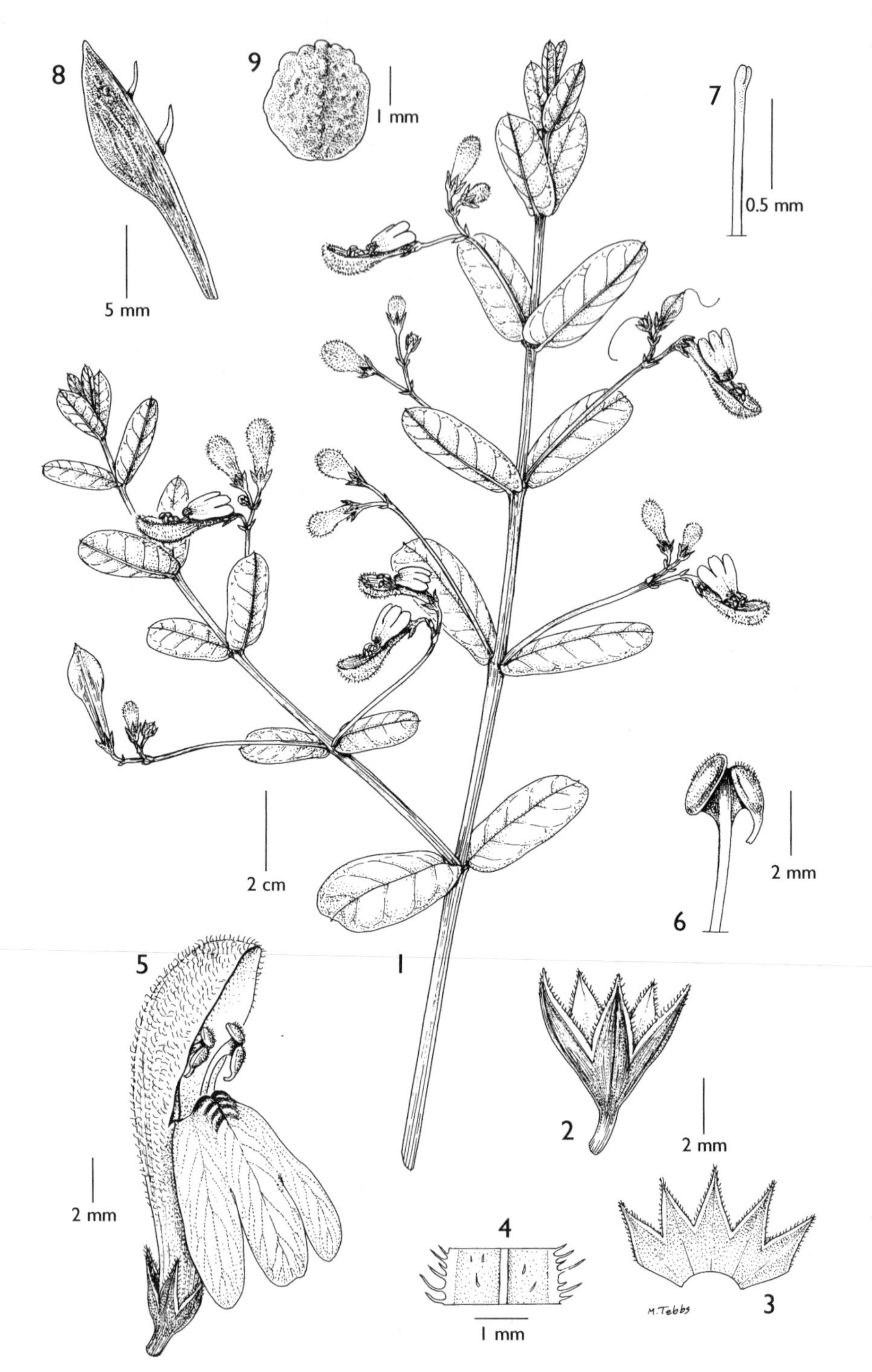

FIG. 68. *JUSTICIA CORDATA* — **1**, habit; **2**, calyx; **3**, calyx opened up; **4**, detail of calyx lobe; **5**, corolla; **6**, apical part of filament and anther; **7**, apical part of style and stigma; **8**, capsule; **9**, seed. 1 from *G. R. Williams* 664, 2–7 from *Karani* 128, 8–9 from *Friis et al.* 6793. Drawn by Margaret Tebbs.

glands, bending backwards to make flower resupinate. Calyx 3–6 mm long, puberulous or sparsely so and with sparse to dense short-stalked capitate glands, lobes lanceolate to narrowly ovate, acute to acuminate. Corolla held upside down, dull purplish brown tinged with green, 13–17 mm long, puberulous and with scattered capitate glands; tube 7–9 mm long and 3–4 mm diameter, slightly gibbose ventrally; upper lip 5–9 mm long, hooded; lower lip deflexed, 5–9 mm long, deeply 3-lobed, lobes lanceolate-oblong. Filaments held in upper lip; thecae 1.5–2 mm long, oblong, parallel, glabrous, lower with broadly ellipsoid rounded appendage ± 0.5 mm long. Capsule pendulous, 17–27 mm long, puberulous and with short capitate glands; seed kidney-shaped, 3.5–4 × 3.5–4 mm, densely tuberculate.

TANZANIA. Korogwe District: Potwe Forest, 29 Dec. 1960, *Semsei* 3138!; Ulanga District: Mahenge Plateau, Mahenge Scarp Forest Reserve, 13 Oct. 1987, *Pócs* 87206/C!; Iringa District: Udzungwa Mts National Park, Sanje Falls, 23 July 1983, *Polhill & Lovett* 5124!

DISTR. **T** 3, 6, 7; not known elsewhere

HAB. Evergreen and semi-evergreen lowland forest and riverine forest; 200–950 m

SYN. *Duvernoia interrupta* Lindau in E.J. 22: 123 (1895)

NOTE. The peculiar "inverted" corolla, which is a result of the bending backwards of the pedicel, also occurs in some – but not all – plants of *J. cordata*, but is otherwise unique in *Justicia*. A similar "inverted" corolla also occurs in the Socotra endemic *Angkalanthus*.

21. **Justicia stachytarphetoides** (*Lindau*) *C.B. Clarke* in F.T.A. 5: 194 (1900); T.T.C.L.: 11 (1949); Vollesen in Opera Bot. 59: 81 (1980); Luke & Robertson, Kenya Coast. For. 2. Checklist Vasc. Pl.: 82 (1993); Vollesen *et al.*, Checklist Mkomazi: 83 (1999). Type: Tanzania, Uzaramo District: Madessa, *Stuhlmann* 8121 (B†, holo.; K!, iso.)

Erect (rarely basally decumbent) shrubby herb or shrub to 1.5 m tall; young branches subglabrous to densely puberulous. Leaves with petiole to 2.5 cm long; lamina ovate to elliptic, largest 4.5–15 × 2.2–6.5 cm, base cuneate to attenuate, apex subacute to shortly acuminate; subglabrous to puberulous on midrib (rarely also on lamina). Flowers all solitary or in 3(–7)-flowered sessile dichasia aggregated into terminal racemiform cymes 2.5–15(–50) cm long, sometimes with lateral cymes from lower nodes thus forming a loose panicle to 25 × 25(–50 × 50) cm; peduncle to 4 cm long; peduncle and rachis sparsely to densely puberulous; bracts and bracteoles lanceolate to elliptic, 4–6(–8) mm long, ciliate-puberulous to densely uniformly puberulous. Calyx 3.5–5 mm long, ciliate-puberulous to uniformly puberulous, lobes lanceolate, acute to acuminate. Corolla greenish white to salmon pink or greyish purple, lower lip white to pale yellow with pink to purple markings, 8–11 mm long, densely puberulous; tube 4–6 mm long and 1–1.5 mm diameter; upper lip 4–5 mm long, strongly hooded; lower lip deflexed, 3–5 mm long, deeply 3-lobed into one broad triangular middle lobe and two strap-shaped recoiled and twisted lateral lobes. Filaments held in upper lip; thecae ± 1 mm long, oblong, at an angle to each other, upper hairy, lower glabrous, lower with curved appendage ± 0.5 mm long. Capsule 14–18 mm long, with longitudinal and transverse ridges, puberulous or densely so; seed reniform, not compressed, ± 2 × 2.5 mm long, reticulately sculptured with central ridge.

KENYA. Lamu District: Mararani, 5 April 1980, *M.G. Gilbert et al.* 5909!; Tana River District: Bfunbe, 4 Aug. 1988, *Robertson & Luke* 5337!; Kilifi District: 6 km S of Kilifi, Takaungu, 23 Dec. 1973, *B.R. Adams* 78!

TANZANIA. Lushoto District: Mkomazi Game Reserve, Umba River, 12 June 1996, *Abdallah et al.* 96/195!; Rufiji District: Selous Game Reserve, Kipalala Hot Springs, 6 Aug. 1993, *Luke* 3723!; Lindi District: 10 km S of Mbemkuru River, 6 Dec. 1955, *Milne-Redhead & Taylor* 7573!

DISTR. **K** 7; **T** 3, 6, 8; Somalia, Mozambique

HAB. Evergreen and semi-deciduous lowland forest, riverine forest and thicket, swamp forest; 5–400 m

SYN. *Duvernoia stachytarphetoides* Lindau in P.O.A. C: 372 (1895) & in E. & P. Pf. IV, 3b: 339 (1895)

NOTE. *Justicia stachytarphetoides* has an inflorescence superficially similar to Sect. *Tyloglossa*, but it differs from this Section in having a corolla with a strongly hooded upper lip.

22. **Justicia francoiseana** *Brummitt* in K.B. 40: 792 (1985); White *et al.*, For. Fl. Malawi: 116 (2001); Mapura & Timberlake, Checklist Zimb. Vasc. Pl.: 14 (2004). Type: Uganda, Toro/Mubende District, Lake Albert, Mizizi River, *Bagshawe* 1332 (BM!, holo.; BM!, iso.)

Shrub or small tree to 5 m tall (rarely scandent); young branches longitudinally striate, glabrous to finely sericeous. Leaves with petiole to 3 cm long; lamina ovate or narrowly so, largest 9–14(–18) × 3–5.8(–9.3) cm, base attenuate, decurrent, apex acuminate or subacuminate; glabrous. Flowers in short axillary and/or terminal subsessile racemiform cymes 1–3(–4.5) cm long; peduncle and rachis finely sericeous-puberulous; bracts and bracteoles triangular or narrowly so, 2–3 mm long, acute to acuminate; pedicels to 1 mm long. Calyx 3–4.5 mm long, puberulous, without capitate glands, lobes triangular, acute. Corolla dull brownish or greenish purple, 10–12 mm long, puberulous; tube 6–7 mm long and 2–3 mm diameter, not gibbose; upper lip 4–6 mm long, strongly hooded; lower lip deflexed, 5–6 mm long, deeply 3-lobed, lobes ovate-elliptic. Filaments held in upper lip; thecae ± 1 mm long, oblong, parallel, glabrous, lower with small appendage ± 0.25 mm long. Capsule 19–27 mm long, finely puberulous; seed reniform, not compressed, ± 4 mm long, verrucose or reticulately sculptured.

UGANDA. Toro/Mubende District: Lake Albert, mouth of Mizizi River, 4 Dec. 1906, *Bagshawe* 1332!
DISTR. **U** 2/4; Malawi, Mozambique, Zimbabwe
HAB. Lowland evergreen forest; 650 m

SYN. *Adhatoda bagshawei* S. Moore in J.B. 45: 333 (1907), *non Justicia bagshawei* S. Moore (1910). Type: as for *Justicia francoiseana*
Justicia bagshawei (S. Moore) Eyles in Trans. Roy. Soc. S. Afr. 5: 486 (1916), *non* S. Moore (1910)

NOTE. This has never been re-collected in Uganda. Despite the large disjunction there are no obvious morphological differences between the type and the material from southern Africa.

23. **Justicia breviracemosa** *Vollesen* **sp. nov.** a *J. interrupta* cymis racemiformibus 5–9 cm longis axillaribus vel terminalibus (nec cymis paniculatis grandibus usque 40 cm longis semper terminalibus) et corolla non resupinata differt. A *J. francoiseana* foliis maioribus 19–26 cm (nec 9–14(–18) cm) longa, corolla maiore ± 16 mm (nec 10–12 mm) longa et glandulis capitatis in axibus inflorescentiarum calyceque dispositis (nec omnino carentibus) differt. Typus: Tanzania, Rufiji District, near WWF Office [8°19'S 38°58'E], *Kibure* 83 (K!, holo.; MO, iso.)

Shrub to 2 m tall; young stems glabrous. Leaves with petiole to 7.5 cm long; lamina ovate to elliptic or broadly so, largest 19–26 × 8–13.5 cm, base attenuate, not decurrent, apex subacuminate with a rounded tip; glabrous. Flowers solitary or with 2 non-developing lateral buds, in axillary and/or terminal subsessile racemiform cymes 5–9 cm long; rachis glabrous to puberulous and with scattered subsessile capitate glands; bracts and bracteoles lanceolate to triangular, 2–3 mm long; pedicels erect, 2–3 mm long, puberulous or sparsely so and with subsessile capitate glands. Calyx 4–5 mm long, puberulous or sparsely so and with sparse short-stalked capitate glands, lobes narrowly ovate to triangular, acute. Corolla held erect, purplish green, ± 16 mm long, puberulous and with scattered capitate glands; tube ± 8 mm long and ± 3.5 mm diameter, not gibbose; upper lip ± 8 mm long, hooded; lower lip deflexed,

± 7 mm long, deeply 3-lobed, lobes lanceolate-triangular. Filaments held in upper lip; thecae oblong, ± 1.5 mm long, at a slight angle to each other, glabrous, lower with broadly ellipsoid rounded appendage ± 0.3 mm long. Immature capsule erect, 19–22 mm long, puberulous and with short capitate glands; seed not seen.

KENYA. Kilifi District: Vipingo, 13 Jan. 1995, *Luke* 4271!
TANZANIA. Rufiji District: near WWF Office [8°19'S 38°58'E], 19 Sept. 1997, *Kibure* 83!
DISTR. **K** 7; **T** 6; not known elsewhere
HAB. Evergreen lowland forest, riverine forest; 10–250 m

NOTE. Known only from these two collections. Differs from *Justicia interrupta* in the short mostly axillary racemoid cymes and the non-resupinate corolla, and from *J. francoiseana* in the larger leaves, larger corolla and presence of capitate glands on inflorescences axes and calyx.

24. **Justicia gendarussa** *Burm. f.*, Fl. Ind.: 10 (1768); Lindau in P.O.A. C: 373 (1895); C.B. Clarke in F.T.A. 5: 203 (1900); T.T.C.L.: 12 (1949); U.O.P.Z.: 318 (1949). Type: India and Indonesia, "Malabar, Amboina and Java", *Herb. Burmann* (? G, not seen)

Shrubby herb or shrub to 1.5(?–3) m tall; young branches purple, glabrous or with a thin band of hairs at nodes. Leaves with petiole to 7 mm long; lamina narrowly ovate or narrowly elliptic, largest 7.3–9.5 × 1.3–1.5 cm, base attenuate, not decurrent, apex obtuse; glabrous. Flowers in 2–3(–7)-flowered cymules at base and single upwards in terminal (or also from upper leaf axils) subsessile racemiform cymes to 12 cm long; rachis glabrous or sparsely puberulous upwards; lower bracts foliaceous, upper bracts and bracteoles linear to lanceolate, to 1 cm (bracts) or 4 mm (bracteoles) long, caducous; pedicels to 0.5 mm long. Calyx 4–7 mm long, sparsely sericeous-puberulous, lobes linear-lanceolate, cuspidate. Corolla white to purple, with darker markings on lower lip, 14–16 mm long, glabrous; tube 9–10 mm long and ± 2 mm diameter; upper lip 5–6 mm long, flat or slightly hooded; lower lip spreading, 7–8 mm long, deeply 3-lobed, lobes oblong-elliptic. Filaments held under upper lip; thecae oblong, ± 1 mm long, parallel, almost superposed, glabrous, lower with small appendage ± 0.25 mm long. Capsule ± 10 mm long, with placentae separating from outer fruit wall and rising elastically, minutely puberulous; immature seed densely reticulate-tuberculate.

UGANDA. Mengo District: Kansira, Singo, July 1949, *Sangster* 1043!; Mubende District: Masaka, 2 km SW of Sunga, 15 May 1972, *Lye* 6874!
KENYA. Malindi District: Jilore, Nov. 1943, *Dale* 2015!; Kilifi District: no locality, Oct. 1979, *Reitsma* 448!
TANZANIA. Tanga District: Tanga, Feb. 1899, *Volkens* 146!; Uzaramo District: 12 km WNW of Dar es Salaam, Kimaro, 16 Dec. 1974, *Wingfield* 2859!; Ulanga District: Msolwa, 7 Jan. 1999, *Mwangulango* 144!; Zanzibar, without locality, 11 Oct. 1889, *Stuhlmann* 698!
DISTR. **U** 4; **K** 7; **T** 3, 4, 6; **Z**; widespread (usually as a cultivated plant or escaped) in India and SE Asia
HAB. Grown as a hedge plant, sometimes persisting in abandoned gardens and fields; near sea level to 1200 m

NOTE. Many authors (e.g. Hansen in Nordic Journ. Bot. 9: 209 (1989) and collectors comment on the fact that ripe fruits are hardly ever seen. The species is invariably propagated by cuttings. It is widely cultivated and has been so for centuries. Its exact area of origin is unknown but is probably somewhere in India or SE Asia. It was almost certainly brought to East Africa with Arab traders many centuries ago.

25. **Justicia gesneriflora** *Rendle* in J.B. 34: 398, as "*gesnerifolia*", corr. p. 414 (1896); C.B. Clarke in F.T.A. 5: 203 (1900); Glover, Check-list Brit. and Ital. Somal.: 66 (1947); E.P.A.: 970 (1964); Hedrén in Symb. Bot. Upsal. 29(1): 118 (1989), all as "*gesnerifolia*"; Hedrén in Fl. Somalia 3: 410 (2006); Ensermu in Fl. Eth. 5: 463, Fig. 166.54 (2006). Type: Ethiopia/ Somalia, Shebelle River, *Donaldson Smith* s.n. (BM!, holo.)

Small compact shrublet to 30 cm tall, often forming cushions; young branches pale yellow, glabrous to sparsely bifariously puberulous or pubescent with curly hairs; older branches pale grey with papery bark peeling in strips. Leaves sessile; lamina obovate-spatulate, largest 2–3.5 × 0.8–1.5 cm, base attenuate, decurrent to stem, apex broadly rounded (with a small apiculus) to emarginate; cilate near base, otherwise glabrous. Flowers solitary (rarely 2 per axil), axillary; peduncle ± 1 mm long; bracts foliaceous, to 1.2(–1.6) cm long; bracteoles oblanceolate-spatulate, 0.8–1.2 cm long, puberulous. Calyx 6–9 mm long, puberulous or sparsely so with stubby non-capitate glandular hairs and with few to many long glossy curly pilose hairs, lobes linear-lanceolate, acuminate. Corolla white, without or with pale purple markings on lower lip, 12–15 mm long, puberulous; tube 7–8 mm long and 2–3 mm diameter; upper lip 5–8 mm long, slightly hooded, tapering upwards; lower lip spreading, 5–8 mm long, shortly 3-lobed, middle lobe wider but shorter than lateral. Filaments 3–4 mm long, held under upper lip; thecae oblong, parallel, ± 1 mm long, 75% overlapping, glabrous, lower with small appendage ± 0.3 mm long. Capsule 8–12 mm long, glabrous; seed circular in outline, 2–2.5 mm diameter, tuberculate.

KENYA. Northern Frontier District: Ramu–Banissa road, 28 km from turning to Banissa, 4 May 1978, *M.G. Gilbert & Thulin* 1405!

DISTR. **K** 1; Ethiopia, Somalia

HAB. Open *Acacia-Commiphora* bushland on grey loamy soil; 575 m

SYN. *J. dodonaeifolia* Chiov., Fl. Somala: 268, tab. 33, 2 (1929); Glover, Check-list Brit. and Ital. Somal.: 66 (1947); E.P.A.: 969 (1964); Kuchar, Pl. Somalia [CRDP Techn. Rep. 16]: 247 (1986). Type: Somalia, Nogal, *Stefanini & Puccioni* 875 (FT, holo.)

NOTE. With its solitary flowers this species would at first sight seem to belong to Sect. *Harnieria*. But the large bracts and bracteoles exclude it from this section and its fits more comfortably into a combined Sect. *Justicia* and Sect. *Rhaphidospora*.

26. **Justicia fittonioides** *S. Moore* in J.B. 16: 134 (1878); C.B. Clarke in F.T.A. 5: 189 (1899); Vollesen in Opera Bot. 59: 80 (1980); Luke & Robertson, Kenya Coast. For. 2. Checklist Vasc. Pl.: 82 (1993); Ruffo *et al.*, Cat. Lushoto Herb. Tanzania: 6 (1996). Type: Kenya, near Mombasa, *Wakefield* s.n. (K!, holo.; K!, iso.)

Acaulescent perennial herb with thick creeping rhizome or with a short puberulous stem to 2 cm long. Leaves in a rosette, often appressed to the ground; petiole 1–7.5 cm long; lamina narrowly oblong to obovate, largest 8–19 × 5–15 cm, base cuneate to cordate (rarely shortly attenuate), usually with one lobe larger than the other, margin often wavy towards base, apex broadly rounded (rarely retuse); puberulous or sericeous-puberulous on petiole, midrib and larger veins. Flowers in 3–15-flowered dichasia; cymes 1.5–8 cm long, sometimes with 2 short lateral cymes from lowermost node; peduncle 4–17 cm long; peduncle and rachis puberulous or retrorsely sericeous-puberulous; main axis bracts broadly elliptic to orbicular or reniform, 6–9 × 3.5–8.5 mm, subacute to broadly rounded and apiculate, puberulous or sparsely so, distinctly 5-veined from base; dichasia bracts and bracteoles lanceolate to elliptic or obovate, 4–6 mm long. Calyx 3–5 mm long, subglabrous to minutely puberulous, lobes linear-lanceolate, acute to cuspidate. Corolla white, with pink markings on lower lip, upper lip greenish outside, 7–11.5 mm long, glabrous to sparsely puberulous; tube 4–6 mm long; upper lip 4–7 mm long; lower lip spreading or deflexed, 4–6 mm long, deeply 3-lobed, middle lobe much wider than laterals. Filaments 2–4 mm long; thecae 1.5–2 mm long, oblong, ± 50% overlapping, upper hairy, lower glabrous, both thecae with acute appendage 0.5 (upper) to 1 (lower) mm long. Capsule 15–20 mm long, glabrous to sparsely and minutely puberulous; seed ovoid, ± 4 mm long, densely verrucose-reticulately sculptured and slightly crested apically. Fig. 69, p. 531.

KENYA. Kilifi District: Mangea Hill, 9 Jan. 1992, *Robertson* 6560!; Mombasa District: near Mombasa, 1877, *Wakefield* s.n.!

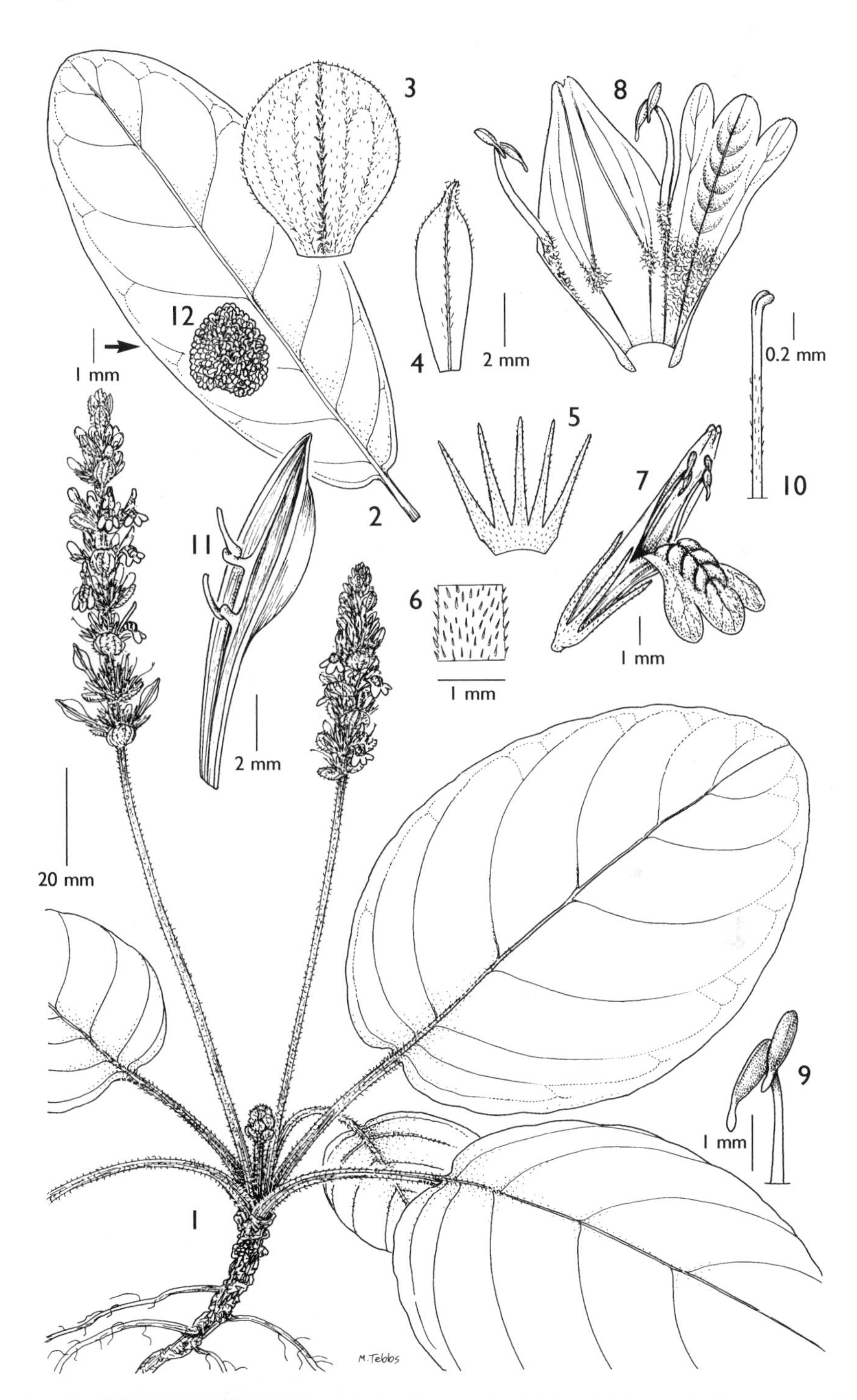

FIG. 69. *JUSTICIA FITTONIOIDES* — **1**, habit; **2**, narrow leaf; **3**, bract; **4**, bracteole; **5**, calyx opened up; **6**, detail of calyx lobe; **7**, corolla; **8**, corolla opened up; **9**, apical part of filament and anther; **10**, apical part of style and stigma; **11**, capsule; **12**, seed. 1 from *Bidgood et al.* 1373, 2 from *Peter* K894, 3–11 from *MacKinder* 40, 12 from *Luke* 4304. Drawn by Margaret Tebbs.

TANZANIA. Lushoto/Handeni District: Korogwe–Handeni road, 14 Jan. 1954, *Faulkner* 1321!; Morogoro District: Nguru Mts, Koruhamba, Nov. 1953, *Paulo* 212!; Lindi District: Rondo Plateau, Rondo Forest Reserve, 6 Feb. 1991, *Bidgood et al.* 1373!
DISTR. **K** 7; **T** 3, 6, 8; Mozambique
HAB. Dry deciduous and semi-evergreen lowland forest and thicket; 150–800(–900) m

SYN. *J. dolichopoda* Mildbr. in N.B.G.B. 12: 721 (1935). Type: Tanzania, Lindi District, Lake Lutamba, *Schlieben* 5847 (B†, holo.; BM!, K!, iso.)

27. **Justicia drummondii** *Vollesen* **sp. nov.** a *J. fittonioide* foliis angustioribus 1.5–3.5(–5) cm (nec 5–15 cm) latibus ad basin attenuatis decurrentibusque (nec cuneatis neque cordatis) atque in regione juxta costam conspicue pallidis, et bracteis axis principalis minoribus 4–5 × 2–3 mm (nec 6–9 × 3.5–8.5 mm) differt. Typus: Kenya, Kwale District, Lungalunga–Msambweni road, between Umba and Mwena Rivers, *Drummond & Hemsley* 3792 (K!, holo.; EA!, K! iso.)

Acaulescent perennial herb with thick creeping rhizome. Leaves in a rosette, appressed to the ground; petiole 1.5–8 cm long; lamina oblong-elliptic or broadly so to ovate-cordiform or reniform, largest 6.5–10(–12) × 1.5–3.5(–5) cm; central area on both sides of midrib conspicuously pale, base attenuate, decurrent, margin wavy towards base, apex broadly rounded to subacute; puberulous or or sparsely so on petiole, midrib and larger veins. Flowers in 3–7-flowered dichasia; cymes 1.5–3.5 cm long; peduncle 3–8 cm long; peduncle and rachis puberulous; main axis bracts ovate to elliptic or slightly obovate, 4–5 × 2–3 mm, acute, subglabrous to sparsely puberulous, distinctly 5-veined from base or not; dichasia bracts and bracteoles linear-lanceolate, 4–5 mm long. Calyx 4–5 mm long, subglabrous to sparsely puberulous, lobes lanceolate, acute to acuminate. Corolla white, with purple markings on lower lip and two purple lines dorsally in throat, 8–10 mm long, glabrous to very sparsely puberulous; tube 4–5 mm long; upper lip 4–5 mm long; lower lip deflexed, 3–4 mm long, deeply 3-lobed, middle lobe wider than laterals. Filaments ± 3 mm long; thecae oblong, ± 1.5 mm long, ± 50% overlapping, upper hairy, lower glabrous, both thecae with acute appendage 0.5 (upper) to 1 (lower) mm long. Immature capsule glabrous; seed not seen.

KENYA. Kwale District: Lungalunga–Msambweni road, between Umba and Mwena Rivers, 14 Aug. 1963, *Drummond & Hemsley* 3792! & Malunganji Forest Reserve, 14 Nov. 1989, *Robertson & Luke* 5985! & Shimba Hills, Mwalugange Forest Reserve, 1 June 1996, *Luke* 4508!
DISTR. **K** 7; not known elsewhere
HAB. Wet evergreen lowland forest; 75–200 m

SYN. *Justicia sp. aff. fittonioides* S. Moore *sensu* Luke & Robertson, Kenya Coast. For. 2. Checklist Vasc. Pl.: 82 (1993)

NOTE. Differs from *J. fittonioides* in the narrower leaves with attenuate and decurrent base and with a conspicuous pale area both sides of midrib. Also the bracts are smaller. The habitat also seems different with this species growing in wetter forest types than *J. fittonioides*.

28. **Justicia flava** (*Vahl*) *Vahl*, Symb. Bot. 2: 15 (1791); C.B. Clarke in F.T.A. 5: 190 (1899); Chiovenda, Fl. Somala: 268 (1929); Lugard in K.B. 1933: 94 (1934); F.P.S. 3: 179 (1956); Heine in F.W.T.A., ed. 2, 2: 427 (1963); E.P.A.: 969 (1964); Binns, Check List Herb. Fl. Malawi: 14 (1968); U.K.W.F.: 604 (1974); Champluvier in Fl. Rwanda 3: 470, fig. 144 (1985); Synnott, Checklist Fl. Budongo For.: 69 (1985); Blundell, Wild Fl. E. Afr.: 393, pl. 401 (1987); Luke & Robertson, Kenya Coast. For. 2. Checklist Vasc. Pl.: 82 (1993); U.K.W.F., ed. 2: 279, pl. 123 (1994); Ruffo *et al.*, Cat. Lushoto Herb. Tanzania: 6: (1996); Vollesen *et al.*, Checklist Mkomazi: 82 (1999); Friis & Vollesen in Biol. Skr. 51(2): 447 (2005); Hedrén in Fl. Somalia 3: 412 (2006); Ensermu in Fl. Eth. 5: 471 (2006); Fischer & Killmann, Pl. Nyungwe National Park Rwanda: 454 (2008). Type: Yemen, Hadie, *Forsskål* 394 (C!, lecto.).

Perennial herb with 1–several erect or scrambling stems from creeping often branched rhizome, sometimes rooting at lower nodes; stems to 1.5(–3 in scrambling plants) m long, subglabrous or sparsely to densely pubescent (rarely tomentose) to retrorsely sericeous. Leaves with petiole to 1.5(–2.5) cm long; lamina narrowly to broadly ovate to elliptic, largest 2–15(–19) × 1–6.5(–7.5) cm, base truncate to attenuate, apex acuminate to obtuse; sericeous to pubescent on midrib and larger veins or also on lamina. Flowers in 1–3-flowered dichasia; cymes 2–25(–35) cm long; peduncle to 9(–17) cm long; peduncle and rachis with indumentum as stems, occasionally with stalked capitate glands; main axis bracts lanceolate to ovate, 7–17 mm long, sparsely puberulous to pubescent and with scattered to dense stalked capitate glands, often conspicuously ciliate with hairs to 2 mm long, persistent; dichasia bracts and bracteoles linear-lanceolate to lanceolate, 7–13 mm long, puberulous and with capitate glands. Calyx 3–6(–8) mm long, sericeous-puberulous, at least on veins and with long pilose ciliae, without (rarely with) capitate glands, lobes narrowly triangular, acute to acuminate. Corolla yellow, lower lip with 3 and upper lip with 2 conspicuous reddish brown streaks (absent in S & C Tanzania), 7–16 mm long, puberulous; tube 4–8 mm long and 3–4 mm diameter; upper lip 3–8 mm long; lower lip spreading or deflexed, 5–12 mm long. Filaments 2–5 mm long, bending out of flower after anthesis; thecae oblong, 1–1.5 mm long, ± 33% overlapping, upper hairy, lower with linear appendage ± 1 mm long. Capsule 6–12 mm long, uniformly puberulous or densely so; seed black, ovoid, 1.5–2 mm long, densely tuberculate. Fig. 70: 1–10, p. 534.

UGANDA. Karamoja District: Nabilatuk, Emoruagaberru, 9 July 1956, *Dyson-Hudson* 20!; Bunyoro District: Budongo Forest, 25 Oct. 1971, *Synnott* 709!; Busoga District: Lake Victoria, Lolui Island, 22 May 1964, *Jackson* U143!

KENYA. Northern Frontier District: Mt Kulal, 20 Nov. 1978, *Hepper & Jaeger* 6928!; North Kavirondo District: Kakamega Forest, 17 April 1965, *Gillett* 16692!; Masai District: Namanga, 16 June 1974, *Faden & Ng'weno* 74/813!

TANZANIA. Musoma District: Klein's Camp, 23 May 1962, *Greenway* 10660!; Arusha District: Monduli, Selela Forest Reserve, 15 April 2000, *Kindeketa et al.* 341!; Njombe District: Livingstone Mts, Mt Masusa, 14 March 1991, *Gereau & Kayombo* 4256!

DISTR. **U** 1–4; **K** 1–7; **T** 1–8; widespread in tropical Africa, South Africa; Arabia

HAB. Forest margins and clearings and in a wide range of woodland, bushland and grassland from high rainfall to semi-desert conditions, often in secondary vegetation, weed; (25–)350–2500 m

SYN. *Dianthera flava* Vahl, Symb. Bot. 1: 5 (1790)

Tyloglossa palustris Hochst. in Flora 26: 72 (1843). Types: Sudan, Cordofan, Arasch-Cool, *Kotschy* 80 (K!, isosyn.); Ethiopia, Gafta, *Schimper* II.1211 (BM!, K!, isosyn.)

T. acuminata Hochst. in Flora 26: 73 (1843). Type: Ethiopia, without locality, *Schimper* s.n. (?TUB, holo.)

T. major Hochst. in Flora 26: 73 (1843). Type: Ethiopia, Tacazze River, *Schimper* II.1251 (BM!, K!, iso.)

T. minor Hochst. in Flora 26: 73 (1843). Type: Ethiopia, Modat, Mai Oui, *Schimper* II.1043 (?TUB, holo.)

Adhatoda major (Hochst.) Nees in DC., Prodr. 11: 397 (1847); A. Richard, Tent. Fl. Abyss. 2: 156 (1850)

A. acuminata (Hochst.) Nees in DC., Prodr. 11: 400 (1847); A. Richard, Tent. Fl. Abyss. 2: 157 (1850)

A. minor (Hochst.) Nees in DC., Prodr. 11: 400 (1847); A. Richard, Tent. Fl. Abyss. 2: 156 (1850); Solms Laubert in Schweinfurth, Beitr. Fl. Aeth.: 103, 242 (1867)

A. flava (Vahl) Nees in DC., Prodr. 11: 401 (1847)

A. palustris (Hochst.) Nees in DC., Prodr. 11: 402 (1847); A. Richard, Tent. Fl. Abyss. 2: 157 (1850); Solms Laubert in Schweinfurth, Beitr. Fl. Aeth.: 103 (1867)

Justicia major (Hochst.) T. Anderson in J.L.S. Bot. 7: 39 (1863); Lindau in P.O.A. C: 373 (1895) & in Ann. Ist. Bot. Roma 6: 83 (1896)

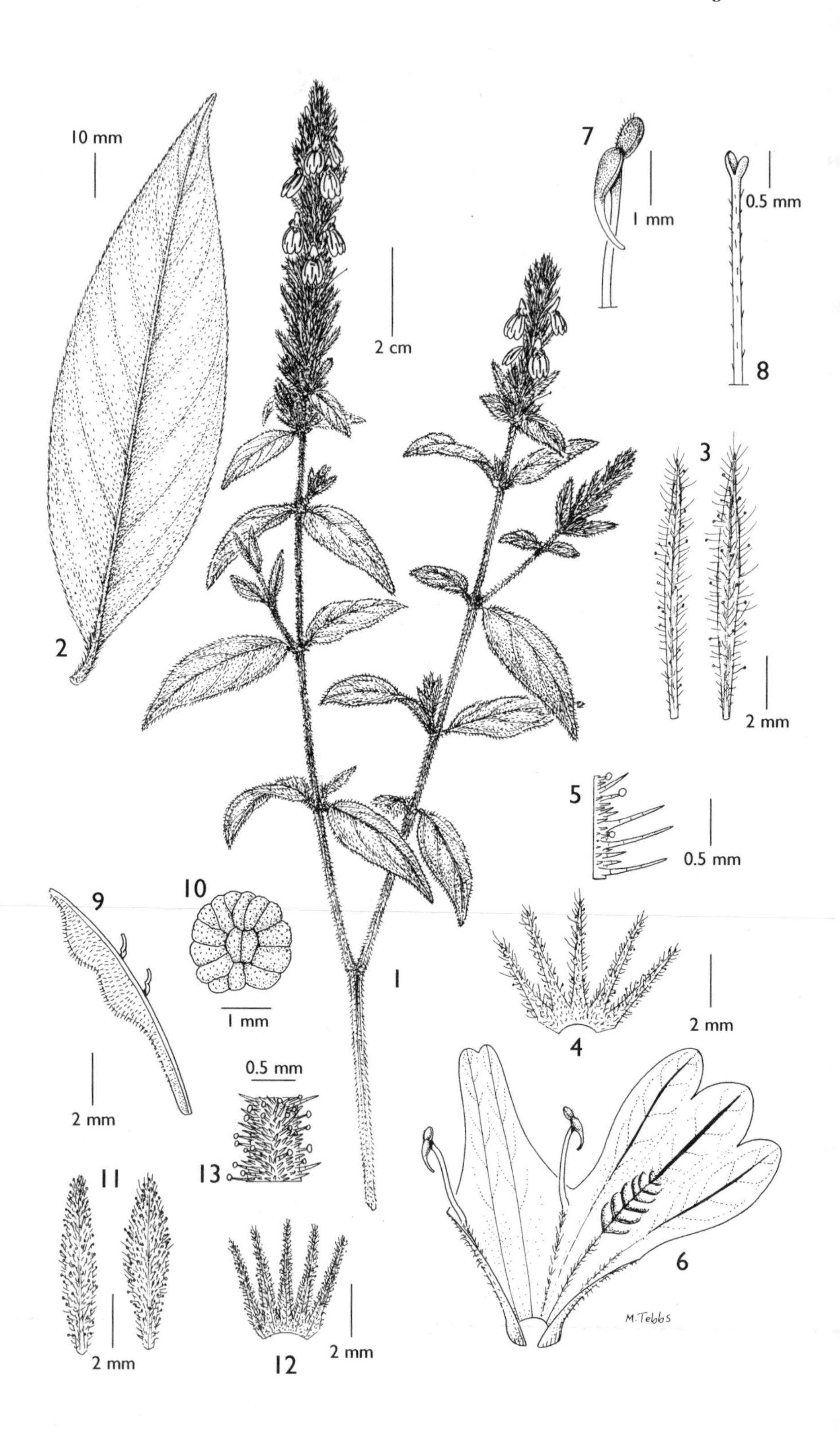
10 mm
2
7
1 mm
0.5 mm
8
2 cm
3
2 mm
5
0.5 mm
9
10
1 mm
1
2 mm
4
2 mm
0.5 mm
13
11
2 mm
12
2 mm
6
M.Tebbs

J. minor (Hochst.) T. Anderson in J.L.S. Bot. 7: 39 (1863); Engler, Hochgebirgsfl. Trop. Afr.: 392 (1892); Lindau in E. & P. Pf. IV, 3b: 349 (1895); Rendle in J.B. 34: 411 (1896)

J. palustris (Hochst.) T. Anderson in J.L.S. Bot. 7: 38 (1863); Oliver in Trans. Linn. Soc., ser. 2, Bot. 2: 345 (1887); Lindau in E. & P. Pf. IV, 3b: 349 (1895) & in P.O.A. C: 373 (1895) & in Ann. Ist. Bot. Roma 6: 82 (1896); C.B. Clarke in F.T.A. 5: 191 (1899); F.P.S. 3: 179 (1956); E.P.A.: 973 (1964); Ensermu in Fl. Eth. 5: 474 (2006)

J. palustris (Hochst.) T. Anderson var. *dispersa* Lindau in E.J. 20: 72 (1894). Types: Uganda, Manganji (not traced), *Stuhlmann* 1367 (not seen); Tanzania, Kilimanjaro, *Johnston* 59 (BM!, K!, isosyn.); Angola, Malandje, *Buchner* 46 (not seen)

J. fruticulosa Lindau in E.J. 20: 75 (1894). Type: Sudan, Erkowit, *Schweinfurth* 256 (B†, holo.; K!, iso.)

J. smithii S. Moore in J.B. 39: 304 (1901). Type: Somalia, Hamaro, *Donaldson Smith* s.n. (BM!, holo.)

J. nelsonioides Fiori in Bull. Soc. Bot. Ital. 1915: 57 (1915). Type: Somalia, Aca-aca, *Paoli* 511 (FT, holo.)

J. sp. aff. nyassana sensu Luke & Robertson, Kenya Coast. For. 2. Checklist Vasc. Pl.: 82 (1993)

NOTE. The material from C and S Tanzania, N & C Mozambique, Malawi and N & E Zambia differs in the absence of the conspicuous reddish brown streaks on the corolla. But no other differences can be found, and at the moment I am keeping all material as one variable taxon, although separating this form out at subspecific rank is tempting. Occasionally faint lines are visible at the base of the lower lip, and a few collections from S Zimbabwe, well outside the area of this "form", also lack the streaks.

Luke et al. TPR10 (**K** 7, Tana River Primate Reserve) is a puzzling plant which may be a hybrid between *J. flava* and *J. nyassana*. It is described as having a "dull beige yellow" corolla. It has the conspicuous streaks on the corolla lips but seems to have pale mauve veins in the throat. The bracts are rather narrow but with the correct indumentum for *J. flava*.

29. **Justicia gilbertii** *Vollesen* **sp. nov.** a *J. flava* caulibus tomentum densum album gerentibus (nec pubescentiam densum neque indumentum retrorse sericeum gerentibus neque subglabris), bracteis caduceis (nec persistentibus), calyce glanduloso et capsula retrorse sericea (nec puberula) differt. Typus: Kenya, Northern Frontier District, 20 km SSW of El Wak on Wajir road, *Gillett* 13380 (K!, holo.; EA!, K!, iso.)

Perennial herb with several erect much-branched brittle stems from woody rootstock; stems to 75 cm long, densely whitish tomentose. Leaves with petiole 1–3 mm long; lamina ovate, largest 3–4 × 1–1.5 cm, base cuneate to rounded, apex acute to obtuse; whitish pubescent (densely so to tomentose when young). Flowers all single or in 1-flowered dichasia; cymes 3–20 cm long; peduncle to 4 cm long tomentose; rachis densely pubescent to tomentose and with stalked capitate glands; upper main axis bracts soon caducous, narrowly elliptic, 9–10 mm long (lower not seen), puberulous and with short- stalked capitate glands, conspicuously ciliate with hairs to 2 mm long; dichasia bracts and bracteoles linear to lanceolate, 8–9 mm long, with similar indumentum, bracts caducous, bracteoles persistent. Calyx 7–9 mm long, minutely puberulous and with short capitate glands, ciliate, lobes linear-lanceolate, cuspidate. Corolla yellow, lower lip with 3 and upper lip with 2 conspicuous reddish brown streaks, 11–14 mm long, puberulous; tube 6–8 mm long and 3–4 mm diameter; upper lip 5–6 mm long; lower lip spreading or deflexed, 7–9 mm long. Filaments 2–3 mm long, bending out of flower after anthesis; thecae oblong, ± 1 mm long, ± 25% overlapping, upper hairy, lower with linear furcate appendage ± 1 mm long. Capsule 9–11 mm long, uniformly retrorsely sericeous; seed black, ovoid, ± 2 mm long, densely tuberculate.

FIG. 70. *JUSTICIA FLAVA* — **1**, habit; **2**, leaf; **3**, bract (right) and bracteole (left); **4**, calyx opened up; **5**, detail of calyx lobe; **6**, corolla opened up; **7**, apical part of filament and anther; **8**, apical part of style and stigma; **9**, capsule; **10**, seed. *J. BARAVENSIS* — **11**, bract (right) and bracteole (left); **12**, calyx opened up; **13**, detail of calyx lobe. 1 & 3–6 from *Drummond & Hemsley* 4122, 2 & 7–8 from *Bidgood et al.* 109, 9–10 from *Gereau* 4256, 11–13 from *Robertson* 6763. Drawn by Margaret Tebbs.

KENYA. Northern Frontier District: 20 km SSW of El Wak on Wajir road, 29 May 1952, *Gillett* 13380! & 12 km S of El Wak, 11 May 1978, *Gilbert & Thulin* 1652!

DISTR. **K** 1; not known elsewhere

HAB. Rich *Acacia-Commiphora-Delonix* bushland on deep sandy soil overlying limestone; 400–450 m

NOTE. Clearly related to *J. flava*, but differs in the dense indumentum, the caducous bracts, the solitary flowers, the glandular calyx and the retrorsely haired capsule.

30. **Justicia baravensis** *C.B. Clarke* in F.T.A. 5: 189 (1899); Fiori in Bull. Soc. Bot. Ital. 1915: 57 (1915); Chiovenda, Fl. Somala 2: 355 (1932); E.P.A.: 968 (1964); Kuchar, Pl. Somalia [CRDP Techn. Rep. Ser. No. 16]: 247 (1986). Type: Somalia, Barava [Brava], *Hildebrandt* 1309 (not seen)

Perennial herb with 1-several erect (rarely), procumbent, ascending or scrambling stems from woody rootstock; stems to 1.25 m long, subglabrous (hairy at nodes) to antrorsely sericeous (rarely puberulous or retrorsely sericeous). Leaves with petiole to 1.5(–2) cm long; lamina ovate to elliptic to subcircular, largest (1–)2–7.5(–15) × 1–4.5(–7) cm, base truncate to attenuate (rarely subcordate), apex acute to broadly rounded; subglabrous to sparsely sericeous-pubescent or puberulous on midrib and larger veins or also on lamina. Flowers in 3–7-flowered dichasia or 1-flowered towards apex; cymes 1–20 cm long, sometimes with short lateral cymes from lowermost nodes; peduncle to 7(–12) cm long; peduncle and rachis with indumentum as stems or slightly denser, often with scattered stalked capitate glands; main axis bracts and dichasia bracts lanceolate to narrowly ovate or narrowly elliptic, 4–8(–9) mm long, puberulous and with usually dense stalked capitate glands and conspicuously ciliate; bracteoles linear to linear-lanceolate, 4–7 mm long, with similar indumentum. Calyx 4–7 mm long, puberulous with stubby shiny moniliform hairs, ciliate, and with scattered to dense stalked capitate glands, lobes linear-lanceolate, cuspidate. Corolla yellow, without streaks on lips, 7–11 mm long, puberulous; tube 4–7 mm long and 2–3 mm diameter; upper lip 3–5 mm long; lower lip spreading or deflexed, 4–8.5 mm long. Filaments 2–4 mm long, bending out of flower after anthesis; thecae oblong, ± 1 mm long, ± 33% overlapping, upper hairy, lower with linear acute or furcate appendage ± 1 mm long. Capsule 6–9 mm long, densely puberulous or retrorsely sericeous-puberulous; seed black, ovoid, ± 1.5 mm long, densely tuberculate. Fig. 70: 11–13, p. 534.

KENYA. Lamu District: Kiunga Point, 24 July 1961, *Gillespie* 41!; Kilifi District: Watamu, Blue Lagoon, 25 Nov. 1992, *Robertson & Brummitt* 6763!; Kwale District: Diani Beach, 6 Dec. 1975, *Kokwaro* 3900!

TANZANIA. Tanga District: Tanga, Ngamiani, 22 April 1955, *Faulkner* 1597! & Sawa, 9 July 1957, *Faulkner* 2019! & 7 Sept. 1965, *Faulkner* 3641!; Zanzibar, Chwaka, 26 Dec. 1933, *Vaughan* 2183!

DISTR. **K** 7; **T** 3; **Z**; Somalia

HAB. Sandy foreshores just above high-water mark, sand dunes, coastal bushland on sand, coral rag, persisting in disturbed places and becoming a weed; near sea level to 15 m

SYN. *J. palustris* (Hochst.) T. Anderson var. *dispersa* Lindau in E.J. 20: 72 (1894), quoad *Hildebrandt* 1940

J. ovalifolia Fiori in Bull. Soc. Bot. Ital. 1915: 57 (1915); E.P.A.: 973 (1964); Kuchar, Pl. Somalia [CRDP Techn. Rep. Ser. No. 16]: 248 (1986). Types: Somalia, Mogadishu, *Stefanini & Paoli* 33, 71, 73 (all FT, syn.)

J. ovalifolia Fiori var. *psammophila* Fiori in Bull. Soc. Bot. Ital. 1915: 57 (1915). Type: Somalia, Mogadishu to Gezira, *Paoli* 61 (FT, holo.)

J. arenaria Hedrén [in Fl. Somalia 3: 413 (2006), *nom. nud.*] & in Willdenowia 36: 751 (2006). Type: as for *J. ovalifolia* var. *psammophila*

NOTE. Plants growing close enough to the sea to be exposed to salt spray develop thick fleshy broad leaves, while those further away have thinner leaves which tend to be relatively larger. Examples of the latter are *Robertson & Brummitt* 6763 and *Faulkner* 3119. This is a phenomenon also known in other East African species, e.g. *Asystasia gangetica* and from various European plants, e.g. *Tussilago farfara* in the Compositae.

Hedrén in Fl. Somalia 3: 410 (2006) considers *J. baravensis* a synonym of *J. flava*, but in the diagnosis C.B. Clarke clearly states: "spikes .... with many gland-tipped hairs". The type collection is also from or near the sea shore. I see no reason – despite not having seen the type – to doubt that this is the correct name.

31. **Justicia bizuneshiae** *Ensermu* in K.B. 54: 445 (1999) & in Fl. Eth. 5: 470, Fig. 166.60 (2006). Type: Ethiopia, Sidamo, Magada, *Ensermu* 2659 (ETH, holo.; K, UPS, WU, iso.)

Perennial herb with scrambling or scandent stems from creeping rhizome; stems to 1.5(–3) m long, sparsely sericeous to puberulous. Leaves with petiole to 2 cm long; lamina ovate-elliptic, largest 7–15 × 2.5–5 cm, base cuneate to attenuate, apex subacuminate with acute to obtuse tip; sparsely sericeous-puberulous on midrib, larger veins and on edges. Flowers in 1–3-flowered dichasia; cymes 2–10 cm long, terminal and axillary; peduncle to 4.5 cm long, sericeous-puberulous; rachis densely pubescent and with long stalked capitate glands; main axis bracts ovate to elliptic, 8–12 mm long, pubescent and with long stalked capitate glands, distinctly ciliate; dichasia bracts and bracteoles linear-lanceolate to narrowly ovate, 7–11 mm long, with similar indumentum. Calyx 5–7 mm long, pubescent and with stalked capitate glands, lobes narrowly triangular, distinctly white edged, cuspidate. Corolla pale mauve to mauve, lower lip with 3 and upper lip with 2 conspicuous dark streaks, 12–15 mm long, puberulous and with capitate glands; tube 6–8 mm long and 3–3.5 mm diameter; upper lip 6–7 mm long; lower lip spreading or deflexed, 9–10 mm long. Filaments 3–5 mm long, bending out of flower after anthesis; thecae oblong, ± 1 mm long, ± 25% overlapping, upper hairy, lower with linear appendage ± 1 mm long. Capsule ± 10 mm long, sericeous; seed (immature) ovoid, ± 2 mm long, tuberculate.

KENYA. Masai District: Olodungora, 13 July 1961, *Glover et al.* 2129! & Nguruman, 5 km SE of Entasekera, 10 Oct. 1977, *Fayad* 268!
TANZANIA. Masai District: Loliondo, Kingarana Forest Reserve, 24 Aug. 1968, *Carmichael* 1479! & 23 March 1995, *Congdon* 426!
DISTR. **K** 6; **T** 2; Ethiopia
HAB. Paths, clearings and margins of dry montane forest; 2100–2400 m

NOTE. A very disjunct distribution but I can find no significant differences between these four collections and the plentiful material from Ethiopia.

32. **Justicia caerulea** *Forssk.*, Fl. Aegypt.-Arab.: 5 (1775); Schwartz, Fl. Trop. Arab.: 258 (1939); Milne-Redhead in Proc. Linn. Soc. 165: 28 (1954); E.P.A.: 968 (1964); U.K.W.F.: 604 (1974); Luke & Robertson, Kenya Coast. For. 2. Checklist Vasc. Pl.: 82 (1993); U.K.W.F., ed. 2: 279, pl. 123 (1994); Ruffo *et al.*, Cat. Lushoto Herb. Tanzania: 6 (1996); Vollesen *et al.*, Checklist Mkomazi: 82 (1999); Friis & Vollesen in Biol. Skr. 51(2): 446 (2005); Hedrén in Fl. Somalia 3: 412 (2006); Ensermu in Fl. Eth. 5: 475 (2006). Type: Yemen, Djebel Melhan, *Forsskål* 383 (C!, holo.)

Perennial or shrubby herb with 1-several erect to decumbent (rarely scrambling) stems from branched rhizome; stems to 0.8(?–1.5) m long, subglabrous to retrorsely sericeous and glabrescent or persistently puberulous to sericeous-puberulous. Leaves with petiole to 1(–2.5) cm long; lamina lanceolate to ovate or elliptic, largest 2–10(–14) × 0.7–3.2(–4.8) cm, base attenuate, decurrent, apex acute to obtuse; glabrous to sparsely sericeous on midrib, larger veins and on edges. Flowers in 1–3-flowered dichasia (often all 1-flowered); cymes (1–)3–30 cm long, terminal, loose at base, denser upwards; peduncle to 11 cm long; peduncle and rachis with indumentum as stem; main axis bracts foliaceous in basal part, bract-like upwards, middle and upper linear-lanceolate to narrowly elliptic, 5–10 mm long, sparsely sericeous and towards apex of bract usually with straight stalked capitate glands with globose head; dichasia bracts and bracteoles similar or narrowly obovate, with similar indumentum. Calyx 7–11(–13) mm long, sericeous or sericeous-puberulous or sparsely so, with sparse to dense stalked straight capitate glands in apical part (rarely without), lobes linear-lanceolate, white-edged, acute. Corolla pale blue to blue, throat white with mauve stripes, 12–17 mm long, puberulous; tube 6–9 mm long and 2.5–3.5 mm diameter; upper lip 6–8 mm long; lower lip deflexed, 8–14 mm long. Filaments 3–5 mm long, bending out of flower after anthesis; thecae oblong, ± 1.5 mm long, hardly overlapping, upper hairy, lower with acute curved appendage 1–1.5 mm long. Capsule 9–12 mm long, glabrous; seed black, round, 2–2.5 mm long, reticulately spinulose.

UGANDA. Karamoja District: Koputh, 6 June 1942, *Dale* U277! & Moroto, 5 Oct. 1952, *Verdcourt* 751! & Pian, 2 km E of Lokapel, 17 Nov. 1968, *Lye & Lester* 465!

KENYA. Northern Frontier District: Sololo Police Post, 2 Aug. 1952, *Gillett* 13665!; Turkana District: Oropoi, Feb. 1965, *Newbould* 7015!; Voi District: Tsavo National Park East, Voi Gate, 5 Jan. 1967, *Greenway & Kanuri* 12968!

TANZANIA. Masai District: 12 km on Longido–Namanga road, 2 Jan. 1962, *Polhill & Paulo* 1022!; Moshi District: 12 km W of Himo, 29 March 1952, *Bally* 8105!; Lushoto District: 5 km NW of Mombo, 29 April 1953, *Drummond & Hemsley* 2271!

DISTR. **U** 1; **K** 1–7; **T** 2, 3; Sudan, Eritrea, Ethiopia, Somalia, Yemen

HAB. Grassland on black cotton soil, *Acacia drepanolobium* grassland and bushland, riverine scrub, roadsides, persisting in cultivations; 100–1550 m

SYN. *J. longecalcarata* Lindau in E.J. 20: 73 (1894) & in P.O.A. C: 373 (1895); C.B. Clarke in F.T.A. 5: 193 (1900). Types: Kenya, Ukamba, *Hildebrandt* 2721 (B†, syn.); Tanzania, Pangani District: Pangani to Himo, *Volkens* 563 (B†, syn.)

J. tanaënsis C.B. Clarke in F.T.A. 5: 193 (1900). Type: Kenya, Tana River, *A.S. Thomas* 38 (not seen)

J. paolii Fiori in Bull. Soc. Bot. Ital. 1915: 58 (1915). Types: Somalia, Uanle Uein, *Paoli* 1189 (FT, syn.) & Juba Valley, *Scassellati & Mazzocchi* s.n. (FT, syn.)

33. **Justicia kulalensis** *Vollesen* **sp. nov.** a *J. caerulea* glandulis capitatis bractearum calycumque capita clavata (nec globosa) gerentibus, calyce breviore 5–6(–8) mm (nec 7–11(–13) mm) longa, corolla glandulas capitatas gerenti (nec glandulis e corolla carentibus) et tubo corollae quam labio superiore distincte longiore (nec labium aequanti) differt. Typus: Kenya, Northern Frontier District, Mt Kulal, *Tweedie* 4278 (K!, holo.)

Perennial or shrubby herb with decumbent or scrambling stems from branched rhizome; stems to 2 m long, antrorsely or retrorsely sericeous or sericeous-puberulous, glabrescent. Leaves with petiole to 1.5(–2.5) cm long; lamina ovate or elliptic, largest 4–10 × 2–4.8 cm, base attenuate, decurrent, apex acute to rounded; sparsely sericeous or sericeous-puberulous on midrib, larger veins and on edges. Flowers in 1–3-flowered dichasia; cymes 1.5–7 cm long, terminal, loose at base, denser upwards; peduncle to 7.5 cm long, with indumentum as stem; rachis densely sericeous and with or without long curly capitate glands with clavate head; main axis bracts foliaceous in basal part, bract-like upwards, middle and upper oblanceolate to obovate, 5–8 mm long, sericeous-puberulous and along the whole length with long curly capitate glands with clavate head; dichasia bracts and bracteoles linear-lanceolate to oblanceolate, with similar indumentum. Calyx (5–)6–8 mm long, sericeous-puberulous and with long curly stalked capitate glands with oblong head along the whole length, lobes lanceolate, white edged, acute. Corolla pale blue to blue, with red venation in throat, 15–17 mm long, sparsely puberulous and with stalked capitate glands; tube 9–10 mm long and 3–4 mm diameter; upper lip 6–7 mm long; lower lip deflexed, 8–11 mm long. Filaments 4–5 mm long, bending out of flower after anthesis; thecae oblong, ± 1 mm long, hardly overlapping, upper hairy, lower with acute curved appendage ± 1 mm long. Capsule and seed not seen.

KENYA. Northern Frontier District: Samburu, Wamba, 28 Nov. 1958, *Newbould* 2929! & Mt Kulal, March 1972, *Tweedie* 4278! & 18 Nov. 1978, *Hepper & Jaeger* 6905!

DISTR. **K** 1; not known elsewhere

HAB. Edges and clearings in dry montane forest, thickets around ponds in *Acacia-Commiphora* bushland; (1200–)1650–2100 m

NOTE. This is basically a high altitude forest form of *J. caerulea*. It differs in the clavate glandular heads on bracts and sepals, corolla with capitate glands, shorter calyx and corolla with tube distinctly longer than upper lip.

34. **Justicia linearispica** *C.B. Clarke* in F.T.A. 5: 192 (1899); Binns, Check List Herb. Fl. Malawi: 14 (1968), as *linearispicata*; Richards & Morony, Check List Mbala: 231 (1969); Cribb & Leedal, Mountain Fl. S. Tanzania: 127 (1982); Ruffo *et al.*, Cat. Lushoto Herb. Tanzania: 6 (1996). Type: Zambia, Stevenson Road, *Scott Elliot* 8267 (K!, holo.; BM!, iso.)

Perennial pyrophytic herb with 1-several erect (rarely scrambling) stems from large often branched woody rootstock; stems to 50 cm long, glabrous to sparsely and finely puberulous below nodes (very rarely uniformly so). Leaves sessile; lamina linear-lanceolate to narrowly elliptic or slightly obovate, largest 1.5–6.7 × 0.3–1.4 cm, base attenuate, decurrent, apex acute to rounded; glabrous (rarely sparsely puberulous on midrib and larger veins). Flowers single (very rarely 2 per bract at lowermost nodes); cymes 1–10(–13.5) cm long, terminal, loose at base, denser upwards, (rarely only 1 developing flower per node, cyme thus becoming 1-sided); peduncle (0.5–)1–7.5 cm long, glabrous (towards base) to sparsely and finely puberulous and towards apex sometimes with subsessile capitate glands; rachis finely puberulous and with (rarely without) subsessile stalked capitate glands; bracts and bracteoles ovate-triangular or narrowly so, 3–6.5 mm long, puberulous and with dense stalked capitate glands. Calyx 4–6 mm long, with similar indumentum, lobes lanceolate to narrowly triangular, acute. Corolla white with pink "herring bones" on lower lip, 7–10 mm long, minutely puberulous and with subsessile glands; tube 3–5 mm long and 2–3 mm diameter; upper lip 4–5 mm long; lower lip spreading to deflexed, 6–8 mm long. Filaments 3–4 mm long, bending out of flower after anthesis; thecae dark purple, oblong, ± 1.5 mm long, only opening for ± $^1/_2$ of their length, ± 33% overlapping, upper hairy, lower with obtuse or furcate appendage 0.5–0.8 mm long. Capsule 10–14 mm long, densely puberulous; seed subcircular, ± 3 mm long, verrucose-spinulose.

TANZANIA. Sumbawanga District: Mbizi Forest Reserve, 2 Dec. 1993, *Gereau et al.* 5218!; Iringa District: Mufindi, Lake Ngwazi, 19 Feb. 1986, *Bidgood* 30!; Mbeya District: Pungaluma Hills, Mshewe Village, 26 Nov. 1989, *Lovett & Kayombo* 3505!

DISTR. **T** 4, 7, 8; Angola, Zambia, Malawi

HAB. Montane grassland, often in seasonally waterlogged soil but also in drier types, *Brachystegia* and *Isoberlinia* woodland; (1000–)1400–2300 m

SYN. *J. "longispica"* C.B. Clarke *sensu* Richards & Morony, Check List Mbala: 231 (1969), *nom. illeg.*

NOTE. *Lovett & Keeley* 1976 is unusually hairy with uniformly puberulous stems and leaves. This also has only one flower developed per node throughout the cyme. The latter condition has been seen partly developed in a few other collections, and a single other collection with a sparsely puberulous (as opposed to the normal minutely puberulous) stem has also been seen.

35. **Justicia nyassana** *Lindau* in E.J. 20: 66 (1894) & in P.O.A. C: 373 (1895); C.B. Clarke in F.T.A. 5: 192 (1899); Milne-Redhead in Mem. N. Y. Bot. Gard. 9: 24 (1954); F.P.S. 3: 179 (1956); Binns, Check List Herb. Fl. Malawi: 14 (1968); Richards & Morony, Check List Mbala: 232 (1969); U.K.W.F.: 604 (1974); Vollesen in Opera Bot. 59: 81 (1980); Luke & Robertson, Kenya Coast. For. 2. Checklist Vasc. Pl.: 82 (1993); U.K.W.F., ed. 2: 280 (1994); Ruffo *et al.*, Cat. Lushoto Herb. Tanzania: 7 (1996); Vollesen *et al.*, Checklist Mkomazi: 83 (1999); Friis & Vollesen in Biol. Skr. 51(2): 448 (2005); Strugnell, Checklist Mt Mulanje: 34 (2006); Ensermu in Fl. Eth. 5: 474 (2006). Type: Malawi, *Buchanan* 290 (B†, holo.; BM!, K! iso.)

Perennial (rarely annual) herb with erect, ascending, scrambling or scandent stems from creeping rhizome or slender rootstock, often rooting at lower nodes; stems to 1 m long, sparsely sericeous to puberulous or pubescent. Leaves with petiole to 2(–3.5) cm long; lamina narrowly to broadly ovate to elliptic, largest 2.5–15(–17) × 1–6(–7) cm, base attenuate and decurrent to cuneate, apex acute to acuminate (rarely obtuse); sparsely sericeous to puberulous or pubescent on midrib and larger veins or also on lamina. Flowers in 1-flowered dichasia or 2–3-flowered in basal part; cymes 1–15 cm long, terminal (sometimes also axillary), dense to very open; peduncle to 9(–14) cm long, indumentum as stem and towards apex with (rarely without) stalked capitate glands; rachis densely whitish to yellowish puberulous, with (rarely without) scattered long pilose hairs and with stalked capitate glands with globose head; main axis bracts linear to lanceolate, 4–13 mm long, puberulous or sparsely so and with stalked capitate glands with gloose head, distinctly ciliate with hairs to 2 mm long; dichasia bracts and bracteoles linear to lanceolate, 4–8(–9) mm long, with similar indumentum. Calyx 3–5 mm long, shorter than bracts and bracteoles, with similar indumentum, lobes filiform to linear, cuspidate. Corolla mauve or pink, with purple "herring bones" on lower lip, 6–13 mm long, sparsely puberulous on lobes, glabrous on whole or basal part of tube and with stalked capitate glands; tube 4–8 mm long and 2–3.5 mm diameter; upper lip 2–5 mm long; lower lip spreading, 4–10 mm long. Filaments 2–4 mm long, usually not bending out of flower after anthesis; thecae narrowly oblong, 0.5–1.5 mm long, ± 25–33% overlapping, upper hairy, lower with linear acute appendage ± 0.5 mm long. Capsule 6–9.5 mm long, puberulous; seed black, subcircular, 1–1.5 mm diameter, densely tuberculate, the tubercles warty to spinulose or hooked.

UGANDA. West Nile District: Zoka Forest, Jan. 1952, *Leggat* 68! & Metu Rest Camp, 20 Sept. 1953, *Chancellor* 301!; Kigezi District: Queen Elizabeth National Park, Birereko track, 9 Dec. 1969, *Lock* 69/429!

KENYA. Northern Frontier District: Marsabit, 4 Aug. 1957, *Verdcourt* 1807!; Fort Hall District: Thika, Blue Posts Hotel, 19 Feb. 1953, *Drummond & Hemsley* 1225!; Kwale District: Marenge Forest, 18 Aug. 1953, *Drummond & Hemsley* 3872!

TANZANIA. Same District: Mkomazi Game Reserve, Ibaya Hill, 1 May 1995, *Abdallah & Vollesen* 95/78!; Morogoro District: Mikumi National Park, 30 April 1968, *Renvoize & Abdallah* 1826!; Songea District: Likonde River, 26 June 1956, *Milne-Redhead & Taylor* 10900!

DISTR. **U** 1, 2; **K** 1–4, 6, 7; **T** 2–8; Central African Republic, Congo-Kinshasa, Burundi, Sudan, Ethiopia, Zambia, Malawi, Mozambique, Zimbabwe

HAB. Lowland to montane forest, usually on edges, on paths and in clearings, riverine forest, damp places in woodland, termite mounds, roadsides; 100–2150 m

SYN. *J. ulugurica* Lindau in E.J. 22: 126 (1895); C.B. Clarke in F.T.A. 5: 194 (1900). Type: Tanzania, Morogoro District, Uluguru Mts, Ukami, *Stuhlmann* 8866 (B†, holo.)

36. **Justicia heterotricha** *Mildbr.* in N.B.G.B. 11: 1088 (1934). Type: Tanzania, Morogoro District, Uluguru Mts, *Schlieben* 2737 (B†, holo.; BM!, iso.)

Perennial or shrubby (? rarely annual) herb, basal part of stems creeping and rooting, apical part erect, ascending or scrambling, to 0.5(?–1) m long, sparsely sericeous to sericeous (rarely sericeous-pubescent). Leaves with petiole to 3 cm long; lamina ovate to elliptic or broadly so or slightly obovate, largest 5.8–11.5 × 2.7–5.5 cm, base attenuate, apex rounded to subacuminate with acute to obtuse tip; sericeous or sparsely so on midrib and larger veins (rarely uniformly sericeous-pubescent). Flowers in 1–3-flowered dichasia; cymes 2–11 cm long, terminal (sometimes also axillary), usually very open; peduncle to 5 cm long; peduncle and rachis whitish to pale yellowish sericeous and with straight moniliform shiny hairs and with long-stalked capitate glands with oblong to clavate head; main axis bracts ovate-spatulate or narrowly so, 3–5(–8) mm long, puberulous or densely so and with long-stalked capitate glands with oblong to clavate head; dichasia bracts and bracteoles oblanceolate to narrowly obovate, 3–7 mm long, with similar indumentum. Calyx 4–7 mm long, often longer than bracts and bracteoles, with similar indumentum, lobes filiform to linear, cuspidate.

Corolla white, with faint pink "herring bones" on lower lip, 7–10 mm long, sparsely puberulous on lobes, glabrous on tube; tube 5–7 mm long and 2–3 mm diameter; upper lip 2–3 mm long; lower lip spreading to deflexed, 4–5 mm long. Filaments 1–2 mm long, usually not bending out of flower after anthesis; thecae oblong, 0.5–1 mm long, ± 33% overlapping, upper hairy, lower with linear acute appendage ± 0.5 mm long. Capsule 6–8 mm long, puberulous in apical half, glabrous towards base; seed black, subcircular, ± 2 mm diameter, densely tuberculate.

KENYA. Kwale District: Shimba Hills, Risley's Ridge, 21 March 1991, *Luke & Robertson* 2770! & Shimba Hills, Mwele, 16 Oct. 1991, *Luke* 2931!

TANZANIA. Lushoto District: E Usambara Mts, Amani to Derema, Hunga Valley, 8 Feb. 1987, *Pócs* 87029/J!; Morogoro District: Mangala Forest Reserve, 17 Aug. 2000, *Mhoro* UMBCP345!; Kilwa District: Tongomba Forest Reserve, 13 July 1995, *G. P. Clarke* 83!

DISTR. **K** 7; **T** 3, 6, 8; not known elsewhere

HAB. Semi-deciduous or wet lowland, intermediate and montane forest; 200–1050(–1650) m

SYN. *Justicia sp.* Sect. *Rostellaria*, Luke & Robertson, Kenya Coast. For. 2. Checklist Vasc. Pl.: 82 (1993)

NOTE. Differs most conspicuously from the closely related *J. nyassana* in the obovate bracts and bracteoles which have glands with oblong or clavate heads; and in the white corolla.

37. **Justicia bridsoniana** *Vollesen* **sp. nov.** a *J. heterotricha* bracteis longioribus (5–)6–8(–12) mm (nec 3–5(–8) mm) longis atque indumentum flavum nec album gerentibus, labio inferiore corollae maiore 5–7 mm (nec 4–5 mm) longo et capsula maiore 8.5–9.5 mm (nec 6–8 mm) longa atque ad basin sericea (nec glabra) differt. Typus: Tanzania, Iringa District, Udzungwa Mts National Park, Sanje Loggers Camp, *Bridson* 644 (K!, holo.; EA!, K!, iso.)

Erect to scrambling shrubby herb to 1.5 m tall; young stems antrorsely sericeous to sericeous-pubescent. Leaves with petiole to 3(–4.5) cm long; lamina ovate to elliptic, largest 7–12 × 3–6 cm, base cuneate to attenuate, apex acute to subacuminate; sericeous or sparsely so on midrib and larger veins or uniformly sericeous-pubescent. Flowers in 1–3-flowered dichasia; cymes 3.5–8.5 cm long, terminal and axillary from upper leaf-axils, open to condensed; peduncle to 6 cm long; peduncle and rachis yellowish puberulous or densely so with straight moniliform shiny hairs, with scattered long bent hairs and with long-stalked capitate glands with oblong to clavate head; main axis bracts ovate-spatulate, (5–)6–8(–12) mm long, densely yellowish puberulous and with long-stalked capitate glands with oblong to clavate head; dichasia bracts and bracteoles similar, (4–)5–8 mm long, with similar indumentum. Calyx 5–7 mm long, with similar indumentum, lobes lanceolate, cuspidate. Corolla white, with blue to purple "herring bones" on lower lip, 9–11 mm long, sparsely uniformly puberulous; tube 6–7 mm long and 2–3 mm diameter; upper lip 3–4 mm long; lower lip spreading to deflexed, 5–7 mm long. Filaments 2–4 mm long, bending out of flower after anthesis; thecae oblong, ± 1 mm long, ± 25% overlapping, upper hairy, lower with bifurcate appendage ± 0.5 mm long. Capsule 8.5–9.5 mm long, puberulous in apical half, sericeous towards base; seed black, subcircular, ± 1.5 mm diameter, tuberculate-verrucose.

TANZANIA. Kilosa District: Mikumi National Park, Malondwe Hill, 25 Oct. 1983, *J. Lovett* 194!; Iringa District: Udzungwa Mts National Park, Sanje Loggers Camp, 8 Sept. 1984, *Bridson* 644! & Mt Luhomero, 24 Sept. 2000, *Luke et al.* 6598!

DISTR. **T** 6, 7; not known elsewhere

HAB. Intermediate and lower montane evergreen forest; 950–1450 m

NOTE. *J. bridsoniana* is known only from five collections. It is closely related to *J. heterotricha* from which it differs in the larger bracts with yellow indumentum, larger differently dimensioned corolla and larger capsule with different indumentum. It also generally occurs at higher altitudes.

38. **Justicia sp. C**

Erect shrubby herb, 1 m tall; young stems antrorsely sericeous. Leaves with petiole to 2.5 cm long; lamina ovate to elliptic or narrowly so, largest ± 15 × 2.5 cm, base attenuate, decurrent, apex subacuminate with an obtuse tip; subglabrous. Flowers in 1–3-flowered dichasia; cymes 4.5–7.5 cm long, terminal and axillary from upper leaf-axils, open to condensed; peduncle to 2.5 cm long; peduncle and rachis densely yellowish puberulous with straight moniliform shiny hairs and with many long-stalked (to 2 mm) capitate glands with oblong to clavate head; main axis bracts oblanceolate-spatulate, 6–10 mm long, densely yellowish puberulous and with long-stalked capitate glands with oblong to clavate head; dichasia bracts and bracteoles similar, (4–)5–8 mm long, with similar indumentum. Calyx 6–7 mm long, with similar indumentum, lobes linear-lanceolate, cuspidate. Corolla white, with blue to purple "herring bones" on lower lip, 15–18 mm long, sparsely uniformly puberulous; tube 6–9 mm long and 3–4 mm diameter; upper lip 8–10 mm long; lower lip to deflexed, 8–10 mm long. Filaments 4–6 mm long, bending out of flower after anthesis; thecae oblong, ± 1 mm long, ± 25% overlapping, upper hairy, lower with bifurcate appendage ± 0.5 mm long. Capsule and seed not seen.

TANZANIA. Iringa District: Mufindi, Uhafiwa-Luhega Forest, 3 Aug. 1989, *Kayombo* 834!
DISTR. **T** 7; not known elsewhere
HAB. Undergrowth in wet evergreen forest; 1100 m

NOTE. Closely related to the preceeding species, from which it mainly differs in the much larger corolla. Further collections may eventually prove that the last three species should all be considered as one.

39. **Justicia migeodii** (*S. Moore*) *V.A.W. Graham* in K.B. 43: 591 (1988). Types: Tanzania, Lindi District: Tendaguru, *Migeod* 137 (BM!, syn.) & 473 (BM!, syn.)

Perennial herb with 1-few erect stems from creeping rhizome or short rootstock; stems to 0.6 m long, antrorsely sericeous or sericeous-puberulous, sometimes with a few stalked capitate glands. Leaves with petiole 0–1 mm long; lamina lanceolate to narrowly ovate, largest 6.5–15 × 0.5–2.3 cm, base attenuate, apex acute to acuminate; subglabrous to sparsely sericeous, densest on midrib and along edges, sometimes with a few stalked capitate glands. Flowers in 1–3-flowered dichasia; cymes 4–21 cm long, sometimes with short lateral cymes from lowermost nodes; peduncle to 7 cm long; peduncle and rachis puberulous or sericeous-puberulous and with many stalked capitate glands; main axis bracts and dichasia bracts sometimes tinged purple, lanceolate, 8–14 mm long, puberulous and with dense stalked capitate glands; bracteoles linear, 5–11 mm long, with similar indumentum. Calyx 5–9 mm long, puberulous and with scattered stalked capitate glands, lobes linear-lanceolate, acuminate to cuspidate. Corolla pure white or with red lines on lower lip, 15–18 mm long, sparsely puberulous; tube 10–13 mm long and 1.5–2 mm diameter; upper lip 4–5 mm long; lower lip spreading or deflexed, 6–10 mm long. Filaments 2–3 mm long, bending out of flower after anthesis; thecae purple, oblong, ± 1 mm long, ± 25% overlapping, both hairy, lower with linear acute appendage ± 1 mm long. Capsule 9–10 mm long, puberulous; seed black, ovoid, ± 2 mm long, tuberculate.

TANZANIA. Kilwa District: 10 km SW of Liwale, 17 Nov. 1988, *Lock* 88/77!; Masasi District: Ndanda Mission, 10 March 1991, *Bidgood et al.* 1890! & 30 km NW of Masasi, 13 March 1991, *Bidgood et al.* 1954!
DISTR. **T** 8; not known elsewhere
HAB. *Brachystegia* woodland on clayey hardpans and on blackish gritty-clayey soils; 150–500 m

SYN. *Siphonoglossa migeodii* S. Moore in J.B. 67: 271 (1929)

40. **Justicia kirkiana** *T. Anderson* in J.L.S. Bot. 7: 39 (1863); Lindau in P.O.A. C: 373 (1895); C.B. Clarke in F.T.A. 5: 192 (1899); Binns, Check List Herb. Fl. Malawi: 14 (1968); Bjørnstad in Serengeti Res. Inst. Publ. 215: 26 (1976). Type: Mozambique, Tette, *Kirk* s.n. (K!, holo.)

Annual herb with a single (but often branched) erect stem from small taproot; stem to 0.8 m long, retrorsely sericeous or sericeous-puberulous or sparsely so (rarely puberulous), upwards often with scattered stalked capitate glands. Leaves with petiole to 1(–1.5) cm long; lamina lanceolate to narrowly ovate or narrowly elliptic (rarely ovate or elliptic), largest 4–17 × 0.7–4.5 cm, base attenuate, decurrent, apex acute to acuminate; sparsely sericeous along on midrib and larger veins, often with stalked capitate glands on margins near base. Flowers in 1–3-flowered dichasia (sometimes all 1-flowered); cymes 1.5–12(–21) cm long, terminal; peduncle to 6(–11.5) cm long; peduncle and rachis puberulous or sericeous-puberulous and often with stalked capitate glands; main axis bracts lanceolate to narrowly obovate, 7–15(–20) mm long, puberulous with moniliform glossy hairs and with long stalked capitate glands, distinctly ciliate with pilose hairs to 2 mm long; dichasia bracts narrowly obovate to narrowly spatulate, 6–11(–13) mm long; bracteoles linear to oblanceolate, slightly shorter; both with similar indumentum. Calyx 3–4 mm long, puberulous and with stalked capitate glands, lobes narrowly triangular, with white edges, cuspidate. Corolla greenish white to pale yellow or pale greyish blue, with distinct blue to purple venation on both lips and with a yellow patch at base of lower lip, 7–11 mm long, sparsely puberulous on lobes, glabrous on tube; tube 4–7 mm long and 2–3 mm diameter; upper lip 3–4 mm long; lower lip spreading or deflexed, 5–8 mm long. Filaments 2–4 mm long, not bending out of flower after anthesis; thecae oblong, ± 1 mm long, ± 25% overlapping, upper hairy, lower with linear acute to obtuse appendage ± 0.5 mm long. Capsule 6–9 mm long, sericeous-puberulous or sparsely so; seed black, round, ± 1.5 mm diameter, tuberculate.

UGANDA. Karamoja District: Loyoro, Aug. 1960, *J. Wilson* 1059!
TANZANIA. Mpanda District: Mpanda road, 13 Feb. 1971, *Sanane* 1535!; Kilosa District: Ruaha Valley, 15 March 1986, *Bidgood & Lovett* 262!; Iringa District: Ruaha National Park, 6 km from Msembe, 24 Feb. 1970, *Greenway & Kanuri* 13945!
DISTR. **U** 1; **T** 1–7; Zambia, Malawi, Mozambique, Zimbabwe, Botswana
HAB. *Acacia*, *Acacia-Commiphora*, *Brachystegia* woodland and bushland, riverine woodland, grassland, rocky hills, on a wide range of soils; (500–)700–1450 m

SYN. *Justicia woodsiae* S. Moore in J.B. 67: 271 (1929). Type: Zambia, Mazabuka, *H.S. Woods* 23 (BM!, holo.)

41. **Justicia ruwenzoriensis** *C.B. Clarke* in F.T.A. 5: 185 (1899); U.K.W.F.: 604 (1974); U.K.W.F., ed. 2: 279 (1994). Type: Tanzania, Bukoba District, Karagwe, *Scott Elliot* 8120 (K!, holo.; BM!, iso.)

Perennial herb with several (up to 20) erect to decumbent stems from woody rootstock, often pyrophytic; stems to 60 (usually less than 30) cm long, glabrous to densely puberulous. Leaves with petiole to 7(–15) mm long; lamina ovate to elliptic or broadly so, largest 1.8–8.5 × 1.2–4.2 cm, base cuneate to attenuate, decurrent, apex rounded to acute; glabrous to puberulous. Spikes terminal, 2–12 cm long, often interrupted towards base; peduncle to 6(–10.5) cm long; rachis subglabrous to densely puberulous; bracts pale to dark green without (more rarely with) dark reticulate venation, sometimes tinged purplish towards apex, without white scarious margin, ovate to elliptic or broadly so, 6–9(–12) × 3–5(–8) mm, glabrous to puberulous, acute to acuminate at apex, cuneate to truncate at base; bracteoles like bracts. Calyx 3–5.5(–6.5) mm long, finely puberulous with short stubby capitate glandular hairs, usually also with longer hairs on edges and midrib, lobes acute to acuminate. Corolla white, without or with faint purple markings at throat, 8.5–11.5 mm

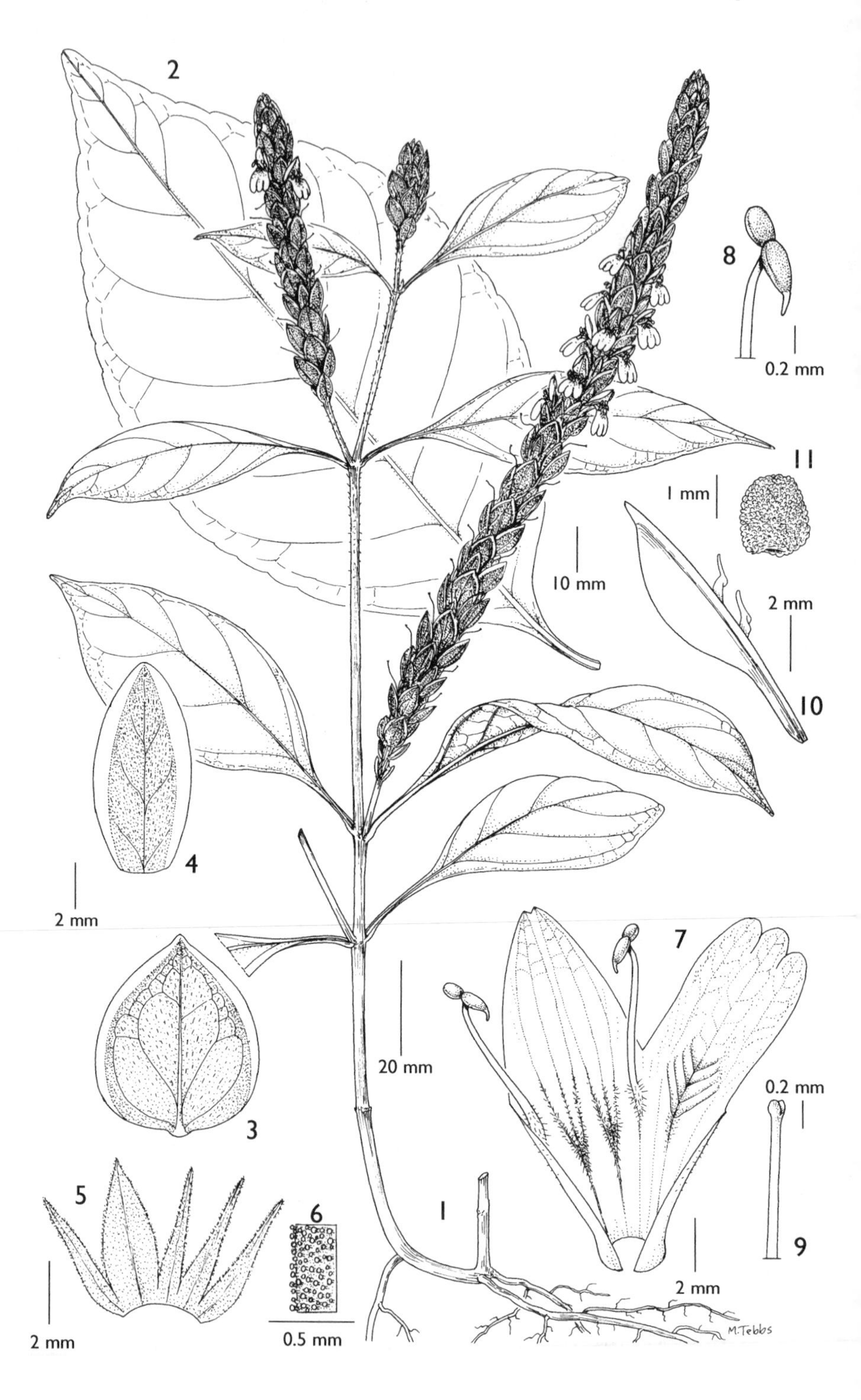
2
8
0.2 mm
11
1 mm
10 mm
2 mm
10
4
2 mm
7
20 mm
0.2 mm
3
5
6
1
9
2 mm
2 mm
0.5 mm
M.Tebbs

long, puberulous or retrorsely sericeous puberulous; tube 4.5–6.5 mm long and 2–3.5 mm in diameter apically; upper lip slightly hooded, 3.5–5 mm long; lower lip deflexed, 3–6 mm long, deeply 3-lobed with middle lobe much wider. Anthers purple; thecae ± 1 (upper) and ± 1.5 (lower) mm long, lower with an acute curved appendage ± 1 mm long, connective glabrous. Capsule 13–15 mm long, densely uniformly puberulous; seed not seen. Fig. 72: 12–13, p. 548.

UGANDA. Acholi District: 2 km NE of Lututuru, 17 Feb. 1969, *Lye & Lester* 2067!; Bunyoro District: Bujenje, Feb. 1943, *Purseglove* 1282!; Busoga District: Bukoli, Bugiri Hill, 18 Feb. 1953, *G.H.S. Wood* 620!

KENYA. North Kavirondo District: Broderick Falls, March 1967, *Tweedie* 3418!; Masai District: Mara Plains, Sept. 1960, *D. Stewart* 324! & Elegesegonyek, Mara River, 15 April 1961, *Glover et al.* 533!

TANZANIA. Biharamulo District: 47 km on Biharamulo–Muleba road, 11 July 2000, *Bidgood et al.* 4901!; Mpanda District: 31 km on Mpanda–Inyonga road, 17 May 1997, *Bidgood et al.* 3994!; Sumbawanga District: 10 km on Sumbawanga–Chala road, 2 March 1994, *Bidgood et al.* 2554!

DISTR. **U** 1–4; **K** 5, 6; **T** 1, 4; Sudan, Ethiopia, Rwanda, Angola, Zambia, ?Namibia (see note)

HAB. Woodland, bushland and grassland, rocky hills, persisting in degraded areas; 950–1900 m

SYN. *Justicia andongensis* C.B. Clarke in F.T.A. 5: 185 (1899). Types: Angola, Pungo Andongo, *Welwitsch* 5111 (K!, syn.; BM!, iso.) & Malange, *Pogge* 301 (not seen); Namibia, Amboland, *Schinz* 36 (not seen)

J. betonica sensu Friis & Vollesen in Biol. Skr. 51(2): 446 (2005) pro parte, *non* L. (1753)

NOTE. Superficially this is very similar to some forms of *J. betonica* and, in my opinion, shows the superficiality of the distinctions between Sect. *Vascia* and Sect. *Betonica* as defined by Graham in K.B. 43: 582 (1988).

I have not seen any collections of *J. ruwenzoriensis* from Namibia, and the *Schinz* syntype of *J. andongensis* cited above might well prove to belong to *J. betonica.*

42. **Justicia pseudorungia** *Lindau* in E.J. 20: 63 (1894) & in P.O.A. C: 374 (1895); C.B. Clarke in F.T.A. 5: 186 (1899); T.T.C.L.: 11 (1949); Iversen in Symb. Bot. Upsal. 29, 3: 161 (1991); Robertson & Luke, Kenya Coast. For. 2. Checklist Vasc. Pl.: 82 (1993); K.T.S.L.: 604 (1994); Ruffo *et al.*, Cat. Lushoto Herb. Tanzania: 7 (1996); Vollesen *et al.*, Checklist Mkomazi: 83 (1999). Types: Tanzania, Usambara Mts, Mashewa, *Holst* 8801 (B†, syn.; K!, iso.); Zanzibar, Kidoti, *Hildebrandt* 982 (B†, syn.; BM!, iso.)

Shrubby herb or shrub to 2(–3) m tall; young stems subglabrous to antrorsely sericeous-puberulous, usually densest in two bands. Leaves with petiole 0–3(–5) cm long; lamina ovate to elliptic, largest (6.5–) 10–25(–27) × (2.5–)4–11(–12.5) cm, base attenuate, decurrent, margin entire to crenate, apex acute to acuminate; glabrous or with a few short hairs on midrib beneath. Spikes terminal and axillary, (1.5–)2.5–23 cm long; peduncle to 5.5 cm long; rachis sericeous-puberulous or sparsely so; bracts green or dark green with (or some without) broad white hyaline margin, ovate to elliptic or broadly so or cordiform, 7–10 × 5–8 mm, glabrous, subacute to broadly rounded at apex, rounded to subcordate at base; bracteoles like bracts but longer, 8–12(–13) × 4–7 mm, acute, folded around the flower. Calyx 6–8 mm long, finely and sparsely puberulous with stubby non-capitate glandular hairs and with subsessile capitate glands, lobes acuminate to cuspidate. Corolla white with faint purple markings at throat, 12–16 mm long, sparsely minutely puberulous; tube 6–9 mm long and 2–4 mm in diameter apically; upper lip strongly hooded, 6–8 mm long; lower lip deflexed, 4–6(–7) mm long, shortly 3-lobed. Thecae 2–2.5 mm long, ± similar, lower with a short stubby triangular appendage less than 0.3 mm long, connective hairy. Capsule 11–16 mm long, glabrous; seed 2–3 mm long, densely reticulately sculptured. Fig. 71, p. 544.

FIG. 71. *JUSTICIA PSEUDORUNGIA* — **1**, habit; **2**, leaf; **3**, bract; **4**, bracteole; **5**, calyx opened up; **6**, detail of calyx lobe; **7**, corolla opened up; **8**, apical part of filament and anther; **9**, apical part of style and stigma; **10**, capsule; **11**, seed. 1 & 2 from *Drummond & Hemsley* 3476, 3–6 from *Tanner* 1999, 7–9 from *Bridson* 646, 10–11 from *Abdallah & Vollesen* 96/99. Drawn by Margaret Tebbs.

KENYA. Teita District: Taita Hills, Ngaongao Forest, 15 May 1989, *Faden et al.* 530!; Kilifi District: 13 km NE of Kaloleni, Cha Simba, 28 July 1971, *Faden & Evans* 71/736!; Kwale District: Golini, Kitsantse Waterfall, 4 Jan. 1988, *Robertson* 5085!

TANZANIA. Moshi District: Rau Forest, 25 Feb. 1953, *Drummond & Hemsley* 1303!; Lushoto District: North Kitivo Forest Reserve, 26 Nov. 1968, *Ruffo* 175!; Morogoro District: Kimboza Forest Reserve, Kibungo Mission, 21 April 1988, *Bidgood et al.* 1244!; Zanzibar, Kizimkazi, 11 Jan. 1931, *Vaughan* 1817!

DISTR. **K** 7; **T** 2, 3, 6, 7; **Z**; not known elsewhere

HAB. Lowland to montane evergreen forest, riverine forest; near sea level to 1900 m

43. **Justicia schimperiana** (*Nees*) *T. Anderson* in J.L.S. Bot. 7: 38 (1863); Engler, Hochgebirgsfl. Trop. Afr..: 392 (1892); Oliver in Trans. Linn. Soc., Ser. 2, Bot. 2: 345 (1887); Lindau in P.O.A. C: 373 (1895); Hedrén in Fl. Somalia 3: 410 (2006) pro parte; Ensermu in Fl. Eth. 5: 468 (2006). Types: Ethiopia, Adua, *Schimper* I.27 (K!, syn.; BM!, iso.) & Axum, *Schimper* III.1549 (K, syn.; BM!, iso.) & Kira, *Quartin-Dillon* 471 (not seen)

Shrubby herb or shrub to 1.5(–2) m tall; young stems bifariously puberulous or retrorsely sericeous-puberulous. Leaves with petiole to 1.5 cm long; lamina ovate to elliptic or narrowly so, largest 5–11(–14) × 1.5–4.5(–5.5) cm, base attenuate, decurrent, margin entire to crenate, apex acute to rounded; glabrous to sparsely puberulous (densest along midrib) beneath. Spikes terminal, 3–16 cm long; peduncle to 3.5(–7) cm long; rachis bifariously puberulous; bracts green with broad white scarious margin, broadly ovate or broadly elliptic, (12–)14–23 × (8–)9–16 mm, puberulous or sparsely so, acute to subacute at apex (ignoring scarious margin), cuneate at base; bracteoles like bracts but narrower, folded around the flower. Calyx (8–)9–12(–13) mm long, densely glandular-puberulous with stalked capitate glands and with longer non-glandular hairs towards tip of lobes, lobes cuspidate, dorsal wider than the others. Corolla white with purple lines on lower lip and purple patch at base of upper lip, 2.5–3.5 cm long, puberulous with stalked capitate glands and non-glandular hairs; tube 9–13 mm long and 4–5 mm in diameter apically; upper lip hooded, 1.5–2 cm long (6–8 mm longer than tube); lower lip oblong, deflexed, 11–16 × 7–11 mm long, with 3 short broad lobes 3–4 mm long. Filaments 12–17 mm long; thecae 3–4 mm long, ± similar, with 80% overlap, lower with a minute tooth but no distinct appendage. Capsule 17–22 mm long, densely uniformly glandular-puberulous, retinacula splitting off from wall at maturity but not rising elastically; seed 3–4 mm long, densely reticulately sculptured.

SYN. *Gendarussa schimperiana* Hochst. in Flora 1, Intell.: 24 (1841), *nom. nud.*
Adhatoda schimperiana Nees in DC., Prodr. 11: 388 (1847); A. Richard, Tent. Fl. Abyss. 2: 155 (1850); Solms Laubert in Schweinf., Beitr. Fl. Aeth.: 242 (1867); C.B. Clarke in F.T.A. 5: 221 (1900) pro parte

subsp. **campestris** *Vollesen* **subsp. nov.** a subsp. *schimperiana* indumento caulorum retrorso (nec patenti) et foliis pilos longos secus costam dispositos gerentibus (nec pilis carentibus) differt. Typus: Kenya, Teita District, Taveta, *Polhill & Paulo* 984 (K!, holo.; EA!, iso.)

Stem-indumentum retrorse; leaves with long spreading hairs along the midrib.

KENYA. Tana River District: 1 km S of Bfumbe, 5 Aug. 1988, *Robertson & Luke* 5328!; Kwale District: Samburu to Mackinnon Road, near Taru, 3 Sept. 1953, *Drummond & Hemsley* 4119!; Teita District: S of Maungu Station, Maungu Hills, 31 May 1970, *Faden* 70/159!

TANZANIA. Pare District: Kiruru, May 1928, *Haarer* 1374! & Suju, Kanziani, Makana, 10 Jan. 2001, *Mlangwa & Mbuso* 1276!; Lushoto District: Mkomazi, Mbalu Hill, 25 Jan. 1948, *Bally* 5756!

DISTR. **K** 4, 6, 7; **T** 3; S Somalia

HAB. *Acacia-Commiphora* and *Acacia-Terminalia-Dobera* bushland, rocky hills, dry riverine forest, coastal bushland; 25–950(–1200) m

SYN. [*Adhatoda schimperiana sensu* C.B. Clarke in F.T.A. 5: 221 (1900) pro parte; Chiovenda, Fl. Somala 2: 356 (1932); T.T.C.L.: 1 (1949); Blundell, Wild Fl. E. Afr.: 386, pl. 795 (1987); Iversen in Symb. Bot. Upsal. 29, 3: 161 (1991); K.T.S.L.: 597 (1994) *non* Nees (1847)]

[*Justicia schimperiana sensu* Robertson & Luke, Kenya Coast. For. 2. Checklist Vasc. Pl.: 82 (1993); Hedrén in Fl. Somalia 3: 410 (2006) quoad distrib. Kenya et Tanzania, *non* (Nees) T. Anderson (1863)]

NOTE. This differs from subsp. *schimperiana* (Ethiopian highlands and N Somalia) in the retrorse (not spreading) stem-indumentum and the long spreading hairs along the midrib of the leaves (absent in subsp. *schimperiana*). I have not been able to find any differences in inflorescence, flowers and fruits, and have therefore decided to recognise two geographical subspecies. Subsp. *schimperiana* grows in montane bushland, grassland and forest margins from 1325 to 2550 m but only rarely below 1500 m.

44. **Justicia grandis** (*T. Anderson*) *Vollesen*, **comb. nov**. Types: Nigeria, Eppah, *Barter* 3274 (K!, syn.); Congo-Kinshasa, *Smith* s.n. (K!, syn.)

Shrubby herb or shrub to 3 m tall (see note); young stems with two (often thin) bands of spreading or slightly antrorse curly hairs. Leaves with petiole to 4 cm long; lamina ovate to elliptic, largest 13–27 × 5–12.5 cm, base attenuate, decurrent, margin entire to slightly crenate, apex with a drawn out obtuse tip; glabrous or with scattered hairs on midrib. Spikes terminal, 3–10.5 cm long, sometimes branched from basal bracts; peduncle to 5 mm long; rachis puberulous or sparsely so; bracts green with narrow to broad white scarious margin, ovate to elliptic, broadly so if margin included, (12–)14–23 × (8–)9–16 mm, finely puberulous or sparsely so, sometimes also with scattered long hairs, not ciliate, acute to acuminate at apex, cuneate at base; bracteoles elliptic to obovate, 8–12 mm long, puberulous. Calyx 7–10 mm long, puberulous, lobes acuminate to cuspidate. Corolla white with pale mauve lines on lower lip near throat, 1.8–2 cm long, puberulous or densely so; tube 9–11 mm long and 4–5 mm in diameter apically; upper lip slightly hooded, tapering upwards, 8–10 mm long; lower lip broadly obovate, spreading to deflexed, 9–11 × 8–12 mm long, with 3 short broad lobes 3–4 mm long. Filaments 5–6 mm long; thecae 2–3 mm long, ± similar, with 25–50% overlap, upper with small appendage, lower with a 1 mm long appendage varying from acute to truncate with several small teeth. Capsule almost circular in outline, 13–18 mm long, with placentae rising elastically from base, densely puberulous; seed ± 3 mm long, densely reticulately sculptured.

UGANDA. Kigezi District: Ishasha Gorge, Feb. 1950, *Purseglove* 3310!; Mengo District: Mawakota, Mpigi, Mpanga Forest, 12 Nov. 1952, *Dawkins* 760! & Entebbe, Aug. 1905, *E. Brown* 303!

DISTR. U 2, 4; Nigeria, Cameroon, Gabon, Central African Republic, Congo-Kinshasa, S Sudan, SW Ethiopia, Angola

HAB. Wet evergreen forest; 1150–1400 m

SYN. *Rungia grandis* T. Anderson in J.L.S. Bot. 7: 46 (1863); C.B. Clarke in F.T.A. 5: 252 (1900); T. Durand, Syll. Fl. Congo: 429 (1909); De Wildeman, Pl. Beq. 4: 30 (1926); F.P.N.A. 2: 308 (1947); Friis & Vollesen in Biol. Skr. 51(2): 454 (2005); Ensermu in Fl. Eth. 5: 490 (2006)

Justicia garckeana Büttn. in Verh. Bot. Ver. Brandenburg 32: 38 (1890). Type: Congo-Kinshasa, Myacca, Quango River, *Büttner* 356 (K!, iso.)

J. grandis (T. Anderson) Lindau in Schlechter, Westafr. Kautschuk.-Exp.: 317 (1900), *nom. nud.*

NOTE. The notes with *Dawkins* 760 say " reaching 15 m high". I am more than a bit dubious about this. It should probably read 15 feet, but even that is rather taller than stated on any other specimen.

45. **Justicia lukei** *Vollesen* **sp. nov.** ab omnibus ceteris speciebus africanis ad Sectionem *Vasciam* pertinentibus foliis grandibus angustes plusquam triplo longioribus quam latioribus, bracteis ad marginem viridibus (nec albo-hyalinus), et corolla cremeo-viridi (nec alba) atque atropurpureo-maculata differt. Typus: Tanzania, Morogoro District, Kimboza Forest Reserve, *Luke et al.* 7645 (K!, holo.; EA!, iso.)

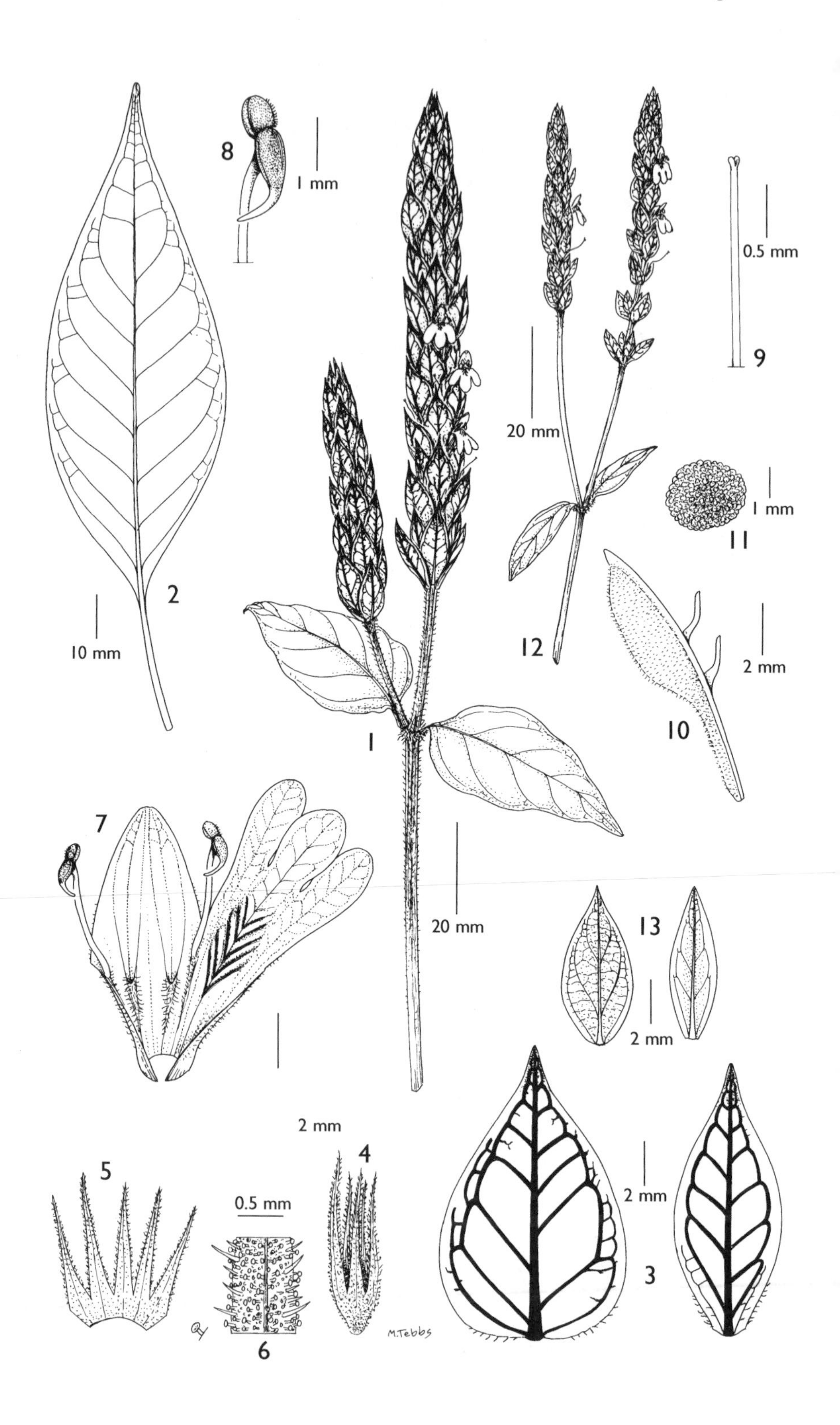
8
1 mm
0.5 mm
9
20 mm
1 mm
11
2
10 mm
12
2 mm
10
1
7
20 mm
13
2 mm
2 mm
5
4
0.5 mm
2 mm
3
6
M.Tebbs

Shrubby herb or shrub to 2.5(–4) m tall; young stems uniformly antrorsely sericeous or densely so with rufous hairs. Leaves with petiole 2–6 cm long with long spreading hairs on the sides; lamina elliptic or narrowly so, largest 20–33 × 6.5–11 cm, base attenuate, decurrent, margin entire to crenate, apex acuminate or with a drawn out obtuse tip; densely rufous-sericeous when young, soon glabrescent or persistently sericeous along main veins and sometimes also sparsely so on lamina. Spikes axillary (rarely 2 per axil), 2.5–12.5 cm long; peduncle to 4.5(–7) cm long; rachis sparsely sericeous; bracts dull green, without white scarious margin, ovate, 14–21 × 7–11 mm, acute obtuse and apiculate at apex, cuneate at base, with 3 strong longitudinal veins and transverse ladder-like veins, sericeous; bracteoles oblanceolate to narrowly obovate, 7–9 × 1.5–3 mm, sericeous with purple hairs. Calyx 5–7 mm long, sericeous with white or purple hairs, lobes acute to acuminate. Corolla creamy-green with dark purple streaks in throat, 13–14 mm long, densely retrorsely sericeous-puberulous with non-glandular hairs; tube 6–8 mm long and 3–4 mm in diameter apically; upper lip slightly hooded, tapering upwards, 6–7 mm long; lower lip deflexed, oblong to obovate, 5–7 × 4–6 mm long, with 3 short broad obtuse lobes 1–2 mm long. Filaments ± 4 mm long; thecae 2–2.5 mm long, ± similar, lower with an acute appendage ± 0.3 mm long. Capsule dark brown, 13–16 mm long, sericeous-puberulous near apex; seed 3–4 mm long, densely reticulately sculptured.

TANZANIA. Lushoto District: Kisiwani, 7 April 1941, *Greenway* 6166! & Manga Forest Reserve, 31 July 1997, *Simon et al.* 11!; Morogo District: Kimboza Forest Reserve, 1 April 1983, *Mwasumbi et al.* 12420!

DISTR. **T** 3, 6; not known elsewhere

HAB. Lowland evergreen forest, swamp forest, riverine forest, sometimes planted as hedge in and around villages; 150–600(–800) m

46. **Justicia betonica** *L.*, Sp. Pl.: 15 (1753); C.B. Clarke in F.T.A. 5: 184 (1899) & in Fl. Cap. 5: 57 (1901); T.T.C.L.: 11 (1949); F.P.S. 3: 178 (1956); E.P.A.: 968 (1964); Binns, Check List Herb. Fl. Malawi: 14 (1968); U.K.W.F.: 603 (1974); Bjørnstad in Serengeti Res. Inst. Publ. 215: 26 (1976); Vollesen in Opera Bot. 59: 80 (1980); Blundell, Wild Fl. E. Afr.: 393, pl. 20 (1987); U.K.W.F., ed. 2: 279, pl. 123 (1994); Ruffo *et al.*, Cat. Lushoto Herb. Tanzania: 6 (1996); Vollesen *et al.*, Checklist Mkomazi: 82 (1999); Friis & Vollesen in Biol. Skr. 51(2): 446 (2005); Ensermu in Fl. Eth. 5: 468 (2006). Type: Rheede, Hort. Mal. 2: t. 21 (1679), iconotype

Perennial (rarely annual) or shrubby herb with 1-several erect or ascending (rarely decumbent) stems from a creeping rhizome or a woody rootstock (rarely taproot); stems to 1.5 m long, glabrous (with band of hairs at nodes) to densely uniformly puberulous. Leaves with petiole 0–2 cm long; lamina lanceolate to ovate or elliptic to broadly so (rarely linear-lanceolate), largest 2–16 × (0.4–)0.7–5.5 cm, base rounded to attenuate (rarely truncate), decurrent, margin entire to crenate, apex rounded to cuspidate; glabrous to puberulous. Spikes terminal, 2–12 cm long, secund; peduncle to 4(–6.5) cm long; rachis sparsely to densely pubescent; bracts white to pale green with dark green venation, more rarely uniformly green or tinged purplish, ovate to elliptic or broadly so, 6–17 × 3–8 mm, glabrous to puberulous, acute to cuspidate and apiculate at apex, cuneate to truncate at base; bracteoles 5–14 × 2–5 mm. Calyx 4–7 mm long, finely puberulous with stubby non-capitate glandular hairs and with longer hairs on edges, at least towards apex, lobes acute to acuminate. Corolla white with 2 pink ridges or spots in throat, 8.5–14 mm long, densely puberulous and with capitate

FIG. 72. *JUSTICIA BETONICA* — **1**, habit; **2**, leaf; **3**, bract (left) and bracteole (right); **4**, calyx; **5**, calyx opened up; **6**, detail of calyx lobe; **7**, corolla opened up; **8**, apical part of filament and anther; **9**, apical part of style and stigma; **10**, capsule; **11**, seed. *J. RUWENZORIENSIS* — **12**, habit; **13**, bract (left) and bracteole (right). 1 from *Richards* 19975, 2, 3 & 7–8 from *Verdcourt* 1675, 4–6 & 9 from *B. J. Harris* 1613, 10 & 11 from *Drummond & Hemsley* 2289, 12 & 13 from *Lye* 2067. Drawn by Margaret Tebbs.

glands; tube 4.5–7.5 mm long and 2.5–4.5 mm in diameter apically; upper lip slightly hooded, 4–7.5 mm long; lower lip deflexed, 4–7 × 5–8 mm, central lobe wider and longer than laterals. Anthers purplish, upper theca 0.5–1.5 mm long, lower 1–2 mm long, with an acute appendage ± 1 mm long, connective hairy. Capsule 11–17(–21) mm long, uniformly puberulous or densely so; seed ± 3 mm long, densely tuberculate. Fig. 72: 1–11, p. 548.

UGANDA. Karamoja District: Mt Morongole, July 1965, *J. Wilson* 1662!; Toro District: Kibale Forest, 16 Dec. 1938, *Loveridge* 257!; Mengo District: 8 km SW of Mpigi, Mpanga Forest Reserve, 11 Oct. 1953, *Drummond & Hemsley* 4738!

KENYA. West Suk District: Sebit, Dec. 1963, *Tweedie* 2958!; North Kavirondo District: Kakamega Forest, 9 Dec. 1956, *Verdcourt* 1675!; Kilifi District: Pangani, 13 Dec. 1990, *Luke & Robertson* 2632!

TANZANIA. Mbulu District: Lake Manyara National Park, Mto wa Chem Chem, 24 March 1964, *Greenway & Kanuri* 11411!; Kigoma District: Mugombasi River, 31 Aug. 1959, *Harley* 9459!; Mbeya District: Songwe Valley, 25 March 1988, *Bidgood et al.* 690!

HAB. In a wide range of grassland, bushland and woodland, riverine forest and scrub, margins of wet forest, also in secondary vegetation and as a weed; 50–2450 m

DISTR. **U** 1–4; **K** 1–7; **T** 1–8; widespread in tropical and South Africa; tropical Asia, introduced in America

SYN. *J. trinervia* Vahl, Enum. Pl. 1: 156 (1804); C.B. Clarke in F.T.A. 5: 185 (1899); F.P.S. 3: 178 (1956). Type: "India orientalis", *Rottler* s.n. (not seen)

Adhatoda betonica (L.) Nees in Wall., Pl. As. Rar. 3: 103 (1832) & in DC., Prodr. 11: 385 (1847)

A. trinervia (Vahl) Nees in Wall., Pl. As. Rar. 3: 103 (1832) & in DC., Prodr. 11: 386 (1847)

A. variegata Nees in DC., Prodr. 11: 385 (1847); Richard, Tent. Fl. Abyss. 2: 154 (1850); Solms Laubert in Schweinfurth, Beitr. Fl. Aeth.: 104 (1867). Type: Ethiopia, Dochli, *Schimper* II.516 (K!, holo.; BM!, iso.)

Justicia variegata (Nees) Martelli in N. Giorn. Bot. Ital. 20: 393 (1888), *non J. variegata* Aubl. (1775)

Nicoteba betonica (L.) Lindau in E.J. 18: 56, 63, t.2 (1894) & in P.O.A. C: 370 (1895)

Justicia betonicoides C.B. Clarke in F.T.A. 5: 184 (1899); Binns, Check List Herb. Fl. Malawi: 14 (1968); Richards & Morony, Check List Mbala: 231 (1969); U.K.W.F.: 604 (1974). Types: Sudan, Jur Ghattas, *Schweinfurth* 1423 (K!, syn.) & Bongo, *Schweinfurth* 2543 (not seen) & Mittu, *Schweinfurth* 2793 (not seen); Kenya, Naivasha District: Gilgil, *Scott Elliot* 6647 (K!, syn.; BM!, iso.); Malawi, Fort Hill, *Whyte* s.n. (K!, syn.)

J. sp., Richards 6166 *sensu* Richards & Morony, Check List Mbala: 233 (1969)

J. sp. nov. Sect. Betonica sensu Luke & Robertson, Kenya Coast. For. 2. Checklist Vasc. Pl.: 82 (1993)

NOTE. A widespread and very variable species. But the variation – sometimes fairly distinct within certain regions – viewed over the total area of the species does not allow for any infraspecific taxa to be separated.

47. **Justicia engleriana** *Lindau* in E.J. 20: 62 (1894) & in P.O.A. C: 374 (1895); Iversen in Symb. Bot. Upsal. 29, 3: 161 (1991); Robertson & Luke, Kenya Coast. For. 2. Checklist Vasc. Pl.: 82 (1993); Ruffo *et al.*, Cat. Lushoto Herb. Tanzania: 6 (1996); Vollesen *et al.*, Checklist Mkomazi: 82 (1999). Types: Tanzania, Lushoto District: Usambara Mts, *Holst* 652 (B†, syn.), Usambara Mts, Mashewa, *Holst* 3491 (B†, syn.; K!, iso.), Kilimanjaro, Ngovi, *Volkens* 506 (B†, syn.; BM!, iso.), without locality, *Fischer* 150 (B†, syn.)

Shrubby herb or shrub to 3(–4) m tall with 1-several stems; young branches quadrangular, sparsely to densely retrorsely sericeous-puberulous on edges or uniformly so, glabrescent. Leaves ovate or elliptic to broadly so (rarely narrowly ovate), largest 5–45 × 4–22 cm, base decurrent to stem and ± distinctly auriculate, more rarely (especially in Uganda) with distinct petiole to 1(–2) cm long, margin slightly crenate, apex acuminate to acute; uniformly pubescent when young, soon glabrescent or persistently puberulous on midrib and veins (sometimes sparsely so on lamina). Spikes axillary and often congested towards tip of stems, 4–23(–30) cm long; peduncle (3–)5–23 cm long, subglabrous to densely puberulous or retrorsely sericeous-puberulous; rachis with similar indumentum; bracts pale green with darker venation, broadly ovate to cordiform, 2–3.8 × 1.3–2.5 cm, acute to acuminate at apex, subcuneate to subcordate at base, sparsely puberulous, usually densest (sometimes only) on veins; bracteoles similar to bracts but narrower. Calyx (7–)8–12 mm long, puberulous or densely so with stalked capitate glands, divided to 1.5–2 mm from base, lobes acute. Corolla white with greenish upper lip, no markings on lower lip, 2–2.5 cm long, sparsely puberulous and with scattered stalked capitate glands; tube 1–1.3 cm long, basal part cylindrical, 5–6 mm in diameter, apical part strongly widened, 10–12 mm in diameter; upper lip strongly hooded, almost sugblobose in side view, 11–13 × 13–17 mm; lower lip deflexed, deeply 3-lobed, 10–16 mm long, central lobe triangular, 8–12 × 6–9 mm, lateral lobes 8–12 × 3.5–5 mm. Thecae oblong, 2.5–3 mm long, lower with short stubby appendage to 0.5 mm long, connective hairy. Capsule 2–3 cm long, glandular-puberulous or densely so and with scattered non-glandular hairs, sometimes breaking irregularly near base; seed ellipsoid, flattened, 6–7 mm long, slightly reticulate sculptured to almost smooth.

UGANDA. Toro District: Bwamba, Sempaya, 26 Oct. 1953, *Dawkins* 810!; Bunyoro District: Masindi, 25 Dec. 1932, *Hazel* 286!; Mengo District: Kampala, 5 July 1927, *Maitland* 35!

KENYA. Northern Frontier District: Ndoto Mts, E of Nguronit Mission, 8 June 1979, *M.G. Gilbert et al.* 5558!; Machakos District: Kibwezi, 3 Aug. 1963, *Verdcourt* 3693!; Kwale District: Pengo Hill, 19 Feb. 1968, *Magogo & Glover* 139!

TANZANIA. Same District: 10 km on Kisiwani–Same road, Makuu River, 3 May 1995, *Abdallah & Vollesen* 95/104!; Lushoto District: Kwashemshi Sisal Estate, Ombeyi Hill to Miembeni, 9 Nov. 1999, *Mwangoka et al.* 1025!; Morogoro District: Wami River to Turiani, 4 Nov. 1947, *Brenan & Greenway* 8273!

DISTR. **U** 2–4; **K** 1, 4, 7; **T** 2, 3, 6–8; Congo-Kinshasa, Mozambique, Zimbabwe.

HAB. Lowland to lower montane evergreen forest (including secondary), riverine forest and thicket, often planted as hedge in and around villages; 150–1500 m

USES. In Uganda this is planted because it is supposed to keep wild animals (especially leopards) away. Used medicinally (not specified) in Tanzania.

SYN. *Adhatoda engleriana* (Lindau) C.B. Clarke in F.T.A. 5: 222 (1900); T.T.C.L.: 1 (1949); Vollesen in Opera Bot. 59: 78 (1980); U.K.W.F., ed. 2: 281 (1994); K.T.S.L.: 597 (1994)

NOTE. Ugandan specimens are always hairy and Tanzanian always glabrous, but the Kenyan material is a mixture.

48. **Justicia paxiana** *Lindau* in E.J. 20: 63 (1894) & in E. & P. Pf. IV, 3b: 350 (1895); T. Durand & De Wildeman, Mat. Fl. Congo 1: 38 (1897); De Wildeman & T. Durand, Contrib. Fl. Congo 2: 49 (1900); De Wildeman, Etud. Fl. Bas- et Moyen Congo 1: 322 (1906); T. & H. Durand, Syll. Fl. Congo: 430 (1909). Type: Cameroon, Buea, *Preuss* 956 (B†, holo.; BM!, K!, iso.)

Perennial or shrubby herb or shrub, usually with several erect stems to 2.5 m long; young stems sparsely to densely antrorsely sericeous, densest in two bands. Leaves with petiole to 3(–6) cm long; lamina ovate to elliptic, largest 8.5–17 × 2.8–7.5 cm, base attenuate, decurrent, margin entire to slightly crenate, apex with a long drawn out acute to obtuse tip; subglabrous to sericeous along midrib and larger veins, glabrous to sparsely sericeous on lamina. Spikes terminal and axillary, 1–5.5 cm long; peduncle to 0.5(–1.8) cm long; rachis sericeous to puberulous; bracts green or dark green with broad white scarious margin, elliptic or broadly so (often with jagged edges), broadly obovate if scarious margin included, 8–10 × 5–7 mm (incuding margin), sericeous to puberulous or sparsely so and conspicuously ciliate, acute to acuminate at apex, cuneate at base; bracteoles ovate or broadly so, 5–8 mm long, acuminate, sericeous. Calyx 3–5 mm long, sericeous, ciliate on edges, lobes acute, dorsal wider than the rest. Corolla white to very pale pink with purple streaks on both lips, 9–11 mm long, glabrous; tube 4–5 mm long and 2–3 mm in diameter apically; upper lip slightly hooded, tapering upwards, 5–6 mm long; lower lip deflexed, obovate, 5–6 × 4–5 mm, with 3 short broad lobes ± 1 mm long, central wider and longer than laterals. Filaments 2–3 mm long; thecae apiculate to rounded, 1–1.5 mm long, lower without appendage. Capsule thin-walled, with placentae rising elastically from base, 8–9 mm long, sparsely puberulous near apex; seed black, 2–3 mm long, densely tuberculate.

UGANDA. Kigezi District: Ishasha Gorge, May 1950, *Purseglove* 3423!; Ankole District: Bushenyi, Kasyoka-Kitomi Forest Reserve, Kyambura River, 12 June 1994, *Poulsen et al.* 576!; Mengo District: Nansagazi, Nakiza Forest, 27 Nov. 1950, *Dawkins* 674!

TANZANIA. Bukoba District: Minziro Forest Reserve, 2 July 2000, *Bidgood et al.* 4786! & 6 July 2000, *Bidgood et al.* 4868! & 12 July 2001, *Festo et al.* 1611!

DISTR. **U** 2–4; **T** 1; Guinée, Ivory Coast, Nigeria, Cameroon, Equatorial Guinea incl. Bioko, Central African Republic, Congo-Kinshasa, S Sudan, Burundi

HAB. Wet evergreen medium altitude forest, swamp forest; 1150–1400 m

SYN. *Rungia buettneri* Lindau in E.J. 20: 46 (1894) & in E. & P. Pf. IV, 3b: 350 (1895); C.B. Clarke in F.T.A. 5: 253 (1900); Friis & Vollesen in Biol. Skr. 51(2): 454 (2005), *non Justicia buettneri* Lindau in E.J. 20: 68 (1894). Type: Congo-Kinshasa, Kasongo Lunda, Ganga River, *Buettner* 456 (B†, holo.)

R. paxiana (Lindau) C.B. Clarke in F.T.A. 5: 253 (1900)

Isoglossa rungioides S. Moore in J.B. 45: 333 (1907). Types: Uganda, Bunyoro District: Lake Albert Edward, Ngusi River, *Bagshawe* 1361 (BM!, syn.) & Unyoro, Hoima, *Bagshawe* 1461 (BM!, syn.)

49. **Justicia sp. D**

Shrubby herb or shrub; young stems densely pubescent to antrorsely sericeous-pubescent with yellow hairs. Leaves with petiole to 2.5 cm long; lamina elliptic or narrowly so, largest 9.5–12 × 3–4 cm, base attenuate, decurrent, margin slightly crenate, apex with a drawn out acute to obtuse tip; sericeous to pubescent along midrib and larger veins beneath. Spikes axillary (rarely 2 per axil), 1.5–3 cm long; peduncle to 0.4 cm long; rachis sericeous; bracts yellowish green with narrow white scarious margin, ovate to elliptic, 6–7 × 2–3 mm (incuding margin), pubescent with pale yellow hairs, acuminate with recurved tip at apex, cuneate at base; bracteoles broadly ovate, ± 5 mm long, with broader white margin. Calyx 4–5 mm long, sparsely strigose-pubescent, glabrous towards base, lobes cuspidate. Corolla white with brown markings on lower lip, ± 6 mm long, hairy apically on lobes, otherwise glabrous; tube ± 3 mm long and ± 1 mm in diameter apically; upper lip slightly hooded, ± 3 mm long; lower lip deflexed, ± 3 mm long. Stamens bending out of corolla during female phase; filaments ± 2 mm long; thecae subequal, 1–1.5 mm long, lower with appendage ± 0.3 mm long. Capsule and seed not seen.

KENYA. North Kavirondo District: Yala River opposite Nature Reserve, Iloro, 26 Jan. 1982, *M.G. Gilbert* 6900!

DISTR. **K** 5; Ivory Coast

HAB. Wet evergreen montane forest; 1650 m

NOTE. A most peculiar distribution. There are a few minor differences between the Kenyan specimen and the collections from Ivory Coast; but nothing more than can be expected from such a vast disjunction.

50. **Justicia sp. E**

Perennial herb, basal part of stems creeping and rooting, apical part ascending to 35 cm long retrorsely sericeous-puberulous with bent curly hairs. Leaves slightly anisophyllous; petiole to 3.5 cm long; lamina ovate, largest ± 6 × 3.5 cm, base cuneate to attenuate, apex subacuminate with acute to rounded tip; sericeous-puberulous on midrib and veins, otherwise glabrous. Spikes axillary (sometimes seemingly terminal), 1–2.5 cm long; peduncle to 4 cm long, indumentum as stem, with 1–2 sterile bracts at base; rachis puberulous with short stubby non-capitate glands; bracts not strobilate, green with white scarious margin, narrowly elliptic to elliptic or obovate, 3.5–4.5 mm long, puberulous; bracteoles filiform to linear, ± 3 mm long, finely puberulous and with a few stalked capitate glands. Calyx ± 4 mm long, with indumentum as bracteoles, lobes filiform, cuspidate. Corolla white, no markings on lower lip, 5–6 mm long, puberulous; tube 3–4 mm long and ± 1.5 mm in diameter; upper lip slightly hooded, 2–3 mm long; lower lip spreading with recurved lobes, 2–3 mm long. Filaments 1–2 mm long; thecae green, oblong, ± 0.8 mm long, lower with appendage ± 0.5 mm long. Capsule and seed not seen.

TANZANIA. Mpanda District: Ntakatta Forest, 12 June 2000, *Bidgood et al.* 4657!
DISTR. **T** 4; not known elsewhere
HAB. Wet evergreen forest along stream; 1100 m

NOTE. Known only from this collection. It is clearly, despite the absence of fruits, a "*Rungia*" and is closest to West and central African *R. congoensis*, from which it differs in the small non-strobilate bracts and small corolla.

51. **Justicia faulknerae** *Vollesen* **sp. nov.** ab omnibus speciebus africanis ad sectionem *Betonicam* pertinentibus bracteis margine inconspicua pallide viridi ornatis atque in fructu brunneis, calyce glanduloso-puberulo et placenta e pariete exteriore capsulae secedenti atque a basin elastice oriente differt. Typus: Tanzania, Pangani District, Segera Forest, *Faulkner* 4076 (K!, holo.)

Shrubby herb or shrub to 2.5 m tall; young stems uniformly antrorsely sericeous-puberulous or densely so. Leaves with petiole to 3 cm long; lamina ovate to elliptic, largest 6.5–10.5 × 2.7–4.2 cm, base attenuate, decurrent, margin entire, apex acute to subacuminate; antrorsely sericeous-puberulous along midrib and larger veins, sparsely so or subglabrous on lamina. Spikes terminal, 2.5–7 cm long, sometimes with a single (rarely two) branch at base (rarely axillary); peduncle to 1(–1.5) cm long; rachis finely and sparsely puberulous; bracts pale green (turning brown) with a rather inconspicuous scarious margin, broadly ovate-cordiform to reniform, 8–10 × 5–7 mm, subglabrous to finely puberulous and finely ciliate, acute to obtuse at apex, cuneate to truncate at base; bracteoles similar to bracts but slightly shorter and narrower. Calyx ± 5 mm long, minutely and sparsely glandular-puberulous and with a few ciliae, lobes acute. Corolla (none available for dissection) white with small purple markings in throat, ± 9 mm long, sparsely puberulous; tube ± 4.5 mm long and ± 2 mm in diameter apically; upper lip slightly hooded, tapering upwards, ± 4.5 mm long; lower lip spreading, broadly obovate, ± 5.5 × 5 mm, with 3 short broad lobes. Stamens not seen. Capsule with placentae rising elastically from base, 8–9 mm long, sparsely sericeous-puberulous in upper half; seed black, ± 2 mm long, densely tuberculate.

KENYA. Kilifi District: Mangea Hill, 27 Dec. 1988, *Luke* 1590! & 25 April 1989, *Luke & Robertson* 1819!
TANZANIA. Pangani District: Segera Forest, 9 Feb. 1968, *Faulkner* 4076!; Bagamoyo District: Zaraninge Forest Reserve, Gongo Village, 27 July 1999, *Abeid* 629!
DISTR. **K** 7; **T** 3, 6; not known elsewhere
HAB. Semi-deciduous lowland forest; 150–450 m

SYN. *Rungia sp. nov. sensu* Robertson & Luke, Kenya Coastal For. 2. Check List Vasc. Pl.: 83 (1993)

52. **Justicia mkungweensis** *Vollesen* **sp. nov.** sectioni *Anselliae* floribus in cymis racemoideis axillaribus unilateralibus similis sed ab omnibus speciebus ad hanc sectionem pertinentibus herba erecta suffruticosa, cymeis in axillis foliorum non singulariter tantum dispositis, foliis maioribus et bracteis bracteolisque obovato-spathulatis (nec subulatis neque linearibus) differt. A speciebus ad sectionem *Betonicam* pertinentibus bracteis parvis non strobilatis praecipue differt. Typus: Tanzania, Morogoro District, Uluguru Mts, Mt Mkungwe, *Schlieben* 3976 (K!, holo.; MO!, iso.)

Shrubby herb; young stems antrorsely sericeous, densest in two bands. Leaves with petiole 3–5 cm long; lamina ovate, largest 13.5–18 × 6–8 cm, base attenuate, decurrent, apex acute to subacuminate with obtuse tip; sparsely sericeous on midrib and veins. Spikes solitary, axillary, 2.5–4 cm long; peduncle and rachis sericeous or (upwards) sericeous-puberulous with a few stalked capitate glands; peduncle 0.5–1.2 cm long; rachis puberulous with short stubby non-capitate glands; bracts and bracteoles not strobilate, obovate-spatulate, 4–5 mm long, without scarious margin, sparsely sericeous and with long-stalked capitate glands. Calyx 4–5 mm long, densely strigose-sericeous, no capitate glands, lobes lanceolate, acuminate. Corolla white with purple lines on lower lip, 8–9 mm long, curly-puberulous; tube 6–7 mm long and ± 2 mm in diameter; upper lip slightly hooded, ± 2 mm long; lower lip spreading, ± 4 mm long. Filaments 1–2 mm long; thecae oblong, ± 1 mm long, hairy, lower with appendage ± 0.5 mm long. Capsule and seed not seen.

TANZANIA. Morogoro District: Uluguru Mts, Mt Mkungwe, 23 May 1933, *Schlieben* 3976! & 28 March 1987, *Pócs & Nsolomo* 87051/V!
DISTR. **T** 6; not known elsewhere
HAB. Medium altitude evergreen forest; 900–1150 m

NOTE. Known only from these two collections made at 50 years interval from the same forest. The one-sided racemoid inflorescence with only one bract per node supporting a flower seems – despite the small non-strobilate bracts – to indicate that the species belongs in Sect. *Betonica*.

53. **Justicia roseobracteata** *Vollesen* **nom. nov.** Type: Tanzania, Morogoro District, Uluguru Mts, *Schlieben* 2846 (B†, holo.; BM!, iso.)

Perennial herb; stems to 35 cm long, sparsely puberulous to retrorsely sericeous-puberulous with broad curly hairs. Leaves with petiole to 1.5 cm long; lamina ovate, largest 5–12 × 2.5–6 cm, base attenuate, decurrent, margin slightly crenate, apex subacute; sparsely sericeous-puberulous along midrib and larger veins beneath. Spikes terminal and axillary, 1.5–4.5 cm long; peduncle to 1.3 cm long; rachis antrorsely sericeous-puberulous; bracts green with broad white scarious margin, tinged purplish towards apex of spikes, ovate to broadly ovate if margins included, green central part lanceolate to oblanceolate, 6–7 × 4–5 mm (sterile 8–11 × 4–5 mm), sparsely sericeous and ciliate, sterile acuminate with scarious margin tapering gradually, fertile with short acute mucro overtopped by the scarious "wings", cuneate at base; bracteoles 5–6 × 2–3 mm. Calyx tinged purplish towards apex of spikes; the 4 long lobes 4–5 mm long, cuspidate, the 5th lobe ± 0.5 mm long, finely ciliate, otherwise glabrous. Corolla greenish white with purple lower lip, ± 6 mm long, hairy on lower lip, otherwise glabrous; tube ± 3.5 mm long and ± 1.5 mm in diameter apically; upper lip flat, ± 2.5 mm long; lower lip spreading, obovate, ± 2.5 mm long, with 3 short ovate lobes. Stamens not bending out during female phase; filaments ± 2 mm long; thecae oblong, ± 1 mm long, upper hairy, lower with obtuse appendage ± 0.5 mm long. Capsule with placentae rising elastically from base, pale brown, ± 4.5 mm long, thin walled, sparsely puberulous near apex; seed ± 1 mm long, finely echinate.

TANZANIA. Morogoro District: Uluguru Mts, NE side, 18 Oct. 1932, *Schlieben* 2846!; Iringa District: Mwanihana Forest Reserve, above Sanje Village, 30 Aug. 1984, *D. W. Thomas* 3625! & Luhega Forest Reserve, 20 Jan. 1997, *Frimodt-Møller et al.* TZ22!
DISTR. **T** 6, 7; not known elsewhere
HAB. Dense evergreen forest; 900–1650 m

FIG. 73. *JUSTICIA TENELLA* — **1**, habit; **2**, bract (right) and bracteole (left); **3**, calyx, whole and opened up; **4**, detail of calyx lobe; **5**, corolla; **6**, corolla opened up; **7**, apical part of filament and anther; **8**, apical part of style and stigma; **9**, capsule; **10**, seed. 1 & 9–10 from *Mhoro* 170, 2–4 & 8 from *Richards* 9851, 5–7 from *Vaughan* 607. Drawn by Margaret Tebbs.

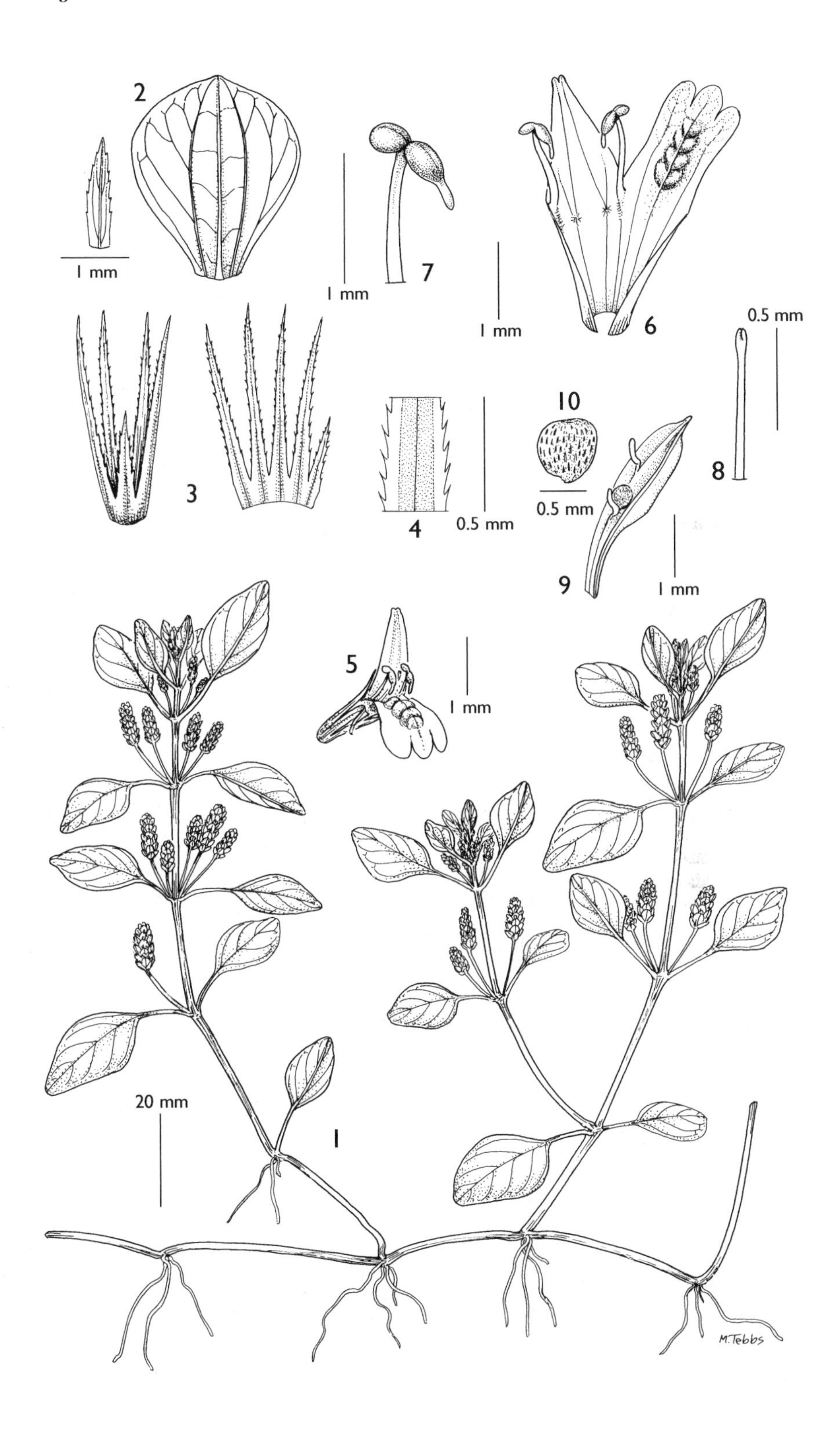
2
1 mm
7
1 mm
1 mm
6
0.5 mm
3
4
0.5 mm
10
0.5 mm
8
9
1 mm
5
1 mm
20 mm
1
M.Tebbs

SYN. *Rungia schliebenii* Mildbr. in N.B.G.B. 11: 1087 (1934), *non Justicia schliebenii* Mildbr. in N.B.G.B. 11: 411 (1932). Type as for *J. roseobracteata*

NOTE. Known only from these three collections. With its 4-lobed calyx (plus one minute tooth) this clearly belongs in Sect. *Anisostachya.*

54. **Justicia tenella** (*Nees*) *T. Anderson* in J.L.S. Bot. 7: 40 (1863); Lindau in Bol. Soc. Brot. 10: 147 (1892); C.B. Clarke in F.T.A. 5: 187 (1899); Benoist, Cat. Pl. Madagascar, Acanthacées: 25 (1939); F.P.S. 3: 178 (1956); Binns, Check List Herb. Fl. Malawi: 14 (1968); Ruffo *et al.*, Cat. Lushoto Herb. Tanzania: 7 (1996); Friis & Vollesen in Biol. Skr. 51(2): 448 (2005). Types: Senegal, Rio Nunez, *Heudelot* 4 (P, syn.); Madagascar, Imerina [Emirna], *Bojer* s.n. (K!, syn.)

Annual herb, basal part of stem erect or (more often) creeping and rooting at nodes, apical part always erect; stems to 0.6 m long, puberulous or sparsely so in two bands with broad curly hairs. Leaves with petiole 0.5–1.5(–3) cm long; lamina ovate to elliptic or broadly so, largest 2–4.5 × 1.3–3 cm, base attenuate, decurrent, margin entire to slightly crenate, apex subacute to broadly rounded; with scattered hairs along veins beneath, above with uniformly scattered hairs. Spikes axillary, 0.5–2(–3) cm long; peduncle to 2(–3.5) cm long; rachis with indumentum as stem; bracts pale green, broadly obcordiform, 3–4.5 × 2.5–4 mm, glabrous to sparsely puberulous-ciliate, truncate to retuse at apex, attenuate at base; bracteoles 1.5–2 mm long. Calyx with the 4 long lobes 2.5–3 mm long, cuspidate, the 5th lobe 0.5–1 mm long, glabrous or finely and sparsely puberulous-ciliate. Corolla white to pale mauve with magenta markings on lower lip, 2.5–3 mm long, glabrous; tube 1.5–2 mm long and ± 1 mm in diameter apically; upper lip ± 1 mm long; lower lip deflexed, ± 1 mm long. Stamens exserted; thecae ± 0.3 mm long, lower with flattened obtuse appendage ± 0.3 mm long hairy. Capsule 2.5–3 mm long, with a few hairs near apex; seed ± 0.8 mm long, sparsely hairy. Fig. 73, p. 555.

TANZANIA. Morogoro District: Turiani, Divue River, 22 Aug. 1981, *Abdallah* 978!; Rungwe District: Lusangu, 21 May 1957, *Richards* 9851!; Lindi District: Nyangedi, 20 March 1935, *Schlieben* 6150!; Pemba, Ngezi Forest, 30 Aug. 1929, *Vaughan* 607!

DISTR. **T** 6–8; **P**; widespread in tropical Africa, Madagascar

HAB. Evergreen forest, secondary forest, riverine forest, weed in cleared forest areas; near sea level to 900(–1500) m

SYN. *Rostellaria tenella* Nees in DC., Prodr. 11: 369 (1847); Bentham in Hooker, Niger Fl.: 482 (1849)
Anisostachya tenella (Nees) Lindau in E. & P. Pf. IV, 3b: 329 (1895)

55. **Justicia anagalloides** (*Nees*) *T. Anderson* in J.L.S. Bot. 7: 42 (1863); Mildbraed in N.B.G.B. 9: 504 (1926); Immelman in Fl. Pl. Afr. 49: pl. 1932 (1986); V.A.W. Graham in K.B. 43: 598 (1988); Ensermu in Symb. Bot. Upsal. 29, 2: 52, fig. 20 & 21 (1990); U.K.W.F. ed. 2: 279, pl. 123 (1994); Hedrén in Fl. Somalia 3: 417 (2006); Ensermu in Fl. Eth. 5: 479 (2006). Type: South Africa, Apjes River, *Burke* s.n. (K!, holo.)

Perennial herb with several unbranched or branched erect, decumbent or creeping (then sometimes rooting) stems from vertical or horizontal woody rootstock; stems to 80 cm long, when young bifariously puberulous or sparsely so to uniformly pubescent (rarely tomentose) with retrorse hairs, sometimes also with longer (to 2 mm) broad curly glossy many-celled hairs with white walls. Leaves with petiole 0–5 mm long; lamina lanceolate to ovate or elliptic or obovate or narrowly so (rarely orbicular), largest 0.7–5(–7) × 0.4–1.9(–3.5) cm, base cuneate to attenuate, margin flat, apex subacute to rounded; glabrous (finely ciliate towards apex) or with scattered to dense long (to 2.5 mm) broad glossy many-celled hairs on margins (espc. towards base) and sometimes also on veins (rarely also on lamina or uniformly pubescent). Spikes (1–)2–5(–7)-flowered; peduncle 0.5–4(–5) cm long, bifariously

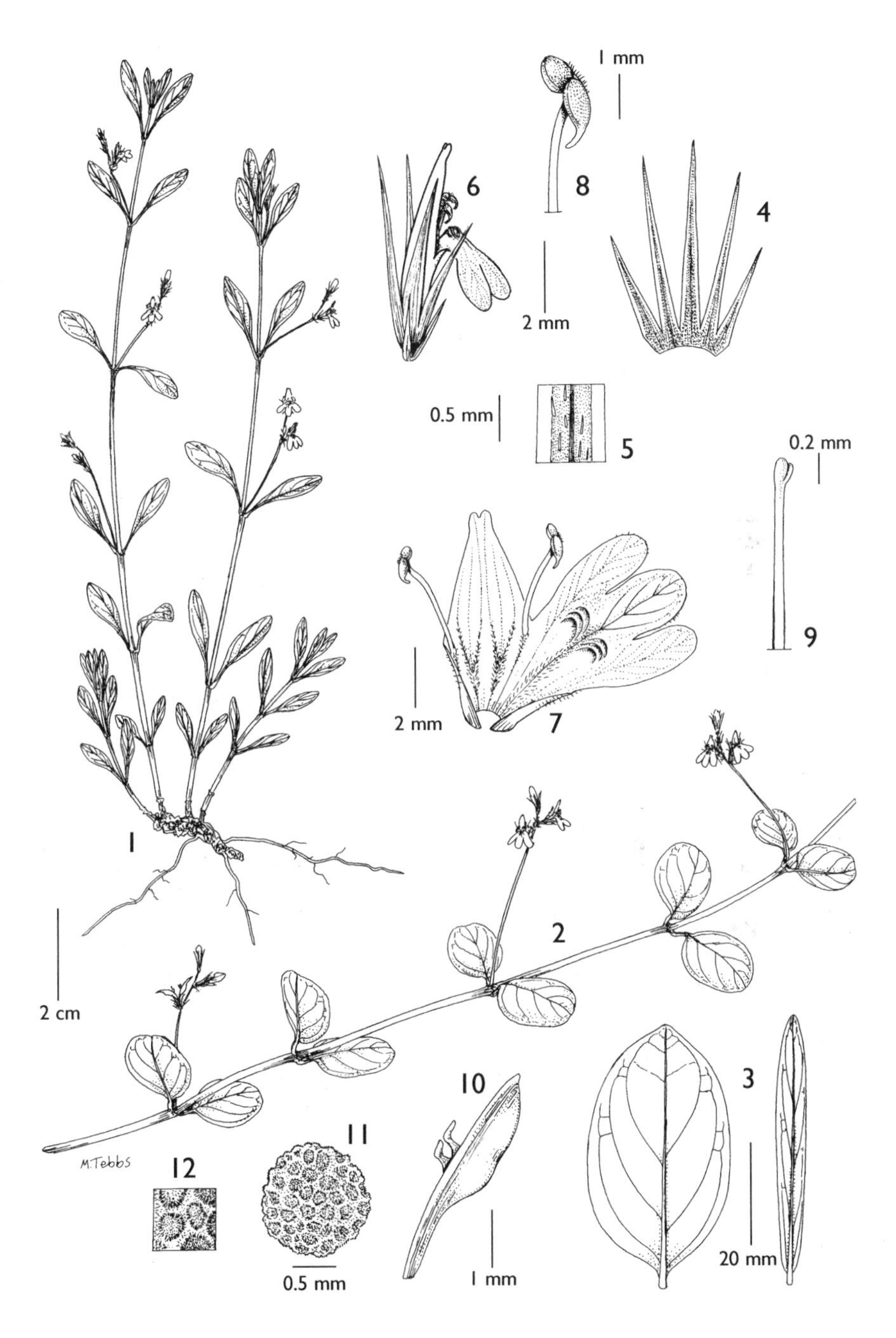

FIG. 74. *JUSTICIA ANAGALLOIDES* — **1**, habit, erect form; **2**, habit, trailing form; **3**, variation in leafshape; **4**, calyx opened up; **5**, detail of calyx lobe; **6**, calyx and corolla; **7**, corolla opened up; **8**, apical part of filament and anther; **9**, apical part of style and stigma; **10**, capsule; **11**, seed; **12**, detail of seed. 1 & 4–5 from *Richards* 24171, 2 from *Faden & Ng'weno* 74/546, 3 (narrow leaf) from *Drummond & Hemsley* 1776 and (broad leaf) *Vollesen* MRC3614, 6–9 from *Faulkner* 507, 10–12 from *Bidgood et al.* 3969. Drawn by Margaret Tebbs.

(rarely uniformly) sericeous-puberulous or sparsely so, sometimes also with scattered long many-celled hairs; rachis 4–18(–30) mm long, with similar indumentum; bracts and bracteoles subulate, 1–4 mm long, glabrous. Calyx minutely hispid-ciliate towards apex (rarely also on midrib), lobes subulate to linear, apiculate, two lower 3–7 mm long, three upper 5–10(–12) mm long. Corolla white with faint pink to mauve "herring bones" on lower lip and two pink lines on upper lip, 6–9(–10) mm long; tube 2–4 mm long; upper lip 3–6 mm long; lower lip 3–6(–7) mm long, middle lobe 2.5–5(–6) × 1.5–3(–4) mm. Filaments 2–4 mm long; anthers 1.5–2 mm long, thecae 0.7–1 mm long. Capsule (5–)6–8 mm long, hairy in upper half, more rarely glabrous; seed ± 1.5 mm in diameter. Fig. 74, p. 557.

UGANDA. Kigezi District: Ruzumbura, April 1939, *Purseglove* 673!; Masaka District: Buddu, Buganga, Feb. 1894, *Scott Elliot* 7454! & Kalungu, N of Mukoko, Kasasa, 5 June 1971, *Lye* 6195!
KENYA. Nakuru District: Eastern Mau Forest Reserve, 9 Sept. 1949, *Maas Geesteranus* 6226!; South Nyeri District: Mwea Irrigation Scheme, 3 June 1976, *Kahurananga & Kibui* 2791!; Masai District: 45 km on Nairobi–Kajiado road, 17 Dec. 1961, *Polhill & Paulo* 1020!
TANZANIA. Lushoto District: 5 km NW of Mombo, 29 April 1953, *Drummond & Hemsley* 2270!; Mpanda District: 10 km on Mpanda–Inyonga road, 15 May 1997, *Bidgood et al.* 3969!; Iringa District: Kitonga Gorge, Mt Image, 9 Dec. 1986, *J. Lovett & Congdon* 1069!
DISTR. **U** 2–4; **K** 1–7; **T** 1–8; Ethiopia, Somalia, Congo-Kinshasa, Rwanda, Burundi, Zambia, Malawi, Mozambique, Zimbabwe, Swaziland, South Africa
HAB. In a wide variety of grassland, bushland, woodland and dry forest; on a wide range of soils; (100–)250–3000 m

SYN. *Adhatoda anagalloides* Nees in DC., Prodr. 11: 403 (1847)
Justicia uncinulata Oliv. in Trans. Linn. Soc. 29: 130, tab. 129A (1875); Lindau in P.O.A. C: 373 (1895); C.B. Clarke in F.T.A. 5: 210 (1900); Mildbraed in N.B.G.B. 9: 503 (1926); Chiovenda in K.B. 1941: 174 (1941); E.P.A.: 976 (1964); U.K.W.F.: 603 (1974); Bjørnstad in Serengeti Res. Inst. Publ. 215: 26 (1976); Champluvier in Fl. Rwanda 3: 468, fig. 145, 2 (1985); Synnott, Checklist Fl. Budongo For.: 69 (1985); Blundell, Wild Fl. E. Afr.: 394, pl. 125 (1987); V.A.W. Graham in K.B. 43: 598 (1988); Ruffo *et al.*, Cat. Lushoto Herb. Tanzania: 7 (1996). Type: Tanzania, Mpwapwa/Kilosa District: Rubeho Mts, *Speke & Grant* s.n. (K!, holo.)
[*J. anselliana sensu* Britten *et al.* in Trans. Linn. Soc. Bot. London, Ser. 2, 4: 32 (1894); C.B. Clarke in F.T.A. 5: 208 (1900), quoad *Buchanan* 483, 876 & 1385; Lugard in K.B. 1933: 94 (1934), *non* (Nees) T. Anderson (1863)]
[*J. matammensis sensu* C.B. Clarke in F.T.A. 5: 209 (1900), quoad *Scott Elliot* 6543 & 7454, *Speke & Grant* s.n., Unyamwezi, *non* (Schweinf.) Oliv. (1875)]
J. crassiradix C.B. Clarke var. *hispida* C.B. Clarke in F.T.A. 5: 210 (1900). Type: Malawi, Shire Highlands, *Buchanan* 461 (K!, lecto.; selected by Ensermu, l.c.)
J. psammophila Mildbr. in N.B.G.B. 11: 1089 (1934); V.A.W. Graham in K.B. 43: 598 (1988). Type: Tanzania, Morogoro, *Schlieben* 3309 (B, lecto.; selected by Ensermu, l.c.; BM!, iso.)
J. sp. C sensu Agnew, U.K.W.F.: 603 (1974) pro parte
J. sp. aff. J. anselliana sensu Bjørnstad in Serengeti Res. Inst. Publ. 215: 26 (1976)
[*J. crassiradix sensu* Vollesen in Opera Bot. 59: 80 (1980), *non* C.B. Clarke (1900)]
J. sp. ? nov. aff. J. nuttii sensu Vollesen in Opera Bot. 59: 81 (1980)
J. sp. ? nov. aff. J. uncinulata sensu Vollesen in Opera Bot. 59: 81 (1980)

NOTE. Ensermu (l.c.) has a long and detailed discussion of the variation within this widespread and very variable species. He concludes that it is possibly heterogenous, but that extensive field studies are needed to prove or disprove this.

56. **Justicia lorata** *Ensermu* in Nordic Journ. Bot. 9: 403, figs. 2 & 3 (1989) & in Symb. Bot. Upsal. 29, 2: 57 (1990); U.K.W.F., ed. 2: 279 (1994). Type: Kenya, 40 km NW of Nairobi, Rarre Ranch, *Boulos* 12202 (B, holo.; ILCA, iso.)

Perennial herb with several branched erect, decumbent or creeping (then often rooting) stems from vertical rootstock; stems to 45 cm long, when young bifariously pubescent or sparsely so with broad curly glossy many-celled hairs with white walls. Leaves with petiole 0–2 mm long; lamina linear to lanceolate, largest 1.3–4 × 0.2–0.5 cm, base attenuate, margin strongly recurved, apex acute to rounded; finely apiculate;

towards base and on petiole with long (to 2 mm) broad glossy many-celled hairs, otherwise glabrous or with a few similar hairs on midrib, finely hispid ciliate towards apex. Spikes 3–4(–5)-flowered; peduncle 5–18 mm long, bifariously sericeous-puberulous or sparsely so; rachis 2–13 mm long, with similar indumentum; bracts and bracteoles subulate to narrowly triangular, 1–2 mm long, glabrous. Calyx minutely hispid-ciliate, at least near apex, lobes subulate to linear, apiculate, two lower 2–4 mm long, three upper 3–6 mm long. Corolla white with or without faint pink to mauve "herring bones" on lower lip, 4–5(–6) mm long; tube 1.5–2.5 mm long; upper lip 2.5–3(–4) mm long; lower lip 2–3.5 mm long, middle lobe 1.5–2.5 × 1–2 mm. Filaments 2–2.5 mm long; anthers 1–1.5 mm long, thecae 0.5–0.8 mm long. Capsule 4–5 mm long, glabrous or with a few hairs at the apex; seed ± 1 mm in diameter.

KENYA. Naivasha District: E side of Lake Naivasha, 18 April 1968, *Mwangangi* 738!; Nairobi District: Nairobi National Park, Mokoyeti Gorge, 31 Dec. 1977, *Gillett* 21647!; Masai District: 40 km on Nairobi–Magadi road, 19 Feb. 1969, *Napper* 1872!

DISTR. **K** 3, 4, 6; not known elsewhere

HAB. *Acacia* bushland on rocky soils, usually volcanic, grassland; (1200–)1500–2100 m

SYN. [*Justicia anselliana sensu* C.B. Clarke in F.T.A. 5: 208 (1900) quoad *Scott Elliot* 6633 & 6810, *non* (Nees) T. Anderson (1863)]

J. uncinulata Oliv. var. *tenuicapsa* C.B. Clarke in F.T.A. 5: 210 (1900) quoad *Scott Elliot* 6521, *non* sensu stricto

J. sp. A sensu Agnew, U.K.W.F.: 603 (1974)

57. **J. ornatopila** *Ensermu* in Nordic Journ. Bot. 9: 399, figs. 1 & 2 (1989) & in Symb. Bot. Upsal. 29, 2: 58 (1990); Luke & Robertson, Kenya Coast. For. 2. Checklist Vasc. Pl: 82 (1993); U.K.W.F., ed. 2: 279 (1994); Vollesen *et al.*, Checklist Mkomazi: 83 (1999); Hedrén in Fl. Somalia 3: 417 (2006); Ensermu in Fl. Eth. 5: 480 (2006). Type: Ethiopia, Harerge, 4 km on Gewane–Awash road, *Ensermu* 1393 (ETH, holo.)

Shrubby perennial herb with trailing (sometimes rooting at nodes) and looping much-branched stems from woody rootstock; stems to 70 cm long, older woody, grey, with corky bark, younger subglabrous to bifariously to uniformly sericeous to puberulous with thin bent hairs with walls not visible. Leaves with petiole to 7 mm long; lamina ovate-oblong or narrowly so (rarely ovate), largest 1–4.5 × 0.3–1.7 cm, base cuneate to attenuate, apex rounded (rarely retuse); subglabrous to sparsely sericeous-puberulous, densest along veins (rarely uniformly puberulous). Spikes 2–3(–4)-flowered; peduncle (0–)0.3–2.5(–3.5) cm long, subglabrous to bifariously sericeous-puberulous with thin hairs; rachis 2–12 mm long, with similar indumentum; bracts and bracteoles linear to narrowly triangular, 1–2 mm long, acuminate, glabrous to puberulous. Calyx glabrous to sericeous or puberulous and finely hispid-ciliate, lobes linear to linear to subulate, two lower 3–5(–6) mm long, three upper 5–8(–9) mm long. Corolla white without (rarely with) pink "herring bones" on lower lip, 4.5–6.5 mm long; tube 2–3 mm long; upper lip 2.5–4 mm long; lower lip 2.5–4 mm long, middle lobe 2–3 × 1–2 mm. Filaments 2–2.5 mm long; anthers 1–1.5 mm long, thecae 0.5–0.8 mm long. Capsule 8–12 mm long, glabrous or sparsely puberulous at the apex (rarely all over); seed 2.5–3.5 mm in diameter.

UGANDA. Karamoja District: Kangole, 22 May 1940, *A.S. Thomas* 3488! & Amudat, Sept. 1949, *Tweedie* 797! & 75 km S of Moroto, 13 Sept. 1956, *Bally* 10818!

KENYA. Northern Frontier District: Furroli, 14 Sept. 1952, *Gillett* 13868!; Turkana District: Kateruk, 2 May 1954, *Popov* 1547!; Baringo District: Kampi ya Samaki, 30 Oct. 1992, *Harvey et al.* 21!

TANZANIA. Lushoto District: 8 km SE of Mkomazi, 2 May 1953, *Drummond & Hemsley* 2388! & Mombo to Same, 2 June 1970, *Mwasumbi et al.* 10732! & Mkomazi Game Reserve, SE of Ndea Hill, 6 May 1995, *Abdallah & Vollesen* 95/167!

DISTR. **U** 1; **K** 1–3, 7; **T** 3; SE Sudan, S Ethiopia, Somalia

HAB. *Acacia-Commiphora* bushland on clayey to sandy or rocky soils, rocky lava hills and lava flows; (25–)150–1400 m

58. **Justicia cufodontii** (*Fiori*) *Ensermu* in Symb. Bot. Upsal. 29, 2: 59, figs. 22 & 23 (1990); Hedrén in Fl. Somalia 3: 418 (2006); Ensermu in Fl. Eth. 5: 480 (2006). Type: Ethiopia, Moyale to Mega, *Cufodontis* 701 (FT, lecto.; selected by Ensermu (l.c.))

Perennial or shrubby perennial herb with woody rootstock, basal part of stems often woody, decumbent, sometimes rooting at nodes, much-branched with many erect to trailing flowering stems; basal decumbent stems to 50 cm long, eventually with grey slightly corky bark, flowering stems to 50(? –100) cm long, bifariously pubescent with broad curly glossy many-celled hairs with white walls. Leaves with petiole to 5(–10) mm long; lamina narrowly ovate-oblong to obovate (rarely ovate), largest 2–4.5(–6) × 0.5–1.3(–2.7) cm, base cuneate to attenuate, apex acute to rounded; subglabrous to sparsely pubescent along veins beneath, glabrous above; lateral veins distinct. Spikes 3–5(–6)-flowered; peduncle 0.5–2.5(–4.5) cm long, sparsely to densely bifariously puberulous to pubescent with broad glossy many-celled hairs; rachis 0.5–1.5(–2.5) cm long, with similar indumentum; bracts and bracteoles linear, 1.5–4 mm long, acuminate, glabrous or finely ciliate. Calyx sparsely sericeous-puberulous on midrib and finely hispid-ciliate (rarely glabrous), lobes linear to linear-lanceolate, two lower 4–7.5 mm long, three upper 7–10 mm long. Corolla white without (rarely with) pink "herring bones" on lower lip, 8–12 mm long; tube 3.5–5 mm long; upper lip 5–7 mm long; lower lip 5–7 mm long, middle lobe 3.5–5 × 2–4 mm. Filaments 3–5 mm long; anthers 1.5–2 mm long, thecae 0.7–1 mm long. Capsule 8–9 mm long, sparsely puberulous at the apex; seed 1.5–2 mm in diameter.

KENYA. Northern Frontier District: 4 km NNW of Marsabit, Gar Jirimi, 24 Nov. 1977, *Carter & Stannard* 679!; Masai District: Kajiado, 9 Jan. 1931, *Napier* 737! & 50 km on Nairobi–Magadi road, *Napper et al.* 1876!

TANZANIA. Masai District: Mto wa Mbu to Katete River, 13 March 1964, *Greenway et al.* 11356! & foot of Mt Longido, 29 Dec. 1968, *Richards* 23512! & Ngaserai Plain, 13 Dec. 1969, *Richards* 24934!

DISTR. **K** 1, 6; **T** 2; Ethiopia, Somalia

HAB. *Acacia drepanolobium* grassland on black cotton soil, volcanic hills, limestone pavements; 900–1700 m

SYN. *Justicia uncinulata* Oliv. var. *cufodontii* Fiori in Miss. Biol. Borana 4: 224, fig. 70 (1939); E.P.A.: 976 (1964)

59. **Justicia crassiradix** *C.B. Clarke* in F.T.A. 5: 210 (1900); Binns, Check List Herb. Fl. Malawi: 14 (1968); Richards & Morony, Check List Fl. Mbala Distr.: 231 (1969); V.A.W. Graham in K.B. 43: 598 (1988); Ensermu in Symb. Bot. Upsal. 29, 2: 61, figs. 24 & 25 (1990). Type: Zambia, Urungu, Fwambo, *Carson* 107 (K!, lecto.; selected by Ensermu, l.c)

Perennial herb with 1–2 erect (rarely decumbent) unbranched (rarely with a single branch) stems from creeping sometimes branched rhizome; stems to 30 cm long, glabrous (with fringe of hairs at nodes) to pubescent with broad glossy many-celled hairs with white walls. Leaves linear to lanceolate, sessile, largest 1.5–6 × 0.2–0.7 cm, base cuneate to attenuate, apex acute to acuminate; glabrous to pubescent with similar hairs; lateral veins absent or very inconspicuous. Spikes 3–4(–5)-flowered; peduncle 1–7.5 cm long, glabrous to pubescent with white hairs; rachis 0.5–3 cm long, with similar indumentum; bracts and bracteoles linear to narrowly triangular, 2–6 mm long, acuminate, glabrous or finely ciliate. Calyx glabrous (rarely sparsely puberulous) and finely hispid-ciliate, lobes subulate to linear, two lower 4–7 mm long, three upper 6–10 mm long. Corolla white with pink to pale mauve "herring bones" on lower lip, 8–12 mm long; tube 3.5–5.5 mm long; upper lip 4–7 mm long; lower lip 6–8 mm long, middle lobe 4–6 × 3–5 mm. Filaments 4–5.5 mm long; anthers 2–3 mm long, thecae 1–1.5 mm long. Capsule 14–16 mm long, glabrous; seed 2.5–3 mm in diameter.

TANZANIA. Ufipa District: Sumbawanga, Mbaa Mt, 16 Nov. 1986, *Brummitt et al.* 18008! & Tatanda, 25 Feb. 1994, *Bidgood et al.* 2455! & 2 km W of Mkowe on Chapota road, 21 Nov. 1994, *Goyder et al.* 3766!
DISTR. **T** 4, 7; Zambia
HAB. Seasonally inundated grassland (dambos) in woodland, valley grassland; 1050–1800 m

60. **Justicia nuttii** *C.B. Clarke* in F.T.A. 5: 210 (1900); Binns, Check List Herb. Fl. Malawi: 14 (1968); Bjørnstad in Serengeti Res. Inst. Publ. 215: 26 (1976); Cribb & Leedal, Mountain Fl. S. Tanzania: 127, pl. 31 (1988); V.A.W. Graham in K.B. 43: 598 (1988); Ensermu in Symb. Bot. Upsal. 29, 2: 66, figs. 28 & 29 (1990); Ruffo *et al.*, Cat. Lushoto Herb. Tanzania: 7 (1996). Type: Tanzania, Ufipa District: between Lake Tanganyika and Lake Rukwa, *Nutt* s.n. (K!, holo.)

Perennial herb with 1-several erect to decumbent unbranched or branched stems from creeping sometimes branched rhizome; stems to 50 cm long, glabrous to densely pubescent with broad glossy many-celled hairs with coloured walls. Leaves with petiole 1–5 mm long, ovate to elliptic or narrowly so, largest 1.5–5.5(–7.5) × 0.7–2 cm, base cuneate to subcordate, apex acute to broadly rounded; glabrous (rarely) to pubescent with similar hairs, above uniformly so, beneath mainly along veins. Spikes (1–)2(–3)-flowered; peduncle (0.5–)1.5–7.5 cm long, glabrous to pubescent with broad hairs with coloured walls; rachis 3–12 mm long, puberulous (sometimes only in a narrow band) to pubescent; bracts and bracteoles subulate to linear, 2–7 mm long, glabrous to puberulous. Calyx glabrous to puberulous, lobes subulate to linear, apiculate, two lower 4–6.5(–10) mm long, three upper 6.5–11.5(–15) mm long. Corolla white, no markings on lips, 8–13 mm long; tube 4–6 mm long; upper lip 4–7 mm long; lower lip 5–8 mm long, middle lobe 3.5–6.5 × 3–5 mm. Filaments 4–6 mm long; anthers 2–3 mm long, thecae 1–1.5 mm long. Capsule 9–12 mm long, pubescent in upper half with broad hairs with coloured walls (rarely glabrous); seed 2.5–3 mm in diameter.

TANZANIA. Ufipa District: Sumbawanga, 15 Jan. 1950, *Bullock* 2258!; Iringa District: Mufindi, Ngwazi, 3 Feb. 1987, *J. Lovett* 1418!; Njombe District: Poroto Mts, Kitulo Plateau, Ndumbi Valley, 21 March 1991, *Bidgood et al.* 2118!
DISTR. **T** 4, 7; Congo-Kinshasa, Zambia, Malawi, Mozambique
HAB. Montane grassland and bushland, seasonally flooded grassland and valley grassland in woodland, rarely in *Brachystegia* woodland; 1350–2800 m

SYN. *J. nuttii* C.B. Clarke var. *blantyrensis* C.B. Clarke in F.T.A. 5: 210 (1900); Binns, Check List Herb. Fl. Malawi: 14 (1968); Richards & Morony, Check List Fl. Mbala Distr.: 232 (1969). Type: Malawi, Shire Highlands, Blantyre, *Buchanan* 20 (K!, lecto.)
J. goetzei Lindau in E.J. 28: 484 (1900); C.B. Clarke in F.T.A. 5: 514 (1900); V.A.W. Graham in K.B. 43: 598 (1988). Type: Tanzania, Iringa District: Uhehe, Ukano Mts, *Goetze* 685 (B†, holo.; BM!, K!, iso.)
J. schliebenii Mildbr. in N.B.G.B. 11: 411 (1932). Type: Tanzania, Njombe District: Lupembe, Likanga, *Schlieben* 458 (B†, holo.)

61. **Justicia calyculata** *Defl.* in Bull. Soc. Bot. France 44: 224 (1896); Ensermu in Symb. Bot. Upsal. 29, 2: 68, fig. 30 & 31 (1990); Robertson & Luke, Kenya Coast. For. 2. Checklist Vasc. Pl.: 82 (1993); U.K.W.F., ed. 2: 279, pl. 123 (1994); Vollesen *et al.*, Checklist Mkomazi: 82 (1999); Hedrén in Fl. Somalia 3: 418 (2006); Ensermu in Fl. Eth. 5: 482 (2006). Type: Yemen, Haifan, Hodjeria, *Deflers* 548 (P, holo.)

Annual or short-lived perennial herb with often several erect, ascending, straggling or decumbent (then often rooting) usually branched (more rarely unbranched) stems to 60(–100) cm long, when young bifariously puberulous to tomentellous and often with few to many long broad curly glossy many-celled hairs to 3 mm long. Leaves with petiole 3–20 mm long, with few to many long broad glossy curly many-celled hairs; lamina linear-lanceolate to ovate (widest below middle), largest 1.7–5.5(–6.5) × 0.2–2.3(–2.8) cm, base cuneate to attenuate, apex acute to rounded; with sparse to dense long broad curly many-celled hairs along midrib (sometimes also lamina) and margins towards base. Spikes 2–5(–7)-flowered; peduncle 0.5–3.3(–4) cm long, bifariously puberulous or sparsely so and towards base also with long broad glossy hairs; rachis 3–20(–30) mm long, bifariously puberulous; bracts and bracteoles subulate, 1–2.5 mm long, glabrous. Calyx glabrous (rarely puberulous-ciliate near base), lobes linear-lanceolate, cuspidate, two lower 2–4 mm long, three upper 3–6(–7) mm long. Corolla white with faint pink to mauve "herring bones" on lower lip and two pink lines on upper lip, 4–5.5 mm long; tube 2–3 mm long; upper lip 2–3 mm long; lower lip 3–4 mm long, middle lobe 1.5–3 × 1–2 mm. Filaments 2–2.5 mm long; anthers 1–1.5 mm long, thecae 0.5–0.8 mm long. Capsule sometimes with purple lines, 4.5–7 mm long, puberulous to below middle; seed 1–1.5 mm in diameter.

UGANDA. Karamoja District: Kangole, 22 May 1940, *A. S. Thomas* 3463! & East Matheniko, Sept. 1958, *Wilson* 623!

KENYA. Northern Frontier District: Dadaab–Wajir road, just N of Lagh Dera, 12 May 1974, *Gillett & Gachati* 20637!; Kitui District: 24 km on Embu–Endau road, 17 Nov. 1979, *Mungai et al.* 79/38!; Kwale District: Kaya Lunguma, 11 Nov. 1992, *Luke* 3355!

TANZANIA. Masai District: Ngorongoro, Olbalbal, 19 Feb. 1966, *Herlocker* 291! & Ngorongoro Crater Rim, 7 Aug. 1989, *Chuwa* 2837!; Pare District: Mkomazi Game Reserve, SE of Ndea Hill, 4 May 1995, *Abdallah & Vollesen* 95/118!

DISTR. **U** 1; **K** 1–4, 6, 7; **T** 2, 3; SE Sudan, S Ethiopia, S Somalia; Yemen

HAB. *Acacia* and *Acacia-Commiphora* woodland and bushland, grassland, glades in lowland forest, rocky outcrops, roadsides, on a wide range of soils; 25–1800(–2350) m

SYN. [*J. matammensis sensu* C.B. Clarke in F.T.A. 5: 209 (1900), quoad *Gregory* s.n., *Hildebrandt* 2840, *Scott Elliot* 6343 & *Volkens* 512; U.K.W.F.: 603 (1974), *non* (Schweinf.) Oliv. (1875)]

J. beguinotii Fiori in Miss. Biol. Borana 4: 222 (1939); E.P.A.: 968 (1964). Type: Ethiopia, Moyale, *Cufodontis* 659 (FT, lecto.; selected by Ensermu, l.c.)

62. **Justicia exigua** *S. Moore* in J.B. 38: 204 (1900); C.B. Clarke in F.T.A. 5: 514 (1900); U.K.W.F.: 603 (1974); V.A.W. Graham in K.B. 43: 598 (1988); Ensermu in Symb. Bot. Upsal. 29, 2: 71, fig. 32 & 33 (1990); Robertson & Luke, Kenya Coast. For. 2. Checklist Vasc. Pl.: 82 (1993); U.K.W.F., ed. 2: 279 (1994); Ruffo *et al.*, Cat. Lushoto Herb. Tanzania: 6 (1996); Friis & Vollesen in Biol. Skr. 51(2): 446 (2005); Ensermu in Fl. Eth. 5: 482 (2006). Type: Zimbabwe, Bulawayo, *Rand* 389 (BM!, holo.)

Annual (rarely biennial) herb with much-branched erect, ascending or decumbent (then often rooting) stems to 60 cm long; stems bifariously puberulous and with or without long broad curly many-celled hairs to 2 mm long. Leaves with petiole 2–10 mm long, with few to many long broad curly many-celled hairs; lamina narrowly to broadly elliptic (rarely lanceolate or ovate), most or all widest near middle, largest 1–4(–5) × 0.3–1.7 cm, base cuneate to attenuate, apex subacute to broadly rounded; glabrous or with scattered to dense long broad curly many-celled hairs along midrib, more rarely also on lamina and edges towards base. Spikes (2–)3–7-flowered; peduncle (0.3–)0.5–3.2 cm long, bifariously puberulous or sparsely so and (espc. towards base) often with long broad curly many-celled hairs; rachis 0.5–2.2 cm long, bifariously puberulous or sparsely so; bracts and bracteoles subulate to linear, 0.5–2 mm long, glabrous. Calyx glabrous, lobes linear, two lower 2–4 mm long, three upper 3–6(–7) mm long. Corolla white with faint pink to mauve "herring bones" on lower lip and two pink lines on upper lip, 4–5 mm long; tube 2–2.5 mm long; upper lip 2–3 mm long; lower lip 2.5–3.5 mm long, middle lobe 1.5–2.5 × 1.5–2 mm. Filaments 1.5–2 mm long; anthers ± 1 mm long, thecae ± 0.5 mm long. Capsule sometimes with purple lines and patches, (3.5–)4–6.5 mm long, with scattered hairs in apical half (rarely almost to base); seed ± 1 mm in diameter.

UGANDA. Acholi District: Murchison Falls National Park, Chobi, 5 Sept. 1967, *Angus* 5844!; Busoga District: Bulamogi, Kaliro, 18 Sept. 1952, *G.H.S. Wood* 390!; Mengo District: Kome Island, Bugombe, 27 Oct. 1968, *Lye* 57!

KENYA. Ravine District: 20 km NE of Eldama Ravine, 21 Aug. 1956, *Bogdan* 4224!; Kiambu District: Thika, 28 July 1967, *Faden* 67/506!; South Kavirondo District: Mbita Point Field Station, 27 Nov. 1981, *Gachati & Opon* 22/81!

TANZANIA. Bukoba District: Minziro Forest Reserve, Kigazi, 21 March 2001, *Festo et al.* 1082!; Arusha District: Ngurdoto Crater, The Rock, 18 March 1966, *Greenway & Kanuri* 12433!; Dodoma District: 3 km on Manyoni–Singida road, 15 April 1988, *Bidgood et al.* 1109!

DISTR. **U** 1–4; **K** 1–7; **T** 1–7; Sudan, Ethiopia, Congo-Kinshasa, Rwanda, Burundi, Angola, Zambia, Mozambique, Zimbabwe, Botswana, Namibia, South Africa

HAB. In a wide variety of dry forest, woodland, bushland and grassland, often in disturbed or secondary vegetation or as a weed in cultivated areas, lawns, on a wide range of soils; 75–2300 m

SYN. [*J. matammensis sensu* C.B. Clarke in F.T.A. 5: 209 (1900), quoad *Elliott* s.n. & *Holub* 1208–1210; Lindau in Wiss. Ergebn. Schwed. Rhod.–Kongo-Exp. 1911–12 2: 309 (1914); Mildbraed in N.B.G.B. 9: 503 (1926); F.P.N.A. 2: 314 (1947); Synnott, Checklist Fl. Budongo For.: 69 (1985); Champluvier in Fl. Rwanda 3: 468, fig. 145 (1985), *non* (Schweinf.) Oliv. (1875)]

J. uncinulata Oliv. var. *tenuicapsa* C.B. Clarke in F.T.A. 5: 210 (1900). Type: Sudan, Upper Nile, *Freeman & Lucas* 79 (K!, lecto.; selected by Ensermu, l.c.)

J. exilissima Chiov. in Ann. Bot. Roma 10: 401 (1912); E.P.A.: 969 (1964). Type: Ethiopia, Lake Zway, *Negri* 815 (FT, holo.)

[*J. anselliana sensu* De Wildeman, Pl. Beq. 4: 34 (1926) pro parte; Benoist in Bol. Soc. Brot., ser. 2, 24: 27 (1950); Immelman in Fl. Pl. Afr. 49: pl. 1932 (1986) pro parte, *non* (Nees) T. Anderson (1863)]

63. **Justicia brevipedunculata** *Ensermu* in Pl. Syst. Evol. 163: 121, figs. 1–3 (1989) & in Symb. Bot. Upsal. 29, 2: 74 (1990). Type: Tanzania, Kondoa District, Mbulu Mts, *Hedrén et al.* 763 (UPS, holo.; NHT, iso.)

Annual herb with erect or ascending (rarely decumbent) much-branched (more rarely unbranched) stems to 35 cm long, when young bifariously (rarely uniformly) puberulous to pubescent (rarely tomentose) with long broad curly glossy many-celled hairs. Leaves held like a V (folded on herbarium sheets); petiole to 4 mm long; lamina narrowly ovate to ovate or elliptic, largest 1.2–7 × 0.5–2.2 cm, base on lower pairs often attenuate to cuneate, otherwise subcordate to cordate, apex acute to rounded; below with sparse to dense broad glossy many-celled hairs along midrib, above with sparse to dense uniformly distributed hairs. Spikes (3–)4–7-flowered; peduncle 2–5(–6) mm long, the basal 1.5–2 mm adnate to petiole, sparsely to densely pilose with long broad glossy hairs; rachis 5–22 mm long, with similar indumentum; bracts and bracteoles subulate, 0.5–1 mm long, glabrous; sterile bracts to 2 mm long. Calyx pilose-ciliate or sparsely so on midrib and margins, lobes linear-lanceolate, cuspidate, two lower 2.5–4 mm long, three upper 4–8 mm long. Corolla white with faint pink to mauve "herring bones" on lower lip and two pink lines on upper lip, 5–6 mm long; tube 2–3 mm long; upper lip 3–4 mm long; lower lip 3–4 mm long, middle lobe 2–3 × 1–2 mm. Filaments 2–3 mm long; anthers 1–1.5 mm long, thecae 0.5–0.8 mm long. Capsule 5.5–8 mm long, with long pilose hairs in upper half; seed 2–2.5 mm in diameter.

TANZANIA. Mbulu District: Serengeti, Endabash, 9 May 1958, *Paulo* 442!; Nzega District: 5 km on Igunga–Singida road, 9 May 1994, *Bidgood & Vollesen* 3311!; Iringa District: Ruaha National Park, Msembi Airstrip, 1 April 1970, *Greenway & Kanuri* 14246!

DISTR. **T** 1, 2, 4, 5, 7; not known elsewhere

HAB. Grassland on grey crumbling usually alkaline clay; 700–1450 m

64. **Justicia anselliana** (*Nees*) *T. Anderson* in J.L.S. Bot. 7: 44 (1863); S. Moore in J.B. 18: 341 (1880); Lindau in P.O.A. C: 373 (1895); C.B. Clarke in F.T.A. 5: 208 (1900) pro parte; De Wildeman, Pl. Beq. 4: 34 (1926); F.P.N.A. 2: 314 (1947); F.P.S. 3: 179 (1956); F.P.U.: 103 (1962); Binns, Check List Herb. Fl. Malawi: 14 (1968); U.K.W.F.: 603 (1974); Champluvier in Fl. Rwanda 3: 468 (1985); V.A.W. Graham in K.B. 43: 598 (1988); Ensermu in Symb. Bot. Upsal. 29, 2: 74, fig. 34 & 35 (1990); U.K.W.F., ed. 2: 279, pl. 123 (1994); Friis & Vollesen in Biol. Skr. 51(2): 446 (2005). Type: Liberia, Cape Palmas, *Ansell* s.n. (K!, lecto., selected by Graham, l.c.; C!, iso.)

Annual or short-lived perennial herb with one to several erect, ascending, straggling or decumbent (then often rooting) unbranched or sparsely branched sometimes inflated and spongy stems to 85 cm long from short creeping rhizome; stems glabrous except for a thin band at nodes (rarely uniformly puberulous or sparsely so). Leaves with petiole 0–5 mm long, sometimes with a few long hairs; lamina linear to lanceolate, largest (2–)3–10.5 × 0.15–1 cm, base cuneate to attenuate, apex acute to acuminate (rarely rounded); minutely hispid-ciliate, otherwise glabrous or with a few hairs at base (rarely also along midrib or uniformly puberulous). Spikes (2–)3–7-flowered; peduncle 0.7–4 cm long, glabrous (very rarely puberulous); rachis (0.5–)1–3.3(–5) cm long, glabrous (very rarely sparsely puberulous); bracts and bracteoles subulate to linear, 1–2.5 mm long, glabrous. Calyx minutely scabrid-ciliate, lobes linear-lanceolate, cuspidate, two lower 2.5–4(–5) mm long, three upper 3.5–6(–7) mm long. Corolla white with faint pink to mauve "herring bones" on lower lip and two pink lines on upper lip, 5.5–8.5 mm long; tube 2.5–4 mm long; upper lip 3–4.5 mm long; lower lip 3.5–5.5 mm long, middle lobe 3–4.5 × 2–3.5 mm. Filaments 2.5–3.5 mm long; anthers 1–1.5(–2) mm long, thecae 0.5–0.8(–1) mm long. Capsule sometimes with purple lines, (7.5–)8–10.5 mm long, glabrous (very rarely hairy at apex); seed ± 2 mm in diameter.

UGANDA. West Nile District: Omnyo Rest Camp, 10 Aug. 1953, *Chancellor* 151!; Teso District: 15 km on Soroti–Moroti road, Arabaka Dam, 30 July 1967, *Kabuye* 98!; Masaka District: Mawagola, 17 km SE of Ntusi, 19 Oct. 1969, *Lye* 4491!

KENYA. Uasin Gishu District: 25 km on Eldoret–Nakuru road, 14 Oct. 1981, *M.G. Gilbert* 6754!; Kericho District: Sotik, 15 June 1953, *Verdcourt* 985!; Masai District: Loita Plains, Olongeri, 29 April 1961, *Glover et al.* 883!

TANZANIA. Bukoba District: Minziro Forest Reserve, 20 March 2001, *Festo et al.* 1065!; Tabora District: Ugalla River, Isimbila, 25 Oct. 1960, *Richards* 13399!; Singida District: 25 km E of Singida, Mgori, 1 May 1962, *Polhill & Paulo* 2271!

DISTR. **U** 1–4; **K** 3, 5, 6; **T** 1, 2, 4, 5; Liberia, Ghana, Togo, Benin, Nigeria, Central African Republic, Sudan, Congo-Kinshasa, Rwanda, Angola, Zambia, Namibia

HAB. *Acacia drepenolobium* bushland and grassland on grey to black clayey soils; 900–2350 m

SYN. *Adhatoda anselliana* Nees in DC., Prodr. 11: 403 (1847)

Justicia anselliana (Nees) T. Anderson var. *angustifolia* Oliv. in Trans. Linn. Soc. 29: 130 (1875); S. Moore in J.B. 18: 342 (1880); Lindau in P.O.A. C: 373 (1895); F.P.N.A. 2: 314 (1947); E.P.A.: 967 (1964). Type: Uganda, West Nile District: Madi, *Speke & Grant* s.n. (K!, lecto., selected by Ensermu, l.c.)

Dianthera anselliana (Nees) Bentham in G.P. 2: 1114 (1876)

Justicia kapiriensis De Wild. in B.J.B.B. 5: 12 (1915). Type: Congo-Kinshasa, Kapiri Valley, *Homblé* 1089 (BR, holo.; BR, iso.)

NOTE. A few collections (*Lavranos & Newton* 17751 (Kenya, Eldoret), *Norman* 69 (Tanzania/Uganda border), *Bax* 332 (Tanzania, Shinyanga)) have uniformly puberulous stems and leaves. Otherwise they have the characteristic habit of this species: thin straggling stems combined with narrow leaves. Their habitat (black cotton soil) is also typical of the species.

65. **Justicia matammensis** (*Schweinf.*) *Oliv.* in Trans. Linn. Soc. 29: 130 (1875); Lindau in P.O.A. C: 373 (1895); C.B. Clarke in F.T.A. 5: 209 (1900); F.P.S. 3: 180 (1956); E.P.A.: 972 (1964); Binns, Check List Herb. Fl. Malawi: 14 (1968); U.K.W.F.: 603 (1974); Bjørnstad in Serengeti Res. Inst. Publ. 215: 26 (1976); Vollesen in Opera Bot. 59: 81 (1980); Blundell, Wild Fl. E. Afr.: 394, pl. 124 (1987); V.A.W. Graham in K.B. 43: 598 (1988); Ensermu in Symb. Bot. Upsal. 29, 2: 78, fig. 36 & 37 (1990); Ruffo *et al.*, Cat. Lushoto Herb. Tanzania: 7 (1996); Vollesen *et al.*, Checklist Mkomazi: 83 (1999); Friis & Vollesen in Biol. Skr. 51(2): 447 (2005); Ensermu in Fl. Eth. 5: 482 (2006). Type: Sudan/ Ethiopia, Matamma, *Schweinfurth* 130C (K!, lecto.; selected by Ensermu, l.c.; BM!, iso.)

Annual herb with usually branched erect, ascending, straggling or decumbent (then often rooting) stems to 45 cm long; stems bifariously puberulous and usually with few to many long glossy curly many-celled hairs to 3 mm long. Leaves with petiole 3–15(–25) mm long, with few to many long glossy curly many-celled hairs; lamina ovate, less often some elliptic, largest 1.5–7(–11) × 0.5–3(–4) cm, base cuneate to attenuate, apex subacute to broadly rounded; with scattered to dense long curly hairs on midrib and lower part of margin and sometimes also on lamina. Spikes 3–6(–7)-flowered; peduncle 0.8–4.5(–7) cm long, bifariously puberulous and (sometimes only in basal part) with few to many long glossy curly many-celled hairs; rachis 0.5–2.8 cm long, bifariously puberulous; bracts and bracteoles subulate, 1–2 mm long, glabrous. Calyx glabrous, lobes linear-lanceolate, two lower 1.5–4 mm long, three upper 2.5–6 mm long. Corolla white with faint pink to mauve "herring bones" on lower lip and two pink lines on upper lip, 4–5.5 mm long; tube 2–2.5 mm long; upper lip 2–3.5 mm long; lower lip 3–4 mm long, middle lobe 1.5–3 × 1.5–2.5 mm. Filaments 2–2.5 mm long; anthers 1–1.5 mm long, thecae 0.5–0.8 mm long. Capsule sometimes with purple lines, 7–10(–11) mm long, hairy in apical half; seed ± 2 mm in diameter.

UGANDA. Karamoja District: Kangole, Aug. 1948, *Wilson* 600!; West Nile District: Maracha, 27 July 1953, *Chancellor* 63!; Kigezi District: Kinkizi, Kambuga, April 1946, *Purseglove* 2043!
KENYA. Turkana District: Napau Pass, 16 Nov. 1953, *Dale* K835!
TANZANIA. Lushoto District: Amani, 13 May 1950, *Verdcourt* 191!; Dodoma District: Great Ruaha River, 6 km upstream from Mtera Bridge, 9 Feb. 1975, *Backéus* 1091!; Chunya District: Mbangala Village to Lake Rukwa, 16 Feb. 1994, *Bidgood et al.* 2308!; Zanzibar, Kisimbani, 27 May 1961, *Faulkner* 2846!
DISTR. **U** 1, 2; **K** 2; **T** 1, 3–7; **Z**; Central African Republic, Congo-Kinshasa, Burundi, Sudan, Ethiopia, Zambia, Malawi, Mozambique, Zimbabwe, Botswana, South Africa
HAB. In a wide range of woodland, bushland and grassland (including secondary types), riverine forest, often a weed in cultivations, on a wide variety of soils; near sea level to 1550 m

SYN. *Adhatoda matammensis* Schweinf. in Verh. Königl. Zool. Bot. Ges. Wien 18: 464 (1868)
Justicia aethiopica Martelli in N. Giorn. Bot. Ital. 20: 393 (1888); Fiori in Miss. Biol. Boran. 4: 224 (1939). Type: Sudan, Fazogl, *Figari* s.n. (not seen). See note.

NOTE. An isotype of *J. aethiopica* is cited by Ensermu (l.c.) as being at Kew, but I have been unable to trace it. This may be a printing error as most other *Figari* collections are at Florence.

66. **Justicia odora** (*Forssk.*) *Lam.*, Encycl. Meth. Bot. 1, 2: 629 (1785); Vahl, Symb. Bot. 2: 11 (1791) & Enum. Pl. 1: 127 (1804); C.B. Clarke in F.T.A. 5: 201 (1900); Glover, Check-list Brit. and Ital. Somal.: 67 (1947); F.P.S. 3: 180 (1956); E.P.A.: 973 (1964); U.K.W.F.: 604 (1974); Kuchar, Pl. Somalia (CRDP Tech. Rep. 16): 248 (1986); Blundell, Wild Fl. E. Afr.: 394, pl. 402 (1987); Hedrén in Symb. Bot. Upsal. 29, 1: 66, fig. 25 (1989); Robertson & Luke, Kenya Coast. For. 2. Checklist Vasc. Pl.: 82 (1993); K.T.S.L.: 603 (1994); U.K.W.F., ed. 2: 280 (1994); Audru *et al.*, Pl. Vasc. Djibouti 2: 734 (1994); Vollesen *et al.*, Checklist Mkomazi: 83 (1999); Hedrén in Fl. Somalia 3: 414 (2006); Ensermu in Fl. Eth. 5: 465 (2006). Type: Yemen, Surdud, *Forsskål* 384 (C!, holo.; K!, photo).

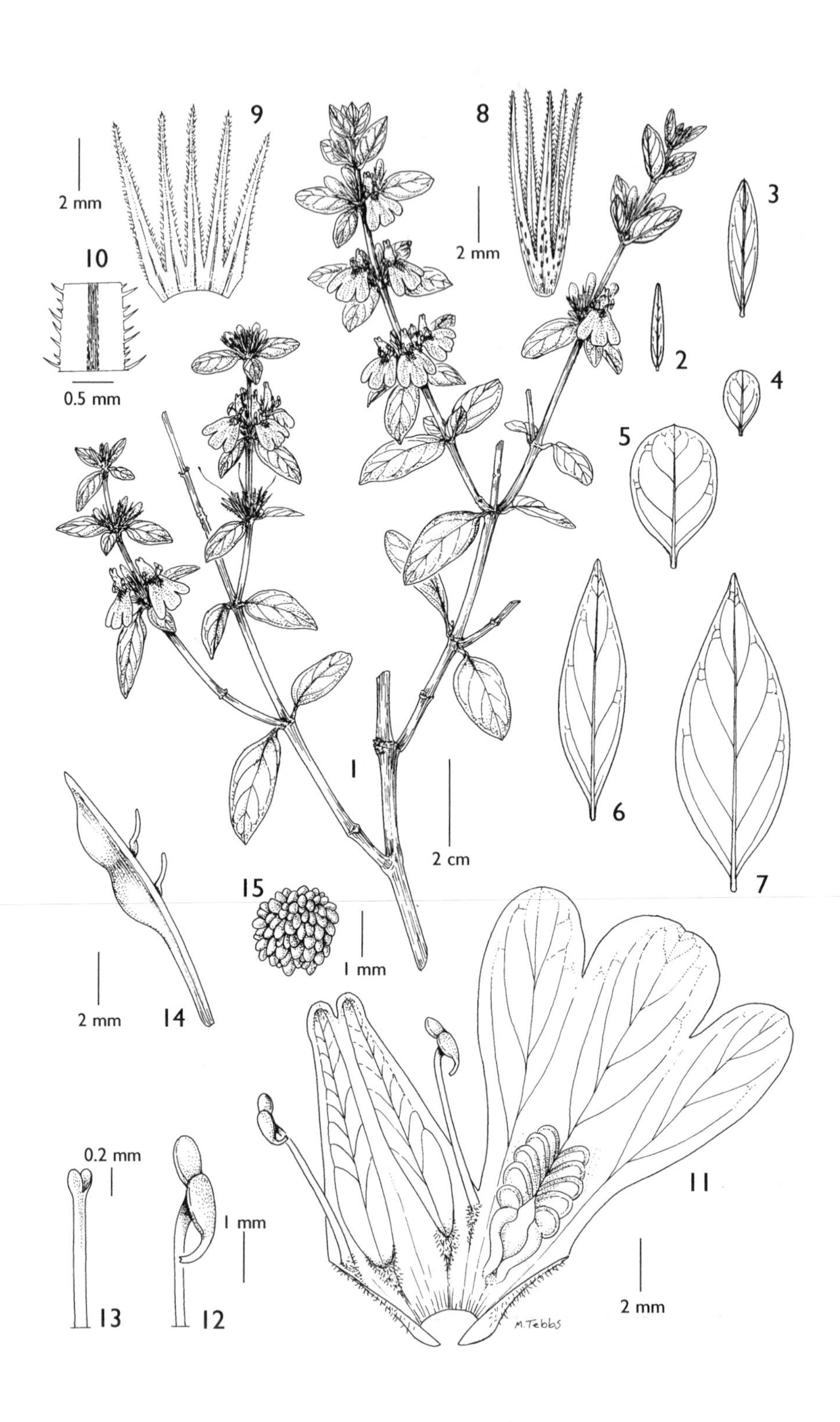
9
8
3
2 mm
2 mm
10
0.5 mm
2
4
5
6
7
1
2 cm
15
1 mm
2 mm
14
0.2 mm
1 mm
11
2 mm
13
12
M.Tebbs

Much-branched shrubby herb or shrub to 1.5 m tall; young stems pale green with straw-coloured ribs, glabrous to puberulous with short hispid hairs, older branches with first dark brownish then pale grey bark. Leaves with petiole 0–3 mm long; lamina lanceolate to broadly ovate, broadly elliptic or obovate, largest (0.7–)1.1–7.5 × (0.2–)0.4–3 cm, base attenuate to cuneate, margin entire, apex acute to emarginate, often apiculate; glabrous to uniformly puberulous with hispid hairs, usually also with longer (to 2 mm) hairs on margins near base, with few to many yellow to orange subsessile capitate glands, often only or mostly along midrib. Flowers single or 2–4 per axil of upper leaves; bracts like leaves, 0.5–1.5 cm long. Calyx sparsely puberulous-ciliate, otherwise glabrous, divided to 1–1.5 mm from base, lobes 4–8 mm long. Corolla lemon yellow to bright yellow, upper lip sometimes with red to purple veins and a purple patch at base, 9–13 mm long; tube 4–5 mm long, 2.5–4 mm diameter apically; upper lip 5–8 mm long (upper lip 1–3 mm longer than tube); lower lip 6–11 × 5–10 mm. Filaments 3.5–7 mm long; thecae yellow, with glandular exudate, 1–1.5 mm long, lower with bifid appendage 0.5–1.5 mm long. Capsule (9–)10–15(–18) mm long, glabrous (rarely with a few hairs near apex); seed 2.5–3 mm in diameter. Fig. 75, p. 566.

UGANDA. Karamoja District: East Matheniko, March 1959, *J. Wilson* 711 & Turkwell Gorge, April 1960, *J. Wilson* 916!

KENYA. Northern Frontier District: Dandu, 13 May 1952, *Gillett* 13163!; Turkana District: Lorengipe, 21 Jan. 1965, *Newbould* 6786!; Masai District: Selengai Game Post, 16 Dec. 1969, *Kibue* 113!

TANZANIA. Mbulu District: Lake Manyara National Park, Msasa River, 4 Dec. 1963, *Greenway & Kanuri* 11129!; Masai District: Mt Kitumbeine, 2 March 1969, *Richards* 24261!; Dodoma District: 25 km on Dodoma–Morogoro road, 12 April 1988, *Bidgood et al.* 1029!

DISTR. **U** 1; **K** 1–4, 6, 7; **T** 2, 3, 5; Sudan, Eritrea, Ethiopia, Djibouti, Somalia, Angola, Zimbabwe, Botswana, Namibia, South Africa; Egypt (Jebel Elba), Saudi Arabia, Yemen, Oman

HAB. *Acacia* and *Acacia-Commiphora* bushland on sandy to loamy or stony soil, rocky hillsides, riverine scrub; near sea level to 1850 m

SYN. *Dianthera odora* Forssk., Fl. Aegypt.-Arab.: 8 (1775); C. Christensen in Dansk Bot. Arkiv 4, 3: 11 (1922)

Adhatoda odora (Forssk.) Nees in DC., Prodr. 11: 399 (1847); Deflers, Voy. Yemen: 183 (1889)

A. hypericum Solms Laub. in Schweinf., Beitr. Fl. Aeth.: 102 (1867). Type: Ethiopia, Mawerr, *Schimper* 136 (not seen)

Justicia urbaniana Lindau in E.J. 20: 64 (1894) & in E. & P. Pf. IV, 3b: 349 (1895); C.B. Clarke in F.T.A. 5: 203 (1900); Glover, Check-list Brit. and Ital. Somal.: 67 (1947); E.P.A.: 976 (1964); Kuchar, Pl. Somalia (CRDP Tech. Rep. 16): 248 (1986). Types: Somalia, Ahl Mts, *Hildebrandt* 860b (B†, syn.) & Meid, *Hildebrandt* 1402 (B†, syn.)

J. fischeri Lindau in E.J. 20: 65 (1894) & in P.O.A. C: 373 (1895); C.B. Clarke in F.T.A. 5: 202 (1900); T.T.C.L.: 11 (1949); Glover, Check-list Brit. and Ital. Somal.: 66 (1947); E.P.A.: 969 (1964); Kuchar, Pl. Somalia (CRDP Tech. Rep. 16): 247 (1986). Types: Tanzania, *Fischer* 490 (B†, syn.; BM!, K!, iso.) & Lushoto District: Kwa Mshuza, *Holst* 8897 (B†, syn.; K!, iso.)

J. shebelensis Rendle in J.B. 34: 409 (1896); C.B. Clarke in F.T.A. 5: 200 (1900); Glover, Check-list Brit. and Ital. Somal.: 67 (1947). Type: Ethiopia, W of Shebele River, *Donaldson Smith* s.n. (BM!, holo.)

J. laetevirens Rendle in J.B. 34: 409 (1896), *nom. illeg., non J. laetevirens* Vahl (1805). Type: Ethiopia/Kenya, between Lake Stefanie and Lake Turkana, *Donaldson Smith* s.n. (BM!, holo.)

J. lorteae Rendle in J.B. 35: 379 (1897); C.B. Clarke in F.T.A. 5: 201 (1900); Glover, Check-list Brit. and Ital. Somal.: 67 (1947); E.P.A.: 972 (1964); Kuchar, Pl. Somalia (CRDP Tech. Rep. 16): 247 (1986). Types: Somalia, Wagga Mts, *Lort Phillips* s.n. (BM!, syn.); Bihin, *Lort Phillips* s.n. (BM!, syn.)

FIG. 75. *JUSTICIA ODORA* — **1**, habit; **2–7**, series of leaves; **8**, calyx; **9**, calyx opened up; **10**, detail of calyx lobe; **11**, corolla opened up; **12**, apical part of filament and anther; **13**, apical part of style and stigma; **14**, capsule; **15**, seed. 1 & 7 from *Bidgood et al.* 1029, 2 from *Greenway* 13828, 3 & 14–15 from *Tweedie* 4019, 4 from *Bally* 14651, 5 & 8–10 from *Richards* 24219, 6 from *Pearce & Vollesen* 942, 11–13 from *Bidgood et al.* 1277. Drawn by Margaret Tebbs.

J. romaniae Schweinf. & Volkens, Liste Pl. Ghika-Comanesti: 17 (1897); C.B. Clarke in F.T.A. 5: 211 (1900); Glover, Check-list Brit. and Ital. Somal.: 67 (1947); E.P.A.: 974 (1964); Kuchar, Pl. Somalia (CRDP Tech. Rep. 16): 248 (1986). Type: Ethiopia, Salul River, *Ghika-Comanesti* s.n. (not seen)

J. fischeri Lindau var. *laetevirens* (Rendle) C.B. Clarke in F.T.A. 5: 202 (1900)

67. **Justicia capensis** *Thunb.*, Prodr. Fl. Cap. 2: 104 (1800); T. Anderson in J.L.S. Bot. 7: 41 (1863); Engler in E.J. 10: 264 (1888); Lindau in E. & P. Pf. IV, 3b: 349 (1895); Hedrén in K.B. 43: 349, fig. 1 (1988) & in Symb. Bot. Upsal. 29, 1: 72 (1989); Robertson & Luke, Kenya Coast. For. 2. Checklist Vasc. Pl.: 82 (1993); U.K.W.F., ed. 2: 280 (1994); Vollesen *et al.*, Checklist Mkomazi: 82 (1999). Type: South Africa, *Herb. Thunberg* 362 (UPS, lecto., selected by Hedrén in K.B. 43: 349 (1988); K!, photo)

Shrubby herb or shrub to 1(–2) m tall, sparsely to densely branched; stems pale yellow to dark green, puberulous or sparsely so with short erect or bent hairs, sometimes with capitate glands near nodes. Leaves with petiole to 3 cm long; lamina ovate to elliptic or obovate, largest 2.5–11(–13) × 1.2–4(–6) cm, base attenuate, margin entire, apex subacute to rounded; glabrous or sparsely hairy on midrib and larger veins and conspicuously ciliate, with few to many yellow to orange subsessile capitate glands. Flowers usually 2 (more rarely some single) per axil of upper leaves; bracts like leaves, 0.3–1(–1.5) cm long. Calyx puberulous or sericeous-puberulous or sparsely so (rarely also with a few stalked capitate glands), lobes 2.5–8(–10.5) mm long. Corolla white to purple with white tube, upper lip without darker veins, (8–)10–17 mm long; tube (4–)5–9 mm long, 1.5–4 mm diameter apically; upper lip (4–)5–9 mm long (tube and lip same length); lower lip (4–)6–11 × 5.5–12 mm. Filaments (3–)5–8 mm long; thecae yellow or brown, ± 1 mm long, lower with bifid appendage 0.5–1 mm long. Capsule 12–19 mm long, glabrous; seed 2–3 mm in diameter.

KENYA. Masai District: Mara Game Reserve, 28 Feb. 1972, *Taiti* s.n.!; Kilifi District: Mangea Hill, 25 March 1989, *Luke & Robertson* 1810!; Kwale District: Kilibasi Hill, 17 Nov. 1989, *Luke & Robertson* 2029!

TANZANIA. Moshi District: Moshi, 6 Feb. 1934, *Schlieben* 4720!; Lushoto District: 3 km E of Mashewa, 29 May 1953, *Drummond & Hemsley* 3101!; Uzaramo District: Mbudya Island, 19 June 1966, *B. J. Harris* 282!; Zanzibar, Mtende, 17 July 1999, *Fakih* 335!

DISTR. **K** 6, 7; **T** 2, 3, 6; **Z**; Mozambique, South Africa

HAB. Coastal and medium altitude to lower montane forest and forest margins, forest-grassland mosaics, coral rag thicket, termite mounds; near sea level to 1800 m

SYN. *Gendarussa capensis* (Thunb.) Nees in Linnaea 15: 366 (1841); Drège in Linnaea 20: 200 (1847)

Adhatoda capensis (Thunb.) Nees in DC., Prodr. 11: 391 (1847)

Justicia sansibarensis Lindau in E.J. 20: 71 (1894) & in P.O.A. C: 373 (1895); C.B. Clarke in F.T.A. 5: 202 (1900); T.T.C.L.: 12 (1949); Ruffo *et al.*, Cat. Lushoto Herb. Tanzania: 7 (1996). Type: Tanzania, Zanzibar, Kidoti, *Hildebrandt* 983 (COR, lecto., selected by Hedrén in K.B. 43: 350 (1988); BM!, iso.)

NOTE. It is interesting that this normally coastal species occurs inland in Kenya and N Tanzania. There are no obvious differences between the inland and coastal specimens.

68. **Justicia euosmia** *Lindau* in E.J. 30: 113 (1901); T.T.C.L.: 11 (1949); Hedrén in K.B. 43: 353, fig. 2 (1988) & in Symb. Bot. Upsal. 29, 1: 73 (1989). Type: Tanzania, Lushoto District: W Usambara Mts, Kwai, *Albers* 42 (B†, syn.) & 72 (B†, syn.). Neotype: Tanzania, W Usambara Mts, Baga Forest Reserve, *Hedrén et al.* 84533 (UPS!, neo., selected by Hedrén in K.B. 43: 353 (1988); K!, UPS!, iso.)

Shrub, 2–3 m tall, sparsely to densely branched, plant with strong smell of cumarin; young stems dark green, puberulous or sparsely so with short bent hairs, rarely with a few capitate glands. Leaves with petiole to 2 cm long, with capitate glands; lamina ovate, elliptic or obovate, largest (on flowering branches) 2–5.3 × 0.8–2.7 cm, on sterile branches to 10 × 5.3 cm, base attenuate to subcuneate, margin entire, apex subacute to truncate, apiculate; sericeous-puberulous or sparsely so on midrib and larger veins, margin ciliate and with stalked capitate glands, lamina with scattered orange subsessile capitate glands. Flowers single or 2 per axil of upper leaves; bracts elliptic or broadly so, 0.5–1.2 cm long. Calyx sparsely puberulous-ciliate, lobes 3–7.5 mm long. Corolla white to pale purple with white tube, upper lip with purple streaks, 8–11 mm long; tube 4–5 mm long, 2.5–4 mm diameter apically; upper lip 4–6 mm long (tube and lip same length or lip slightly longer); lower lip 5–7 × 5.5–8.5 mm. Filaments 3–5 mm long, thecae yellow, ± 1 mm long, lower with bifid appendage ± 0.5 mm long. Capsule (8.5–)16–18 mm long, glabrous or puberulous; seed ± 3 mm in diameter.

TANZANIA. Lushoto District: West Usambara Mts, Mkusu Valley, between Mkusi and Kifungilo, 14 April 1953, *Drummond & Hemsley* 2095! & Magamba–Mkuzi road, 7 June 1953, *Drummond & Hemsley* 2875! & Baga Forest Reserve, above Mabanda, 3 March 1984, *Hedrén et al.* 84533!

DISTR. **T** 3; not known elsewhere

HAB. Evergreen montane forest and forest margins, grassland-forest mosaics; 1600–1900 m

NOTE. There is really very little to separate this species from the more widespread *J. capensis*. In the West Usambaras it grows at higher altitudes, but in other areas *J. capensis* grows at similar high altitudes. Recent collections from the Kenya coast have eroded the difference in corolla size given by Hedrén (l.c.). The difference in corolla coloration might just be local variation and variation in colouration has not been studied in populations of *J. capensis*. It is true that the small-flowered Kenyan collections of *J. capensis* have a slenderer corolla tube (± 1.5 mm diameter) and for that reason and the streaked upper lip I am keeping them separate.

69. **Justicia brevipila** *Hedrén* in K.B. 43: 356, fig. 3 (1988) & in Symb. Bot. Upsal. 29, 1: 72 (1989); Luke & Robertson, Kenya Coast. For. 2. Checklist Vasc. Pl.: 82 (1993); Hedrén in Fl. Somalia 3: 415 (2006). Type: Kenya, Kwale District, Samburu to Mackinnon Road, *Drummond & Hemsley* 4043 (K!, holo.; EA!, K!, iso.)

Shrubby herb or shrub to 1(–2) m tall, sparsely to densely branched; stems pale yellow to dark green, finely retrorsely sericeous or densely so with appressed hairs. Leaves with petiole to 1.5(–3) cm long, with scattered capitate glands; lamina ovate to elliptic, oblong or slightly obovate, largest 1.5–5.7 × 1–3.5 cm, base attenuate to cuneate, margin entire, apex truncate to retuse; uniformly puberulous or sparsely so on lamina, denser on larger veins, with scattered stalked capitate glands. Flowers usually 2(–3) per axil of upper leaves (rarely a few single); bracts like leaves, 0.5–1.3 cm long, with many capitate glands. Calyx finely antrorsely sericeous-puberulous, usually also with scattered stalked capitate glands, lobes 2–4(–6) mm long. Corolla white to pale mauve with white tube, upper lip without darker veins, 10–15 mm long; tube 5–7 mm long, 2–3 mm diameter apically; upper lip 5–8 mm long (tube and lip same length or lip slightly longer); lower lip 7–12 × 5–10 mm. Filaments 5–6 mm long; thecae yellow, ± 1 mm long, lower with bifid appendage 0.5–1 mm long. Capsule 11–14 mm long, uniformly puberulous; seed 2.5–3 mm in diameter.

KENYA. Tana River District: Tana River Primate Reserve, Mulondi Camp, 22 March 1990, *Luke et al.* TRP785!; Kilifi District: Kaya Dagamra to Kaya Bura, 23 April 1989, *Robertson & Luke* 5728!; Kwale District: 15 km beyond Marikani on Mombasa–Nairobi road, 3 Dec. 1961, *Polhill & Paulo* 904!

DISTR. **K** 7; Somalia

HAB. *Acacia-Commiphora* bushland on sandy to loamy soil or on rocky hills, *Acacia-Dobera* grassland; 25–350 m

70. **Justicia sp. F**

Shrubby herb to 75 cm tall, loosely branched; stems dark green, densely retrorsely sericeous-puberulous. Leaves with petiole to 1.5 cm long; lamina ovate to elliptic, largest ± 8 × 4 cm, base attenuate, margin entire, apex subacute with rounded tip; uniformly puberulous on lamina, denser on midrib and larger veins, with scattered stalked capitate glands. Flowers single or 2 per axil of upper leaves; bracts like leaves, 0.5–0.7 cm long, with dense capitate glands. Calyx uniformly puberulous, lobes 3–4 mm long. Corolla "pink", upper lip without darker veins, lower lip with poorly developed "herring bones", ± 7 mm long; tube ± 4 mm long, ± 1.5 mm diameter apically; upper lip ± 3 mm long (tube 1 mm longer than lip); lower lip ± 4 × 5 mm. Filaments ± 3 mm long; thecae yellow, ± 1 mm long, lower with bifid appendage ± 0.5 mm long. Capsule apparently 2-seeded by abortion, 9–12 mm long, glabrous; seed not seen.

KENYA. Kwale District: Maluganji Forest Reserve, 14 Nov. 1989, *Robertson & Luke* 5989!
DISTR. **K** 7; not known elsewhere
HAB. Semi-deciduous lowland forest with *Cynometra* and *Scorodophloeos*; 200–300 m

NOTE. This puzzling collection combines characters from *J. capensis* and *J. brevipila* and may be a hybrid between the two. Both species occur in the general area. The densely puberulous stems and leaves indicate *J. brevipila*, but the subacute leaf-apex and glabrous capsule is *J. capensis*. The corolla is smaller than in any specimen of either species.

The hybrid origin theory is supported by the small flowers and the apparent development of only two seeds. But the anthers seem to be normally developed and have obviously contained fully developed pollen.

71. **Justicia phillipsiae** *Rendle* in J.B. 35: 378 (1897); C.B. Clarke in F.T.A. 5: 201 (1900); Glover, Check-list Brit. and Ital. Somal.: 67 (1947); Kuchar, Pl. Somalia (CRDP Tech. Rep. 16): 248 (1986); Hedrén in Symb. Bot. Upsal. 29, 1: 77, fig. 32 (1989) & in Fl. Somalia 3: 415 (2006); Ensermu in Fl. Eth. 5: 475 (2006). Types: Somalia, Wagga Mts, *Lort Phillips* s.n. (BM!, syn.); Upper Sheik, *Lort Phillips* s.n. (BM!, syn.) and between Dobar and Hammer, *Lort Phillips* s.n. (BM!, syn.). See note below

Loosely branched twiggy shrub to 1 m tall, sometimes heavily browsed and then densely branched, plant with strong smell of cumarin; young stems pale yellow, whitish sericeous with antrorse hairs, older branches pale grey. Leaves flat; petiole 2–7(–12) mm long; lamina ovate to elliptic or broadly so, largest 0.7–2.2 × 0.5–1.1(–1.9) cm, base cuneate to truncate, margin entire, apex subacute to truncate, apiculate; subglabrous to whitish sericeous, densest on veins and with scattered stalked capitate glands, sometimes also with long (to 1 mm) curly pilose hairs on margins and petiole. Flowers in 3-flowered dichasia or single with undeveloping lateral buds in axils of upper leaves; bracts like leaves, to 0.6 cm long; bracteoles 1–2 mm long, finely sericeous. Calyx sericeous-puberulous or sparsely so, sometimes with scattered stalked capitate glands, lobes 2.5–6 mm long. Corolla bright purple, upper lip with darker veins, 12–16 mm long; tube 7–9 mm long, 2–3 mm diameter apically; upper lip 5–7 mm long (tube ± 2 mm longer than upper lip); lower lip 8–13 × 9–15 mm. Filaments 5–6 mm long; thecae yellow or brown, ± 1 mm long, lower with bifid appendage ± 0.5 mm long. Capsule 9–12 mm long, glabrous to finely retrorsely sericeous-puberulous all over; seed 1.5–2 mm in diameter.

KENYA. Northern Frontier District: Suguta, 8 Jan. 1956, *J. Adamson* 511! & Loriu Plateau, 1 June 1970, *B. Mathew* 6505! & 1 km S of South Horr on Baragoi road, 4 June 1979, *M.G. Gilbert et al.* 5506!
DISTR. **K** 1; Ethiopia, Somalia
HAB. Dry *Acacia-Delonix* bushland on rocky hills or lava flows; 450–1250 m

SYN. *Justicia gillettii* Chiov. in K.B. 2: 172 (1941); Glover, Check-list Brit. and Ital. Somal.: 66 (1947); Kuchar, Pl. Somalia (CRDP Tech. Rep. 16): 247 (1986). Type: Somalia/Ethiopia boundary, *Gillett* 4110 (K!, holo.; K!, iso.)

J. scabrula Chiov. in K.B. 2: 174 (1941); Glover, Check-list Brit. and Ital. Somal.: 67 (1947); Kuchar, Pl. Somalia (CRDP Tech. Rep. 16): 248 (1986); Hedrén in Symb. Bot. Upsal. 29, 1: 75, fig. 30 (1989) & in Fl. Somalia 3: 415 (2006). Type: Somalia, Buramo, Dumuk, *Gillett* 4881 (K!, holo.; K!, iso.)

J. luteocinerea Hedrén in Symb. Bot. Upsal. 29, 1: 79, fig. 34 (1989). Type: Kenya, Northern Frontier District, Loriu Plateau, *B. Mathew* 6505 (K!, holo.)

NOTE. Hedrén in Symb. Bot. Upsal. 29, 1: 77 (1989) says: "Type: Somalia, Wagga Mt, Upper Sheik and between Dobar and Hammer, VI.1895, *Lort Phillips* s.n. (BM lectotype, selected here, K isotype)". The same citation is given by Ensermu (l.c.). A closer study of the sheet at BM shows that three different collections are mounted on the same sheet thus making the lectotypification invalid. The Kew sheet is a fourth collection and thus not a type.

72. **Justicia elliotii** *S. Moore* in J.B. 38: 466 (1900); U.K.W.F.: 604 (1974); Hedrén in Symb. Bot. Upsal. 29, 1: 81, fig. 36 (1989); U.K.W.F., ed. 2: 280, pl. 124 (1994). Types: Kenya, Masailand, *Scott Elliot* 6593 (BM!, syn.) & 6637 (BM!, syn.)

Much branched shrubby herb or shrub to 75 cm tall, often ± decumbent and cushion-forming; young stems whitish sericeous or densely (rarely sparsely) so with antrorse hairs, older branches with pale brownish bark. Leaves with recurved margins; petiole 0–2 mm long; lamina linear to narrowly ovate (rarely ovate), largest 0.5–2(–3.5) × 0.1–0.8(–1.3) cm, base attenuate to cuneate (rarely rounded), margin entire, apex acute to obtuse; sparsely to densely whitish sericeous (very rarely subglabrous). Flowers single in axils of upper leaves; bracts absent; bracteoles 1.5–3(–4) mm long, densely whitish sericeous (rarely sparsely so or subglabrous), occasionally also with stalked capitate glands. Calyx densely whitish sericeous to tomentellous and usually with stalked yellow capitate glands, lobes 2.5–4 mm long. Corolla mauve to purple (rarely white), upper lip with darker veins, 7–12 mm long; tube 3–6 mm long, 1.5–2.5 mm diameter apically; upper lip 3–6 mm long (upper lip and tube same length); lower lip 5–7 × 6–10 mm. Filaments 4–5 mm long; thecae purplish brown, 0.7–1.5 mm long, lower with bifid appendage 0.5–0.8 mm long. Capsule 7–8.5 mm long, retrorsely sericeous-puberulous almost to base; seed ± 2 mm in diameter.

KENYA. Naivasha District: Gilgil, Aug. 1929, *Humbert* 9128!; Masai District: Morijo Loita, 14 July 1961, *Glover et al.* 2232! & W of Ngong Hills, 29 Jan. 1969, *Napper* 1838!

TANZANIA. Musoma District: 54 km on Seronera–Soitayai road, 28 March 1961, *Greenway* 9920!; Masai District: Ngorongoro Crater Floor, 18 July 1989, *Chuwa* 2823! & Losimingori Hills, Essimingori Forest Reserve, 14 April 2000, *Mollel* 168!

DISTR. **K** 3, 4, 6; **T** 1, 2; not known elsewhere

HAB. Open grassland and *Acacia* bushland, often in disturbed places or in secondary vegetation, rocky hillsides; (1350–)1550–2500 m

SYN. *J. rooseveltii* Standley in Smithsonian Misc. Coll. 68(5): 18 (1917). Type: Kenya, Sotik Country, Nguaso Nyeri River, *Mearns* 721 (US!, holo.)

J. friesiorum Mildbr. in N.B.G.B. 9: 502 (1926). Type: Kenya, Naivasha, *R. & T. Fries* 2764 (B†, lecto.; K!, iso.)

73. **Justicia pinguior** *C.B. Clarke* in F.T.A. 5: 197 (1900); Lugard in K.B. 1933: 94 (1934); U.K.W.F.: 605 (1974); Hedrén in Symb. Bot. Upsal. 29, 1: 84, fig. 40 (1989); U.K.W.F., ed. 2: 280, pl. 123 (1994); Friis & Vollesen in Biol. Skr. 51(2): 448 (2005). Types: Uganda, Toro District, Kivata, *Scott Elliot* 7656 (K!, syn.; BM!, iso.); Kenya, Laikipia, *Thomson* s.n. (K!, syn.). See note below

Single- or few-stemmed perennial herb with with large woody rootstock (? always), stems erect or scrambling, sparsely branched, to 1.25 m long, when young glabrous to puberulous or pubescent. Leaves with petiole to 1 cm long, pubescent; lamina lanceolate to ovate, largest 2.8–8 × 1.2–3 cm, base shortly attenuate to truncate, margin entire, apex subacuminate to rounded; sericeous-pubescent or sparsely so, densest on midrib and veins. Flowers 2–4 in axils of upper leaves (rarely some single), upper internodes often contracted forming a "pseudo-inflorescence"; bracts lanceolate to narrowly ovate or narrowly elliptic (rarely elliptic), to 1.6 cm long, often with long pilose hairs on edges and sometimes with stalked capitate glands; bracteoles to 2.5 mm long. Calyx divided to 1–1.5 mm from base, puberulous-ciliate or sparsely so and sometimes with scattered capitate glands, lobes 6–10 mm long. Corolla mauve to purple or magenta (very rarely white), upper lip usually not with dark veins, puberulous and with scattered capitate glands, 15–22 mm long; tube 8–12 mm long, 2–4 mm diameter in middle, very distinctly widened from ± $^1/_2$ way up; upper lip distinctly hooded, 6–10(–12) mm long (tube usually ± 2 mm longer than upper lip, rarely same length), teeth ± 1 mm long; lower lip 8–14 × 7–13 mm. Filaments 8–11 mm long; thecae yellow, 1.5–2 mm long, lower with entire acute appendage 0.5–0.8 mm long. Capsule (9–)10–14 mm long, glabrous (rarely sparsely puberulous near apex); seed 2–2.5 mm in diameter.

UGANDA. West Nile District: Paida, Ayoda River, 27 Aug. 1953, *Chancellor* 191!; Ruwenzori Mts, Minimba Camp, 21 Jan. 1962, *Loveridge* 367!; Musaka District: NW side of Lake Nabugabo, 9 Oct. 1953, *Drummond & Hemsley* 4725!

KENYA. Nandi District: Kaimosi, Yala River, 2 June 1933, *Gilbert Rogers* 733!; Trans-Nzoia District: Kitale, 28 July 1951, *G. R. Williams* 279!; North Kavirondo District: W of Kakamega Forest, 10 July 1960, *Paulo* 519!

DISTR. **U** 1–4; **K** 3, 5; S Sudan, NE Congo

HAB. Margins and glades of montane forest, montane grassland, bushland and *Combretum* woodland, old fields; (900–)1050–2700 m

SYN. [*J. leikipiensis sensu* C.B. Clarke in F.T.A. 5: 197 (1900), quoad *Speke & Grant* s.n., *non* S. Moore (1892)]

J. scaettae Mildbr. in B.J.B.B. 14: 360 (1937). Type: Congo-Kinshasa, Kivu, Mt Muhede, *Scaetta* 31M (BR, holo.)

[*J. whytei sensu* F.P.S. 3: 180 (1956) pro parte, *non* S. Moore (1894)]

NOTE. Hedrén (l.c.) cites the Kenyan syntype as *Thomson* 9184, but what the label actually says is 9/84 meaning September 1884.

74. **Justicia ladanoides** *Lam.*, Tabl. Encycl. Met. 1, 1: 42 (1791); Hedrén in Symb. Bot. Upsal. 29, 1: 88, fig. 45 (1989); Friis & Vollesen in Biol. Skr. 51(2): 447 (2005); Hedrén in Fl. Somalia 3: 415 (2006); Ensermu in Fl. Eth. 5: 476 (2006). Type: Unknown collector in Herb. Lamarck (P-LA, holo.)

Single- or few-stemmed annual or perennial herb with taproot or short creeping rhizome, basal part of stems often creeping and rooting, apical part erect or scrambling, sparsely branched, to 1 m long, when young glabrous to puberulous or pubescent (rarely densely so). Leaves with petiole to 1 cm long, pubescent; lamina linear-lanceolate to ovate or narrowly elliptic, largest 3.8–10.5(–12) × 0.3–3.5 cm, base attenuate to cuneate, margin entire, apex acuminate to acute; puberulous to pubescent or sparsely so, densest on midrib. Flowers single or 2–4 in axils of upper leaves, upper internodes not contracted; bracts (absent in single flowers) lanceolate to narrowly ovate, to 1.6 cm long, often with long (to 2 mm) pilose hairs on edges and sometimes with stalked capitate glands; bracteoles to 1.5(–2.5) mm long. Calyx divided to 1–1.5 mm from base, glabrous to puberulous-ciliate and sometimes with scattered capitate glands, lobes 4–9(–12) mm long. Corolla purple to deep purple (rarely white), upper lip pale to almost white, with purple to dark

FIG. 76. *JUSTICIA LADANOIDES* — **1–2**, habit, long- and short-tubed plants; **3**, broad leaf; **4**, bract; **5**, calyx opened up; **6**, detail of calyx lobe; **7**, short-tubed corolla; **8**, long-tubed corolla; **9**, corolla opened up; **10**, apical part of filament and anther; **11**, apical part of style and stigma; **12**, capsule; **13**, seed. 1, 4, 8 & 13 from *Verdcourt* 778, 2 from *Kerfoot* 3794, 3 from *Eggeling* 5919, 5–7 & 9–11 from *Tweedie* 2557, 12 from *Wilson* 193. Drawn by Margaret Tebbs.

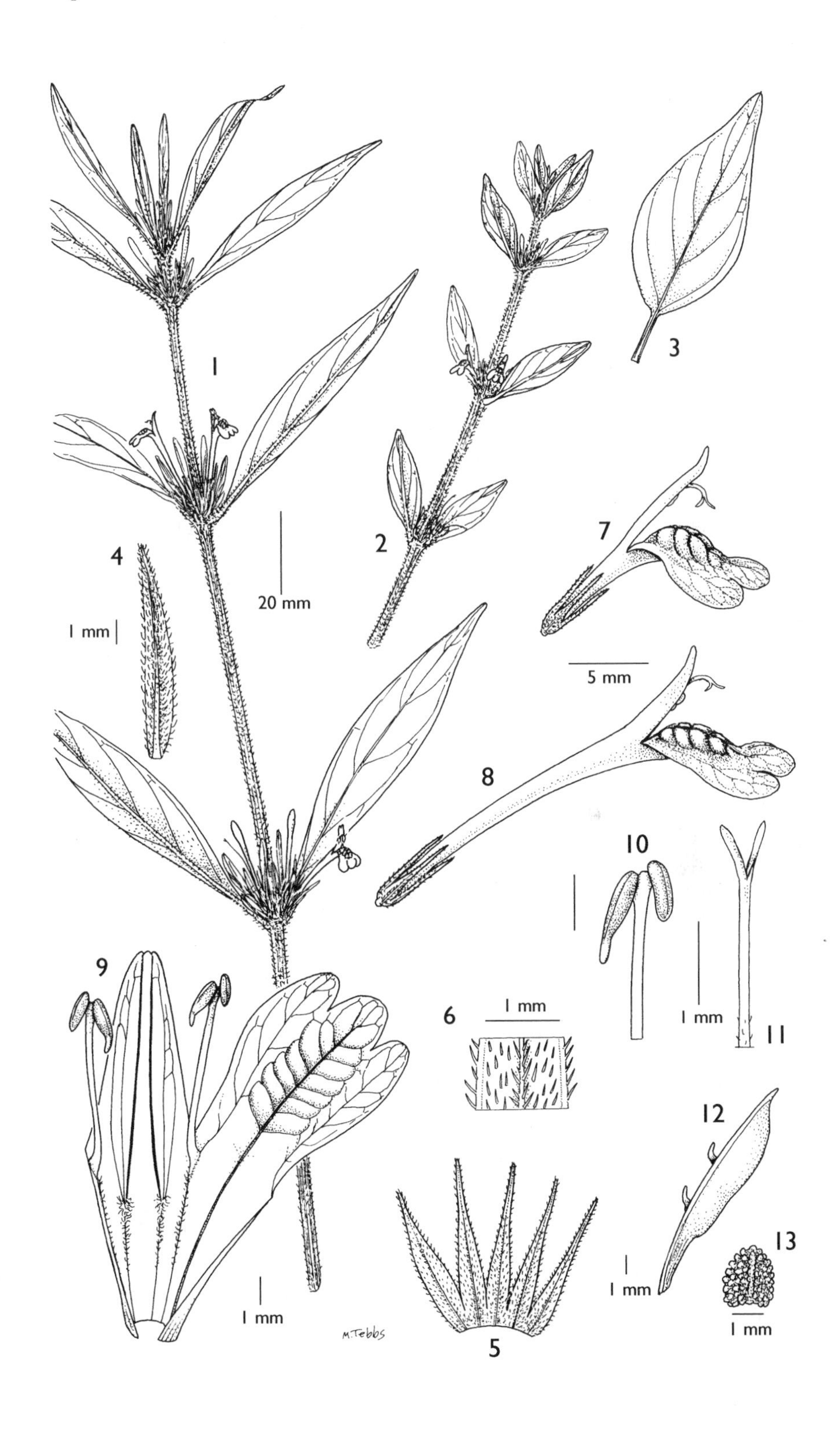
1
2
3
4
1 mm
20 mm
7
5 mm
8
10
9
1 mm
6
1 mm
11
12
5
1 mm
13
1 mm
1 mm
M. Tebbs

purple veins, 13–27 mm long, puberulous and usually with capitate glands; tube 9–22 mm long, 1–2 mm diameter in middle, only expanded in the apical 2–3 mm; upper lip 3–7 mm long (tube 6–15 mm longer than upper lip); lower lip 4–12 × 4–14 mm. Filaments 3–9.5 mm long; thecae yellow (more rarely purple), 1–1.5(–2) mm long, lower with entire acute appendage ± 0.5 mm long. Capsule 7–10 mm long, glabrous (more rarely sparsely puberulous in apical half); seed ± 2 mm in diameter. Fig. 76, p. 573.

UGANDA. Karamoja District: base of Mt Moroto, 6 Oct. 1952, *Verdcourt* 788!; Acholi District: Murchison Falls National Park, Chobi, 20 Sept. 1967, *Angus* 6028!; Mbale District: Bukedi, Tororo Hill, 15 Dec. 1968, *Kimani* 157!

KENYA. Northern Frontier District: Moyale, 29 Sept. 1952, *Gillett* 13969!; West Suk District: Sigorr, Sept. 1971, *Tweedie* 4261!; Baringo District: Upper Kerio Valley, Chebloch, Sept. 1962, *Tweedie* 2441!

DISTR. **U** 1–4; **K** 1–3, 5; Gambia, Guinea-Bissau, Guinea, Senegal, Sierra Leone, Liberia, Ivory Coast, Togo, Benin, Nigeria, Cameroon, Central African Republic, Chad, Congo-Kinshasa, Sudan, Eritrea, Ethiopia, Somalia

HAB. *Combretum-Terminalia* woodland, wetter places in *Acacia* bushland, tall grassland, montane woodland and bushland, riverine thicket; (650–)800–1850(–2400) m

SYN. *J. lithospermifolia* Jacq., Hort. Schoenbrun. 1: 3, t.4 (1800); Vahl, Enum. Pl. 1: 151 (1805). Type: Plate 4 in Jacq. (l.c.), iconotype

Tyloglossa schimperi Hochst. in Flora 26: 73 (1843). Type: Ethiopia, Adua, *Schimper* I.106 (BM!, lecto.; K!, iso.; selected by Morton in K.B. 32: 435 (1978)

T. kotschyi Hochst. in Flora 26: 74 (1843). Type: Sudan, Kordofan, *Kotschy* 293 (K!, iso.)

Adhatoda kotschyi (Hochst.) Nees in DC., Prodr. 11: 397 (1847)

A. lithospermifolia (Jacq.) Nees in DC., Prodr. 11: 398 (1847)

A. rostellaria Nees in DC., Prodr. 11: 397 (1847). Type: as for *Tyloglossa schimperi*

A. rostellaria Nees var. *humilis* Nees in DC., Prodr. 11: 397 (1847). Type: Ethiopia, Djeladjeranne, *Schimper* III.1657 & 1659 (K!, holo.). See Note

Justicia neglecta T. Anderson in J.L.S. Bot. 7: 40 (1863). Type: as for *Tyloglossa schimperi*

J. rostellaria (Nees) Lindau in P.O.A. C: 373 (1895) & in Ann. Ist. Bot. Roma 6: 82 (1896)

J. calcarata C.B. Clarke in F.T.A. 5: 195 (1900); Fiori in N. Giorn. Bot. Ital. 20: (1913) & in N. Giorn. Bot. Ital. 47: 41 (1940), *nom. illeg., non J. calcarata* Wall. (1831)

J. galeopsis C.B. Clarke in F.T.A. 5: 196 (1900); Lindau in Z.A.E.: 309 (1911); F.P.S. 3: 180 (1956). Type: Nigeria, Nupe, *Barter* 1038 (K!, lecto.; selected by Morton in K.B. 32: 437 (1978)

J. leikipiensis sensu C.B. Clarke in F.T.A. 5: 197 (1900), quoad *Wilson* 91, *non* S. Moore (1892)

J. sexangularis C.B. Clarke in F.T.A. 5: 198 (1900), *nom. illeg., non J. sexangularis* L. (1753)

J. calcarata C.B. Clarke forma *parviflora* Fiori in N. Giorn. Bot. Ital. 20: (1913). Type: Eritrea, Filfil, *Fiori* 772 (FT, holo.)

J. fistulosa S. Moore in J.B. 58: 47 (1920). Type: Congo-Kinshasa, Bokala, *Sparano* 26 (BM!, holo.)

J. calcarata C.B. Clarke var. *baddabunae* Fiori in N. Giorn. Bot. Ital. 47: 41 (1940). Type: Ethiopia, Jimma, Baddabuna, *Saccardo* s.n. (FT, holo.)

[*J. insularis sensu* F.P.N.A. 2: 316 (1947), *non* T. Anderson (1863)]

J. kotschyi (Hochst.) Dandy in F.P.S. 3: 180 (1956); E.P.A.: 971 (1964); Adam in Mem. Mus. Nat. Hist. Nat., Ser. B., Bot. 25: 1451 (1975)

J. schimperi (Hochst.) Dandy in F.P.S. 3: 180 (1956); E.P.A.: 974 (1964); Morton in K.B. 32: 435, fig. 1 (1978); Synnott, Checklist Fl. Budongo For.: 69 (1985)

J. schimperi (Hochst.) Dandy var. *kotschyi* (Hochst.) Morton in K.B. 32: 437 (1978)

J. schimperi (Hochst.) Dandy subsp. *fistulosa* (S. Moore) Morton in in K.B. 32: 439 (1978)

NOTE. The label of the type specimen of *Adhatoda rostellaria* Nees var. *humilis* Nees bears two numbers. C.B. Clarke in F.T.A. 5: 196 (1900) quite rightly points out that the sheet is a mixture of *Justicia* and *Asystasia*. But Nees' description is quite clearly of the *Justicia* ("floribus ..... in verticillo") and it is therefore quite clear which part should be considered the type.

Hedrén (l.c.) gives a lenghty and thorough discussion of the variation within this widespread and variable species pointing out that closer studies may well show it to consist of several taxa.

75. **Justicia leikipiensis** *S. Moore* in J.B. 32: 137 (1892); Lindau in P.O.A. C: 373 (1895); C.B. Clarke in F.T.A. 5: 197 (1900) pro parte; U.K.W.F.: 605 (1974); Hedrén in Symb. Bot. Upsal. 29, 1: 87, fig. 43 (1989); U.K.W.F., ed. 2: 280 (1994). Type: Kenya, Laikipia District, Gopolal Mwaru, *Gregory* s.n. (BM!, holo.)

Perennial herb with several decumbent, ascending or scrambling (? sometimes erect) stems from woody rootstock; stems sparsely branched, to 50 cm long, when young glabrous or sparsely puberulous at nodes. Leaves with petiole 0–2 mm long; lamina linear to lanceolate, largest 1.8–6.5 × 0.2–0.5 cm, base attenuate, margin entire, apex acute to obtuse; glabrous to sparsely sericeous-puberulous and scabrid-ciliate, upper often with stalked capitate glands near base. Flowers single or some or all in pairs in axils of upper leaves; bracts absent (single flowers) or linear, to 5 mm long; bracteoles to 1.5 mm long. Calyx subglabrous to sparsely puberulous, densest on margins, and with stalked capitate glands, lobes (3–)5–9 mm long. Corolla pale mauve to magenta, upper lip with darker veins, 12–16 mm long, puberulous and with stalked capitate glands; tube not distinctly expanded in upper part, 7–10 mm long, 1–1.5 mm diameter in middle; upper lip 5–7 mm long (tube 2–3 mm longer than upper lip), teeth ± 1 mm long; lower lip 4–6 × 4–6 mm. Filaments 3–5 mm long; thecae yellow, ± 1 mm long, lower with entire acute appendage ± 0.5 mm long. Capsule sometimes 2-seeded, 7–10 mm long, glabrous. Mature seed not seen.

UGANDA. Karamoja District: Pian, Namalu, 20 Aug. 1965, *J. Wilson* 1704!
KENYA. Northern Frontier District: Poror Plateau, Lerogi Forest, 15 Nov. 1977, *Carter & Stannard* 436!; West Suk District: Kapenguria, 13 May 1932, *Napier* 1914!; Rumuruti District: Ngeleshwa, 9 July 1931, *Napier* 1220!
DISTR. **U** 1; **K** 1–3; not known elsewhere
HAB. Grassland and bushland, mostly on rocky hillsides, secondary grassland after forest clearing; (1200–)1850–2450 m

NOTE. Differs most conspicuously from the widespread *J. ladanoides* in the narrow subglabrous leaves and the smaller corolla with relatively (in relation to upper lip) shorter corolla tube.

76. **Justicia toroensis** *S. Moore* in J.B. 48: 255 (1910); Hedrén in Symb. Bot. Upsal. 29, 1: 97, fig. 51 (1989). Type: Uganda, Toro District: Fort Portal, *Bagshawe* 1268 (BM!, holo.)

Annual or short-lived perennial herb, basal part of stems creeping and rooting, apical part erect or scrambling, sparsely branched, to 50 cm long, sparsely and uniformly or bifariously sericeous-puberulous with curly or bent hairs. Leaves with petiole to 3.5 cm long, subglabrous to sericeous; lamina ovate, largest 4–9.5 × 2–5.2 cm, base attenuate to cuneate, margin entire to slightly crenate, apex acuminate to acute; with scattered hairs on midrib, otherwise glabrous or with a few long broad hairs on lamina above. Flowers single (rarely 2) in axils of upper leaves (rarely congested on short axillary branches); bracts absent; bracteoles 1.5–3(–4) mm long, densely whitish sericeous (rarely sparsely so or subglabrous), occasionally also with stalked capitate glands. Calyx with scattered long straight hairs towards tip of lobes, otherwise glabrous (rarely sparsely sericeous-puberulous), lobes 5–7 mm long. Corolla white to mauve, upper lip without darker veins, 16–24 mm long, puberulous, without capitate glands; tube 10–16 mm long, 1–1.5 mm diameter in middle; upper lip 6–8 mm long (tube 4–8 mm longer than upper lip); lower lip 8–12 × 8–14 mm. Filaments 5–7 mm long; thecae yellow, 0.5–1 mm long, lower with entire acute appendage ± 0.3 mm long. Capsule 9–11 mm long, glabrous or with scattered hairs near apex; seed not seen.

UGANDA. Ankole District: Bunganguma, July 1939, *Purseglove* 860!; Bunyoro District: Bugoma, Jan. 1920, *Dummer* 4366! & Budongo Forest, Masindi, Nyabisabo River, 8 Dec. 1996, *Poulsen* 1255!
KENYA. North Kavirondo District: Kakamega Forest, Yala River, 25 Nov. 1969, *Bally* 13681!

TANZANIA. Kigoma District: Gombe Stream National Park, Kakombe Valley, 15 May 1999, *Gobbo et al.* 316!
DISTR. **U** 2; **K** 5; **T** 4; E Congo-Kinshasa, Burundi
HAB. Medium altitude and montane evergreen forest and forest margins; 800–1500 m

SYN. *Justicia pseudoruellia* Mildbr. in B.J.B.B. 17: 91 (1943); F.P.N.A. 2: 315 (1947). Type: Congo-Kinshasa, Rutshuru, *de Witte* 1667 (BR, holo.)

77. **Justicia striolata** *Mildbr.* in N.B.G.B. 11: 411 (1932); Richards & Morony, Check List Fl. Mbala Distr.: 232 (1969); Hedrén in Symb. Bot. Upsal. 29, 1: 99, fig. 53 (1989); Strugnell, Checklist Mt Mulanje: 35 (2006). Type: Tanzania, Njombe District, Lupembe, Ruhudje, *Schlieben* 678 (B†, holo.; BM!, EA!, K!, iso.)

Annual or short-lived perennial herb, basal part of stems creeping and rooting, apical part erect or ascending, sparsely branched, to 1(?–1.5) m long, subglabrous to uniformly or bifariously sericeous-pubescent with curly or bent hairs. Leaves with petiole to 6 cm long, sericeous, often also with long pilose hairs to 2 mm long and with scattered capitate glands; lamina ovate to elliptic, largest 2.5–10.5 × 1.3–5 cm, base attenuate to rounded, margin entire, apex acuminate to acute; subglabrous or with scattered hairs on midrib and larger veins (occasionally also on lamina). Flowers single (rarely) in axils of upper leaves or 2–4 per leaf in short spiciform cymes on short axillary branches; bracts (absent in solitary flowers) ovate to elliptic or broadly so, to 1.1 cm long, conspicuously ciliate with long white hairs to 1.5 mm long. Calyx subglabrous or with scattered to dense long straight hairs, often only towards apex but sometimes all over, sometimes with stalked capitate glands, lobes 4–6 mm long. Corolla white to pale mauve, upper lip with darker veins, 12–20 mm long, puberulous and with capitate glands; tube 9–13 mm long, 1–1.5 mm diameter in middle; upper lip 3–7 mm long (tube 4–6 mm longer than upper lip), teeth 1.5–2 mm long; lower lip 6–10 × 7–13 mm. Filaments 4–5 mm long; thecae white to pale yellow, 0.5–1 mm long, lower with entire acute appendage ± 0.3 mm long. Capsule 6.5–8.5 mm long, with scattered hairs near apex; seed 1.5–2 mm in diameter.

TANZANIA. Mpanda District: Mt Livandabe [Lubalisi], 30 May 1997, *Bidgood et al.* 4195!; Iringa District: Mufindi, Luhega Forest Reserve, Uhafiwa, 10 June 1989, *J. Lovett et al.* 3259!; Songea District: Matengo Hills, Luiri Kitesi, 24 May 1956, *Milne-Redhead & Taylor* 10427!
DISTR. **T** 4, 7, 8; SE Congo-Kinshasa, Zambia, Malawi, Mozambique
HAB. Medium altitude to montane moist evergreen forest and forest margins, persisting in plantation and in old fields; 850–1900 m

SYN. [*Justicia phyllostachys sensu* C.B. Clarke in F.T.A. 5: 188 (1899), quoad *Whyte* s.n., Mt Chiradzula, *non sensu stricto*]

78. **Justicia ukagurensis** *Hedrén* in Nordic Journ. Bot. 6: 301, fig. 2 (1986) & in Symb. Bot. Upsal. 29, 1: 99 (1989). Type: Tanzania, Kilosa District, Ukaguru Mts, Mt Mnyera, *Thulin & Mhoro* 2830 (UPS, holo.; K!, iso.)

Perennial or shrubby perennial herb, basal part of stems creeping and rooting, apical part wiry, often somewhat woody in basal part, erect, ascending or scrambling, unbranched or sparsely branched, to 0.6(–1) m long, when young sparsely to densely puberulous with short straight hairs and usually with stalked capitate glands. Leaves with petiole to 2.5 cm long, finely puberulous, sometimes also with long pilose hairs to 2 mm long and scattered capitate glands; lamina ovate or broadly so, largest 2.8–6.5 × 1–3.6 cm, base shortly attenuate to truncate, margin entire to slightly crenate, apex acuminate to acute; puberulous or sparsely so on midrib, otherwise glabrous or with scattered long straight hairs above. Flowers single in axils of upper leaves (rarely 2 per leaf); bracts absent; bracteoles 1–1.5 mm long; pedicels 1–1.5 mm long. Calyx puberulous and with longer straight hairs to 1.5 mm long and usually with stalked capitate glands, lobes 7–9 mm long. Corolla white to pale mauve, upper lip with many purple spots and patches along the edges sometimes merging to give an almost uniformly purple upper lip, 19–25 mm long, puberulous and with dense capitate glands; tube 11–14 mm long, 2–3 mm diameter in middle; upper lip 7–10 mm long (tube 3–4 mm longer than upper lip), teeth 3–4 mm long; lower lip 13–20 × 18–24 mm. Filaments 6–8 mm long; thecae dark purple, 1–1.5 mm long, lower with bifid appendage ± 0.5 mm long. Capsule (fide Hedrén) ± 15 mm long, uniformly puberulous all over; seed (fide Hedrén) ± 2.5 mm in diameter.

TANZANIA. Kilosa District: Ukaguru Mts, Mamiwa Forest Reserve, near Mandege Forest Station, 2 Aug. 1972, *Mabberley* 1355!; Ifakara/Iringa District: Mwanihana Forest Reserve, above Sanje Village, 17 June 1986, *J. Lovett et al.* 855!; Iringa District: Luhomero Massif, Upper Lafia Valley, 18 Aug. 1985, *Rodgers & Hall* 4528!

DISTR. **T** 5–7; not known elsewhere

HAB. Montane forest, often on steep rocky slopes or in crevices, *Erica* woodland above forest; 1400–2200 m

79. **Justicia afromontana** *Hedrén* in Nordic Journ. Bot. 8: 161, fig. 1 (1988) & in Symb. Bot. Upsal. 29, 1: 101 (1989); Friis & Vollesen in Biol. Skr. 51(2): 445 (2005). Type: Sudan, Imatong Mts, Mt Kinyeti, *Myers* 11641 (K!, holo.)

Wiry shrubby herb with one to several erect to decumbent stems, stems sparsely to densely branched, to 75 cm long, when young sparsely to densely puberulous to pubescent; older stems woody, dark brown. Leaves with petiole to 4 mm long, pubescent; lamina narrowly to broadly ovate, largest 1.2–2.3(–3.5) × 0.8–1.3(–1.8) cm, base shortly attenuate to truncate, margin entire, apex subacute to rounded; on both sides sparsely uniformly pubescent. Flowers single or 2(–4) in axils of upper leaves, upper internodes usually contracted forming a "pseudo-inflorescence"; bracts lanceolate to ovate or elliptic, to 0.8(–1.2) cm long, sparsely to densely pubescent; bracteoles lanceolate or narrowly ovate-elliptic, 3–6 mm long (rarely linear and 1.5–2 mm). Calyx pubescent-ciliate or sparsely so (hairs to 1 mm long), lobes 4–9 mm long. Corolla purple or scarlet, lower lip usually with poorly developed "herring bones", upper lip sometimes with dark veins, 13–17 mm long, puberulous and with scattered capitate glands; tube 8–9 mm long, ± 2 mm diameter in middle, very distinctly widened from ± $^1/_2$ way up; upper lip distinctly hooded, 5–8 mm long (tube 1–3 mm longer than upper lip), teeth ± 1 mm long; lower lip 8–12 × 10–12 mm. Filaments 4–6 mm long; thecae yellow, 1–1.5 mm long, lower with entire acute appendage ± 0.5 mm long. Capsule 8–10 mm long, sparsely puberulous in upper half; seed 1.5–2 mm in diameter.

UGANDA. Karamoja District: Mt Moroto, Feb. 1936, *Eggeling* 2915! & Oct. 1945, *Dale* 456! & Mt Morongole, 11 Nov. 1939, *A. S. Thomas* 3299!

DISTR. **U** 1; S Sudan

HAB. High altitude montane grassland, *Erica* scrub; (2450–)2700–3050 m

SYN. [*Justicia whytei sensu* F.P.S. 3: 180 (1956) pro parte, *non* S. Moore (1894)]

NOTE. This is basically a high altitude form of *J. pinguior*. It is more shrubby, has smaller leaves with a different indumentum, smaller corolla and smaller capsule. It is known only from the Imatong Mts in the Sudan and Moroto and Morongole Mts in NE Uganda.

80. **Justicia sulphuriflora** *Hedrén* in Nordic Journ. Bot. 6: 303, fig. 3 (1986) & in Symb. Bot. Upsal. 29, 1: 102 (1989). Type: Tanzania, Lushoto District, Usambara Mts, Magamba Peak, *Drummond & Hemsley* 2818 (K!, holo.; EA!, K!, iso.)

Perennial herb, basal part of stems creeping and rooting, apical part scandent or scrambling, sparsely branched, to 1(–2) m long, when young subglabrous to densely pubescent or retrorsely sericeous with straight pale yellow hairs, towards apex sometimes with stalked capitate glands. Leaves with petiole to 1(–2) cm long, sparsely to densely sericeous-pubescent with pale yellow hairs; lamina linear-lanceolate to ovate, largest 4–7(–9.5) × 0.6–2.5(–3) cm, base shortly attenuate to rounded, margin entire to crenate, apex acuminate to cuspidate (rarely acute); on both sides with sparse appressed pale yellow hairs to 1.5 mm long. Flowers single or in 2's in axils of upper leaves; bracts lanceolate to ovate, to 1.2 cm long, conspicuously ciliate; bracteoles 0.5–1.5 mm long. Calyx subglabrous to sericeous-puberulous on margins, lobes (2–)3–5 mm long. Corolla bright yellow, upper lip with dark purple veins to almost uniformly dark purple, 9–16 mm long, puberulous and with or without capitate glands; tube 6–9 mm long, 2–3.5 mm diameter in middle, very distinctly widened from ± $^1/_2$ way up; upper lip flat, 4–7 mm long (tube 2–3 mm longer than upper lip), teeth 1.5–2 mm long; lower lip 7–12 × 9–14 mm. Filaments 4–5 mm long; thecae purple, 1–1.5 mm long, lower with square-cut appendage ± 0.5 mm long. Capsule 10–13 mm long, acuminate, glabrous; seed ± 2 mm in diameter.

TANZANIA. Lushoto District: W Usambara Mts, Shagayu Forest Reserve, 22 Oct. 1986, *Borhidi et al.* 86122!; Morogoro District: Uluguru Mts, Mgeta River, Hululu Falls, 19 March 1953, *Drummond & Hemsley* 1686!; Iringa District: Mufindi, Luisenga Stream, 24 Aug. 1984, *Bridson* 566!

DISTR. **T** 3, 6, 7; not known elsewhere

HAB. Understorey in evergreen montane forest, often near or along streams; 1500–2400 m

NOTE. Hedrén (1989) mentions 1-seeded indehiscent fruits for this species, but the material available to me shows none.

81. **Justicia diclipteroides** *Lindau* in E.J. 20: 65 (1894) & in P.O.A. C: 373 (1895); C.B. Clarke in F.T.A. 5: 203 (1900); U.K.W.F.: 605 (1974); Blundell, Wild Fl. E. Afr.: 393, pl. 703 (1987); Hedrén in Bull. Mus. Hist. Nat., Ser. 4, 10. Sect. B, Adansonia 4: 353, figs. 6–13 (1988) & in Symb. Bot. Upsal. 29, 1: 102 (1989); Luke & Robertson, Kenya Coast. For. 2. Checklist Vasc. Pl.: 82 (1993); U.K.W.F., ed. 2: 280 (1994); Vollesen *et al.*, Checklist Mkomazi: 82 (1999); Ensermu in Fl. Eth. 5: 476 (2006). Type: Kenya, Machakos District, Kibwezi, *Scheffler* 420 (W, neo., selected by Hedrén (1988); C!, iso.)

Perennial or shrubby herb with several suberect, ascending, decumbent or scandent stems from erect to creeping rhizome, basal part of stem sometimes creeping and rooting, stems unbranched to much branched, to 60 cm long in erect plants, to 2(–3) m long in scandent ones, when young sparsely to densely sericeous to puberulous or pubescent, sometimes also with long pilose hairs to 2(–3) mm long, with or without capitate glands. Leaves with petiole to 2.5(–5) cm long, sericeous to pubescent; lamina narrowly to broadly ovate, largest (0.8–)1.8–10(–11) × (0.6–)0.8–5(–6.2) cm, base attenuate to truncate (very rarely subcordate), margin entire to crenate, apex acute to acuminate; subglabrous to pubescent on both sides, densest on midrib and larger veins. Flowers single or 2–4 in axils of upper leaves; bracts absent or ovate to broadly so, to 1.5 cm long; bracteoles 1–2 mm long, glabrous to pubescent; pedicels to 1(–2) mm long. Calyx puberulous or sparsely so (hairs to 0.5 mm long), lobes 3–8(–9) mm long. Corolla bright purple (very rarely white), lower lip with or without "herring bone" pattern, upper lip often with dark purple veins sometimes merging to give an almost uniformly dark purple lip, 8–19 mm long, with short (to 0.5 mm) straight to slightly curly hairs, with or without capitate glands; tube (4–)5–10 mm long, ± 1.5 mm diameter in middle, widening immediately above base; upper lip flat or slightly hooded, (3–)4–9 mm long (tube and upper lip same length or tube to 2(–3) mm longer), teeth 1–2 mm long; lower lip 6–18 × 7–17 mm. Filaments 2–7 mm long; thecae yellow, 1–1.5 mm long, lower with acute appendage ± 0.5 mm long. Capsule 6–12(–14) mm long, acute to acuminate, hairy in apical $^1/_2$ to $^2/_3$ (rarely only near apex); seed 1–1.5 mm in diameter. Spiny indehiscent 1-seeded fruits 3–4 mm long occasionally present.

UGANDA. Karamoja District: Karasuk, Kenailmet, 16 June 1959, *Symes* 599!
KENYA. Baringo District: Kamasia, Katimok Forest, Oct. 1930, *Dale* 2439!; Nairobi District: Thika Road House, 15 April 1951, *Verdcourt* 486!; Teita District: 1.5 km W of Voi Gate, 19 Dec. 1966, *Greenway & Kanuri* 12782!
TANZANIA. Masai District: Engaruka Gorge, 25 Feb. 1970, *Richards* 25530!; Pare District: Mkomazi Game Reserve, Kiholo, 30 April 1995, *Abdallah & Vollesen* 95/57!; Iringa District: Mufindi, 27 April 1986, *Congdon* 70!
DISTR. **U** 1; **K** 1–4, 6, 7; **T** 1–5, 7; Ethiopia, E Congo-Kinshasa, Rwanda, Burundi
HAB. Montane evergreen forest, particularly in glades and on margins, montane grassland and bushland, lowland and intermediate bushland, rocky hills, riverine forest; (250–)550–2600 m

SYN. *J. taylori* S. Moore in J.B. 39: 303 (1901). Type: Tanzania, Moshi, *W. E. Taylor* s.n. (BM!, holo.)
J. praetervisa Lindau in E.J. 33: 192 (1902). Type: Tanzania, Kilimanjaro, *Volkens* 1903 (HBG, lecto., selected by Hedrén (1988))
J. nierensis Mildbr. in N.B.G.B. 9: 502 (1926). Type: Kenya, Nyeri Forest, *Fries & Fries* 200 (UPS, holo.). See Note
J. sp. E sensu U.K.W.F.: 605 (1974)
J. diclipteroides Lindau subsp. *praetervisa* (Lindau) Hedrén in Bull. Mus. Hist. Nat., Ser. 4, 10. Sect. B, Adansonia 4: 356, fig. 7 (1988) & in Symb. Bot. Upsal. 29, 1: 104 (1989); U.K.W.F., ed. 2: 280 (1994)
J. diclipteroides Lindau subsp. *usambarica* Hedrén in Bull. Mus. Hist. Nat., Ser. 4, 10. Sect. B, Adansonia 4: 359, fig. 10 (1988) & in Symb. Bot. Upsal. 29, 1: 104 (1989). Type: Tanzania, Lushoto District: W Usambara Mts, *Hedrén et al.* 84831 (UPS!, holo.; K!, iso.)
J. diclipteroides Lindau subsp. *nierensis* (Mildbr.) Hedrén in Bull. Mus. Hist. Nat., Ser. 4, 10. Sect. B, Adansonia 4: 363, fig. 12 (1988) & in Symb. Bot. Upsal. 29, 1: 105 (1989); U.K.W.F., ed. 2: 280 (1994); Ensermu in Fl. Eth. 5: 478 (2006)

NOTE. An exceedingly variable species, but – as opposed to Hedrén – I see no way of separating it into meaningful infraspecific taxa. The characters used by Hedrén (flower size, indumentum) all intergrade to an extent larger than indicated by him as do the geographical distributions. A large number of recent collections from N Tanzania in particular blur the distinction between his subsp. *praetervisa* and subsp. *nierensis* in **T** 2 and between subsp. *praetervisa* and subsp. *usambarica* in **T** 3.

The type collection of *J. nierensis* is cited as *Fries & Fries* 260 by Mildbraed but as 200 by Hedrén.

82. **Justicia kiborianensis** (*Hedrén*) *Vollesen*, **comb. nov**. Type: Tanzania, Mpwapwa District, Mt Kiboriani, *B.D. Burtt* 4543 (K!, holo.)

Perennial herb with several erect or suberect stems to 60 cm long, when young densely pubescent with long (to 1.5 mm) straight spreading hairs, no capitate glands. Leaves with petiole to 1.5 cm long, densely pubescent; lamina broadly ovate, largest 4–6.8 × 3.5–5.5 cm, base truncate to subcordate, margin crenate, apex obtuse; on both sides pubescent (densely so on midrib). Flowers single or in 2's in axils of upper leaves; bracts lanceolate, to 1.5 cm long, conspicuously ciliate; bracteoles 1–2 mm long, densely pubescent. Calyx densely pubescent (hairs to 1 mm long), lobes 5–6.5 mm long. Corolla bright purple, lower lip without coloured "herring bone" pattern, upper lip with dark purple veins, 13–14 mm long, with long (to 1 mm) curly hairs, no capitate glands; tube 6–7 mm long, ± 1.5 mm diameter in middle, widening immediately above base; upper lip flat or slightly hooded, 6–7 mm long (tube and upper lip same length), teeth ± 1 mm long; lower lip 9–11 × 10–13 mm. Filaments ± 3 mm long; thecae yellow, ± 1 mm long, lower with acute appendage ± 0.5 mm long. Capsule ± 7 mm long, acute, with scattered hairs in apical half; seed ± 1.5 mm in diameter.

TANZANIA. Mpwapwa District: Mt Kiboriani, 26 Feb, 1933, *B.D. Burtt* 4543! & 27 Feb. 1933, *Hornby* 497!
DISTR. **T** 5; not known elsewhere
HAB. Margins of evergreen montane forest; 1825 m

SYN. *J. diclipteroides* Lindau subsp. *kiborianensis* Hedrén in Bull. Mus. Hist. Nat., Ser. 4, 10. Sect. B, Adansonia 4; 362, fig. 11 (1988) & in Symb. Bot. Upsal. 29, 1: 104 (1989), as "*kibarianensis*"

NOTE. Known only from these two collections but is probably more widespread in similar habitats in the Ukaguru Mts, a rather poorly known region. The forest on Mt Kiboriani is still in pretty good shape, so there also ought to be good chances of re-collecting the species here.
Hedrén considers this a subspecies of the variable *J. diclipteroides* but it differs in its subcordate leafbase and different calyx and corolla indumentum.

83. **Justicia petterssonii** (*Hedrén*) *Vollesen*, **comb. nov**. Type: Tanzania, Arusha, *Hedrén & Pettersson* 1106 (UPS, holo.; K!, iso.)

Perennial herb with several erect or ascending unbranched or sparsely branched stems from a small woody rootstock, stems to 50 cm long, when young densely puberulous with or without capitate glands. Leaves with petiole to 0.5 cm long; lamina ovate or narrowly so, largest 3.2–7 × 1–2.7(–3.5) cm, base attenuate, margin entire, apex subacute to obtuse; sparsely pubescent, densest on midrib and margins, sometimes with capitate glands on margins towards base. Flowers single or 2–4(–6) in axils of upper leaves, sometimes only one leaf of a pair carrying flowers; bracts absent or linear-lanceolate, to 1 cm long, conspicuously ciliate; bracteoles 1–2 mm long. Calyx puberulous and with many stalked capitate glands, lobes 3–4 mm long, with basal part slightly thickened and spongiform in fruit. Corolla mauve, lower lip without coloured "herring bone" pattern, upper lip with dark purple veins to almost uniformly dark purple, 7–8.5 mm long, puberulous and with capitate glands; tube 4.5–5 mm long, widening immediately above base; upper lip 2.5–3.5 mm long (tube 1.5–2 mm longer than upper lip), teeth less than 0.5 mm long; lower lip ± 6 × 7 mm. Filaments ± 3 mm long; thecae yellow, 0.5–1 mm long, lower with acute appendage ± 0.3 mm long. Capsule 7–9 mm long, uniformly puberulous all over; mature seed not seen; indehiscent fruits not seen.

TANZANIA. Masai District: 53 km SW of Arusha, Lolkisale, March 1967, *Beesley* 238!; Arusha District: Arusha Town, 7 March 1985, *Hedrén & Pettersson* 1106; Dodoma District: Bereko, Salanga Forest, 7 Feb. 1973, *Richards* 28561!
DISTR. **T** 2, 5; not known elsewhere
HAB. Grassland, including secondary grassland after forest clearing, weed; 1150–2100 m

SYN. *J. heterocarpa* T. Anderson subsp. *petterssonii* Hedrén in Symb. Bot. Upsal. 29, 1: 106 (1989), *nom. nud.* & in Nordic Journ. Bot. 10: 379, fig. 27C & 28O-Q (1990)

NOTE. Hedrén (l.c.) considers this to be a subspecies of the polymorphic *J. heterocarpa* but in my opinion it is distinct enough to be recognised at species level. It differs most conspicuously in being a perennial herb and in the larger corolla and larger capsule.

84. **Justicia heterocarpa** *T. Anderson* in J.L.S. Bot. 7: 41 (1863); Lindau in P.O.A. C: 373 (1895); C.B. Clarke in F.T.A. 5: 200 (1900); F.P.S. 3: 181 (1956); E.P.A.: 971 (1964); U.K.W.F.: 605 (1974); Hedrén in Symb. Bot. Upsal. 29, 1: 105 (1989) & in Nordic Journ. Bot. 10: 374, fig. 27 & 28 (1990); Robertson & Luke, Kenya Coast. For. 2. Checklist Vasc. Pl.: 82 (1993); U.K.W.F., ed. 2: 280 (1994); Ruffo *et al.*, Cat. Lushoto Herb. Tanzania: 6 (1996); Vollesen *et al.*, Checklist Mkomazi: 83 (1999); Hedrén in Fl. Somalia 3: 416 (2006); Ensermu in Fl. Eth. 5: 478 (2006). Type: Ethiopia, Gageros, *Schimper* in *Hohenacker* 2300 (K!, lecto.; selected by Hedrén (1990); BM!, iso.)

Annual or short-lived perennial herb with a single unbranched to much-branched stem (often branched from near base to give impression of having several stems) from a taproot or short creeping rhizome, stems erect, decumbent, scrambling or scandent, to 1(–2) m long, when young glabrous to densely puberulous, sometimes with long (1–2 mm) pilose hairs, with or without capitate glands. Leaves with petiole to 4 cm long; lamina narrowly to broadly ovate or elliptic (rarely lanceolate or slightly obovate), largest 1.2–10.5 × 0.4–6.8 cm, base attenuate to truncate, margin entire or slightly crenate, apex acuminate to rounded; sparsely puberulous or sparsely pubescent, densest on midrib and margins towards base and here usually also with stalked capitate glands. Flowers single or 2–6 in axils of upper leaves; bracts absent or lanceolate to suborbicular or reniform, to 0.8 cm long, sometimes with long (to 2(–3) mm) glossy curly hairs, with or without stalked capitate glands. Calyx sparsely to densely puberulous with hairs to 1.5 mm long, often with stalked capitate glands, lobes 2–6.5(–7) mm long, basal part sometimes spongiform in fruit. Corolla white to pale or dark mauve, lower lip with or without faint to distinct "herring bone" pattern, upper lip with dark veins to uniformly dark purple, 3–8(–9) mm long; tube 1.5–5 mm long, widening immediately above base; upper lip 1.5–3(–4.5) mm long (tube same length as upper lip or to 2 mm longer), teeth to 1 mm long; lower lip 2–7 × 3–8 mm. Filaments 1–2 mm long; thecae yellow, ± 0.5 mm long, lower with acute appendage 0.2–0.3 mm long. Capsule 3.5–6 mm long, glabrous to uniformly puberulous all over; seed 1–1.5 mm long. Indehiscent fruits with entire to (usually) dissected wings, 3–4 mm long.

1. Young stems with long (1–2 mm) pilose hairs and with stalked capitate glands; corolla 3.5–7 mm long .... c. subsp. *dinteri*
   Young stems usually without long pilose hairs, if pilose hairs present then capitate glands absent 2
2. Corolla 3–5.5(–6) mm long, tube 1.5–3.5 mm long, same length as or to 1 mm longer than upper lip, lower lip 2–4 × 3–5 mm .... a. subsp. *heterocarpa*
   Corolla 6–8(–9) mm long, tube 3.5–4.5 mm long, 1–2 mm longer than upper lip, lower lip 4–7 × 5–8 mm .... b. subsp. *praetermissa*

a. subsp. **heterocarpa**; Hedrén in Symb. Bot. Upsal. 29, 1: 106 (1989) & in Nordic Journ. Bot. 10: 375, fig. 27A & 28A-G (1990); Ensermu in Fl. Eth. 5: 478 (2006)

Annual herb, often branched from near base; stems sparsely to densely puberulous with hairs to 1 mm long, with or without stalked capitate glands (rarely with a few hairs to 1.5 mm long but then without stalked capitate glands). Largest leaf 1.5–7(–10.5) × 0.4–3.7(–5.5) cm, base attenuate, apex acute to rounded. Flowers 2–6 in axils of upper leaves. Calyx lobes 2–6(–7) mm long, basal part sometimes spongiform in fruit. Corolla 3–5.5(–6) mm long, tube 1.5–3.5 mm long, same length as or to 1 mm longer than upper lip, lower lip 2–4 × 3–5 mm. Capsule 3.5–5(–6) mm long.

Uganda. Karamoja District: base of Mt Moroto, 6 Oct. 1952, *Verdcourt* 789! & Moroto township, Sept. 1956, *J. Wilson* 255!

Kenya. Northern Frontier District: Isiolo, 9 March 1952, *Gillett* 12516!; West Suk District: Lokwien, 14 Aug. 1978, *Lye* 9100!; Teita District: Taita Hills, SE end of Mwandango Forest, N of Msau, 17 June 1998, *Mwachala* EW959!

Tanzania. Musoma District: Serengeti National Park, Seronera River, Banagi, 30 March 1961, *Greenway* 9941!; Lushoto District: Lushoto, 1 March 1976, *Shabani*, 1098!; Dodoma District: 33 km on Itigi–Chunya road, 26 April 1964, *Greenway & Polhill* 11739!

Distr. **U** 1; **K** 1–4, 6, 7; **T** 1–5, 7; Sudan, Eritrea, Ethiopia, Somalia, Socotra; Egypt (Jebel Elba), Yemen, Saudi Arabia, Oman, Pakistan (Sind)

Hab. *Acacia*- and *Acacia-Commiphora* bushland, rocky hills and lava flows, riverine thicket, riverbeds, forest margins and secondary grassland, roadsides, weed of cultivation; 300–1700 m

SYN. *Harnieria dimorphocarpa* Solms Laub. in Sitzber. Ges. Naturf. Berlin 1864: 21 (1864) & in Schweinfurth, Beitr. Fl. Aeth. 1: 110 (1867). Type: as for *J. heterocarpa*

Justicia leptocarpa Lindau in E.J. 20: 70 (1894) & in P.O.A. C: 373 (1895); C.B. Clarke in F.T.A. 5: 200 (1900); E.P.A.: 971 (1964). Type: Tanzania, Lushoto District, Mashewa, *Holst* 8799 (K!, lecto.; selected by Hedrén (1990)

J. sp. Sect. *Rostellularia sensu* Robertson & Luke, Kenya Coast. For. 2. Checklist Vasc. Pl.: 82 (1993)

b. subsp. **praetermissa** *Hedrén* in Symb. Bot. Upsal. 29, 1: 106 (1989), *nom. nud.* & in Nordic Journ. Bot. 10: 379, fig. 27J-N (1990). Type: Uganda, Masaka District, Lake Nabugabo, *Drummond & Hemsley* 4733 (K!, holo.; EA!, K!, iso.)

Annual or short-lived perennial herb, often branched from near base; young stems glabrous to puberulous (rarely densely so) with hairs to 1 mm long, no capitate glands. Largest leaf 1.2–6(–7.5) × 0.4–2.8 cm, base attenuate, apex acute to obtuse. Flowers single or 2–4 in axils of upper leaves. Calyx lobes 2–4 mm long, basal part not spongiform in fruit. Corolla 6–8(–9) mm long, tube 3.5–4.5 mm long, 1–2 mm longer than upper lip, lower lip 4–7 × 5–8 mm. Capsule (4.5–)5–6 mm long.

UGANDA. Kigezi District: Kachwekano Farm, May 1949, *Purseglove* 2846!; Mengo District: Kitamiro, 26 Sept. 1949, *Dawkins* 402!; Mubende District: Wattuba, 13 April 1970, *Katende* 106!

KENYA. Trans-Nzoia District: Kitale, 19 May 1969, *Napper & Tweedie* 2103!; Kisumu/Londiani District: Londiani–Fort Ternan road, Mt Limutit, 26 Sept. 1953, *Drummond & Hemsley* 4452!; Masai District: Oldebesi Lemoko, 28 April 1961, *Glover et al.* 794!

TANZANIA. Mwanza District: Mwanza, Capri Point, 17 June 1951, *Tanner* 408!; Kigoma District: Kasye Forest, 19 March 1994, *Bidgood et al.* 2830!; Dodoma District: 25 km on Manyoni–Singida road, 15 April 1988, *Bidgood et al.* 1151!

DISTR. **U** 2–4; **K** 3, 5, 6; **T** 1, 4, 5; E Congo-Kinshasa, Rwanda, Burundi

HAB. A wide range of woodland and bushland on sandy, stony or clayey soils, forest margins and glades, secondary forest, termite mounds; 800–2600 m

SYN. *J. striata* (Klotzsch) Bullock subsp. *melampyrum* (S. Moore) J. K. Morton in K.B. 32: 443 (1978) pro parte

[*J. striata sensu* Champluvier in Fl. Rwanda 3: 466 (1985) pro parte, *non* (Klotzsch) Bullock (1932)]

J. sp. cf. J. striata sensu Synnott, Checklist Fl. Budongo For.: 69 (1985)

c. subsp. **dinteri** (*S. Moore*) *Hedrén* in Symb. Bot. Upsal. 29, 1: 106 (1989) & in Nordic Journ. Bot. 10: 378, fig. 27B & 28H-I (1990). Type: Namibia, Otjitua, *Dinter* 87 (BM, holo.; K!, iso.)

Annual herb, not branched from near base; young stems with long (1–2 mm) pilose hairs and with stalked capitate glands. Largest leaf (3–)4–10.5 × 1.3–6.8 cm, base attenuate to truncate, apex acuminate to acute. Flowers 2–4 in axils of upper leaves. Calyx lobes 2–5(–6.5) mm long, basal part whitish and spongiform in fruit. Corolla 3.5–7 mm long, tube 2.5–4.5 mm long, 1–2 mm longer than upper lip, lower lip 3–7 × 3–8 mm. Capsule 4–6 mm long.

TANZANIA. Dodoma District: 15 km SW of Dodoma on Bihawana road, 27 Jan. 1962, *Polhill & Paulo* 1267!; Kilosa District: Ruaha Valley, near Mbuyni, 15 March 1986, *Bidgood & J. Lovett* 254!; Iringa District: Ruaha National Park, Mwayenye, Ruaha River Camp, 6 Jan. 1987, *J. Lovett* 1325!

DISTR. **T** 5–7; Angola, Zambia, Malawi, Zimbabwe, Botswana, Namibia, South Africa

HAB. *Acacia-Commiphora* woodland and bushland, often on rocky hills and slopes, riverine forest, Brachystegia woodland and grassland; 500–1250 m

SYN. *Justicia dinteri* S. Moore in J.B. 57: 246 (1919); Meyer in Prodr. Fl. Südw. Afr. 130: 35 (1968)

NOTE. I have with some hesitation decided to keep the three subspecies. There are large overlaps in the distributions, and – especially in central Tanzania – the distinction between subsp. *heterocarpa* and subsp. *praetermissa* is not very clear-cut. Subsp. *dinteri* seems on the whole slightly more distinct but it overlaps with the other two in **T** 5. The ecology is pretty similar for all three in central Tanzania except subsp. *dinteri* seems to prefer *Brachystegia* woodland while the other two are more likely to be found on rocky outcrops.

85. **Justicia unyorensis** *S. Moore* in J.B. 49: 308 (1911); Rendle in J.B. 70: 162 (1932); Hedrén in Symb. Bot. Upsal. 29, 1: 107 (1989) & in Nordic Journ. Bot. 10: 380, fig. 30 (1990); U.K.W.F., ed. 2: 280 (1994); Ensermu in Fl. Eth. 5: 479 (2006). Type: Uganda, Toro District, Kibale, *Bagshawe* 1228 (BM!, lecto.; selected by Morton in K.B. 32: 444 (1978))

Short-lived perennial herb with one to several sparsely branched stems from short erect or creeping rhizome; stems erect, ascending, decumbent or scrambling (if decumbent usually rooting at lower nodes), to 1 m long, when young subglabrous to densely retrorsely to spreadingly pubescent with pale yellow hairs, usually densest just below nodes, without capitate glands. Leaves with petiole to 1(–2.5) cm long; lamina narrowly ovate to ovate or elliptic, largest (0.8–)1.5–5(–7) × 0.5–2.5(–3) cm, base attenuate to rounded, margin entire or slightly crenate, apex subacuminate to broadly rounded; subglabrous to pubescent (rarely densely so), densest on midrib and margins, no stalked capitate glands. Flowers single or 2–4 in axils of upper leaves; bracts absent(in single flowers) or ovate to elliptic, to 0.8 cm long, never with long glossy curly hairs, with or without stalked capitate glands. Calyx subglabrous to densely puberulous with straight hairs to 0.5 mm long, sometimes with stalked capitate glands, lobes (2.5–)3–5.5(–6) mm long. Corolla white to mauve, lower lip with or without distinct "herring bone" pattern (if without then sometimes with dark patch), upper lip with dark veins to uniformly purple, 6–9.5 mm long; tube 3–5.5 mm long, widening immediately above base; upper lip 2.5–4.5 mm long (tube same length as upper lip or to 2 mm longer), teeth to 1 mm long; lower lip 4–8 × 5–9 mm. Filaments 2–3 mm long; thecae yellow or purple, 0.5–1 mm long, lower with acute appendage ± 0.4 mm long. Capsule 5–7 mm long, acute, glabrous; seed ± 1 mm long. Indehiscent fruits absent.

UGANDA. Kigezi District: Mt Mgahinga, June 1951, *Purseglove* 3696!; Mbale District: Bugishu, Budadiri, Jan. 1932, *Chandler* 499!; Mengo District: Mabira Forest, 7 Nov. 1938, *Loveridge* 5!

KENYA. Trans-Nzoia District: NE Elgon, Oct. 1969, *Tweedie* 3717!; Kiambu District: 1.5 km N of Uplands turning on Njabini road, 25 Jan. 1969, *Napper* 1816!; North Kavirondo District: Kakamega Forest, 10 Dec. 1956, *Verdcourt* 1688!

TANZANIA. Bukoba District: Maruku, no date, *Panayotis* 94! & Minziro Forest Reserve, Kele Hill, 20 April 1994, *Congdon* 355! & 20 Jan. 1999, *Festo* 124!

DISTR. **U** 1–4; **K** 3–6; **T** 1; Nigeria, Cameroon, Ethiopia, Congo-Kinshasa, Rwanda, Burundi

HAB. Intermediate and montane evergreen forest, often on margins, paths and in glades, montane grassland, wetter types of woodland and bushland, hedges, roadsides, old fields; 1150–2900 m

SYN. *J. keniensis* Rendle in J.B. 70: 162 (1932); U.K.W.F.: 605 (1974); Blundell, Wild Fl. E. Afr.: 394, pl. 611 (1987). Type: Kenya, N Nyeri District, Nanyuki, *Rendle* 589 (BM!, holo.)

J. striata (Klotzsch) Bullock in K.B. 1932: 502 (1932), quoad *Lugard* 242, *non* sensu stricto

[*J. striata sensu* Lugard in K.B. 1933: 94 (1934); Champluvier in Fl. Rwanda 3: 466 (1985), *non* (Klotzsch) Bullock (1932)]

J. wittei Mildbr. in B.J.B.B. 17: 92 (1943); F.P.N.A. 2: 315 (1947). Type: Congo-Kinshasa, Kamatembe, *de Witte* 1531 (BR, holo.)

J. striata (Klotzsch) Bullock subsp. *melampyrum* (S. Moore) J. K. Morton in K.B. 32: 443 (1978) pro parte, *non* sensu stricto

J. unyorensis S. Moore var. *keniensis* (Rendle) Hedrén in Symb. Bot. Upsal. 29, 1: 108 (1989) & in Nordic Journ. Bot. 10: 382 (1990)

NOTE. The plants from higher altitudes in Kenya (*J. keniensis*, sensu stricto) tend to have more slender stems, more rounded leaves and slightly larger corolla. But the differences are more like a grade and there are no clear-cut differences which justify recognising distinct taxa.

86. **Justicia striata** (*Klotzsch*) *Bullock* in K.B. 1932: 502 (1932); Milne-Redhead in Mem. N. Y. Bot. Gard. 9, 1: 24 (1954); Binns, Check list Herb. Fl. Malawi: 14 (1968); Richards & Morony, Check List Mbala: 232 (1969); U.K.W.F.: 605 (1974); Moriarty, Wild Fl. Malawi: 84, pl. 42 (1975); Morton in K.B. 32: 441 (1978); Hedrén in Symb. Bot. Upsal. 29, 1: 110 (1989) & in Nordic Journ. Bot. 10: 385, fig. 35 (1990); Robertson & Luke, Kenya Coast. For. 2. Checklist Vasc. Pl.: 82 (1993); U.K.W.F., ed 2: 280, pl. 124 (1994); Ruffo *et al.*, Cat. Lushoto Herb. Tanzania: 7 (1996); Vollesen *et al.*, Checklist Mkomazi: 83 (1999); Strugnell, Checklist Mt Mulanje: 34 (2006); Hedrén in Fl. Somalia 3: 416 (2006); Ensermu in Fl. Eth. 5: 479 (2006). Type: Mozambique, Zambesia, *Torre* 4994 (LISC, neo. selected by Hedrén (1990)

Annual or perennial herb with erect, ascending (sometimes rooting at lower nodes) or scrambling stems from taproot (when annual) or from creeping often branched rhizome (rarely a scrambling subshrub with woody basal stems); stems to 0.75 m long in erect plants, to 2 m when scrambling, when young subglabrous to sericeous, puberulous, pubescent or tomentose. Leaves with petiole to 3.5(–5.5) cm long; lamina narrowly ovate to ovate (rarely elliptic), towards tip of stems often lanceolate, largest 2–8(–11.5) × 1–4.5(–5.5) cm, base attenuate to subcuneate, margin entire to crenate, apex acuminate to subacute; subglabrous to sparsely sericeous, puberulous or pubescent, densest on midrib and margins, with or without stalked capitate glands on margins towards base and on petioles. Flowers 2–6 in axils of upper leaves (rarely some single); bracts lanceolate to broadly ovate or broadly elliptic, to 1.1 cm long, with or without long (to 1.5 mm) pilose hairs, with or without stalked capitate glands. Calyx puberulous or sparsely so with straight hairs to 0.5 mm long, sometimes with stalked capitate glands, lobes 3–6(–7) mm long. Corolla white to mauve or purple, lower lip with distinct "herring bone" pattern, upper lip without dark veins, 5–13 mm long; tube 3–7 mm long, widening immediately above base; upper lip (2–)3–6(–8) mm long (tube same length as upper lip or to 2 mm longer, very rarely with lip to 2 mm longer than tube), teeth to 1 mm long; lower lip 4–10 × 5–10 mm. Filaments 2–6 mm long; thecae yellow with brown pigment patches at apex and base of upper theca and at apex of lower theca, 0.5–1 mm long, lower with acute appendage 0.3–0.5 mm long. Capsule 6–9.5 mm long, acute, glabrous to sparsely puberulous in upper half; seed 1–1.5 mm long. Indehiscent fruits 3–4 mm long, 4-winged. Fig. 77, p. 585.

KENYA. Machakos District: 4 km N of Nunguni, Kibungu, 10 June 1967, *Mwangangi* 39!; Masai District: Cis Mara Masai, Chepkorobotik Forest, 28 March 1961, *Glover et al.* 33!; Kwale District: 12 km WSW of Gazi, Buda, Mafisine Forest, 16 Aug. 1953, *Drummond & Hemsley* 3829!

TANZANIA. Arusha District: Korongo, E of Legumishwa Hill, 25 April 1994, *Grimshaw* 94/419!; Ufipa District: 2 km on Tatanda–Mbala road, 24 April 1997, *Bidgood et al.* 3388!; Masasi District: NE of Masasi, Pangani Hill, 11 March 1991, *Bidgood et al.* 1914!

DISTR. **K** 1, 4, 6, 7; **T** 2–8; Ethiopia, S Somalia, SE Congo-Kinshasa, Angola, Zambia, Malawi, Mozambique (see notes)

HAB. Lowland to montane forest, usually in grassy glades or on edges, riverine forest, grassland, bushland and woodland (including *Acacia* and *Brachystegia*), rocky hills, roadsides, often becoming weedy or semi-weedy and surviving in plantations and disturbed vegetation; near seal level to 2300 m

SYN. *Adhatoda striata* Klotzsch in Peters, Reise Mossamb.: 216 (1861)

Justicia filifolia Lindau in E.J. 20: 70 (1894) & in P.O.A. C: 373 (1895); C.B. Clarke in F.T.A. 5: 198 (1900) pro parte; Binns, Check list Herb. Fl. Malawi: 14 (1968). Type: Malawi, Shire Highlands, *Last* s.n. (K!, lecto.; selected by Hedrén (1990)

FIG. 77. *JUSTICIA STRIATA* — **1**, habit; **2**, branch with many-flowered nodes; **3**, bract; **4**, calyx whole and calyx opened up; **5**, detail of calyx lobe; **6**, corolla opened up; **7**, apical part of filament and anther; **8**, apical part of style and stigma; **9**, capsule; **10**, indehiscent fruit; **11**, seed. 1, 9 & 10 from *Drummond & Hemsley* 3829, 2, 4–6 & 8 from *Bidgood et al.* 703, 3 from *Faulkner* 3995, 7 from *Tweedie* 1614, 11 from *Milne-Redhead & Taylor* 7281. Drawn by Margaret Tebbs.

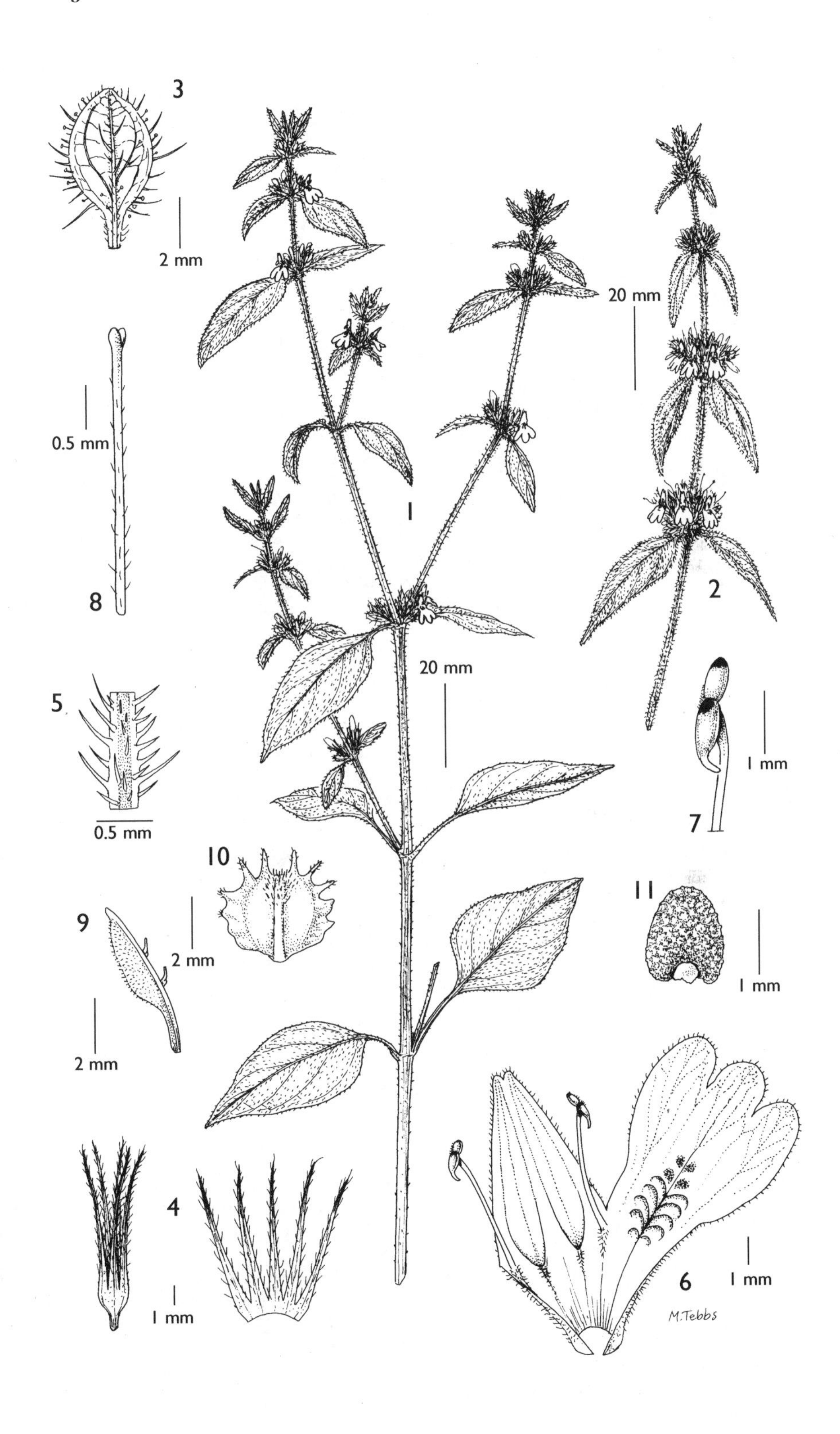
3
2 mm
0.5 mm
8
5
0.5 mm
9
2 mm
10
2 mm
4
1 mm
1
20 mm
20 mm
2
1 mm
7
11
1 mm
6
1 mm
M.Tebbs

J. rostellaria (Nees) Lindau in E.J. 20: 70 (1894) & in P.O.A. C: 373 (1894) pro parte, *non* *Adhatoda rostellaria* Nees (1847)

J. melampyrum S. Moore in Trans. Linn. Soc. London, ser. 2, Bot. 4: 32 (1894); C.B. Clarke in F.T.A. 5: 199 (1900). Type: Malawi, Mt Mulanje, *Whyte* 135 (BM!, holo.)

[*J. heterocarpa sensu* Lindau in P.O.A. C: 373 (1895) pro parte, *non* T. Anderson (1863)]

J. dyschoristeoides C.B. Clarke in F.T.A. 5: 197 (1900). Type: Tanzania, Kilimanjaro, *Johnston* s.n. (K!, holo.; BM!, iso.)

[*J. whytei sensu* C.B. Clarke in F.T.A. 5: 198 (1900), quoad *Buchanan* 304, *non* S. Moore (1894)]

J. lindaui C.B. Clarke in F.T.A. 5: 199 (1900). Type: Tanzania, Lushoto District, Usambara, *Holst* 4324 (K!, lecto.; selected by Hedrén (1990)

J. kaessneri S. Moore in J.B. 40: 345 (1902). Type: Kenya, Kwale District, Gadu, *Kässner* 409 (BM!, holo.; K!, iso.)

J. forbesii S. Moore in J. Bot. 44: 219 (1906); Benoist, Cat. Pl. Madagascar, Acanthacées: 24 (1929). Type: "Madagascar", *Forbes* 22 (BM!, holo.). See notes

J. infirma Mildbr. in N.B.G.B. 11: 410 (1932). Type: Tanzania, Njombe District, Ruhudje, Massagati, *Schlieben* 1145 (B†, holo.; BM!, iso.)

J. sp. Sect. *Calophanoides sensu* Richards & Morony, Check List Mbala: 231 (1969) pro parte

J. striata (Klotzsch) Bullock subsp. *melampyrum* (S. Moore) J. K. Morton in K.B. 32: 443 (1978) pro parte; Vollesen in Opera Bot. 59: 81 (1980)

J. striata (Klotzsch) Bullock subsp. *striata* var. *filifolia* (Lindau) J. K. Morton in K.B. 32: 443 (1978)

[*J. pinguior sensu* Cribb & Leedal, Mountain Fl. S. Tanzania: 127 (1988); Ruffo *et al.*, Cat. Lushoto Herb. Tanzania: 7 (1996) pro parte, *non* C.B. Clarke (1900)]

J. striata (Klotzsch) Bullock subsp. *striata* var. *dyschoristeoides* (C.B. Clarke) Hedrén in Symb. Bot. Upsal. 29, 1: 111 (1989) & in Nordic Journ. Bot. 10: 390 (1990)

J. striata (Klotzsch) Bullock subsp. *austromontana* Hedrén in Symb. Bot. Upsal. 29, 1: 111 (1989), *nom. nud.* & in Nordic Journ. Bot. 10: 390 (1990). Type: Tanzania, Mbeya, *Hedrén et al.* 248 (UPS, holo.; K!, iso.)

J. sp. nr. striata sensu Robertson & Luke, Kenya Coast. For. 2. Checklist Vasc. Pl.: 82 (1993)

NOTE. Hedrén (1990) divides the Eastern and Southern African material of *J. striata* into two subspecies and one of these into two varieties. I have decided not to keep up these infraspecific taxa here. There are many intermediates between the ± glabrous small-flowered lowland form (subsp. *striata* var. *striata*) and the more hairy large-flowered northern highland form (subsp. *striata* var. *dyschoristeoides*). It is true that some specimens of the southern glabrous to hairy large-flowered highland form (subsp. *austromontana*) look very distinct with their large purple corollas, but closer study shows a total gradation towards the coastal form.

The type of *J. forbesii* is said to be from Madagascar, but – as explained by Hedrén (1990) – Forbes also collected in Mozambique. As no species from this section is known from Madagascar it seems reasonable to assume that this collection was made in Mozambique and wrongly labelled. The specimen is typical of coastal *J. striata.*

Subsp. *occidentalis* from Ghana to Central African Republic and W Congo has not been considered here. It may or may not be worthy of subspecific rank.

87. **Justicia phyllostachys** *C.B. Clarke* in F.T.A. 5: 188 (1900); Binns, Check List Herb. Fl. Malawi: 14 (1968); Hedrén in Symb. Bot. Upsal. 29, 1: 115 (1989) & in Nordic Journ. Bot. 10: 396, fig. 39 (1990). Type: Malawi, Nyika Plateau, *Whyte* 118 (K!, lecto.; selected by Hedrén (1990); K!, iso.)

Perennial herb (often pyrophytic) with one to several erect, ascending, decumbent or scrambling usually unbranched or little-branched stems from woody rootstock, basal part of stems sometimes subshrubby; stems to 1 m long in erect plants, to 2(?–3) m when scrambling, when young subglabrous to sericeous, puberulous or pubescent with white or pale yellow hairs. Leaves with petiole to 0.8 cm long; lamina narrowly to broadly ovate, towards tip of stems and in "inflorescence" often lanceolate, largest (1–)2–6.3 × (0.4–)0.8–2.8 cm (2–3 times as long as wide), base shortly attenuate to truncate, margin entire to indistinctly crenate, apex subacuminate to rounded; subglabrous to pubescent, densest on midrib and margins, without stalked capitate glands. Flowers 2–4 in axils of upper leaves or some single, upper internodes usually contracted thus creating a pseudo-racemose inflorescence; bracts held erect (usually covering flowers), ovate to elliptic or broadly so (rarely lanceolate), to 1.2(–1.5) cm long, often ciliate with long (to 2 mm) pilose hairs, without stalked capitate glands. Calyx ciliate or sparsely so (hairs to 1 mm) on edges and midrib, more rarely subglabrous or finely puberulous, lobes 4–8 mm long. Corolla from pale pink or pale mauve to dark purple (very rarely white), lower lip with distinct "herring bone" pattern, upper lip with dark veins, 9–14 mm long; tube 5–7 mm long, widening immediately above base; upper lip 4–7 mm long (tube same length as upper lip or to 2 mm longer), teeth to 1 mm long; lower lip 5–10(–12) × 5–12(–15) mm. Filaments 2–5 mm long; thecae yellow, without brown pigment patches, 0.5–1 mm long, lower with acute appendage 0.3–0.5 mm long. Capsule 6.5–9.5 mm long, acute, glabrous to sparsely puberulous in upper half; seed 1–1.5 mm in diameter. Indehiscent fruits absent.

TANZANIA. Ufipa District: 5 km on Namanyere–Karonga road, 4 March 1994, *Bidgood et al.* 2612!; Iringa District: Dabaga Highlands, Kilolo, 9 Feb. 1962, *Polhill & Paulo* 1402!; Mbeya District: Mt Mbeya, 13 May 1956, *Milne-Redhead & Taylor* 10330!

DISTR. **T** 1, 4–7; Congo-Kinshasa, Burundi, Angola, Zambia, Malawi, Mozambique, Zimbabwe

HAB. Montane grassland and bushland (including secondary types), forest margins and glades, higher altitude wetter *Brachystegia* and *Uapaca* woodland, weed in old fields and in plantations, roadsides, termite mounds, rarely in seasonally wet valley grassland at lower altitudes; (1100–)1500–2450 m

SYN. [*J. whytei sensu* C.B. Clarke in F.T.A. 5: 198 (1900), quoad *Whyte* s.n., Masuku Plateau; Lindau in Wiss. Ergebn. Schwed. Rhod.-Kongo-Exp. 1911–1912: 308 (1914) pro parte, *non* S. Moore (1894)]

[*J. melampyrum sensu* C.B. Clarke in F.T.A. 5: 199 (1900), quoad *Johnson* s.n. & *Whyte* s.n., Masuku Plateau, *non* S. Moore (1894)]

J. umbratilis S. Moore in J.B. 51: 216 (1913). Type: Congo/Zambia, Lake Mweru [Moero], Lukongolawa, *Kässner* 2804 (BM!, holo.)

J. sp. Sect. *Calophanoides sensu* Richards & Morony, Check List Mbala: 231 (1969) pro parte

[*J. striata sensu* Morton in K.B. 32: 443 (1978) pro parte, *non* (Klotzsch) Bullock (1932)]

[*J. pinguior sensu* Ruffo *et al.*, Cat. Lushoto Herb. Tanzania: 7 (1996) pro parte, *non* C.B. Clarke (1900)]

88. **Justicia chalaensis** *Vollesen* **sp. nov.** a *J. phyllostachyae* foliis minoribus latioribus minus quam duplo longioribus quam latioribus et corolla alba vel pallide flava (nec pallide rosea nec pallide malvina neque atropurpurea) differt. Species omnes ceterae flaviflorae ad sectionem *Harnieriam* pertenentes vel habitu erecto frutescentique (*J. odora*) vel foliis floribusque maioribus (*J. sulphuriflora*) distinguenda sunt. Typus: Tanzania, Ufipa District, 29 km on Sumbawanga–Chala road, *Bidgood et al.* 5399 (K!, holo.; BR!, C!, CAS!, DSM!, EA!, K!, MO!, NHT!, P!, UPS!, iso.)

Perennial pyrophytic herb with several trailing or ascending unbranched or little-branched stems from woody rootstock; stems to 60 cm long, when young puberulous (hairs pale yellow, to 1 mm), densest on two sides. Leaves with petiole to 0.5 cm long; lamina broadly ovate, towards tip of stems and in "inflorescence" narrowly ovate or narrowly elliptic, largest 1.7–2.5 × 1.2–1.5 cm (less than twice as long as wide), base cuneate to rounded, margin entire, apex rounded; sparsely pubescent on midrib and margins, no stalked capitate glands. Flowers 2 in axils of upper leaves, upper internodes contracted thus creating a pseudo-racemose inflorescence; bracts ovate to elliptic, held erect and covering flowers, to 1 cm long, ciliate with long (to 1.5 mm) pilose hairs, no stalked capitate glands. Calyx ciliate (hairs to 1 mm) on edges, lobes 4–6 mm long. Corolla white to pale yellow, lower lip with distinct mauve and white "herring bone" pattern, upper lip with dark veins, 9–10 mm long; tube ± 5 mm long, widening immediately above base; upper lip 4–5 mm long (tube same length as upper lip), teeth ± 0.5 mm long; lower lip ± 7 × 8 mm. Filaments ± 4 mm long; thecae yellow, without brown pigment patches, 0.5–1 mm long, lower with acute appendage ± 0.5 mm long. Capsule ± 7 mm long, acute, glabrous; seed ± 1.5 mm in diameter, black, reticulate.

TANZANIA. Ufipa District: 29 km on Sumbawanga–Chala road, 1 May 1997, *Bidgood et al.* 3604! & 9 April 2006, *Bidgood et al.* 5399!

DISTR. **T** 4; not known elsewhere

HAB. Seasonally wet grassland on floor of broad flat valley, dark grey sandy-peaty soil; 1800 m

NOTE. Known only from these two collections. Superficially very similar to *J. phyllostachys* (which occurs in the same general area); but none of the many collections of this species have yellow flowers. Yellow flowers are rare in Sect. *Harnieria* and otherwise only occur in *J. odora* and *J. sulphuriflora*.

89. **Justicia lithospermoides** *Lindau* in Wiss. Ergebn. Schwed. Rhod.–Kongo-Exp. 1: 308 (1916); Hedrén in Symb. Bot. Upsal. 29, 1: 114 (1989) & in J.L.S. 103: 272, fig. 7 (1990). Type: Zambia, Fort Roseberry, *R. E. Fries* 532b (UPS, lecto.; selected by Hedrén (1990))

Perennial herb with several erect (in flower) to decumbent (in fruit) unbranched stems from woody rootstock; stems to 15 cm long in flowering and to 25 cm long in fruiting specimens, when young subglabrous (hairy below nodes) to puberulous or pubescent. Leaves often only partly developed when flowering; petiole 0–1 mm long; lamina narrowly elliptic to narrowly obovate (rarely lanceolate), largest 1–2.8(–4.7) × 0.2–0.6 cm, base attenuate, margin entire, apex subacute to broadly rounded; glabrous to pubescent. Flowers single in axils of upper leaves; bracts absent; bracteoles 2–4 mm long; pedicels 1–1.5 mm long. Calyx puberulous or densely so, sometimes with stalked capitate glands, lobes 4–7(–10 in fruit) mm long. Corolla bluish or reddish purple, mouth almost closed with stamens hidden under upper lip, lower lip with distinct "herring bone" pattern, upper lip with dark veins, 8–12 mm long; tube 5–6 mm long, widening immediately above base; upper lip 3–6 mm long (tube same length as upper lip or to 2 mm longer), teeth ± 0.5 mm long; lower lip 5–10 × 5–10 mm. Filaments 2–4 mm long; thecae yellow or tinged purple, ± 1 mm long, lower with appendage ± 0.5 mm long, broadly ellipsoid in outline, bent 90° relative to theca, obtuse. Capsule 6.5–8.5 mm long, puberulous near apex only or in upper half; seed ± 1.5 mm in diameter.

TANZANIA. Kigoma District: 57 km S of Uvinza, 31 Aug. 1950, *Bullock* 3295!; Ufipa District: 2 km W of Mkowe on Chapota road, 21 Nov. 1994, *Goyder et al.* 3768!; Mbeya District: Mbogo, 17 Oct. 1932, *Geilinger* s.n.!

DISTR. **T** 4, 7; Congo-Kinshasa, Burundi, Zambia, Malawi, Mozambique

HAB. Appearing shortly after burning in short-grass dambos and valley grassland, on sandy-peaty to clayey soils; 1200–1800 m

SYN. [*Justicia anselliana sensu* C.B. Clarke in F.T.A. 5: 209 (1900), quoad *Whyte* s.n., Fort Hill, *non* (Nees) T. Anderson (1863)]

[*Peristrophe usta* C.B. Clarke in F.T.A. 5: 244 (1900), quoad *Carson* s.n., *non* sensu stricto]

90. **Justicia** sp. **G**

Perennial herb with a single erect or ascending unbranched stem from a short creeping rhizome; stems to 65 cm long, when young glabrous to very sparsely puberulous and with band of hairs at nodes. Leaves sessile, lamina lanceolate to narrowly ovate, largest 3–4.2 × 0.4–0.8 cm, base attenuate to cuneate, margin entire, apex subacute, finely apiculate; glabrous or very sparsely ciliate near base. Flowers single in axils of upper leaves; bracts absent; bracteoles 2–7 mm long; pedicels 1–1.5 mm long. Calyx puberulous, lobes 5–6 mm long. Corolla dark purple, mouth almost closed with stamens hidden under upper lip, lower lip with distinct mauve and white "herring bone" pattern, upper lip with dark veins, 9–12 mm long; tube 5–6 mm long, widening immediately above base; upper lip 4–6 mm long (tube same length as upper lip or to 1 mm longer), teeth ± 0.5 mm long; lower lip 7–9 × 8–10 mm. Filaments ± 4 mm long; thecae purple, ± 1 mm long, lower with appendage ± 0.3 mm long, ellipsoid in outline, bent 90° relative to theca, obtuse. Capsule and seed not seen.

TANZANIA. Mpanda District: 19 km on Mpanda–Uvinza road, 14 May 1997, *Bidgood et al.* 3936!
DISTR. **T** 4; not known elsewhere
HAB. Seepage area in tall *Brachystegia-Julbernardia* woodland, grey sandy-peaty soil; 1100 m

NOTE. Known only from this collection. Close to *J. lithospermoides*, from which it mainly differs in having only a single stem from a creeping rootstock, in the larger leaves and in the differently shaped anther appendage.

91. **Justicia mollugo** *C.B. Clarke* in F.T.A. 5: 200 (1900); Milne-Redhead in Mem. N. Y. Bot. Gard. 9: 24 (1954); Binns, Check List Herb. Fl. Malawi: 14 (1968); Richards & Morony, Check List Mbala: 232 (1969); Hedrén in B.J.B.B. 58: 140, fig. 5 (1988) & in Distrib. Pl. Afr. 35: map 1166 (1988) & in Symb. Bot. Upsal. 29, 1: 117 (1989); Strugnell, Checklist Mt Mulanje: 34 (2006). Type: Malawi, *Buchanan* s.n. (K!, lecto.; selected by Hedrén (1988); BM!, iso.)

Single-stemmed annual herb, stem erect (rarely decumbent), unbranched or sparsely branched from near base (side branches leaving under angles of less than 45°), to 20(–35) cm long, but usually less than 10 cm, when young bifariously sericeous-puberulous or sparsely so. Leaves with petiole to 6(–10) mm long; lamina ovate to elliptic or narrowly so, sometimes lanceolate towards apex, largest 0.5–2.7 × 0.2–1.3 cm (usually 2–3 times as long as wide), base attenuate to cuneate, margin entire, apex acute to obtuse; sparsely puberulous on midrib and margins, densest towards base, often with stalked glands. Flowers 2–4 in axils of upper leaves or some single; bracts lanceolate to narrowly ovate or narrowly elliptic, to 7 mm long, sparsely puberulous, not pilose-ciliate; bract fused to pedicel making it appear as if flower attached at apex of petiole; pedicels 1–2(–3) mm long. Calyx sparsely puberulous-ciliate, lobes yellowish green to green, 2–4(–5 in fruit) mm long. Corolla white to mauve, lower lip with indistinct to distinct "herring bone" pattern with raised ribs but not with crenulate edges, upper lip with faintly darker veins, 3–5 mm long; tube white, 2–3 mm long, widening immediately above base; upper lip 1–2 mm long (tube ± 1 mm longer than lip), teeth minute; lower lip 1.5–3 × 2–3.5 mm. Filaments 1–2 mm long; anthers exserted, thecae with a purplish band on sides, 0.2–0.5 mm long, lower with acute appendage ± 0.2 mm long. Capsule 3–4.5(–5) mm long, acute, sparsely puberulous near apex, more rarely glabrous; seed ± 1 mm in diameter.

TANZANIA. Kigoma District: Kasye Forest, 21 March 1994, *Bidgood et al.* 2882!; Iringa District: Sao Hill Sawmill, 2 May 1975, *Hepper & Field* 5248!; Songea District: 40 km on Songea–Njombe road, 22 March 1991, *Bidgood et al.* 2093!
DISTR. **T** 4, 7, 8; Congo-Kinshasa, Burundi, Angola, Zambia, Malawi, Mozambique
HAB. *Brachystegia* woodland and grassland in open sandy places or on loamy to stony soil, rocky hills, also in disturbed woodland, roadsides, pine plantations, weed; 900–2000 m

SYN. [*Justicia leptocarpa sensu* Lindau in E.J. 20: 70 (1894), quoad *Buchanan* s.n., *non* sensu stricto]
[*J. heterocarpa sensu* C.B. Clarke in F.T.A. 5: 200 (1900), quoad *Nutt* s.n., *non* T. Anderson (1863)]

NOTE. This species must be a strong contender for the "smallest Acanthaceae in the World" title! I have seen fully developed flowering and fruiting specimens smaller than a match stick.

92. **Justicia boaleri** *Hedrén* in B.J.B.B. 58: 146, fig. 8 (1988) & in Distrib. Pl. Afr. 35: map 1167 (1988) & in Symb. Bot. Upsal. 29, 1: 117 (1989). Type: Tanzania, Chunya District, Lupa Forest Reserve, *Boaler* 513 (K!, holo.)

Single-stemmed annual herb, stem erect (rarely scrambling), unbranched to densely branched, to 1 m long, when young subglabrous to sericeous-puberulous. Leaves with petiole to 5 mm long; lamina linear-lanceolate to narrowly ovate or narrowly elliptic, largest 1.5–9.5 × (0.1–)0.3–1.5 cm (more than 3 (and usually more than 5) times as long as wide), base attenuate, margin entire, apex acute to obtuse; subglabrous to sericeous-puberulous, usually densest on midrib and edges, hairs to 1 mm long, rarely with stalked glands. Flowers 2–4 in axils of upper leaves or some single; bracts lanceolate to narrowly elliptic, to 8 mm long, puberulous or sparsely so, sometimes with capitate glands, not pilose-ciliate. Calyx sparsely puberulous-ciliate, lobes uniformly dark green, 2–3.5(–5 in fruit) mm long. Corolla white to mauve, lower lip with distinct "herring bone" pattern with raised ribs but not with crenulate edges, upper lip with dark veins and darker towards apex, 5–8.5 mm long; tube white, 2.5–4.5 mm long, widening immediately above base; upper lip 2.5–4 mm long (tube 0–1 mm longer than lip), teeth ± 0.5 mm long; lower lip 4–7 × 5–8 mm. Filaments 1.5–3 mm long; anthers exserted, thecae yellow, 0.5–0.8 mm long, lower with slightly or sharply bent acute appendage ± 0.3 mm long. Capsule 3.5–6 mm long, acute to obtuse, glabrous to puberulous all over; seed 1–1.5 mm in diameter.

TANZANIA. Kigoma District: 43 km on Uvinza–Mpanda road, 11 March 1994, *Bidgood et al.* 2764!; Dodoma District: 10 km W of Bagamoyo, 22 Jan. 1969, *Mawalla* 4993!; Chunya District: Muzibini Village, 23 March 1965, *Richards* 19827!

DISTR. **T** 4, 5, 7; Congo-Kinshasa, Burundi

HAB. On dry sandy to stony soil in *Brachystegia*, *Combretum* and *Acacia* woodland and bushland, seasonally wet grassland, roadside ditches; 850–1800(–2000) m

SYN. *Justicia sp. A*; Hedrén in Symb. Bot. Upsal. 29, 1: 108 (1989) & in Nordic Journ. Bot. 10: 383 (1990)

NOTE. This species is very widespread and common in West and Central Tanzania, often occurring in large stands in late rainy to early dry season.

93. **Justicia obtusicapsula** *Hedrén* in B.J.B.B. 58: 151, fig. 10 (1988) & in Distrib. Pl. Afr. 35: map 1169 (1988) & in Symb. Bot. Upsal. 29, 1: 118 (1989). Type: Zambia, Mungwi to Kasama, *Richards* 16419 (K!, holo.)

Erect single-stemmed annual herb, much branched from ± 10 cm above ground level, lateral branches spreading at 45–60° to main stem; stems to 60 cm long, when young uniformly sericeous-puberulous. Leaves with petiole to 3 mm long; lamina lanceolate to narrowly elliptic or narrowly obovate, often linear towards apex, largest 3–5.5 × 0.3–1.1 cm (4–10 times as long as wide), base attenuate, margin entire, apex subacute to obtuse; finely and sparsely pubescent, densest on midrib and margins towards base. Flowers 2–4 in axils of upper leaves, upper internodes not contracted; bracts lanceolate to narrowly elliptic or oblanceolate, to 4(–8) mm long, not pilose-ciliate. Calyx finely uniformly puberulous and sometimes with subsessile capitate glands, lobes 2–3 mm long. Corolla pale mauve or bright purple, lower lip with indistinct "herring bone" pattern with raised ribs but not with crenulate edges, upper lip with darker veins, 4–5 mm long; tube white, 2.5–3 mm long, widening immediately above base; upper lip 1.5–2 mm long (tube ± 1 mm longer than lip), teeth ± 0.3 mm long; lower lip 2–3 × 2.5–4 mm. Filaments 1–2 mm long; thecae yellow, ± 0.5 mm long, lower with linear straight acute appendage ± 0.3 mm long. Capsule 3–4 mm long, obtuse, puberulous all over; seed ± 1 mm in diameter.

TANZANIA. Ufipa District: Kalambo River, Tanzania side, 9 April 1969, *Sanane* 585!

DISTR. **T** 4; Zambia

HAB. In rough grass on rocky slope in *Brachystegia* woodland; 1225 m

SYN. *J. sp.* Sect. *Calophanoides sensu* Richards & Morony, Check List Mbala: 231 (1969) pro parte

94. **Justicia mariae** *Hedrén* in B.J.B.B. 58: 135, fig. 3 (1988) & in Distrib. Pl. Afr. 35: map 1164 (1988) & in Symb. Bot. Upsal. 29, 1: 117 (1989). Type: Zambia, Mporokoso, Mweru Wantipa, *Richards* 9173 (K!, holo.; K!, iso.)

Single-stemmed annual herb, basal part of stem creeping and rooting, apical part erect; stem unbranched or sparsely branched, to 35 cm long, when young puberulous or sparsely so. Leaves with petiole 0–2 mm long; lamina narrowly obovate, largest 1.5–3.5 × 0.3–0.8 cm (more than 3 times as long as wide), base attenuate, margin entire, apex subacute to obtuse; subglabrous to puberulous on midrib and edges, densest towards base, hairs to 1 mm long. Flowers single or 2 in axils of upper leaves, upper internodes contracted; bracts lanceolate, to 1 cm long, pilose-ciliate with broad glossy hairs to 1.5 mm long and sometimes with stalked capitate glands. Calyx uniformly puberulous, lobes dark green towards apex, yellowish towards base, 3.5–5(–6 in fruit) mm long. Corolla a rich bluish purple, lower lip with distinct "herring bone" pattern with raised crenulate edges, upper lip with dark veins, ± 7 mm long; tube white, ± 3.5 mm long, widening immediately above base; upper lip ± 3.5 mm long (tube and lip same length), teeth ± 0.5 mm long; lower lip 4–5 × 5–6 mm. Filaments ± 3 mm long; thecae yellow, ± 0.5 mm long, lower with appendage ± 0.3 mm long, broadly ellipsoid in outline, straight, obtuse. Capsule ± 5.5 mm long, obtuse, sparsely puberulous near apex; seed ± 1 mm in diameter.

TANZANIA. Mpanda District: Kibwesa Point, 20 Aug. 1956, *Newbould & Jefford* 1668!
DISTR. **T** 4; Zambia
HAB. In water in small sedge swamp in woodland; 925 m

NOTE. This species has a broadly ellipsoid obtuse anther appendage like *J. lithospermoides*, but here the appendage is straight and not bent 90° relative to the theca.

95. **Justicia acutifolia** *Hedrén* in Nordic Journ. Bot. 10: 268, fig. 5 (1990). Type: Tanzania, Njombe District, Kitulo [Elton] Plateau, *Richards* 14014 (K!, holo.; BR, iso.)

Perennial herb with one to several erect unbranched or sparsely branched stems from creeping rhizome; stems to 30 cm long, glabrous apart from band of short hairs at nodes (rarely sparsely uniformly puberulous). Leaves held erect; petiole 0.5–1.5 mm long; lamina linear to lanceolate, largest 1–3 × 0.15–0.5 cm, base attenuate to cuneate, margin entire, raised, yellow, apex acute, apiculate (rarely subacute); glabrous (rarely scabrid-puberulous on margins); with strong midrib and strong lateral veins running parallel to midrib to apex. Flowers single, only one per node on alternating sides of stem; bracts foliaceous, 0.7–2 cm long, indumentum as leaves; bracteoles absent. Calyx glabrous to puberulous-ciliate, usually only towards apex, lobes 4–7 mm long. Corolla mauve to reddish purple, lower lip with distinct "herring bone" pattern and with two dark purple longitudinal streaks, upper lip with dark veins, 8–12 mm long; tube 4–6 mm long, widening immediately above base; upper lip 4–6 mm long (tube same length as upper lip), teeth 0.5–1 mm long; lower lip 6–8 × 7–11 mm. Filaments 3–4 mm long; thecae yellow, brown on sides, ± 1 mm long, lower with linear appendage ± 0.3 mm long. Capsule ± 1 cm long, glabrous or with a few hairs at apex; seed ± 2.5 mm in diameter.

TANZANIA. Njombe District: Kipengere Mts, Mtorwi Peak, 13 Jan. 1957, *Richards* 7722! & 14 Jan. 1957, *Richards* 7757! & Livingstone Mts, Mt Msalaba, above Luana, 23 Nov. 1992, *Gereau et al.* 5136!
DISTR. **T** 7; not known elsewhere
HAB. Montane grassland, often rocky; 2100–2700 m

NOTE. I am not at all certain that this and the following species belong in Sect. *Harnieria*. It only develops a single flower at each node on alternating sides of the stem and lacks bracteoles. But it does not naturally fit into any other group, and at the moment it is probably best considered an advanced species here.

96. **Justicia alterniflora** *Vollesen* **sp. nov.** ab omnibus ceteris speciebus ad sectionem *Harnieriam* pertinentibus (*J. acutifolia* excepta) flore unico tantum e omne nodo evoluto atque floribus secus caulem alternatim dispositis (nec floribus ad omnem nodem binatim dispositis) differt. A *J. acutifolia* foliis maioribus 4.5–5 cm (nec 1–3 cm) longis atque venis lateralibus parallelis ad costam currentibus carentibus, bracteis grandibus foliaceis carentibus et bracteolis praesentibus differt. Typus: Tanzania, Kigoma District, 35 km on Uvinza road from Kigoma–Kasulu road, *Bidgood & Vollesen* 3211 (K!, holo.; C!, NHT!, iso.)

Perennial pyrophytic herb with a solitary erect or scrambling unbranched stem from short creeping rhizome; stems to 75 cm long, glabrous. Leaves held erect; petiole 0–1 mm long; lamina linear, largest 4.5–5 × 0.2–0.3 cm, base attenuate, margin entire, apex acuminate; glabrous; with strong midrib and no lateral veins. Flowers single, only one per node on alternating sides of stem; bracts absent; bracteoles present, minute. Calyx very sparsely puberulous-ciliate, otherwise glabrous, lobes 3.5–5 mm long. Corolla mauve, lower lip with distinct "herring bone" pattern, upper lip with dark veins, 10–11.5 mm long; tube 5–6 mm long, widening immediately above base; upper lip 5–6 mm long (tube same length as upper lip), teeth ± 0.5 mm long; lower lip 8–10 × 7–9 mm. Filaments ± 4 mm long; thecae yellow, tinged purple on sides, ± 1 mm long, lower with linear bifurcate appendage ± 0.5 mm long. Capsule and seed not seen.

TANZANIA. Kigoma District: 35 km on Uvinza road from Kigoma–Kasulu road, 1 April 1994, *Bidgood & Vollesen* 3035! & 26 April 1994, *Bidgood & Vollesen* 3211!

DISTR. **T** 4; not known elsewhere

HAB. Wet seepage grassland in transition zone between *Brachystegia* woodland and valley grassland, grey sandy-loamy soil; 1000 m

NOTE. Known only from these two collections. Superficially this is very similar to narrow-leaved forms of *J. lithospermoides* but has a completely different anther appendage. It is probably closer to *J. acutifolia* which also only develops a single flower at each node.

97. **Justicia ciliata** *Jacq.*, Hort. Vindob. 2: 47, t. 104 (1772). Type: Table 104 in Hort. Vindob. 2 (1772), from plant cultivated in Vienna (iconotype)

Erect to decumbent often much-branched annual herb; stems to 60 cm long, sparsely to densely pubescent or pilose. Leaves with petiole to 8 mm long; lamina linear-lanceolate to narrowly elliptic, largest 5–10.5 × 0.5–1 cm, at least 5 times longer than wide, base cuneate to attenuate, not decurrent, apex subacute to rounded; sparsely to densely pubescent or pilose on veins and margins. Flowers solitary in axils, upwards usually congested into terminal pseudo-racemes, sometimes with short lateral branches; bracts foliaceous, as leaves or slightly smaller, distinctly white-ciliate in basal part with ciliae (1.5–)2–4 mm long; bracteoles linear or linear-lanceolate, (1–)1.2–2 cm long, with similar ciliae. Calyx 8–13 mm long, with similar ciliae, lobes linear, acuminate, with a single broad central vein and with broad hyaline white margins. Corolla: lower lip white with purple to mauve lines, upper lip white to pale green with brownish veins, 7–8 mm long, sparsely sericeous, shorter than calyx; tube 3–4 mm long, with a conspicuous central pouch and two smaller lateral pouches; upper lip ± 4 mm long, hooded; lower lip ± 5 mm long, broadly obovate, lobes ± 2 mm long. Filaments ± 2.5 mm long; anthers hairy, thecae superposed, upper smaller and held at angle to lower, ± 0.5 (upper) and 1 (lower) mm long, lower with bifurcate appendage ± 1 mm long. Capsule 9–10 mm long, glabrous; seed black, 3–3.5 × 4.5–5 mm, with two large tufts of thick moniliform hairs.

UGANDA. West Nile District: 8 km N of Metu, 17 Sept. 1953, *Chancellor* 280!; Busoga District: Bugaya, 30 Nov. 1949, *Jameson* 81!; Mengo District: Nakasongola, Sungira, 11 July 1956, *Langdale-Brown* 2197!

TANZANIA. Kigoma District: 6 km S of Kigoma, Kitwe Point, 19 April 1994, *Bidgood & Vollesen* 3136!; Chunya District: 50 km NW of Chunya, Mkwajuni, 6 May 1981, *Mwasumbi* 12011!; Songea District: 3 km NW of Gumbiro, 9 May 1956, *Milne-Redhead & Taylor* 10131!

DISTR. **U** 1, 3, 4; **T** 4, 6–8; Gambia, Senegal, Guinea, Sierra Leone, Mali, Burkina Faso, Ivory Coast, Ghana, Togo, Niger, Nigeria, Cameroon, Chad, Central African Republic, Congo-Kinshasa, Rwanda, Burundi, Sudan, Ethiopia, Zambia, Malawi

HAB. Grassland on black cotton soil, *Combretum* woodland on clay, *Julbernardia* woodland on gravelly soil, sometimes a weed of cultivation; 250–1200 m

SYN. *J. ciliaris* L. f., Suppl. Pl.: 84 (1781); Vahl, Symb. Bot. 2: 15 (1791); Willd., Sp. Pl. 1: 90 (1798). Type: as for *J. ciliata*

Monechma hispidum Hochst. in Flora 24: 375 (1841) & in Flora 26: 76 (1843); C.B. Clarke in F.T.A. 5: 213 (1900); Broun & Massey, Fl. Sudan: 347 (1929); Binns, Checklist Herb. Fl. Malawi: 15 (1968). Type: Sudan, Kordofan, Mt Kohn, *Kotschy* 239 (K!, iso.)

Pogonospermum ciliare (L. f.) Hochst. in Flora 27, Beil.: 6 (1844)

P. hispidum (Hochst.) Hochst. in Flora 27, Beil.: 6 (1844)

Schwabea ciliaris (L. f.) Nees in DC., Prodr. 11: 384 (1847); Bentham in Hooker, Niger Fl.: 482 (1849); A. Richard, Tent. Fl. Abyss. 2: 154 (1851); T. Anderson in J.L.S. Bot. 7: 45 (1863); Solms-Laubert in Schweinfurth, Beitr. Fl. Aeth.: 113 (1867); Oliver in Trans. Linn. Soc. 29: 130 (1875); Lindau in E.J. 18: 64, t. 2 fig. 98 (1894) & in P.O.A. C: 372 (1895)

Monechma ciliatum (Jacq.) Milne-Redh. in K.B. 1934: 304 (1934) & in K.B. 5: 381 (1951); F.P.S. 3: 184 (1956); Friis & Vollesen in Biol. Skr. 51(2): 450 (2005); Ensermu in Fl. Eth. 5: 485 (2006)

NOTE. In Suppl. Pl. the type is said to come from Ceylon, but there is nothing to indicate that this species occurs outside Africa. Jacquin says its origin is unknown and as his plate is a perfect match for the species it can be used as the type. The seeds from which the plant illustrated was grown almost certainly came from West Africa.

98. **Justicia bracteata** (*Hochst.*) *Zarb*, Cat. Spec. Bot. Pfund: 32 (1879); Hedrén in Fl. Somalia 3: 420 (2006). Type: Sudan, Kordofan, Tejara, *Kotschy* 261 (K!, iso.)

Erect (rarely ascending to decumbent) unbranched or little-branched annual herb; stems to 1(–1.25) m long, when young finely sericeous with downwardly directed appressed hairs. Leaves with petiole to 2 cm long; lamina narrowly ovate or narrowly elliptic (rarely ovate to elliptic), largest 3–12(–15) × 0.7–4(–5.5) cm, usually more than 4 times as long as wide (rarely 2–3 times), base attenuate, decurrent, apex acute or subacute; sericeous-puberulous or sparsely so along veins, with sparse yellow subsessile glands. Flowers solitary or paired, in many axillary racemoid cymes, long and many-flowered downwards, gradually shorter upwards, sometimes ± fusing to form an elongated terminal pseudo-raceme; main axis bracts foliaceous, raceme bracts imbricate, oblong-elliptic to orbicular (or lowermost reniform), 6–13 × 4–9 mm in middle of racemes, acute to subacute with recurved tip, lower with cuneate to rounded base, finely puberulous and with scattered stalked yellow capitate glands, distinctly ciliate with with broad glossy ciliae (1–)1.5–2 mm long; bracteoles linear or acicular, 1–1.5 mm long. Calyx 4–7(–8) mm long, puberulous with a mixture of broad non-glandular hairs, short non-capitate glands and scattered stalked yellow capitate glands, lobes linear to filiform, cuspidate, with a single central vein. Corolla white to purple (upper lip white to yellowish green), without or with purple lines on lower lip, 6–10 mm long, puberulous and with scattered capitate glands; tube 4–6 mm long; upper lip 2–4 mm long, flat or slightly hooded; lower lip 3–5 mm long, broadly obovate, lobes 1–2 mm long. Filaments ± 2 mm long, not bending out after anthesis; anthers with a few hairs, thecae ± 0.5 mm long, lower with simple or bifurcate appendage ± 0.3 mm long. Capsule 4–9 mm long, densely sericeous-puberulous (hairs spreading near apex, deflexed near base); seed mottled grey and black, 2–2.5 × 2.5–3 mm, glabrous. Fig. 78, p. 594.

UGANDA. Karamoja District: Kangole, July 1957, *J. Wilson* 382!; Acholi District: Adilang, Nov. 1937, *Tothill* 2647!; Teso District: Serere, Aug. 1932, *Chancellor* 874!

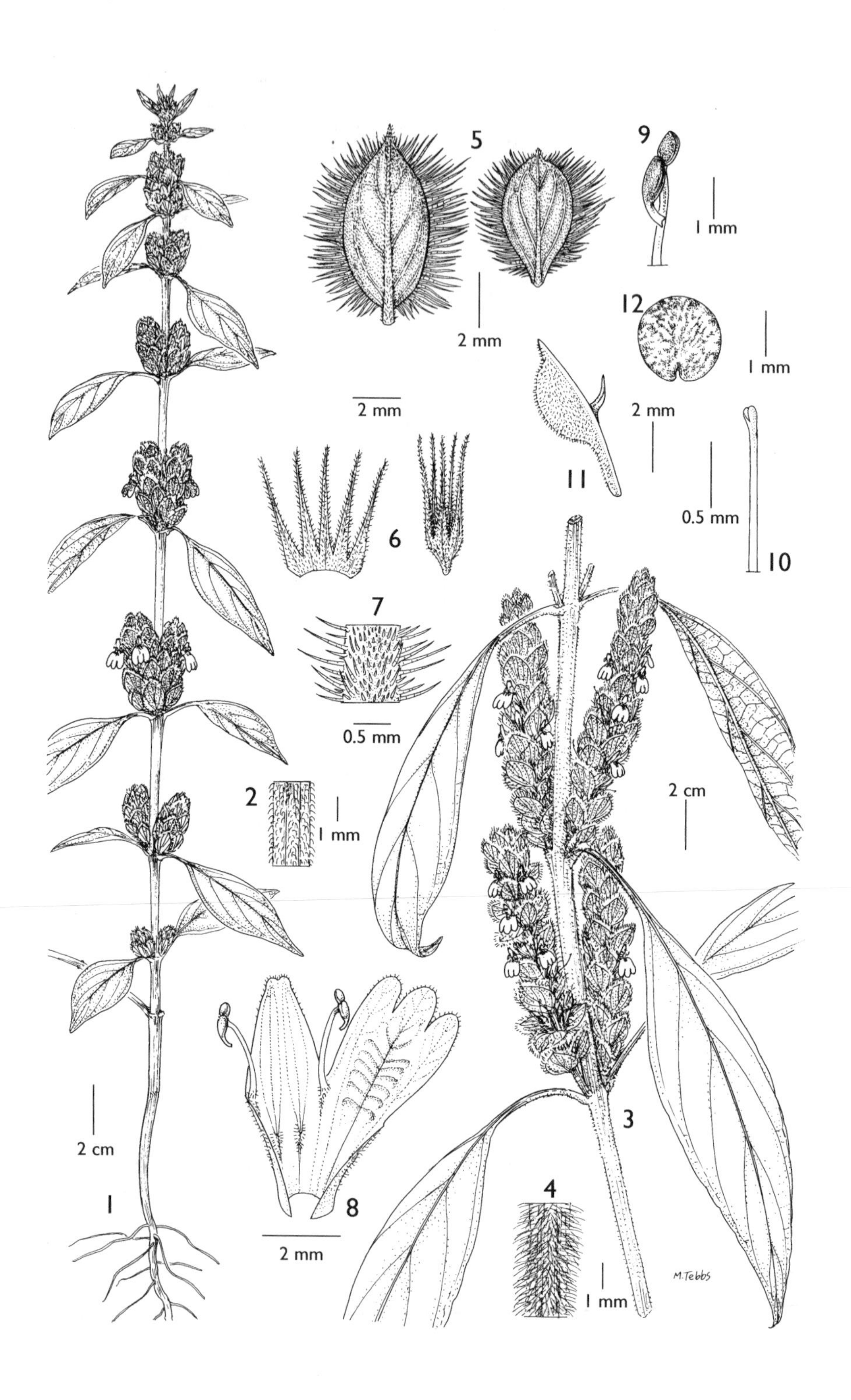
1
2 cm
2
1 mm
3
2 cm
4
1 mm
5
2 mm
6
2 mm
7
0.5 mm
8
2 mm
9
1 mm
10
0.5 mm
11
2 mm
12
1 mm
M.Tebbs

KENYA. Northern Frontier District: Loriu Plateau, 1 June 1970, *Mathew* 6520!; Machakos District: 120 km on Nairobi–Kibwezi road, 21 April 1969, *Napper & Kanuri* 2018!; Teita District: Tsavo East National Park, Kandala Swamp, 27 Feb. 1977, *Hooper & Townsend* 1077!

TANZANIA. Mbulu District: Endabash Plain, 26 May 1965, *Greenway & Kanuri* 12083!; Dodoma District: 8 km on Manyoni-Singida road, 15 April 1988, *Bidgood et al.* 1123!; Iringa District: Ruaha National Park, Msembe, 16 May 1968, *Renvoize & Abdallah* 2181!

DISTR. **U** 1, 3; **K** 1–7; **T** 1–8; Sudan, Eritrea, Ethiopia, Somalia, SE Congo-Kinshasa, Angola, Zambia, Malawi, Mozambique, Zimbabwe, Botswana, South Africa, Namibia; Yemen, Oman, India (?)

HAB. Grassland on black clay or sandy loam, alluvial *Acacia* bushland, *Acacia* and *Combretum* bushland on sandy soil or on rocky hills, coastal bushland, in Tanzania also in *Brachystegia* woodland, common in old cultivations, on roadsides and as a weed of cultivation, seems to thrive in disturbed areas; near sea-level to 1900(–2300) m

SYN. *Monechma bracteatum* Hochst. in Flora 24: 375 (1841) & in Flora 26: 75 (1843); Nees in DC., Prodr. 11: 411 (1847); Solms-Laubert in Schweinfurth, Beitr. Fl. Aeth.: 112 & 244 (1867); C.B. Clarke in F.T.A. 5: 214 (1900)

M. affine Hochst. in Flora 26: 76 (1843); Nees in DC., Prodr. 11: 411 (1847); Richard, Tent. Fl. Abyss. 2: 158 (1851). Type: Ethiopia, *Schimper* s.n. [? II.759] (BM!, K!, iso.). See note

M. angustifolium Nees in DC., Prodr. 11: 412 (1847). Type: South Africa, Natal, Durban, *Drège* s.n. (K!, holo.)

Justicia heterostegia T. Thoms. in Speke, Journ. Disc. Source Nile Append.: 643 (1863). Type: Tanzania, Bukoba District, Karagwe, *Speke & Grant* 433 (K!, holo.)

J. debilis (Forssk.) Vahl var. *angustifolia* (Nees) Oliv. in Trans. Linn. Soc. 29: 129 (1873); C.B. Clarke in F.T.A. 5: 215 (1900); Binns, Check List Herb. Fl. Malawi: 15 (1968)

J. ukambensis Lindau in E.J. 20: 69 (1894) & in P.O.A. C: 373 (1895). Type: Kenya, "Ukamba", *Hildebrandt* 2725 (B†, holo.)

Monechma scabrinerve C.B. Clarke in F.T.A. 5: 215 (1900); E.P.A.: 977 (1964); Kuchar, Pl. Somalia [CRDP Techn. Rep. Ser. No. 16]: 249 (1980). Type: Kenya, Nakuru District, Elmenteita, *Scott Elliot* 6676 (K!, holo.; BM!, iso.)

M. bracteatum Hochst. var. *non-strobilifera* C.B. Clarke in F.T.A. 5: 215 (1900). Type: Malawi, Kondowa to Karonga, *Whyte* s.n. (K!, syn.)

M. bracteatum Hochst. var. *angustifolium* (Nees) C.B. Clarke in F.T.A. 5: 215 (1900); Binns, Check List Herb. Fl. Malawi: 15 (1968)

M. ukambense (Lindau) C.B. Clarke in F.T.A. 5: 220 (1900)

[*M. debile sensu* Schinz in Mem. Herb. Boiss. 10: 64 (1900); F.P.S. 3: 184 (1956) pro parte; E.P.A.: 976 (1964) pro parte; Binns, Checklist Herb. Fl. Malawi: 15 (1968); Richards & Morony, Check List Mbala: 233 (1969); U.K.W.F.: 605 (1974); Bjørnstad in Serengeti Res. Inst. Publ. 215: 26 (1976); Kuchar, Pl. Somalia [CRDP Techn. Rep. Ser. No. 16]: 249 (1980) pro parte; Vollesen in Opera Bot. 80: 81 (1980); Robertson & Luke, Kenya Coast. For. 2. Checklist Vasc. Pl.: 83 (1993); U.K.W.F., ed. 2: 281, pl. 124 (1994); Ruffo *et al.*, Cat. Lushoto Herb. Tanzania: 8 (1996) pro parte; Vollesen *et al.*, Checklist Mkomazi: 83 (1999) quoad *B. J. Harris* 791; Friis & Vollesen in Biol. Skr. 51(2): 451 (2005); Ensermu in Fl. Eth. 5: 485 (2006) pro parte, *non* (Forssk.) Nees (1847)]

NOTE. The number of the *Schimper* collection is not specified in the original description of *Monechma affine*, but the only *Schimper* collection from Ethiopia at that time of any species in the *J. debile*-complex is *Schimper* II.759.

Material from Uganda, Kenya and N Tanzania (partly) has small (6–7 mm long) corollas and usually small (up to 7 mm long) capsules. Most Tanzanian material has larger (up to 10 mm long) corollas and larger (up to 9 mm long) capsules. But the overlap in corolla size is considerable in N Tanzania and some material from C Kenya ("*Monechma scabrinerve*") also has capsules up to 9 mm long.

FIG. 78. *JUSTICIA BRACTEATA* — **1**, habit; **2**, detail of stem indumentum; **3**, branch with many-flowered inflorescences; **4**, detail of stem indumentum; **5**, bract (left) and bracteole (right); **6**, calyx whole and calyx opened up; **7**, detail of calyx lobe; **8**, corolla, opened up; **9**, apical part of filament and anther; **10**, apical part of style and stigma; **11**, capsule; **12**, seed. 1 & 2 from *Richards* 20425, 3 & 4 from *Bullock* 3923, 5–12 from *Greenway & Kanuri* 12083. Drawn by Margaret Tebbs.

99. **Justicia eminii** *Lindau* in E.J. 20: 68 (1894) & in P.O.A. C: 373 (1895); C.B. Clarke in F.T.A. 5: 187 (1900); Hedrén in Symb. Bot. Upsal. 29, 1: 73, fig. 29 (1989). Type: Uganda, Ankole District, Mpororo, Ruhanga, *Stuhlmann* 2086 (B†, holo.; K!, iso.)

Brittle-stemmed shrubby perennial herb to 2 m tall, usually sparsely branched; young stems sericeous-puberulous with upwardly directed (or spreading and bent up apically) hairs, more rarely puberulous or densely so with straight spreading hairs. Leaves with petiole to 1 cm long; lamina ovate to elliptic, largest 3–7.5 × 1.2–3 cm, 2–3 times as long as wide, base attenuate, decurrent, apex acute to rounded; sparsely to densely sericeous-puberulous, densest on veins. Flowers solitary or in 3(–5)-flowered cymules aggregated into many terminal and axillary racemoid cymes; main axis bracts foliaceous, raceme bracts ± imbricate, ovate to cordiform or broadly so, 5–12 × 4–8 mm in middle of racemes, subacute to rounded and apiculate (rarely truncate), finely puberulous and with scattered to dense subsessile or short-stalked capitate glands, not or indistinctly ciliate with ciliae to 0.5 mm long; bracteoles minute, acicular to triangular, ± 1 mm long. Calyx 4–8 mm long, puberulous and with broad long-stalked capitate glands, lobes linear, acuminate to cuspidate, with a single strong central vein. Corolla purple or dark purple with white lines on lower lip (rarely white with purple lines), (9–)10–14 mm long, puberulous and with scattered capitate glands; tube 5–7 mm long; upper lip 4–7 mm long, slightly hooded; lower lip (5–)6–10 mm long, broadly obovate, lobes 1–4 mm long. Filaments 2–4 mm long, bending out after anthesis; anthers hairy, thecae ± 1 mm long, lower with simple or bifurcate appendage 0.7–1 mm long. Capsule 7–9 mm long, densely puberulous; seed mottled grey and black, 2.5–3 × 3–3.5 mm, glabrous.

UGANDA. Karamoja District: Mt Moroto, Dec. 1963, *Wilson* 1591!; Kigezi District: Kabale, May 1949, *Purseglove* 2896!; Mbale District: NW Elgon, Tulel Valley, Oct. 1958, *Tweedie* 1723!

TANZANIA. Bukoba District: Minziro Forest Reserve, Muhangu, 20 May 2001, *Festo* 1403!; Ufipa District: 6 km on Tatanda–Sumbawanga road, 26 April 1997, *Bidgood et al.* 3486!; Iringa District: 20 km on Iringa–Mafinga road, 5 April 1988, *Bidgood et al.* 930!

DISTR. **U** 1–3; **T** 1, 4, 7; Congo-Kinshasa, Rwanda, Burundi, Zambia, Malawi

HAB. *Brachystegia* woodland and bushland on rocky hills, montane grassland and forest margins, riverine woodland, thicket clumps and termite mounds in *Brachystegia* woodland; 1200–1850(–2100) m

SYN. *Monechma bracteatum* Hochst. var. *eciliata* C.B. Clarke in F.T.A. 5: 215 (1900). Types: Malawi, Nyika Plateau, *Whyte* 118b (K!, syn.) & *Whyte* 135 (K!, syn.)

M. debile (Forssk.) Nees var. *eciliata* (C.B. Clarke) Chiov., Nuov. Giorn. Bot. Ital. 26: 161 (1919)

[*M. debile sensu* Champluvier in Fl. Rwanda 3: 476, fig. 147 (1985); Fischer & Killmann, Pl. Nyungwe National Park Rwanda: 514 (2008), *non* (Forssk.) Nees (1847)]

NOTE. The material from N and E Uganda (**U** 1 & 3) differs from typical material in having conspicuously ciliate bracts and a white corolla with pink markings; also the stem hairs are spreading and bending upwards apically.

The third collection (*Schweinfurth & Riva* 2195, Eritrea, Ghinda) cited by C.B. Clarke under *Monechma bracteatum* var. *eciliata* seems to be a distinct taxon.

100. **Justicia debilis** (*Forssk.*) *Vahl*, Symb. Bot. 2: 15 (1791) & Enum. Pl. 1: 135 (1805); T. Anderson in J.L.S. Bot. 7: 43 (1863); Oliver in Trans. Linn. Soc. 29: 129 (1875) & in Trans. Linn. Soc., ser. 2, Bot. 2: 345 (1886); Engler, Hochgebirgsfl. Trop. Afr.: 393 (1892); Lindau in P.O.A. C: 373 (1895) & in Ann. Ist. Bot. Roma 6: 83 (1896); Hedrén in Fl. Somalia 3: 418 (2006). Type: Yemen, Taizz, *Forsskål* 391 (C!, lecto.)

Erect or straggling usually much-branched perennial or shrubby (rarely annual) herb; stems to 1.25 m long, when young puberulous or sericeous-puberulous to pubescent or densely (rarely sparsely) so with spreading or downwardly curved hairs. Leaves with petiole to 1(–2) cm long; lamina narrowly ovate to ovate or elliptic, largest 1.5–7.5 × 0.5–2.7(–3) cm, base attenuate, decurrent, apex subacute to rounded; puberulous or sparsely (rarely densely) so along veins. Flowers solitary or paired, in many axillary and terminal racemoid cymes, long and many-flowered downwards, gradually shorter upwards; main axis bracts foliaceous, raceme bracts usually imbricate (but occasionally racemes with long internodes and non-imbricate bracts), ovate to elliptic or slightly obovate or lower cordiform, 6–13 × 3–9 mm in middle of racemes, acute to rounded with small straight or recurved tip, lower with cuneate to subcordate base, subglabrous to densely puberulous and with scattered (rarely dense) stalked capitate glands, distinctly (rarely indistinctly) ciliate with broad glossy ciliae to 1 mm long; bracteoles like bracts, slightly smaller, usually with denser glands. Calyx 4–8 mm long, minutely puberulous with non-capitate glands and scattered short capitate glands, also with (rarely without) scattered (rarely dense) long glossy hairs and/or glands, lobes linear to filiform, acuminate to cuspidate, with a single central vein. Corolla white to mauve or bright purple, with white lines on lower lip or (if white) with red lines, 6–10 mm long, puberulous; tube 3–6 mm long; upper lip 2–5 mm long, flat or slightly hooded; lower lip 3–7 mm long, broadly obovate, lobes 1–2 mm long. Filaments 2–3 mm long, bending out after anthesis; anthers hairy, thecae ± 0.8 mm long, lower with acute or bifurcate appendage ± 0.5 mm long. Capsule 5–7.5 mm long, densely sericeous-puberulous (hairs spreading near apex, deflexed near base); seed mottled grey and black, ± 2.5 × 2.5 mm, glabrous.

KENYA. Embu District: 36 km on Embu–Meru road, Karurumo, 14 Nov. 1979, *Mungai et al.* 79/5!; Masai District: 45 km on Nairobi–Magadi road, 28 May 1995, *Vollesen* 95/212!; Teita District: Tsavo East National Park, 7 Sept. 1968, *Kokwaro* 1518!

TANZANIA. Musoma District: Serengeti National Park, Engari Nanyuki Springs, 3 June 1962, *Greenway & Turner* 10684!; Moshi District: SW of Ngari Nairobi, Rongai, 22 April 1957, *Greenway* 9192!; Pangani District: Korogwe, Segera Forest, 24 Sept. 1968, *Faulkner* 4139!

DISTR. **K** 1, 3, 4, 6, 7; **T** 1–3, 6; Sudan, Eritrea, Ethiopia, Djibouti, Somalia; Yemen, Saudi Arabia

HAB. *Acacia-*, *Combretum-*, *Acacia-Commiphora* bushland, montane grassland and bushland, often disturbed or secondary, roadsides, weed of cultivation, often on stony soil; near sea level to 2100 m

SYN. *Dianthera debilis* Forssk., Fl. Aegypt.-Arab.: 9 (1775)

Gendarussa debilis (Forssk.) Nees in Linnaea 16: 302 (1842)

Monechma debile (Forssk.) Nees in DC., Prodr. 11: 411 (1847); Solms-Laubert in Schweinfurth, Beitr. Fl. Aeth.: 112 (1867); F.P.S. 3: 184 (1956) pro parte; E.P.A.: 976 (1964); Brummitt *et al.* in K.B. 38: 444 (1983); Luke & Robertson, Kenya Coastal For. 2. Checklist Vasc. Pl.: 83 (1993); U.K.W.F., ed. 2: 281 (1994); Ruffo *et al.*, Cat. Lushoto Herb. Tanzania: 8 (1996) pro parte; Vollesen *et al.*, Checklist Mkomazi: 83 (1999) quoad *Abdallah & Vollesen* 95/63; Ensermu in Fl. Eth. 5: 485 (2006) pro parte

Justicia gregorii S. Moore in J.B. 32: 138 (1894). Type: Kenya, Naivasha to Baringo, Inhuyuni [not traced], *Gregory* s.n. (BM!, holo.)

Monechma bracteatum Hochst. var. *hirsutior* C.B. Clarke in F.T.A. 5: 215 (1900). Types: Kenya, Teita District, Maungu, *Johnston* s.n. (K!, syn.) & Taita [Teita], Ndara, *Hildebrandt* 2397 (K!, syn.; BM!, iso.) & Ukamba, *Scott Elliot* 6749 (K!, syn.; BM!, iso.); Tanzania, Lushoto District, Usambara, Kwa Mshuza, *Holst* 8991 (K!, syn.)

M. bracteatum Hochst. var. *stricta* C.B. Clarke in F.T.A. 5: 215 (1900). Types: Eritrea, Keren, *Steudner* 1499 (not seen); Ethiopia, *Schimper* 207 (BM!, syn.), 511 (BM!, syn.), 876 (not seen), *Steudner* 1521 (not seen)

M. troglodytica Chiov. in K.B. 1941: 174 (1941); Glover, Checklist Brit. & Ital. Somal.: 68 (1947); E.P.A.: 977 (1964); Kuchar, Pl. Somalia [CRDP Techn. Rep. ser. No. 16]: 249 (1986). Type: Somalia, Hargeisa, *Gillett* 4255 (K!, holo.; K!, iso.)

M. sp. (= Hucks 796) *sensu* Luke & Robertson, Kenya Coastal For. 2. Checklist Vasc. Pl.: 83 (1993)

M. sp. B sensu Agnew & Agnew, U.K.W.F., ed. 2: 281 (1994)

NOTE. Some of the material from coastal areas (**K** 7; **T** 3, 6) has generally larger wider bracts ((7–)9–13 × (4–)5–9 versus 6–10(–12) × 3–6(–8) mm) with truncate to subcordate (not cuneate to rounded) base in basal part of inflorescence. This material also often has a white corolla and is often straggling. But similarly large bracted forms also occasionally occur inland and some of the coastal material has pink flowers.

101. **Justicia sp. H**

Perennial herb with 1 or few erect unbranched (or branched from near base) stems from woody rootstock; stems to 25 cm long, densely puberulous. Leaves with petiole to 2 mm long; lamina narrowly ovate-elliptic, largest ± 3 × 1 cm, base cuneate, apex subacute; with sparse hairs along midrib and veins; venation of 5 strong raised longitudinal veins from base to apex, occasionally also a few off midrib in upper half. Flowers in ± 4 cm long dense racemoid cymes, terminal and with short branches from lower nodes; peduncle not clearly defined; lowermost pair of bracts foliaceous, upwards differentiated, without hyaline margins, middle and upper elliptic-obovate, 10–12 × 4–5 mm, whitish pubescent and ciliate, with 3(–5) strong longitudinal veins; bracteoles similar in length but only ± 2 mm wide, with similar indumentum. Calyx subequally 5-lobed, ± 11 mm long, with similar indumentum, lobes narrowly elliptic (distinctly widened above base), without hyaline margins, with three strong longitudinal veins from base to apex, subacute. Corolla white, without markings on lower lip, ± 14 mm long, sparsely puberulous on lobes and apically on tube; tube ± 8 mm long; upper lip ± 6 mm long, slightly hooded; lower lip ± 6 mm long, broadly obovate, lobes ± 2.5 mm long. Filaments ± 4 mm long, bending out after anthesis; anthers black, glabrous or with a few hairs, thecae ± 2 mm long, lower with acute appendage ± 1 mm long. Capsule and seed not seen.

TANZANIA. Mbeya District: Umalila, Isuto, 18 March 1977, *Leedal* 4092!
DISTR. **T** 7; not known elsewhere
HAB. Probably montane grassland; 1675 m

NOTE. Morphologically very similar to *J. varians* from the Nyika Plateau in N Malawi. It differs most conspicuously in the very characteristic leaf, bract and calyx venation and in the larger calyx and corolla.

102. **Justicia tricostata** *Vollesen* **sp. nov.** a *J. rigida* bracteis atque bracteolis lobisque calycis venas tres inter se similimas omnes robustas costiformes (nec venam centralem robustem venas duas laterales tenues) gerentibus differt. Typus: Tanzania, Sumbawanga District, Tatanda Mission, *Bidgood et al.* 2431 (K!, holo.; BR!, C!, CAS!, DSM!, EA!, K!, MO!, NHT!, P!, WAG!, iso.)

Brittle-stemmed shrubby herb or shrub; stems to 0.5 m long, puberulous. Leaves with ill-defined petiole, to 5 mm long; lamina narrowly ovate-elliptic, largest 3.5–6 × 0.8–1.2 cm, base attenuate, decurrent, apex acute to rounded; sparsely puberulous on midrib and veins; venation of 1–2 pairs of strong veins from below middle curving up and running to near apex. Flowers single in 2–9 cm long racemoid cymes, terminal and often also from upper leaf axils, dense or interrupted in basal part; peduncle not clearly defined; lowermost pair of bracts foliaceous, middle and upper pale greenish yellow, narrowly ovate, 7–10 × 2–3 mm, acute with straight or recurved tip, without broad hyaline margins, sparsely sericeous-puberulous, not ciliate, with 3 strongly raised dark green rib-like longitudinal veins from base to apex; bracteoles similar in shape, colour, venation and indumentum, longer than bracts, 8–11 mm long. Calyx subequally 5-lobed, greenish yellow, 9–13 mm long, with similar indumentum, lobes lanceolate, with 3 strong rib-like dark green veins from base to apex. Corolla tube white, lower lip pale mauve, without darker lines on lower lip, ± 18 mm long, sparsely puberulous; tube ± 9 mm long; upper lip ± 9 mm long, slightly hooded; lower lip ± 12 mm long, obovate, lobes ± 2 mm long. Filaments ± 9 mm long, bending out after anthesis; anthers glabrous, thecae ± 2 mm long, lower with bifurcate appendage ± 1.5 mm long. Capsule 10–13 mm long, glabrous; seed mottled grey and black, ± 3.5 × 4 mm, glabrous.

TANZANIA. Sumbawanga District: Tatanda Mission, 24 Feb. 1994, *Bidgood et al.* 2431! & 25 April 1997, *Bidgood et al.* 3450! & 15 km on Tatanda–Mbala road, 24 April 2006, *Bidgood et al.* 5669!

DISTR. **T** 4; Zambia
HAB. *Brachystegia* woodland on large rocky outcrops; 1550–1800 m

NOTE. Related to the following species and to *J. rigida* from Angola. It differs from both in the distinctive bract, bracteole and calyx venation.

103. **Justicia attenuifolia** *Vollesen* **sp. nov.** a *J. subsessile* foliis ad basin attenuatis decurrentibusque (nec truncatis nec subcordatis), corolla maiore 17–19 mm (nec 11–16 mm) longa et labio dorsali corollae quam tubo plus quam duplo longiore (nec tubum aequanti) et capsula maiore ± 17 mm (nec 10–14 mm) longa differt. Typus: Tanzania, Tunduru District, 95 km on Masasi–Tunduru road, *Richards* 17968 (K!, holo.)

Perennial herb with several erect or ascending stems from woody rootstock; stems to 75 cm long, sericeous-puberulous when young. Leaves with petiole to 3 mm long; lamina narrowly ovate or narrowly elliptic, largest 8–9 × 1.7–2.2 cm, more than 3 times longer than wide, base attenuate, decurrent, apex subacute to rounded; sparsely sericeous-puberulous on midrib, otherwise glabrous. Flowers in 4–9 cm long racemoid cymes, terminal and from upper leaf axils, dense or interrupted basally; peduncle clearly defined, 3–6.5 cm long; lowermost bracts usually foliaceous, upwards differentiated, middle and upper green, without hyaline margins, narrowly elliptic, 18–25 × 4–6 mm, sericeous-puberulous and with scattered subsessile pale yellow capitate glands, distinctly ciliate; bracteoles similar or narrower, 20–25 mm long, with similar indumentum. Calyx 5-lobed with subequal lobes or 5th lobe $^3/_4$ the other four, 17–19 mm long, puberulous, lobes linear-lanceolate, cuspidate, with strong central and 2 weaker lateral veins. Corolla cream or yellowish green, without (?) pink lines on lower lip, 17–20 mm long, sparsely puberulous; tube 5–6 mm long; upper lip 12–14 mm long, slightly hooded; lower lip 10–12 mm long, oblong-obovate, lobes ± 2 mm long. Filaments 8–9 mm long, bending out after anthesis; anthers glabrous, thecae ± 2.5 mm long, lower with bifurcate appendage ± 2 mm long. Capsule ± 17 mm long, densely and finely sericeous-puberulous (hairs spreading near apex, deflexed near base); seed not seen.

TANZANIA. Songea District: Songea–Tunduru road, Nsamala Baraga, 19 April 1950, *Tanner* 130!; Tunduru District: 95 km on Masasi–Tunduru road, 19 March 1963, *Richards* 17968!
DISTR. **T** 8; Mozambique
HAB. Grassland along road in *Brachystegia* woodland on sandy-gritty soil; 900 m

NOTE. Known from these two and one collection from N Mozambique. Differs from *J. subsessilis* in the large attenuate leaves, large corolla with long upper lip and large capsules. *J. scabrida* from W Zambia and Angola has similar leaves but corolla and capsule are smaller than *J. subsessilis*.

104. **Justicia subsessilis** *Oliv.* in Trans. Linn. Soc. 29: 129, t.129B (1875); Lindau in P.O.A. C: 373 (1895). Type: Tanzania, Bukoba District, Karagwe, *Speke & Grant* 213 (K!, holo.)

Perennial herb with several erect or ascending stems from large woody rootstock; stems to 0.75(–1) m long, sericeous to puberulous or pubescent often with a mixture of short appressed and longer spreading hairs. Leaves sometimes only partly developed when flowering; petiole to 3(–5) mm long; lamina ovate to broadly so or elliptic or narrowly so, largest 2.5–5.2(–10) × 1.2–3.5 cm, base truncate to subcordate (rarely rounded), apex subacute to rounded; subglabrous to pubescent or sericeous-pubescent, densest along veins. Flowers in terminal 2–10(–13) cm long terminal racemoid cymes, dense or interrupted in basal part; peduncle not clearly defined; lowermost bracts foliaceous, upwards differentiated, middle and upper green, without or with indistinct hyaline margins, elliptic or narrowly so (rarely ovate), 12–20 × 3–8 mm, subglabrous to puberulous or sericeous and with scattered to dense subsessile pale yellow capitate glands, distinctly ciliate; bracteoles similar or narrower, 11–18 mm long, with similar indumentum. Calyx 5-lobed with 5th lobe $^1/_2$ to $^2/_3$ the other four, (10–)12–15(–18) mm long, with similar indumentum, lobes linear-lanceolate, cuspidate, with strong central and 2 weaker lateral veins. Corolla white, pale yellow or yellow, with or without pink lines on lower lip, 11–16 mm long, puberulous; tube 5–7 mm long; upper lip 6–9 mm long, hooded; lower lip 5–10 mm long, oblong-obovate, lobes ± 2 mm long. Filaments 4–7 mm long, bending out after anthesis; anthers glabrous, thecae ± 2 mm long, lower with bifurcate appendage ± 1.5 mm long. Capsule 10–14 mm long, densely and finely sericeous-puberulous (hairs spreading near apex, deflexed near base); seed mottled grey and black, 3–3.5 × 4–4.5 mm, whitish sericeous or densely so on edges, sparser on flanks, glabrescent.

UGANDA. Ankole District: Mbarara, 29 Sept. 1970, *Katende* 565!; Masaka District: Katera, 1 Oct. 1953, *Drummond & Hemsley* 4495! & SW of Lake Nabugabo, Bugabo, 1 Feb. 1969, *Lye et al.* 1776!
KENYA. Machakos District: Mt Donyo Sabuk, 25 July 1954, *Bally* 9804!; Embu District: Embu, Siakago, 3 April 1968, *Kayu* 545!; South Nyeri District: Kirinyaga, Murinduko, 9 Jan. 1972, *Robertson* 1690!
TANZANIA. Bukoba District: Minziro Forest Reserve, Bulembe Hill, 17 Nov. 1999, *Sitoni & Simon* 925!; Buha District: 20 km NE of Kibondo, Keza Mission, 3 May 1994, *Bidgood & Vollesen* 3256!; Iringa District: 25 km on Mafinga–Madibira road, 21 March 1988, *Bidgood et al.* 625!
DISTR. **U** 2, 4; **K** 4; **T** 1, 4, 7; E Congo-Kinshasa, Rwanda, Burundi, Zambia, Zimbabwe
HAB. Grassland, *Combretum* and *Brachystegia* woodland and bushland with low grasscover, on a wide range of soils; 900–1700(–1850) m

SYN. *J. simplicispica* C.B. Clarke in F.T.A. 5: 188 (1900). Type: "East Africa", *Scott Elliot* s.n. (K!, holo.)
Monechma subsessile (Oliv.) C.B. Clarke in F.T.A. 5: 216 (1900); Richards & Morony, Check List Mbala: 233 (1969); U.K.W.F.: 605 (1974); U.K.W.F., ed. 2: 281 (1994); Ruffo *et al.*, Cat. Lushoto Herb. Tanzania: 8 (1996)
M. nemoralis S. Moore in J.B. 47: 296 (1909). Type: Congo-Kinshasa, Niembe River, *Kässner* 3010 (BM!, holo.; K!, iso.)

NOTE. The occurrence of this species in C Kenya is very disjunct from the rest of the area of this species, but – apart from being slightly more hairy – the specimens are typical.
Newbould & Harley 1178 (K) from **T** 4, Mpanda District, Mahali Mts has some characters reminiscent of *Justicia scabrida* from W Zambia and Angola; particularly the large leaves (maximum length in description) and absence of distinct ciliae on bracts and bracteoles. But it differs from *J. scabrida* in the larger calyx and corolla and in the absence of a conspicuous hyaline margin to bracts and bracteoles.

105. **Justicia tetrasperma** *Hedrén* in Nordic Journ. Bot. 10: 151, fig. 1 (1990). Type: Congo-Kinshasa, Shaba, Marungu, *Vilain* 60 (BRLU, holo.)

Shrubby perennial herb with several erect unbranched stems from woody rootstock; stems to 2 m long, densely pubescent. Leaves with petiole to 8 mm long; lamina ovate, largest ± 6.5 × 3 cm, base attenuate, apex subacute to obtuse; pubescent on midrib and veins sparsely so on lamina. Flowers few together in 3–4 cm long dense racemoid cymes, terminal and from upper leaf axils; peduncle not clearly defined; lowermost bracts foliaceous, upwards differentiated, middle and upper green, without hyaline margins, ovate, 14–17 × 7–10 mm, sparsely puberulous, with longer hairs on veins and distinctly ciliate; bracteoles small, narrowly triangular, ± 2 × 0.5 mm long. Calyx subequally 5-lobed or 5th lobe ± $^2/_3$ of the other four, ± 16 mm long, puberulous and ciliate, lobes linear-lanceolate, cuspidate, with 1 strong central vein. Corolla purple, without (?) darker markings on lower lip, ± 19 mm long, puberulous; tube ± 10 mm long; upper lip ± 9 mm long, slightly hooded; lower lip ± 13 mm long, broadly obovate, lobes ± 5 mm long, middle much wider than laterals. Filaments ± 9 mm long, bending out after anthesis; anthers glabrous, thecae ± 2 mm long, lower with bifurcate appendage ± 1 mm long. Capsule 4-seeded, 11–13 mm long, densely puberulous; seed mottled grey and black, ± 3 × 3 mm, glabrous.

TANZANIA. Mpanda District: Mwese, 20 May 1975, *Kahurananga et al.* 2582!
DISTR. **T** 4; Congo-Kinshasa, Zambia
HAB. Scrub grassland, riverine forest edges; 1525 m

NOTE. Hedrén (l.c.) discusses at length the affinities of this species and how it breaks down the characters which have been used to differentiate between *Justicia* and *Monechma*.

41. RHINACANTHUS

Nees in Wall. Pl. As. Rar. 3: 76 (1832) & in DC., Prodr. 11: 442 (1847); C.B. Clarke in F.T.A. 5: 224 (1900); Darbyshire & Harris in K.B. 61: 401–418 (2006)

Perennial herbs or subshrubs, evergreen or rarely deciduous; cystoliths many, ± conspicuous. Stems 6-angular or terete, often with short pale retrorse and/or antrorse hairs, sometimes glabrate. Leaves opposite-decussate, pairs equal to somewhat anisophyllous; blade margin subentire to obscurely repand. Cymes axillary and/or terminal, fasciculate or laxly paniculate; bracts and bracteoles inconspicuous, linear to lanceolate; flowers subsessile, in clusters or solitary. Calyx divided almost to base; lobes 5, subequal, linear-lanceolate. Corolla bilabiate, pubescent outside except towards base of tube where glabrous; tube narrowly long-cylindrical, usually longer than limb, floor rarely ventricose in upper half, roof with 2 lines of hairs within extending onto upper lip; upper lip ovate to lanceolate, apically notched or entire; lower lip divided in distal half or third into 3 subequal rounded to oblong lobes. Stamens 2; filaments attached near mouth of corolla tube, shortly exserted; anthers bithecous, thecae slightly to strongly offset, ± oblique, lower theca mucronate or sometimes minutely tailed; staminodes absent. Disk shallowly cupular, with a V-shaped slit. Ovary 2-locular, 2 ovules per locule; style filiform; stigma positioned at same level as anthers, bifid, lobes unequal, linear. Capsule clavate, long-stipitate, apex shortly beaked. Seeds held on retinacula, lenticular, with a hilar excavation, brown to black, ± tuberculate.

A genus of ± 25 species, confined to the Old World tropics and subtropics. In addition to the 11 taxa recorded from East Africa, the recently described *R. mucronatus* Ensermu from Sidamo, Ethiopia, is to be expected in *Acacia-Commiphora* bushland in northernmost Kenya. Of the taxa in our region, several from E and SE Tanzanian (**T** 6–8) are known from only one or two collections and further material is highly desirable to fully delimit the taxa there.

1. Inflorescences axillary, fasciculate or verticillate or if narrowly paniculate then shorter than the mature leaves, these ovate or ovate-lanceolate, green 2
   Inflorescences terminal and often also axillary, laxly paniculate, usually much longer than the leaves and widely spreading, or if shorter than the leaves then these elliptic or obovate, somewhat glaucous 4
2. Plant procumbent; mature leaves obovate (-elliptic); bracts, bracteoles and calyx lobes with a hyaline margin; corolla 13–19 mm long; capsule glabrous .... 1. *R. ndorensis*
   Plant erect or straggling; mature leaves broadly ovate to ovate-lanceolate; bracts, bracteoles and calyx lobes lacking a hyaline margin; corolla 20 mm long or more, capsule hairy 3
3. Leafy stems rounded to subangular; mature leaves 4.5–16 cm long, base attenuate; corolla (in our region) mauve-pink, tube 15–17 mm long, lower lip 5–7 mm; lower anther theca shortly tailed .... 2. *R. virens*
   Leafy stems strongly 6-angular; mature leaves 1–5 cm long, base rounded to subcordate; corolla white, tube 31–34 mm long, lower lip 11–13 mm; lower anther theca not tailed .... 3. *R. rotundifolius*

4. Lower lip of corolla 20–30 mm long; anther thecae separated by 1–2 mm; plants often lacking mature leaves at flowering .... 11. *R. pulcher*
   Lower lip of corolla 3–14 mm long; anther thecae overlapping or separated by up to 0.6 mm; plants with mature leaves at flowering 5
5. Corolla 25–40 mm long 6
   Corolla 9–21 mm long 8
6. Peduncles with multicellular spreading hairs to 1.7 mm long; capsule 17–21 mm long; seeds 3.2–3.8 mm in diameter, largely smooth .... 6. *R. sp. A*
   Peduncles lacking long multicellular spreading hairs; capsule 13–19 mm long; seeds 2–2.8 mm in diameter, tuberculate 7
7. Stems strongly 6-angular with prominent pale ridges where hairs most dense; mature leaves narrowly elliptic to lanceolate, usually with sparse multicellular hairs above; corolla tube 16–24 mm long; lower lip (5.5–)7–8.5 mm long, lobes rounded-ovate .... 4. *R. angulicaulus*
   Stems subangular, not pale-ridged, hairs more evenly distributed; mature leaves obovate to elliptic, glabrous above except on the veins; corolla tube (20–)22–28 mm long; lower lip (7–)10–14 mm long, lobes oblong-ovate 5. *R. dichotomus*
8. Anther thecae slightly overlapping or immediately superposed, (sub)oblique but the upper theca not almost patent to the filament; corolla 19–21 mm long; upper leaf surface glabrous 9
   Anther thecae separated by 0.2–0.6 mm, highly oblique, the upper theca held almost patent to the filament; corolla 9–15(–17) mm long; upper leaf surface sparsely hairy at least towards margin 10
9. Stems pale, angular but without prominent ridges; upper corolla lip lanceolate, ± 1.8 mm wide; style glabrous; leaves obovate with acute or obtuse apex .... 7. *R. sp. B*
   Stems prominently pale-ridged; upper corolla lip triangular-ovate, ± 3 mm wide; style sparsely pubescent in lower half; leaves elliptic with attenuate apex ...... 8. *R. sp. C*
10. Base of lower leaves acute, cuneate or attenuate; anther thecae separated by 0.3–0.6 mm, lower theca 0.9–1.2 mm long; plants of submontane forest .... 9. *R. submontanus*
    Base of lower leaves obtuse; anther thecae separated by 0.2–0.3 mm, lower theca 0.6–0.8 mm long; plants of lowland forest .... 10. *R. selousensis*

1. **Rhinacanthus ndorensis** *Schweinf.* in Höhnel, Rudolph & Stephanie-See, Appendix: 7 (1892) & in Disc. Teleki Rudolf & Stefanie 2: 357 (1894); C.B. Clarke in F.T.A. 5: 225 (1900); Schweinfurth & Mildbraed in N.B.G.B. 9: 504 (1926); U.K.W.F. ed. 2: 277 (1994). Type: Kenya, North Nyeri District, Ndoro, foot of Mt Kenya, *Höhnel* s.n. (B†, holo.; BM!, iso.)

Procumbent herb, branching widely from woody base. Young stems ridged, glabrous except for tufts of hairs at nodes. Leaves fleshy, obovate (-elliptic), 0.8–1(–1.4) cm long, 0.4–0.6(–0.8) cm wide, base cuneate, margin entire, apex obtuse or rounded, surfaces glabrous; lateral veins 3(–4) pairs; petiole to 1.5 mm long. Inflorescences axillary, fasciculate, (1–)2–3-flowered; peduncle 0–3 mm long, glabrous; bracts linear-lanceolate, 6–9 mm long, margin hyaline, ciliate with hairs of variable length, midrib prominent outside; bracteoles as bracts but 5–6.5 mm long; flowers subsessile. Calyx lobes 4–5.5 mm long, margin narrowly hyaline, ciliate, surfaces with sparse short appressed hairs. Corolla bright pink to mauve, (13–)15–19 mm long, rather densely retrorse-pubescent outside; tube (8–)10–11 mm long, 1–1.5 mm in diameter centrally, widening somewhat towards mouth, lines of hairs on the roof barely extending onto upper lip; upper lip suberect, ovate, 5–7 mm long, 3–3.7 mm wide, apex shortly bilobed; lower lip pendant, 6–7.5 mm long, glabrous within, lobes oblong, 2.5–3 mm long, 1.4–1.8 mm wide, apex obtuse or rounded, median lobe slightly longer and narrower than lateral pair. Staminal filaments 1.2–2.3 mm long, glabrous; anther thecae only slightly offset, 1–1.2 mm long. Ovary glabrous; style ± sparsely appressed-pubescent. Capsule 8–10 mm long, glabrous or largely so; seeds blackish, 1.3–1.8 mm in diameter, tuberculate.

KENYA. Laikipia District: Rumuruti, Nov. 1978, *Hepper & Jaeger* 6626!; North Nyeri District: Nyeri, Amboni R., Jan. 1933, *Napier* 2579! & ± 40 km S of Nanyuki, Apr. 1975, *Hepper & Field* 4838!

DISTR. **K** 3, 4; not known elsewhere

HAB. Grassland, open woodland and roadside verges, sometimes on seasonally flooded black clays; 1700–2150 m

USES. None recorded on herbarium specimens

CONSERVATION NOTES. This species has a highly restricted range and is recorded from two discrete areas, the larger between Nyeri and Nanyuki in the rain shadow of Mt Kenya and the smaller centred around Rumuruti ± 70 km to the west. Only 11 specimens are known to the author. This is despite extensive collection in central Kenya, and suggests that the species is scarce, though it is possibly overlooked due to its small, procumbent habit. Its grassland habitat is perhaps threatened by increased intensity of pastoral agriculture, though label data (*Moreau* 34; *Napier* 2579) suggests that this species is tolerant of some grazing. It is provisionally assessed as Near Threatened (NT).

2. **Rhinacanthus virens** (*Nees*) *Milne-Redh.* in Exell, Suppl. Cat. Vasc. Pl. S. Tomé: 37 (1956); Heine in F.W.T.A. ed. 2, 2: 425 (1963) excl. var. *obtusifolius* & in Fl. Gabon 13: 201, pl. 42 (1966); Darbyshire & Harris in K.B. 61: 414–415 (2006); Hawthorne & Jongkind, Woody Pl. W. Afr. Forests: 446, 447 (2006). Type: Gabon, *Middleton* s.n. (K!, holo.)

Erect or straggling herb, 50–120 cm tall; stems rounded or subangular, with dark longitudinal striations, young stems antrorse-pubescent, soon glabrescent. Leaves ovate to ovate-lanceolate, (4.5–)8.5–16 cm long, (1.7–)3–6 cm wide, base attenuate, margin subentire to obscurely repand, apex acuminate, surfaces with sparse multicellular hairs, most dense on margins, sometimes absent beneath; lateral veins 6–7 pairs, pale and prominent and with short antrorse hairs beneath; petiole 0.5–2.5 cm, pubescent. Inflorescences axillary, fasciculate or extended and shortly verticillate or narrowly paniculate, up to 3.5 cm long, flowers in subsessile clusters of 2–4 or solitary; peduncles antrorse-pubescent and with scattered stalked glands; bracts and bracteoles linear-lanceolate, 2–3.5 mm long. Calyx lobes 4.5–5.5 mm long, with mixed antrorse and spreading short hairs and stalked glands outside. Corolla mauve-pink, 20–24 mm long, tube sometimes whitish, spreading-pubescent outside; tube 15–17 mm long, 1.2–1.7 mm in diameter, lines of hairs on roof becoming longer and more dense towards base; limb only slightly spreading; upper lip triangular-ovate, 3.5–4.5 mm long and wide, apex notched; lower lip 5–7 mm long, glabrous within, lobes rounded, 2.3–4.3 mm long, 2–4 mm wide, median lobe slightly larger than lateral pair. Staminal filaments 2.5–3.5 mm long, glabrous; anther thecae offset and somewhat oblique, overlapping, ± 1 mm long, lower theca slightly larger and with a short pale tail. Ovary largely glabrous; style pubescent, glabrous towards apex. Capsule 15–19 mm long, pubescent, hairs spreading distally, becoming deflexed towards base; seeds brown, 2–3.3 mm in diameter, densely tuberculate.

UGANDA. Busoga District: Lake Victoria, Butembe Bunya, S end of Nainavi Island, Jan. 1953, *Wood* 682!; Masaka District: Malabigambo forest, 3 km SSW of Katera, Oct. 1953, *Drummond & Hemsley* 4588!; Mengo District: Namutamba, Mityana, introduced from Entebbe, Jan. 1968, *Musaka & Ferreira* in NC 213!

TANZANIA. Bukoba District: Minziro Forest Reserve, July 2000, *Bidgood et al.* 4763!

DISTR. **U** 3–4; **T** 1; widespread in the forests of West and Central Africa, from Sierra Leone to Congo-Kinshasa and W Zambia

HAB. Forest including seasonally flooded areas, forest margins; 1100–1250 m

USES. Cultivated in **U** 4 (*Musaka & Ferreira* NC 213), although the purpose is unknown

CONSERVATION NOTES. Widespread and common in the forested regions of West and Central Africa, though apparently becoming much more localised at the eastern end of its range: Least Concern (LC).

SYN. *Leptostachya virens* Nees in DC., Prodr. 11: 378 (1847)
Rhinacanthus dewevrei De Wild. & T. Durand in B.S.B.B. 38: 105 (1899). Type: Congo-Kinshasa, Bokakata, Feb. 1896, *Dewevre* 804 (BR!, holo.; K!, iso.)
[*R. parviflorus* T. Anderson, *nom. nud.*; De Wild. & T. Durand in Compte Rendu Soc. Bot. Belge 38: 106 (1899), in obs.]
R. subcaudatus C.B. Clarke in F.T.A. 5: 225 (1900). Syntypes: Sierre Leone, near the Scarcies R., Wallia, *Scott-Elliot* 4276 (K!, syn.) & near Sasseni, Scarcies, *Scott-Elliot* 4428 (K!, syn.); without location, *Afzelius* s.n. (not traced, syn.)
Siphonoglossa rubra S. Moore in J.B. 44: 88 (1906). Type: Uganda, Mengo District, Entebbe, *Bagshawe* 750 (BM!, holo.), **syn. nov**.
[*R. communis sensu* C.B. Clarke in F.T.A. 5: 224 (1900) pro parte, & *sensu* Heine in F.W.T.A. ed. 1, 2: 266 (1931) pro parte quoad spec. ex West Africa, *non* Nees]

NOTE. The description here is based upon material from East Africa and E Congo-Kinshasa. Further west this species becomes much more variable, with flowers commonly in the range 15–20 mm long, often white or pale yellow, and often borne on longer, lax axillary panicles. The indumentum of the inflorescence axes also varies, with many specimens having predominantly short, spreading hairs or abundant stalked glands. However, there are also specimens from across the species' range which resemble the East African material, for example in the Ashanti region of Ghana, and no clear geographic trends are discernable.

3. **Rhinacanthus rotundifolius** *C.B. Clarke* in F.T.A. 5: 225 (1900); Meeuse in Bothalia 7: 412 (1960); Hedrén in Fl. Somalia 3: 401 (2006). Syntypes: Kenya, Lamu District, Witu, *A.S. Thomas* 7 (B†, BM!, K!, syn.) & Tana River, *A.S. Thomas* 89 (B†, syn.)

Erect or straggling herb or subshrub, 30–60(–100) cm tall; stems woody towards base, flowering shoots 6-angular with conspicuous pale ridges, with pale short retrorse hairs, often most dense on ridges. Leaves ovate to broadly so, 1–5 cm long, 0.7–3.2 cm wide, base rounded to subcordate, margin subentire, apex obtuse to rounded, often apiculate, surfaces puberulous, most dense on veins and margin; lateral veins 4–6 pairs, tertiary venation reticulate, ± prominent; petiole 2–5 mm, pubescent. Inflorescences axillary, fasciculate, 2–4-flowered; peduncle 1–4 mm long, with dense pale retrorse hairs and occasional short-stalked glands; bracts and bracteoles lanceolate, 1.5–2 mm long; flowers subsessile. Calyx lobes linear, 5–7 mm long, with dense spreading multicellular hairs and capitate stalked glands outside. Corolla white, 38–44 mm long, sometimes with purple markings on lower lip, pubescent outside; tube 31–34 mm long, 0.5–1 mm in diameter above somewhat swollen base; upper lip erect, slightly reflexed distally, lanceolate, 8.5–11 mm long, 2 mm wide, apex notched; lower lip pendant, 11–13 mm long, lobes subequal, ovate, 6–7.5 mm long, 4.5–6 mm wide, apex rounded. Staminal filaments 2–2.3 mm long, glabrous or with scattered short hairs; anther thecae superposed, slightly overlapping, 1.2–1.4 mm long, lower theca slightly larger. Ovary glabrous; style sparsely pubescent towards base. Capsule 14–18 mm long, densely puberulous, hairs spreading distally, becoming deflexed towards base, with scattered capitate stalked glands; seeds black, 1.8–2.1 mm in diameter, tuberculate.

KENYA. Northern Frontier Province: Korokoro, N bank of Garissa, June 1960, *Paulo* 457!; Tana River District: Garissa to Malindi road, 16 km N of junction for Garsen, Jan. 1972, *Gillett* 19527! & Tana River National Primate Reserve, Mar. 1990, *Luke et al.* in TPR 640!

DISTR. **K** 1, 7; SE Somalia

HAB. Open dry bushland; 0–350 m

USES. None recorded on herbarium specimens

CONSERVATION NOTES. This species occupies a small range, centred upon the lower Tana R. of Kenya. It is known from few collections, 11 having been seen at Kew. However, it is noted as locally common in collections from both Kenya and Somalia, and in the latter country has been recorded in grazed bushland (*Bally* 9508), hence it may be tolerant of some disturbance. Human population density is low throughout this species' range with the exception of areas around Garissa and Garsen. It therefore appears largely unthreatened at present: Least Concern (LC).

4. **Rhinacanthus angulicaulis** *I. Darbysh.* in K.B. 61: 404, fig. 1L–P (2006). Type: Kenya, Northern Frontier District, Marsabit, *Oteke* 36 (K!, holo.; EA!, iso.)

Straggling subshrub, 60–100(–200) cm tall; stems 6-angular, with pale antrorse and retrorse hairs mainly on the prominent pale ridges, shorter suberect hairs elsewhere, nodes usually with a line of long ascending hairs. Leaves narrowly elliptic to lanceolate, 7.5–13.5 cm long, 1.5–5 cm wide, base acute to attenuate, becoming obtuse in uppermost leaves, margin subentire to obscurely repand, apex acute to subacuminate, upper surface with sparse multicellular hairs, rarely glabrous, veins shortly antrorse-pubescent, longer on midrib above; lateral veins 4–6(–7) pairs; petiole 4–10 mm long, furrowed above, antrorse-pubescent. Inflorescences terminal, laxly paniculate, to 30 cm long, widely branched, often with additional shorter axillary panicles; flowers subsessile, solitary or in clusters of 2–3; peduncles with short, blunt-tipped, spreading hairs and increasing numbers of capitate stalked glands distally; bracts and bracteoles 1.5–3 mm long. Calyx lobes 4–6(–7) mm long, with short, blunt-tipped spreading hairs and many capitate stalked glands outside. Corolla 25–31 mm long, pubescent and with scattered stalked glands outside; tube white or pink, 16–24 mm long, 1–1.5 mm in diameter, slightly swollen at base, shallowly curved; limb white, cream or tinged pink, with occasional stalked glands within, lips widely spreading; upper lip lanceolate, 6–7.5 mm long, 2–3 mm wide, apex shallowly notched; lower lip (5.5–)7–8.5 mm long, speckled purple only in throat or not at all, lobes subequal, rounded-ovate, 2.5–4 mm long, 2.5–3.5 mm wide. Staminal filaments 3–3.3 mm long, glabrous; anther thecae offset and oblique, barely overlapping, lower theca 1.2–1.4 mm long, base apiculate, upper theca 0.7–1 mm long. Ovary glabrous; style sparsely pubescent, becoming glabrous towards apex. Capsule 13.5–19 mm long, pubescent, hairs spreading distally, becoming deflexed towards base, and with scattered capitate stalked glands; seeds black, 2–2.8 mm in diameter, tuberculate. Fig. 79/11–14, p. 606.

KENYA. Northern Frontier District: Ndoto Mts, track up from Ngurunit Mission, June 1979, *Gilbert, Kanuri & Mungai* 5623!; Kitui District: Kibwezi to Kitui road, Yatta Gap, Apr. 1969, *Napper & Kanuri* 2028!; Teita District: Tsavo East National Park, Feb. 1969, *Hucks* 1111!

TANZANIA. Mpwapwa District: Mpwapwa, Apr. 1929, *H.E. Hornby* 111! & *idem*, June 1948, *R.M. Hornby* 3004!

DISTR. **K** 1, 4, 6, 7; **T** 5; not known elsewhere

HAB. Dense or open *Acacia-Commiphora* bushland, dry semi-evergreen forest, sometimes on riverbanks; 450–1550 m

USES. "Eaten by livestock" (**T** 5, *Hornby* 111)

CONSERVATION NOTES. Assessed as of Least Concern (LC) by Darbyshire (l.c.).

SYN. [*R. communis sensu* C.B. Clarke in F.T.A. 5: 224 (1900) pro parte quoad *Johnston* s.n. & ?*Stuhlmann* 249 (based on locality), *non* Nees]
[*R. gracilis sensu* U.K.W.F. ed. 2: 277 (1994), *non* Klotzsch]

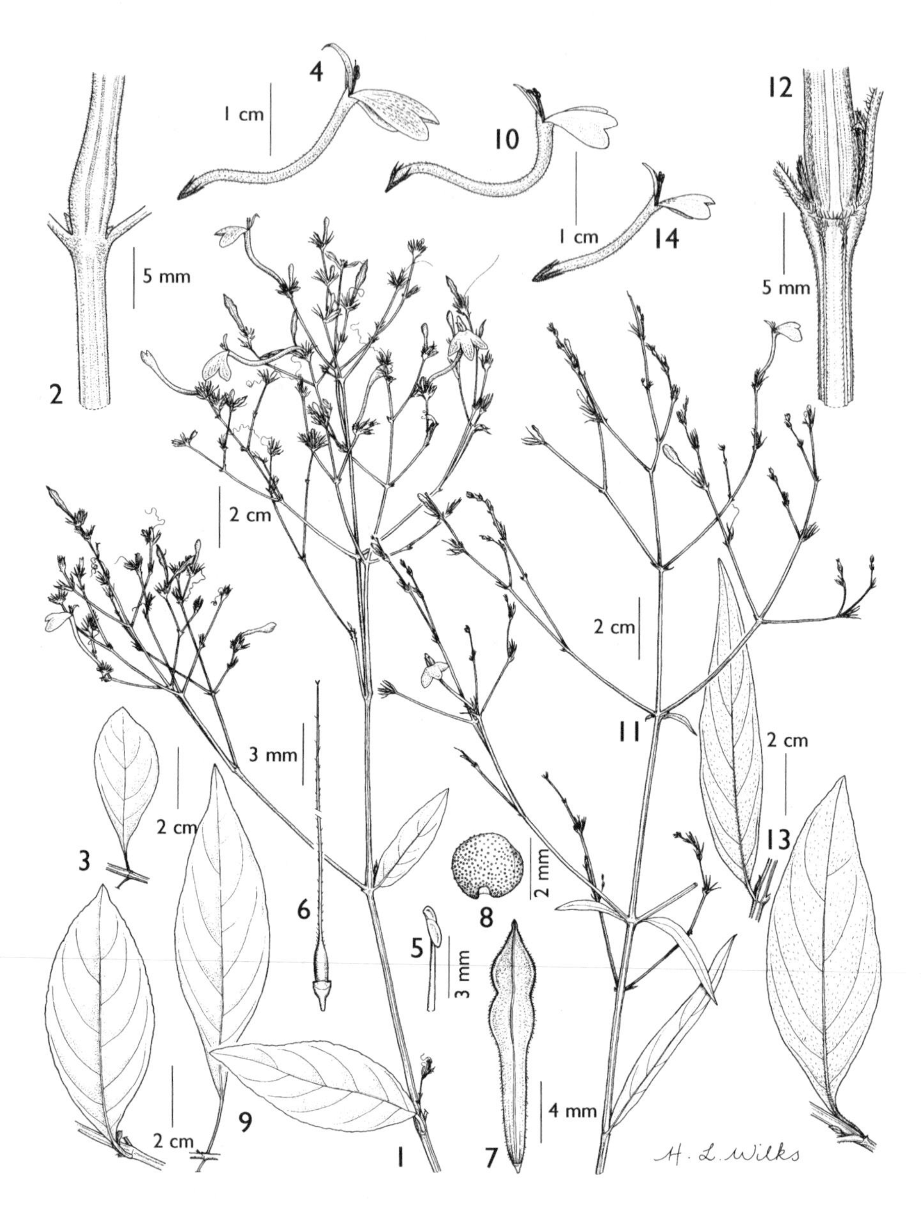

FIG. 79. *RHINACANTHUS DICHOTOMUS* var. *DICHOTOMUS* — **1**, habit, flowering and fruiting stem with upper cauline leaves; **2**, detail of stem at leaf node; **3**, variation in lower cauline leaf shape; **4**, corolla and calyx; **5**, stamen; **6**, gynoecium within floral disc, style length reduced; **7**, mature capsule; **8**, mature seed. *R. DICHOTOMUS* var. *EMACULATUS* — **9**, lower cauline leaf; **10**, corolla and calyx. *R. ANGULICAULIS* — **11**, habit, flowering stem with upper cauline leaves; **12**, detail of stem at leaf node; **13**, variation in lower cauline leaf shape; **14**, corolla and calyx, lateral view. 1, 2, 3 (elliptic leaf), 7 & 8 from *Leach & Brunton* 10216; 3 (obovate leaf) from *Drummond & Hemsley* 3757; 4–6 from *Faulkner* 3826; 9 from *Carmichael* 561; 10 from *Faulkner* 1893; 11, 12 & 13 (elliptic variant) from *Oteke* 36; 13 (lanceolate variant) from *Napper & Kanuri* 2028; 14 from *Luke* 4342 & *Oteke* 36. All drawn by Hazel Wilks. Reproduced from Kew Bulletin 61: 405 (2006) with permission of the Trustees of the Royal Botanic Gardens, Kew.

NOTE. *R. angulicaulus* differs from the similar *R. gracilis* Klotzsch, of Malawi and Mozambique south to Zimbabwe, principally in having angular stems with prominent ridges, those of *R. gracilis* more closely resembling *R. dichotomus*, and in lacking the rounded to cordate leaf bases of the upper cauline leaves characteristic of *R. gracilis*. In addition, the corolla of *R. angulicaulus* is smaller, with a shorter, more curved tube, and shorter, more rounded lobes to the lower lip.

Variation within *R. angulicaulus* is minimal; however, specimens from the Mpwapwa area of Tanzania have a more densely hairy and more strongly curved corolla tube than specimens from Kenya.

5. **Rhinacanthus dichotomus** (*Lindau*) *I. Darbysh.* in K.B. 61: 406, fig. 1A–K (2006); Luke in Journ. E. Afr. Nat. Hist. Soc. 94: 93 (2005). Type: Tanzania, Lushoto District, Usambara Mts, Mashewa, *Holst* 8782 (K!, lecto., chosen by Darbyshire, l.c.)

Erect or straggling herb or subshrub, 30–100(–150) cm tall; stems subangular with dark longitudinal striations, pale antrorse- and retrorse-pubescent. Leaves obovate to elliptic, (2.7–)4.5–9(–13) cm long, (1.6–)2.2–4.5(–5) cm wide, base cuneate or attenuate, becoming obtuse in uppermost leaves, margin obscurely repand, apex obtuse to attenuate, surfaces glabrous except for short antrorse hairs on the veins; lateral veins 4–6 pairs; petiole 3–20 mm long, furrowed above, antrorse-pubescent. Inflorescences terminal, laxly paniculate, to 26 cm long, widely branched, often with additional shorter axillary inflorescences; flowers subsessile, in clusters of 2–3 or solitary; peduncles with short, blunt-tipped, spreading hairs and increasing numbers of capitate stalked glands distally; bracts and bracteoles 1–3 mm long. Calyx lobes 3–4.5(–5) mm long, with short, blunt-tipped, spreading hairs and many capitate stalked glands outside. Corolla 29–40 mm long, pubescent and with scattered stalked glands outside; tube pink or red, (20–)22–27 mm long, 1.2–1.7 mm in diameter, slightly swollen at base, declinate to the centre then ± strongly upcurved towards the mouth; limb white to pale yellow or pink, with scattered stalked glands within, lips widely spreading; upper lip lanceolate, 5.5–7.5 mm long, 2–3 mm wide, apex shallowly notched; lower lip (7–)10–14 mm long, speckled purple, sometimes only in the throat, lobes oblong-ovate, (3.5–)5.5–8.5 mm long, 2.5–4(–6) mm wide, median lobe slightly longer than lateral pair, apex rounded or obtuse. Staminal filaments 3–3.5 mm long, glabrous; anther thecae offset and oblique, barely overlapping, lower theca 1.1–1.4 mm long, base apiculate, upper theca 0.8–1 mm long. Ovary glabrous; style sparsely pubescent, glabrous towards apex. Capsule 13–16(–18) mm long, pubescent, hairs spreading distally, becoming deflexed towards base, and with scattered capitate stalked glands; seeds black, 2.2–2.5 mm in diameter, tuberculate.

a. var. **dichotomus**

Stems pubescent. Mature leaves 2.7–9 cm long, 1.6–4.5 cm wide, base cuneate-attenuate, apex obtuse to subattenuate; petiole of mature leaves 3–12(–20) mm long. Corolla tube strongly, or rarely shallowly, upcurved towards mouth, (20–)22–27 mm long; lower lip speckled purple. Fig. 79/1–8, p. 606.

KENYA. Lamu District: Mararani, Boni Forest, Sept. 1961, *Gillespie* 310!; Kilifi District: 43 km from Mombasa on Nairobi road, Aug. 1965, *Gillett* 16863!; Kwale District: 1.6 km N of Lungalunga, July 1960, *Leach & Brunton* 10216!

TANZANIA. Lushoto District: W slopes of E Usambara Mts, below Ubili, Oct. 1973, *Magogo* 1570!; Pangani District: Segera Forest, July 1966, *Faulkner* 3826!; Uzaramo District: near Msorwa [Msolwa], 72 km from Dar es Salaam towards Morogoro, Nov. 1969, *Batty* 702!

DISTR. **K** 7; **T** 3, 6; not known elsewhere

HAB. Bushland, coastal thicket, margins of cultivation and roadsides, rarely in dry forest; 0–850 m

USES. None recorded on herbarium specimens

CONSERVATION NOTES. Assessed as of Least Concern (LC) by Darbyshire (l.c.).

SYN. *Pseuderanthemum dichotomum* Lindau in E.J. 20: 40 (1894) pro parte excl. *Bachmann* 1270 ex S Africa
[*Rhinacanthus communis sensu* C.B. Clarke in F.T.A. 5: 224 (1900) pro parte quoad *Wakefield* s.n. & *Holst* 3289, 8782 & 3902A, *non* Nees]

NOTE. In *Batty* 702 from **T** 6 the corolla tube is less strongly curved and the lower lip is somewhat shorter, with shorter and more rounded lobes than typical.

b. var. **emaculatus** *I. Darbysh.* in K.B. 61: 407, fig. 1J–K (2006). Type: Tanzania, Tanga District, Steinbruch gorge, *Carmichael* 561 (K!, holo.; DSM!, EA, iso.)

Stems sparsely pubescent. Mature leaves 6–10.5 cm long, 2–3.6 cm wide, base long-cuneate, apex (sub-)attenuate; petiole of mature leaves (6–)10–20 mm long. Corolla tube strongly upcurved towards mouth, 25–28 mm long; lower lip unspeckled. Fig. 79/9, 10.

TANZANIA. Tanga District: Steinbruch gorge, July 1956, *Faulkner* 1893! & Amboni cave near Tanga, Jan. 1969, *Batty* 895! & *idem*, Dec. 1969, *Botany students* in DSM 1336!; Zanzibar, near Kinyasini, Aug. 1935, *Vaughan* 2263!
DISTR. **T** 3; **Z**; not known elsewhere
HAB. Coastal forest and thicket; near sea level
USES. None recorded on herbarium specimens
CONSERVATION NOTES. This variety was assessed as Data Deficient (DD) by Darbyshire (l.c.) but is highly likely to be threatened by habitat loss or disturbance in the vicinity of Tanga town and on Zanzibar.

6. **Rhinacanthus sp. A** (= *Ward* U70)

Herb; stems and leaves not seen. Inflorescence terminal, widely branched, laxly paniculate, to 23 cm long; flowers mainly in subsessile clusters of 3–4 at the ends of the branches; peduncles with mixed antrorse and spreading short hairs, sparse multicellular hairs to 1.7 mm long and dense capitate stalked glands throughout; bracts and bracteoles 1.5–2 mm long. Calyx lobes 4–4.5 mm long, with short blunt-tipped spreading hairs and capitate staked glands outside. Corolla colour unknown, 36 mm long, pubescent outside; tube 25–26 mm long, 1.5 mm in diameter, shallowly curved; limb with scattered stalked glands within; upper lip lanceolate, ± 6 mm long, apex notched; lower lip 11 mm long, lobes oblong-ovate, 5–6 mm long. Staminal filaments ± 3.3 mm long, glabrous; anther thecae offset but overlapping, oblique, lower theca 1.5 mm long, base apiculate, upper theca 1 mm long. Ovary glabrous; style sparsely pubescent, glabrous towards apex. Capsule 17–21 mm long, pubescent and with many capitate stalked glands towards apex; seeds 3.2–3.8 mm in diameter, largely smooth.

TANZANIA. Iringa District: Usagara to Mazombe, Aug. 1936, *Ward* U70!
DISTR. **T** 7; not known elsewhere
HAB. Not recorded; ? 750–2000 m
USES. None recorded on herbarium specimen
CONSERVATION NOTES. Data Deficient (DD): more information on the status of this taxon is required prior to further assessment.

SYN. [*R.* sp. 2 *sensu* Darbyshire & Harris in K.B. 61: 407 (2006), in notes]

NOTE. Known only from the single specimen cited, which is notable for its large fruits, large smooth seeds and long spreading hairs on the peduncles. The corolla form is close to that of *R. dichotomus*, of which it may be a variant. However, the *Ward* collection is geographically isolated from the closest collections of *R. dichotomus*, being recorded on the edge of the Iringa Plateau in the rainshadow of the Eastern Arc Mts, away from the coastal Swahelian vegetation element. Further collections, including vegetative parts, are required to delimit this taxon.

7. **Rhinacanthus sp. B** (= *Flock* 364)

Erect unbranched herb to 10 cm tall; stem pale, 6-angular, antrorse-pubescent. Leaves glaucous, obovate, 3–3.7 cm long, 1.5–2 cm wide, base cuneate, margin obscurely repand, apex acute to obtuse, surfaces glabrate; lateral veins 4 pairs, pale and prominent beneath; petiole 5 mm long, furrowed above, antrorse-pubescent. Inflorescence terminal, single-branched, 1.5 cm long, flowers in terminal pairs; peduncle with short blunt-tipped spreading hairs and capitate stalked glands; bracts and bracteoles 1–1.5 mm long. Calyx lobes ± 2.5 mm long, with short blunt-tipped spreading hairs and sparse capitate stalked glands outside. Corolla white, ± 21 mm long, limb speckled purple, pubescent and with scattered stalked glands outside; tube 14 mm long, 1 mm wide, shallowly curved; limb widely spreading; upper lip lanceolate, ± 4 mm long, 1.8 mm wide, apex barely notched; lower lip 6 mm long, lobes subequal, ovate, ± 4 mm long, 2.5 mm wide. Staminal filaments ± 2.5 mm long, glabrous; anther thecae offset, ± 0.8 mm long, lower theca with base apiculate. Pistil glabrous. Capsule ± 8 mm long, sparsely pubescent and with many stalked glands towards apex; seeds black, ± 1.5 mm in diameter, tuberculate.

TANZANIA. Uzaramo District: Kibaha, 40 km W of Dar es Salaam, July 1972, *Flock* 364!
DISTR. **T** 6; not known elsewhere
HAB. In deep shade on termite mound; 150 m
USES. None recorded on herbarium specimen
CONSERVATION NOTES. Data Deficient (DD): more information on the status of this taxon is required prior to further assessment.

SYN. [*R.* sp. 1 *sensu* Darbyshire & Harris in K.B. 61: 407 (2006), in notes]

NOTE. Known only from the single specimen cited. The leaf and flower morphology resemble that of a dwarf form of *R. dichotomus*. However, in addition to the much smaller habit, it differs in having a smaller corolla with a less strongly curved tube, a glabrous style and pale, angular stems. The short habit and short, few-branched inflorescence closely resemble *R. humilis* Benoist from Madagascar but the leaves are more ovate and the style pubescent in that species. Further material is required to determine the status of this taxon.

8. **Rhinacanthus sp. C** (= *Busse* 2992)

Slender erect or decumbent herb, ± 40 cm tall; stems 6-angular with prominent pale ridges, glabrous except for few inconspicuous antrorse hairs on the ridges immediately below the nodes. Leaves rather glaucous, elliptic, 3.3–8.3 cm long, 1.2–3 cm wide, base and apex attenuate, margin entire, surfaces glabrous except for few short antrorse hairs on veins beneath; lateral veins 4–5 pairs; decurrent into petiole to 7 mm long. Inflorescences terminal and axillary, shortly paniculate, 2–12.5 cm long, few-branched, branching partially monochasial; flowers sessile, clustered at apex of branches; peduncles glabrous proximally, with increasing numbers of short, blunt-tipped, spreading hairs and capitate stalked glands distally; bracts and bracteoles 1–1.5 mm long. Calyx lobes 2–2.5 mm long, with short, blunt-tipped, spreading hairs and capitate stalked glands outside. Corolla colour unknown, ± 19 mm long, eglandular- and glandular-pubescent outside; tube ± 13.5 mm long, 0.8 mm in diameter, upcurved in distal half; limb with scattered stalked glands within; upper lip triangular-ovate, 3.5–4.5 mm long, 3 mm wide, apex acute; lower lip 5–5.5 mm long, lobes rounded, 2–2.5 mm long and wide, median lobe slightly longer than lateral pair. Staminal filaments ± 2.5 mm long, glabrous; anther thecae superposed and oblique, lower theca 0.8 mm long, base minutely tailed, upper theca 0.4–0.5 mm long. Ovary largely glabrous; style sparsely pubescent in lower half. Capsule and seeds not seen.

TANZANIA. Lindi District: Kitulu hill, June 1903, *Busse* 2992! & *idem*, *Busse* 2994!
DISTR. **T** 8; not known elsewhere
HAB. Not recorded; 150 m

USES. None recorded on herbarium specimens

CONSERVATION NOTES. Known only from the two specimens cited. As habitat notes are lacking, and with no information on the current status of the population at the single known site, this taxon must be assessed as Data Deficient (DD) but will probably prove to be threatened.

NOTE. This taxon appears close to *R. angulicaulis* in the conspicuously angular stems (which separate it from other small-flowered species in our region) but is smaller in all parts, notably in flower size, and is less hairy on the stems and leaves. Further material is needed.

9. **Rhinacanthus submontanus** *T. Harris & I. Darbysh.* in K.B. 61: 411, fig. 2A–F (2006). Type: Malawi, Chitipa District, Musitu Forest, 21 km SE of Chisenga, *Brummitt* 11987 (K!, holo.; LISC, MAL, SRGH, iso.)

Decumbent or erect perennial herb, 15–50(–90) cm tall; stems 6-angular, shortly pubescent particularly on the angles, hairs mainly antrorse but sometimes retrorse or spreading, sometimes with occasional longer hairs. Leaves elliptic or ovate to narrowly so, 3.5–14 cm long, 1–3.7 cm wide, base acute, cuneate or attenuate, becoming ± rounded in uppermost leaves, margin subentire to obscurely repand, apex acute to shortly attenuate, upper surface with sparse long, multicellular hairs, veins and margin shortly antrorse-pubescent; lateral veins 4-7 pairs; petiole 2.5–15(–30) mm long, antrorse-pubescent. Inflorescences axillary and (sub)terminal, the latter extending beyond the uppermost leaves, laxly paniculate, 3–19 cm long, few–several-branched; flowers subsessile, solitary or paired, few flowers open at any given time; peduncles shortly antrorse-pubescent towards base, with short spreading blunt-tipped hairs and increasing numbers of capitate stalked glands distally; bracts lanceolate, 1.5–2.5 mm long, bracteoles somewhat shorter. Calyx lobes linear-lanceolate, 2–3.5 mm long, with dense spreading blunt-tipped hairs and many capitate stalked glands outside. Corolla white with pink to purple speckling on lower lip, (9–)12–15(–17) mm long, spreading-pubescent and with scattered glandular hairs outside; tube (6–)7.5–9(–10) mm long, (0.7–)1–1.5 mm in diameter, straight; upper lip ovate, 3–4.5 mm long, apex notched; lower lip 3.5–7 mm long, lobes (oblong-) rounded, 1.5–3.5 mm long, median lobe up to 0.5 mm longer than lateral pair. Staminal filaments 1.8–2.5 mm long, glabrous; thecae highly oblique and separated by 0.3–0.6 mm, lower theca 0.9–1.2 mm long, parallel to filament, upper theca 0.6–0.8 mm long, held almost patent to filament. Ovary glabrous; style pubescent towards base or glabrate. Capsule 12–14 mm long, sparsely pubescent and with short stalked-glands towards apex; seeds black, 2–2.7 mm in diameter, tuberculate.

TANZANIA. Pare District: S Pare Mts, Chome Forest Reserve, E side above Gonja, Nov. 1999, *Massawe et al.* 500!

DISTR. **T** 3; N Malawi & NE Zambia

HAB. Submontane forest; 1380 m

USES. None recorded on herbarium specimens

CONSERVATION NOTES. Assessed as Vulnerable (VU B2ab(iii)) by Darbyshire & Harris (l.c.). This assessment still stands despite the recent discovery of an outlying population in the Pare Mts; submontane forest is highly fragmented and threatened in this mountain range.

NOTE. The single specimen cited clearly matches the highly disjunct populations from Malawi and Zambia, particularly in the stamen morphology and in the rather narrow leaves with a long cuneate to attenuate base. The Tanzanian material has a larger, more branched inflorescence and a more deeply lobed lower lip than the southern populations. Both these characters bring this specimen close to *B. selousensis* (sp. 10 below), from which it is however separated by the larger stamens with the anthers more clearly offset, the different leaf shape and the differing habitat. Further collections may, however, ultimately prove these to be a single species.

10. **Rhinacanthus selousensis** *I. Darbysh.* in K.B. 61: 413, fig. 2G–M (2006). Type: Tanzania, Ulanga District, Magombera Forest Reserve, *Vollesen* in MRC 4159 (C!, holo.; EA!, K!, iso.)

Suberect or decumbent perennial herb to 40 cm tall; stems 6-angular, shortly pubescent, hairs concentrated on the angles, mainly antrorse. Leaves elliptic, 5.7–7.2 cm long, 2–3.2 cm wide, base obtuse, becoming rounded in reduced uppermost leaves, margin obscurely undulate, apex shortly attenuate, upper surface sparsely pubescent with long, multicellular hairs most dense at margin, midrib above and veins beneath shortly antrorse-pubescent; lateral veins 5–7 pairs; petiole 8–10 mm, antrorse-pubescent, uppermost leaves subsessile. Inflorescences axillary and terminal, widely and laxly paniculate, to 22 cm long; flowers paired or rarely solitary, subsessile; peduncles spreading-pubescent, hairs mainly short and blunt-tipped, with occasional longer hairs, and with increasing numbers of capitate stalked glands distally; bracts lanceolate, 1.4–2 mm long, bracteoles somewhat shorter. Calyx lobes linear-lanceolate, 2.5–3 mm long, with short spreading hairs and capitate stalked glands outside. Corolla white with violet spots on lower lip, 10–12 mm long, retrorse- to spreading-pubescent with scattered glandular hairs outside; tube 7.5–8 mm long, 1 mm in diameter, straight; limb widely spreading; upper lip ovate, 3–3.5 mm long, apex notched or rounded; lower lip 3.5–4 mm long, lobes obovate, 2.2–2.5 mm long, median lobe somewhat larger than lateral pair. Staminal filaments 1.5–1.8 mm long, glabrous; anther thecae highly oblique and slightly separated, up to 0.3 mm apart, lower theca 0.6–0.8 mm long, upper theca 0.5–0.6 mm long. Pistil glabrous. Capsule 12–13 mm long, sparsely pubescent, hairs becoming deflexed towards base, and with scattered short glandular hairs; seeds black, 2–2.4 mm in diameter, tuberculate.

TANZANIA. Ulanga District: Magombera Forest Reserve, Sept. 1979, *Vollesen* in MRC 4159! (type)
DISTR. **T** 6; not known elsewhere
HAB. Lowland forest; 250 m
USES. None recorded on herbarium specimen
CONSERVATION NOTES. Assessed as Critically Endangered (CR B1ab(iii) 2ab(iii)) by Darbyshire (l.c.); known only from the type collection.

SYN. [*Rhinacanthus virens sensu* Vollesen in Opera Bot. 59: 81 (1980), *non* (Nees) Milne-Redh.]

11. **Rhinacanthus pulcher** *Milne-Redh.* in K.B.: 283 (1935). Type: Kenya, Northern Frontier District, Tana River, between Hameyi and Dakacha, *Sampson* 4 (K!, holo.)

Spreading subshrub, 30–200 cm tall; flowering shoots terete, glaucous, longitudinally striate, with dense pale retrorse hairs. Leaves often immature or absent at flowering, narrowly ovate or lanceolate, 3–5 cm long, 1.5–2.5 cm wide, base acute to obtuse, margin entire, apex acute, surfaces with sparse antrorse hairs most dense on veins and margin; lateral veins 5(–7) pairs; petiole to 6 mm long. Inflorescence terminal, laxly paniculate, to 40 cm long, flowers clustered in subsessile pairs or rarely solitary; peduncles with mixed retrorse and spreading blunt-tipped hairs, and with increasing numbers of capitate stalked glands distally; bracts and bracteoles lanceolate, 1.5–2.5 mm long. Calyx lobes linear-lanceolate, (3–)6–8 mm long, with ± dense capitate stalked glands and short spreading hairs outside. Corolla white or cream, 36–50 mm long, limb sometimes tinged pink or blue outside, pubescent outside; tube 15–23 mm long, 1.5 mm in diameter at base, floor ventricose in upper 10–14 cm, where 2.5–3 mm in diameter, lines of multicellular hairs on the roof within most dense distally; limb with scattered stalked glands within; upper lip recurved distally, lanceolate, 20–25 mm long, 3 mm wide, apex notched; lower lip protruding, (20–)24–30 mm long, lobes subequal, ovate, 12–16 mm long, 7–10 mm wide, apex obtuse. Staminal filaments 8–9 mm, pubescent at base; anther thecae separated by 1–2 mm, lower theca 1.5–1.8 mm long, upper theca 1.2–1.5 mm. Ovary largely glabrous; style densely pubescent at base, glabrous towards apex. Capsule 15–25 mm long, puberulous outside, hairs spreading distally, becoming deflexed towards base; seeds dark brown, ± 3.5 mm in diameter, tuberculate.

KENYA. Northern Frontier District: 27 km S of Mado Gashi, Jan. 1972, *Bally & Smith* 14973! & Isiolo to Garba Tula road, June 1970, *Gillett & Newbould* 19156!; Tana River District: Kora Game Reserve, area between Tana R. K4/7 boundary and Mwitamyisi watercourse, May 1977, *Gillett* 21100!

TANZANIA. Pare District: Nyumba ya Mungu, Pangani R., Aug. 1974, *Mhoro* 2195!

DISTR. **K** 1, 4, 7, **T** 3; not known elsewhere

HAB. *Acacia-Commiphora* bushland, occasionally at the margins of gallery forest and in riverine thicket, on sandy or stony ground; 300–700 m

USES. None recorded on herbarium specimens

CONSERVATION NOTES. Although localised, several collectors have noted this species as locally frequent to common. It also occurs within a region of low human population pressure, therefore disturbance may be limited, although overgrazing may cause losses of its favoured bushland habitat. It is currently assessed as of Least Concern (LC).

NOTE. This species is deciduous, with few mature leaves having been observed on fertile herbarium specimens. They may therefore grow to a considerably larger size than that documented here during non-flowering periods.

The single specimen seen from Tanzania is in fruit only and, whilst largely matching the Kenyan material, the calyx lobes are much shorter (3 mm versus 6–8 mm); flowering material from that locality is desirable.

42. ISOGLOSSA

Oerst. in Vidensk. Medd. Dansk. Naturhist. Foren. Kjobenh.: 155 (1854), *nom. cons.*; C.B. Clarke in F.T.A. 5: 227 (1900); Brummitt in K.B. 40: 785–791 (1985); B. Hansen in Nordic Journ. Bot. 5: 5 (1985); Kiel *et al.* in Taxon 55: 683–694 (2006); Darbyshire in K.B. 64: 401–427 (2009)

Rhytiglossa Lindl. in Introd. Nat. Syst., ed. 2: 444 (1836), *nom. rejic.*, Brummitt in Taxon 23: 440 (1974); McVaugh in Taxon 24: 247–248 (1975)
Ecteinanthus T. Anderson in J.L.S. 7: 45 (1864)
Homilacanthus S. Moore in J.B. 32: 129 (1894)
Schliebenia Mildbr. in N.B.G.B. 12: 99 (1934)
Plagiotheca Chiov. in Racc. Bot. Miss. Consol. Kenya: 96 (1935)

Annual or perennial herbs or shrubs, evergreen or deciduous, sometimes monocarpic and cyclically mass-flowering (plietesial); cystoliths present, linear, ± conspicuous. Stems (sub-) 4- or 6-angular, often sulcate between the angles when young. Leaves opposite-decussate, pairs equal to anisophyllous, blade margin entire, undulate or shallowly crenate, lower leaves petiolate or subsessile, uppermost leaves and those subtending the inflorescences often sessile with base rounded or cordate. Inflorescences terminal and/or axillary in upper leaf axils, rarely on older, leafless branches; spiciform or laxly paniculate thyrses, branching dichasial and/or monochasial; bract pairs ± equal, free, shape and size variable, sometimes foliaceous, sometimes rapidly becoming smaller in size along main inflorescence axis; bracteoles as bracts or much-reduced. Calyx usually divided almost to the base into 5 subequal lobes, often lengthening in fruit. Corolla bilabiate; indumentum variable but always glabrous towards base of tube outside; tube cylindrical, campanulate or inflated; upper lip ± hooded, apex 2-lobed, lobes often reflexed; lower lip oblong, elliptic or obovate, apex 3-lobed, median lobe broader than lateral pair, palate and throat often upraised with a longitudinal central furrow and with "herring-bone"-like transverse ridging. Stamens 2; filaments attached at various heights within the corolla tube, exserted or held at the mouth; anthers bithecous, thecae strongly offset or at a subequal height, parallel to highly oblique, upper theca larger than lower theca or the pair subequal in size, basally muticous; staminodes absent. Disk shallowly cupular. Ovary (oblong-) ovoid, largely glabrous, 2-locular, 2 ovules per locule; style filiform; stigma shortly bilobed. Capsule stipitate, placental base inelastic. Seeds 4 per capsule or 2 by abortion, held on retinacula, lenticular, compressed-ellipsoid or subdiscoid, with a hilar excavation, variously tuberculate, rugose or smooth.

A genus of ± 50–70 species with a palaeotropical and -subtropical distribution, most diverse within our region where it is particularly associated with high montane habitats. Several species are **plietesial**, i.e. they display population-level periodic mass-flowering over a life cycle of several years, often becoming dominant in the forest understorey during maturity.

The generic circumscription here follows B. Hansen (in Nordic Journ. Bot. 5: 5 (1985)), who expanded the Asian *Isoglossa* to include taxa with prolate tricolporate pollen grains (spangenpollen and rahmenpollen) in addition to those with the girdled bi-porate grains (gürtelpollen) traditionally assigned to *Isoglossa*. The former pollen type is represented in our region by the recently described *I. faulknerae* and *I. variegata*. The principal diagnostic characters of the expanded genus are the combination of a bilabiate, rugulate corolla and 2 bithecous anthers, the thecae slightly to strongly offset and basally muticous.

1. Inflorescences with cymules (sub)sessile, shortly to widely spaced along a main axis, spiciform or several-branched, sometimes strobilate; each cymule 1–several flowered 2
   Inflorescences ± laxly paniculate, the flowers terminating well-developed dichasial and/or monochasial branches, rarely reduced to a simple cyme of 1–3 flowers 16
2. Corolla 7–17 mm long*, tube 3.5–9.5 mm long 3
   Corolla 19–35 mm long, tube 8.5–26 mm long 9
3. Upper anther theca with a line of hairs either side of the connective and at the apex 4
   Anther thecae glabrous 5
4. Leaves (ovate-) elliptic, base attenuate or cuneate, upper surface uniformly green; cymule bracts lanceolate, 0.5–1 mm wide .... 1. *I. faulknerae* p.618
   Leaves broadly ovate, base cordate, upper surface variegated, paler around the midrib and lateral veins; cymule bracts obovate, 1.5–3.5 mm wide ... 2. *I. variegata* p.619
5. Leaves with 17–25 pairs of lateral veins, tertiary venation scalariform, conspicuous; capsule 29–38 mm long with seeds 5–6 mm wide; inflorescence glandular-villose .... 18. *I. multinervis* p.635
   Leaves with 3–13 pairs of lateral veins, tertiary venation not conspicuously scalariform; capsule 6.5–14 mm long with seeds 1–2.5 mm wide; if inflorescence glandular then not villose 6
6. Seeds with elongate, hooked tubercles; bract margin glabrous or minutely ciliate; corolla throat and palate largely lacking "herring-bone" patterning 7
   Seeds rugose; bract margin conspicuously ciliate (rarely sparsely so); corolla throat and palate with prominent "herring bone" patterning 8
7. Cymule bracts 1.5–4.5(–6) mm long, triangular-ovate; inflorescences 2-sided, often branched, axes straight with well-spaced cymules, bracts not imbricate; calyx lacking pilose hairs and only rarely with glandular hairs; corolla white or rarely pink 3. *I. punctata* p.620
   Cymule bracts (3.7–)4.5–14 mm long, ovate, elliptic or somewhat obovate; inflorescences wholly or partially unilateral, spiciform or few-branched, axes often curved, cymules usually shortly spaced and with imbricate bracts distally, more rarely lax; calyx often with long pilose hairs and/or glandular hairs; corolla pink, mauve or rarely white .... 4. *I. gregorii* p.621

*Corolla measurements in the key and descriptions are of the tube plus lower lip when straightened

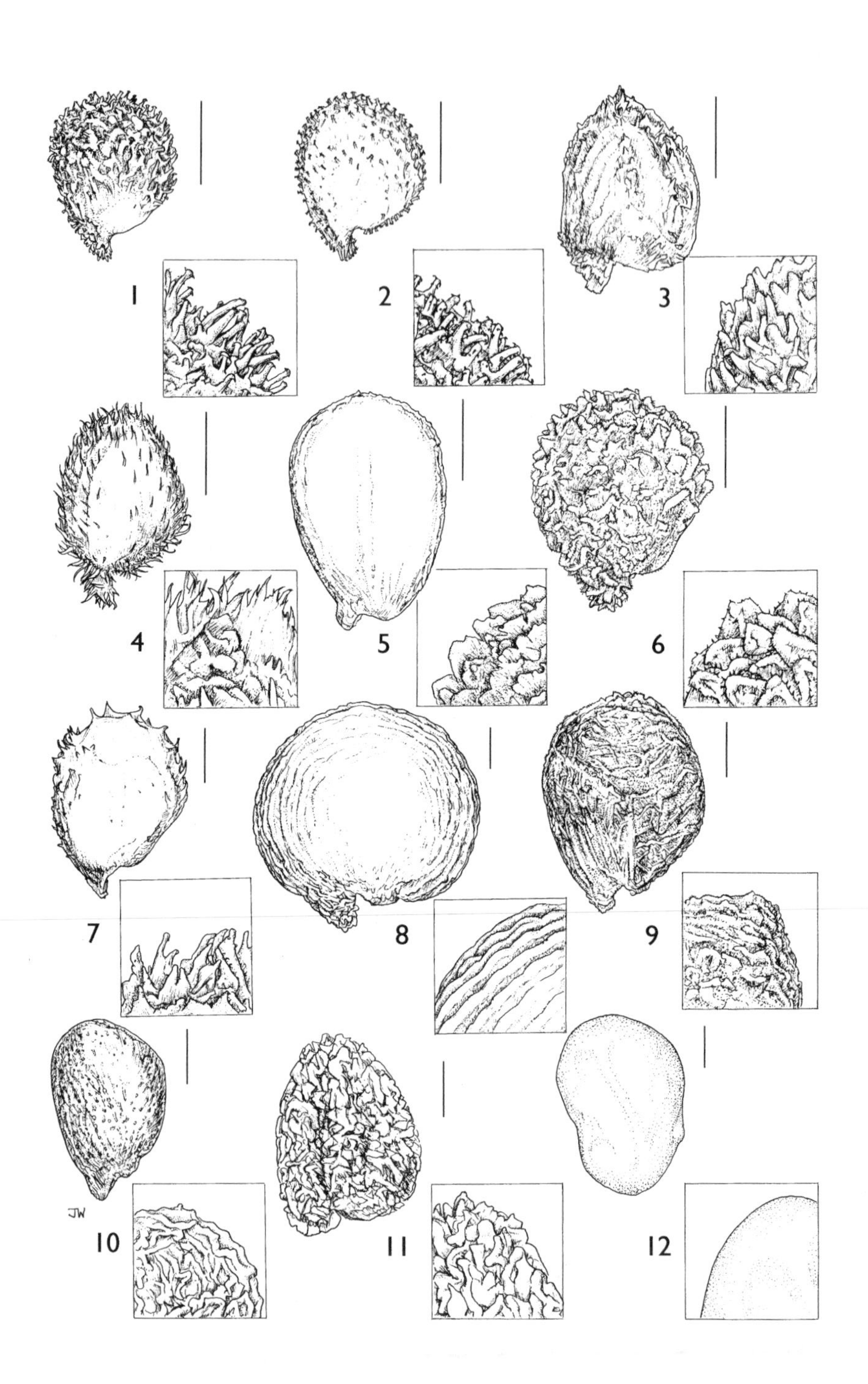
1
2
3
4
5
6
7
8
9
10
11
12
JW

8. Mature leaves (if present at flowering) 1.5–14 cm long, ovate or elliptic, lateral veins 3–7 pairs; bracts not reflexed towards apex (or if somewhat so when immature then straightening with maturity, except sometimes the acumen) . . . . 20. *I. substrobilina* p.637
   Mature leaves 14.5–22 cm long, elliptic or the upper pairs somewhat obovate, lateral veins 7–13 pairs; bracts reflexed towards apex when immature, sometimes remaining so at maturity . . . . 21. *I. bracteosa* p.638
9. Largely glabrous shrubs lacking glandular hairs; leaves somewhat coriaceous; anther thecae strongly offset, one or both held almost patent to the filament 10
   Perennial herbs or subshrubs with glandular hairs at least on the calyx, often dense throughout the inflorescence; leaves thin; anther thecae not or barely offset, parallel or somewhat oblique, never almost patent to the filament 11
10. Inflorescences axillary, often from the old, leafless axils; flowers on distinct pedicels (1.5–)3–6.5 mm long; corolla pink to red . . . . 22. *I. candelabrum* p.639
   Inflorescence terminal (additional inflorescences sometimes present in the upper axils); pedicels 0–2 mm long; corolla white . . . . 23. *I. asystasioides* p.640
11. Corolla glabrous outside, tube markedly longer than limb, declinate, 17–21 mm long; capsule with sparse long glandular hairs . . . . 24. *I. ixodes* p.641
   Corolla sparsely to densely pubescent outside at least on the upper lip, tube shorter than limb, upcurved, 8.5–15 mm long; capsule glabrous 12
12. Inflorescence with pale patent eglandular hairs to 1.5–3 mm long in addition to the shorter glandular indumentum, conspicuous and dense at least on the calyx; seeds rugose . . . . 26. *I. floribunda* p.642
   Inflorescence lacking long patent eglandular hairs, indumentum predominantly glandular with inconspicuous eglandular hairs only; seeds smooth or rugose 13
13. Seeds rugose; staminal filaments free for 14–17 mm 27. *I. grandiflora* p.645
   Seeds smooth; staminal filaments free for (7.5–) 9–12.5 mm 14

FIG. 80. *ISOGLOSSA* seeds with detail of surface sculpturing — **1**, *I. PUNCTATA* (*G. Williams* 301); **2**, *I. MEMBRANACEA* subsp. *SEPTENTRIONALIS* (*Verdcourt* 1696); **3**, *I. LACTEA* subsp. *LACTEA* (*Thulin & Mhoro* 3182); **4**, *I. PAUCINERVIS* (*Polhill et al.* 4983); **5**, *I. STRIGOSULA* (*Richards* 6674); **6**, *I. BRUCEAE* (*Verdcourt & Newbould* 2276); **7**, *I. LAXA* (*Richards* 24659); **8**, *I. MULTINERVIS* (*Pawek* 10137); **9**, *I. CANDELABRUM* (*Luke* 4137); **10**, *I. SUBSTROBILINA* subsp. *SUBSTROBILINA* (whole seed: *Glover et al.* 2504; detail from young seed: *Irwin* 479); **11**, *I. GRANDIFLORA* (*Lovett & Kayombo* 4680); **12**, *I. UFIPENSIS* (*Mwangulango* 868). Whole seeds at × 12, except 7 & 9–11 at × 8, 8 & 12 at × 6. Detail of seed sculpture: 1–4 at × 40, 5 & 6 at × 30; 7 at × 20; 8, 9 & 12 at × 10, 10 & 11 at × 16. Scale bars (for whole seeds) = 1 mm. Drawn by Juliet Williamson.

14. Calyx lobes 5–8 mm long in flower; inflorescence unbranched or with up to 3 pairs of simple branches; corolla pale blue to pale mauve, rarely purple; capsule stipe (2.5–)3.5–6 mm long ...... 30. *I. ufipensis* p.647
    Calyx lobes 2.5–5 mm long in flower; inflorescence with 3–9 pairs of branches, these sometimes with secondary branching, or if unbranched or largely so then corolla deep purple; capsule stipe (4.5–)5.5–8.5 mm long 15
15. Inflorescence with 3–9 pairs of branches from the main axis, these up to 23 cm long; seeds 4–5 mm long ... 28. *I. mbalensis* p.645
    Inflorescence unbranched or with 1–2 branches to 5 cm long; seeds 3.5–3.8 mm long (incompletely known species) .... 29. *I. sp. C* p.646
16. Leaves anisophyllous, the larger of each pair lanceolate, to 1.8 cm wide; corolla 3.5–5 mm long; capsule 5–7 mm long .... 6. *I. anisophylla* p.623
    Leaf pairs subequal to somewhat anisophyllous, ovate or elliptic, 1–13 cm wide; corolla 5–33 mm long; capsule 7.5–26 mm long 17
17. Corolla magenta or purple; capsule 23–26 mm long; calyx 8–11.5 mm long in flower, this and the upper inflorescence axes glabrous .... 19. *I. humbertii* p.636
    Corolla white to rose-pink; capsule 7.5–20 mm long; calyx less than 8 mm long in flower or if longer then glandular-hairy; inflorescence indumentum variable, often glandular- and/or eglandular-hairy, rarely glabrous 18
18. Anther thecae barely offset; upper lip of corolla pubescent within .... 25. *I. rubescens* p.641
    Anther thecae offset by at least half their length, often fully superposed; upper lip of corolla glabrous within 19
19. Seeds with elongate, hooked tubercles; corolla 7.5–10 mm long; capsule 8.5–11.5 mm long; upper anther thecae 0.5–0.8 mm long; mature leaves often with a narrowly cuneate wing along the upper petiole .... 5. *I. membranacea* p.622
    Seed sculpture variable, if tubercles elongate then lacking hooks (but sometimes with minute hair-like processes); upper anther thecae 0.8–2.4 mm long; mature leaves lacking a narrowly cuneate extension along the upper petiole 20
20. Young stems and petioles with slender white multicellular hairs lacking conspicuous cell walls; staminal filaments free for 7–8.5 mm; style strigulose or rarely glabrous .... 12. *I. bruceae* p.631
    Young stems and petioles with broader hairs with conspicuously darker cell walls, or stems largely glabrous; staminal filaments free for 1–6 mm; style glabrous 21
21. Inflorescence conspicuously glandular-pilose in addition to the shorter eglandular indumentum, at least some of the glandular hairs 1–4 mm long 22
    Inflorescence glabrate, eglandular-hairy, or if glandular hairs present then sparse and/or less than 1 mm long 28

22. Calyx lacking short hairs on the margin of the lobes; corolla palate pilose .... 11. *I. vulcanicola* p.629
    Calyx with short hairs on the margin of the lobes; corolla palate glabrous 23
23. Corolla tube shorter than lower lip, ± abruptly ventricose on the floor in the upper half; capsule glabrous; leaves with scattered hairs above ...... 10. *I. ventricosa* p.629
    Corolla tube (where known) longer than lower lip, gradually campanulate or inflated from the base; capsule with sparse pilose glandular hairs towards the apex (but glabrescent); leaves with hairs restricted to the midrib above 24
24. Calyx lobes 4.5–14 mm long in fruit; corolla 13.5–33 mm long 25
    Calyx lobes 3–5.5 mm long in fruit; corolla 5–15 mm long 27
25. Corolla pink or rarely white, with purple markings on palate; tube slender, 1.5–2.5 mm in diameter at the base, gradually expanded to 3–5.5 mm at the mouth 13. *I. laxa* p.632
    Corolla white with purple (-black) markings on palate; tube inflated from base, 5–10 mm in diameter for much of its length 26
26. Calyx lobes 3.5–4.5 mm long in flower, 6–8.5 mm in fruit; bracteoles triangular-lanceolate, 1–2 mm long; corolla pilose outside .... 16. *I. bondwaensis* p.634
    Calyx lobes 8–12 mm long in flower, to 14 mm in fruit; bracteoles linear-lanceolate, 4.5–7.5 mm long; corolla glabrous outside .... 17. *I. oreacanthoides* p.635
27. Corolla 5–7 mm long; inflorescence branches widely divergent (incompletely known species) .... 14. *I. sp. A* p.633
    Corolla 13–15 mm long; inflorescence more slender with less divergent, ascending branches (incompletely known species) .... 15. *I. sp. B* p.633
28. Corolla (rose-) pink or rarely white, with purple markings on palate, tube longer than lower lip, slender and gradually expanded .... 13. *I. laxa* p.632
    Corolla white or tinged pink, usually with pink or purple markings on palate, tube subequal in length to the lower lip, often either inflated from the base or with the floor ventricose in the upper half 29
29. Mature leaves 1.5–5.3 cm long, with 3–5 pairs of lateral veins; branching of inflorescence largely monochasial, sometimes reduced to 1–2 flowers; seeds with elongate appressed tubercles, these lacking hair-like processes .... 8. *I. paucinervis* p.627
    Mature leaves 3–16 cm long, with 5–12 pairs of lateral veins; branching of inflorescence at least in part dichasial; seeds rugose with shorter, broader tubercles or if these elongate then with minute, curved hair-like processes 30
30. Corolla tube subcampanulate or the floor ventricose in the upper half 31
    Corolla tube inflated from the base 33

31. Young stems inconspicuously antrorsely or appressed-ascending pubescent in two opposite lines or largely glabrous . . . . 7. *I. lactea* p.624
    Young stems conspicuously retrorse-pubescent in two opposite lines, often soon glabrescent 32
32. Corolla floor gradually declinate; seeds rugose, becoming rather smooth with maturity; glandular hairs usually present at least on the minor inflorescence branches and calyx, less than 1 mm long or subsessile and with a conspicuous, broadly capitate gland-tip . . . . 9. *I. strigosula* p.628
    Corolla floor ± abruptly ventricose in the distal half; seeds with elongate tubercles with minute curved hair-like processes; glandular hairs on the inflorescence either absent or 1–3 mm long, gland-tip minute . . . . 10. *I. ventricosa* p.629
33. Calyx lobes lacking short marginal hairs; anther thecae held almost patent to the filament; lower main axis bracts often ovate and foliaceous, gradually to rapidly narrowing upwards . . . . 11. *I. vulcanicola* p.629
    Calyx lobes with short marginal hairs; anther thecae subparallel or oblique to the filament but not almost patent; main axis bracts all triangular- to linear-lanceolate . . . . 7. *I. lactea* p.624

1. **Isoglossa faulknerae** *I. Darbysh.* in K.B. 64: 416, figs. 1D & 5K–R (2009). Type: Tanzania, Tanga District, Steinbruch Gorge, *Faulkner* 2128 (K!, holo.; BR!, iso.)

Straggling or decumbent perennial herb, 10–35 cm tall, often rooting at lower nodes; stems slender, pale green with darker longitudinal stripes, retrorse-pubescent in two opposite lines, glabrescent. Leaves (ovate-) elliptic, 3–7.5 cm long, 1.3–3.3 cm wide, base attenuate or cuneate, apex rounded, acute or bluntly acuminate, surfaces largely glabrous; lateral veins 4–6 pairs; petiole 3–11 mm long. Inflorescences terminal but sometimes overtopped by lateral shoots, 1(–3) per branch, spiciform or more rarely with 1–2 branches at the lowest inflorescence node, main axis 3–8 cm long, cymules distant, sessile, 1–2 flowers per bract; axes with many patent glandular hairs and shorter eglandular hairs, the latter ± concentrated in two opposite lines; bracts lanceolate, 1.5–3 mm long, 0.5–1 mm wide, glandular-pubescent; bracteoles (elliptic-) lanceolate, 1–3 mm long, glandular pubescent and eglandular-ciliate, margin often paler. Calyx lobes lanceolate, 2–2.5(–3) mm long in flower, to 3.5 mm long in fruit, indumentum as bracteoles. Corolla white or tinged purple, 7–11 mm long, with mauve or purple markings on palate, glabrous outside except for short hairs towards apex of lobes; tube subcylindrical, 5–7.5 mm long, 1–2.5 mm in diameter at base, 1.5–3 mm at mouth, pubescent within particularly on floor below insertion point of stamens; upper lip hooded, 1.5–2.5 mm long, lobes 0.3–0.6 mm long; lower lip 2–3.5 mm long, lobes 1–1.5 mm long, palate glabrous, with prominent "herring-bone" patterning. Stamens attached in upper half of corolla tube; filaments free for ± 1.5 mm, with scattered glandular hairs; anther thecae offset, overlapping for half their length, slightly oblique, each 1–1.2 mm long, upper theca with broad hairs on each side above connective and at apex, lower theca glabrous. Style strigulose in lower half. Capsule 8.5–9 mm long, glandular- and eglandular-pubescent; seeds ± 1.7 mm wide, rugose.

TANZANIA. Lushoto District: between Amani and Monga, May 1950, *Verdcourt* 211! & W part of Kwamgumi Forest Reserve, June 2000, *Mwangoka & Maingo* 1384!; Tanga District: Kange Gorge, June 1956, *Faulkner* 1870!

DISTR. **T** 3; not known elsewhere

HAB. Coastal forest and mid-altitude moist forest; 0–900 m

USES. None recorded on herbarium specimens

CONSERVATION NOTES. Assessed as Endangered (EN B1ab(iii), 2ab(iii)) by Darbyshire (l.c.).

SYN. [*I. bracteosa sensu* Iversen, Integr. Usambara Rain For. Proj., App. 1: 9 (1988) pro parte quoad *Verdcourt* 211, *non* Mildbr.]
[*Justicia* sp. nov. (= *Verdcourt* 211) *sensu* Iversen, Integr. Usambara Rain For. Proj., App. 1: 10 (1988)]

2. **Isoglossa variegata** *I. Darbysh.* in K.B. 64: 415, figs. 1C & 4 (2009). Type: Tanzania, Lushoto District, E Usambara Mts, Amani, Mt Bomole, *Verdcourt* 271 (K!, holo.; EA!, K!, iso.)

Creeping, decumbent perennial herb, 10–30 cm tall, rooting at lower nodes; stems pale grey (-green), pale pubescent in two opposite furrows, hairs largely retrorse, sometimes with few glandular hairs on upper internodes. Leaves variegated, paler along midrib and lower portion of the principal veins above, these darker and ?purplish beneath, broadly ovate, 1.8–7.3 cm long, 1–5.5 cm wide, base cordate, apex obtuse to subacuminate, upper surface with ± conspicuous pale strigulose hairs between the veins and finer antrorse hairs on margin and veins, margin also with sparse glandular hairs; lateral veins 4–6 pairs; petiole 3–45 mm long. Inflorescences terminal, 1–2(–3) per branch, spiciform or with a single branch developing in the lower half, main axis 5–14 cm long, cymules distant, sessile or pedunculate for up to 2.5 mm, flowers 1(–2) per bract; axes densely patent glandular-pubescent, usually also with shorter patent or antrorse eglandular hairs; bracts obovate, 2–6 mm long, 1.5–3.5 mm wide, base attenuate, apex rounded or shallowly emarginate, with a minute acumen, surfaces glandular-pubescent, bracts subtending inflorescence branch (if present) larger and ovate; bracteoles lanceolate, 0.8–2.5 mm long, glandular pubescent and shortly eglandular-ciliate. Calyx lobes lanceolate, 2–3 mm long in flower, indumentum as bracteoles. Corolla white with pale purple markings on throat and palate and purple lines along dorsal portion of tube, 9–11 mm long; shortly eglandular- and glandular-pubescent outside particularly on upper tube; tube dorso-ventrally flattened, expanded laterally in upper half, 5.5–6.5 mm long, pubescent within, most dense below and between insertion of stamens; upper lip hooded, 2.8–4 mm long, lobes 0.8–1.8 mm long; lower lip 3.5–4.5 mm long, lobes 1.8–3 mm long, palate glabrous, with prominent "herring-bone" patterning. Stamens attached in upper half of corolla tube; filaments free for 1.3–2 mm, with short glandular hairs; anther thecae offset overlapping for half their length, slightly oblique, upper theca 1.2–1.5 mm long, with broad blunt hairs on each side above connective and at apex, lower theca 1.1–1.4 mm long, blunt hairs restricted to apex. Style strigulose towards base. Capsule and seeds not seen.

TANZANIA. Lushoto District: 0.8 km W of Amani, July 1953, *Drummond & Hemsley* 3474!; Tanga District: Muheza, between Amani Research Station and Greenway's forest house, Oct. 1986, *Borhidi, Iversen & Steiner* 86131!; Morogoro District: Nguru Mts, Chazi River valley above Mhonda Mission, Feb. 1993, *Manktelow & Gustafsson* 678!

DISTR. **T** 3, 6; not known elsewhere

HAB. Moist forest, forest margins and tracks, roadsides and areas of secondary regrowth; 800–1100 m

USES. None recorded on herbarium specimen

CONSERVATION NOTES. Assessed as Endangered (EN B1ab(iii), B2ab(iii)) by Darbyshire (l.c.).

SYN. [*Justicia* sp. nov. (= *Drummond & Hemsley* 3474) *sensu* Iversen, Integr. Usambara Rain For. Proj., App. 1: 10 (1988)]

3. **Isoglossa punctata** (*Vahl*) *Brummitt & J.R.I. Wood* in K.B. 38: 448 (1983); Blundell, Wild Fl. E. Afr.: 393, pl. 801 (1987); Friis & Vollesen in Biol. Skr. 51 (2): 445 (2005) pro parte excl. *Friis & Vollesen* 744; Ensermu in Fl. Eth. 5: 493 (2006). Type: Yemen, Hadie [Hadiyah], *Forsskål* 389 (C holo., microfiche 38: II. 1–2, K!, photo.; LD, iso., K!, photo.)

Straggling or decumbent perennial herb, 10–150(–300) cm tall, often rooting at lower nodes; stems pubescent with predominantly retrorse hairs in two opposite lines, glabrescent. Mature leaves ovate, 2.5–15(–20) cm long, 1–8.5 cm wide, base abruptly to gradually attenuate, apex acuminate, margin and midrib pubescent, upper surface and veins beneath often sparsely pilose; lateral veins (5–)6–9 pairs; petiole 7–90 mm long. Inflorescences terminal, spiciform or more usually with 1–5 pairs of branches, these rarely with secondary branching, main axis 3.5–27 cm long, cymules distant, (sub)sessile; axes with two opposite lines of dense, often retrorse hairs, rarely with scattered glandular hairs; bracts sometimes tinged purple at apex, triangular-ovate, 1.5–4.5(–6) mm long, 0.7–2.5 mm wide, largely glabrous, margin pale, bracts subtending the branches often larger and foliaceous; bracteoles as bracts but lanceolate to triangular, 0.7–4 mm long; flowers sessile or pedicels to 1.5 mm long; reduced inflorescences often present in upper leaf axils. Calyx sometimes tinged purple, lobes linear-lanceolate, 2.5–4.5 mm long in flower, extending to 3–8.5 mm in fruit, shortly ciliate, rarely with scattered longer glandular hairs. Corolla white or rarely pink, 7–16 mm long, always with pink to purple markings on palate; largely glabrous outside except for short hairs at limb apex or rarely more widespread on upper tube and limb, glabrous within; tube subcylindrical or narrowly campanulate, 4.5–8 mm long, 1–2.5 mm in diameter at base, 1.5–4 mm at mouth; upper lip hooded, 2.5–7 mm long, lobes often reflexed, 0.7–2 mm long; lower lip 3–7.5 mm long, lobes 1–3 mm long, palate with "herring-bone" patterning inconspicuous. Staminal filaments attached in upper half of corolla tube, free for 3–5 mm, glabrous; anther thecae widely separated, each 0.6–1.2 mm long. Style glabrous. Capsule 8.5–14 mm long, glabrous or with minute eglandular hairs, occasionally with scattered longer glandular hairs; seeds 1.5–2.4 mm wide, tuberculate particularly towards rim, tubercles elongate and hooked. Fig. 80/1, p. 614.

Uganda. Ruwenzori, above Minimba Camp, Jan. 1962, *Loveridge* 359!; Kigezi District: Luhizha, June 1951, *Purseglove* 3673!; Mbale District: Sebei, Kyosoweri, Oct. 1955, *Norman* 304!

Kenya. Northern Frontier District: Marsabit, Feb. 1953, *Gillett* 15115!; Trans-Nzoia District: Suam Bridge, Sept. 1959, *Tweedie* 1896!; North Kavirondo District: Kakamega Forest, June 1961, *Lucas* 96!

Tanzania. Mbulu District: Great North Road, Pienaars Heights, 195 km S of Arusha, May 1962, *Polhill & Paulo* 2333!; Moshi District: Lyamungu, S slope, Aug. 1932, *Greenway* 3064! & Marangu, Dec. 1963, *Archbold* 372!

Distr. **U** 2–4; **K** 1–5; **T** 2; Sudan, Ethiopia, Congo-Kinshasa, Rwanda, Burundi; Yemen

Hab. Montane forest undergrowth, particularly along margins and pathways, also in secondary bushland; 1300–2750 m

Uses. None recorded on herbarium specimens

Conservation notes. Widespread and locally common in suitable habitat. It favours more open areas in forest and may therefore be tolerant of moderate habitat disturbance: Least Concern (LC).

Syn. [*Dianthera b americana sensu* Forssk., Fl. Aegypt.-Arab.: 9 (1775)]

Dianthera punctata Vahl in Symb. Bot. Upsal. 1: 4 (1790)

Justicia punctata (Vahl) Vahl in Symb. Bot. Upsal. 2: 15 (1791)

Isoglossa runssorica Lindau in E.J. 20: 56 (1894); C.B. Clarke in F.T.A. 5: 231 (1900); Champluvier in Fl. Rwanda, Sperm. 3: 462 (1985). Type: Uganda, Ruwenzori [Runssoro], *Stuhlmann* 2474 (B†, holo.), **syn. nov.**

[*I. oerstediana sensu* C.B. Clarke in F.T.A. 5 (1900) pro parte quoad *Schimper* ex Ethiopia & *Volkens* 1387 & 1852 (pro parte, mixed coll. with *I. gregorii*) ex Tanzania; *sensu* U.K.W.F. ed. 1: 599, 600 (fig.) (1974); *sensu* U.K.W.F. ed. 2: 278, pl. 122 (1994); *sensu* Brummitt & J.R.I. Wood in K.B. 38: 447–448 (1983) pro parte, *non* Lindau]

Note. Plants of *I. punctata* from tropical Africa have regularly been misnamed *I. oerstediana* Lindau in the past, an error stemming from Clarke's application of that name in F.T.A. (1900). *I. oerstediana*, described from the W Usambara Mts, is referable to *I. gregorii* (see below).

Plants from Mt Marsabit (**K** 1) are superficially distinct in having large corollas (12–16 mm long, compared to 7–12.5 mm elsewhere in our region), more broadly ovate leaves and in usually having glandular hairs on the inflorescence axes and calyx lobes. These plants reach a maximum [height] of only 55 cm and tend to have a simple spiciform or few-branched inflorescence. Large-flowered forms of *I. punctata* are also recorded from the Kefa region of southwestern Ethiopia; these populations also tend to have a simple spike but lack glandular hairs. A glandular indumentum is however not entirely restricted to the Marsabit populations; plants from Burundi often display glandular hairs, matched in our region by *Lye* 25509 from Mt Elgon and *B.D. Burtt* 2368 from northern Tanzania. The Marsabit populations are therefore best considered an extreme variant of a single, variable taxon.

Populations from the Ruwenzori Mts were previously separated under *I. runssorica*, although Brummitt & J.R.I. Wood (in K.B. 38: 448 (1983)) noted that they are probably conspecific. Plants named *I. runssorica* at K and EA clearly fall within the range of variability of *I. punctata*, hence it is reduced to synonymy here.

4. **Isoglossa gregorii** (*S. Moore*) *Lindau* in P.O.A. C: 372 (1895); C.B. Clarke in F.T.A. 5: 232 (1900); Mildbraed in N.B.G.B. 9: 505 (1926); U.K.W.F. ed. 1: 600, 608 (fig.) (1974); U.K.W.F. ed. 2: 278, pl. 122 (1994). Type: Kenya, Mt Kenya, *Gregory* s.n. (BM!, holo.)

Straggling or trailing perennial herb, 10–200 cm tall, rooting at lower nodes; stems pubescent with mainly retrorse hairs in two opposite lines, glabrescent. Mature leaves ovate, 1.5–12 cm long, 0.8–7.5 cm wide, base attenuate to rounded, apex (sub)acuminate, margin and midrib pubescent, upper surface and veins beneath sparsely pilose; lateral veins (4–)5–7(–8) pairs; petiole 4–55(–90) mm long. Inflorescences terminal, spiciform or more rarely few-branched, sometimes with reduced inflorescences in upper leaf axils, main axis 2–24 cm long, often curved, cymules subsessile, clustered with bracts imbricate towards apex or throughout, sometimes distantly spaced towards base, wholly or partially 1-sided, one bract of each pair often sterile; axes with two opposite lines of dense, retrorse hairs, sometimes with scattered glandular and/or long eglandular hairs; bracts purple towards apex or throughout, elliptic to somewhat obovate or basal pairs often ovate, (3.5–)4.5–14 mm long, 0.7–6 mm wide, apex attenuate to acuminate, glabrous except for short marginal hairs, occasionally with scattered glandular hairs; bracteoles as bracts but elliptic-lanceolate, 3–7 mm long; flowers sessile. Calyx tinged purple towards apex, lobes subulate, 3.5–6.5(–7.5) mm long in flower, extending to 5.5–11 mm in fruit, with short eglandular hairs particularly along margin, often also with scattered glandular hairs and/or long eglandular hairs. Corolla pink, mauve or rarely white, 7.5–17 mm long, with pink to purple markings on palate and throat, largely glabrous or pubescent outside on limb and upper tube, glabrous within; tube subcylindrical, (4–)5–9.5 mm long, 1–1.5 mm in diameter at base, 1–2.7 mm at mouth; upper lip hooded, 2.5–6 mm long, lobes often reflexed, 0.5–1.5 mm long; lower lip 3–7 mm long, lobes 1.3–2.5 mm long, palate with "herring-bone" patterning inconspicuous. Staminal filaments attached in upper third of corolla tube, free for 1.5–3.5 mm, glabrous; anther thecae immediately superposed or widely separated, each 0.65–1.1 mm long. Style glabrous or sparsely hairy towards base. Capsule 7–12.5 mm long, glabous or with minute eglandular hairs, occasionally with scattered glandular and/or long eglandular hairs; seeds 1–2 mm wide, tuberculate particularly towards rim, the tubercles elongate and hooked.

Uganda. Karamoja District: Moroto Mt, Oct. 1958, *J. Wilson* 563!; Mbale District: Elgon, Sasa Trail, near Mudangi Cliff, Dec. 1996, *Wesche* 589! & Mt Elgon National Park, Tutum Cave along path from Forest Exploration Centre to the Caldera, Feb. 2002, *Lye* 25497!

KENYA. Northern Frontier District: Mathews Range, Uaraguess, Aug. 1962, *Archer* 362!; Nakuru District: Eburru Forest Reserve, July 2002, *Luke, Hoft & Ndeche Mwatsuma* 8933!; Embu District: Mt Kenya Forest, vicinity of Castle Forest Station, Jan. 1967, *Perdue & Kibuwa* 8398!
TANZANIA. Arusha District: Leopard Point, Ngurdoto Crater, Oct. 1965, *Greenway & Kanuri* 11952!; Lushoto District: Madala, July 1961, *Semsei* 3256!; Rungwe District: Poroto Mts, foot of ridge S of Ngozi Crater, June 1992, *Gereau, Mwasumbi & Kayombo* 4624!
DISTR. **U** 1, 3; **K** 1, 3, 4, 6, 7; **T** 2–4, 6, 7; NE Congo-Kinshasa, Malawi, E Zimbabwe
HAB. Forest, particularly margins and clearings, often near water, rarely in ericaceous montane bushland or abandoned cultivation; 1700–2900 m
USES. "Browsed by livestock" (**K** 1; *Bytebier et al.* 58)
CONSERVATION NOTES. *I. gregorii* is most common in the mountains of central and S Kenya where it can dominate the forest understorey. Elsewhere it appears to be more localised, being apparently absent from several suitable highlands regions such as those of N Malawi, but may have been overlooked there particularly if it does not flower annually. It appears to tolerate substantial disturbance and probably benefits from low-level disturbance along forest margins and pathways: Least Concern (LC).

SYN. *Homilacanthus gregorii* S. Moore in J.B. 32: 129, t. 343 (1894)
Isoglossa oerstediana Lindau in E.J. 20: 56 (1894); C.B. Clarke in F.T.A. 5: 232 (1900) pro parte quoad *Scott Elliot* 6767 ex Kenya & *Johnston* 11a; *Holst* 523 & *Volkens* 1852 (pro parte, mixed coll. with *I. punctata*) ex Tanzania. Type: Tanzania, Lushoto District, Usambara, *Holst* 523 (B†, holo.), **syn. nov.**
Justicia nyassae Lindau in N.B.G.B. 8: 424 (1923). Type: Tanzania, Rungwe District, Kyimbila, N of Lake Nyasa, Rungwe, *Stolz* 1565 (B†, holo.; BM!, C!, K!, iso.) **syn. nov.**
[*Isoglossa punctata sensu* Iversen, Integr. Usambara Rain For. Proj., App. 1: 10 (1988), quoad *Peter* 60386, *non* (Forssk.) Brummitt & J.R.I. Wood]

NOTE. Inflorescence structure and indumentum are rather variable in this species. The typical form, common in the mountains of central and S Kenya, has strongly unilateral inflorescences and the cymule bracts are imbricate at least towards the apex; such populations often have glandular and/or eglandular-pilose hairs on the inflorescence axes and calyx lobes, and pubescent corollas. Plants from N Tanzania, particularly Mt Meru and the W Usambara, tend to have more lax, branched inflorescences with widely spaced cymules at least towards the base; such plants also have less strictly unilateral inflorescences and the indumentum is more variable. These specimens can easily be confused with *I. punctata*. Indeed, *I. oerstediana*, described from the W Usambara, has previously been placed in synoymy with *I. punctata* (Brummitt & J.R.I. Wood in K.B. 38: 448 (1983)). The type specimen of this taxon has been destroyed but the description matches the lax form of *I. gregorii* from that mountain range. Mildbraed (in N.B.G.B. 9: 506 (1926)), who saw the type, noted the close similarity with *I. gregorii*, stating that they only differed in the bracts being slightly smaller in relation to the calyx in the former. The basionym *Homilacanthus gregorii* S. Moore was described several months prior to *I. oerstediana* and so has nomenclatural priority. A few specimens (e.g. *Verdcourt* 2238 from Mt Kulal, **K** 1; *Faden* 70/52 and *Sequeira* 18 from Donyo Sabuk, **K** 4) prove difficult to key out and it is quite possible that hybridisation occurs between *I. gregorii* and *I. punctata* in regions of range overlap.

High altitude populations on Kilimanjaro are unusual in having narrowly elliptic bracts. Intermediate specimens between this and the typical form with broader bracts occur at lower altitudes.

5. **Isoglossa membranacea** *C.B. Clarke* in F.T.A. 5: 230 (1900); Darbyshire in K.B. 64: 410 (2009). Type: Malawi, Misuku Hills, *Whyte* s.n. (K!, holo.)

Straggling or decumbent perennial herb or subshrub, 20–200 cm tall; stems largely glabrous. Mature leaves (ovate-) elliptic to broadly so, 6–22 cm long, 3–13 cm wide, base of largest leaves narrowly cuneate, often forming a wing along the upper petiole, then abruptly to gradually attenuate, apex acuminate, midrib antrorse-pubescent above, margin with or without multicellular hairs becoming shorter upwards, surfaces glabrous or with scattered hairs above; lateral veins 6–11 pairs; petiole 12–65 mm long. Inflorescences terminal, laxly paniculate, ± pyramidal in outline, 10–36 cm long, much-branched, dichasial; axes densely puberulous and sparsely to densely glandular-pubescent, the latter with hairs patent, 0.2–1 mm long, gland tips conspicuous, sometimes with additional scattered long eglandular hairs to 1.5 mm long; bracts and bracteoles minute, (triangular-) lanceolate, 0.4–2.5 mm long; flowers subsessile or on pedicels to 1 mm long. Calyx lobes lanceolate, 1.7–3 mm long in flower, 2.5–3.5 mm long in fruit, indumentum as axes. Corolla white, pale pink or pale violet, 7.5–12.5 mm long, often with pink to purple markings on palate of lower lip, pubescent outside, glabrous within; tube 4–5.7 mm long, subcampanulate, floor declinate, 1–1.5 mm in diameter at base, 2.5–3.5 mm at mouth; upper lip broadly triangular, hooded, reflexed distally, 2.5–5.5 mm long, lobes 1–2.5 mm long; lower lip protruding, 3.5–7 mm long, lobes 1–3 mm long, palate with shallow "herring-bone" patterning. Staminal filaments attached ± midway along corolla tube, free for 2–2.5 mm, glabrous; anther thecae offset, slightly overlapping or separated by up to 0.7 mm, each 0.5–1.1 mm long. Style glabrous. Capsule 8.5–11.5 mm long, indumentum variable (see subspecies); seeds 1.6–1.9 mm wide, tuberculate particularly towards rim, tubercles elongate, hooked.

subsp. **septentrionalis** *I. Darbysh.* in K.B. 64: 410 (2009). Type: Kenya, Kakamega Forest, *Verdcourt* 1696 (K!, holo.; BR!, EA!, iso.)

Leaves and petiole largely glabrous, margin sometimes with minute, inconspicuous hairs. Corolla 7.5–10 mm long. Anther thecae superposed and separated by up to 0.7 mm, 0.5–0.8 mm long. Capsule with scattered long eglandular and occasional glandular hairs, with or without additional short eglandular hairs, glabrescent. Fig. 80/2, p. 614.

UGANDA. Bunyoro District: Budongo Forest, Feb. 1935, *Taylor* 3311! & *idem*, Nov. 1936, *Sangster* 207!; Ankole District: Kasyoha-Kitomi Forest Reserve, NE of Kyambura R., June 1994, *Poulsen* 550!

KENYA. North Kavirondo District: Yala R. area of Kakamega Forest, Dec. 1956, *Verdcourt* 1696! (type) & Kaimosi, W of iron bridge over R. Yala on Kapsabet–Kisumu road, Nov. 1971, *Tweedie* 4151! & Kakamega Forest along Kubiranga Stream at N end of forest, Mar. 1977, *R.B. & A.J. Faden* 77/848!

DISTR. **U** 2; **K** 5; S Sudan, E Congo-Kinshasa

HAB. Forest understorey, sometimes along streams or pathsides, forest margins; 1050–1750 m

USES. None recorded on herbarium specimens

CONSERVATION NOTES. Assessed as Near Threatened (NT) by Darbyshire (l.c.).

SYN. [*I. punctata sensu* Friis & Vollesen in Biol. Skr. 51 (2): 445 (2005) pro parte quoad *Friis & Vollesen* 744, *non* (Vahl) Brummitt & J.R.I. Wood]

NOTE. Subsp. *membranacea* C.B. Clarke, apparently endemic to the Misuku Hills of N Malawi, differs in having conspicuous multicellular hairs along the petiole and proximal margins of the leaf blade, larger corollas (11–13 mm long), larger anther thecae (0.85–1.1 mm long) which overlap slightly and eglandular- and glandular-puberulous capsules. The apparent disjunction in distribution is unusual, although this species may have been overlooked in the under-botanised mountains of W Tanzania, the few collecting trips perhaps not having coincided with flowering years. Subsp. *membranacea* should also be sought for in the mountains of southernmost Tanzania.

6. **Isoglossa anisophylla** *Brummitt* in K.B. 40: 788 (1985). Type: Tanzania, Tanga District, Kange Gorge, *Faulkner* 1848 (K!, holo.; BR!, EA!, K!, LISC!, iso.)

Decumbent, probably annual herb, 15–60 cm tall, rooting at lower nodes; stems weak, antrorse-pubescent, hairs mainly arranged in two opposite lines, most dense below the nodes. Leaves subsessile, anisophyllous, blade of larger of a pair lanceolate, 1.5–8.5 cm long, 0.3–1.8 cm wide, the smaller ovate to lanceolate, 0.4–2.7(–5.3) cm long, 0.2–1 cm wide, base asymmetric, one side rounded or obtuse, the other cuneate, apex acute, surfaces glabrous except for inconspicuous antrorse hairs on some of the veins; lateral veins 4–8 pairs in the larger of each leaf pair. Inflorescences axillary and terminal, paniculate, 1–5.2 cm long, largely dichasial or the branches sometimes monochasial; axes shortly antrorse-pubescent, with or without patent glandular hairs; bracts lanceolate, 0.7–2 mm long, those subtending the lowermost branches on the main axis sometimes ovate, to 3.5 mm long; bracteoles lanceolate, 0.5–1.2 mm long; flowers subsessile or pedicels to 1.5 mm long. Calyx lobes lanceolate, 1.5–2.5 mm long in flower, extending to 2–3.5 mm in fruit, indumentum as axes outside. Corolla white or pale mauve, 3.5–5.5 mm long, sparsely pubescent outside; tube cylindrical, 2–2.8 mm long, 0.8–1.2 mm in diameter, shallowly waisted; upper lip ovate, held erect, 1.5–2 mm long, lobes 0.2–0.5 mm long; lower lip 1.8–2.8 mm long, lobes 0.7–1.6 mm long, mouth and throat pubescent, palate with prominent "herring-bone" patterning. Staminal filaments attached in upper half of corolla tube, free for 0.4–0.8 mm, glabrous; anther thecae offset, slightly oblique, each 0.6–0.8 mm long. Style glabrous or with occasional hairs near base. Capsule 5–7 mm long, glabrous; seeds 0.9–1.3 mm wide, rugose.

TANZANIA. Tanga District: Steinbruch gorge, Apr. 1958, *Faulkner* 2129!; Pangani District: Kiama, Mfumoni, Madanga, May 1956, *Tanner* 2809! & Kibubu, Mkuzi Katani, Madanga, July 1957, *Tanner* 3623!

DISTR. **T** 3, 6; not known elsewhere

HAB. Lowland semi-evergreen forest and damp shaded areas, persisting in cultivated sites; 0–850 m

USES. None recorded on herbarium specimens

CONSERVATION NOTES. Currently known only from five collections, with a highly limited distribution in coastal NE and E Tanzania. It may form large stands at these sites and may therefore be locally common. However, lowland forest is threatened in this region of Tanzania. *I. anisophylla* is therefore likely to have suffered from habitat loss, although it may be able to persist in cultivated sites, as evidenced by *Tanner* 3623. *Faulkner* 2129 noted that colonies growing on limestone ledges were susceptible to being washed away during heavy rains. Local subpopulation losses through such stochastic events, together with the threat from human encroachment upon its habitat, render this subspecies threatened: Endangered (EN B1ab(iii)+2ab(iii)).

NOTE. A collection from the Dindili Forest Reserve, Morogoro District (*Pócs* 89182/X, K!) differs from the Tanga District material in being larger (to 60 cm tall, not 15–35 cm) with only some of the leaf pairs being strongly anisophyllous, the larger of the pair having 5–8, not 4–6 pairs of lateral veins which are more conspicuous, in having more-glandular hairs on the inflorescence, and in having longer fruiting calyx lobes. This location is isolated both geographically and ecologically from the Tanga sites, being recorded from higher altitude forest (450–850 m, not 0–100 m), and possibly represents a distinct taxon at least at the subspecific rank. Further collections are however required to confirm the consistency of these differences.

7. **Isoglossa lactea** *Lindau* in E.J. 20: 55 (1894); C.B. Clarke in F.T.A. 5: 230 (1900) pro parte excl. *Scott-Elliot* 7664 & *Stuhlmann* 8822; U.K.W.F. ed. 1: 599 (1974); U.K.W.F. ed. 2: 278 (1994); Darbyshire in K.B. 64: 406, 407, fig. 3L (2009). Type: Tanzania, Lushoto District, Usambara Mts, Ngwelo, *Holst* 2279 (K!, lecto., chosen here)

Perennial herb or subshrub, 30–250 cm tall; stems glabrous or with two opposite lines of antrorse or appressed-ascending hairs when young, rarely spreading to somewhat retrorse immediately below the nodes, soon glabrescent. Mature leaves often somewhat anisophyllous; blade ovate (-elliptic), (4.5–)6–16 cm long, (2–)4–7 cm wide, base ± asymmetric, attenuate or cuneate, apex acuminate, ascending-pubescent on midrib above and often on margin, upper surface sometimes with scattered hairs, glabrous beneath; lateral veins 6–12 pairs, prominent; petiole 10–75 mm long. Inflorescences terminal on principal branches and particularly on short axillary shoots, laxly paniculate, 3.5–21 cm long, few- to much-branched, up to 6 orders of branching, branches sometimes partially monochasial; axes eglandular-puberulous or glabrous particularly towards base, hairs spreading, antrorse or somewhat retrorse, with or without scattered glandular hairs to 1 mm long, these rarely more dense; bracts lanceolate, 0.8–3(–5) mm long; bracteoles as bracts but 0.5–1.5 mm long; flowers subsessile or on pedicels to 1.5 mm long. Calyx lobes linear-lanceolate,

1.8–4.5(–6) mm long in flower, 2.5–5 mm in fruit, minutely ciliate or puberulous throughout, with or without scattered longer glandular hairs. Corolla white or tinged pink with pink to purple markings on palate, 7.5–21 mm long, glabrous to pubescent outside; tube 3.5–10 mm long, inflated from base or the floor ± abruptly ventricose midway, 2–5 mm in diameter at mouth, pubescent within mainly on the floor and most dense towards base, rarely glabrous; upper lip triangular, 3–9 mm long, lobes reflexed, 1.3–4 mm long; lower lip 4–11 mm long, lobes 1.5–6 mm long, palate sparsely pubescent or glabrous, palate with prominent "herring-bone" patterning. Staminal filaments attached ± midway along corolla tube, free for 1–6 mm, glabrous or with scattered hairs towards base; anther thecae offset, overlapping for half their length or superposed, somewhat oblique, each 0.8–1.8 mm long. Style glabrous. Capsule 11–18.5(–20) mm long, glabrous; seeds 1.9–2.5 mm wide, rugose-tuberculate, tubercles short, with or without minute hair-like processes.

a. subsp. **lactea**

Inflorescence with minor branches often monochasial, slender ascending branches often developing. Corolla 7.5–14.5 mm long; tube campanulate, floor ± ventricose in upper half; limb with lobes often oblong, ratio of lip length/lobe length 1.5–3/1. Filaments free for 1–3.5 mm; anther thecae 0.8–1.3 mm long. Seeds with minute hair-like processes visible at × 40 magnification. Figs. 80/3, p. 614 & 81/5–8, p. 626.

TANZANIA. Lushoto District: Amani, road to Monga, June 1970, *Kabuye* 177!; Morogoro District: Liwale R., Aug. 1951, *Greenway & Farquhar* 8636!; Iringa District: Udzungwa Scarp Forest Reserve, Dec. 1997, *Frimodt-Møller, Jøker & Ndangalasi* TZ561!

DISTR. **T** 3, 6, 7; not known elsewhere

HAB. Undergrowth and margins of moist forest, including disturbed areas, along path margins and streamsides; 600–1700 m

USES. "Root used for syphilis" (**T** 3; *Koritschoner* 1370)

CONSERVATION NOTES. Subsp. *lactea* is widespread and locally abundant in the Eastern Arc Mts of Tanzania, often dominating the understorey of mid-altitude forest and forest margins in the E Usambara and Uluguru ranges. It appears adaptable to both primary forest and somewhat disturbed habitats. Therefore, although large areas of forest within its altitudinal range have been cleared, it is not considered threatened at present: Least Concern (LC).

SYN. *I. flava* Lindau in E.J. 22: 125 (1895). Type: Tanzania, Morogoro District, E Uluguru Mts, *Stuhlmann* 9032 (B†, holo.)

[*I. dichotoma sensu* B. Hansen in Nordic Journ. Bot. 5: 10 (1985) pro parte, *non* (Hassk.) B. Hansen]

NOTE. Several specimens from the Nguru Mts are notable for having rather densely glandular-pubescent calyx lobes in comparison to the sparsely glandular peduncles (e.g. *Semsei* 1941). However, this variation is inconsistent; *Semsei* 1485 (K!) from the same location lacks glandular hairs on the calyx.

Lindau listed two syntypes in the protologue, *Holst* 2279 and 3252, both from the Usambara Mts. However, the latter specimen is highly unusual for the species in having flowering calyx lobes to 6 mm long (these measuring up to only 3.5 mm in all the many other gatherings of subsp. *lactea* seen). For this reason, the more representative *Holst* 2279 is here chosen as a lectotype. *Holst* 3252 is otherwise a good match for subsp. *lactea* as circumscribed here.

b. subsp. **saccata** *I. Darbysh.* **subsp. nov.** a subspecie typica praecipue tubo corollae magis inflato et lobis labii inferioris pro rata brevioribus atque magis rotundatis differt. Type: Kenya, Teita District, Taita Hills, Chawia Forest, Bura bluff, *Drummond & Hemsley* 4373 (K!, holo.; BR!, EA!, K!, iso.)

Inflorescence with minor branching at least in part dichasial, more widely spreading. Corolla (9.5–)11–21 mm long; tube ± inflated from base, floor not ventricose in upper half; limb with lobes (oblong-) rounded lobes, ratio of lower lip length/lobe length (2.2–)2.7–4.5/1. Staminal filaments free for 3–6 mm; anther thecae 1–1.8 mm long. Seeds without visible hair-like processes at × 40 magnification.

FIG. 81. *ISOGLOSSA BRUCEAE* — **1**, habit, × $^2/_3$; **2**, detail of hairs on young stem × 12; **3**, flower, lateral view, × 2; **4**, detail of anther, × 6. *I. LACTEA* subsp. *LACTEA* — **5**, habit, × $^2/_3$; **6**, detail of hairs on young stem, × 20; **7**, flower, lateral view, × 3; **8**, detail of anther, × 12. 1 & 2 from *Martin* 234; 3 & 4 from *Verdcourt* 660; 5 & 7 from *Faulkner* 2036; 6 from *Verdcourt* 95; 8 from *Greenway & Farquhar* 8636. Drawn by Juliet Williamson.

KENYA. Northern Frontier District: Mathews Range, Mantachien, Dec. 1960, *Kerfoot* 2558!; Kiambu District: Kikuyu Escarpment Forest, Gatamayu R., Jan. 1969, *Napper & Stewart* 1822!; Embu District: near Thiba R. above fishing camp, Dec. 1966, *Gillett* 18052!

TANZANIA. Pare District: S Pare Mts, Chome Forest Reserve, ± 1.5 km SW of Kanza Village, Feb. 2001, *Gereau et al.* 6584!; Lushoto District: W Usambara Mts, Shagayu Forest Reserve, June 1951, *Eggeling* 6157! & Lushoto–Gare path, Aug. 1959, *Semsei* 2882!

DISTR. **K** 1, 4, 7; **T** 3; not known elsewhere

HAB. Undergrowth and margins of moist forest, sometimes in disturbed areas and streamsides; 1250–2350 m

USES. "Used as animal feed" (**T** 3; *Mlangwa et al.* 1484)

CONSERVATION NOTES. This subspecies is locally abundant in habitats similar to (though often at higher elevations than) subsp. *lactea*. Although some losses of forest have occurred within its altitudinal range, much montane forest remains intact. It is therefore not considered threatened at present: Least Concern (LC).

SYN. [*I. dichotoma sensu* B. Hansen in Nordic Journ. Bot. 5: 10 (1985) pro parte, *non* (Hassk.) B. Hansen]

NOTE. The largest and most inflated corollas within this subspecies (and hence those most easily distinguished from subsp. *lactea*) are recorded from the Taita Hills and W Usambara Mts, where it is notably recorded at higher elevations than the populations of subsp. *lactea* from the adjacent E Usambara. Plants from elsewhere in Kenya and the Pare Mts of Tanzania have smaller, more campanulate corollas. They are, however, separable from subsp. *lactea* by having proportionally smaller corolla lobes, lacking a ventricose upper half to the corolla tube, having larger stamens and a more spreading inflorescence, all of which place them close to the Taita - W Usambara plants.

Luke 4129 from 1200 m altitude at Kasigau (**K** 7) appears rather intermediate between the two subspecies. It has the more spreading, 2-sided inflorescence of subsp. *saccata* but the flowers are small and with somewhat oblong lobes to the lips, and the seeds have minute hair-like processes on the tubercles which would place it closer to subsp. *lactea*.

I. somalensis, to be expected in the Ugandan extension of the Imatong Hills as it is recorded just across the border in Sudan, is rather similar to subsp. *saccata* but has larger flowers and fruits.

8. **Isoglossa paucinervis** *I. Darbysh.* in K.B. 64: 411, fig. 2J–R (2009). Type: Tanzania, Kilosa District, Ukaguru Mts, Matandu Mt, 3–5 km WNW of Mandege Forest Station, *Thulin & Mhoro* 2966 (K!, holo.; EA!, DSM!, MO!, UPS, iso.)

Creeping or decumbent perennial herb, 20–200 cm tall, rooting at lower nodes; stems retrorse-pubescent, rather dense particularly in two opposite lines, hairs with conspicuous cell walls. Mature leaves ovate to broadly so, 1.5–5.3 cm long, 1–3 cm wide, base truncate or obtuse to attenuate, apex (sub)acuminate, midrib above and margin antrorse-pubescent, surfaces sometimes pubescent or pilose; lateral veins 3–5 pairs; petiole 2–25 mm long. Inflorescences terminal and in upper leaf axils, ± laxly paniculate, 0.5–11.5 cm long, usually few-branched, sometimes reduced to a simple 1–2-flowered cyme, branching largely monochasial; axes puberulous to pubescent, hairs largely antrorse or spreading, becoming smaller in length upwards, glandular hairs occasionally present, sometimes glabrous towards base or throughout; main axis bracts often foliaceous towards base, ovate to narrowly elliptic, ± soon becoming linear-lanceolate upwards, 1–11.5 mm long, 0.3–6 mm wide in central portion of axis, glabrous or pubescent; bracteoles linear-lanceolate, 1–3 mm long; flowers subsessile or on pedicels to 2 mm long. Calyx lobes subulate, 3–6 mm long in flower, to 4–8.5 mm in fruit, minutely ciliate, with or without scattered spreading glandular and/or ascending eglandular hairs on the outer surface. Corolla white or tinged pink, 10–20 mm long, with pink to purple markings on palate of lower lip, pubescent outside; tube subcampanulate, sometimes inflated, 4.5–9 mm long, pubescent on floor within, hairs most dense or restricted to near the base; upper lip triangular, 4.5–6.5 mm long, lobes reflexed, 1–2.3 mm long; lower lip 5–11 mm long, lobes 1.5–4 mm long, palate glabrous or sparsely pubescent, with prominent "herring-bone" patterning. Staminal filaments attached ± midway along corolla tube, free for 2.5–4 mm, glabrous; anther thecae offset, usually superposed and separated by 0.5–1.5 mm, oblique, each 0.65–1.8 mm long. Style glabrous. Capsule 7.5–14.5 mm long, glabrous; seeds 1.8–2 mm wide, with appressed elongate tubercles extending towards rim. Fig. 80/4, p. 614.

TANZANIA. Moshi District: Kilimanjaro, Mweka route, July 1968, *Bigger* 2090!; Lushoto District: Shagayu Forest Reserve, Aug. 1955, *Semsei* 2203!; Kilosa District: Ukaguru Mts, Mamiwa Forest Reserve, summit to E of Ikwamba Peak, Aug. 1972, *Mabberley & Salehe* 1464!

DISTR. **T** 2, 3, 6, 7; not known elsewhere

HAB. Montane forest including open elfin forest; 1700–2750 m

USES. None recorded on herbarium specimens

CONSERVATION NOTES. Assessed as of Least Concern (LC) by Darbyshire (l.c.).

NOTE. This species is found within four rather isolated populations, all sharing the small, few-nerved leaves, largely monochasial, few-flowered inflorescences, small capsules and seeds with a characterisitic unhooked appressed-tuberculate sculpture which together readily separate this species from other closely related taxa. There is however considerable variation in leaf and inflorescence indumentum and flower and fruit size which may warrant formal taxonomic recognition with further analysis; this variation is fully documented in the protologue.

9. **Isoglossa strigosula** *C.B. Clarke* in F.T.A. 5: 231 (1900); P. Winter in Pl. Nyika Plateau: 51 (2005) as *I. strigulosa*; Darbyshire in K.B. 64: 406, 407 (2009). Type: Malawi, Misuku Hills [Masuku Plateau], *Whyte* 288 (K!, holo.)

Straggling perennial herb, 20–300 cm tall, lower stems creeping and rooting at nodes; stems retrorse-pubescent in two opposite lines, hairs with conspicuous cell walls, glabrescent. Mature leaves ovate to narrowly so, 3–11.5 cm long, 1.5–5.5 cm wide, base attenuate or cuneate, apex acuminate, margin, midrib above and principal veins beneath pubescent, upper surface sometimes with scattered hairs; lateral veins 5–8(–9) pairs; petiole 6–30(–45) mm long. Inflorescences terminal and in upper leaf axils, laxly paniculate, 2–12(–19) cm long, few- to much-branched, branching largely dichasial but minor branching often monochasial; axes eglandular-puberulous, sparse or glabrous towards base, more dense on minor branches, these with additional capitate short-stalked glands; bracts linear-lanceolate, 2.5–9 mm long; bracteoles as bracts but 1.5–3 mm long; flowers subsessile or on pedicels to 2 mm long. Calyx lobes linear-lanceolate, 3.5–6 mm long in flower, to 4.5–7 mm in fruit, eglandular-puberulous particularly along margin, with scattered capitate short-stalked or subsessile glands. Corolla white or tinged pink, 13–20 mm long, with red or purple markings on palate and throat, rather densely pubescent outside; tube 7–9.5 mm long, somewhat inflated, subcampanulate, 1.5–3 mm in diameter at base, 4.5–7 mm at mouth, floor pubescent within, hairs most dense or restricted to near the base; upper lip broadly triangular, hooded, 4.5–7 mm long, lobes reflexed, 1–2.2 mm long; lower lip protruding, 6–11 mm long, lobes 2–3 mm long, palate sparsely pubescent or glabrous, with prominent "herring-bone" patterning. Staminal filaments attached ± midway along corolla tube, free for 2.5–4 mm, glabrous; anther thecae offset, separated by 1–1.8 mm, subparallel, upper theca 1.6–2.1 mm long, lower theca 1.2–1.6 mm long. Style glabrous or nearly so. Capsule 14–18 mm long, glabrous; seeds 1.75–2 mm wide, rugose, inconspicuously so at maturity. Fig. 80/5, p. 614.

TANZANIA. Mbeya District: Poroto Mts, May 1957, *Richards* 9776b!; Rungwe District: Rungwe Mt, ± 4 km NNE of Rungwe Peak and 5 km SSE of Kiwira Forest Station, June 1992, *Mwasumbi* 16247!; Njombe District: Mtorwi Mt, Ikuwo side, Sept. 1991, *Congdon* 315!

DISTR. **T** 7; N Malawi, NE Zambia

HAB. Undergrowth and margins of moist montane forest including stands of bamboo; 1500–2600 m

USES. None recorded on herbarium specimens

CONSERVATION NOTES. This is a locally common species, with several collectors noting its abundance or dominance in the understorey of montane forest. Some habitat loss has occurred through forest clearance by man, for example on the Nyika Plateau. However, significant areas of montane forest remain within its range and as this species appears adaptable to both the forest understorey and marginal habitats, it is not considered threatened: Least Concern (LC).

SYN. [*I. dichotoma sensu* B. Hansen in Nordic Journ. Bot. 5: 10 (1985) pro parte, *non* (Hassk.) B. Hansen]

10. **Isoglossa ventricosa** *I. Darbysh.* in K.B. 64: 413, figs. 1A & 2A–H (2009). Type: Tanzania, Iringa District, Lulando [Lulanda] Village, S of Fufu forest patch, *Mwangoka & Mduvike* 715 (MO!, holo.; K!, iso.)

Perennial herb, 45–200 cm tall; stems retrorse-pubescent in two opposite lines, hairs with conspicuous cell walls, glabrescent. Mature leaves ovate (-elliptic) to broadly so, 6–10 cm long, 2.5–4.5 cm wide, base ± asymmetric, attenuate, apex acuminate, midrib and margin antrorse-pubescent, upper surface ± sparsely pubescent, glabrous beneath or pubescent along principal veins; lateral veins 7–10 pairs; petiole 12–60 mm long. Inflorescences terminal on principal and short lateral branches, ± laxly paniculate, 5.5–15 cm long, rather slender, branching dichasial or partially monochasial; axes eglandular-puberulous to -pubescent, hairs concentrated in two opposite lines, retrorse or spreading, often also sparsely to densely glandular-pilose, these hairs 1–2(–3) mm long, gland tips minute; bracts lanceolate, 1–2.5 mm long, ciliate, lowermost pair on main axis rarely foliaceous; bracteoles triangular-lanceolate 0.9–1.5 mm long; flowers subsessile or on pedicels to 2 mm long. Calyx often tinged red-purple, lobes subulate, 3–5 mm long in flower, to 5–6 mm in fruit, pale-ciliate, hairs ascending, sometimes also with long glandular hairs outside. Corolla white or tinged pink, 12.5–17 mm long, palate with red or purple markings, sparsely to densely pubescent outside; tube 5–6.5 mm long, subcampanulate, base 1.5–2 mm in diameter, the floor ± abruptly ventricose in upper half, mouth 4–5 mm in diameter, rather densely pubescent 1.5–2 mm from base within and with sparse hairs extending onto floor of throat; upper lip hooded, 5.5–7.5 mm long, lobes reflexed, 1.5–2 mm long; lower lip protruding, 8–10 mm long, lobes 1.5–3 mm long, palate glabrous, with prominent "herring-bone" patterning. Staminal filaments attached ± midway along corolla tube, free for 3.5–4.5 mm, glabrous; anther thecae offset, separated by 1–1.5 mm, subparallel, each 1.1–1.5 mm long. Style glabrous. Capsule 12–14 mm long, glabrous; seeds ± 2 mm wide, rugose and with elongate tubercles towards rim, these with minute curved hair-like processes, most conspicuous when immature.

TANZANIA. Iringa District: Lulando Forest, May 1987, *J. & J. Lovett* 2257! & Lulando Forest Reserve, Fufu forest patch, July 1999, *Kayombo et al.* 2522! & Mufindi Scarp Forest Reserve, Aug. 1987, *Congdon* 175!
DISTR. **T** 7; not known elsewhere
HAB. Submontane and montane forest, sometimes by streams; 1400–1800 m
USES. None recorded on herbarium specimens
CONSERVATION NOTES. Assessed as Endangered (EN B2ab(iii + iv)) by Darbyshire (l.c.).

NOTE. *Congdon* 175 is a rather more hairy plant throughout than the Lulando gatherings with more pubescence on the leaves, inflorescences and corollas, but otherwise matches that material and is clearly conspecific.

11. **Isoglossa vulcanicola** *Mildbr.* in B.J.B.B. 17: 88 (1943); Robyns, F.P.N.A. 2: 304 (1947); Champluvier in Fl. Rwanda, Sperm. 3: 462, fig. 143.3 (1985); Fischer & Killmann, Illus. Field Guide Pl. Nyungwe N.P. Rwanda: 354 (2008); Darbyshire in K.B. 64: 406, 407 (2009). Type: Congo-Kinshasa, Kabara (Mikeno), *de Witte* 1687 (BR!, holo.)

Subshrub or perennial herb, sometimes scandent, 30–200 cm tall; stems glabrous or retrorse-pubescent in two opposite lines, hairs with conspicuous cell walls, glabrescent. Mature leaves ovate, 5–16 cm long, 2.5–7.5 cm wide, base shortly attenuate, apex acuminate, surfaces ± sparsely pubescent, sometimes only on margin and midrib; lateral veins 6–11 pairs; petiole 6–55 mm long. Inflorescences terminal on principal and short lateral branches, ± laxly paniculate, 5.5–30 cm long, (few- to) much-branched, dichasial; axes eglandular-pubescent, hairs retrorse and concentrated in opposite lines towards base, often patent or antrorse and more evenly distributed elsewhere, sometimes also densely glandular-pilose, these hairs 1–2.5(–4) mm long, gland tips conspicuous, rarely glabrous; main axis bracts foliaceous and broadly ovate towards the base, gradually narrowing and elliptic to lanceolate upwards, 3.5–23 mm long, 1–14 mm wide in central portion of axis; secondary bracts narrowly elliptic (-obovate) or lanceolate, 1–7 mm long; bracteoles lanceolate, 0.5–2 mm long; flowers subsessile, pedicels to 0.5–3 mm long in fruit. Calyx lobes subulate, 2.5–5 mm long in flower, to 3.5–8.5 mm in fruit, glandular-pilose or glabrous. Corolla white with pink or purple markings on palate, 11–18 mm long, sparsely pubescent on dorsal side of tube outside or largely glabrous; tube 5.5–8.5 mm long, inflated from base, mouth 4–6 mm in diameter, pilose on floor within particularly towards base; upper lip broadly triangular, hooded, 3–5 mm long, lobes reflexed, 0.5–1.2 mm long; lower lip protruding, 5.5–10.5 mm long, lobes 1–2.5 mm long, palate pilose, with prominent "herring-bone" patterning. Staminal filaments attached ± midway along corolla tube, free for 3–4.5 mm, glabrous; anther thecae offset, separated by 0.8–1.7 mm, both held almost patent to filament, each 0.8–1.4 mm long. Style glabrous. Capsule 11–20 mm long, glabrous or either glandular-pilose or eglandular-puberulous towards apex; seeds 1.9–2.6 mm wide, rugose-tuberculate particularly when young, tubercles with minute hair-like processes.

UGANDA. Kigezi District: Impenetrable Forest, May 1939, *Purseglove* 730! & Kinkizi, Rutanga, Mar. 1947, *Purseglove* 2344! & Luhizha, June 1951, *Purseglove* 3671!

TANZANIA. Morogoro District: Nguru Mts, Mesumba Mt, July 1933, *Schlieben* 4171!; Kilosa District: Rubeho Mts, Ukwiva Forest Reserve, May 2005, *Mwangoka et al.* 3876!; Njombe District: Livingstone Mts, Isalala R., SE of Bumbigi, Mar. 1991, *Gereau & Kayombo* 4158!

DISTR. **U** 2; **T** 5–7; E Congo-Kinshasa, Rwanda, Burundi

HAB. Undergrowth of montane forest including degraded areas, swamp or river margins, roadsides; 1750–2600 m

USES. None recorded on herbarium specimens

CONSERVATION NOTES. *I. vulcanicola* has a highly limited distribution, being largely restricted to the Ruwenzori and Virungu Mts with isolated populations in the mountains of E and SE Tanzania. In the latter country it is apparently scarce, having been collected only four times. However, it appears much more common in the Ruwenzori where it has been recorded as a "carpet plant" (*Synge* 1405). It is likely to be a periodic mass-flowering species, which may have resulted in under-recording from poorly collected areas. It may therefore also be present in the little-known mountains of W Tanzania (e.g. the Mahali Mts). Its habitat within its core range is largely unthreatened due to the high altitude: Least Concern (LC).

SYN. *I. vulcanicola* Mildbr. var. *eglandulosa* Mildbr. in B.J.B.B. 17: 89 (1943); Robyns, F.P.N.A. 2 Sympétales: 305 (1947). Type: Congo-Kinshasa, Nyiragongo [Ninagongo], *Humbert* 7973 (BR!, holo.) **syn. nov.**

[*I. lactea sensu* C.B. Clarke in F.T.A. 5: 230 (1900) pro parte quoad *Scott-Elliot* 7664 ex Kivata, Uganda]

NOTE. Mildbraed recognised two varieties, var. *vulcanicola* having glandular-pilose inflorescences and var. *eglandulosa* having eglandular-pilose axes and glabrous calyces. Both forms are present in Uganda. Plants of the latter tend to have shorter bracteoles and calyx lobes, and either glabrous or eglandular-puberulous capsules (those of var. *vulcanicola* often being glandular pilose). However, there is much overlap and *Purseglove* 209 from Ruwenzori, Uganda is entirely intermediate between the two, having the glabrous calyx of var. *eglandulosa* but having glandular-pilose inflorescence axes. As inflorescence indumentum is highly variable in several species of *Isoglossa* (both *I. laxa* and *I. ventricosa* from our region displaying similar glandular and eglandular forms) it is not deemed of great taxonomic significance and the varieties are not maintained here.

The K sheets of *Poulsen* 709 from Kashoya-Kitomi forest, **U** 2, have been labelled "*I. gracilis* Champl." by Dominique Champluvier but this name has never been published. This specimen is matched by *Purseglove* 2003 from Ishasha Gorge, both having very lax inflorescences with widely divergent, glabrous branches. The flower morphology is however identical to that of *I. vulcanicola* and there are intermediates between those specimens and eglandular forms of this species. They are therefore regarded here as an extreme variant of *I. vulcanicola*.

12. **Isoglossa bruceae** *I. Darbysh.* in K.B. 64: 414, fig. 3M (2009). Type: Kenya, Nairobi District, near Nairobi, *Whyte* s.n. (K!, holo.)

Erect or straggling perennial herb, 30–100(–200) cm tall; stems white-retrorse- to patent-pubescent in two opposite lines, glabrescent. Mature leaves ovate to broadly so, 4–14 cm long, 1.7–8 cm wide, base rounded, obtuse or shortly attenuate, apex acuminate, surfaces sparsely pubescent, most dense on margin and principal veins; lateral veins (5–)6–8 pairs, ± conspicuous beneath; petiole 8–55 mm long. Inflorescences terminal on principal and short lateral branches, laxly paniculate, 3–16 cm long, few–much-branched, largely dichasial but minor branching sometimes monochasial; axes densely glandular-pubescent to -pilose, hairs to 2 mm long, gland tips conspicuous, also eglandular-puberulous concentrated in opposite lines and sometimes with eglandular-pilose hairs 2–3.5 mm long; main axis bracts foliaceous and broadly ovate towards base, soon becoming smaller to linear-lanceolate upwards, 2–18 mm long, 0.3–10 mm wide in central portion of axis, pubescent; secondary bracts linear-lanceolate 2–6 mm long; bracteoles 2–3.5 mm long; flowers subsessile. Calyx lobes linear-lanceolate, (3–)4–6(–8.5) mm long in flower, to 4.5–9 mm in fruit, indumentum as axes outside, shortly eglandular ciliate. Corolla white or tinged pink towards apex of limb, (13.5–)15–20(–23) mm long, with pink to purple markings on palate, pubescent outside; tube 7–11 mm long, campanulate and inflated in upper half, 1.5–3 mm in diameter at base, 4.5–6.5 mm at mouth, with an interrupted ring of hairs within 2.5–3 mm from base; upper lip hooded, 8–9.5(–12) mm long, lobes 0.5–1 mm long; lower lip 6.5–10(–14) mm long, lobes 0.7–2 mm long, palate glabrous, with prominent "herring-bone" patterning. Staminal filaments attached ± midway along corolla tube, free for 7–8.5 mm, glabrous; anther thecae offset, separated by 0.7–1.3 mm, upper or both thecae held almost patent to filament, upper theca 1.6–2.4 mm long, lower theca 1.4–2 mm long. Style strigulose in lower half or rarely glabrous. Capsule (10.5–)12–17 mm long, glabrous; seeds 2.2–3 mm wide, rugose-tuberculate, tubercles reticulate when young and with conspicuous hair-like processes. Figs. 80/6, p. 614, & 81/1–4, p. 626.

UGANDA. Karamoja District: Moroto township, Oct. 1958, *J. Wilson* 602!

KENYA. Northern Frontier District: Moyale, July 1952, *Gillett* 13532!; Trans-Nzoia District: Suam Gorge, Nov. 1961, *Tweedie* 2245!; Nairobi District: Karura Forest, Oct. 1967, *Mwangangi & Abdalla* 228!

TANZANIA. Mbulu District: NW of Msasa near crest of W Rift Wall, Mar. 1964, *Greenway & Kanuri* 11412!; Moshi District: W end of Lake Chala, July 1968, *Bigger* 1995!; Nzega District: 44 km on Tabora–Mambili road, Jan. 2009, *Bidgood, Leliyo & Vollesen* 7533!

DISTR. **U** 1; **K** 1–4, 6, 7; **T** 1–4; S Ethiopia, ?Rwanda (see note)

HAB. Undergrowth and margins of forest, particularly drier forest-types, *Acacia*-woodland, dry bushland on hillslopes, riverine thicket, large termite mounds in *Brachystegia* woodland; 850–2150 m

USES. None recorded on herbarium specimens

CONSERVATION NOTES. Widespread and locally abundant in a variety of habitats, therefore although suitable locations may be decreasing through removal of woodland and forest by man, it is not considered threatened: Least Concern (LC).

SYN. *I. laxa* Oliv. var. *pilosa* Schweinf. in Höhnel, Rudolph & Stephanie-See, 2: 357 (1892). Type: Kenya, Laikipia District, upper course of Ewaso Narok, *Höhnel* s.n. (B†, holo.), **syn. nov.**
I. ovata E.A. Bruce in K.B.: 100 (1932), *nom. illegit.*, *non* (Nees) Lindau. Type as for *I. bruceae*
[*I. laxa sensu* Agnew, U.K.W.F. ed. 1: 599, 600 (fig.) (1974) & ed. 2: 278, pl. 122 (1994); *sensu* Ensermu in Fl. Eth. 5: 493 (2006) pro parte excl. type, *non* Oliv.]

NOTE. The white stem hairs with inconspicuous cell walls clearly separate this species from closely related taxa, as does the characteristic seed ornamentation. Inflorescence indumentum is variable, ranging from having rather sparse and short glandular hairs to very dense long glandular hairs and with or without longer, wispy eglandular hairs. It is this latter form that most likely accounts for *I. laxa* var. *pilosa*, although the type of that taxon has not been seen and there is no description in the protologue; the name is tentatively synoymised here based principally upon geography.

Troupin 6921 from Kibungu, Rwanda (BR!, K!), recorded as *Isoglossa* sp. A by Champluvier in Fl. Rwanda, Sperm. 3: 462 (1985), appears to match this species but the flowers are immature.

13. **Isoglossa laxa** *Oliv.* in Johnston, Kilim. Exped. Append.: 344 (1886) & in Trans. Linn. Soc., Bot. Ser. II, 2: 345 (1887); C.B. Clarke in F.T.A. 5: 229 (1900) pro parte excl. var. *pilosa* Schweinf.; Darbyshire in K.B. 64: 406, 407, fig. 3K (2009). Type: Tanzania, Kilimanjaro, *Johnston* 11 (K!, holo., BM!, iso.)

Perennial herb or subshrub, 30–200 cm tall; stems retrorse-pubescent largely in two opposite lines, hairs with conspicuous cell walls, soon glabrescent. Mature leaves (ovate-) elliptic, 9–15 cm long, 3–6 cm wide, base cuneate, apex long-acuminate, surfaces glabrous except for antrorse hairs on midrib above, sometimes sparsely ciliate; lateral veins 8–17 pairs; petiole 20–55 mm long. Inflorescences terminal on principal and short lateral branches, laxly paniculate, (3.5–)8–25 cm long, branching dichasial; axes usually glandular-pilose, hairs often purple, 1.5–3.5 mm long, gland tips minute, additionally or rarely exclusively eglandular-puberulous, hairs patent to retrorse, with conspicuous purplish cell walls; main axis bracts often foliaceous, narrowly elliptic to (linear-) lanceolate, 1.5–18 mm long, 0.7–4 mm wide in central portion of axis, largely glabrous; secondary bracts lanceolate, 1–6 mm long; bracteoles triangular-lanceolate, 0.5–2 mm long; pedicels 0.5–4 mm long. Calyx lobes lanceolate, 3–5.5 mm long in flower, to 4.5–9 mm in fruit, indumentum as axes outside, rarely glabrous except for short marginal hairs. Corolla (rose-) pink or rarely white, 13.5–28 mm long, with purple markings on palate, floor of tube and throat sometimes paler, pilose outside, glabrous within; tube 8–15 mm long, slender, gradually campanulate, floor somewhat declinate in upper half, 1.5–2.5 mm in diameter at base, 3–5.5 mm at mouth; upper lip broadly triangular, hooded, reflexed distally, 5–10.5 mm long, lobes 1–2 mm long; lower lip protruding, 5.5–13 mm long, lobes 2–4.5 mm long, palate with "herring-bone" patterning. Staminal filaments attached in upper half of corolla tube, free for 3.5–5 mm, glabrous; anther thecae strongly offset, separated by 0.7–2 mm, oblique, upper theca 1.1–1.7 mm long, lower theca 0.9–1.5 mm long. Style glabrous. Capsule 13–17 mm long, with scattered long glandular hairs, glabrescent; seeds 2–2.5 mm wide, rugose and with elongate tubercles towards rim, these most conspicuous when immature. Fig. 80/7, p. 614.

KENYA. Masai District: Chyulu-South, June 1938, *Bally* 1204c!

TANZANIA. Arusha District: Mt Meru, Aug. 1970, *Richards* 25742!; Moshi District: Kilimanjaro, S slope between Umbwe and Weru Weru, Aug. 1932, *Greenway* 3027! & Kilimanjaro, Mweka route, Sept. 1968, *Bigger* 2212!

DISTR. **K** 6; **T** 2; not known elsewhere

HAB. Montane forest understorey, clearances and secondary growth, often dominant; 1650–2750 m

USES. None recorded on herbarium specimens

CONSERVATION NOTES. *I. laxa* has a highly restricted distribution but is locally abundant. It has been recorded as periodically mass-flowering on both Mt Meru (*Richards* 25742, K!, MO!) and Kilimanjaro (*Grimshaw* 93375, K!), the former believed to be on a cycle of 7 years. At such times it is dominant in the understorey. Habitat loss may have resulted in some population reduction but it appears tolerant of some disturbance and so is probably still locally common within its range: Least Concern (LC).

SYN. *I. volkensii* Lindau in E.J. 20: 55 (1894); C.B. Clarke in F.T.A. 5: 230 (1900). Type: Tanzania, Kilimanjaro, Marangu, *Volkens* 990 (B†, holo., BM!, iso.) **syn. nov.**

[*I. oerstediana sensu* C.B. Clarke in F.T.A. 5: 232 (1900) pro parte quoad *Volkens* 1936 ex Kilimanjaro (B†), fide Mildbraed in N.B.G.B. 9: 506 (1926) as *I. volkensii*]

NOTE. The corolla is usually pink in this species but in *Vesey-Fitzgerald* 6788 from Arusha National Park, Jekukumia, it is recorded as white (all plants have purple markings on the palate).

Greenway & Kanuri 13443 from **T** 2, Ngurdoto Crater, is atypical in largely lacking glandular hairs on the inflorescence, these are here restricted to the pedicels. The axes have two lines of retrorse hairs in this specimens and the calyx is largely glabrous. The corolla has the characteristic long, slender tube of *I. laxa*. Density of glandular hairs on the inflorescence is highly variable in several other taxa of *Isoglossa* (see note under *I. vulcanicola*).

The isotype of *I. volkensii* at BM is a rather scant specimen but the inflorescence indumentum and elongate corolla support its placement within *I. laxa*, as does the collecting locality.

14. **Isoglossa sp. A** (= *Rounce* 520)

Herb to 90 cm tall; stems retrorse-pubescent in two opposite lines, hairs with conspicuous cell walls, soon glabrescent. Mature leaves sometimes absent at flowering; blade ovate-elliptic, 9–?12 cm long, 3.5–5.5 cm wide, base attenuate or cuneate, apex acuminate, surfaces glabrous except for inconspicuous antrorse hairs on midrib above; lateral veins 10 pairs; petiole 15–27 mm long. Inflorescences terminal, laxly paniculate, 9.5–30 cm long, much-branched, widely spreading, largely dichasial but minor branching sometimes monochasial; axes glandular-pilose, hairs to 2 mm long, gland tips minute, cell walls purple-tinged, also eglandular-puberulous; main axis bracts narrowly elliptic to linear, 4.5–18 mm long, 1–5 mm wide in central portion of axis; secondary bracts linear-lanceolate 1–3 mm long; bracteoles lanceolate 0.8–1.5 mm long; pedicels extending to 1.5–2.5 mm long in fruit. Calyx lobes linear-lanceolate, 1.8–2.8 mm long in flower, to 3.5–5.5 mm in fruit, indumentum as axes. Corolla pink (where colour known), 5–7 mm long, pilose outside; tube 3–4.5 mm long, slender; lips 2–2.7 mm long, palate of lower lip with prominent "herring-bone" venation. Staminal filaments ± 1 mm long; anther thecae offset but overlapping, each ± 0.9 mm long. Style glabrous. Capsule 11.5–14.5 mm long, with scattered long glandular hairs, glabrescent; seeds ± 2 mm wide, rugose-tuberculate, tubercles elongate towards rim, with or without minute hair-like processes.

TANZANIA. Morogoro District: Uluguru Mts, without date, *Rounce* 520! & Nguru Mts, Maskati–Mhonda path, Dec. 1966, *Robertson* 405!

DISTR. **T** 6; not known elsewhere

HAB. Bamboo forest & forest glades; 1500–1850 m

USES. None recorded on herbarium specimens

CONSERVATION NOTES. Clarification of the taxonomic status is required prior to a conservation assessment: Data Deficient (DD).

NOTE. The two specimens cited are mainly in fruit but also bear a few minute corollas. They clearly fall within the *I. laxa* - *I. bondwaensis* group, sharing the long glandular hairs with minute gland tips on the inflorescence, largely glabrous mature leaves and sparsely glandular-pilose capsules. The tiny flowers are diagnostic but it is possible that these are aberrant specimens. Unusually small-flowered plants have been recorded in other plietesial species, for example in the genus *Strobilanthes* (Wood in Edinburgh J. Bot. 51: 175–273 (1994)), particularly associated with plants flowering outside the normal cycle and apparently under stress. As *I. laxa* and allies are believed to be plietesial, it is quite possible that this is the case here too. The flowers of *Robertson* 405 do not appear well-developed and are barely open at the mouth and may be cleistogamous. However, no other specimens of the *I. laxa* group have been seen from the Nguru Mts; clearly more material is needed from both this range and the Ulugurus.

15. **Isoglossa sp. B** (= *Luke et al.* 6561)

Perennial herb, 65–100 cm tall; stems retrorse-pubescent in two opposite lines, hairs with conspicuous cell walls, soon glabrescent. Mature leaves elliptic, 6–12.2 cm long, 1.8–4.3 cm wide, base attenuate or cuneate, apex acuminate, surfaces glabrous except for inconspicuous antrorse hairs on midrib above and sometimes on margin;

lateral veins 7–14 pairs; petiole 23–40 mm long. Inflorescences terminal, laxly paniculate, 6–27 cm long, slender, branching largely dichasial; axes glandular-villose, hairs 1.5–3.5 mm long, gland tips minute, also eglandular-puberulous, hairs mainly patent, more dense upwards; main axis bracts narrowly elliptic to linear-lanceolate, 3–8 mm long, 0.5–1.5 mm wide in central portion of axis; secondary bracts linear-lanceolate 1–2 mm long; bracteoles triangular-lanceolate 0.5–1 mm long; pedicels to 1–4 mm long in fruit. Calyx lobes linear-lanceolate, 2.5–3.5 mm long in flower, to 3–5.5 mm in fruit, indumentum as axes. Immature corolla white with purple markings on palate,13–15 mm long, pilose outside; tube ± 7.5 mm long; upper lip ± 5 mm long; lower lip ± 5.5 mm long. Staminal filaments free for ± 3.5 mm; anther thecae strongly offset, separated by ± 2 mm, each 1–1.2 mm long, upper theca held almost patent to filament. Style glabrous. Capsule 12.5–15 mm long, with scattered long glandular hairs towards apex, glabrescent; seeds ± 2.2 mm wide, rugose-tuberculate, tubercles with minute hair-like processes.

TANZANIA. Iringa District: Ndunduru Forest Reserve, Sept. 2000, *Luke et al.* 6561! & Ndunduru, above camp 232, Sept. 2001, *Luke et al.* 8022! & Udzungwa Mts National Park, Nov. 2005, *Mwangoka & Festo* 4517!

DISTR. **T** 7; known only from the Udzungwa Mts

HAB. Montane forest; 1200–1850 m

USES. None recorded on herbarium specimens

CONSERVATION NOTES. Clarification of the taxonomic status is required prior to a conservation assessment: Data Deficient (DD).

NOTE. This is quite probably a white-flowered form of *I. laxa* but I have only seen immature corollas from which it is impossible to work out what their final shape and size will be. The range disjunction would be unusual but not improbable. Mature flowering material is required before any further conclusions can be made.

16. **Isoglossa bondwaensis** *I. Darbysh.* in K.B. 64: 414, fig. 3A–J (2009). Type: Tanzania, Morogoro District, Uluguru N Catchment Forest Reserve, *Jannerup & Mhoro* 181 (K!, holo.; C, iso.)

Perennial herb, 80–150 cm tall; stems retrorse-pubescent largely in two opposite lines, hairs with conspicuous cell walls, soon glabrescent. Mature leaves ovate, 9–11 cm long, 3.3–4.7 cm wide, base attenuate, apex long-acuminate, surfaces glabrous except for inconspicuous antrorse hairs on midrib above and margin; lateral veins 9–12 pairs; petiole 20–35 mm long. Inflorescences terminal on principal and short lateral branches, laxly paniculate, 4.5–18 cm long, minor branching sometimes monochasial; axes glandular-pilose, hairs 1–2 mm long, gland tips minute, also eglandular-puberulous, hairs patent to retrorse; bracts linear or lanceolate, 1.5–5 mm long in central portion of axis, basal pairs of main axis often foliaceous, narrowly ovate, to 12 mm long; bracteoles triangular-lanceolate, 1–2 mm long; flowers subsessile, pedicels extending to 2.5–5 mm long in fruit. Calyx lobes linear-lanceolate, 3.5–4.5 mm long in flower, to 6–8.5 mm in fruit, glandular-pilose and with minute eglandular hairs along margin and towards base. Corolla white with purple markings on palate of lower lip, 21–26 mm long, pilose outside; tube strongly inflated, 12–16 mm long, 7–9 mm in diameter at mouth, with a ring of pilose hairs at base within; upper lip broadly triangular, hooded, ± 7 mm long, lobes reflexed, ± 1.5 mm long; lower lip protruding, 9–10 mm long, lobes 2–3 mm long, palate glabrous, "herring-bone" patterning rather inconspicuous. Staminal filaments attached ± midway along corolla tube, free for 5–6 mm, glabrous; anther thecae offset, separated by 1–1.5 mm, oblique, upper theca held almost patent to filament, 1.5–1.7 mm long, lower theca 1.4–1.5 mm long. Style glabrous. Capsule 12–13 mm long, with scattered long glandular hairs towards apex; seeds 1.6–1.9 mm wide, rugose and with elongate tubercles towards rim, most conspicuous when immature.

TANZANIA. Morogoro District: Uluguru Mts, Bondwa Peak, Oct. 1969, *Pócs & Gibbon* 6051/J! & Uluguru N Catchment Forest Reserve, on path Tegetero–Luhongo, just W of ridge from Bondwa Peak to Nziwane, Jan. 2001, *Jannerup & Mhoro* 142! & E of path Tegetero–Luhungo, W of ridge from Bondwa Peak to Nziwane, Jan. 2001, *Jannerup & Mhoro* 181! (type)
DISTR. **T** 6; known only from the Uluguru Mts
HAB. Montane mist forest, including on rocks in streambeds and waterfalls; 1650–1950 m
USES. None recorded on herbarium specimens
CONSERVATION NOTES. Assessed as Critically Endangered (CR B1ab(iii) + B2ab(iii)) by Darbyshire (l.c.).

NOTE. Differs from the closely related *I. laxa* in having an inflated, not narrowly campanulate, corolla tube with a ring of hairs at the base within, not glabrous, and with the filaments attached further down the tube and longer.

17. **Isoglossa oreacanthoides** *Mildbr.* in N.B.G.B. 11: 1086 (1934); Darbyshire in K.B. 64: 406, 407 (2009). Type: Tanzania, Morogoro District, Uluguru Mts, Lupanga Mt, *Schlieben* 2983 (B†, holo.; BM!, BR!, LISC!, M!, iso.)

Perennial herb, 40–250 cm tall, woody towards base; stems retrorse-pubescent in two opposite lines, hairs with ± conspicuous cross-walls, soon glabrescent. Mature leaves drying blackish; blade ovate or ovate-lanceolate, 4.5–13 cm long, 1.5–4.5 cm wide, base obtuse to attenuate, apex long-acuminate, apiculate, surfaces largely glabrous except for antrorse hairs on midrib above and margin; lateral veins 6–10 pairs, conspicuous beneath; petiole 4–40 mm long. Inflorescences terminal, ± laxly paniculate, often pyramidal in outline, 8.5–20 cm long, lateral branches short, to 2(–6.5) cm long at base of inflorescence; axes densely glandular- and eglandular-pilose, the former spreading, evenly distributed, the latter spreading or retrorse and concentrated in two opposite lines; main axis bracts foliaceous, ovate to elliptic, 5.5–25 mm long, (1.5–)4–14 mm wide in central portion of axis, becoming smaller in size and with increasingly dense glandular hairs upwards; secondary bracts linear-lanceolate to narrowly ovate-elliptic, 6–9.5 mm long, glandular-pilose; bracteoles as bracts but linear-lanceolate, 4.5–7.5 mm long; flowers subsessile. Calyx lobes linear, 8–12 mm long in flower, to 14 mm in fruit, densely glandular-pilose outside, hairs to 1.5 mm long. Corolla white with purple-black markings on throat and palate of lower lip, 22–29 mm long, glabrous; tube 13–17 mm long, inflated from base, 5–10 mm in diameter; upper lip broadly triangular, hooded, 5–6.5 mm long, lobes reflexed, 1.5–2 mm long; lower lip 7–13.5 mm long, median lobe 3.5–6 mm long, lateral lobes 2.5–4 mm long, palate with "herring-bone" patterning inconspicuous. Staminal filaments attached ± midway along corolla tube, free for 4.5–6 mm, glabrous; anther thecae strongly offset, separated by ± 1 mm, oblique, upper theca almost patent to filament, 1.7–2 mm long, lower theca 1.5–1.8 mm long. Style glabrous. Capsule 11–11.5 mm long, glandular-pilose towards apex; only immature seeds seen, rugose.

TANZANIA. Morogoro District: NE Uluguru Mts, Magali, Aug. 1933, *Schlieben* 4222! & Kinole Ridge, Sept. 1970, *Harris et al.* 5103! & Uluguru N Forest Reserve, Bondwa Ridge, July 1972, *Mabberley* 1161!
DISTR. **T** 6; known only from the Uluguru Mts
HAB. Submontane forest, montane mist forest and elfin forest; (1000–)2100–2200 m
USES. None recorded on herbarium specimens
CONSERVATION NOTES. Restricted to the northern Uluguru Mts where it appears scarce, only five herbarium collections having been seen. It is usually found at high altitudes where human activity is likely to be minimal. However, the lower altitude populations are likely to be threatened by the widespread deforestation which continues in this mountain range: Endangered (EN B1ab(iii)+2ab(iii)).

18. **Isoglossa multinervis** *I. Darbysh.* in K.B. 64: 422, fig. 6 (2009). Type: Tanzania, Rungwe District, Ngozi Crater, *Kayombo* 1105 (MO!, holo.; K!, iso.)

Shrub, 200–400 cm tall; stems glabrous. Mature leaves ovate, (10.5–)17–26 cm long, (7–)10.5–13.5 cm wide, base truncate or rounded, with a 1–2.5 cm long cuneate extension along upper petiole, or more evenly attenuate, apex acuminate to caudate; surfaces glabrous except for inconspicuous antrorse hairs on principal veins above; lateral veins 17–25 pairs, these and the scalariform tertiary venation conspicuous; petiole 45–110 mm long, glabrous. Inflorescences terminal, 1–3 developing from the apex of the branch, each spiciform or with 1–3 pairs of simple branches, main axis 9–28 cm long, cymules sessile, densely clustered towards branch apex, more widely spaced towards base; axes densely glandular-villose, hairs yellow, to 2–3.5 mm long, with a shorter secondary indumentum; bracts and bracteoles minute and obscured by the indumentum, triangular to lanceolate, 1–1.5 mm long; flowers subsessile, pedicels sometimes extending to 1.5 mm long in fruit. Calyx lobes lanceolate, 3.5–6.5 mm long in flower, barely extending in fruit, densely glandular-villose. Corolla white or yellow, 13.5–18 mm long, with deep red markings on palate, glabrous; tube strongly inflated from base, 6.5–8.5 mm long, 5–8 mm in diameter at mouth; upper lip broadly triangular, hooded, 6–8.5 mm long, lobes 0.8–1.5 mm long; lower lip 7–9 mm long, lobes 1.5–2 mm long, palate with conspicuous "herring-bone" patterning. Staminal filaments attached in upper half of the corolla tube, free for 4–6.5 mm, glabrous; anther thecae widely separated, somewhat oblique, upper theca 2.4–3 mm long, lower theca 1.9–2.2 mm long. Style glabrous. Capsule 29–38 mm long, glabrous; seeds 5–6 mm wide, rugose when immature, rather smooth with maturity except along rim. Fig. 80/8, p. 614.

TANZANIA. Mbeya District: Umalila, Sept. 1975, *Leedal* 2798! & Mbalizi to Jojo road, June 1983, *Mhoro & Mashala* 3347!; Rungwe District: slopes of Ngozi, June 1975, *Aleljung* 536!
DISTR. **T** 7; N Malawi
HAB. Montane forest; 1830–2500 m
USES. None recorded on herbarium specimens
CONSERVATION NOTES. Assessed as Endangered (EN B2ab(iii + iv)) by Darbyshire (l.c.).

19. **Isoglossa humbertii** *Mildbr.* in B.J.B.B. 14: 358 (1937). Type: Congo-Kinshasa, W of Lake Kivu, between Tshibinda and Kanziba, *Humbert* 7445 (B†, holo.; BR!, P!, iso.)

Straggling perennial herb, 100–200 cm tall; stems appressed-pubescent or crisped-pilose when young, soon glabrescent. Mature leaves elliptic (-obovate) to narrowly so, 8–14.5 cm long, 2–6.5 cm wide, base asymmetric, long-cuneate or attenuate, apex acuminate, upper surface, margin and veins beneath pubescent to pilose; lateral veins 12–16 pairs, conspicuous beneath; petiole 5–30 mm long. Inflorescences terminal on principal and short lateral branches, laxly paniculate, main axis 7.5–14 cm long, branching at least partially monochasial; axes largely glabrous, basal portion of primary peduncle sometimes sparsely pubescent or pilose; bracts (ovate-) lanceolate, 2.5–5 mm long, 0.7–2.2 mm wide, glabrous; bracteoles as bracts but 2–4.5 mm long, pairs cupping the immature calyx; flowers sessile or on pedicels to 2 mm long in flower, to 4.5 mm in fruit. Calyx lobes linear-lanceolate, 8–11.5 mm long in flower, to at least 12 mm long in fruit, glabrous. Corolla magenta or purple, (23–)28–33 mm long, glabrous outside, pilose within towards base; tube (14–)18–20 mm long, cylindrical in lower half where 2–2.5 mm in diameter, somewhat constricted above ovary, upper half expanded, 4–7.5 mm in diameter at mouth; upper lip hooded, 9.5–12 mm long, lobes minute; lower lip 9–13.5 mm long, lobes minute, palate with sparse pale strigulose hairs, "herring-bone" patterning largely undeveloped. Staminal filaments attached ± midway along corolla tube, free for ± 10–15 mm, glabrous; anther thecae strongly offset, partially overlapping, parallel, upper theca 3.5–4.2 mm long, lower theca 2.8–3.2 mm long. Style glabrous. Capsule 23–26 mm long, glabrous; seeds not seen.

UGANDA. Toro District: Nyamagasani Valley, near Musalombe Forest, Aug. 1952, *Ross* 918! & Ruwenzori, Butahu valley, July 1932, *Humphreys* 1362!

DISTR. **U** 2; E Congo-Kinshasa
HAB. Upper forest margins, bamboo forest including streamsides; 2250–2750 m
USES. None recorded on herbarium specimen
CONSERVATION NOTES. Recorded from a highly restricted range in the Congo-Uganda borderlands and known from rather few herbarium collections. Its high montane habitat is however unlikely to receive significant disturbance and it appears tolerant of some natural disturbance and forest fragmentation, being recorded from marginal habitats and from more open bamboo forest: Least Concern (LC).

NOTE. *Ross* 918 has the largest and broadest leaves seen for this species, the Congo material seen having leaves to only 11.5 × 3 cm. It is also atypical in having long spreading, crisped hairs on the stems and petioles, these being shorter and appressed on the Congo material. These vegetative characters in fact agree more closely with the related *I. laxiflora* Lindau of Rwanda and E Congo but that species has cream, pale yellow or greenish corollas with a shorter, more broadly campanulate tube (11–13.5 mm long and 6–11 mm in diameter at the mouth in material at K). Dominique Champluvier (*pers. comm.*) also records *I. humbertii* as having less distant thecae on each anther (1 mm versus 2–4 mm); the Ross gathering more closely agrees with *I. humbertii* in this respect. The two species are therefore maintained at present based upon these floral differences but may prove worthy of only subspecific separation following further study.

20. **Isoglossa substrobilina** *C.B. Clarke* in F.T.A. 5: 232 (1900); U.K.W.F. ed. 1: 600 (fig.), 601 (1974); U.K.W.F. ed. 2: 278, pl. 122 (1994); Darbyshire in K.B. 64: 418 (2009). Type: Kenya, Kericho/Masai District: Mau, *Scott Elliot* 6769 (K!, holo.; BM!, iso.)

Trailing or straggling perennial herb or shrub, 15–300 cm tall; stems retrorse- or appressed-pubescent in two opposite lines. Leaves sometimes absent at flowering; blade ovate to elliptic, 1.5–14 cm long, 1–5.2 cm wide, base attenuate, apex attenuate to acuminate, margin, midrib and often upper surface and veins beneath pubescent; lateral veins 3–7 pairs; petiole 4–32 mm long. Inflorescences terminal on principal and short lateral branches, spiciform or rarely with up to 3 pairs of branches, 2–10 cm long, with decussately arranged imbricate bracts, each supporting 1(–3) flowers, sometimes unilateral particularly when young, axis then curved; axis with two opposite lines of dense multicellular hairs, with or without short patent glandular hairs towards apex; bracts broadly obovate, elliptic or suborbicular, 3–11(–15) mm long, 2–8 mm wide, base attenuate, apex subattenuate to acuminate, margin ciliate, surfaces eglandular-pubescent particularly within or largely glabrous, upper bracts often with short patent glandular hairs outside; bracteoles lanceolate or narrowly elliptic, 2.5–8.5 mm long, indumentum as bracts but glandular hairs often more numerous; flowers sessile. Calyx lobes lanceolate, 3–6.5 mm long in flower, to 5–10.5 mm in fruit, with patent glandular hairs outside, shortly ciliate. Corolla white or pale mauve, 9–12.5 mm long, palate with pink to purple markings, sparsely puberulous on limb outside; tube subcylindrical, 3.5–5.5 mm long, 0.8–2.5 mm in diameter at base, 1.8–3 mm at mouth, with an interrupted ring of short hairs 1.5–3 mm from base within; upper lip triangular, hooded, 5.5–8.5 mm long, lobes 0.3–1.5 mm long; lower lip 4.5–8 mm long, lobes 0.5–1.5 mm long, palate minutely papillose, upraised and with prominent "herring-bone" patterning. Staminal filaments attached in upper half of corolla tube, free for 3.5–6 mm, glabrous; anther thecae widely separated, ± strongly oblique, one or both held almost patent to filament, upper theca 1–1.5 mm long, lower theca 0.8–1.3 mm long. Style glabrous. Capsule (6.5–)8–12 mm long, glabrous; seeds 1.7–2.7 mm wide, rugose, inconspicuously so when mature.

a. subsp. **substrobilina**; Darbyshire in K.B. 64: 420 (2009)

Leaves with upper surface and veins beneath at least sparsely pubescent. Inflorescence with bracts often remaining imbricate at maturity; blade broadly obovate to elliptic, apex (sub)attenuate to broadly and shortly acuminate, margin often densely ciliate, surfaces eglandular-pubescent particularly within, upper bracts often also with glandular hairs outside, venation ± conspicuous. Bracteoles with apex acute. Fig. 80/10, p. 614.

UGANDA. Mbale District: Sebei, Kapkwata, Dec. 1969, *Hamilton* 1380! & ridge top, 5 km W of Suam, Apr. 1993, *van Heist & Sheil* 1963!
KENYA. Trans-Nzoia District: Suam Sawmills, May 1969, *Napper & Tweedie* 2128!; Kiambu District: Muguga, Sept. 1957, *Verdcourt* 1840!; Masai District: Narosura, Aug. 1961, *Glover et al.* 2504!
TANZANIA. Arusha District: Meru Mt, Jekukumia, Mar. 1971, *Richards & Arasululu* 26696!; Moshi District: Kilimanjaro, May 1994, *Grimshaw* 94515!; Iringa District: Udzungwa Mts National Park, Karenga Peak, Nov. 2005, *Luke, Mwangoka & Festo* 11392!
DISTR. **U** 3; **K** 3–7; **T** 2, 5–7; not known elsewhere
HAB. Montane forest undergrowth, particularly in glades, margins and disturbed areas; (1000–)1750–2600 m
USES. "grazed by all domestic stock" (**K** 6; *Glover et al.* 2199 & 2504)
CONSERVATION NOTES. Assessed as of Least Concern (LC) by Darbyshire (l.c.).

SYN. *I. schliebenii* Mildbr. in N.B.G.B. 12: 98 (1934). Type: Tanzania, Morogoro District, Nguru Mts, Kombola, *Schlieben* 4102 (B†, holo.; BR!, BM!, LISC!, M!, iso.)
Plagiotheca fallax Chiov., Racc. Bot. Miss. Consol. Kenya: 96 (1935). Types: Kenya, Nyeri, Mt Kenya, *Balbo* 601 (TOM, syn.) & *idem, Balbo* 828 (TOM, K!, syn.)

b. subsp. **tenuispicata** *I. Darbysh.* in K.B. 64: 421, fig. 5A–J (2009). Type: Tanzania, Lushoto District, 1.6 km N of Lushoto, *Leach & Brunton* 10196 (K!, holo.; BR!, SRGH, iso.)

Leaves with upper surface and veins beneath sparsely pubescent or largely glabrous. Inflorescence with bracts at first imbricate but often becoming clearly spaced at maturity; blade suborbicular, apex with a conspicuous narrow acumen, margin ciliate or inconspicuously so, surfaces largely glabrous or with scattered eglandular hairs, upper bracts sometimes with additional glandular hairs outside, venation inconspicuous. Bracteoles with apex (sub)acuminate.

TANZANIA. Lushoto District: Usambara Mts, Shume Forest Reserve, July 1924, *Maber* 147! & Baga–Bumbuli road, 2 km NE of Sakarani, May 1953, *Drummond & Hemsley* 2416!; Morogoro District: Uluguru Mts, Morningside, Dec. 1987, *Mhoro* 4517!
DISTR. **T** 3, 6; known only from the Usambara and Uluguru Mts
HAB. Moist (sub-)montane forest understorey, dense undergrowth or grassland at forest margins; 1350–1950 m
USES. None recorded on herbarium specimens
CONSERVATION NOTES. Assessed as Vulnerable (VU B2ab(iii)) by Darbyshire (l.c.).

21. **Isoglossa bracteosa** *Mildbr.* in N.B.G.B. 11: 1086 (1934). Type: Tanzania, Morogoro District, Uluguru Mts, Kinole, *Schlieben* 2881 (B†, holo.; BM!, BR!, G!, iso.)

Perennial herb or shrub, 100–200 cm tall; stems retrorse- or appressed-pubescent in two opposite lines or sometimes more widespread on young stems, hairs with conspicuous, brown (-orange) cell walls, glabrescent. Leaf blade elliptic or somewhat obovate, 14.5–22 cm long, (3.5–)5–8.5 cm wide, base long-attenuate or cuneate, apex acuminate; surfaces pubescent or largely glabrous; lateral veins 7–13 pairs; petiole 18–53 mm long. Inflorescences terminal on principal and short lateral branches, spiciform or with a single branch towards base, main axis 2–12.5 cm long, with decussately arranged bracts, imbricate when young, becoming more distant towards base of spike at maturity, each supporting a single flower; axes villose in two opposite lines, sometimes also with patent glandular hairs towards apex; bracts (for colour see Note) broadly ovate to elliptic, 5–10 mm long, 4–7 mm wide, base attenuate, apex attenuate or acuminate, upper margins reflexed at least when young, ± densely pilose along margins and on inner surface towards apex, outer surface largely glabrous or with ± dense glandular hairs; bracteoles lanceolate or narrowly elliptic, 4–7.5 mm long; flowers sessile. Calyx lobes lanceolate, 4.5–7.5 mm long in flower, to 7–9 mm in fruit, shortly ciliate and with patent glandular hairs outside. Corolla white with pink or brown markings on palate, 10.5–15 mm long, sparsely puberulous towards apex of lips outside; tube subcylindrical, 4–5 mm long, ± 1.5 mm in diameter at base, 2–2.5 mm at mouth, with an interrupted ring of long hairs 2–3 mm from base within or these shorter and restricted to the floor; upper lip narrowly triangular, hooded, 7.5–10 mm long, lobes 0.5–1.5 mm long; lower lip 7–10 mm long, lobes 0.5–1.2 mm long, palate minutely papillose, upraised and with prominent "herring-bone" patterning. Staminal filaments attached in upper half of corolla tube, free for 3.5–5.5 mm, glabrous; anther thecae widely separated, ± strongly oblique, one or both held almost patent to filament, upper theca 1–1.7 mm long, lower theca 1–1.5 mm long. Style glabrous. Capsule ± 11 mm long, glabrous; only immature seeds seen, rugose.

TANZANIA. Iringa District, ridge above Sanje Falls, July 1983, *Polhill, J.C. & J.M. Lovett* 5146!; Rungwe District: Rungwe Mt, ± 4 km NNE of Rungwe Peak and 5 km SSE of Kiwira Forest Station, June 1992, *Mwasumbi* 16236!; Njombe District: Livingstone Mts, Bumbigi to Kitulo, Feb. 1991, *Gereau & Kayombo* 4108!

DISTR. **T** 6, 7; N Malawi, NE Zambia

HAB. Montane and submontane forest including areas with abundant bamboo thicket; 1050–2500 m

USES. None recorded on herbarium specimens

CONSERVATION NOTES. This species appears to be scarce, currently being known from four sites in Tanzania, one in the Misuku Hills of N Malawi and two on the Nyika Plateau. Several of these sites have only recently been discovered and are rather isolated from one another. Whilst its high altitude populations may be undisturbed at present, lower altitude sites are likely to be threatened by human encroachment. It is therefore considered Vulnerable (VU B2ab(iii)).

SYN. *I. imbricata* Brummitt in K.B. 40: 790 (1985). Type: Tanzania, Iringa District, ridge above Sanje Falls, *Polhill, J.C. Lovett & J.M. Lovett* 5150 (K!, holo.; DSM!, EA!, MO!, iso.) **syn. nov.**

[*I.* aff. *substrobilina sensu* Dowsett-Lemaire in B.J.B.B. 55: 317, 378 (1985); *I. substrobilina sensu* Phiri, Checkl. Zambian Vasc. Pl.: 19 (2005); *I. substrobilina* or aff. *sensu* Winter in Burrows & Willis, Pl. Nyika Plateau: 51 (2005), *non* C.B. Clarke]

NOTE. The types of *I. bracteosa* and *I. imbricata* differ in the former having more elongate, less strobilate inflorescences, and in the bracts having a less strongly reflexed apex and being densely glandular-pubescent on the outer surface away from the margin, not largely glabrous. The first two differences are likely a product of maturity. *Mwasumbi* 16236 is found to be intermediate between the two with regard to bract indumentum, having glandular hairs on the bracts in the upper portion of the inflorescence whilst the lower bracts are largely glabrous outside. *I. bracteosa* and *I. imbricata* are therefore considered conspecific.

The bracts of this species are apparently discolorous, being green within and yellow-green with darker venation outside, therefore the reflexed tips contrast with the lower portion of each bract when young. However, the bracts (and inflorescence as a whole) often dry blackish.

22. **Isoglossa candelabrum** *Lindau* in E.J. 30: 113 (1901); Iversen in Symb. Bot. Upsal. 28: 242 (1988). Type: Tanzania, Lushoto District, Usambara Mts, Ngwelo, *Scheffler* 52 (B†, holo.; BM!, iso.)

Straggling shrub, 120–350(–500) cm tall; stems glabrous. Leaf blade coriaceous, oblong-elliptic or somewhat obovate, 9.5–24 cm long, 2.5–10 cm wide, base attenuate, apex shortly acuminate, glabrous; lateral veins (5–)6–8(–9) pairs; petiole 0.5–2.5(–6) cm long. Inflorescences axillary, solitary or clustered, often on the old, bare or leafy branches, spiciform or several-branched, main axis 0.4–5.5 cm long, cymules subsessile, shortly spaced; axes largely glabrous; bracts triangular to lanceolate, 1.5–4.5 mm long, shortly ciliate; bracteoles as bracts but to 2 mm long; pedicels (1.5–)3–6.5 mm long, with short, broad-based ascending hairs. Calyx with distinct tube 1–3.5(–4.5) mm long, lobes (oblong-) lanceolate, 3.5–6.5 mm long in flower, to 6–7.5 mm in fruit, largely glabrous outside, puberulous within particularly towards margin and apex. Corolla bright pink or red, 25–35 mm long, outer surface unevenly pilose, glabrous within; tube curved, 19–26 mm long, 1.5–3.5 mm in diameter at base, constricted above ovary then gradually expanded to 4.5–7.5 mm at mouth; upper lip broadly ovate, hooded, 7–10 mm long, lobes minute; lower lip 7–10.5 mm long, lobes 0.7–1.2 mm long, "herring-bone" patterning inconspicuous. Staminal filaments attached in upper half of corolla tube, free for 10–14.5 mm, glabrous; anther thecae offset, both held almost patent to filament, upper theca 1.75–2.2 mm long, lower theca 1.4–1.8 mm long. Style glabrous. Capsule 14–17 mm long, glabrous; seeds 2.3–3.3 mm wide, rugose. Fig. 80/9, p. 614.

KENYA. Teita District: Kasigau Hill, 64 km S of Voi, W slopes, June 1968, *Archer* 570! & Kasigau, Nov. 1994, *Luke* 4137!

TANZANIA. Lushoto District: Kwamkuyo/Dodwe Falls, Amani, Dec. 1956, *Verdcourt* 1745!; Tanga District: 6 miles from Amani on road to Marvera above International Business Combine Saw Mill, July 1969, *Magogo* 1265! & E Usambara, Mlinga Peak area, below the S summit, Nov. 1986, *Borhidi et al.* 86490!

DISTR. **K** 7; **T** 3; not known elsewhere

HAB. Moist lowland and mid-altitude forest; (250–)700–1300 m

USES. None recorded on herbarium specimens, but cultivated at Amani and of high horticultural potential

CONSERVATION NOTES. Most of the collections of this species are from the E Usambaras, where *Greenway* (7020) recorded it as "very locally dominant in small rare patches". Its more recent discovery on the Kasigau inselberg in SE Kenya extends its known range significantly, but it is clearly rare there, never having been recorded from the nearby Taita Hills. Lowland and mid-altitude forest has been severely depleted in the Usambaras thus it is highly likely that several former sites have been lost. With over 10 sites known historically, this species is currently assessed as Near Threatened (NT) but may well prove to be Vulnerable following an assessment of the status of the historic sites.

23. **Isoglossa asystasioides** *I. Darbysh. & Ensermu* in K.B. 62: 618, fig. 1 (2007). Type: Tanzania, Morogoro District, Kimboza Forest Reserve, ± 48–50 km Morogoro–Matombo road, *Rodgers, Hall & Mwasumbi* 2512 (K!, holo.; DSM, iso.)

Shrub, 50–120 cm tall; stem surface often longitudinally striate, largely glabrous or with inconspicuous antrorse hairs when young. Leaves subcoriaceous, oblong-elliptic, 10–23 cm long, 3.2–7.7 cm wide, base acute, cuneate or attenuate, apex shortly acuminate or acute, glabrous; lateral veins (5–)6–7 pairs; petiole (3–)8–23 mm long. Inflorescences terminal, spiciform, sometimes with additional spikes in upper leaf axils, main axis 1.5–5.5 cm long, cymules subsessile, ± dense, shortly-spaced; axes largely glabrous or with scattered pale antrorse hairs; bracts triangular to lanceolate, 1.5–7 mm long, becoming shorter towards apex of spike where imbricate, shortly ciliate; bracteoles as bracts but triangular, 1–3 mm long; flowers subsessile or pedicels 0.5–2 mm long. Calyx lobes lanceolate, 2–7 mm long in flower, extending somewhat in fruit, shortly ciliate and with scattered short hairs and subsessile glands outside, puberulous within towards margins and apex. Corolla white, 21–35 mm long, outer surface unevenly pilose, glabrous within; tube 10.5–19 mm long, 1.5–3.5 mm in diameter at base, centrally waisted then abruptly expanded laterally, somewhat dorso-ventrally compressed, to 5.5–10 mm wide at mouth; upper lip broadly ovate, hooded, 10.5–14.5 mm long, lobes ± 1 mm long; lower lip 10–16 mm long, lobes ± 1.5 mm long, palate with "herring-bone" patterning. Staminal filaments attached in upper half of corolla tube, free for 7.5–13.5 mm, glabrous; anther thecae offset, highly oblique or both held almost patent to filament, upper theca 2–2.5 mm long, lower theca 1.5–2 mm long. Style glabrous. Capsule (8.5–)16–19 mm long, glabrous; seeds 2.7–3.2 mm wide, rugose.

TANZANIA. Morogoro District: Uluguru Mts, Kimboza Forest Reserve, ± 48–50 km on Morogoro to Matombo road, Apr. 1983, *Mwasumbi, Rodgers & Hall* 12438! & 40 km from Morogoro on Kisaki road, Sept. 1969, *Harris & Pócs* 3212! & Kimboza Forest Reserve, Sept. 2001, *Luke et al.* 7631!

DISTR. **T** 6; known only from the Uluguru Mts

HAB. Lowland rainforest and swamp forest, particularly associated with exposed karstic marble; 200–400 m

USES. None recorded on herbarium specimens

CONSERVATION NOTES. Assessed as Endangered (EN B1ab(iii) B2ab(iii)) by Darbyshire & Ensermu (l.c.).

NOTE. When first published, only malformed, single-seeded fruits had been seen for this species, these being 8.5–11.5 mm long. Mature, 4-seeded capsules have since been seen at EA (*Luke et al.* 7631; *Pócs et al.* 6801/C) resulting in the measurements recorded here.

24. **Isoglossa ixodes** *Lindau* in E.J. 22: 124 (1901); C.B. Clarke in F.T.A. 5: 234 (1900). Type: Tanzania, Morogoro District, Uluguru Mts, near Lukwangule, *Stuhlmann* 9174 (B†, holo.)

Straggling or scandent perennial herb or subshrub, 150–300 cm tall, laxly branched; stems with two opposite lines of sparse, mainly retrorse hairs when young, soon glabrescent. Mature leaves ovate, 3–10 cm long, 1.5–4.8 cm wide, base ± asymmetric, attenuate, apex acuminate or caudate, upper surface with hairs on midrib and margin particularly towards base, lower surface sparsely pubescent; lateral veins 5–7 pairs; petiole 6–35 mm long. Inflorescences terminal, solitary or 2–3 terminating the branches, spiciform or with few short branches towards base, main axis (1.5–) 3–11.5 cm long, often curved, cymules subsessile, 1–several-flowered, often unilateral; axes shortly eglandular-pubescent and with many long pale crisped hairs with inconspicuous gland tips; main axis bracts foliaceous, ovate or lanceolate, 5–14.5 mm long, 1–7 mm wide in central portion of axis, gradually becoming smaller upwards, apex (sub)attenuate, surface glabrous or with long glandular hairs, ciliate; secondary bracts and bracteoles (linear-) lanceolate, 4–8 mm long, the latter often clasping the pedicel towards base; pedicels 0.8–2.5 mm long. Calyx lobes often tinged purple, linear-lanceolate, 5–8 mm long in flower, to 9.5–11 mm in fruit, with crisped glandular hairs outside towards apex, shortly ciliate, each lobe with 3 conspicuous parallel veins. Corolla red-purple or scarlet, 25–30 mm long, outer surface glabrous; tube 17–21 mm long, 2–2.5 mm in diameter at base, gradually expanded to 5.5–7.5 mm at mouth, dorsally declinate, ventral side abruptly bent above base then ± straight, pubescent towards base within; upper lip triangular, hooded, 7.5–9.5 mm long, lobes 0.7–1.2 mm long; lower lip 7.5–10 mm long, lobes 1.5–3 mm long, palate glabrous, "herring-bone" patterning largely undeveloped. Staminal filaments attached in upper half of corolla tube, free for 9.5–15 mm, glabrous; anther thecae partially offset, slightly oblique, upper theca 2.1–2.6 mm long, lower theca 1.75–2.4 mm long. Style glabrous. Capsule 8.5–11 mm long, apex with sparse long glandular hairs; only immature seeds seen, rugose.

TANZANIA. Morogoro District: Uluguru Mts, [Kienzema] Chenzema-Lukwangule Plateau, Aug. 1951, *Greenway & Eggeling* 8679! & Lukwangule Plateau, above Chenzema Mission, Mar. 1953, *Drummond & Hemsley* 1524! & above Chenzema towards Lukwangule Plateau, Jan. 1975, *Polhill & Wingfield* 4659!

DISTR. **T** 6; known only from the Uluguru Mts

HAB. Montane forest, including in clearings; 2200–2250 m

USES. None recorded on herbarium specimens

CONSERVATION NOTES. Apparently restricted to the Lukwangule Plateau and adjacent areas of the Uluguru Mts, currently known from only six specimens. It is probably locally common, *Drummond & Hemsley* recording it as "one of the most common constituents" of the forest field layer and *Greenway & Eggeling* recording it as "very locally dominant". As it is recorded at such high altitudes, its habitat may receive little human disturbance. However, with such a restricted distribution, this species is threatened by local stochastic events such as forest fires or by future human disturbance, and so is assessed as Vulnerable (VU D2).

25. **Isoglossa rubescens** *Lindau* in E.J. 20: 57 (1894); C.B. Clarke in F.T.A. 5: 233 (1900). Syntypes: Congo-Kinshasa, Ruwenzori, *Stuhlmann* 2397 & 2424a (both B†, syn.)

Erect, straggling or decumbent perennial herb or subshrub, 40–200 cm tall, sometimes rooting at lower nodes. Young stems glandular- and eglandular-pilose, gland tips reddish, older stems glabrescent. Mature leaves ovate, 1.7–8.5 cm long, 0.8–4.5 cm wide, base attenuate, apex acuminate, surfaces pilose, hairs most dense on upper surface and margin, sparse on veins beneath; lateral veins 6–10 pairs; petiole 3.5–14 mm long. Inflorescence a narrow, often large terminal panicle (2–)5–35 cm long, interrupted by reduced, often lanceolate leaves, lateral branches ascending, 1.5–15 cm long, with or without secondary branching, flowers clustered at apices of branches; axis indumentum as uppermost stems, dense; main axis bracts foliaceous, narrowly elliptic to lanceolate, 5–11 mm long; secondary bracts linear-lanceolate, 3–6 mm long; bracteoles as secondary bracts but 2.5–5 mm long, often obscured by the dense indumentum; flowers sessile or on pedicels to 3 mm long. Calyx lobes linear, (4–)5–8 mm long in flower, extending to 11.5–18 mm in fruit, densely glandular-pilose outside, shortly eglandular-ciliate. Corolla white, sometimes tinged pink outside, 19–33 mm long, with or without mauve markings on palate of lower lip, pilose outside, pubescent towards base of tube and on limb within; tube 13.5–20 mm long, slender, 1.8–3(–4) mm in diameter at base, gradually expanded to 4.5–6.5(–8) mm at mouth; upper lip triangular, upcurved, 5–9.5 mm long, lobes 2–2.5 mm long; lower lip 5.5–12.5 mm long, deeply lobed, these 3–7.5 mm long, palate flat and largely lacking "herring-bone" patterning. Staminal filaments attached 3–4 mm below corolla mouth, free for 2–5 mm, upcurved, glabrous; anther thecae barely offset, parallel or slightly oblique, each 1.5–2.2 mm long. Style with sparse hairs towards base. Capsule 13.5–16 mm long, sparsely glandular-pilose towards apex; seeds 1.7–2 mm wide, rugose.

UGANDA. Ruwenzori, near Nyamuleju, Bujuku valley, Aug. 1933, *Eggeling* 1266! & Mubuku valley, Apr. 1948, *Adamson* 44! & near Nyamuleju Hut, Dec. 1968, *Lye* 1245!

DISTR. U 2; Ruwenzori of Congo-Kinshasa

HAB. High montane ericaceous belt and open areas in upper forest margins, bamboo forest, waterfall margins; 2250–3350 m

USES. None recorded on herbarium specimens

CONSERVATION NOTES. *I. rubescens* has a highly restricted distribution, being largely confined to the ericaceous belt of the high Ruwenzori. It may, however, be locally common here (*Noble* 3) or even locally dominant (*Synge* 1412), possibly as a periodically mass-flowering species. There are no known threats from human activity at this altitude. It is therefore not considered threatened at present: Least Concern (LC).

26. **Isoglossa floribunda** *C.B. Clarke* in F.T.A. 5: 233 (1900); Darbyshire in K.B. 64: 424 (2009). Type: Mozambique, Chupanga, lower Zambesi, *Kirk* s.n. (K!, holo.)

Erect or decumbent perennial herb or subshrub, 35–200 cm tall. Uppermost internodes eglandular- and glandular-pubescent, with or without occasional longer eglandular hairs, soon glabrescent. Mature leaves (ovate-) elliptic, 4.8–15 cm long, 1.7–7 cm wide, base attenuate, apex acuminate, surfaces largely glabrous except for inconspicuous hairs on principal veins; lateral veins 5–7 pairs; petiole 7–25 mm long. Inflorescences terminal on principal and short lateral branches, compounded into large pseudopanicles to 45 cm long interrupted by ± reduced leaves; partial inflorescences unilateral thyrses, unbranched to several-branched, main axis 3–19 cm long, often curved, cymules densely clustered or more distant, sometimes strobilate; axes patent glandular- and eglandular-pubescent with scattered long patent eglandular hairs to 3 mm long; cymule bracts lanceolate to (oblong-) obovate, 1.5–9 mm long, 1–6 mm wide, apex acute, acuminate or rounded-apiculate, outer surface glandular-pubescent and eglandular-puberulent, with or without long patent eglandular hairs particularly on margins, bracts subtending the branches often foliaceous, ovate; bracteoles lanceolate to oblanceolate, 2–10.5 mm long, 0.5–3.5 mm wide, otherwise as bracts; flowers

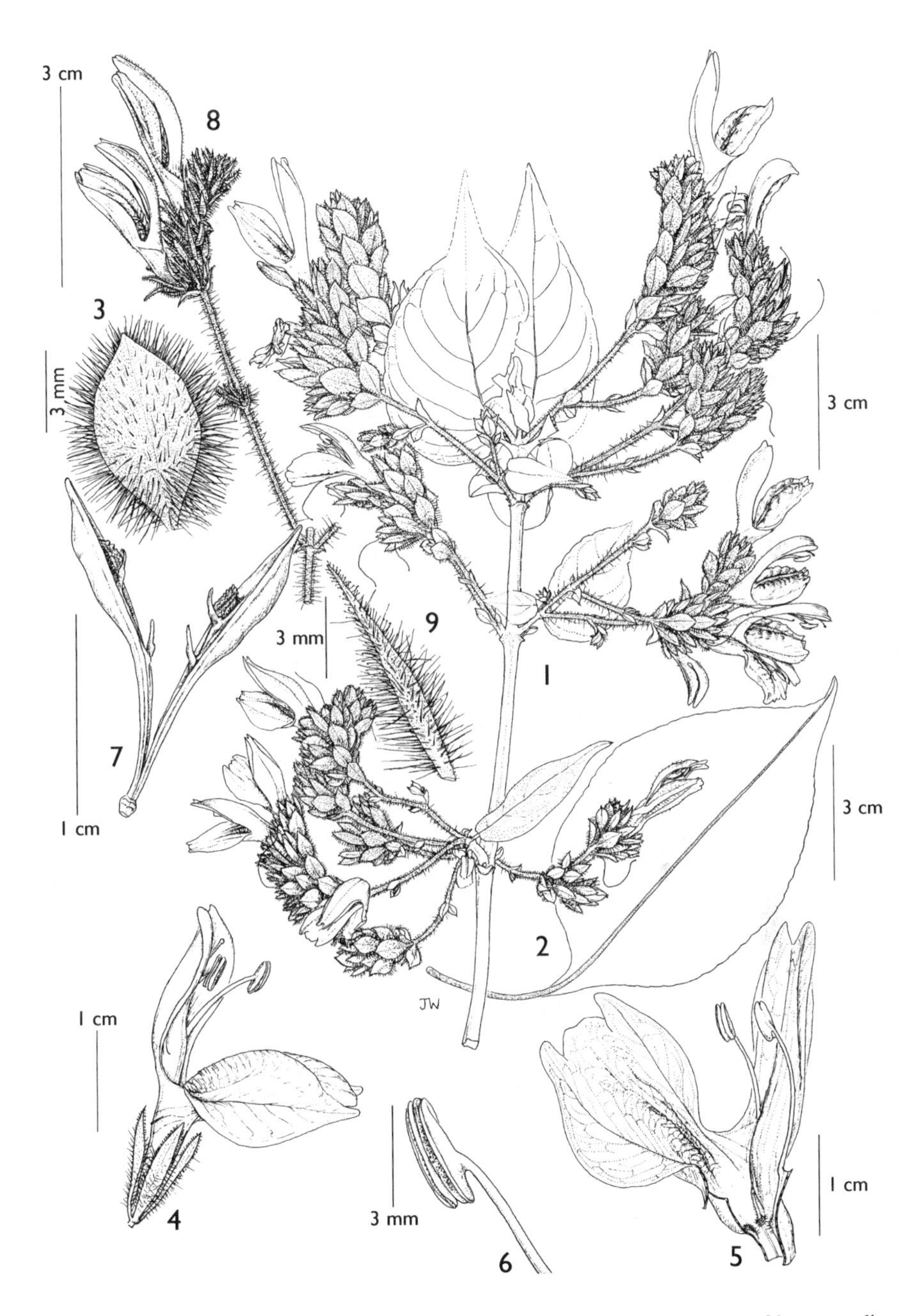

FIG. 82. *ISOGLOSSA FLORIBUNDA* subsp. *SALVIIFLORA* — **1**, habit; **2**, outline of larger cauline leaf; **3**, cymule bract; **4**, corolla within calyx; **5**, dissected corolla showing attachment of stamens; **6**, detail of anther; **7**, capsule with immature seed; subsp. *FLORIBUNDA* — **8**, inflorescence, variant with congested flowers for comparison with subsp. *salviiflora*,; **9** calyx lobe. 1 and 3 from *Mwangulango & Leliyo* 364; 2, 5 & 6 from *de Nevers & Charnley* 3300; 4 from *de Nevers & Charnley* 3402; 7 from *Luke et al.* 7660; 8 & 9 from *Vollesen* in MRC 2323. Drawn by Juliet Williamson. Reproduced from Kew Bulletin 64: 425 (2009) with permission of the Trustees of the Royal Botanic Gardens, Kew.

subsessile. Calyx lobes linear-lanceolate to narrowly oblanceolate, 3–10.5 mm long in flower, extending to 4.5–14 mm in fruit, indumentum as bracts. Corolla (blue-) purple, palate often paler and with or without darker mauve-blue lines, the whole flower rarely white, 19–34 mm long, outer surface sparsely pubescent, sometimes restricted to upper lip; tube 8.5–15 mm long, 2–2.5 mm in diameter at base, slightly upturned centrally then expanded to 3.5–5.5 mm in diameter at mouth, with a raised and interrupted ring of hairs 3–4 mm from base within; upper lip arcuate, hooded, 13.5–23 mm long, lobes minute; lower lip broad, 10.5–19 mm long, lobes 1–3 mm long, palate upraised and with prominent "herring-bone" patterning. Staminal filaments attached 3.5–5 mm below corolla mouth, free for 11–17 mm, arcuate, glabrous; anther thecae parallel, barely offset, upper theca 2.5–5 mm long, lower theca 2.3–4.5 mm long. Style glabrous. Capsule 11.5–17 mm long, glabrous; seeds 1.8–2.7 mm wide, rugose.

a. subsp. **floribunda**, Darbyshire in K.B. 64: 424, fig. 7H & J (2009)

Inflorescence rather lax with clearly spaced cymules at least towards the base, or more dense and clustered throughout; cymule bracts lanceolate to oblanceolate, 1.5–6 mm long, 1–2 mm wide. Calyx lobes 3–10 mm long in flower. Corolla 19–29 mm long. Fig. 82/8 & 9, p. 644.

TANZANIA. Iringa District: Idodi, June–July 1936, *Ward* I30!; Kilwa District: Kingupira, May 1975, *Vollesen* in MRC 2323!; Lindi District: Tendaguru, July 1929, *Migeod* 569!
DISTR. **T** 6–8; Zambia, Malawi, Mozambique, Zimbabwe
HAB. Riverine thicket, wooded grassland; 100–1000 m
USES. None recorded on herbarium specimens
CONSERVATION NOTES. Assessed as Data Deficient (DD) by Darbyshire (l.c.).

SYN. [*Schliebenia salviiflora sensu* Vollesen in Opera Bot. 59: 81 (1980), *non* Mildbr.]
[*Isoglossa salviiflora sensu* Brummitt in K.B. 40: 786 (1985) pro parte quoad *Vollesen* in MRC 2323 & 3715]

b. subsp. **salviiflora** (*Mildbr.*) *I. Darbysh.* in K.B. 64: 426, fig. 7A–G (2009). Type: Tanzania, Morogoro District, Uluguru Mts south, *Schlieben* 4243 (BR!, lecto., EA! & K!, photo.; BM!, M!, P!, PRE!, isolecto.)

Inflorescence with densely clustered cymules, strobilate; cymule bracts (oblong-) obovate, 4.5–9 mm long, 2.5–6 mm wide. Calyx lobes 7.5–10.5 mm long in flower. Corolla 24–34 mm long. Fig. 82/1–7, p. 644.

TANZANIA. Mpwapwa District: Mpwapwa, May 1929, *H.E. Hornby* 139a!; Kilosa District: Mikumi National Park, Apr. 1984, *de Nevers & Charnley* 3300!; Ulanga District: Udzungwa Mt National Park, Mang'ula village, Sept. 1999, *Mwangulango & Leliyo* 364!
DISTR. **T** 5, 6, 8; not known elsewhere
HAB. Lowland woodland and forest, particularly riparian shade; 200–500 m
USES. "Eaten by stock" (**T** 5; *Hornby* 139)
CONSERVATION NOTES. Assessed as Vulnerable (VU B2ab(iii)) by Darbyshire (l.c.).

SYN. *Schliebenia salviiflora* Mildbr. in N.B.G.B. 12: 99 (1934); T.T.C.L.: 16 (1949)
S. secunda Mildbr. in N.B.G.B. 12: 521 (1935); T.T.C.L.: 16 (1949). Type: Tanzania, Lindi District: Lake Lutamba, *Schlieben* 5394 (B†, holo.; BM!, BR!, HBG!, LISC!, M!, P!, iso.)
Isoglossa salviiflora (Mildbr.) Brummitt in K.B. 40: 786 (1985), excl. *Vollesen* in MRC 2323 & 3715

NOTE. S Tanzania represents the geographical boundary between the two subspecies and some gatherings are inevitably somewhat intermediate. In particular, plants of subsp. *floribunda* from our region (with the exception of *Ward* I30 from the isolated, higher altitude Idodi population) display more densely clustered cymules than those most commonly found elsewhere in its distribution, and *Vollesen* in MRC 3715 also has oblanceolate bracts, thus approaching the more broadly obovate bracts of subsp. *salviiflora* (bracts lanceolate in other material of subsp. *floribunda*).

27. **Isoglossa grandiflora** *C.B. Clarke* in F.T.A. 5: 233 (1900); [S. Moore in Trans. Linn. Soc., Bot., Ser. 2, 4: 34 (1894), *nom. nud. in obs.*]. Syntypes: Malawi, Chikwawa [Shibisa] to Tshinsunze [Tshinmuze], *Kirk* s.n. & Manganja Hills, Mt Sochi, *Meller* s.n. (both K!, syn.)

Straggling subshrub, 50–300 cm tall; stems sparsely pubescent, hairs mainly antrorse or appressed, uppermost stems also with short patent glandular hairs, glabrescent. Mature leaves (ovate-) elliptic, 6.4–22 cm long, 3.2–10.5 cm wide, base attenuate or cuneate, apex acuminate, upper surface and veins beneath sparsely pubescent; lateral veins 6–9 pairs, prominent beneath; petiole 11–54 (–83) mm long. Inflorescences terminal, thyrsoid, with 1–3(–5) pairs of lax branches towards base, these sometimes with secondary branching, main axis 7.5–38 cm long, cymules ± distant, sessile, 1–several-flowered, sometimes unilateral along the lateral branches; axes densely patent glandular-pubescent and with appressed eglandular hairs; bracts ovate to lanceolate, 2.5–5.5(–7.5) mm long, indumentum as axes; bracteoles as bracts but narrower; flowers subsessile; reduced inflorescences often present in uppermost leaf axils. Calyx lobes linear, (5–)6–9 mm long in flower, extending to 6.5–12 mm in fruit, glandular-pubescent and with scattered eglandular hairs outside. Corolla pink, purple or blue, 25–34 mm long, outer surface pubescent; tube 10.5–14.5 mm long, 2–2.5(–3) mm in diameter at base, abruptly upturned below the centre then expanded to 4–6 mm in diameter at mouth, dorsally pubescent within and with a ring of hairs 2.5–3.2 mm from base; upper lip arcuate and hooded, 15–20 mm long, lobes minute; lower lip 15–19 mm long, lobes 0.5–1.5 mm long, palate upraised with prominent "herring-bone" patterning. Staminal filaments attached 3.5–4.5 mm below corolla mouth, free for 14–17 mm, arcuate, glabrous; anther thecae barely offset, parallel, upper theca 2.6–3.4 mm long, lower theca 2.3–3 mm long. Style glabrous. Capsule (11.5–)14.5–21 mm long, glabrous; seeds 2.5–3.1 mm wide, rugose. Fig. 80/11, p. 614.

TANZANIA. Ufipa District: Milepa, gorge of R. Mburu, July 1935, *Michelmore* 1142!; Chunya District: Mbeya–Chunya road km 50, 1 km before Salangwe village, June 1996, *Faden et al.* 96/477!; Mbeya District: Songwe, Sept. 1990, *J. Lovett & Kayombo* 4763!

DISTR. **T** 4, 7; E Zambia, Malawi

HAB. Riverine forest and woodland, particularly along rocky stream margins; 1150–1750 m

USES. None recorded on herbarium specimens

CONSERVATION NOTES. Although rather limited in distribution, *I. grandiflora* appears locally common within its range, particularly in Malawi where it is recorded in both the northern and southern highlands. Several collectors have recorded it as locally abundant in suitable habitat. It appears to favour some exposure to light, and has been recorded from somewhat disturbed environments including roadside banks: Least Concern (LC).

SYN. [*Ecteinanthus grandiflorus* T. Anderson, *nom. nud.*; C.B. Clarke in F.T.A. 5: 233 (1900), in syn.]

28. **Isoglossa mbalensis** *Brummitt* in K.B. 40: 786 (1985). Type: Zambia, Mbala [Abercorn], *Fanshawe* 5640 (K!, holo.; LISC!, iso.)

Erect perennial herb, 60–160 cm tall, from a woody rootstock; stems pubescent, hairs antrorse and retrorse, with or without scattered longer hairs, upper stems also with short spreading glandular hairs, lowermost stems glabrescent. Mature leaves ovate, 7–12.5 cm long, 4.3–7 cm wide, base shortly attenuate, apex acuminate; surfaces sparsely pubescent, mainly on veins beneath, uppermost leaves also with glandular hairs towards base; lateral veins 5–6(–7) pairs, prominent beneath; petiole 3–21 mm long. Inflorescences terminal, thyrsoid, with 3–9 pairs of lax branches to 23 cm long, becoming shorter towards apex, sometimes with secondary branching, one of a pair sometimes barely developing, main axis 18–55 cm long, cymules distant, sessile; axes densely patent glandular- and eglandular-pubescent; main axis bracts lanceolate, 5–12 mm long; cymule bracts ovate to lanceolate, 3–5.5(–7) mm long, sometimes tinged purple, margin and midrib slightly paler than blade; bracteoles as bracts but to 4 mm long; flowers sessile or on pedicels to 2 mm long. Calyx lobes lanceolate, (2.5–)3.5–5 mm long in flower, barely extending in fruit, glandular- and eglandular-pubescent. Corolla deep blue-purple to mauve, 23–29 mm long, palate of lower lip sometimes tinged brownish, outer surface glabrous or pubescent; tube 11–13 mm long, somewhat inflated at base where 1.5–2.5(–4) mm in diameter, abruptly upturned below the centre then campanulate to 4–6.5 mm in diameter at mouth, with an interrupted raised ring of hairs 2.5–3.5 mm from base within; upper lip arcuate and hooded, 14.5–19 mm long, lobes 0.8–1.5 mm long; lower lip deflexed distally, 12.5–16 mm long, lobes ± 1.5 mm long, palate upraised and with prominent "herring-bone" venation. Staminal filaments attached at or just below corolla mouth, free for 10–12.5 mm, arcuate, glabrous; anther thecae barely offset, parallel, 3–4 mm long. Style glabrous. Capsule (12.5–)15–25 mm long, of which stipe (4.5–)5.5–8.5 mm long, glabrous; only immature seeds seen, flattened-ellipsoid, 4–5 mm long, 2.8–3.1 mm wide, smooth.

TANZANIA. Ufipa District: Tatanda, Apr. 2006, *Bidgood et al.* 5698!; Mbeya District: near Songwe hotsprings, May 1991, *P. Lovett & Kayombo* 161!; Mbeya/Rungwe District: Ikuwo road, Chimala–Makete, Apr. 1992, *Congdon, P. Lovett & Bampton* 330!

DISTR. **T** 4, 7; NE Zambia

HAB. Woodland including *Brachystegia* miombo, marginal scrub and wooded grassland; 1200–1650 m

USES. None recorded on herbarium specimens

CONSERVATION NOTES. *I. mbalensis* has a highly limited distribution in NE Zambia and SW Tanzania and is currently known from fewer than ten localities. Data on abundance are unavailable for the Tanzanian sites at present, except at Tatanda where only a single plant was found. However, it has been noted as locally abundant in the Mbala-Lunzua-Chilongowelo region of Zambia, forming large colonies in some localities. Although appearing to favour more open bushland and thicket, it is unlikely to tolerate high levels of disturbance. With population pressure in the Tanzania-Zambia border area increasing, *I. mbalensis* is therefore provisionally assessed as Near Threatened (NT).

NOTE. Plants from the Mbala region of NE Zambia differ from the Tanzanian material in the corollas usually being glabrous, not pubescent outside, although some specimens have sparse hairs on the upper tube (notably *Richards* 11288 from Chisungu Bush). They often additionally have conspicuous pale multicellular hairs scattered on the stems (though these are notably lacking in the type) and somewhat fewer glandular hairs in the inflorescence than the Tanzanian plants. Further material from Tanzania is required to determine the consistency of such differences; these populations may warrant subspecific recognition.

This species is very close to *Isoglossa verdickii* (De Wild.) Champl. ined. (*Duvernoia verdickii* De Wild.) from the Shaba Plateaux of S Congo-Kinshasa, which differs primarily in having longer calyx lobes (5–8 mm long in flower) and larger fruits (28–30 mm long).

29. **Isoglossa sp. C** (= *Bjørnstad* 1757)

Erect perennial herb, 75–100 cm tall, 1–several-branched from a woody rootstock; stems pubescent, hairs largely spreading, of a variable length, upper stems also with short spreading glandular hairs, lowermost stems glabrescent. Mature leaves poorly preserved in the material seen; blade ovate, 4.5–8.5 cm long, 2.8–4.3 cm wide, base rounded to subcordate, apex subattenuate, surfaces sparsely pubescent mainly on veins beneath; lateral veins 4–6 pairs, prominent beneath; petiole to 7 mm long. Inflorescences terminal, thyrsoid, spiciform or with up to 2 lateral branches to 5 cm long, main axis 30–60 cm long, cymules distant, sessile, dense; axes densely glandular- and eglandular-pubescent, hairs short, spreading; bracts ovate to lanceolate, 3.5–6.5 mm long, glandular- and eglandular-pubescent; bracteoles as bracts but to 4 mm long; flowers sessile or on pedicels to 2.5 mm long. Calyx lobes lanceolate, 3–4.5 mm long, glandular- and eglandular-pubescent outside, the latter particularly along the margin. Corolla poorly preserved in the material seen, deep purple, ± 23 mm long, outer surface pubescent; tube ± 11.5 mm long, 2.5 mm in diameter at base, abruptly upturned below the centre then campanulate to the mouth, with an interrupted raised ring of hairs 3 mm from base within; upper lip arcuate and hooded, ± 12 mm long; lower lip deflexed distally, ± 12 mm long, palate upraised with prominent "herring-bone" venation. Stamens with filaments attached ± 3.5 mm below corolla mouth, glabrous; anthers not seen. Style glabrous. Capsule 21–22 mm long, including stipe 6–7.5 mm long, glabrous; seeds flattened-ellipsoid, 3.5–3.8 mm long, 2.5–2.8 mm wide, smooth.

Tanzania. Mbeya District: Ruaha National Park, foot of Magangwe Hill, May 1972, *Bjørnstad* 1757!
Distr. **T** 7; not known elsewhere
Hab. *Brachystegia* woodland; 1450 m
Uses. None recorded on herbarium specimens
Conservation notes. With the current uncertainty over its status, this taxon must be considered Data Deficient (DD) at present, but if it does prove to be a distinct species it will almost certainly prove to be threatened.

Note. Brummitt (in K.B. 40: 787 (1985)) noted that this collection falls within the *I. mbalensis* – *I. ufipensis* group of species (*Salvia*-like corolla, smooth seeds) and may represent a further, undescribed species but the material is too poor for full determination. It is closest to *I. mbalensis*, sharing the short calyx lobes and purple flowers of that species. It however differs in having a largely or wholly unbranched inflorescence and smaller seeds. It is somewhat isolated geographically from the currently known populations of *I. mbalensis*. Further flowering and fruiting material is required to confirm the status of this population.

30. **Isoglossa ufipensis** *Brummitt* in K.B. 40: 787 (1985). Type: Tanzania, Ufipa District, Ilemba Gap, road to Rukwa, *Richards* 11192 (K!, holo.)

Erect perennial herb, 35–175 cm tall, from a creeping woody rootstock; stems pubescent, hairs often retrorse, young stems usually also glandular-pilose. Mature leaves ovate (-elliptic), 5–15 cm long, 2.2–8.2 cm wide, base attenuate, apex acuminate; surfaces eglandular-pubescent particularly on veins beneath, uppermost leaves with additional glandular hairs towards base; lateral veins (4–)5–6(–8) pairs, prominent beneath; petiole 3–42 mm long. Inflorescences terminal, thyrsoid, spiciform or with up to 3 pairs of simple branches to 33 cm long, main axis 8.5–37 cm long, extending to 60–70 cm in fruit, cymules distant, sessile; axes and peduncles densely glandular-pilose and with shorter eglandular hairs; bracts lanceolate, 3.5–8 mm long, margin and midrib paler than blade; bracteoles as bracts but to 6.5 mm long; flowers sessile or on pedicels to 3 mm long. Calyx lobes lanceolate, 5–8 mm long in flower, barely extending in fruit, densely glandular-pilose outside and with shorter eglandular hairs particularly on margin. Corolla pale blue to pale mauve, rarely purple, 21–26 mm long, tube and palate of lower lip paler, the latter sometimes orange or brownish at base, outer surface pilose; tube 9–11.5 mm long, somewhat inflated at base where 2–3 mm in diameter, abruptly upturned below the centre then campanulate to 4.5–5.5 mm in diameter at mouth, with an interrupted raised ring of hairs 2.5–3 mm from base within; upper lip arcuate and hooded, 13.5–16 mm long, lobes 0.7–1.3 mm long; lower lip deflexed, 12.5–14.5 mm long, lobes 1.5–2 mm long, palate upraised with prominent "herring-bone" patterning. Staminal filaments attached at corolla mouth, free for (7.5–)9–9.5 mm, arcuate, glabrous; anther thecae parallel to the filament, barely offset, 3–4 mm long. Style glabrous. Capsule (14–)16–24 mm long, including stipe (2.5–)3.5–6 mm long, glabrous; seeds flattened-ellipsoid, 3–4.5 mm long, 2.2–3.2 mm wide, smooth. Fig. 80/12, p. 614.

Tanzania. Mpanda District: ± 13.5 km from Mpanda on Sumbawanga road, June 1980, *Hooper & Townsend* 1937! & 32 km on Mpanda–Inyonga road, May 1997, *Bidgood, Sitoni, Vollesen & Whitehouse* 3999! & 4 km on Sitalike–Kapapa road, Mar. 2009, *Bidgood, Leliyo & Vollesen* 7994!
Distr. **T** 4; not known elsewhere
Hab. *Brachystegia* woodland, riverine woodland, roadsides; 800–1500 m
Uses. None recorded on herbarium specimens

CONSERVATION NOTES. *I. ufipensis* is currently known from only six herbarium collections. Four of the collecting localities are in close proximity to Mpanda. This region is under-botanised and suitable habitat remains widespread here, so *I. ufipensis* may well be under-recorded. The type collection records it as "growing in quantities from 1200–1500 m on bank beside road", hence it may be locally common and tolerant of some disturbance. However, with such a limited distribution it may be threatened by gradual habitat encroachment as human populations increase in SW Tanzania. It is provisionally assessed as Near Threatened (NT).

EXCLUDED SPECIES

Isoglossa violacea *Lindau* in E.J. 22: 125 (1895). Type: Tanzania, Morogoro District, Uluguru Mts, *Stuhlmann* 8822 (B†, holo.)

In the protologue this taxon is separated from *I. lactea* Lindau by the more extended inflorescences and from *I. flava* Lindau (= *I. lactea*) by having narrower leaves, the inflorescence axes having two distinct lines of hairs and a glabrous calyx. These differences seem rather insignificant in view of the variability recorded within *I. lactea*, but of greater significance is the fact the corollas were recorded as "blue-violet" and with a tube only 2.5 mm long, the corolla of *I. lactea* being white or at most flushed pink and with a longer tube (3.5–6 mm in subsp. *lactea*). Clarke in F.T.A. 5: 230 (1900) synonymised *I. violacea* within *I. lactea*, stating that the corolla tube of the type appeared scarcely shorter; he did not however address the difference in corolla colour. The type specimen is believed to have been destroyed in Berlin during World War II, and no further material matching Lindau' s description of *I. violacea* has since come to light; its placement therefore remains uncertain.

Isoglossa heterophylla *sensu* Iversen, Integr. Usambara Rain For. Proj., App. 1: 9 (1988) quoad *Peter* 60382 & 60584

These specimens, both held at K, are referable to *Justicia anisophylla* Lindau. The name *Isoglossa heterophylla* has never been published.

43. **CHLAMYDOCARDIA**

Lindau in E.J. 20: 39 (1894)

[*Genus nov. sensu* Vollesen & Darbyshire in Cheek *et al.* (eds.), Pl. Kupe, Mwanenguba & Bakossi Mts, Cameroon: 231 (2004)]

Weak-stemmed perennial herbs, evergreen; cystoliths many, linear. Leaves opposite-decussate, petiolate or sessile. Inflorescences terminal, solitary or in clusters of 2–3, sometimes also in upper leaf axils, spiciform, unbranched or with 1–2 branches at lowermost nodes; bracts highly modified, conspicuous, decussately arranged along the axis; bracteoles linear (-lanceolate), exceeding the calyx; flowers usually one per bract, sessile. Calyx divided almost to base; lobes 5, subequal, linear-lanceolate. Corolla bilabiate; tube narrowly cylindrical, longer than the limb; lips reflexed with maturity, upper lip oblong-lanceolate, emarginate; lower lip deeply divided into 3 oblong-lanceolate lobes; rugula absent. Stamens 2; filaments attached at mouth of corolla tube, exserted; anthers bithecous, thecae barely offset, slightly oblique, unequal in size, basally muticous; staminodes absent. Ovary 2-locular, 2 ovules per locule; style filiform; stigma capitate, bilobed. Capsule 4-seeded (or 2 by abortion), stipitate, placental base inelastic; retinacula strong, hook-like. Seeds lenticular, with a hilar excavation, surface rugose-tuberculate and with minute glochidia.

A genus of 2 species confined to West and Central Africa and just extending into our region in W Uganda. The 6-zonocolporate pollen grains are an additional diagnostic character for this genus.

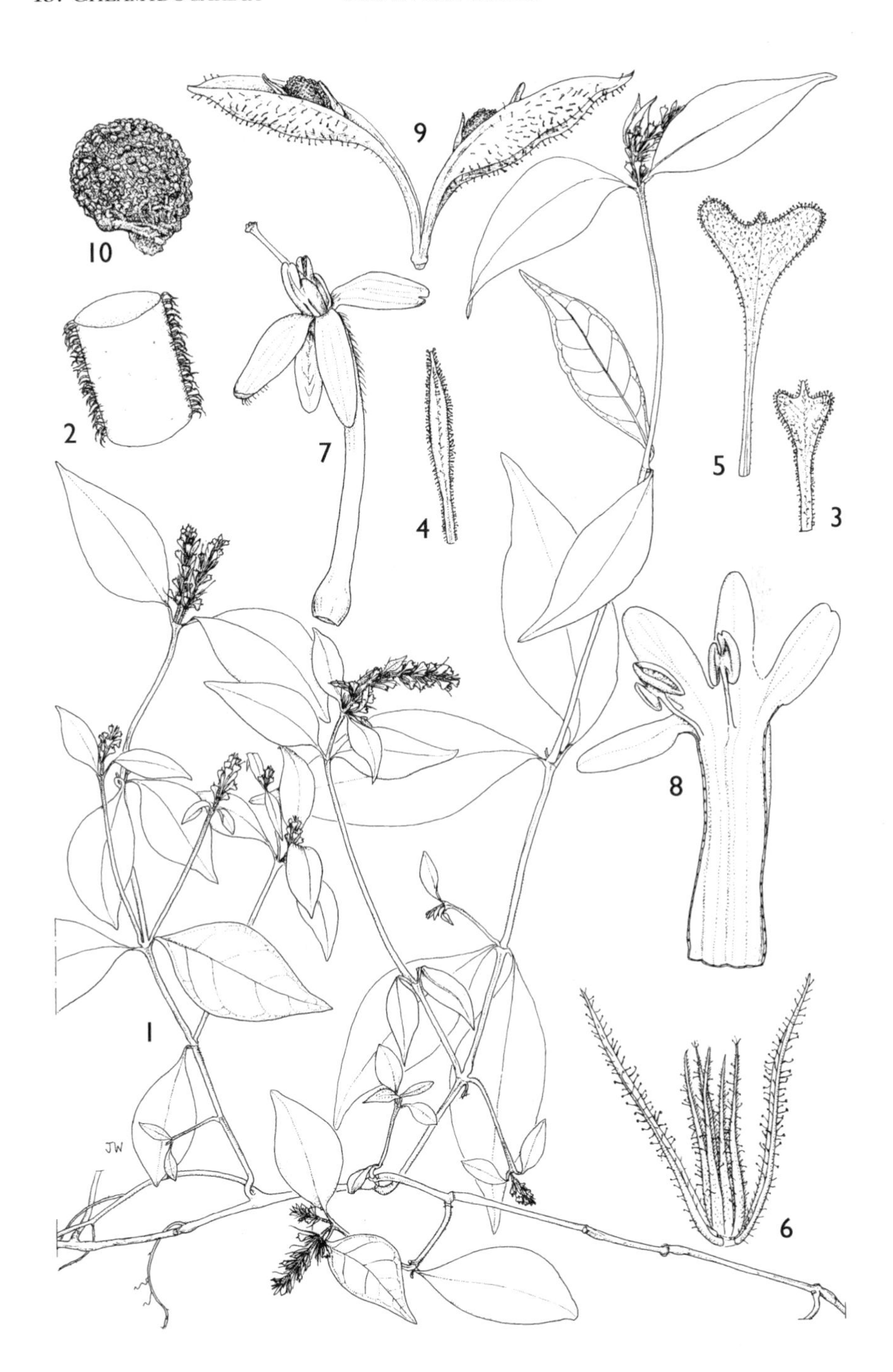

FIG. 83. *CHLAMYDOCARDIA BUETTNERI* — **1**, habit, × $^{2}/_{3}$; **2**, detail of stem indumentum × 12; **3**, bract, outer surface, × 4; **4**, bract, outer surface, narrow extreme, × 4; **5**, bract, outer surface, large West African form, × 4; **6**, bracteoles and calyx, × 8; **7**, corolla with stamens and pistil, × 6; **8**, dissected corolla with stamens, × 6; **9**, mature capsule with seeds, × 5; **10**, mature seed, × 12. 1–3 & 8–10 from *G. Taylor* 3312; 4 from *Eilu* 450; 5 from *Cheek* 7162 ex Cameroon; 6 & 7 from *Brenan* 8455 ex Nigeria. Drawn by Juliet Williamson.

Chlamydocardia buettneri *Lindau* in E.J. 20: 39 (1894); Heine in F.W.T.A. ed. 2, 2: 423 (1963) pro parte excl. *C. subrhomboidea* Lindau; Heine in Fl. Gabon 13: 185, pl. 38: 1–9 (1966); Vollesen & Darbyshire in Cheek *et al.* (eds.), Pl. Kupe, Mwanenguba & Bakossi Mts, Cameroon: 224 (2004); Hawthorne & Jongkind, Woody Pl. W. Afr. Forests: 438, 446, 447 (2006). Type: Gabon, Sibange Farm, *Büttner* 469 (B†, holo.)

Trailing, decumbent or erect herb, 15–60 cm tall, often rooting at lower nodes; stems pale green, retrorse-pubescent in 2 opposite lines. Leaves ovate (-elliptic), 3.8–13.5 cm long, 1.5–5.8 cm wide, base attenuate, cuneate or rarely rounded, margin subentire, apex gradually acuminate to attenuate, surfaces glabrate; lateral nerves (4–)5–6(–7) pairs, prominent beneath; sessile or petiole to 25 mm long. Inflorescence 1.5–8 cm long, slender, axes glandular-pubescent to sparsely so in addition to short eglandular hairs; bracts variable, obovate, spatulate or narrowly oblanceolate, 4–11 mm long, base attenuate to cuneate, gradually or abruptly expanded towards apex, 0.7–6.5 mm wide at widest point, apex often truncate or emarginate and with a prominent acumen, less abruptly narrowed in specimens with the narrowest bracts, surfaces glandular-pubescent, most dense on or restricted to the margin, with or without long marginal eglandular hairs; bracteoles linear, 4–13 mm long, to 0.4 mm wide. Calyx lobes 2–3.5(–4.5) mm long, shortly eglandular-ciliate and with scattered glandular hairs. Corolla greenish-white, with or without purple markings in the throat, 9–10.5 mm long, glabrous or sparsely eglandular-pubescent outside; tube 6–7 mm long; upper lip 2.5–3.5 mm long, 1–1.5 mm wide, emarginate, lobes of lower lip each 2.5–3 mm long, median lobe 1–1.3 mm wide, lateral lobes slightly narrower. Staminal filaments ± 1.5 mm long, glabrous; larger theca of each anther 1–1.5 mm long, smaller theca 0.8–1.2 mm. Ovary minutely glandular or glabrous; style glabrous. Capsule 8–12 mm long, sparsely glandular-pubescent or rarely glabrous; seeds 1.5–2 mm in diameter. Fig. 83, p. 649.

UGANDA. Bunyoro District: Budongo Forest, Feb. 1935, *G. Taylor* 3312! & *idem*, Dec. 1935, *Eggeling* 3335! & *idem*, Kaniyo-Pabidi block, Jan. 1996, *Eilu* 450!

DISTR. **U** 2; Ivory Coast, Nigeria to Congo-Kinshasa

HAB. Moist semi-deciduous and evergreen forest; ± 1000 m

USES. None recorded on herbarium specimens

CONSERVATION NOTES. *C. buettneri* appears locally common in West and Central Africa where it is recorded in low- to mid-altitude primary and secondary forest. It appears absent from much of the Congo Basin, the populations from NE Congo and adjacent Uganda being highly disjunct. In Uganda it is known from only a single site, though the Budongo Forest is an extensive area and is afforded reasonable protection. Therefore, although some areas of its habitat have inevitably been lost through deforestation, it is not currently considered threatened: Least Concern (LC).

SYN. *C. lanciformis* Lindau in Z.A.E.: 300, pl. 32: G–J (1911). Type: Congo-Kinshasa, Beni, Muera, *Mildbraed* 2225 (B†, holo.), **syn. nov**.

NOTE. Bract size and shape and bracteole length are very variable in this species. The Ugandan specimens have bracts (4–6 × 0.7–2.5 mm) and bracteoles (4–6 mm long) at the smaller end of the size range, though with significant overlap with smaller specimens from Cameroon and Nigeria. The most narrow bracts in the Ugandan material (*Eilu* 450) clearly match those illustrated in the protologue of *C. lanciformis* Lindau; here the bracts are narrowly oblanceolate and lack a pronounced acumen. However, bract shape in the Ugandan material is variable, with *Taylor* 3312 having broader, spatulate-attenuate bracts with a pronounced acumen. This clearly matches the common bract form in West African specimens of *C. buettneri*. *C. lanciformis* is therefore considered a variant of *C. buettneri* and is synonymised here.

Bract shape is also very variable in west Africa, with some populations having broader, obovate bracts. This form approaches the second species in the genus, *C. subrhomboidea* Lindau, confined to Cameroon and N Congo-Kinshasa, which differs chiefly in having broader, more ovate to elliptic bracts with an acute to attenuate apex. It also tends to have larger leaves with more lateral veins and somewhat longer calyx lobes and corollas.

44. ANISOTES

Nees in DC., Prodr. 11: 424 (1847); Lindau in E. & P. Pf. IV, 3b: 351 (1895) & in P.O.A. C: 374 (1895); C.B. Clarke in F.T.A. 5: 226 (1900); Baden in Nordic Journ. Bot. 1: 623 (1981)

Himantochilus T. Anderson in G.P. 2, 2: 1117 (1876)
Symplectochilus Lindau in E.J. 20: 45 (1894)
Macrorungia C.B. Clarke in F.T.A. 5: 254 (1900) & in Fl. Cap. 5: 89 (1901); Lindau in E. & P. Pf., Nachtr. 3: 324 (1908); Baden in Nordic Journ. Bot. 1: 143 (1981)
Chlamydostachya Mildbr. in N.B.G.B. 12: 101 (1934)
Metarungia Baden in K.B. 39: 638 (1984)

Shrubby herbs, shrubs or small trees; cystoliths conspicuous or not; young branches, petioles, bracts and bracteoles sometimes with sessile peltate scale-like glands. Leaves opposite, equal, entire to crenate. Plants often flowering when leafless, flowers in dorsiventral (only one bract per node supporting a flower) racemoid cymes with bracts imbricate or not or in 1–12-flowered axillary dichasia or short cymes with minute bracts; bracts persistent, from large and showy to minute; bracteoles small. Calyx variably divided into 5 equal segments. Corolla glandular-puberulous and sometimes also hairy (rarely glabrous) on the outside, basal part of tube short, cylindrical, widening slightly upwards; throat short, very indistinct; limb very distinctly 2-lipped with the lips usually more than twice as long as tube and throat combined; upper lip hooded; lower lip narrowly oblong, usually pendent and coiled at anthesis, without rugula. Stamens 2, glabrous, no staminodes, inserted at top of tube; anthers bithecous, thecae subequal, one from only a little below the other to ± 50% superposed, parallel, both mucronate or lower with minute appendage at base. Style filiform, glabrous or hairy; stigma minutely bifid with 2 ellipsoid lobes. Capsule 2–4-seeded, clavate with a solid basal stalk, placenta solid and not splitting from capsule wall or splitting from base or wall splitting longitudinally. Seed sphaeroid to discoid, compressed or not, smooth to rugose or reticulate-tuberculate.

24 species in tropical Africa, Arabia and Madagascar. The centre of diversity is from S Ethiopia and S Somalia through E Kenya to Tanzania and to a lesser degree onwards through Mozambique to E Zimbabwe and northern South Africa.

1. Flowers in elongated dorsiventral (only one bract per node supporting a flower) racemoid cymes; peduncle (0.3–)1–5.5 cm long; bracts large and conspicuous, 0.8–2.9 cm long 2
   Flowers in 1–12-flowered axillary dichasia or in short racemoid cymes, sessile or with peduncle to 4 mm long; bracts usually minute, rarely up to 1.2 cm long 8
2. The two bracts at each node fused in basal half and forming an ochrea round the axis .... 7. *A. spectabilis*
   Bracts not fused in basal half, not forming an ochrea 3
3. Bracts and calyx with broad pale red to dark red margins; corolla glabrous; capsule with placentae rising elastically from capsule wall at maturity .... 3. *A. pubinervis*
   Bracts and calyx without differently coloured margin or with pale green margin; corolla glandular and with or without non-glandular hairs (glabrous in sp. 1); placentae not rising elastically at maturity 4
4. Flowers with pedicels 4–7 mm long .... 6. *A. tangensis*
   Flowers sessile or with pedicels up to 1 mm long 5

5. Corolla 4.5–4.8 cm long along upper lip, glabrous; calyx 4–5 mm long; anther thecae ± 5 mm long; capsule glabrous . . . . 1. *A. macrophyllus*
Corolla 2.2–3.5 cm long, glandular-puberulous and with or without non-glandular hairs; calyx 5–10 mm long; anther thecae 1.5–3 mm long; capsule with mixture of hairs and glands 6
6. Corolla purple or dark purple; seed reticulate-tuberculate 5. *A. nyassae*
Corolla white, greenish white or various shades of yellow to brownish orange; seed smooth to slightly rugose 7
7. Corolla 2.2–2.5 cm long, lower lip with 2–3 mm long lobes; bracts pale or yellowish green with dark venation and pale crinkly margin . . . . 2. *A. bracteatus*
Corolla (2.7–)3–3.5 cm long, lower lip divided almost to the base; bracts uniformly green, not with pale crinkly margin . . . . 4. *A. umbrosus*
8. Young branches and leaves covered (usually densely) with sessile peltate glands; flowers clearly pedicellate, in (1–)2–12-flowered condensed axillary dichotomous or shortly racemoid cymes; lobes in lower corolla lip 0.5–2 cm long 9
Young branches and leaves without sessile peltate glands; flowers sessile, in 1–2-flowered cymules (but sometimes 2–3 per axil); lobes in lower corolla lip 1–3 mm long 11
9. Bracts and bracteoles 5–8 mm long; calyx 5–8 mm long, sparsely puberulous along veins and on edges . . . . . . 8. *A. ukambensis*
Bracts and bracteoles 1–3(–4) mm long; calyx 2–4 mm long, uniformly puberulous or glandular-puberulous 10
10. Largest leaf 10–14 × 3.7–4.2 cm; bracts densely finely sericeous or tomentellous all over; calyx 3–4 mm long; corolla glandular-puberulous . . . . 9. *A. dumosus*
Largest leaf 4–7 × 1.2–2.4 mm; bracts glabrous or with scattered hairs; calyx 2–3 mm long; corolla glandular-puberulous and with long non-glandular hairs . . . . . . 10. *A. galanae*
11. Calyx 5–7 mm long, the lobes with conspicuous white edges and rib-like midrib; bracts with conspicuous white edges; capsule glabrous . . . . 13. *A. parvifolius*
Calyx 2.5–4.5 mm long, the lobes not with conspicuous white edges nor with rib-like midrib; bracts not white-edged; capsule hairy 12
12. Flowers in a 1-flowered sessile cymule; bracts and bracteoles 2–3.5 mm long; calyx with appressed indumentum; largest leaf 5–14 cm long . . . . 12. *A. sessiliflorus*
Flowers in a 2-flowered cymule; peduncle 0.5–1.5 mm long; bracts and bracteoles 5–12 mm long; calyx with spreading indumentum; largest leaf 2–7.5 cm long . . . 11. *A. tanensis*

1. **Anisotes macrophyllus** (*Lindau*) *Heine* in Fl. Gabon 13: 189, pl. 39 (1966); Baden in Nordic Journ. Bot. 1: 640 (1981). Type: Congo-Kinshasa, Issonge-Semliki, *Stuhlmann* 2938 (B†, holo.)

Erect shrubby herb or shrub to 2 m tall; young branches indistinctly tetragonous, finely antrorsely sericeous, glabrescent. Leaves with petiole 4–9.5 cm long; lamina elliptic or narrowly so, largest 24–33 × 7–13.5 cm, base cuneate to attenuate, apex acuminate with obtuse tip; glabrous or sparsely sericeous along basal part of midrib. Flowers in axillary dorsiventral 3–6 cm long racemoid cymes from upper part of branches, single, subsessile (pedicel under 0.5 mm long); peduncle 1–3 cm long, minutely antrorsely sericeous, rachis with similar indumentum; bracts not or slightly imbricate, folded round flowers, uniformly green, broadly ovate-elliptic, 1–1.7 × 0.8–1.2 cm, finely antrorsely sericeous, apex acute to triangular, base attenuate or cuneate near base of cyme; bracteoles similar in colour and indumentum, ovate-triangular, 1–2 mm long. Calyx 4–5 mm long, divided to 1.5–2 mm from base, with a few antrorse hairs on veins, lobes narrowly triangular, acute. Corolla orange or orange-red (? or white), 4.5–4.8 cm long, glabrous; tube 1.2–1.4 cm long, 3–4 mm in diameter; upper lip 3.3–3.5 cm long; lower lip 3.1–3.3 cm long, with 2–3 mm long lobes. Filaments 3–3.3 cm long, thecae ± 5 mm long. Capsule 2–4-seeded, 2.4–2.6 cm long, glabrous; seed discoid, ± 5 × 4 mm, reticulate-tuberculate, with a very inconspicuous ridge on the inside.

UGANDA. Bunyoro District: Budongo Forest, Nov. 1939, *Eggeling* 3831! & 12 Dec. 1970, *Synnott* 488! & Rabongo Forest, 19 Feb. 1964, *H. E. Brown* 2036!

DISTR. **U** 2, 3; Cameroon, Gabon, Central African Republic, Congo-Brazzaville, Congo-Kinshasa

HAB. Wet evergreen lowland and intermediate forest; 950–1150 m

SYN. *Himantochilus macrophyllus* Lindau in E.J. 20: 60 (1894) & in P.O.A. C: 372 (1895)
Macrorungia macrophylla (Lindau) C.B. Clarke in F.T.A. 5: 255 (1900); Benoist in Mem. Soc. Linn. Soc. Norm., N. S., Sect. Bot., 1(3): 48 (1928)
Himantochilus sereti De Wild., Ann. Mus. Congo, Bot., Ser. 5, 3: 274, t. 47 (1910). Type: Congo-Kinshasa, Nadi River, *Seret* 458 (BR, lecto.; selected by Baden (l.c.))
Macrorungia batesii Wernh. in J.B. 54: 229 (1916). Type: Cameroon, Bitye, Messe, *Bates* 687 (Z, lecto., selected by Baden (l.c.); BM!, iso.)

2. **Anisotes bracteatus** *Milne-Redh.* in Hooker, Ic. Pl.: t. 3268 (1935); T.T.C.L.: 1 (1949); Milne-Redh. in Proc. Linn. Soc. 165: 30 (1954); F.F.N.R.: 381 (1962); Björnstad, Check-list Ruaha National Park: 25 (1976); Baden in Nordic Journ. Bot. 1: 644 (1981); Ruffo *et al.*, Cat. Lushoto Herb. Tanzania: 1 (1996); Mapura & Timberlake, Checklist Zimb. Vasc. Pl.: 13 (2004); Phiri, Checklist Zamb. Vasc. Pl.: 18 (2005). Type: Tanzania, Mpwapwa District, Gulwe, *Greenway* 2407 (EA!, lecto., selected by Baden (l.c.); K!, iso.)

Shrub to 3 m tall; young branches with 4 ridges running down from base of petioles, densely and finely whitish antrorsely sericeous, soon glabrescent. Leaves with petiole to 3.5 cm long; lamina narrowly to broadly ovate or elliptic, largest 8.5–16 × 2.5–9 cm, base attenuate, decurrent, apex acute to subacuminate; glabrous when mature, finely sericeous when young, densely so on petiole. Flowers in axillary dorsiventral 2.5–10 cm long racemoid cymes from upper part of branches, often aggregated to a large pseudo-panicle, single or in 3-flowered cymules (or 1 plus 2 aborted buds) towards base of cyme; peduncle 1–2.5 cm long, densely and finely antrorsely sericeous, rachis with similar indumentum; bracts imbricate, pale green or yellowish green with conspicuously darker green venation and with a well-defined paler margin, ovate or broadly so, 1.5–2.8 × 1–1.6 cm, sparsely sericeous-puberulous, densest on veins, base from cordate at base of cyme to truncate or cuneate towards tip, margin distinctly crinkly, apex acuminate to cuspidate; bracteoles similar in colour and indumentum, ovate, 1.5–2.3 × 0.5–1 cm. Calyx 5–7 mm long, divided to ± 2 mm from base, finely uniformly antrorsely sericeous or sericeous-puberulous, lobes narrowly triangular, acute. Corolla greenish yellow, dull yellow, pale yellow, dirty orange or brownish yellow, 2.2–2.5 cm long, retrorsely sericeous with sharply bent hairs and with stalked capitate glands; tube 0.7–1 cm long, ± 3 mm in diameter; upper lip 1.5–1.7 cm long; lower lip 1.3–1.7 cm long, lobes 2–3 mm long. Filaments 1.2–1.5 cm long, thecae 2.5–3 mm long. Capsule 2–4-seeded, 1.8–2.4 cm long, finely glandular-puberulous, sometimes also with appressed non-glandular hairs; seed discoid, 6–7 × 5–6 mm, smooth and glabrous, with a weak ridge on the inside.

TANZANIA. Mpwapwa District: 15 km S of Gulwe on Kibakwe track, 9 April 1988, *Bidgood et al.* 984!; Singida District: Mkalama, Usasi, Wimbari Steppe, Sept. 1935, *B.D. Burtt* 5221!; Iringa District: Ruaha National Park, 5 km NE of Msembe, 7 Aug. 1970, *Thulin & Mhoro* 632!

DISTR. **T** 5, 7; Zambia, Zimbabwe

HAB. *Combretum* woodland and bushland, *Acacia-Commiphora-Cordyla* thickets, often on rocky hills or poor stony soil, riverine scrub, alluvial *Acacia* grassland; 800–1200 m

3. **Anisotes pubinervis** (*T. Anderson*) *Heine* in Fl. Gabon 13: 189 (1966); da Silva *et al.*, Prelim. Checklist Vasc. Pl. Mozamb.: 18 (2004). Type: Malawi, Mt Tohiradzovu, *Kirk* s.n. (K!, holo.)

Shrub or small tree to 3(–4) m tall (? rarely to 6 m); young branches dark green, subtetragonous, sparsely to densely antrorsely whitish sericeous (rarely puberulous or densely so), soon glabrescent, older branches brownish. Mature leaves with petiole to 5 cm long; lamina ovate to elliptic or narrowly so, largest 9–26 × 3.8–10.5 cm, base attenuate, decurrent, apex acuminate to cuspidate (rarely acute or obtuse); subglabrous to puberulous or sericeous, densest on midrib and large veins. Flowers in axillary (or from old branches below leaves) dorsiventral 1–6.5 cm long racemoid cymes, solitary (very rarely with 1–2 aborted buds at lowermost node), subsessile; peduncle 0.3–1.5 cm long (to lowermost bract with red margin), with a number of small sterile bracts, puberulous to sericeous, rachis with similar indumentum; fertile bracts imbricate or not, green with a broad pale red to claret margin, ovate to elliptic or narrowly so, 0.8–1.5 × 0.3–0.7 cm, subglabrous to sericeous, base truncate or cuneate, apex acute or subacute; bracteoles absent (commonly) or filiform to narrowly elliptic and bract like, to 0.5(–1.2) cm long. Calyx green with broad pale red to claret margins, 8–11 mm long, fused to above middle, subglabrous to sparsely puberulous or sericeous, densest along the broad midrib, ciliate, lobes triangular, acute. Corolla dark claret to crimson (very rarely white), 2.9–3.8 cm long, glabrous; tube 0.9–1.2 cm long, 2.5–3.5 mm in diameter; upper lip 2–2.9 cm long; lower lip 1.9–2.7 cm long, with 3 minute lobes ± 1 mm long. Filaments 1.8–2.8 cm long, thecae 2.5–3.5(–4) mm long. Capsule with placentae rising elastically from base at maturity and splitting from outer wall and each valve usually splitting longitudinally from base, 4-seeded, 1.4–1.9 cm long, glabrous or sparsely puberulous near apex; seed triangular-conical in shape, ± 4 × 3 mm, when young densely reticulate-verrucose and with a transverse fold, at maturity ± smooth and without fold. Fig. 84, p. 655.

UGANDA. West Nile District: Gulu, Zoka Forest, Nov. 1941, *Eggeling* 4687! & 17 Nov. 1941, *A.S. Thomas* 4029; Mbale District: Bugishu, Bulago, 9 Dec. 1938, *A.S. Thomas* 2584!

KENYA. Meru District: Nyambeni Hills, Ngaia Forest, 24 May 2004, *Luke et al.* 10271!; North Kavirondo District: Kakamega Forest, Yala River, 10 Dec. 1956, *Verdcourt* 1702!; Masai District: Emali Forest, 16 March 1940, *van Someren* 131!

TANZANIA. Arusha District: Ngurdoto Crater, Aug. 1965, *Beesley* 162!; Same District: Mkomazi Game Reserve, Maji Kununua Ridge, 7 June 1996, *Abdallah et al.* 96/94!; Mpanda District: Ntakatta Forest, 10 June 2000, *Bidgood et al.* 4636!

DISTR. **U** 1–3; **K** 4–6; **T** 2–7; Nigeria, NE Congo-Kinshasa, Burundi, S Sudan, SW Ethiopia, Zambia, Malawi, Mozambique, Zimbabwe

HAB. Intermediate and montane evergreen forest, mist forest, often in drier forest types (in Kenya and N Tanzania in *Croton megalocarpus* forest with *Diospyros* and *Nuxia*, in W Tanzania in *Pterygota-Newtonia-Pseudospondias* forest), riverine forest; (900–)1100–1850 m

SYN. *Rungia pubinervia* T. Anderson in J.L.S. Bot. 7: 46 (1863)

Himantochilus marginatus Lindau in E.J. 20: 60 (1894) & in E. & P. Pf. IV, 3b: 346 (1895). Type: Tanzania, Lushoto District, Usambara Mts, Kwa Mshuza, *Holst* 9063 (B†, holo.; BM!, K!, iso.)

H. pubinervius (T. Anderson) Lindau in P.O.A. C: 373 (1895)

Macrorungia pubinervia (T. Anderson) C.B. Clarke in F.T.A. 5: 255 (1900); T.T.C.L.: 13 (1949); K.T.S.: 17 (1961); Binns, Checklist Herb. Fl. Malawi: 15 (1968); U.K.W.F.: 507 (1974); Baden in Nordic Journ. Bot. 1: 148, fig. 2 (1981); K.T.S.L.: 605 (1994); U.K.W.F., ed. 2: 276 (1994); Mapura & Timberlake, Checklist Zimb. Vasc. Pl.: 14 (2004); Phiri, Checklist Zamb. Vasc. Pl.: 19 (2005).

FIG. 84. *ANISOTES PUBINERVIS* — **1**, habit; **2**, bracts; **3**, bracteole; **4**, corolla with bracts and calyx; **5**, calyx; **6**, corolla tube, opened up; **7**, corolla limb; **8**, anther; **9**, ovary, style and stigma; **10**, capsule; **11**, dehisced capsule; **12**, seeds. 1, 4, 6–8 from *Perdue & Kibuwa* 9475, 2, 3 & 5 from *Whellan* 261, 9–12 from *Fanshawe* 7081. Drawn by Victoria Gordon-Friis. Reprinted with permission from Nordic Journal of Botany 1: 147 (1981).

Metarungia pubinervia (T. Anderson) C. Baden in K.B. 39: 638 (1984); Vollesen in Checklist Pl. Mkomazi: 83 (1999); White *et al.*, Evergreen For. Fl. Malawi: 116 (2001); Friis & Vollesen, Fl. Sudan-Uganda border area. 2. Cat. Vasc. Pl., part 2: 450 (2005)

NOTE. The genus *Metarungia* is normally separated from *Anisotes* by the placentae rising elastically from base at maturity and splitting from the outer capsule wall. This is the only character separating the two genera and in recent years this has been found to be an unreliable character in other pairs of genera (*Justicia/Rungia, Dicliptera/Peristrophe*) where it has also been the only separating character.

In some collections from the Flora Zambesiaca area, e.g. *Fanshawe* 7086 from Zambia, the placentae do not rise elastically or only rise partly in some capsules although the lateral capsule walls are thin and breaking irregulary. This has not been observed in any material from East Africa.

4. **Anisotes umbrosus** *Milne-Redh.* in Hooker, Ic. Pl.: t. 3267 (1935); T.T.C.L.: 2 (1949); Baden in Nordic Journ. Bot. 1: 645 (1981); Ruffo *et al.*, Cat. Lushoto Herb. Tanzania: 1 (1996). Type: Tanzania, Mpwapwa, Tubugwe Valley, *B.D. Burtt* 4776 (EA!, lecto., selected by Baden (l.c.); EA!, K!, iso.)

Soft-stemmed erect or scrambling shrubby herb or shrub to 2.5 m tall; young branches distinctly quadrangular, sparsely to densely pubescent or slightly retrorsely sericeous-pubescent, soon glabrescent. Leaves with petiole to 4 cm long; lamina narrowly elliptic to elliptic or ovate (rarely broadly ovate), largest 19–31 × 5–14.5 cm, base attenuate, decurrent, apex shortly acuminate; subglabrous to puberulous, densest on veins. Flowers in axillary dorsiventral 3–8(–16) cm long racemoid cymes from upper part of branches, sometimes aggregated apically, single or 1 plus 2 aborted buds towards base of cyme; peduncle 1–4(–8) cm long, pubescent or sparsely so, rachis with similar indumentum; bracts imbricate, uniformly green, ovate or broadly so, 1.8–2.9 × 0.9–1.5 cm, pubescent or sparsely so, distinctly ciliate, base from cordate near base of cyme to truncate or cuneate towards tip, apex acute to subacuminate; bracteoles similar in colour and indumentum, ovate or narrowly so, 0.8–1.5 × 0.2–0.6 cm. Calyx 7–10 mm long, divided to 1.5–2 mm from base, puberulous with mixture of capitate glands and non-glandular hairs, lobes narrowly ovate, acuminate. Corolla white, greenish white or pale yellowish green, (2.7–)3–3.5 cm long, glandular-puberulous, with or without intermixed non-glandular hairs; tube 0.9–1.5 cm long, ± 4 mm in diameter; upper lip (1.7–)1.9–2.3 cm long; lower lip straight at anthesis, eventually coiled and pendent, (1.7–)1.9–2.3 cm long, divided almost to the base. Filaments (1.2–)1.7–2 cm long, thecae 2.5–3 mm long. Capsule 4-seeded, ± 2 cm long, with mixture of capitate glands and non-glandular hairs; seed discoid, ± 6.5 × 5.5 mm, slightly rugose, with a weak ridge on the inside.

TANZANIA. Handeni District: 7 km S of Handeni, Kideleko, 7 July 1974, *Archbold* 1837!; Mpwapwa District: Upper Tubugwe valley, 3 Aug. 1933, *B.D. Burtt* 4776!; Morogoro District: Mikumi National Park Headquarters Village, 26 June 1977, *Wingfield & Mhoro* 3881!

DISTR. **T** 3, 5, 6; not known elsewhere

HAB. Dry semi-evergreen forest and thicket, riverine forest; 500–1100 m

NOTE. Superficially this looks very similar to *Justicia engleriana* and when not in flower the two are easily confused. They are most easily separated by the subamplexicaul leafbases in *J. engleriana* versus the attenuate leafbases in this species.

5. **Anisotes nyassae** *Baden* in Nordic Journ. Bot. 1: 36 (1981) & in Nordic Journ. Bot. 1: 647, fig. 14 (1981); White *et al.*, Evergreen For. Fl. Malawi: 113 (2001). Type: Malawi, Misuku Hills, *Müller* 1675 (K!, holo.; SRGH, iso.)

Erect or scrambling shrubby herb or shrub to 4(–5) m tall; young branches tetragonous, sparsely to densely antrorsely sericeous to strigose, soon glabrescent. Leaves with petiole to 3 cm long; lamina narrowly elliptic to elliptic (rarely ovate or slightly obovate), largest 10–32 × 3.2–13.5 cm, base attenuate, decurrent, apex acuminate; subglabrous to antrorsely sericeous to strigose along major veins (rarely also on lamina). Flowers in axillary dorsiventral 2.5–5 cm long racemoid cymes from near apex of branches, single or 2 towards base of cyme; peduncle 2–5.5 cm long, antrorsely sericeous or sparsely so (rarely pubescent), rachis with similar indumentum; bracts imbricate or not, uniformly green, elliptic or narrowly so or ovate, (1–)1.5–2.3(–2.7) × 0.3–1.1(–1.4) cm, sparsely antrorsely sericeous on veins (rarely uniformly puberulous), base truncate to attenuate, apex subacuminate; bracteoles similar in colour and indumentum, lanceolate to narrowly ovate, 5–10 × 1–2.5 mm. Calyx whitish, 6–9 mm long, divided to ± 2 mm from base, puberulous or sparsely so, with sessile gland dots, lobes narrowly ovate to narrowly triangular, acuminate. Corolla purple or dark purple, 3–3.5 cm long, glandular-puberulous; tube 1.3–1.5 cm long, 2 (base) to 4 (apex) mm in diameter; upper lip 1.7–2 cm long; lower lip 1.5–2 cm long, divided almost to the base. Filaments 1.7–2 cm long, thecae 1.5–2 mm long. Capsule 4-seeded, 2–3.2 cm long, densely pubescent with non-glandular hairs and with scattered short capitate glands; seed discoid, 5–6 × 3–4 mm, reticulate-tuberculate, with a very weak ridge on the inside.

TANZANIA. Iringa District: Kawemba Forest Reserve, 8 Dec. 1995, *Kisena & Mmari* 1731! & Lulanda Village, Magwila Forest, 3 July 2003, *Mwangoka & Kisonga* 3093!; Songea District: Luwira-Kitega Forest Reserve, 10 Oct. 1956, *Semsei* 2529!
DISTR. **T** 7, 8; Zambia, Malawi
HAB. Evergreen montane forest; 1300–1750 m

SYN. *Anisotes* sp 1; White *et al.*, Evergreen For. Fl. Malawi: 113 (2001)

6. **Anisotes tangensis** *Baden* in Nordic Journ. Bot. 1: 36 (1981) & in Nordic Journ. Bot. 1: 642, fig. 12 (1981). Type: Tanzania, Handeni District, Nguu Mts, Lulago, *Parry* 150 (EA!, holo.; K!, iso.)

Erect or scrambling shrubby herb to 3 m tall; young branches indistinctly tetragonous, finely antrorsely sericeous on two sides, soon glabrescent. Leaves with petiole to 10 cm long; lamina ovate or elliptic, largest 25–37 × 10.5–21 cm, base cuneate to shortly attenuate, apex obtuse; glabrous. Flowers in axillary dorsiventral (rarely all bracts supporting flowers) 4–9 cm long racemoid cymes from near apex of branches, single; peduncle 1–2.5 cm long, minutely antrorsely sericeous, rachis with similar indumentum; bracts not imbricate, uniformly green, narrowly ovate-elliptic, 1–1.8 × 0.3–0.6 cm, subglabrous to sparsely minutely antrorsely sericeous or puberulent, base truncate to cuneate, apex acute to subacuminate; bracteoles less than 1 mm long, filiform; pedicels 4–7 mm long, minutely puberulous. Calyx 7–9 mm long, divided to ± 2 mm from base, finely puberulous, lobes narrowly triangular, acuminate. Corolla creamy yellowish green with purple specks, 3.8–4.2 cm long, puberulous with mixture of capitate glands and hairs or with glands only; tube 1.4–1.5 cm long, 3–4 mm in diameter; upper lip 2.5–2.7 cm long; lower lip 2.5–2.7 cm long, divided almost to the base. Filaments ± 2 cm long, thecae ± 2.5 mm long. Capsule and seed not seen.

TANZANIA. Lushoto District: W Usambara Mts, Mazumbai Forest Reserve, 15 Sept. 1983, *J. Lovett* 173! & Sept. 1983, *J.B. Hall* MA2287!; Handeni District: Nguu Mts, Lulago, 15 Aug. 1952, *Parry* 150!
DISTR. **T** 3; not known elsewhere
HAB. Evergreen montane forest; 1200–1450 m

NOTE. Known only from these three collections. This species almost certainly has periodic mass flowering. In 2007 (pers. obs.) it was common in much of Mazumbai Forest Reserve. No specimen in flower or showing signs of having flowered for several years. None of the forest rangers with us could remember having seen this species in flower.

7. **Anisotes spectabilis** (*Mildbr.*) *Vollesen*, **comb. nov**. Type: Tanzania, Morogoro District, Uluguru Mts, *Schlieben* 3755 (B†, holo.; BM!, K!, iso.)

Shrub to 3(? –5) m tall; young branches subtetragonous, green or pale green, glabrous, older branches purplish brown. Leaves with petiole to 4 cm long; lamina elliptic or slightly obovate, largest 19–30 × 5.2–10 cm, base attenuate, decurrent, apex acuminate to cuspidate; glabrous. Flowers in axillary dorsiventral 4.5–9(–12) cm long racemoid cymes from upper part of branches, in 2–5-flowered subsessile dichasia; peduncle 2–5 cm long, glabrous, rachis glabrous; bracts imbricate in dried material but spreading out in living plants, pale green to bright red, the two at each node fused in basal half and forming an ochrea around the stem, orbicular or transversely elliptic, 2–3 × 2.5–4 cm, glabrous, apex acute or subacute and with a small recurved apiculus; bracteoles broadly triangular, 1–2.5 mm long; pedicels ± 1 mm long. Calyx 9–11 mm long, divided to ± 4 mm from base, sparsely glandular-puberulous and lobes sparsely ciliate, lobes ovate, 2–3 mm wide, acuminate. Corolla white to cream (sometimes tinged pink) or pink, 3.2–4 cm long, sparsely glandular-puberulous; tube 1.3–1.7 cm long, ± 3 mm in diameter at base, 4–5.5 mm at apex; upper lip 1.9–2.3 cm long; lower lip straight at anthesis, eventually deflexed but not coiled, 2–2.4 cm long, divided to the base into 3 linear lobes. Filaments 1.8–2.2 cm long, thecae 3–3.5 mm long. Capsule 4-seeded, ± 2.5 cm long, densely glandular-puberulous and with scattered long non-glandular hairs; seed discoid, ± 5.5 × 4.5 mm, densely reticulate-tuberculate, with a ridge on the inside.

TANZANIA. Morogoro District: Uluguru Mts, Lupanga valley, 28 Sept. 1988, *Pócs & Knox* 88192/Y! & Shikurufumi Forest Reserve, 25 Aug. 2000, *Mhoro* UMBCP421!; Iringa District, Udzungwa Mts National Park, Mt Luhomero, Ruipa River, 5 Oct. 2000, *Luke et al.* 7020!
DISTR. **T** 6, 7; not known elsewhere
HAB. Montane evergreen forest; 1300–1700 m

SYN. *Chlamydostachya spectabilis* Mildbr. in N.B.G.B. 12: 101 (1934); T.T.C.L.: 6 (1949); Ruffo *et al.*, Cat. Lushoto Herb. Tanzania: 3 (1996)

NOTE. This species really only differs from other species of *Anisotes* in its fused bracts. Flowers, fruits and seeds are exactly like in the rest of the genus. A closer inspection also shows that in *A. involucratus* from Somalia the bracts are slightly fused at the base.

This species normally has pale green bracts and white corollas but *Luke et al.* 7020 has bright red bracts and pink corollas. It is also the only collection from **T** 7. But on the label of *Thulin & Mhoro* 3074 from the Nguru Mts the corolla is described as "white with a pink tinge".

8. **Anisotes ukambensis** *Lindau* in E.J. 49: 409 (1913); K.T.S.: 17 (1961); U.K.W.F.: 607 (1974); Baden in Nordic Journ. Bot. 1: 638, fig. 9 (1981); Luke & Robertson, Kenya Coast. For. 2. Checklist Vasc. Pl.: 80 (1993); K.T.S.L.: 598 (1994); U.K.W.F., ed. 2: 281 (1994). Type: Kenya, Kibwezi, *Scheffler* 455 (E, lecto., selected by Baden, l.c.; BM!, K!, iso.)

Shrub or small tree to 3(–5) m tall; young branches pale green to pale yellow, subtetragonous with longitudinal furrows, glabrous apart from band of hairs at nodes, with many sessile peltate glands (sometimes covering surface), older branches pale to greyish brown. Leaves sessile or with ill-defined petiole to 5 mm long; lamina narrowly ovate or narrowly elliptic (rarely ovate), largest 7–14 × 1.7–4.5 cm, more than 3 times longer than wide, base attenuate, decurrent to stem, apex acuminate to cuspidate; with scattered hairs on midrib and edges and with usually dense (often covering surface) sessile peltate glands. Flowers in 3–12-flowered condensed axillary dichotomous or shortly racemoid cymes; peduncle 1–3(–4) mm long, puberulous or sparsely so and covered with sessile peltate glands; bracts uniformly green, lanceolate or narrowly triangular, 5–8 mm long, subacute to obtuse, with similar indumentum; bracteoles like bracts; pedicels (1–)2–4 mm long, sparsely puberulous, without or with scattered sessile peltate glands. Calyx yellowish green, 5–8 mm long, divided to ± 1 mm from base, sparsely puberulous along midribs, towards base also with scattered sessile and stalked capitate glands, lobes lanceolate to narrowly triangular, acuminate, with faint white edges. Corolla pale yellow or pale yellowish green, 4–5 cm

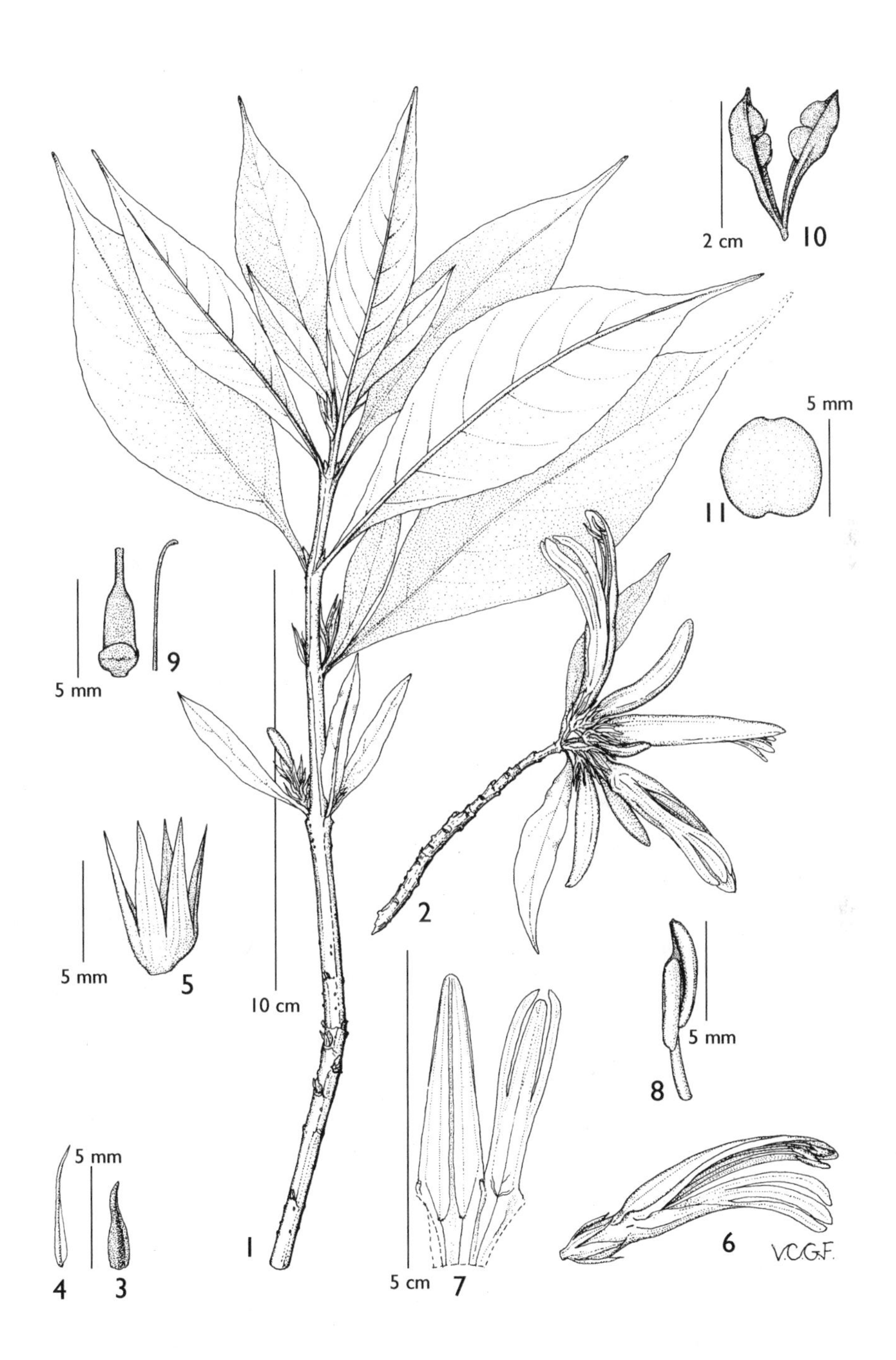

FIG. 85. *ANISOTES UKAMBENSIS* — **1**, habit; **2**, flowering branch; **3**, bract; **4**, bracteole; **5**, calyx; **6**, corolla with calyx; **7**, corolla limb opened up; **8**, anther; **9**, ovary and style; **10**, capsule; **11**, seed. 1 & 2 from *Bally* s.n., 3–11 from *Verdcourt* 1844. Drawn by Victoria Gordon-Friis. Reprinted with permission from Nordic Journal of Botany 1: 639 (1981).

long, very finely glandular-puberulous (rarely also with a few hairs); tube 0.8–1.2 cm long, funnel-shaped with a conspicuous ventral pouch, 4–5 mm in diameter just below pouch; upper lip 3.2–4.3 cm long; lower lip 3–4 cm long, lobes 1.4–2 cm long. Filaments 3.2–3.8 cm long, thecae 3.5–4 mm long. Capsule (fide Baden) 2.5–3 cm long, glabrous; seed (fide Baden) ± 5 × 4 mm, smooth. Fig. 85, p. 659.

KENYA. Machakos District: Kibwezi, 3 Sept. 1957, *Verdcourt* 1844! & 15 Sept. 1961, *Polhill & Paulo* 465! & between Athi and Kibwezi Rivers, 16 Sept. 1969, *Dyson* 601!
DISTR. **K** 4, 6, 7; not known elsewhere
HAB. *Acacia-Commiphora* scrub on rocky lava flows, *Acacia-Combretum* woodland; 750–1000 m

SYN. *Anisotes ukambanensis* Lindau in E.J. 57: 23 (1920). Type: Kenya, Kibwezi, *Scheffler* 181 (E, lecto., selected by Baden, l.c.; BM!, K!, iso.)

NOTE. When Lindau described *A. ukambanensis* in 1920 he made no mention of his *A. ukambensis* of 1913. The two type collections are from the same locality and are so similar they could have been made from the same plant.

9. **Anisotes dumosus** *Milne-Redh.* in K.B. 1936: 487 (1936); T.T.C.L.: 2 (1949); Baden in Nordic Journ. Bot. 1: 636, fig. 7 (1981); Ruffo *et al.*, Cat. Lushoto Herb. Tanzania: 1 (1996). Type: Tanzania, Shinyanga District, Makumba River Flats, *B.D. Burtt* 5144 (EA!, lecto., selected by Baden, l.c.; EA!, K!, iso.)

Shrub to 1.75 m tall; young branches yellowish brown, terete with longitudinal furrows, glabrous to finely antrorsely sericeous, soon glabrescent, densely covered with sessile peltate glands, older branches dark grey. Leaves sessile or with ill-defined petiole to 1 cm long; lamina ovate to elliptic or narrowly so, largest 10–14 × 3.7–4.2 cm, 2.5–3.5 times longer than wide, base attenuate, decurrent to stem, apex acute; glabrous or with scattered hairs on midrib, with scattered sessile peltate glands. Flowers in 2–5-flowered condensed axillary dichotomous cymes; peduncle absent or 1(–2) mm long, densely and finely antrorsely sericeous to tomentellous and with sessile peltate glands; bracts uniformly green, narrowly triangular, subacute to obtuse, 2–3(–4) mm long, with similar indumentum or puberulous; bracteoles like bracts; pedicels 1–2 mm long, sparsely puberulous and with stalked capitate glands. Calyx green, 3–4 mm long, divided to ± 1 mm from base, densely glandular-puberulous all over and with scattered non-glandular hairs, lobes narrowly triangular, acute, with faint white edges. Corolla creamy white, 3.5–4.5 cm long, finely glandular-puberulous; tube 0.7–1.2 cm long, funnel-shaped with a conspicuous ventral pouch, 4–5 mm in diameter just below pouch; upper lip 2.8–3.3 cm long; lower lip 2.6–3 cm long, lobes 0.5–1 cm long. Filaments 2.8–3.2 cm long, thecae 3.5–4.5 mm long. Capsule and seed not seen.

TANZANIA. Shinyanga District: Old Shinyanga, Mnyanga River Flats, 27 June 1931, *B.D. Burtt* 3438! & Shinyanga, Makumba River Flats, July 1935, *B.D. Burtt* 5144! & Shinyanga, Semuye, July 1951, *Eggeling* 6266!
DISTR. **T** 1, 4, 5; not known elsewhere
HAB. *Acacia drepanolobium* grassland on clay hardpans, alluvial mud flats, secondary grassland after cultivation; 1150–1250 m

10. **Anisotes galanae** (*Baden*) *Vollesen*, **comb. et stat. nov**. Type: Kenya, Teita District, Tsavo National Park East, Melka Faya, *Leuthold* 85 (EA!, holo.; K!, iso.)

Shrub or small tree to 2.5(–4) m tall; young branches yellowish brown, terete with longitudinal furrows, glabrous apart from thin band of hairs at nodes, densely covered with sessile peltate glands, older branches dark grey. Leaves sessile or with ill-defined petiole to 2 mm long; lamina narrowly ovate to ovate or narrowly elliptic, largest (? immature) 4–7 × 1.2–2.4 cm, 2.5–3.5 times longer than wide, base attenuate, decurrent to stem, apex rounded; glabrous or with scattered hairs on midrib, with dense sessile peltate glands. Flowers in (1–)2–7-flowered condensed axillary dichotomous or shortly racemoid cymes; peduncle absent or to 0.5 mm long; bracts uniformly green, triangular, 1–3 mm long, subacute to obtuse, glabrous or with scattered spreading hairs, with dense sessile peltate glands; bracteoles like bracts; pedicels (1–)2–3 mm long, sparsely puberulous and with scattered stalked capitate glands. Calyx green, 2(–3) mm long, divided to ± 0.5 mm from base, finely uniformly puberulous and with usually dense stalked capitate glands, lobes triangular (rarely ovate), subacute, with white edges. Corolla lemon yellow or mauve pink to dull purple, 3.8–5 cm long, glandular-puberulous and with few to many non-glandular hairs; tube 1–1.3(–1.5) cm long, cylindrical to slightly funnel-shaped, ventral pouch inconspicuous, 3–4 mm in diameter; upper lip 2.8–3.7 cm long; lower lip 2.8–3.3 cm long, lobes 0.5–0.8 cm long. Filaments 3–3.5 cm long, thecae 3.5–4 mm long. Capsule and seed not seen.

KENYA. Northern Frontier District: Tana River, Galmagalla, no date, *Bally* 6037!; Tana River District: Maziwa road, 3 Aug. 1988, *Robertson* 5284!; Teita District: Galana River, near Lugards Falls, no date, *Bally* 13382!

DISTR. **K** 1, 4, 7; not known elsewhere

HAB. Alluvial *Acacia* and *Acacia-Commiphora* bushland; 25–550 m

SYN. *Anisotes dumosus* Milne-Redh. subsp. *galanae* Baden in Nordic Journ. Bot. 1: 36 (1981) & in Nordic Journ. Bot. 1: 638, fig. 7, L-Q (1981); Luke & Robertson, Kenya Coast. For. 2. Checklist Vasc. Pl: 80 (1993); K.T.S.L.: 597 (1994)

11. **Anisotes tanensis** *Baden* in Nordic Journ. Bot. 1: 36 (1981) & in Nordic Journ. Bot. 1: 655, Fig. 20 (1981); Luke & Robertson, Kenya Coast. For. 2. Checklist Vasc. Pl.: 80 (1993), as "*tanaensis*"; K.T.S.L.: 598 (1994). Type: Kenya, Northern Frontier District, Garissa, Mado Gashi, *Gillett* 19180 (EA!, holo.; K!, iso.)

Shrub to 3(?–5) m tall; young branches dark brown, rounded and longitudinally striate, sparsely to densely finely puberulous or sericeous-puberulous, older branches dark grey. Leaves sessile or with ill-defined petiole to 5 mm long; lamina ovate to elliptic or broadly so, largest 2–7.5 × 1.5–4 cm, base attenuate, decurrent to stem, apex rounded to obtuse or slightly emarginate; subglabrous to sparsely (densely when young) puberulous or sericeous-puberulous. Flowers in axillary 2-flowered dichasia; peduncle absent or 0.5–1.5 mm long, densely puberulous; bracts uniformly green, narrowly ovate to obovate or orbicular, 5–12 mm long, whitish sericeous-puberulous or densely puberulous; bracteoles 2–5 mm long, acicular to narrowly triangular, with similar indumentum. Calyx 2.5–4 mm long, divided to 0.5–1 mm from base, with similar indumentum, lobes narrowly triangular, acute. Corolla tube yellow, limb orange to orange-red, orange brown or reddish brown, (3.5–)4.3–5.5(–6.5) cm long, densely glandular-puberulous; tube (0.5–)1–1.5 cm long, 2.5–4 mm in diameter; upper lip (3–)3.3–4(–5) cm long; lower lip (2.5–)3–4(–4.5) cm long, lobes 2–3 mm long. Filaments 2.5–3.8(–4) cm long, thecae 2.5–3 mm long. Capsule 4-seeded, 2–2.5 cm long, minutely antrorsely sericeous with sharply bent hairs; seed ± 6 × 5 mm, densely reticulate-tuberculate.

KENYA. Northern Frontier District: Daua Valley, Yalichu, 23 May 1952, *Gillett* 13285! & 50 km W of Ramu, Lagh Olla, 25 Jan. 1972, *Bally & Radcliffe-Smith* 14948!; Tana River District: Thika–Garissa road, 4 km towards Garissa after Namorumat Drift, 10 June 1974, *Faden* 74/781!

DISTR. **K** 1, 4, 7; Ethiopia, Somalia

HAB. *Acacia-Commiphora* bushland on sandy or sandy-loamy soil or on limestone escarpments; 250–600 m

SYN. [*Anisotes parvifolius sensu* Dale & Greenway, K.T.S.: 17 (1961) quoad *Gillett* 13285, *non* Oliv. (1886)]

[*A. involucratus sensu* Baden in Nordic Journ. Bot. 1: 655 (1981); K.T.S.L.: 597 (1994), quoad distrib. Kenya, *non* Fiori (1915)]

NOTE. The Kenyan record of *A. involucratus* in Baden (l.c.) and Beentje (K.T.S.L.) is based on *Bally & Radliffe-Smith* 14948. This in my opinion is a very hairy specimen of *A. tanensis* with broad orbicular bracts. Similar specimens have been collected in S Ethiopia. The bract shape as well as bract and corolla indumentum are wrong for *A. involucratus*.

There are, however, Ethiopian collections of *A. involucratus* from as far west as Moyale, so the species might well occur in N Kenya. It has a corolla indumentum of glands and long hairs and much larger bracts with sparse appressed indumentum.

12. **Anisotes sessiliflorus** (*T. Anderson*) *C.B. Clarke* in F.T.A. 5: 226 (1900); Milne-Redhead in Mem. N. Y. Bot. Gard. 9: 26 (1954); Binns, Checklist Herb. Fl. Malawi: 12 (1968); da Silva *et al.*, Prelim. Checklist Vasc. Pl. Mozamb.: 18 (2004). Type: Malawi, Shire River, Shibisa, *Meller* s.n. (K!, holo.; K!, iso.)

Shrub to 3 m tall; young branches pale yellowish brown, subtetragonous and longitudinally ridged, sparsely antrorsely sericeous on two sides, very quickly glabrescent, older branches grey to dark grey or purplish grey. Leaves sessile or with ill-defined petiole to 2 mm long; lamina ovate to elliptic or broadly so, largest 5–14 × 2.5–6 cm, base attenuate, decurrent to stem, apex subacuminate to obtuse; sparsely antrorsely sericeous along midrib and larger veins. Flowers in a 1-flowered cymule enclosed by a pair of bracts and a pair of bracteoles, sometimes 2(–3) cymules per axil; peduncle absent; bracts uniformly green, ovate to triangular, 2–3.5 mm long, subacute to obtuse, densely minutely antrorsely sericeous; bracteoles triangular or ovate-triangular, 2–3.5 mm long, with similar indumentum. Calyx 3–4.5 mm long, divided to ± 1 mm from base, with similar indumentum, lobes narrowly triangular, subacute, with faint white edges and conspicuous rib-like midrib. Corolla tube yellow, limb bright orange, orange-red or bright red, 3.5–5 cm long, glandular-puberulous and with long pilose hairs, densest at base; tube 1.2–1.5 cm long, 2.5–4 mm in diameter; upper lip 2.3–4 cm long; lower lip 2.5–3.5 cm long, lobes 2–3 mm long. Filaments 2.5–3.5 cm long, thecae 2.5–3.5 mm long. Capsule and seed not seen (but see note).

TANZANIA. Kilosa District: Ruaha River Gorge, April 1966, *Procter* 3307! & Ruaha River Gorge, 48 km W of Mikumi, 16 May 1990, *Carter et al.* 2244!; Iringa District: Lukosi River, Mtandika, 8 May 1986, *Lovett & Congdon* 719!

DISTR. **T** 6, 7; Malawi, Mozambique, Zimbabwe

HAB. *Acacia-Commiphora* bushland, often on rocky slopes, dry riverine scrub; 500–900 m

SYN. *Himantochilus sessiliflorus* T. Anderson in G.P. 2, 2: 1117 (1876); Lindau in P.O.A. C: 372 (1895), as *sessilifolius*

Anisotes sessiliflorus (T. Anderson) C.B. Clarke subsp. *iringensis* Baden in Nordic Journ. Bot. 1: 36 (1981) & in Nordic Journ. Bot. 1: 660 (1981). Type: Tanzania, Kilosa District, Ruaha River Gorge, *Mhoro* 1236 (UPS, holo.; EA!, K!, iso.)

NOTE. I do not see any justification for maintaining subsp. *iringensis*. The overlap in corolla size is larger than indicated by Baden (l.c.) and there is also an overlap in calyx size. The differences are smaller than the ones between the northern and southern populations of *A. bracteatus* which have a very similar distribution but was kept as one taxon by Baden.

There are mis-shapen capsules on *Hall-Martin* 1235 from Malawi. They are ± 3 cm long and sparsely antrorsely sericeous with no glandular hairs.

13. **Anisotes parvifolius** *Oliv.* in Hooker, Ic. Pl.: t. 1527 (1886) & in Trans. Linn. Soc. Ser. 2, 2: 345 (1887); Lindau in P.O.A. C: 374 (1895); C.B. Clarke in F.T.A. 5: 227 (1900); T.T.C.L.: 2 (1949); K.T.S.: 17 (1961); Baden in Nordic Journ. Bot. 1: 660 (1981); Blundell, Wild Fl. E. Afr.: 385 (1987); Luke & Robertson, Kenya Coastal For. 2. Checklist Vasc. Pl.: 80 (1993); K.T.S.L.: 597 (1994). Type: Kenya, "40–60 miles inland from the Mombasa Coast", *Johnston* s.n. (K!, lecto., selected by Baden, l.c.)

Shrub to 2.5 m tall; young branches pale yellowish brown, subtetragonous and longitudinally ridged, glabrous or with indistinct band of hairs at nodes, older branches grey. Leaves sessile or with ill-defined petiole to 3 mm long; lamina elliptic to obovate or broadly so, largest 2.2–6(–9) × 1–3.2(–4.5) cm, base attenuate, decurrent to stem, apex acute to obtuse; glabrous or sparsely antrorsely sericeous on midrib. Flowers in a single 1-flowered cymule, more rarely 2–3 cymules per axil or in a 2–3-flowered dichasium; peduncle 1–2(–3) mm long, finely antrorsely sericeous or densely so; bracts green with conspicuous white edges and rib-like midrib, ovate to triangular, 5–9 mm long, acute, finely antrorsely sericeous or densely (rarely sparsely) so; bracteoles similar in size, colour and indumentum. Calyx 5–7 mm long, divided to 1–1.5 mm from base, similar in colour and indumentum, lobes triangular, acute. Corolla tube yellowish green, limb red to scarlet, 4–6.3 cm long, sparsely glandular-puberulous, sometimes also with scattered long pilose hairs; tube 1.2–1.8 cm long, 3–4 mm in diameter; upper lip 2.8–4.5 cm long; lower lip 2.5–3.8 cm long, lobes 1–3 mm long. Filaments 2.8–3.5 cm long, thecae 3–4 mm long. Capsule 2–4-seeded, ± 2.5 cm long, glabrous; seed not seen.

KENYA. Northern Frontier District: S end of Lake Turkana, Mt Nyiru [Nyero], Sept. 1936, *Jex-Blake* 19!; Kwale District: Samburu to Mackinnon Road, Taru, 11 Sept. 1953, *Drummond & Hemsley* 4265!; Teita District: Maungu Hills, Marunga Hill, 30 May 1970, *Archer* 636!

DISTR. **K** 1, 7; not known elsewhere

HAB. *Acacia-Terminalia* and *Acacia-Commiphora* bushland; 25–650 m

SYN. [*Anisotes sessiliflorus sensu* C.B. Clarke in F.T.A. 5: 226 (1900) quoad *Hildebrandt* 2375 & *Wakefield* s.n., *non* (T. Anderson) C.B. Clarke (1900)]

NOTE. There must be some doubt as to the exact provenance of the *Jex-Blake* collection cited above. This is the only collection not from **K** 7 and is a long way outside the rest of the species' distribution area.

Cultivated in and around Nairobi and in Tanga area.

45. ECBOLIUM

Kurz in Journ. As. Soc. Beng. 40: 75 (1871); C.B. Clarke in F.T.A. 5: 235 (1900); Vollesen in K.B. 44: 638 (1989)

Justicia Kuntze, Rev. Gen.: 491 (1891), *non Justicia* L. (1753)

Erect perennial or shrubby herbs or shrubs; stems articulated (with transverse lines at nodes), usually swollen above nodes and with many longitudinal ribs. Leaves opposite, with short often inconspicuous cystoliths, margin entire, usually recurved. Flowers usually sessile, solitary or 3(–7) per bract, in dense strobilate terminal spiciform cymes, axes often flattened; bracts papery, usually caducous, pale to yellowish green (rarely purplish), with a straight non-pungent tip and usually with inconspicuous to almost invisible venation; bracteoles 2, subulate to narrowly triangular. Calyx deeply divided into 5 equal lanceolate to narrowly triangular acuminate to cuspidate lobes. Corolla livid yellowish to bluish green or turquoise green (rarely white), upper part of tube usually pubescent outside; tube long, cylindric, slightly widening upwards, straight; lower lip with three horizontal narrowly elliptic to circular rounded lobes, middle one widest; upper lip with one linear-lanceolate 2-veined bifurcate lobe held erect or curved back. Stamens 2, held erect and parallel under upper lip, inserted near base of lower lip, glabrous, usually 1–3 mm long; anthers usually 2–3 mm long, bithecous, medifixed, held parallel with filament, thecae oblong, subequal, curved, rounded at both ends. Style filiform, glabrous or hairy near base (rarely for whole length); stigma lobes equal, broadly elliptic, rounded, erect; ovary with two ovules per locule. Capsule 2–4-seeded, club-shaped, apiculate; retinacula strong. Seed discoid, cordiform in outline, from smooth to densely tuberculate, usually with a broad raised rim with entire to jagged edge.

22 species; 12 in eastern and southern Africa (10 endemic); 7 on Madagascar and the Comoro Islands (all endemic); 3 in the southern part of Arabia (1 endemic); 3 in India (2 endemic) one of which extends to Malaysia (probably introduced).

1. Corolla white, tube 1.5–1.7 cm long .... 1. *E. albiflorum*
   Corolla livid yellowish to bluish green or turquoise green, tube 2–3.5 cm long 2
2. Corolla glabrous on the outside (rarely with a few scattered hairs); capsule glabrous .... 6. *E. subcordatum*
   Corolla pubescent or densely so on the outside 3
3. Leaves sessile or subsessile, the base cordate to auriculate (rarely truncate); capsule puberulous 4
   Leaves petiolate, the base attenuate to cuneate (rarely rounded) 5
4. Leaves pandurate to elliptic-pandurate; inflorescence axis sericeous-puberulous with downwardly pointing hairs .. 5. *E. amplexicaule*
   Leaves ovate or narrowly so; inflorescence axes puberulous with spreading hairs .... 2. *E. viride*
5. Bracteoles 10–13 mm long and calyx (10–)12–15 mm long; stems 4-angular with 4 longitudinal furrows; seed smooth on one side and densely tuberculate on the other; capsule puberulous or densely so .... 4. *E. boranense*
   Bracteoles 1–4 mm long and calyx 3–7(–9) mm long; stems rounded with many longitudinal ribs; seed from almost smooth to tuberculate on both sides 6
6. Seed 9–12 mm long, smooth to slightly rugose; capsule glabrous or minutely sparsely puberulous; leaves and bracts with distinctly raised venation .... 3. *E. tanzaniense*
   Seed 8–10 mm long, sparsely to densely tuberculate; capsule puberulous or sparsely so; leaves and bracts without distinctly raised venation .... 2. *E. viride*

1. **Ecbolium albiflorum** *Vollesen* in K.B. 44: 645 (1989); Lebrun & Stork, Enum. Pl. Afr. Trop. 4: 480 (1997); Ensermu in Fl. Eth. 5: 454 (2006). Type: Ethiopia, Sidamo, Filtu–Bokol Mayo road, *M.G. Gilbert et al.* 7706 (K!, holo.; C!, ETH!, UPS!, iso.)

Shrub to 1.25 m tall; young branches 4-angular with 4 longitudinal furrows on edges, finely puberulous. Leaves with petiole 2–5 mm long; lamina ovate, largest 2.5–4.3 × 1.5–2.2 cm, base cuneate to truncate, often oblique, apex subacute to rounded; subglabrous to sparsely sericeous-pubescent, hairs curly. Inflorescences 1–3 cm long (but only immature ones seen), single; flowers solitary or in threes at lower nodes; peduncle 3–8 mm long; axes finely sericeous towards base, puberulous towards apex and here also with scattered subsessile capitate glands; bracts broadly ovate or broadly elliptic to orbicular, 10–14 × 9–11 mm, subacute to truncate with a mucro up to 4 mm long, upper puberulous and with dense yellowish capitate glands; bracteoles linear-lanceolate, 1.5–7 mm long. Calyx 5–9 mm long, puberulous and with dense yellowish capitate glands. Corolla white; tube 15–17 mm long; middle lobe in lower lip up to 8 × 4 mm, lateral lobes ± 5 × 3 mm, upper lip up to 6 × 1 mm. Anthers purplish tinged. Capsule and seed not seen.

KENYA. Northern Frontier District: 110 km SW of El Wak, 17 km E of Tarbaj Hill, 16 Dec. 1971, *Bally & Radcliffe-Smith* 14652!
DISTR. **K** 1; Ethiopia
HAB. Dense *Acacia-Commiphora* woodland and bushland on rocky limestone slopes; 500 m

2. **Ecbolium viride** (*Forssk.*) *Alston* in Trimen, Handb. Fl. Ceylon 6: 229 (1931); Milne-Redhead in K.B. 1941: 175 (1941); Glover, Check-list Brit. and Ital. Somal.: 64 (1947); F.P.S. 3: 174 (1956); Täckholm, Students Fl. Egypt, ed. 2: 502 (1974); Migahid, Fl. Saudi Arabia, ed. 2: 517 (1978); Wood *et al.* in K.B. 38: 445 (1983); Collenette, Ill. Fl. Saudi Arabia: 30 (1985); Vollesen in K.B. 44: 655 (1989); Robertson & Luke, Kenya Coastal Forests Checklist Vasc. Pl.: 82 (1993); Lebrun & Stork, Enum. Pl. Afr. Trop. 4: 480 (1997); Thulin in Fl. Somalia 3: 406 (2006); Ensermu in Fl. Eth. 5: 456 (2006). Type: Yemen, Kossajf, Surdud, *Forsskål* s.n. (BM!, lecto.; selected by Wood *et al.*, l.c.)

Shrubby herb or shrub to 2 m tall; young branches glabrous and glaucous to puberulous. Leaves fleshy; petiole 0–10 mm long; lamina ovate to elliptic or broadly so, largest 4–18 × 1.8–7.5 cm, base attenuate to cuneate, usually decurrent, apex acute to broadly rounded; glabrous to puberulous. Inflorescences 3–15 cm long, single or 3(–5) together at upper nodes; flowers solitary or in threes at lower nodes; peduncle 3–20 mm long; axes puberulous or densely so and also with scattered to dense stalked capitate glands; bracts ovate to elliptic or orbicular, 12–27 × 8–20 mm, rounded (basal) to acute with a straight mucro 1–2 mm long, upper puberulous and with many (rarely sparse) capitate glands, indistinctly ciliate, lateral veins indistinct; bracteoles linear-lanceolate, 1–3.5 mm long. Calyx 3–7 mm long, finely puberulous and usually with stalked capitate glands. Corolla tube 2–2.8 cm long; middle lobe in lower lip 13–18 × 6–12 mm, lateral lobes up to 15 × 8 mm, upper lip 8–13 × 1–2 mm. Capsule 1.7–2.6 cm long, puberulous; seed 8–10 × 7.5–9 mm, surfaces sparsely to densely tuberculate, tubercles sometimes merging into ribs near edges, rim distinctly raised. Fig. 86, p. 666.

KENYA. Northern Frontier District: Koya River, Dec. 1956, *J. Adamson* 609! & Marsabit, between Milgis and Siriwa River, 21 May 1970, *Magogo* 1444!; Lamu District: Kitwa Pembe Hill, 15–16 July 1974, *Faden* 74/1114!

DISTR. **K** 1, 7; Sudan, Eritrea, Ethiopia, Djibouti, Somalia; Saudi Arabia, Yemen, Oman, India

HAB. *Acacia-Commiphora* bushland, river-beds, coastal bushland on sand dunes, coral rag thicket; near sea level to 600 m

SYN. *Justicia viridis* Forssk., Fl. Aegypt.-Arab.: 5 (1775)

J. rotundifolia Nees in Wallich, Pl. As. Rar. 3: 108 (1832) & in DC., Prodr. 11: 427 (1847); Drury, Hand. Ind. Pl. 2: 470 (1866). Type: India, *Herb. Wallich* 2432L (K!, lecto.; selected by Vollesen, l.c.)

Ecbolium linneanum Kurz var. *rotundifolium* (Nees) C.B. Clarke in Fl. Brit. Ind. 4: 545 (1884)

[*E. linneanum sensu* Baker in K.B. 1894: 338 (1894); Lindau in P.O.A. C: 371 (1895); C.B. Clarke in F.T.A. 5: 236 (1900); Fiori, Boschi e piante legn. Eritrea: 356 (1912); Chiovenda, Miss. Stefanini-Paoli 1: 141 & 217 (1916); Broun & Massey, Fl. Sudan: 347 (1929); Chiovenda, Fl. Somala 2: 357 (1932); Schwartz, Fl. Arabien: 256 (1939); Glover, Check-list Brit. and Ital. Somal.: 64 (1947); E.P.A.: 963 (1964), Migahid & Hammouda, Fl. Saudi Arabia: 284 (1974), *non* Kurz (1871)]

[*E. barlerioides sensu* Lindau in Ann. R. Ist. Bot. Roma 6: 81 (1896); C.B. Clarke in F.T.A. 5: 238 (1900), excl. type; Fiori, Boschi e piante legn. Eritrea: 356 (1912); Chiovenda, Fl. Somala: 270 (1929); E.P.A.: 963 (1964), excl. type, *non* (S. Moore) Lindau (1894)]

E. viride (Forssk.) Alston var. *rotundifolium* (Nees) Raizada in Ind. For. 84: 482 (1958), *nom. illeg.* & in Ind. For. Rec. 5: 16 (1958), *nom. illeg.*

3. **Ecbolium tanzaniense** *Vollesen* in K.B. 44: 659 (1989); Lebrun & Stork, Enum. Pl. Afr. Trop. 4: 480 (1997). Type: Tanzania, Kilosa District, Ruaha Valley, *Bidgood & Lovett* 260 (K!, holo.; C!, DSM!, EA!, K!, MO!, NHT!, WAG!, iso.)

Shrubby herb or shrub to 1 m tall; young branches glaucous, with two bands of hairs on uppermost node, otherwise glabrous. Leaves with petiole 1–6 mm long; lamina elliptic, largest 4–10.5 × 1.5–4.8 cm, base attenuate to rounded, not decurrent, apex subacute to broadly rounded; venation distinctly raised; glabrous

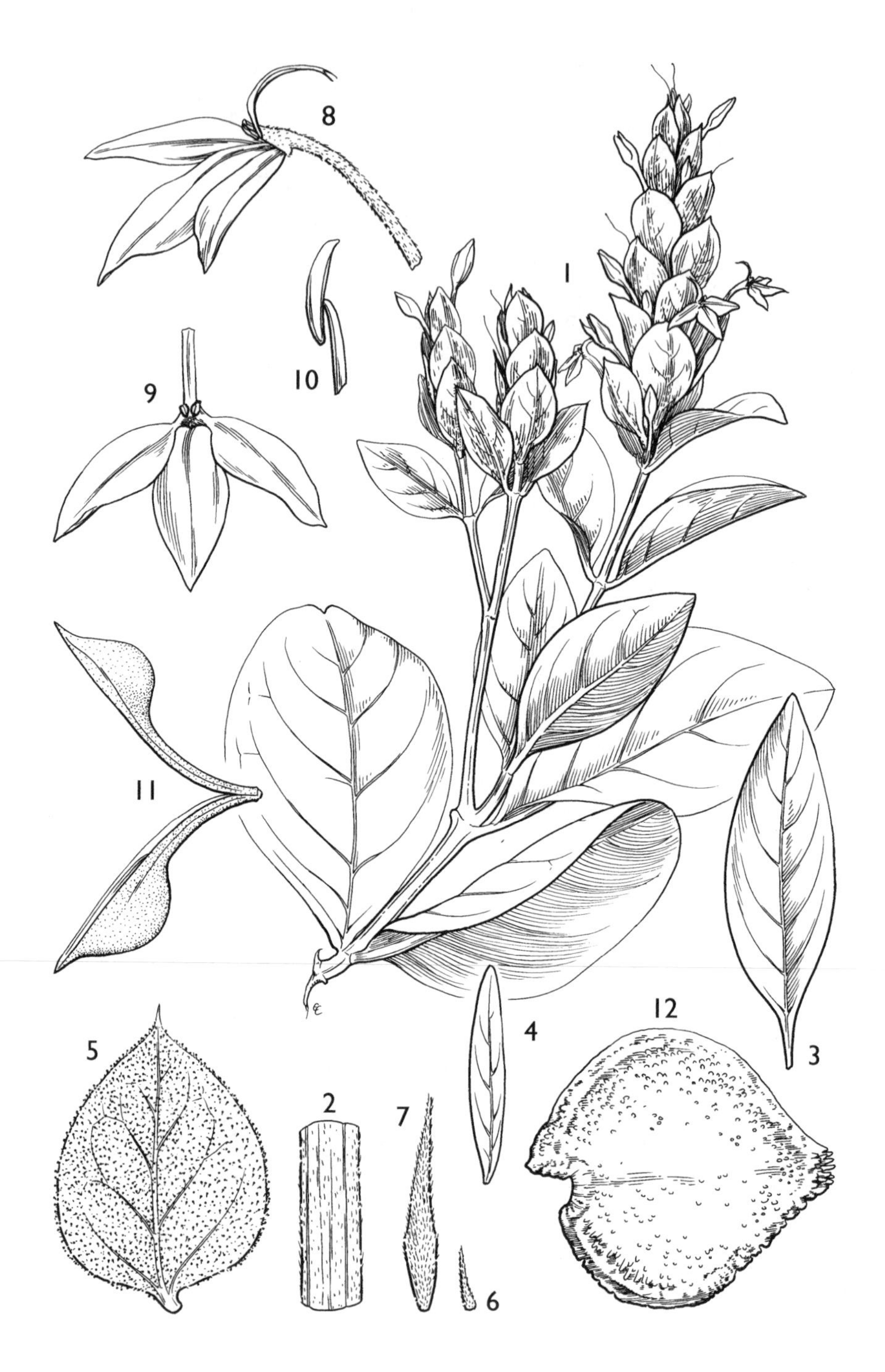

FIG. 86. *ECBOLIUM VIRIDE* — **1**, flowering branch, × ²⁄₃; **2**, detail of young stem, × 6; **3** & **4**, variation in leaf, × ²⁄₃; **5**, bract, × 2; **6**, bracteole, × 6; **7**, calyx lobe, × 6; **8**, corolla, × 2; **9**, corolla, frontal view, × 2; **10**, anther, × 6; **11**, capsule, × 2; **12**, seed, × 5. 1, 2 & 6–10 from *Gilbert et al.* 8186, 3 from *Collenette* 1047, 4 from *Gillett* 23478, 5 from *Magogo* 1444, 11–12 from *Kirk* s.n. Drawn by Eleanor Catherine. Reproduced from Kew Bull. 44: 657 (1989).

apart from hairy midrib, sometimes slightly ciliate near base. Inflorescences 6–23 cm long, single or 3(–5) together at upper nodes; flowers solitary (rarely in 3's at lower nodes); peduncle 5–15 mm long; axes finely puberulous and also with scattered stalked capitate glands; bracts ovate or broadly so, 12–20 × 8–16 mm, acute with a straight mucro up to 3 mm long, upper puberulous and with dense stalked capitate glands, indistinctly ciliate, lateral veins conspicuously raised; bracteoles linear-lanceolate, 2–3 mm long. Calyx 4–7 mm long, base glabrous, lobes finely puberulous and with stalked capitate glands. Corolla tube ± 2.5 cm long; middle lobe in lower lip ± 2 × 1 cm, lateral lobes up to 15 × 6 mm, upper lip ± 15 × 1.5 mm. Capsule 1.8–2.4 cm long, glabrous to minutely and sparsely puberulous; seed 9–12 × 8–9.5 mm, surfaces smooth or slightly rugose, rim distinctly raised, from smooth to verrucose.

TANZANIA. Mpwapwa District: near Winza, 30 July 1937, *Hornby* 867!; Kilosa District: 60 km on Mikumi–Iringa road, Ruaha Gorge, 20 March 1988, *Bidgood et al.* 567!; Iringa District: Iringa–Chemali road, 26 Jan. 1967, *Richards* 22063!
DISTR. **T** 5–7; not known elsewhere
HAB. *Acacia-Commiphora* bushland and thicket on red sandy soil and on rocky slopes; 500–1500 m

4. **Ecbolium boranense** *Vollesen* in K.B. 44: 661 (1989); Lebrun & Stork, Enum. Pl. Afr. Trop. 4: 480 (1997); Ensermu in Fl. Eth. 5: 456 (2006). Type: Ethiopia, Sidamo, Arero, 20 km E of Teltele, *Puff & Ensermu* 821221–4/7 (K!, holo.; ETH!, iso.)

Shrub to 2 m tall; young branches 4-angular with 4 longitudinal furrows on edges, sparsely to densely puberulous to pubescent. Leaves with petiole 3–8 mm long; lamina ovate to elliptic, largest 3–10 × 1.6–4 cm, base attenuate, decurrent, apex acute to rounded; sparsely to densely pubescent, densest below. Inflorescences 4–17 cm long, single or up to 11 clustered at upper nodes; flowers solitary or in threes at lower nodes; peduncle 3–15 mm long; axes puberulous and with stalked capitate glands; bracts elliptic, 12–22 × 5–12 mm, rounded to acute with a straight mucro up to 2 mm long, upper puberulous and with dense stalked capitate glands, indistinctly ciliate; bracteoles 10–13 mm long. Calyx (10–)12–15 mm long, puberulous and with dense stalked capitate glands. Corolla tube 2.5–3 cm long; middle lobe in lower lip 13–16 × 8–12 mm, lateral lobes up to 15 × 6 mm, upper lip ± 11 × 2 mm. Capsule (1.5–)1.7–2.3 cm long, puberulous or densely so; seed 8–9 × 6.5–8 mm, inner surface densely tuberculate, outer completely smooth or very slightly tuberculate near base, rim prominent on inner side, absent on outer.

KENYA. Northern Frontier District: Marsabit, Gulmi, 14 Oct. 1977, *Sato* 381! & Ndoto Mts, Ngurunit Mission, 11 June 1979, *M.G. Gilbert et al.* 5633!
DISTR. **K** 1; Ethiopia
HAB. *Acacia-Commiphora* bushland and thicket on red sandy soil and on rocky slopes; 750 m

5. **Ecbolium amplexicaule** *S. Moore* in J.B. 32: 136 (1894); Lindau in P.O.A. C: 371 (1895); C.B. Clarke in F.T.A. 5: 237 (1900), for type; Vollesen in K.B. 44: 663 (1989); Iversen in Symb. Bot. Upsal. 29(3): 161 (1991); Robertson & Luke, Kenya Coastal Forests Checklist Vasc. Pl.: 82 (1993); Lebrun & Stork, Enum. Pl. Afr. Trop. 4: 480 (1997). Type: Kenya, Sabaki River, *Gregory* s.n. (BM!, holo.)

Perennial or shrubby herb to 2 m tall; young branches subglabrous to sericeous-pubescent. Leaves sessile (rarely a few with up to 5(–15) mm long petiole); lamina pandurate to elliptic-pandurate, largest 8–22 × 2–8.5 cm, base auriculate, clasping the stem, apex acuminate to subacute; subglabrous to puberulous. Inflorescences 3–22(–33) cm long; flowers solitary or in threes at lower nodes; peduncle 4–25 mm long; axes sericeous-puberulous and with few to many stalked capitate glands; bracts ovate (lower) to elliptic, 13–30 × 7–22 mm, acute to acuminate (or lowermost rounded) with a straight mucro 1–2 mm long, upper puberulous or densely so and with usually dense stalked capitate glands, indistinctly ciliate; bracteoles 1.5–3.5(–6.5) mm long. Calyx (4–)5–9(–11) mm long, puberulous and with stalked capitate glands. Corolla tube 2.5–3.5 cm long; middle lobe in lower lip 13–18 × 7–11 mm, lateral lobes up to 17 × 7 mm, upper lip 10–14 × 1–2 mm. Filaments pale green, anthers dark blue. Capsule 1.7–2.3 cm long, puberulous; seed 7.5–10 × 6.5–9 mm, both surfaces densely tuberculate with tubercles denser towards edge, rim prominent.

KENYA. Kilifi District: Watamu, 18 Feb. 1968, *Bally* 13065!; Mombasa District: Bamburi Quarry, 6 Sept. 1989, *Robertson* 5902!; Kwale District: Shimoni, 20 Aug. 1953, *Drummond & Hemsley* 3921!
TANZANIA. Pangani District: Mkwaja Ranch, 13 Sept. 1955, *Tanner* 2184!; Uzaramo District: Dar es Salaam University Campus, 14 Aug. 1968, *Mwasumbi* 10375!; Mikindani District: Mtwara, 11 March 1963, *Richards* 17847!; Zanzibar: Chuaka, 7 Sept. 1959, *Faulkner* 2352!
DISTR. **K** 7; **T** 3, 6, 8; **Z**; Mozambique
HAB. Coastal forest and thicket on brown to black loamy to clayey soil, riverine forest, coral rag thicket, sand dunes, abandoned sisal plantations; near sea level to 400 m

SYN. *Ecbolium auriculatum* C.B. Clarke in F.T.A. 5: 237 (1900); Milne-Redh. in Bot. Mag. 176: t.516 (1967); Vollesen in Opera Bot. 59: 80 (1980), *nom. illeg.*, *non E. auriculatum* (Nees) O. Ktze., Rev. Gen.: 980 (1891). Type: Tanzania, without locality, *Hannington* s.n. (K!, lecto.; selected by Vollesen in K.B. 44: 663 (1989)

6. **Ecbolium subcordatum** *C.B. Clarke* in F.T.A. 5: 237 (1900); Vollesen in K.B. 44: 670 (1989); Robertson & Luke, Kenya Coast. For. 2. Checklist Vasc. Pl.: 82 (1993); Lebrun & Stork, Enum. Pl. Afr. Trop. 4: 480 (1997); Thulin in Fl. Somalia 3: 406 (2006). Type: Kenya, Nyika, Jono Plains, *Gregory* s.n. (BM!, lecto.; selected by Vollesen. l.c.)

Perennial or shrubby herb to 1.5 m tall; young branches glabrous or with two lines of hairs on uppermost node or uniformly pubescent. Leaves sessile or with petiole up to 2 mm long; lamina obovate-pandurate, largest 4.5–16 × 2.2–6.5 cm, base auriculate, clasping the stem, apex subacute to broadly rounded (rarely retuse); glabrous or pubescent. Inflorescences 5–33 cm long; flowers solitary or in threes at lower nodes or almost all in threes; peduncle 2–7(–22) mm long; axes glabrous or with scattered hairs and with subsessile capitate glands or puberulous to pubescent and with capitate glands; bracts elliptic-obovate to broadly so or orbicular (or lower ovate), 10–25 × 8–20 mm, rounded with a straight mucro to 0.5 mm long, upper glabrous but for scattered sessile capitate glands or puberulous to pubescent or sparsely so and with usually dense capitate glands, indistinctly ciliate; bracteoles 1–2 mm long. Calyx 3.5–6 mm long, glabrous apart from finely ciliate margin (rarely uniformly puberulous). Corolla glabrous (rarely with a few scattered hairs); tube 2.5–3.5 cm long; middle lobe in lower lip up to 20 × 15 mm, lateral lobes up to 23 × 10 mm, upper lip 10–17 × 1–2.5 mm. Capsule 1.7–2.3 cm long, glabrous; seed 7–10 × 6.5–8.5 mm, tuberculate, often denser on inner surface, with tubercles denser towards edge and merging into ribs here, rim prominent.

SYN. [*Ecbolium amplexicaule sensu* C.B. Clarke in F.T.A. 5: 237 (1900), quoad *Fischer* 489; T.T.C.L.: 9 (1949); Milne-Redh. in Bot. Mag. 176: t. 516 (1967); Blundell, Wild Fl. E. Afr.: 391 (1987), *non* S. Moore (1894)]

a. var. **subcordatum**

Stems and leaves uniformly pubescent. Bracts uniformly puberulous to pubescent or sparsely so.

KENYA. Machakos District: Tsavo West National Park, Tsavo Gate Campsite, 27 Feb. 1977, *Hooper & Townsend* 1073!; Tana River District: Tana River National Primate Reserve, 19 March 1990, *Luke et al.* TPR 644!; Kwale District: Samburu Station, 2 Feb. 1961, *Greenway* 9818!
HAB. Lowland *Acacia* and *Acacia-Commiphora* bushland on sandy to loamy soil, riverbanks and river-beds; (15–)50–1050(–1200) m
DISTR. **K** 4, 7; Somalia

b. var. **glabratum** *Vollesen* in K.B. 44: 671 (1989); Iversen in Symb. Bot. Upsal. 29(3): 161 (1991); Robertson & Luke, Kenya Coastal Forests Checklist Vasc. Pl.: 82 (1993); Lebrun & Stork, Enum. Pl. Afr. Trop. 4: 480 (1997). Type: Kenya, Kilifi District, 40 km on Malindi–Garsen road, *Polhill & Paulo* 729 (K!, holo.; B!, EA!, FT!, iso.).

Stems glabrous or with two lines of hairs on uppermost internode. Leaves glabrous. Bracts glabrous apart from indistinctly ciliate margins.

KENYA. Tana River District: Tana River National Primate Reserve, 14 March 1990, *Luke et al.* TPR 387!; Teita District: Voi, 6 May 1931, *Napier* 960! & Tsavo National Park East, Lake Kanderi, 6 Dec. 1966, *Greenway & Kanuri* 12669!

TANZANIA. Same District: Lake Kalimawe, 10 Jan. 1967, *Richards* 21921!; Lushoto District: 8 km SE of Mkomazi, 30 April 1953, *Drummond & Hemsley* 2298! & 11 km on Mkomazi–Mombo road, 28 April 1988, *Bidgood & Vollesen* 1264!

HAB. Lowland *Acacia* and *Acacia-Commiphora* bushland on sandy to loamy soil, riverbanks and river-beds; (15–)50–1050(–1200) m

DISTR. **K** 7; **T** 3; not known elsewhere

46. MEGALOCHLAMYS

Lindau in E.J. 26: 345 (1899); C.B. Clarke in F.T.A. 5: 240 (1900); Vollesen in K.B. 44: 605 (1989)

Ecbolium Kurz subgen. *Choananthus* C.B. Clarke in F.T.A. 5: 236 (1900), excl. *E. striatum* and *E. barlerioides.*

Erect shrubby herbs or shrubs; stems 4-angular with 4 longitudinal furrows on the edges (except *M. tanaensis*). Leaves opposite, usually with conspicuous cystoliths and sometimes with dark sessile glands, usually entire. Flowers solitary, sessile, in dense strobilate terminal spiciform cymes; bracts persistent, usually glossy, usually with a straight or recurved pungent tip and usually with conspicuously raised venation; bracteoles 2, subulate. Calyx deeply divided into 5 equal linear to lanceolate or narrowly triangular lobes. Corolla pale blue to bright sky-blue (rarely white), puberulous outside; tube cylindric, not or very slightly widening upwards, straight or very slightly curved; lower lip with three horizontal narrowly elliptic to elliptic subequal lobes; upper lip with one narrowly elliptic 2-veined entire (rarely retuse) lobe. Stamens 2, spreading straight out between lower and upper lip and diverging, inserted near base of lower lip, usually with short downwardly directed hairs along their whole length; anther usually 2–3 mm long, bithecous, medifixed, held at right angle to filament, thecae oblong, subequal, straight, rounded at both ends. Style filiform, hairy in lower part or for whole length (rarely glabrous); stigma lobes equal, oblong-elliptic, flat, rounded, usually recurved; ovary with two ovules per locule. Capsule 1–2(–4)-seeded, club-shaped, apiculate; retinacula strong. Seed discoid, broadly elliptic to orbicular in outline with a slightly raised usually entire rim, smooth and shiny or rugose with dense glandular-glochidiate hairs or tuberculate with finely glandular-glochidiate tubercles.

10 species in eastern and southern Africa; two of which extend to the southern part of the Arabian Peninsula.

1. Corolla white; bracts indistinctly veined; seed tuberculate ..... 6. *M. trinervia*
   Corolla pale blue to bright sky-blue; bracts distinctly reticulately veined 2
2. Stem rounded, with many longitudinal ribs, swollen at nodes; seed tuberculate ..... 7. *M. tanaensis*
   Stem 4-angular with 4 longitudinal furrows on angles, not swollen at nodes, without longitudinal ribs; seed smooth and shiny or with glandular-glochidiate hairs 3

3. Lower pair of bracts completely covering the inflorescence; flowers in (1–)2–3 pairs; corolla tube distinctly longer than lobes; capsule 1-seeded; largest leaf 1–5 mm wide ... 5. *M. linifolia*
   Lower pair of bracts covering only a small part of the inflorescence; flowers in many pairs; corolla tube shorter than or ± same length as lobes; capsule 2(–4)-seeded; largest leaf 5–57 mm wide ... 4
4. Seed smooth and shiny; bracts with short straight tip ... 5
   Seed glandular-glochidiate; bracts with recurved pungent tip ... 6
5. Young branches and inflorescence axes finely sericeous; calyx puberulous all over; largest leaf 1.2–6.5 cm long; corolla lobes 6–13 mm long; filaments 4–7 mm long; capsule 8–12 mm long; seed 4–6 × 3–4.5 mm ... 1. *M. violacea*
   Young branches glabrous to pubescent; inflorescence axes pubescent, basal part of calyx glabrous; largest leaf 6–14 cm long; corolla lobes 13–19 mm long; filaments 8–12 mm long; capsule 12–14 mm long; seed 6–7.5 × 4.5–5.5 mm ... 2. *M. tanzaniensis*
6. Hairs on young branches appressed or if somewhat spreading then pointing straight downwards ... 3. *M. revoluta*
   Hairs on young branches spreading or bending down near tip, usually ± curly ... 7
7. Bracts with 0.5–1 mm long capitate glands, clasping the ripening capsule; inflorescence usually loose with internodes showing, longest (6–)8–17 cm long ... 4. *M. kenyensis*
   Bracts without or with short (less than 0.5 mm long) capitate glands, not clasping the ripening capsule; inflorescence usually dense, internodes not showing, longest 2–8 cm long ... 3. *M. revoluta*

NOTE. Vollesen (1989) divides *Megalochlamys* into three sections:
Sect. *Laevigatae* with smooth and shiny seeds. Species 1 and 2.
Sect. *Megalochlamys* with glandular-glochidiate seeds. Species 3, 4 and 5.
Sect. *Verrucosae* with tuberculate seeds. Species 6 and 7.

1. **Megalochlamys violacea** (*Vahl*) *Vollesen* in K.B. 44: 608 (1989); Robertson & Luke, Kenya Coastal Forests Checklist Vasc. Pl.: 83 (1993); U.K.W.F. ed. 2: 277 (1994); Lebrun & Stork, Enum. Pl. Afr. Trop. 4: 494 (1997); Thulin in Fl. Somalia 3: 403 (2006); Ensermu in Fl. Eth. 5: 457 (2006). Type: Yemen, Jebel Melhan, *Forsskål* 390 (C!, lecto.; selected by Wood *et al.* in K.B. 38: 446 (1983); BM!, iso.)

Shrub to 1(–1.5) m tall; young branches, petioles and peduncles finely whitish sericeous or densely so. Leaves with petiole 1–10 mm long; lamina lanceolate to ovate, largest 1.2–6.5 × 0.5–3 cm, base cuneate to truncate, apex acute to rounded, usually apiculate; subglabrous to densely whitish puberulous on both sides or sparser above. Inflorescences 1–8 cm long; peduncle to 8 mm long; axes finely sericeous or densely so, without or with short (less than 0.5 mm long) spreading non-capitate glands; bracts ovate to elliptic, 8–15 × 2.5–8 mm, acute to acuminate with a straight tip, subglabrous to densely puberulous with non-glandular and/or non-capitate glandular hairs and with (rarely without) scattered stalked capitate glands, usually conspicuously ciliate with shiny hairs to 1.5 mm long; bracteoles 1–3 mm long. Calyx 3–5 mm long, puberulous or densely so, no capitate glands. Corolla pale blue to bright sky-blue; tube 4–8 mm long, lobes 6–13 mm long, middle in lower lip 2–4 mm wide. Filaments 4–7 mm long. Capsule 8–12 mm long, pubescent all over or near tip only; seed brown or yellowish brown, 4–6 × 3–4.5 mm, smooth and shiny without any ornamentation.

KENYA. Northern Frontier District: Dandu, 13 May 1952, *Gillett* 13152! & Samburu National Park, Uaso Nyiro, 5 Dec. 1978, *Brenan* 14867!; Turkana District: Oropoi, March 1965, *Newbould* 7337!

DISTR. **K** 1–4, 6, 7; Sudan, Eritrea, Ethiopia, Somalia; Yemen, Saudi Arabia, Oman

HAB. *Acacia* and *Acacia-Commiphora* bushland on limestone or basement rock or on sandy soil, grassland on black cotton soil, lavaflows, river-beds; (50–)250–1400 m

SYN. *Dianthera violacea* Vahl, Symb. Bot. 1: 6 (1790)
Justicia violacea (Vahl) Vahl, Symb. Bot. 2: 15 (1791)
Monechma violacea (Vahl) Nees in DC., Prodr. 11: 411 (1847)
Justicia anisacanthus Schweinf. in Verh. Zool.-Bot. Ges. Wien 18: 678 (1868). Type: Sudan, Suakin, Gebel Uaratab, *Schweinfurth* 130 (BM!, iso.)
Schwabea anisacanthus (Schweinf.) Lindau in E.J. 20: 59 (1894), *nom. nud.* & in E. & P. Pf. IV, 3b: 346 (1895)
Ecbolium anisacanthus (Schweinf.) C.B. Clarke in F.T.A. 5: 238 (1900); Fiori, Boschi e piante Legn. Eritrea: 356 (1912); Broun & Massey, Fl. Sudan: 347 (1929); Chiovenda, Fl. Somala: 270 (1929) & Fl. Somala 2: 358 (1932); Milne-Redhead in K.B. 1941: 175 (1941); Glover, Check-list Brit. And Ital. Somal.: 64 (1947); F.P.S. 3: 174 (1956); E.P.A.: 963 (1964)
E. violaceum (Vahl) Hillcoat & Wood in K.B. 38: 446 (1983); Collenette, Ill. Fl. Saudi Arabia: 30 (1985); Kuchar, Pl. Somalia in CRDP Tech. Rep. No. 16: 246 (1986)

2. **Megalochlamys tanzaniensis** *Vollesen* in K.B. 44: 611 (1989); Lebrun & Stork, Enum. Pl. Afr. Trop. 4: 493 (1997). Type: Tanzania, Kilosa District, Ruaha Valley, *Bidgood & Lovett* 170 (K!, holo.; C!, EA!, K!, MO!, NHT!, WAG!, iso.)

Shrubby herb or shrub to 1.5 m tall; young branches, petioles and peduncles glabrous or pubescent in two longitudinal bands, hairs straight or curly. Leaves with petiole 0.5–3 cm long; lamina ovate to elliptic, largest 6–14 × 2.5–5.7 cm, base attenuate, decurrent on petiole, apex acuminate; subglabrous to pubescent along veins, lamina glabrous. Inflorescences 3–11 cm long; peduncle 2–8 mm long; axes pubescent with spreading and downwardly directed non-glandular hairs and with sessile or short capitate glands towards apex; bracts glossy, ovate to elliptic or broadly so, 12–20 × 7–15 mm, subacute to retuse with a straight tip less than 1 mm long, puberulous or sparsely so with stalked capitate glands, with longer non-glandular hairs along midrib and distinctly ciliate with long shiny hairs; bracteoles 2–4 mm long. Calyx 4–7 mm long, basal part glabrous or subglabrous, lobes ciliate with non-glandular hairs or with a few capitate glands intermixed. Corolla bright sky-blue; tube 7–11 mm long, lobes 13–19 mm long, middle in lower lip 3–4 mm wide. Filaments 8–12 mm long. Capsule 2(–3)-seeded, 12–14 mm long, glabrous (rarely with a few hairs near apex or finely glandular-puberulous); seed yellowish brown, 6–7.5 × 4.5–5.5 mm, smooth and shiny without any ornamentation.

TANZANIA. Shinyanga District: New Shinyanga, Kisumbi Hill, 17 March 1932, *B.D. Burtt* 3730!; Dodoma District, 11 km on Kilimatinde–Dodoma road, 16 April 1988, *Bidgood et al.* 1174!; Iringa District: Ruaha National Park, Msembi–Kimirimatonge track, 2 April 1970, *Greenway & Kanuri* 14247!

DISTR. **T** 1, 2, 5–7; not known elsewhere

HAB. *Acacia* and *Combretum* woodland and bushland, usually on loamy or clayey soils, rocky hills with *Commiphora* thicket, riverine forest and thicket; 500–1400 m

3. **Megalochlamys revoluta** (*Lindau*) *Vollesen* in K.B. 44: 616 (1989); Iversen in Symb. Bot. Upsal. 29(3): 161 (1991); Robertson & Luke, Kenya Coastal Forests Checklist Vasc. Pl.: 83 (1993); U.K.W.F., ed. 2: 277 (1994); Lebrun & Stork, Enum. Pl. Afr. Trop. 4: 493 (1997). Type: Tanzania, Lushoto District, Lake Manka, *Procter* 3652 (K!, neo., selected by Vollesen, l.c.; EA!, FT!, iso.)

Shrubby herb or shrub to 1.25 m tall; young branches, petioles and peduncles pubescent or finely sericeous. Leaves with petiole 1–8(–15) mm long; lamina narrowly ovate to ovate or narrowly elliptic to elliptic, largest 2–8(–12) ×

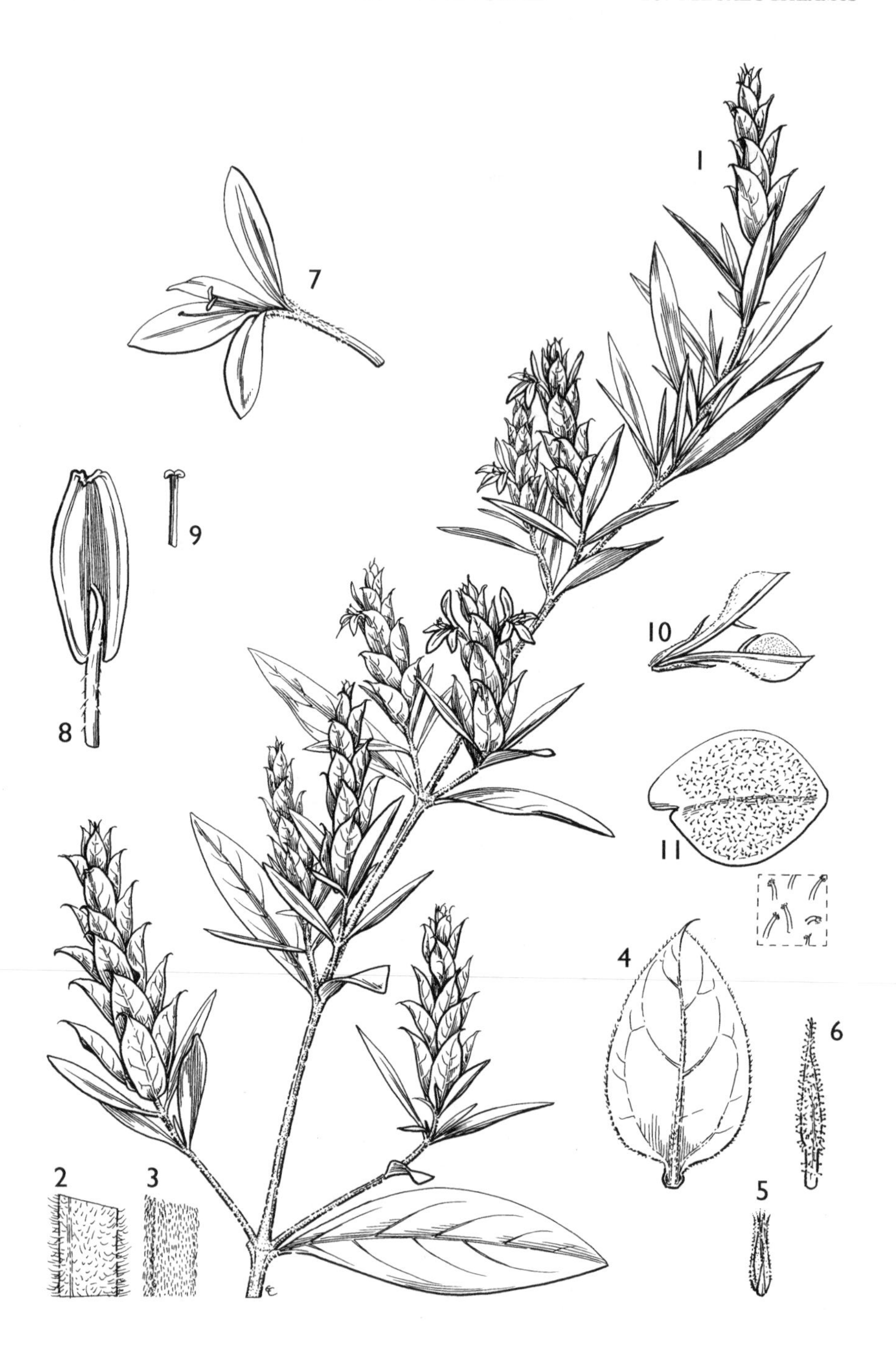

FIG. 87. *MEGALOCHLAMYS REVOLUTA* subsp. *REVOLUTA* — **1**, flowering branch, × $^2/_3$; **2**, detail of young stem, × 3; **4**, bract, × 2; **5**, bracteole and calyx, × 2; **6**, calyx lobe, × 4; **7**, corolla, × 2; **8**, filament and anther, × 15; **9**, upper part of style and stigma, × 6; **10**, capsule, × 2; **11**, seed, × 5 with surface detail much enlarged. *M. REVOLUTA* subsp. *NYANZAE* — **3**, detail of young stem, × 3. 1 & 4–11 from *Napper* 1969, 2 from *Gilbert & Thulin* 1726A, 3 from *Grabner* 284. Drawn by Eleanor Catherine. Reproduced from Kew Bull. 44: 618 (1989).

0.5–2.3(–5.2) cm, base attenuate to truncate, apex acute to retuse, apiculate; pubescent or sparsely so, often densest on veins and margins. Inflorescences 1–8 cm long; peduncle 1–8 mm long; axes finely sericeous to puberulous with non-glandular hairs and with non-capitate glandular hairs, scattered capitate glands sometimes present; bracts green or with purplish veins, ovate or broadly so or elliptic, 7–20 × 3–14 mm, with a sharp recurved acumen 1–6 mm long, subglabrous to puberulous or pubescent with non-glandular hairs or non-capitate glandular hairs, short (less than 0.5 mm long) capitate glands present or not, ciliate or not; bracteoles 1–4 mm long. Calyx 5–8 mm long, puberulous to pubescent with non-capitate glands, usually also with stalked capitate glands. Corolla bright sky-blue; tube 6–13 mm long, lobes 8–14 mm long, middle in lower lip 2–4.5 mm wide. Filaments 5–8 mm long. Capsule 9–14 mm long, puberulous (mainly along sutures with central part of valves usually glabrous); seed greyish, 4.5–6 × 3.5–5 mm, rugose with many glandular-glochidiate hairs, rim entire. Fig. 87, p. 672.

a. subsp. **revoluta**; Vollesen in K.B. 44: 617 (1989)

Young branches pubescent (sometimes only in two lines), hairs spreading straight out or bending down apically. Leaves narrowly elliptic (rarely elliptic), length/width-ratio (3–)3.6–5.2. Fig. 87/1–2 & 4–11, p. 672.

KENYA. Machakos District: Kibwezi–Mtito Andei, Nthawe Hill, 20 March 1969, *Napper & Jones* 1969!; Kitui/Tana River District: 92 km on Garissa–Nairobi road, Katumba Hill, 14 May 1978, *M.G. Gilbert & Thulin* 1726A!; Teita District: Tsavo National Park East, Irima Hill, 16 June 1971, *Kokwaro* 2624!

TANZANIA. Masai/Pare District: Nyumba ya Mungu Dam, 6 June 1970, *Mwasumbi* 10834!; Same District: Mkomazi Game Reserve, Mbulu area, 27 April 1995, *Abdallah & Vollesen* 95/15!; Lushoto District: 8 km on Mkomazi–Mombo road, 28 April 1988, *Bidgood & Vollesen* 1265!

HAB. *Acacia* and *Acacia-Commiphora* woodland, bushland and thicket on sandy, stony or loamy soil or on rocky slopes, grassland on black cotton soil and grey hardpan soils, riverine thicket; 250–1600(–1800) m

DISTR. **K** 4, 6, 7; **T** 2, 3; not known elsewhere

SYN. *Schwabea ecbolioides* Lindau in E.J. 20: 58 (1894). Types: Kenya, Teita District, Ndera, *Hildebrandt* 2436 (B†, syn.); "Ostafrika", *Fischer* 296 (B†, syn.; HBG!, iso.)
S. revoluta Lindau in E.J. 20: 59 (1894)
Ecbolium revolutum (Lindau) C.B. Clarke in F.T.A. 5: 239 (1900); T.T.C.L.: 9 (1949); Blundell, Wild Fl. E. Afr.: 391 (1987)
E. hamatum sensu C.B. Clarke in F.T.A. 5: 239 (1900), quoad *Volkens* 552; T.T.C.L.: 9 (1949), *non* sensu stricto
[*E. revolutum sensu* Blundell, Wild Fl. Kenya: 105 (1982) pro parte]

b. subsp. **nyanzae** *Vollesen* in K.B. 44: 621 (1989); Lebrun & Stork, Enum. Pl. Afr. Trop. 4: 493 (1997). Type: Kenya, Masai District, Magadi–Nairobi road, *Verdcourt* 3763 (K!, holo.; EA!, iso.)

Young branches finely sericeous or densely so with appressed hairs. Leaves narrowly elliptic to elliptic or narrowly ovate to ovate, length/width-ratio 2.2–3.7(–4.1). Fig. 87/3, p. 672.

KENYA. Masai District: Ngong Hills, 6 March 1938, *Bally* 7142! & Ngong Hills to Ol Lorgosailic, 24 Oct. 1955, *Milne-Redhead & Taylor* 7143! & 20 km on Ol Tukai–Namanga road, 15 Dec. 1959, *Verdcourt* 2581!

TANZANIA. Masai District: Ngorongoro Conservation Area, Kakessio, 7 April 1961, *Newbould* 5821! & Mt Longido, 26 March 1970, *Richards* 25686!; Dodoma District: 23 km on Dodoma–Kondoa road, 19 April 1988, *Bidgood et al.* 1215!

HAB. *Acacia* and *Acacia-Commiphora* woodland, bushland and thicket on sandy, stony or loamy soil or on rocky slopes, grassland on black cotton soil and grey hardpan soils, riverine thicket; 250–1600(–1800) m

DISTR. **K** 6; **T** 1, 2, 4, 5; not known elsewhere

SYN. *Neuracanthus subuncinatus* G. Taylor in J.B. 68: 84 (1930). Type: Tanzania, Arusha/Mbulu District, Arusha to Mbulu, *Piemeisel & Kephart* 478 (BM!, holo.)

[*Ecbolium revolutum sensu* Agnew, U.K.W.F.: 599 (1974); Blundell, Wild Fl. Kenya: 105 (1982) pro parte]

NOTE. A few collections are ± intermediate between the two subspecies. They occur in the Kondoa–Dodoma area and have a mixture of appressed and spreading hairs on the stems. Examples are *Burtt* 1508, 2149 and *Bidgood et al.* 1206. Typical subsp. *nyanzae* occurs in the same general area (*Burtt* 2069, *Bidgood et al.* 1215) but not subsp. *revoluta*.

A third subspecies (subsp. *cognata* (N.E. Br.) Vollesen) occurs in Southern Africa and South Africa.

4. **Megalochlamys kenyensis** *Vollesen* in K.B. 44: 625 (1989); U.K.W.F., ed. 2: 277 (1994); Lebrun & Stork, Enum. Pl. Afr. Trop. 4: 493 (1997). Type: Kenya, near Athi River on Kibwezi–Kitui road, *Napper & Jones* 1956 (K!, holo.; EA!, FT!, iso.)

Shrubby herb or shrub to 1 m tall; young branches and petioles whitish pubescent (sometimes only in two bands) or densely sericeous. Leaves with petiole 2–13 mm long; lamina narrowly ovate to ovate or narrowly elliptic to elliptic, largest 3–7 × 1–3.3 cm, base attenuate, apex acute to rounded; sparsely pubescent, mainly on veins, lamina often glabrous. Inflorescences (2–)5–17 cm long; peduncle 3–8 mm long, whitish pubescent; axes pubescent or densely so with non-glandular shiny hairs and stalked capitate glands; bracts greenish with red to purple veins, elliptic to obovate or broadly so, folded around the capsules, 10–15(–20) × 7–13 mm, with a sharp recurved acumen 1–2 mm long, pubescent or sparsely so with non-glandular shiny hairs and long-stalked (0.5–1 mm long) capitate glands, these most conspicuous along edges, usually not ciliate; bracteoles 1–2 mm long. Calyx (6–)7–10 mm long, pubescent or densely so with non-glandular shiny hairs and long-stalked (0.5–1 mm long) capitate glands. Corolla pale blue to bright sky-blue or deep royal blue; tube 10–14 mm long, lobes 12–14 mm long, middle in lower lip 3–4 mm wide. Filaments 6–8 mm long. Capsule 10–15 mm long, puberulous in upper half; seed greyish, 4–5.5 × 3.5–4 mm, almost smooth to rugose, with many glandular-glochidiate hairs, rim entire to crenulate.

UGANDA. Karamoja District: Mathenko County, Magosi Hills, Oct. 1958, *J. Wilson* 572! & Karasuk, Atot River, Aug. 1962, *J. Wilson* 1279!

KENYA. West Suk District: Sigorr, Sept. 1971, *Tweedie* 4265!; Machakos District: Mtito Andei to Voi, Kitani Hill, 20 March 1969, *Napper & Jones* 1976!; Teita District: Voi River, 5 Feb. 1953, *Bally* 8739!

DISTR. **U** 1; **K** 2–4, 7; not known elsewhere

HAB. *Acacia* and *Acacia-Commiphora* wooded grassland and bushland, *Combretum* bushland, riverine thicket, usually on loamy soils; 450–1050(–1450) m

SYN. [*Ecbolium hamatum sensu* Agnew, U.K.W.F.: 599 (1974); Blundell, Wild Fl. Kenya: 105 (1982), *non* C.B. Clarke (1900)]

NOTE. Our material belongs to subsp. *kenyensis* which has pubescent branches. Subsp. *australis* from South Africa has sericeous branches.

5. **Megalochlamys linifolia** (*Lindau*) *Lindau* in E.J. 26: 346 (1899); C.B. Clarke in F.T.A. 5: 241 (1900); Chiovenda, Fl. Somala: 270 (1929) & Fl. Somala 2: 358 (1932); Schwartz, Fl. Trop. Arabien: 256 (1939); Glover, Check-list Brit. and Ital. Somal.: 68 (1947); E.P.A.: 965 (1964); Lebrun & Stork, Enum. Pl. Afr. Trop. 4: 493 (1997); Thulin in Fl. Somalia 3: 404 (2006); Ensermu in Fl. Eth. 5: 459 (2006). Type: Kenya, Northern Frontier District, Beila [Bela] Mt along the Dawa R., *Riva* 847[1616]1462 (FT!, lecto.; selected by Vollesen, l.c.)

Dwarf shrub to 50 cm tall, often intricately branched; young branches glabrous or with thin lines of curly hairs running down from the nodes. Leaves with petiole 0–1.5 mm long; lamina linear to lanceolate (rarely narrowly ovate), largest 1–2.5 ×

0.1–0.5 cm, base cuneate, margin strongly recurved, apex apiculate, but often recurved to make apex look rounded; glabrous to sparsely crisped-puberulous. Flowers in (1–)2–3 pairs; peduncle 1–2 mm long, glabrous to puberulous; axes ± 1 mm long, puberulous; bracts green, lower pair completely covering inflorescence, reniform, 8–15 × 10–12 mm, with a straight or recurved acumen ± 1 mm long, glabrous to sparsely puberulous and then also with a few sessile capitate glands, upper pair(s) lanceolate to narrowly ovate; bracteoles 1–2 mm long. Calyx 2.5–4 mm long, puberulous with non-capitate glands. Corolla bright sky-blue; tube ± 10 mm long, lobes 3.5–5 mm long. Filaments 2–3 mm long, anthers ± 1.5 mm long. Capsule 1-seeded, 6–8 mm long, glabrous to sparsely puberulous; seed greyish, 3–4 × 2.5–3 mm, rugose, with many glandular-glochidiate hairs.

KENYA. Northern Frontier Province: Dawa River, Mt Beila [Bela], 12 June 1893, *Riva* 847 [1616]1462!

DISTR. **K** 1; Ethiopia, Somalia; Yemen

HAB. *Acacia-Commiphora* bushland on sandy to stony soil overlying limestone or on rocky limestone slopes; no altitude given

SYN. *Dicliptera* (?) *linifolia* Lindau in Ann. R. Ist. Bot. Roma 6: 80 (1896)

NOTE. The type is still the only material ever collected in the Flora area.

6. **Megalochlamys trinervia** (*C.B. Clarke*) *Vollesen* in K.B. 44: 635 (1989); Robertson & Luke, Kenya Coastal Forests Checklist Vasc. Pl.: 83 (1993); Lebrun & Stork, Enum. Pl. Afr. Trop. 4: 494 (1997); Thulin in Fl. Somalia 3: 404 (2006). Type: "Ostafrika", *Fischer* 284 (B†, holo.; HBG!, iso.)

Shrubby herb or shrub to 2 m tall; young branches finely whitish sericeous. Leaves with petiole 3–15(–20) mm long; lamina ovate to elliptic or braodly so, largest 2–7(–9) × 1.3–4(–5.2) cm, base cuneate (rarely rounded), apex acute to retuse; subglabrous to sparsely puberulous beneath and with sessile orange glands, above glabrous to sparsely pilose. Inflorescences 2–12(–17) cm long; peduncle to 2(–4.5) cm long, whitish sericeous-puberulous, with scattered sessile orange glands; axes sericeous-puberulous, towards apex with short-stalked capitate glands; bracts green or yellowish green, sometimes with purplish tinged veins, broadly elliptic to circular, 7–12 × 4–10 mm, with a straight or recurved acumen 1–2 mm long, densely glandular-puberulous with stalked capitate glands, with sparse long glossy hairs on edges; bracteoles 1–2.5 mm long. Calyx 4–5.5 mm long, glandular-puberulous or sparsely so with capitate and non-capitate glands, sometimes sparsely ciliate. Corolla pure white (rarely very faintly lilac tinged); tube 5–8(–12) mm long, lobes 10–17 mm long, middle in lower lip 2–3.5 mm wide; filaments 5–10 mm long. Capsule 2(–4)-seeded, 10–14 mm long, sparsely puberulous; seed dark brown, 6–7 × 4–5.5 mm, densely tuberculate all over, the tubercles finely glandular-glochidiate, rim entire to jagged.

KENYA. Northern Frontier/Tana River District: 33 km on Garissa–Hagadera road, 27 Nov. 1978, *Brenan et al.* 14774!; Lamu District: Kipini, Mlango ya Simba, 6 Nov. 1957, *Greenway & Rawlins* 9459!; Tana River District: Garissa–Garsen road, 6 km from Tula, Gelmat Drift, 7 July 1974, *Faden* 74/984!

DISTR. **K** 1, 7; Somalia

HAB. *Acacia*, *Acacia-Commiphora* and *Acacia-Dobera* woodland and bushland on sandy to clayey alluvials and on black cotton soil, usually near rivers or in seasonally wet depressions; 25–200(? –400) m.

SYN. *Schwabea ecbolioides* Lindau var. *tomentosa* Lindau in E.J. 20: 59 (1894). Type: as for *M. trinervia*

Ecbolium trinervium C.B. Clarke in F.T.A. 5: 239 (1900); T.T.C.L.: 9 (1949)

[*E. striatum sensu* C.B. Clarke in F.T.A. 5: 238 (1900); Chiovenda, Result. Sci. Miss. Stef.-Paoli, Coll. Bot.: 142 (1916); E.P.A.: 964 (1964), excl. type, *non* Balf. f. (1884)]

7. **Megalochlamys tanaensis** *Vollesen* in K.B. 44: 637 (1989); Robertson & Luke, Kenya Coastal Forests Checklist Vasc. Pl.: 83 (1993); Lebrun & Stork, Enum. Pl. Afr. Trop. 4: 494 (1997). Type: Kenya, Northern Frontier District, 10 km NE of Garsen, 3 km N of Wema, *Gillett et al.* 19922 (K!, holo.; EA!, iso.)

Shrubby herb to 50 cm tall; young branches swollen above nodes, with many longitudinal ribs, glabrous apart from line of hairs at nodes and a few hairs just above nodes. Leaves with petiole 1.5–4 cm long; lamina ovate, largest 10.5–14 × 6–7 cm, base rounded, margin crenate, apex acute; glabrous apart from hairy midrib above. Inflorescences 6–12 cm long; peduncle of central inflorescence 0.7–1 cm long, of lateral 0.5–4.5 cm long, with two longitudinal bands of spreading hairs; axes puberulous, towards apex with short non-capitate glands; bracts greenish at base to dark purple towards apex, ovate to elliptic, 11–16 × 7–10 mm, with a straight acumen 1–3 mm long, finely puberulous with non-capitate glands and with scattered sessile capitate glands, conspicuously ciliate with long shiny many-celled hairs; bracteoles ± 5 mm long. Calyx 6–7 mm long, indumentum as bracts (but no capitate glands), densely ciliate. Corolla pale lavender; tube 6–7 mm long, lobes 7–8 mm long, middle in lower lip ± 2 mm wide; filaments ± 6 mm long; style glabrous with erect stigma lobes. Capsule 2(–3)-seeded, ± 11 mm long, glabrous; seed brown, 4–6 × 3–4.5 mm, densely tuberculate all over, the tubercles finely glandular-glochidiate, rim strongly crested-jagged all round or basally and apically only.

KENYA. Tana River District: 10 km NE of Garsen, 3 km N of Wema, 15 July 1972, *Gillett et al.* 19922! & Tana Delta Irrigation Project, Wema East, 27 Nov. 2004, *Luke* 10727!

DISTR. **K** 7; not known elsewhere

HAB. Lowland evergreen *Cynometra* forest on deep dark grey alluvial clay; 25 m

NOTE. *M. tanaensis* is known only from these two collection. It shows several characters unique to *Megalochlamys* and could with some justification also be placed in *Ecbolium*. It is the only species with a longitudinally ribbed stem which is swollen at the nodes. It also has tuberculate seeds (a character shared with *M. trinervia*). But the persistent bracts, corolla with equal tube and lobes and four subequal lobes, long stamens spreading straight out and anthers at right angle to filament all point to it being better placed in *Megalochlamys*. The pollen is exactly like in all other species of *Megalochlamys*; see Furness in K.B. 44: 688 (1989).

47. CEPHALOPHIS

Vollesen **gen. nov.** ad partem gregis *Justicieae* Afro-orientale (qui e *Ecbolio*, *Megalochlamyde* et *Trichaulace* constat) pertinet. Ea quattor genera omnia staminibus duobus bithecis proviso atque thecis ad altitudinem eandem affixis distinguenda sunt. *Cephalophis Trichaulace* ob semina persimiliter glochidiata probabiliter proxima, sed inflorescentia corollaque differt. Semina ad ea *Megalochlamyde* etiam aliquantum similia sunt sed ab eo genera inflorescentia corollaque etiam differt. A generibus eis tribus corolla valde bilabiata, labio superiore corollae angusto lateraliter compresso valde falcato-curvato et labio inferiore lato longitudinaliter plicato differt. Typus: *Cephalophis lukei* Vollesen

Herb. Leaves opposite, equal, crenate, with conspicuous cystoliths. Flowers solitary or in 3-flowered cymules in a narrow racemiform panicle; bracts often foliaceous at basal nodes, upwards bract-like, persistent; bracteoles present, large. Calyx deeply divided into 5 subequal often recurved segments. Corolla outside with long (± 1 mm) curly glossy non-glandular hairs all over; basal tube linear, long, widening into a short indistinct throat; limb distinctly 2-lipped, upper lip strongly falcately curved and laterally flattened, with two minute apical teeth, lower lip subcircular in outline, longitudinally plicate with two lateral parts folded over and ± covering central part, with three minute lobes, without conspicuous pattern of

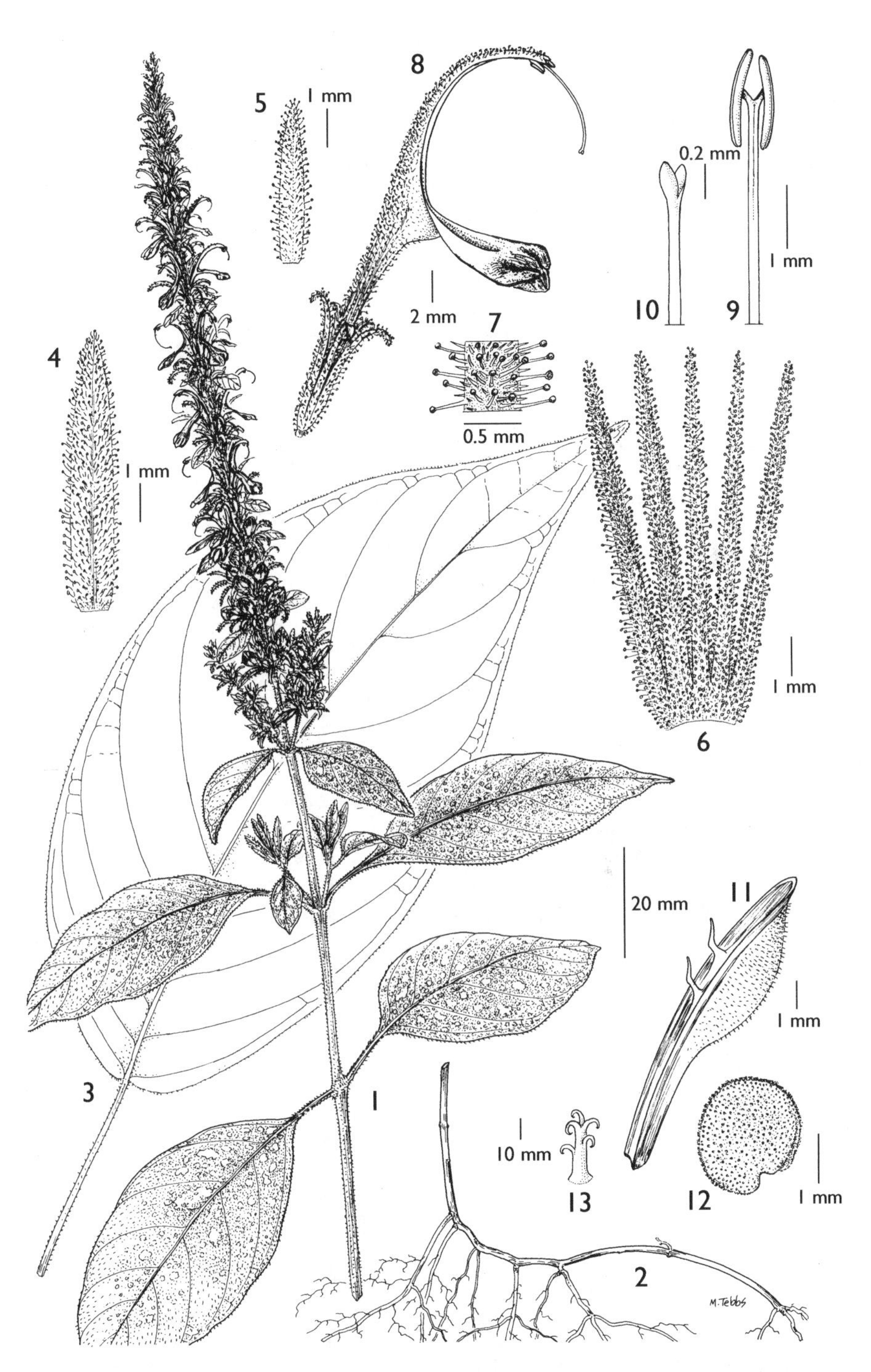

FIG. 88. *CEPHALOPHIS LUKEI* — **1**, habit; **2**, creeping and rooting basal part of stem; **3**, large leaf; **4**, bract; **5**, bracteole; **6**, calyx opened up; **7**, detail of calyx indumentum; **8**, calyx and corolla; **9**, stamen and anther; **10**, stigma; **11**, capsule; **12**, seed; **13**, glochidia from seed surface. 1 from *Torre* 4088; 2 from *Harvey & Vollesen* 41; 3–10 from *Mendonça* 3929; 11–13 from *Luke* 3377. Drawn by Margaret Tebbs.

transverse differently coloured lines ("herringbones"). Stamens 2, no staminodes, attached just inside tube; filaments strongly flattened, strap-shaped, enclosed in upper lip; anthers bithecous, medifixed; thecae equal or one very slightly longer, held at same height and parallel, rounded. Style filiform, held between stamens; stigma with two erect ovoid rounded lobes. Capsule 4-seeded, clavate, with a sterile basal stalk ± equal in length to fertile part; retinacula strong, curved. Seed discoid, circular in outline, without raised rim, densely covered all over with retrorsely barbed glochidiae.

A single species in Kenya and Mozambique.

Cephalophis lukei *Vollesen* **sp. nov.** corolla valde bilabiata, labio superiore corollae angusto valde curvato, labio inferiore lato longitudinaliter plicato partibus lateralibus labii inferioris implicatis partem centralem fere obtegentibus, labio etiam ad apicem lobulis minutis tribus ornato, staminibus duobus in labio superiore inclusis, thecis duobus ad altitudinem eandem affixis, semine discoideo omnino glochidiis retrorse barbatis obtecto atque margine promenenti carenti distinguenda est. Typus: Kenya, Kwale District, Gongoni Forest Reserve, *Harvey & Vollesen* 41 (K!, holotype; CAS!, DSM!, EA!, K!, NHT!, isotypes)

Perennial herb, basal part of stem creeping and rooting, apical part erect, to 0.5(–1) m tall; young stems finely puberulous or sparsely so with antrorsely curved hairs, upwards sometimes also with scattered stalked capitate glands. Leaves with petiole to 6.5 cm long; lamina ovate to elliptic or narrowly so, largest 10–20 × 3.8–9 cm, base cuneate to truncate, slightly to distinctly unequal-sided, margin slightly crenate, apex subacute to shortly acuminate; on both sides antrorsely or spreadingly puberulous along midrib and sparsely so on lateral veins, sometimes also with longer hairs to 1 mm, glabrous on lamina. Inflorescence (5–)10–22(–30) cm long, sometimes with short lateral inflorescences from 1(–2) lower pair of nodes; peduncle 0.5–3(–5) cm long; peduncle and rachis puberulous or densely so, with or without stalked capitate glands, these if present denser upwards; flowers solitary or in 3-flowered cymules towards base of inflorescence (rarely two cymules together), solitary towards apex; main axis bracts often foliaceous at basal nodes, with purplish apical part, upwards narrowly oblong to narrowly obovate, 7–15 × 1.5–3 mm, sparsely to densely puberulous and with scattered to dense stalked capitate glands (rarely glandular hairs only); cymule bracts and bracteoles linear-lanceolate, 0.5–1 cm long, with similar indumentum. Calyx 7–11 mm long, divided to ± 1 mm from base, puberulous with broad glossy capitate glands and minute non-glandular hairs, lobes purple towards apex, linear, acute, often recurved. Corolla 2.2–2.8 cm long along upper lip, basal part of tube greenish, upper part of tube whitish, lower and upper lips purple, dorsal side of upper lip with white area along midrib and weakly rugulate with transverse white bands; basal tube cylindrical, 0.8–1.1 cm long; throat indistinct, 2–3 mm long; upper lip strongly falcately curved and laterally flattened, 1.2–1.6 cm long, with two minute apical teeth; lower lip 7–9 × 6–8 mm, with three minute lobes ± 1 mm long, outside with long (± 1 mm) curly glossy non-glandular hairs all over. Filaments 1.1–1.5 cm long, glabrous; thecae 1.5–2 mm long, linear-oblong. Style 2.1–2.7 cm long, glabrous. Capsule 1–1.3 cm long, finely uniformly puberulous all over; seed 3–4 mm in diameter. Fig. 88, p. 677.

KENYA. Kwale District: Gongoni Forest Reserve, 21 Oct. 1991, *Luke* 2945! & 10 Nov. 1992, *Harvey & Vollesen* 41! & Shimba Hills, Mwele Forest, 13 Nov. 1992, *Luke* 3377!
DISTR. **K** 7; Mozambique
HAB. Wet evergreen lowland forest; 25–400 m

48. **TRICHAULAX**

Vollesen in K.B. 47: 613 (1992); Furness in K.B. 47: 619 (1992)

Shrubby herb; stems articulated (with transverse lines at nodes), swollen above nodes, with two longitudinal furrows (decussate on successive nodes). Leaves opposite, with short often inconspicuous cystoliths, margin crenulate. Flowers sessile, solitary, in dense strobilate distinctly 4-angled terminal spiciform cymes; bracts papery, persistent, with a straight non-pungent tip, with distinct midrib and inconspicuous to almost invisible venation; bracteoles 2, filiform. Calyx deeply divided into 5 equal narrowly triangular acuminate lobes. Corolla white to pale or pinkish mauve, puberulous on the outside; tube cylindric, hardly widening upwards, straight; lower lip with three recurved subequal lanceolate lobes; upper lip with one linear-lanceolate entire or notched recurved lobe. Stamens 2, spreading straight out between lower and upper lip and diverging, inserted near base of lower lip, finely puberulous; anther bithecous, medifixed, held at right angle to filament, thecae oblong, subequal, straight, rounded at both ends. Style filiform, hairy near base; stigma lobes equal, broadly elliptic, rounded, spreading; ovary with two ovules per locule. Capsule 2-seeded, club-shaped, apiculate, puberulous all over, the valves splitting lengthwise from base at maturity; retinacula strong; seed discoid, cordiform in outline, densely verrucose on both sides and on lateral rim, verrucae finely glochidiate.

One species on the East African coast.

Trichaulax mwasumbii *Vollesen* in K.B. 47: 613 (1992); Robertson & Luke, Kenya Coast. For. 2. Checklist Vasc. Pl.: 84 (1993); Lebrun & Stork, Enum. Pl. Afr. Trop. 4: 507 (1997). Type: Tanzania, Uzaramo District, Pande Forest Reserve, *Mwasumbi* 14238 (K!, holo.; BR!, C!, CAS!, DSM!, EA!, NHT!, K!, P!, iso.)

Shrubby herb to 60 cm tall; young branches glabrous apart from a thin line of spreading and downwardly curved hairs in each furrow, soon glabrous. Leaves often tinged purplish; petiole 0.2–3 cm long; lamina ovate to elliptic, largest 3.5–12 × 1.5–4.3 cm, base attenuate, often oblique, apex acute to acuminate or drawn out into a rounded to acuminate tip; glabrous beneath or with a few minute hairs on midrib, above sparsely and minutely puberulous on midrib, ciliate towards base. Inflorescences 1–9(–11.5) cm long; peduncle 3–8 mm long, finely puberulous and sometimes with scattered stalked capitate glands; axes with stalked capitate glands, densest towards apex; bracts green with purple edges, ovate to cordiform or reniform, 5–17 × 3–14 mm, base cordate, apex acute to rounded with a straight acumen to 0.5 mm long, finely puberulous and with many stalked capitate glands, distinctly ciliate; bracteoles 1–4 mm long. Calyx 3–7 mm long, finely puberulous and with stalked capitate glands, finely ciliate towards tip of lobes. Corolla white to pale or pinkish mauve; tube 5–7 mm long, lobes lanceolate, 6–9 mm long, mid-part of lower lip 1.5–2 mm wide. Filaments pink to purple, ± 6 mm long; anthers pink to purple, ± 1.5 mm long. Style white. Capsule 7–9 mm long; seed dark brown when ripe, 3.5–4 × 3–3.5 mm. Fig. 89, p. 680.

KENYA. Kilifi District: Rabai, 2 Apr. 1963, *Verdcourt* 3607! & Sokoke Forest, near Jilore Forest Station, 28 Aug. 1971, *Faden* 71/800! & Dida to Tezo, Nyari, 27 Aug. 1989, *Luke* 1947!

TANZANIA. Lushoto District: Segoma Forest Reserve, 18 June 1966, *Faulkner* 3822!; Uzaramo District: Pugu Hills, 29 Aug. 1982, *Hawthorne* 1668!; Bagamoyo District: Kiono Forest Reserve, 20 Aug. 1989, *Rulangaranga et al.* 175!

DISTR. **K** 7; **T** 3, 6; not known elsewhere

HAB. Undergrowth in dry evergreen or semi-evergreen lowland *Brachylaena-Cynometra-Trachylobium* and *Albizia-Manilkara-Scorodophloeos* forest with undergrowth rich in *Rubiacaeae* and *Euphorbiaceae*, coastal thicket, on red loamy soil; 50–200(–500) m.

FIG. 89. *TRICHAULAX MWASUMBII* — **1**, habit, × $^2/_3$; **2**, small plant, × $^2/_3$; **3**, detail of young stem, × 4; **4**, large leaf, × $^2/_3$; **5**, bract, × 1; **6**, corolla, × 4; **7**, filament and anther, × 6; **8**, stigma, × 20; **9**, capsule, × 4; **10**, seed, × 6; **11**, detail of verrucae on seed surface, × 60. 1, 5 & 9–11 from *Faulkner* 3822, 2 from *Faden* 71/800, 3 & 7–8 from *Mwasumbi* 14238, 4 from *Verdcourt* 3607, 6 from *Frontier Tanzania* 2299. Drawn by Eleanor Catherine. Reproduced from Kew Bull. 47: 615 (1992).

49. DICLIPTERA

Juss. in Ann. Mus. Natl. Hist. Nat. 9: 267 (1807), *nom. cons.*; C.B. Clarke in F.T.A. 5: 256 (1900); K. Balkwill, Getliffe-Norris & M.-J. Balkwill in K.B. 51: 1–61 (1996); Darbyshire & Vollesen in K.B. 62: 119–128 (2007); Darbyshire in K.B. 63: 361–383 (2009)

Diapedium C.W. König in Ann. Bot. 2: 189 (1805); Steud. Nom. ed. 2, 1: 501, 504 (1840), in syn.

Peristrophe Nees in Wall., Pl. As. Rar. 3: 112 (1832); C.B. Clarke in F.T.A. 5: 242 (1900); K. Balkwill in Bothalia 26: 83 (1996)

Annual or perennial herbs or subshrubs, evergreen or deciduous, sometimes suffruticose and pyrophytic; cystoliths linear, many and ± conspicuous. Stems (sub-)6-angular, often ridged. Leaves with pairs equal or anisophyllous; margin entire or obscurely repand. Inflorescences axillary and/or terminal, comprising a series of monochasial cymules, usually umbellately arranged, more rarely solitary, umbels solitary to several in each axil, sometimes compounded and paniculate or in dense terminal heads, rarely spiciform; main axis bracts paired, pairs equal to somewhat unequal, free, often much-reduced, rarely foliaceous; cymule bracts paired, pairs subequal to highly unequal, free, usually conspicuous; bracteoles narrower than the cymule bracts and often much-reduced. Calyx shortly tubular, five-lobed, lobes subequal, lanceolate, indumentum variable outside, appressed-pubescent within. Corolla bilabiate, variously white to pink or purple, usually with darker guidelines on a paler palate, hairy outside except towards base of tube where glabrous; resupinate, the tube twisted through 180°, tube cylindrical, somewhat widened towards mouth, rarely strongly so; lip held in upper position oblong, ± recurved distally, apex shortly three-lobed; lip held in lower position unlobed, apex obtuse or shallowly emarginate. Stamens two; filaments arising from mouth of corolla tube or rarely from within the tube, exserted; anthers bithecous, thecae elliptic, offset, superposed or slightly overlapping. Ovary (oblong-) ovoid, glabrous or puberulous, two-locular, two ovules per locule; style filiform; stigma bifid, exserted. Capsule short- to long-stipitate, placental base elastic, the placental base and thin capsule walls tearing from the thickened raphes at dehiscence, or inelastic. Seeds 4 per capsule, rarely 2 by abortion, held on retinacula, lenticular, discoid or compressed-ellipsoid, with a hilar excavation, smooth or tuberculate, tubercles sometimes hooked.

Under the broadened generic concept applied here, with the inclusion of those species with an inelastic placental base at fruit dehiscence previously separated as the genus *Peristrophe*, *Dicliptera* comprises approximately 175 species with a pantropical and subtropical distribution. The genus is in need of a full revision.

The inflorescence structure in both *Dicliptera* and *Hypoestes* is rather complex. Each cyme is divided into discrete monochasial inflorescence units (cymules), sometimes reduced to a single flower. The branching of each cymule peduncle is subtended by a pair of bracts (the "main axis bracts") which are positioned at the terminus of the primary peduncle when the cymules are umbellately arranged (this being the most common arrangement in *Dicliptera*; e.g. fig. 92: 3, p. 700) or along the inflorescence axis when the cymules are spicately arranged (fig. 90: 8e, p. 682). These main axis bracts are rarely (in *D. pumila* and *D. brevispicata*) absent. Each individual cymule is subtended by a pair of bracts (cymule bracts) that are often showy and typically larger and broader than the main axis bracts, usually enclosing the bracteoles, calyx and lower portion of the developed corolla. It is the cymule bracts that are of greatest taxonomic interest within *Dicliptera* (see figs. 90, p. 682 and 91, p. 685). In several species, they become pale towards the base due to a concentration of sclereids but with the midrib and lateral veins remaining darker and opaque, giving the bract a "windowed" effect (e.g. Figs. 90: 7, 8, 10–12, p. 682; 92: 3, p. 700). Several flowers can develop within each cymule, each flower being subtended by a pair of bracteoles. In both *D. pumila* and *D. brevispicata* the cymule can become shortly spiciform or verticillate at maturity, each flower being subtended by one bract and a pair of bracteoles (figs. 91: 9e, p. 685 & 93: 3, p. 715); in these species the cymule bracts are subequal to the bracts subtending each flower, thus the inflorescence is atypical for the genus.

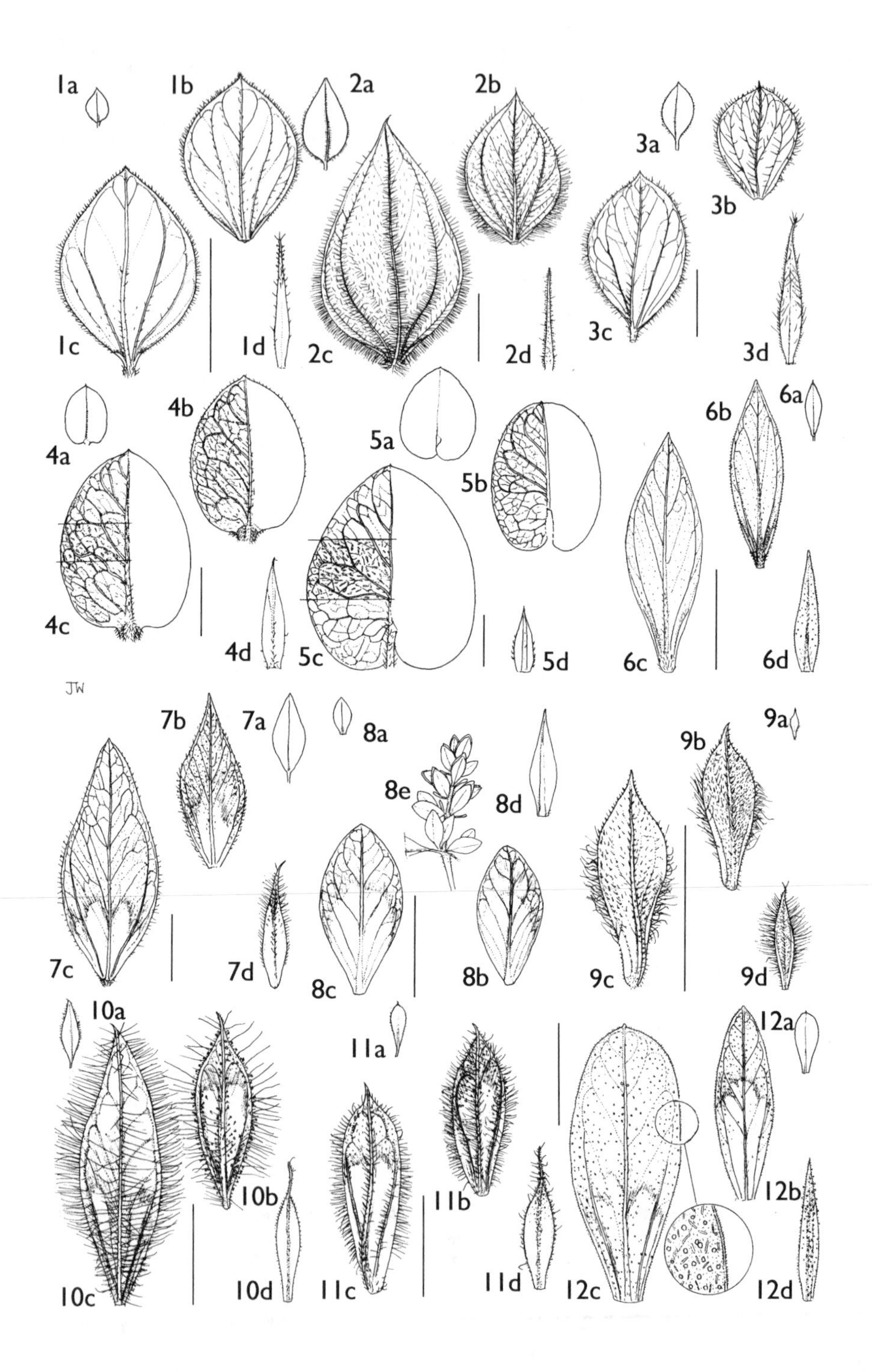
1a
1b
2a
2b
3a
3b
1c
1d
2c
2d
3c
3d
4a
4b
5a
5b
6a
6b
4c
4d
5c
5d
6c
6d
JW
7a
7b
8a
8e
8d
9a
9b
7c
7d
8c
8b
9c
9d
10a
11a
12a
10b
11b
12b
10c
10d
11c
11d
12c
12d

Identification of *Dicliptera* taxa is most easily achieved by comparison to named material. Strong characters for separation of species are few, it usually being the combination of a suite of characters that enables species delimitation. Such differences are difficult to convey in a key, which is therefore necessarily descriptive in places. Geography and ecology can be helpful.

1. Corolla tube 18 mm long or more, clearly expanded and broadly cylindrical above the 180° twist; stamens inserted 7–9 mm below the corolla mouth; scandent herb of (sub)montane forest, **T** 6 .... 10. *D. grandiflora* p.696
   Corolla tube up to 11 mm long, only slightly expanded above the twist, not broadly cylindrical; stamens inserted at or immediately below the corolla mouth 2
2. Cymules arranged in 2–10 pairs along a spike (fig. 90: 8e, p. 682); **U** 2 .... 11. *D. alternans* p.697
   Cymules solitary or usually umbellately arranged, sometimes compounded into loose panicles or congested into globose, conical or cylindrical heads 3
3. Corolla minute, 4.5–7.5 mm long, limb barely extending beyond the cymule bracts; inflorescences predominantly axillary 4
   Corolla larger, 9–30 mm long, clearly exposed beyond the bracts, or if corolla smaller (to 7.5 mm long in *D. vollesenii*) then inflorescences exclusively terminal, in a dense globose or conical head 6
4. Stems strongly 6-angular with prominent pale ridges on the angles; cymule bracts elliptic, obovate or oblanceolate, to 3 mm wide .... 12. *D. verticillata* p.697
   Stems (sub)angular, sulcate when young but not prominently ridged on the angles; cymule bracts broadly ovate to broadly elliptic, 4–10.5 mm wide 5
5. Plants erect, decumbent or trailing; cymule bracts ovate, minutely apiculate, margin pilose (rarely shortly so); inflorescences dense, (2–)3–5(–6) umbels of (2–)3–5(–6) cymules per axil; Congolian forest: **U** 2, 4, **K** 5, **T** 1 .... 1. *D. elliotii* p.688
   Plants creeping; cymule bracts (ovate-) elliptic, mucronate, margin shortly ciliate; inflorescences less dense, 1–2 umbels of (1–)2–3(–4) cymules per axil; coastal forest: **K** 7, **T** 3, 6 .... 2. *D. inconspicua* p.689

FIG. 90. *DICLIPTERA* cymule bracts and bracteoles — **1**, *D. ELLIOTII* (*Bidgood et al.* 4780); **2**, *D. HETEROSTEGIA* (*Magogo & Glover* 137); **3**, *D.* sp. A (*Lovett & Niblett* 2309); **4**, *D. NAPIERAE* (*Verdcourt* 1575); **5**, *D. CORDIBRACTEATA* (*Gillett* 13672); **6**, *D. LAXATA* (*Faden* 74/917); **7**, *D. GRANDIFLORA* (*Mhoro* in UMBCP 491); **8**, *D. ALTERNANS* (*Poulsen* 681); **9**, *D. VERTICILLATA* (*Bidgood & Vollesen* 1214); **10**, *D. MACULATA* subsp. *MACULATA* (*Bidgood et al.* 4894); **11**, *D. MACULATA* subsp. *USAMBARICA* (*Glover et al.* 2088); **12**, *D.* sp. D (*Richards* 13060). For each species: **a**, scaled outline of larger cymule bract; **b**, cymule bract, smaller of each pair, outer surface; **c**, cymule bract, larger of each pair, outer surface; **d**, bracteole, outer surface; **8e**, arrangement of cymules in *D. ALTERNANS* (× $^2/_3$). Scale bar = 5 mm for all cymule bracts. Bracteoles: **1d** × 16; **2d** × 6; **3d**, **5d** × 4; **4d**, **6d**, **9–12d** × 3; **8d** × 8. Drawn by Juliet Williamson.

6. Cymule bracts broadly ovate, broadly elliptic or orbicular (fig. 90: 2–5, p. 682), green throughout or somewhat paler towards base but lacking "windows", sometimes pale-scarious in fruit; bracteoles pale throughout 7
   Cymule bracts variously shaped but usually narrower, if broadly elliptic then conspicuously "windowed" and/or bracteoles with a clearly darker midrib and apex, pale only towards the margin 11
7. Cymule bracts with principal venation 3-veined or palmate, base obtuse, rounded or subcordate (fig. 90: 2/b–c, & 3/b–c, p. 682); capsule 4.5–6.5 mm long 8
   Cymule bracts with venation pinnate-reticulate, base shallowly to deeply cordate (figs. 90: 4/b–c & 5/b–c, p. 682); capsule 7–9 mm long 10
8. Cymule bracts puberulent on the outer surface and margin, pairs subequal, each with apex obtuse or rounded and apiculate; **T** 7 . . . . 5. *D. sp. B* p.692
   Cymule bracts long-pilose at least on the margin where often dense, pairs clearly unequal in size or if subequal then apex acute or attenuate and conspicuously mucronate 9
9. Cymule bracts densely pilose outside at least on margin, larger of each pair with apex conspicuously mucronate; filaments glabrous or with sparse hairs towards base; (1–)2–5(–7) umbels of 2–5(–7) cymules per axil (widespread) 3. *D. heterostegia* p.690
   Cymule bracts rather sparsely pilose, mainly on the margin and midrib, apices minutely apiculate; staminal filaments pubescent on upper surface; 1 inflorescence of 1–3 cymules per axil; **T** 7 . . . . 4. *D. sp. A* p.691
10. Cymule bract pairs subequal or slightly unequal, the larger bract 6–13.5 mm long, base shallowly cordate; main axis bracts ovate (often caducous); capsule glabrous; leaves with apex obtuse or rounded . . . . 6. *D. napierae* p.693
   Cymule bract pairs strongly unequal, the larger bract 11–23 mm long, base deeply cordate; main axis bracts linear-lanceolate; capsule puberulous; leaves with apex acute or acuminate . . . . 7. *D. cordibracteata* p.694

FIG. 91. *DICLIPTERA* cymule bracts and bracteoles — **1**, *D. LATIBRACTEATA* (*Brodhurst Hill* 666); **2**, *D. ALBICAULIS* (*Verdcourt* 1556); **3**, *D. MINUTIFOLIA* (*Sebsebe & Ensermu* 2715); **4**, *D. MELLERI* (*Richards & Arasululu* 25827A); **5**, *D. CAPITATA* (*Bidgood et al.* 3725); **6**, *D. CARVALHOI* subsp. *CARVALHOI* (*Milne-Redhead & Taylor* 10907); **7**, *D. CARVALHOI* subsp. *ERINACEA* (*Lovett & Kayombo* 468); **8**, *D. NILOTICA* (*Bogdan* 3693); **9**, *D. PUMILA* (*Goyder et al.* 3761); **10**, *D.* sp. F (*Lovett & Kayombo* 159); **11** *D. PANICULATA* (*Verdcourt* 1830); **12** *D. ACULEATA* (*Greenway & Kanuri* 13803). For each species: **a**, scaled outline of larger cymule bract (except **11a**); **b**, cymule bract, smaller of each pair, outer surface except **7b** inner surface; **3**, cymule bract, larger of each pair, outer surface except **7c** inner surface; **4** bracteole, outer surface; **9e** arrangement of flowers within cymule in *D. PUMILA* (× 1); **11a** typically bent cymule peduncle in *D. PANICULATA* (× $^{2}/_{3}$). Scale bar = 5 mm for all cymule bracts. Bracteoles: **1d**, **10d** × 2; **2d**, **6d** × 3; **3d** × 8; **4d**, **5d**, **7d**, **11d**, **12d** × 4; **8d**, **9d** × 5. Drawn by Juliet Williamson.

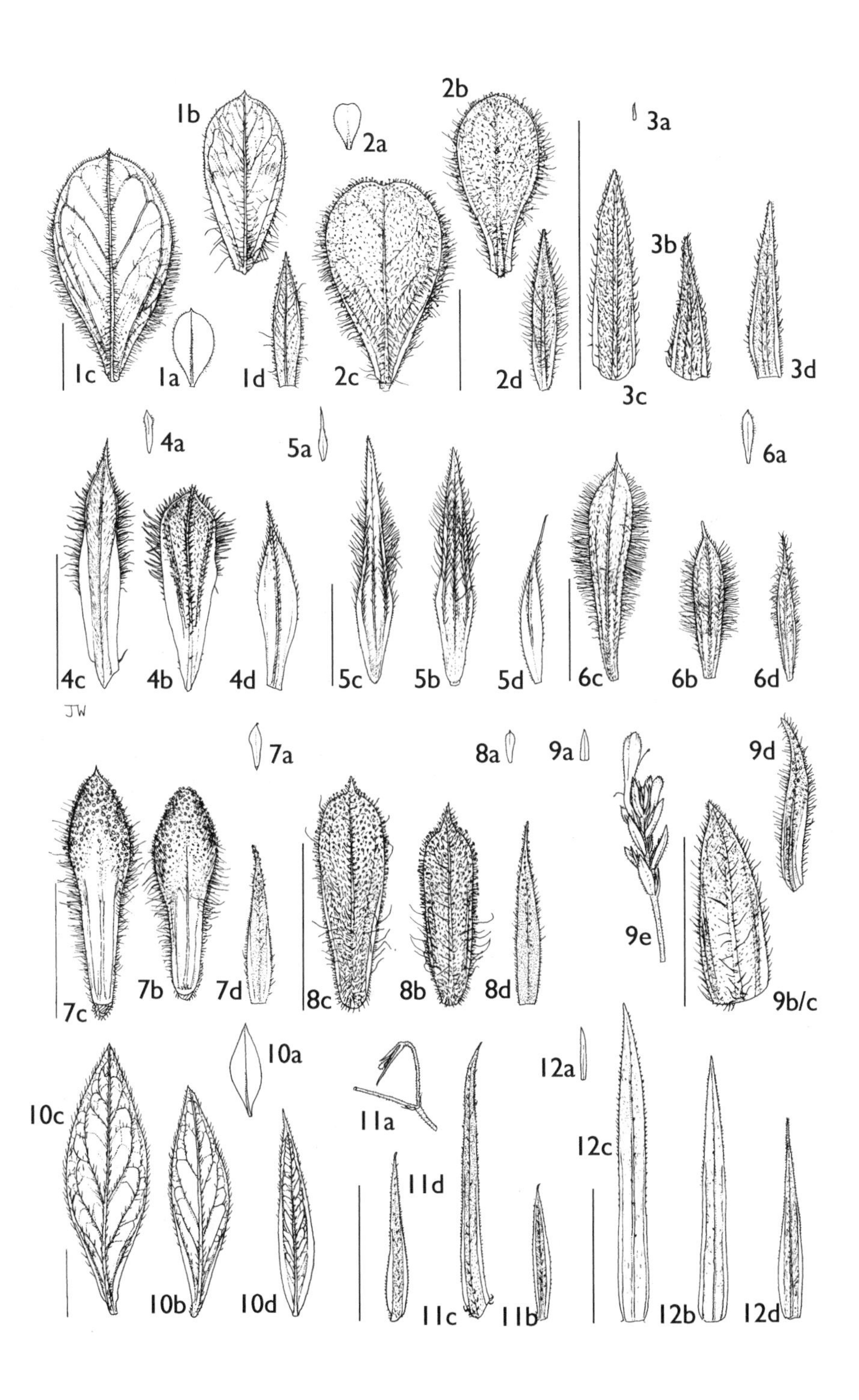
1b
2b
2a
3a
3b
1c
1a
1d
2c
2d
3c
3d
4a
5a
6a
4c
4b
4d
5c
5b
5d
6c
6b
6d
JW
7a
8a
9a
9d
9e
7b
7d
7c
8c
8b
8d
9b/c
10a
10c
11a
12a
12c
11d
10b
10d
11c
11b
12b
12d

11. Cymule bracts with venation pinnate-reticulate, surface initially green but soon scarious, then pinkish-brown in colour and with a hyaline margin, larger bract of a pair 4.5–7 mm wide, elliptic or rhombic (fig. 91: 10/b–c, p. 685); **T** 7 .... 27. *D. sp. F* p.716
    Cymule bracts with principal venation palmate, 3-veined or only midrib prominent, surface not as above or if so then larger bract of a pair to 2(–3) mm wide, (linear-) lanceolate 12
12. Stems soon woody (or softly so), pale grey or whitish 13
    Stems herbaceous or if woody towards base then not pale grey or whitish, leafy stems (blackish-) green or brown-green, sometimes with paler alternating stripes 15
13. Cymule bracts inconspicuous, lanceolate, the larger of each pair up to 4.5 × 0.8 mm; **K** 1 .... 18. *D. minutifolia* p.706
    Cymule bracts conspicuous, obovate, the larger of each pair 7–14 × 2.5–8 mm 14
14. Bare sections of stems with prominent leaf scars between short internodes, leaves and inflorescences usually conspicuously clustered towards stem apices; corolla 19–27 mm long; bracteoles 7.5–12 mm long; ovary conspicuously puberulous particularly towards apex; **K** 6 .... 17. *D. cicatricosa* p.705
    Bare sections of stems not bearing many prominent leaf scars, leaves and inflorescences less clustered; corolla 13–19(–21) mm long; bracteoles 5.5–8 mm long; ovary largely glabrous or with sparse hairs towards apex; **K** 4, 6, **T** 1, 2 .... 16. *D. albicaulis* p.704
15. Perennial herbs and shrubs of moist lowland to montane forest and riverine woodland, not suffruticose 16
    Wiry annual herbs or suffruticose perennials, the latter producing annual stems from a woody base and rootstock, often pyrophytic; plants typically of drier, often fire-prone habitats but also in riverine thicket and woodland 20
16. Cymule bracts largely glabrous except for inconspicuous short hairs towards the base and sometimes on the midrib; seeds with tubercles most dense towards or restricted to the rim 17
    Cymule bracts with a more dense indumentum, often conspicuously ciliate and with glandular hairs; seeds with tubercles more evenly distributed 18
17. Corolla 10.5–16 mm long, white or cream, rarely pink or pale purple; hairs on peduncles antrorse to appressed-ascending .... 8. *D. laxata* p.694
    Corolla 22–25 mm long, purple; hairs on peduncles retrorse .... 9. *D. sp. C* p.695
18. Cymule bracts broadly obovate, length/width ratio less than 2/1, paler towards base but without conspicuous "windows"; main axis bracts usually broadly elliptic to obovate, rarely narrowly elliptic; calyx lobes (5.5–)6.5–8.5 mm long .... 15. *D. latibracteata* p.703

Cymule bracts oblong- to (ovate-) elliptic or if obovate then narrower, length/width ratio 2–5/1, often conspicuously "windowed"; main axis bracts linear-lanceolate or rarely narrowly ovate; calyx lobes 3–6 mm long 19

19. Inflorescences variously pubescent and/or pilose, sometimes interspersed with shorter hairs but not puberulous; capsules with predominantly glandular hairs . . . . 13. *D. maculata* p.699

Inflorescences puberulous throughout (fig. 90: 12c, p. 682); capsules with predominantly eglandular hairs . . . . 14. *D. sp. D* p.703

20. At least the smaller bract of each cymule bract pair prominently 3-veined, sometimes "windowed"; seeds smooth, discoid, 2.2–3 mm in diameter 21

Cymule bracts not prominently 3-veined, either all veins inconspicuous, midrib only prominent or venation reticulate; seeds (where known) tuberculate or if smooth then only subflattened and up to 1.7 mm in diameter 23

21. Pyrophytic suffrutex, producing few to many short flowering shoots from a woody base and rootstock 19. *D. melleri* p.707

Slender annual herbs 22

22. Cymule bracts conspicuously long-ciliate, the smaller bract of each pair lanceolate or rarely elliptic, apex acute to subattenuate . . . . 20. *D. capitata* p.707

Cymule bracts very shortly hairy, not conspicuously ciliate, the smaller bract of each pair elliptic or obovate, apex obtuse or rounded . . . . 21. *D. vollesenii* p.708

23. Placental bases of capsules rising elastically at dehiscence, walls thinner than and tearing from the thickened raphes; cymule bracts often paler towards base, never scarious 24

Placental bases inelastic or largely so at dehiscence, walls not clearly thinner than the raphes and remaining (largely) intact; cymule bracts unicolorous except for the abrupt narrow hyaline margin, sometimes scarious at maturity 28

24. Slender annual or perennial herbs with wiry, laxly branched stems 30–150 (–200) cm tall 25

Pyrophytic suffruticose perennials, producing few to many flowering shoots 5–35 cm tall from a woody base and rootstock, often with precocious flowering 27

25. Inflorescences congested into a terminal globose, cylindrical or verticillate synflorescence, umbels usually many but rarely reduced to 1–2 terminal umbels in small plants, usually with additional umbels in the uppermost leafy axils; each umbel subsessile or shortly and inconspicuously pedunculate . . . . 22. *D. carvalhoi* p.709

Inflorescences more lax, 1–2 umbels in the upper, often leafless, axils, not forming a terminal synflorescence; at least some umbels conspicuously pedunculate, primary peduncle up to 16(–22) mm long 26

26. Larger cymule bract of each pair 6.5–10(–13) mm long, rather abruptly narrowed into the mucro, outer surface with pilose hairs sparse or many on margin and midrib only; capsule 5.5–6.5 mm long; **T** 4 . . . . 22. *D. carvalhoi* p.709
    Larger cymule bract of each pair 11.5–16 mm long, narrowed gradually into the apical mucro, outer surface densely white-pilose; capsule 7.5–8 mm long; **T** 1 . . . . 23. *D. sp. E* p.711
27. Cymules (potentially) several-flowered, subcapitate or shortly spicate, flowers each subtended by a single bract and pair of bracteoles (fig. 91: 9e, p. 685); axillary inflorescences of solitary cymules (terminal inflorescences can appear umbellate), 1–2 per axil; peduncle 7–30(–95) mm long; corolla 13–17 mm long . . . . 25. *D. pumila* p.713
    Cymules with a single pair of bracts enclosing 1–2 flowers, each subtended by a pair of bracteoles only; all inflorescences of 2–5 umbellately arranged cymules; umbels sessile or primary peduncle to 5(–15) mm long; corolla 11–13.5 mm long . . . . 24. *D. nilotica* p.712
28. Cymules solitary or in umbels of 2(–3); each cymule potentially several-flowered, subcapitate or shortly spicate, flowers each subtended by a single bract and pair of bracteoles (fig. 93, 3 & 4, p. 715); capsule lacking glandular hairs . . . . . 26. *D. brevispicata* p.714
    Cymules arranged in umbels, usually of 3–4 cymules, or compounded into large panicles on the bare upper portions of the branches; each cymule 1–2-flowered, with a pair cymule bracts and flower(s) subtended by a pair of bracteoles only; capsule with at least some glandular hairs 29
29. Plants annual; corolla (9–)11–13.5(–17) mm long; seeds with hooked tubercles; peduncles with dense short eglandular hairs and sparse to dense glandular hairs . . . . 28. *D. paniculata* p.716
    Plants perennial; corolla 15–23(–30) mm long; seeds with unhooked tubercles; peduncles glabrous or sparsely hispid, rarely with subsessile glands . . . . 29. *D. aculeata* p.718

1. **Dicliptera elliotii** *C.B. Clarke* in F.T.A. 5: 258 (1900); Heine in F.W.T.A. ed. 2, 2: 425 (1963) & in Fl. Gabon 13: 193 (1966); Darbyshire in K.B. 63: fig. 1H, table 1 (2009). Syntypes: Sierra Leone, Scarcies R., Wallia, *Scott-Elliot* 4632 (K!, syn.); Congo-Kinshasa, *Smith* 8 & 64 (both BM!, syn.)

Erect, decumbent or trailing, annual or perennial herb, 15–100 cm tall; stems weak, sulcate when young, sparsely antrorse- and retrorse-pubescent, soon glabrescent. Leaves ovate, 2–8.5 cm long, 1–4 cm wide, base attenuate, apex attenuate to acuminate, glabrous except for sparse hairs on margin and midrib, upper surface occasionally with few longer hairs; lateral veins (3–)4–5 pairs; petiole (2–)7–18(–28) mm long. Inflorescences axillary, (1–)2–4(–5) umbels of (2–)3–5(–6) cymules per axil; primary peduncle 1.5–12.5 mm long, spreading- or antrorse-pubescent; main axis bracts linear-lanceolate, 1.5–5 mm long, margin and midrib sparsely pubescent; cymule peduncles 2–9 mm long; cymule bracts broadly ovate, pairs unequal, larger bract (5.5–)7–9.5(–11) mm long, (4–)5–8(–10.5) mm wide, apex obtuse or shortly attenuate, apiculate, surfaces sparsely to densely pale-pilose mainly on margin, midrib and towards apex of larger bract within, occasionally only ciliate, palmately 5–7-nerved; bracteoles pale throughout, lanceolate, 1–2 mm long, with sparse short glandular and/or eglandular hairs outside. Calyx lobes 1.3–2 mm long, puberulent. Corolla white or cream, 5.5–7.5 mm long, pubescent outside; tube sometimes curved, 4–5 mm long; lip held in upper position oblong, 2–3 mm long, 1–1.5 mm wide; lip held in lower position rounded-obovate, 1.5–2.5 mm long and wide. Staminal filaments 1–2 mm long, with occasional hairs near base; anther thecae ± 0.4 mm long, superposed. Ovary glabrous or sparsely puberulous towards apex; style sparsely strigulose. Capsule 4.5–5 mm long, densely pubescent on raphes, hairs mainly eglandular, placental base elastic; seeds (1.5–)2 mm in diameter, minutely tuberculate, tubercles hooked. Fig. 90: 1/a–d, p. 682.

UGANDA. Bunyoro District: Budongo Forest, Dec. 1938, *Loveridge* 159!; Mengo District: Kikubamiti, shore of Lake Victoria "a few miles" from Kampala, Dec. 1935, *Chandler* 1477! & Mpanga Forest, Jan. 1960, *Lind* 2693!

KENYA. North Kavirondo District: Kakamega Forest, Kibiri Block, S side of Yala R., *Faden et al.* 70/12! & Kakamega, Buyangu, Feb. 2005, *Malombe & Morawetz* 950!

TANZANIA. Bukoba District: Minziro Forest Reserve, Nyakabanga, Aug. 1999, *Sitoni, Festo & Bayona* 766! & Minziro Forest Reserve, July 2000, *Bidgood et al.* 4780!

DISTR. **U** 2, 4; **K** 5; **T** 1; Sierre Leone to Congo-Kinshasa and Cabinda

HAB. Moist forest, particularly along margins and pathways; 1000–1650 m

USES. None recorded on herbarium specimens

CONSERVATION NOTES. Widespread in the Guineo-Congolian forests of West and Central Africa. It appears commoner in the western part of its range, becoming more scattered from Nigeria eastwards. It appears to tolerate or even favour some forest disturbance and may benefit from fragmentation by human activity: Least Concern (LC).

SYN. *D. microchlamys* S. Moore in J.B. 48: 255 (1910). Type: Uganda, Bunyoro/Mubende District, Nkusi [Ngusi] R., E of Lake Albert, *Bagshawe* 1387 (BM!, holo.), **syn. nov.**

D. silvicola Lindau in Z.A.E. 2: 302 (1914). Type: Tanzania, Bukoba District, Minziro Forest [Budduwald], *Mildbraed* 147 (B†, holo.; BR!, iso.)

D. silvicola Lindau var. *elongata* Lindau in Z.A.E. 2: 302 (1914). Type: Cameroon, Beni, near Muera, *Mildbraed* 2365 (B†,holo.; BR!, iso.)

NOTE. West African specimens have rather densely hairy cymule bracts. In Uganda, Tanzania and Congo bract indumentum becomes more variable, with the more sparsely hairy form from Uganda having previously been separated as *D. microchlamys* S. Moore. However, a range of intermediate specimens, as well as those closely matching the types of *D. elliotii*, are also found in our region indicating that *D. microchlamys* falls within the variation of *D. elliotii*.

2. **Dicliptera inconspicua** *I. Darbysh.* in K.B.: 63: 362, fig. 1A–G (2009). Type: Tanzania, Tanga District, Kiwanda, 15 km NW of Muheza, *Archbold* 2048 (K!, holo.; DSM!, EA!, iso.)

Creeping herb, often rooting at nodes of principal stems; stems 6-angular, sulcate, ridges sparsely pale antrorse- or retrorse-pubescent when young, soon glabrescent. Leaves ovate, 1–4(–5.5) cm long, 0.5–2(–2.5) cm wide, base shortly attenuate, somewhat asymmetric, apex obtuse or rounded, glabrous or midrib and margin with inconspicuous short hairs, upper surface occasionally with few longer hairs; lateral veins 3–4 pairs; petiole 2–13(–25) mm long. Inflorescences axillary and terminal, 1–2 umbels of (1–)2–3(–4) cymules per axil; primary peduncle 1.5–5 mm long, extending to 7.5 mm in fruit, antrorse- or retrorse-pubescent; main axis bracts linear-lanceolate, 1–3 mm long, margin and midrib with sparse antrorse hairs; cymule peduncles 1.5–4.5 mm long, extending to 8 mm in fruit, antrorse-pubescent; cymule bracts becoming scarious in fruit, broadly (ovate-) elliptic, pairs somewhat unequal, larger bract 6–9.5 mm long, 4–6 mm wide, apex acute or shortly attenuate, mucronate, midrib antrorse-pubescent, with scattered hairs elsewhere outside, margin shortly ciliate, largely glabrous within, palmately 5–7-nerved; bracteoles pale throughout, lanceolate, 1–2.5 mm long, with sparse short hairs outside, sometimes glandular. Calyx lobes 1.3–2 mm long, sparsely puberulent. Corolla white with purple markings on lip held in upper position or purple throughout, 5.5–6.5 mm long, pubescent outside; tube 4–4.5 mm long; lip held in upper position oblong, 2–2.5 mm long, 1–1.5 mm wide; lip held in lower position broadly and obtusely trullate, 1.5–2 mm long, ± 2.3 mm wide. Staminal filaments 1–1.5 mm long, glabrous or with sparse hairs towards base; anther thecae ± 0.4 mm long, superposed. Ovary glabrous; style largely glabrous. Capsule 3.5–4.5 mm long, ± densely pubescent on raphes; placental base elastic; seeds ± 1.5 mm in diameter, minutely tuberculate, tubercles hooked.

KENYA. Kwale District: Jombo Hill, Feb. 1989, *Robertson et al.* in MDE 231!; Kilifi District: Marafa, Nov. 1990, *Luke & Robertson* 2493!; Tana River District: Tana River National Primate Reserve, Mnazini South, Mar. 1990, *Luke et al.* in TPR 459!

TANZANIA. Lushoto District: Kwamtili to Segoma, Oct. 1918, *Peter* 60707! & Segoma Forest, Aug. 1968, *Faulkner* 4191!; Morogoro District: near Mkata, Oct. 1968, *Faulkner* 4143!

DISTR. **K** 7; **T** 3, 6; not known elsewhere

HAB. Lowland moist semi-deciduous forest and riverine forest; 0–450 m

USES. None recorded on herbarium specimens

CONSERVATION NOTES. Assessed as Near Threatened (NT) by Darbyshire (l.c.).

NOTE. A specimen from **T** 6, Nyamuete Forest Reserve (*Frontier Tanzania* 3075) appears somewhat intermediate between this species and *D. heterostegia*; the cymule bract shape and indumentum resemble the latter but the inflorescences are less dense and the capsules (4.5 mm) are smaller than in that species and closer to *D. inconspicua*. More specimens, including flowers, are desirable from this site.

3. **Dicliptera heterostegia** *Nees* in DC., Prodr. 11: 478 (1847); K. Balkwill *et al.* in K.B. 51: 24 (1996); Luke in Journ. E. Afr. Nat. Hist. Soc. 94: 92 (2005); Darbyshire in K.B. 63: table 1 (2009). Type: South Africa, Transkei, between Umzimvubu [Omsamwubo] and Umsikaba [Omsamcaba], *Drège* s.n. (B†, holo.; G-DC, K! [see note], iso.)

Erect or scrambling, ?annual or perennial herb or subshrub, 25–100(–200) cm tall; stems (sub-)6-angular, often woody at base, antrorse-pubescent or densely silky white-pilose at least when young, rarely largely glabrous. Leaves sometimes undeveloped at flowering, ovate to elliptic, 3.5–16 cm long, 2–7 cm wide, base (cuneate-) attenuate, apex attenuate to acuminate, upper surface and veins beneath sparsely pilose or largely glabrous, margin and midrib above more densely pubescent; lateral veins 5–8 pairs; petiole 1–6 cm long. Inflorescences axillary, (1–)3–5(–7) umbels of 2–5(–7) cymules per axil; primary peduncle 2–7(–10) mm long, antrorse-pubescent or pilose; main axis bracts linear-lanceolate, 2.5–6(–7.5) mm long, margin and midrib sparsely hairy; cymule peduncles 1–7(–9.5) mm long, longest on central cymule of each umbel; cymule bracts becoming scarious in fruit, broadly ovate or ovate-elliptic, pairs slightly to strongly unequal, larger bract (9–)12–25 mm long, (6–)7.5–14.5 mm wide, margin sometimes revolute, apex acute or attenuate and mucronate in the larger bract, often obtuse or rounded in the smaller bract, surfaces and particularly margin white-pilose, also glandular- and eglandular-puberulent at least towards the apex within, palmately 5–7-nerved; bracteoles pale, lanceolate, 2–4(–7.5) mm long, with short glandular hairs and occasional longer eglandular hairs. Calyx lobes 2.5–4(–5.5) mm long, glandular-puberulent, sometimes with few longer eglandular hairs towards margin. Corolla pink, purple or white, 14–19(–26) mm long, eglandular-pubescent outside; tube 8–11.5 mm long; lip held in upper position oblong-obovate, 6–10.5(–15) mm long, 3–5 mm wide, palate sparsely puberulous towards mouth; lip held in lower position broadly elliptic-obovate, 6–8.5(–14.5) mm long, 4–6 mm wide. Staminal filaments 4.5–8.5(–18) mm long, glabrous or with sparse hairs towards base; anther thecae 1–1.2 mm long, superposed and somewhat oblique. Ovary glabrous; style sparsely strigulose in upper half. Capsule 5–6.5 mm long, puberulous to pubescent particularly on raphes, hairs glandular and/or eglandular; placental base elastic; only immature seeds seen, ± 1.5 mm in diameter, tuberculate, tubercles hooked. Fig. 90: 2/a–d, p. 682.

KENYA. Lamu District: Mararani, Boni Forest, Sept. 1961, *Gillespie* 342!; Kilifi District: Kaya Rabai, Kombeni, Feb. 1993, *Luke* 3510!; Kwale District: Pengo Hill area on the road to Shimba Hills settlement, Feb. 1968, *Magogo & Glover* 137!

TANZANIA. Mpwapwa District: E Mpwapwa, Oct. 1933, *R.M. & H.E. Hornby* 539!; Morogoro District: Uluguru Mts, Kimboza Forest Reserve, Sept. 2001, *Luke* 7639!; Ulanga District: Matundu Forest Reserve, Ifakara, Sept. 1994, *Kisena* 1459!

DISTR. **K** 7; **T** 3, 5–8; E Congo-Kinshasa, Malawi and Mozambique to South Africa; Madagascar

HAB. Coastal and riverine forest and forest margins, undergrowth of secondary woodland and thicket; 0–700(–1150) m

USES. None recorded on herbarium specimens

CONSERVATION NOTES. This species has a wide though somewhat disjunct distribution and can be locally common in suitable habitat. It appears to prefer shaded areas and is therefore unlikely to tolerate substantial habitat disturbance, though it can be found along forest edges. It is not considered threatened: Least Concern (LC).

SYN. *Justicia heterostegia* E. Mey. in Drège, Cat. Pl. Afr. Austral. 1: 3 (1837) & in Drège, Zwei Pflanzengeogr. Doc.: 152, 159 & 195 (1843), *nom. nud.*

Dicliptera mossambicensis Klotzsch in Peters, Reise Mossamb., Bot. 1: 220 (1861); C.B. Clarke in F.T.A. 5: 258 (1900). Type: Mozambique Island and mainland, *Peters* s.n. (B†, holo.), cited by C.B. Clarke (1900) as "Mozambique, *Klotzsch*" in error

D. insignis Mildbr. in N.B.G.B. 12: 521 (1935). Type: Tanzania, Lindi District, Lake Lutamba, ± 40 km W of Lindi, *Schlieben* 5240 (B†, holo.; BM!, BR!, LISC!, M!, iso.), **syn. nov.**

NOTE. *D. heterostegia* varies considerably in terms of stem indumentum, leaf development at flowering and the relative and absolute size of the cymule bract pairs, these characters appearing to vary independently of one another. Specimens from South Africa, including the type material, closely match both the coastal Kenyan material and a robust form from submontane and riverine forest in Malawi, in largely lacking pilose hairs on the stems, having well-developed leaves at flowering and having highly unequal cymule bract pairs. Elsewhere in the Flora Zambesiaca region, plants are densely pilose on the stems and often lack mature leaves at flowering; this form extends to the Mpwapwa region of Tanzania. Plants from E Tanzania often have only slightly unequal cymule bract pairs but vary considerably in stem indumentum and leaf development at flowering.

Specimens from **T** 8 in the Ruvumu catchment (*Kirk* s.n., *Bally* 16931) have unusually large corollas with long stamens, the lip held in upper position up to 15 mm long (max. 7(–10.5) mm elsewhere in our region), and correspondingly long bracteoles (5–7.5 mm versus 2–4 mm) and calyx lobes (4.5–5.5 mm versus 2–4 mm). However, intermediate populations are recorded in N Mozambique. Moreover, Balkwill *et al.* (1996) record corollas in *D. heterostegia* from South Africa with the lip held in upper position up to 20 mm long.

Of the Asian and American species listed by K. Balkwill *et al.* (l.c.) as possibly conspecific with *D. heterostegia*, *D. foetida* (Forssk.) Blatter, described from Yemen, appears closest. It however differs in having fewer umbels per axil (1–2), the cymule bracts and peduncles lacking a pilose indumentum, the corolla being smaller (11–13 mm long), the staminal filaments being more densely pubescent and the leaf bases being acute to rounded. *D. foetida* occurs in Africa only in the Ethiopian highlands.

The two specimens at K believed to be type material of *D. heterostegia* are confused by labelling of the locality by C.B. Clarke as "between St. John's R. and Umtsikaba River" which does not exactly match that written on the isotype in de Candolle's herbarium (as cited by K. Balkwill *et al.* 1996), but the year of collection (1837) matches that cited in the protologue by Nees.

4. **Dicliptera sp. A** (=*J. & J. Lovett & Niblett* 2309)

Decumbent herb to 50 cm tall, rooting at lower nodes; stems 6-angular, somewhat sulcate, ridges antrorse-pubescent immediately below nodes, elsewhere glabrous. Leaves ovate, 3–6 cm long, 1.3–2.3 cm wide, base shortly attenuate, apex subacuminate, surfaces sparsely pubescent, more dense on margin and midrib above; lateral veins 4–5 pairs; petiole 4–18 mm long. Inflorescences axillary, 1 per axil, each with 1–3 cymules; primary peduncle to 1.5 mm long, largely glabrous; main axis bracts linear-lanceolate, 2–4 mm long, midrib with occasional hairs; cymule peduncles 1.5–3.5 mm long, sparsely pilose; cymule bracts broadly elliptic, pairs highly unequal, larger bract 9–13 mm long, 5–8.5 mm wide, apiculate, margin and (sparsely) midrib pale-pilose, elsewhere largely glabrous except towards apex within on larger bract where pubescent, inconspicuously 5–7-nerved from base; bracteoles pale, lanceolate, 4–6 mm long, with minute glandular hairs, margin pilose. Calyx lobes 3.5–4.5 mm long, glandular-puberulent, margin pilose. Corolla pale purple, 14.5–16 mm long, eglandular-pubescent and with occasional glandular hairs outside; tube 6.5–9 mm long; lip held in upper position oblong, 8–9 mm long, 2.5–3.5 mm wide; lip held in lower position obovate, 7–8 mm long, 3.5–4.5 mm. Staminal filaments 5.5–7 mm long, pubescent mainly above; anther thecae 0.8–1 mm long, slightly overlapping. Ovary glabrous; style sparsely strigulose. Immature capsules only seen, 4.5–6 mm long, raphes densely pilose, some hairs minutely gland-tipped; placental base elastic; seeds not seen. Fig. 90: 3/a–d, p. 682.

TANZANIA. Iringa District: Luisenga Stream, Oct. 1987, *J. & J. Lovett & Niblett* 2309!
DISTR. **T** 7; not known elsewhere
HAB. Submontane moist forest; 1750 m
USES. None recorded on herbarium specimen
CONSERVATION NOTES. If this proves to be a good species, it is clearly scarce and highly localised. Although forest remains intact along the Luisenga Stream, much of the forest of the surrounding Mufindi Escarpment has been cleared for tea plantations and shambas; this species will therefore probably prove to be threatened.

NOTE. This incompletely known taxon is close to *D. heterostegia* Nees but differs in having fewer inflorescences with fewer cymules per axil, elliptic, not ovate (-elliptic) cymule bracts with a less dense indumentum, and more dense hairs on the filaments of the stamens. See also note to sp. B. Further collections are required.

5. **Dicliptera sp. B** (= *Mwangoka et al.* 4493)

Subshrub, 60–100 cm tall; stems 6-angular, antrorse-pubescent when young, hairs yellowish-brown, soon glabrescent. Leaves ovate, 4.5–14 cm long, 2–6.5 cm wide, base attenuate, apex acuminate, midrib above, margin and veins beneath shortly antrorse-pubescent at least when young; lateral veins 5–6 pairs; petiole 0.8–4.7 cm long. Inflorescences axillary, 1–4 umbels of (2–)4–7 cymules per axil; primary peduncle 2.5–15 mm long on the most developed umbel of each axil, antrorse-pubescent; main axis bracts ± caducous, linear-lanceolate, 3–5.5 mm long, margin and midrib sparsely hairy; central cymule of each umbel on peduncle 2.5–10.5 mm long, shorter on lateral cymules; cymule bracts scarious in fruit, broadly (ovate-) elliptic to orbicular, pairs subequal, 9.5–11.5 mm long, 7–9 mm wide, apex obtuse or rounded, apiculate, outer surface ± sparsely puberulent and with occasional short antrorse hairs on principal veins, glabrous within, palmately 5–9-nerved; bracteoles pale, lanceolate, 1.5–3.5 mm long, glandular-puberulent, sometimes with occasional longer eglandular hairs. Calyx lobes 2–3 mm long, glandular-puberulent. Corolla white, ± 17 mm long, eglandular-pubescent outside; tube ± 10 mm long; lip held in upper position ± 7 mm long, 3.5 mm wide; lip held in lower position poorly preserved in material seen. Staminal filaments ± 6 mm long, with scattered hairs; anther thecae ± 0.9 mm long, superposed and somewhat oblique. Pistil glabrous. Capsule 5.5–6.5 mm long, pubescent particularly on the raphes, hairs mixed glandular and eglandular or mainly the latter; placental base elastic; only immature seeds seen, tuberculate principally around the rim, tubercles hooked.

TANZANIA. Iringa District: Udzungwa Mts National Park, Oct. 2005, *Mwangoka, Festo & Luke* 4437! & *idem*, Oct. 2005, *Mwangoka, Monongi & Festo* 4493!
DISTR. **T** 7; not known elsewhere
HAB. Submontane moist forest; 1450–1500 m
USES. None recorded on herbarium specimens

CONSERVATION NOTES. Apparently restricted to the Udzungwa Mts where it is known from only two collections made during the same expedition; it was then recorded as "occasional". Although the forests of the Udzungwa Mts have been subject to less disturbance than elsewhere in the Eastern Arc chain, this species is still likely to be threatened due to its extremely limited distribution and scarcity.

NOTE. This is a further species within the *D. heterostegia* complex, appearing closest to the robust forest forms of that species. It is, however, readily separable by lacking pilose hairs on the cymule bracts, the posticous bract having an obtuse or rounded (not acute or attenuate) apex which lacks the conspicuous mucro of *D. heterostegia*. In addition, the antrorse hairs on the stems and young leaves are darker, yellowish-brown in colour. *D.* sp. A is also close but has more unequal cymule bract pairs which are long-ciliate and has longer bracteoles and calyx lobes. More material, particularly of flowers and mature seeds, is required before this species can be formally described.

6. **Dicliptera napierae** *E.A. Bruce* in K.B.: 99 (1932); U.K.W.F. ed. 1: 609 (1974); U.K.W.F. ed. 2: 276 (1994). Type: Kenya, Masai District, Ngong, *Napier* 539 (K!, holo.; EA!, iso.)

Erect herb or subshrub, 10–50 cm tall, sometimes caespitose; stems branching from towards base where woody, bark pale; herbaceous stems subangular, shallowly sulcate, pale retrorse-pubescent. Leaves sometimes immature or absent at flowering, ovate or rounded, 1–2.5 cm long, 0.5–2 cm wide, base rounded, apex obtuse or rounded, apiculate, surfaces shortly antrorse-pubescent, most dense on margins; lateral veins 3–5 pairs; petiole 0.5–3 mm long. Inflorescences axillary, 1 umbel of (1–)2–3 cymules per axil; primary peduncle 3–40 mm long, extending in fruit, retrorse-pubescent; main axis bracts soon turning scarious and ± caducous, ovate, 4.5–10 mm long, 3–7.5 mm wide, base shallowly cordate, apex acute, margin and midrib sparsely pubescent; cymule peduncles 3–15 mm long, extending in fruit, retrorse- or spreading-pubescent; cymule bracts turning scarious in fruit, ovate-elliptic, pairs subequal or slightly unequal, 6–13.5 mm long, 3.5–10 mm wide, base shallowly cordate, apex apiculate, indumentum as main axis bracts, venation pinnate-reticulate; bracteoles pale, lanceolate-acuminate, 2.5–6 mm long, sparsely ciliate. Calyx scarious in fruit, lobes 2.5–3 mm long, puberulent. Corolla mauve to lilac with a paler tube and palate to lip held in upper position, the latter streaked purple, 19–24 mm long, retrorse-pubescent outside; tube 6.5–9 mm long; lip held in upper position oblong, 13–14 mm long, 3–4 mm wide, with two lines of short hairs at mouth; lip held in lower position ovate-elliptic, 9.5–12.5 mm long, 7–9 mm wide. Staminal filaments 6–7 mm long, with retrorse hairs mainly on upper side; anther thecae ± 1.5 mm long, superposed and separated by up to 0.5 mm. Ovary glabrous; style sparsely strigulose. Capsule 7.5–9 mm long, glabrous; placental base elastic; seeds 2.3–3 mm in diameter, minutely and sparsely tuberculate. Fig. 90: 4/a–d, p. 682.

KENYA. North Nyeri District: near Timau R., Oct. 1960, *Verdcourt & Polhill* 2924!; Meru District: 16 km on Isiolo–Nanyuki road, Aug. 1963, *Heriz-Smith* s.n.!; Masai District: Magadi Road, top of Ngong Hills, Sept. 1956, *Verdcourt* 1575!

DISTR. **K** 4, 6; not known elsewhere

HAB. Grassland in open, rocky bushland, roadsides; 1700–2700 m

USES. None recorded on herbarium specimens

CONSERVATION NOTES. Of the eleven collections seen of this species, eight are from the Ngong Hills where it is locally common and recorded both on the peaks and on the lower slopes towards the Rift Valley. This area is rather heavily populated, being close to Nairobi, and so experiences some disturbance, particularly through over-grazing. Such conditions may not, however, disfavour this species which has been recorded as a pioneer of disturbed grassland (*Verdcourt* 1575). Beyond the Ngong Hills it is clearly scarce, with very few collections despite this area of Kenya being well botanised. It is, however, recorded as "apparently common" in the Isiolo area (*Heriz-Smith* s.n.). It is provisionally assessed as of Least Concern (LC) but further information on its ecology and current status of the populations is desirable.

7. **Dicliptera cordibracteata** *I. Darbysh.* in K.B. 63: 364, fig. 2 (2009). Type: Kenya, Meru District, Ura R., Kinna area, *J. Adamson* in EAH 11835 (K!, holo.; EA!, iso.)

Erect herb, 15–80 cm tall; stems 6-angular, ± prominently ridged, ridges sparsely pale-pubescent, more dense immediately below the nodes, hairs variously antrorse, retrorse or spreading. Leaves sometimes absent or immature at flowering, ovate, 2–5.5 cm long, 1–3 cm wide, base abruptly narrowed into petiole, apex acute to acuminate, upper surface and veins beneath sparsely pale-pilose, with somewhat shorter, more dense hairs along margin; lateral veins 5–6 pairs; petiole 9–15 mm long. Inflorescences axillary, 1–2 lax umbels of (1–)3 cymules per axil; primary peduncle (1–)5–25 mm long, sparsely antrorse- and retrorse-pubescent; main axis bracts turning scarious, linear-lanceolate, 3.5–7 mm long, midrib and margin sparsely antrorse-pubescent or largely glabrous; cymule peduncles 9–40 mm long; cymule bracts becoming scarious in fruit, the smaller bract then paler than the larger, cordiform, pairs highly unequal, larger bract 11–23 mm long, 10–19 mm wide, apex obtuse or rounded, minutely apiculate, surfaces glabrous or with sparse hairs on midrib and margin, venation pinnate-reticulate; bracteoles hyaline, lanceolate, 1.5–3 mm long, glabrous or sparsely ciliate. Calyx scarious in fruit, lobes 2–3 mm long, outer surface largely glabrous, margin sparsely appressed-pubescent. Corolla pink to mauve, 16–21 mm long, eglandular-pubescent outside; tube 6.5–7.5 mm long; lip held in upper position oblong, 9.5–14.5 mm long, 2.5–4.5 mm wide; lip held in lower position elliptic, 9.5–13 mm long, 5.5–9 mm wide. Staminal filaments 6.5–7.5 mm long, pilose; anther thecae ± 1 mm long, somewhat overlapping. Ovary and style glabrous or the latter sparsely strigulose at base. Capsule 7–7.5 mm long, puberulous; placental base elastic; seeds 2.2–2.5 mm in diameter, minutely tuberculate, tubercles hooked. Fig. 90: 5/a–d, p. 682.

KENYA. Northern Frontier District: Sololo police post, Aug. 1952, *Gillett* 13672!; Meru District: Ura R., Kinna area, Feb. 1960, *J. Adamson* in EAH 11835! (type); Embu District: Tana R., Emberre, June 1932, *Graham* 1710!

DISTR. **K** 1, 4; not known elsewhere

HAB. Open areas near rivers, "submontane scrub"; 670–? 1350 m

USES. None recorded on herbarium specimens

CONSERVATION NOTES. Assessed as Data Deficient (DD) by Darbyshire (l.c.).

8. **Dicliptera laxata** *C.B. Clarke* in F.T.A. 5: 258 (1900); Mildbraed in N.B.G.B. 9: 507 (1926); Heine in F.W.T.A. ed. 2, 2: 425 (1963) pro parte excluding *Guinea* 2298 ex Fernando Po; U.K.W.F. ed. 1: 607, 608 (fig.) (1994); U.K.W.F. ed. 2: 276, pl. 121 (1994); Friis & Vollesen in Biol. Skr. 51 (2): 441 (2005); Ensermu in Fl. Eth. 5: 443 (2006). Syntypes: Kenya, Central Kavirondo District, Samia, *Scott-Elliot* 7098 (K!, BM!, syn.); Malawi, [Masuku Plateau] Misuku Hills, *Whyte* s.n. (K!, syn.)

Erect or straggling perennial herb or subshrub, 30–120(–200) cm tall; stems (sub-)6-angular, shortly antrorse-pubescent mainly on the angles, glabrescent. Leaves ovate to elliptic, (3–)7–12.5(–16) cm long, 1.5–5.5(–7) cm wide, base attenuate or cuneate, apex acuminate, largely glabrous or with sparse hairs on upper surface, midrib above antrorse-pubescent; lateral veins (5–)6–7(–8) pairs, conspicuous; petiole 0.5–5.5 cm long. Inflorescences axillary, (1–)2–3(–4) umbels of (3–)4–5(–7) cymules per axil; subsessile or primary peduncle to 13(–34) mm long, shortly antrorse-pubescent; main axis bracts linear-lanceolate, (1–)2–4.5 mm long, with sparse antrorse hairs; cymule peduncles 1–4(–9) mm long; cymule bracts elliptic (-obovate), pairs slightly to strongly unequal, larger bract 7–17 mm long, 2–9 mm wide, apex acute or shortly attenuate, minutely apiculate, surfaces largely glabrous except for sparse antrorse hairs at base, with or without minute subsessile glands towards apex, inconspicuously 3–5-nerved from base, sometimes faintly "windowed"; bracteoles linear-lanceolate, 3.5–9 mm long, with minute glandular and antrorse eglandular hairs, margin broadly hyaline. Calyx lobes

2.5–5 mm long, glandular-puberulent. Corolla white, cream or rarely pink or purple, 10.5–16 mm long, with or without purple markings on lips, pubescent and with interspersed glandular hairs outside; tube 4–5.5 mm long; lip held in upper position oblong, 6–10 mm long, 2–3.5 mm wide, palate puberulous towards mouth; lip held in lower position elliptic, 6.5–9.5 mm long, 3–4.5 mm wide. Staminal filaments (3.5–)5.5–8 mm long, pubescent on upper side; anther thecae 0.6–0.85 mm long, superposed. Ovary with or without short hairs towards apex; style sparsely strigulose. Capsule 5.5–7 mm long, ± densely pubescent; placental base elastic; seeds 2–3 mm in diameter, tuberculate, tubercles more dense towards rim, hooked. Fig. 90: 6/a–d, p. 682.

UGANDA. Toro District: Bundibugyo, Bwamba, Dec. 1938, *Loveridge* 262!; Mengo District: Kipayo, Oct. 1914, *Dummer* 1206! & Entebbe, Jan. 1938, *Chandler* 2134!

KENYA. Northern Frontier District: top of Mt Kulal, July 1958, *Verdcourt* 2241!; Meru District: Upper Imenti Forest, June 1974, *Faden & Faden* 74/917!; North Kavirondo District: near Kakamega Forest Station, Nov. 1971, *Magogo* 1532!

TANZANIA. Bukoba District: Minziro Forest Reserve, 3 km E of Kabwoba village, Nov. 1999, *Sitoni et al.* 972!; Lushoto District: Kitivo Forest Reserve, Sept. 1970, *Shabani* 623!; Iringa District: Ipafu Hill, beyond Supeme Estate, Sept. 1971, *Perdue & Kibuwa* 11480!

DISTR. **U** 2, 4; **K** 1, 3–6; **T** 1, 3, 7; Nigeria to Ethiopia and to N Malawi

HAB. Submontane and montane moist forest and forest margins; 1000–2350(–2750) m

USES. "Used to make a blue dye; the plants are dried and powdered, put in a hole in the ground lined with banana leaves and soaked in water thus making a dye which is used in colouring grasses for weaving purposes" (**U** 2; *Loveridge* 262); in Cameroon, the leaves of this species are used to produce the drink "Bakossi Tea"

CONSERVATION NOTES. Widespread and locally abundant in suitable habitat, where it can be dominant in the forest herb layer. It is tolerant of some disturbance, persisting in disturbed and secondary forest patches. The assessment of Least Concern (LC) made in Cheek *et al.* (Pl. Kupe, Mwanenguba & Bakossi Mts, 2004: 224) is maintained here.

SYN. [*Hypoestes phaylopsoides sensu* C.B. Clarke in F.T.A. 5: 248 (1900) pro parte quoad *Whyte* s.n. ex Misuku Hills [Masuka Plateau], *non* S. Moore]

Dicliptera humbertii Mildbr. in B.J.B.B. 14: 356 (1937). Type: Congo-Kinshasa, Lubero to Libongo, *Humbert* 8740 (?B†, holo.; BR!, P!, iso.), **syn. nov.**

NOTE. Collections from **T** 7 and N Malawi have somewhat longer bracteoles (5–9 mm, not 3.5–7 mm long) and calyx lobes ((3–)3.5–5 mm, not 2.5–3.5(–4) mm long) than the Guineo-Congolian forest material, though the overlap is considerable. In our region, they are additionally separable by having consistently large and proportinately broad cymule bracts, length/width ratio 0.4–0.8/1, versus 0.2–0.5/1 in the Ugandan, Kenyan and N Tanzanian material. The bracts are also more clearly "windowed" in these plants. However, in West Africa (particularly Nigeria and Cameroon), bract size and shape and the prominence of "windows" are more variable, with overlap between the two East African forms.

9. **Dicliptera sp. C** (= *Bidgood et al.* 4166)

Perennial herb to 100 cm tall; stems subangular, shortly retrorse-pubescent, becoming sparse with maturity. Leaves ovate, 6–11.5 cm long, 3.5–5 cm wide, base attenuate, apex acuminate, apiculate, margin, midrib and veins beneath shortly antrorse-pubescent, with sparse longer hairs on upper surface; lateral veins 6–7 pairs, conspicuous beneath; petiole 12–30 mm long. Inflorescences axillary, (1–)2 umbels of 3–5 cymules per axil; primary peduncle 2.5–5 mm long, shortly retrorse-pubescent; main axis bracts linear to narrowly elliptic, 3–6.5 mm long, margin and midrib sparsely hairy; cymule peduncles 2.5–5.5 mm long, the central cymule of each umbel with peduncle extending to 14–25 mm long, retrorse-pubescent; cymule bracts somewhat paler towards base, elliptic, pairs unequal, larger bract 13–16 mm long, 4.5–7 mm wide, apiculate, surfaces with short antrorse hairs mainly towards base and on midrib; bracteoles lanceolate, 5.5–7 mm long, with minute glandular hairs mainly towards margin, this broadly hyaline. Calyx lobes 2.5–3.5 mm long, glandular-puberulent. Corolla purple with a white tube and palate to lip held in upper position, the latter streaked purple, 22–25 mm long, pubescent outside with mixed eglandular and glandular hairs, the latter rather dense on the limb; tube 8.5–9 mm long; lip held in upper position oblong, 14–16 mm long, 4.5–5 mm wide, palate puberulous towards mouth; lip held in lower position elliptic-rounded, 10.5–14.5 mm long, 7–8 mm wide. Staminal filaments 9.5–12.5 mm long, with sparse hairs mainly on upper side; anther thecae ± 1 mm long, slightly overlapping. Ovary glabrous; style sparsely strigulose. Capsule 6.5–7 mm long, rather densely pubescent; placental base elastic; seeds 2.5–2.8 mm in diameter, tuberculate, tubercles more dense towards rim, hooked.

TANZANIA. Kigoma District: Mt Livandabe [Lubalisi], May 1997, *Bidgood et al.* 4166!
DISTR. **T** 4; not known elsewhere
HAB. Closed moist forest; 1200–1300 m
USES. None recorded on herbarium specimens
CONSERVATION NOTES. This species is currently known only from the above cited collection.

NOTE. This distinctive taxon is closest to *D. laxata*, but differs in the much larger purple flowers, with more numerous stalked glands on the lips and in the retrorse (not antrorse) stem and peduncle indumentum. Further collections are desirable from this site and possibly also from the adjacent Mahali Mts to confirm its status.

10. **Dicliptera grandiflora** *Gilli* in Ann. Naturhist. Mus. Wien 77: 49 (1973); Darbyshire in K.B. 63: 373 (2009). Type: Tanzania, Morogoro District, Uluguru Mts, Chenzema, *Gilli* 521 (W!, holo.; K!, iso.)

Scandent perennial herb, 100–200 cm tall or more; stems slender, 6-angular, ± sulcate, sparsely pilose on the angles, more dense immediately below the nodes, soon glabrescent. Leaves anisophyllous, ovate, 3.5–8.5 cm long, 1.5–4 cm wide, base rounded, abruptly narrowed into petiole, apex acuminate, upper surface and veins beneath sparsely pilose; lateral veins 4–5 pairs; petiole slender, 6–37 mm long. Inflorescences axillary, 1–2 umbels of 2–4 cymules per axil; primary peduncle 3–22 mm long, pilose and with or without glandular hairs; main axis bracts linear or rarely foliaceous and narrowly ovate, 3.5–9 mm long, margin and midrib sparsely pilose; cymule peduncles 2–9 mm long; cymule bracts inconspicuously "windowed", ovate- to elliptic-rhombic, pairs unequal, larger bract 9.5–21 mm long, 4–9.5 mm wide, apex mucronulate, surface sparsely pubescent to pilose mainly on margin and midrib, with or without short glandular hairs, 3-veined; bracteoles linear-lanceolate, 7.5–10 mm long, indumentum as cymule bracts, margin broadly hyaline. Calyx lobes 6–7 mm long, ciliate and with or without scattered short glandular hairs. Corolla purple, darker in the throat, (25–)28–31(–35) mm long, pubescent outside, hairs glandular on limb; tube 18–22(–25) mm long, strongly widened above the twist; lip held in upper position ovate, (8–)10–10.5 mm long, 6–7 mm wide, palate with minute broad hairs in mouth and throat; lip held in lower position ovate, (7–)9–10 mm long, 6–6.5 mm wide. Staminal filaments 13–14 mm long, arising from 7–9 mm below the corolla mouth, with short hairs on upper side particularly towards base; anther thecae 1–1.5 mm long, superposed and highly oblique, upper theca almost patent to filament. Pistil glabrous. Capsule ± 6.5 mm long, sparsely glandular-pubescent; placental base elastic; seeds ± 2 mm in diameter, tuberculate, tubercles squat and unhooked. Fig. 90: 7/a–d, p. 682.

TANZANIA. Morogoro District: Nguru Mts, Mesumba Peak, July 1933, *Schlieben* 4170! (EA sheet only; BR sheet of 4170 is of *Mimulopsis solmsii* Schweinf.) & Uluguru Mts, Uluguru S Catchment Forest Reserve, path from N'gungulu village to Lukwangule plateau, May 2000, *Jannerup & Mhoro* 49! & Bunduki Forest Reserve, Aug. 2000, *Mhoro* in UMBCP 491!
DISTR. **T** 6; known only from the Uluguru and Nguru Mts
HAB. Moist submontane and montane forest; 1200–2300 m
USES. None recorded on herbarium specimens

CONSERVATION NOTES. Assessed as Endangered (EN B2ab(iii)) by Darbyshire (l.c.).

SYN. [*D. sp. aff. umbellata* Juss. (*D. sp. 2* of CFS) *sensu* Luke in Journ. E. Afr. Nat. Hist. Soc. 94: 92 (2005) pro parte quoad *Schlieben* 4170]

11. **Dicliptera alternans** *Lindau* in E.J. 20: 47 (1894); C.B. Clarke in F.T.A. 5: 258 (1900); Heine in F.W.T.A. ed. 2, 2: 426 (1963). Type: Cameroon, W of Buea, *Preuss* 604 (B†, holo.; K!, iso.)

Decumbent herb, 15–50(–150) cm tall, rooting at lower nodes; stems slender, sulcate, sparsely antrorse-pubescent on ridges or largely glabrous. Leaves ovate to elliptic, 4–10 cm long, 2–4.5 cm wide, base shortly attenuate, apex acuminate, largely glabrous; lateral veins 4–6 pairs; petiole 8–26 mm long. Inflorescences terminal, occasionally also in uppermost leaf axils, laxly spiciform, 1–4(–8) cm long, 2–10 pairs of cymules per spike, opposite pairs sometimes developing at different times; primary peduncle antrorse-pubescent, sometimes with additional short glandular hairs, rarely pilose; main axis bracts linear, 2–5(–7.5) mm long, apex blunt, glabrous or with short glandular hairs; cymules subsessile, bracts "windowed", elliptic, pairs subequal or slightly unequal, larger bract 6.5–9 mm long, 2.5–5 mm wide, apex apiculate, surfaces largely glabrous or with short glandular hairs sometimes many, occasionally long-ciliate; bracteoles hyaline, linear-lanceolate, 1–3 mm long, glabrous or with few short glandular hairs. Calyx lobes hyaline, 2.5–3.5 mm long, indumentum as bracteoles. Corolla whitish to pale violet, 8–11.5 mm long; limb sparsely pubescent outside, most dense towards apex of lip held in upper position; tube 4.5–6 mm long; lip held in upper position narrowly oblong, 3.5–5.5 mm long, 0.7–2 mm wide; lip held in lower position narrowly elliptic, 3.5–4.5 mm long, 1–1.5 mm wide. Staminal filaments 3.5–4 mm long, with short hairs on upper side; anther thecae 0.5–0.8 mm long, superposed. Pistil glabrous. Capsule 5–6 mm long, sparsely puberulous, hairs eglandular and/or glandular; placental base elastic; only immature seeds seen, tuberculate, tubercles hooked. Fig. 90: 8/a–e, p. 682.

UGANDA. Toro District: Kibale National Park, near Kanyawara, July 1994, *Poulsen* 681! (pro parte, mixed with *D. maculata* Nees subsp. *maculata*)

DISTR. **U** 2; Cameroon, Congo-Brazzaville & Congo-Kinshasa

HAB. Closed moist forest, more rarely in disturbed forest; 1400 m

USES. None recorded on herbarium specimens

CONSERVATION NOTES. Although widely distributed in central Africa, *D. alternans* appears scarce, being known from very few collections. Its ecology, particularly the extent to which it tolerates disturbance, is not fully understood, although it seems to prefer intact forest patches. It may consequently be threatened by forest loss across its range and is provisionally assessed as Vulnerable (VU B2ab(iii)).

NOTE. The Ugandan material has almost glabrous inflorescences, whilst the type specimen has a glandular indumentum with additional long pilose hairs on the margins of the cymule bracts. However, material from elsewhere in Cameroon and Congo is intermediate. Further collections from across this taxon's range are desirable to gain a more complete understanding of its variability and its relationship to the closely related *D. laxispica* Lindau, recorded from Ivory Coast to Cameroon, which may represent a dwarf form of *D. alternans*.

12. **Dicliptera verticillata** (*Forssk.*) *C. Chr.* in Dansk Bot. Arkiv. 4 (3): 11 (1922); F.P.S. 3: 173 (1956); Heine in F.W.T.A. ed. 2, 2: 425 (1963) pro parte; Raynal in Adansonia sér. 2, 7: 304 (1967); U.K.W.F. ed. 1: 607 (1974); U.K.W.F. ed. 2: 276 (1994); Thulin in Fl. Somalia 3: 424 (2006); Ensermu in Fl. Eth. 5: 445 (2006), excl. fig. 166.45. Type: Yemen, *Forsskål* 393 (C, microfiche 39: I. 3–4!, lecto., chosen by J.R.I. Wood, Hillcoat & Brummitt in K.B. 38: 450 (1983), K!, photo.; BM!, isolecto.)

Erect to decumbent annual herb, 10–50(–75) cm tall; stems strongly 6-angular, pale-ridged, largely glabrous or sparsely retrorse-pubescent. Leaves often immature or absent at flowering, ovate, 1.5–7(–9) cm long, 1–4(–5.5) cm wide, base shortly attenuate, apex subattenuate, apiculate, surfaces sparsely antrorse-pubescent, soon glabrescent except for short hairs on margin; lateral veins 4–6 pairs, conspicuous beneath; petiole 5–25(–45) mm long. Inflorescences axillary, developing even at the lowermost nodes, 2–3 umbels of 3–5 cymules per axil, umbels becoming compounded and forming dense clusters; primary peduncle 1–5(–20) mm long, shortly pubescent; main axis bracts linear-lanceolate, 2.5–7.5 mm long, ciliate and with antrorse hairs on midrib; cymules subsessile, bracts elliptic, obovate or oblanceolate, pairs slightly to strongly unequal, larger bract (5.5–)7–10.5(–12) mm long, 1.5–2.5(–3) mm wide, apex attenuate into a curved mucro, densely pilose-ciliate proximally, hairs shorter towards apex, surface with hairs of variable length, longest on prominent midrib, margin hyaline towards base; bracteoles linear-lanceolate, 4–6 mm long, indumentum as cymule bracts or with additional short glandular hairs, margin hyaline. Calyx lobes 2.5–3.5 mm long, sparsely pubescent and with minute glandular hairs outside. Corolla pink to purple, 4–6 mm long, pubescent outside; tube 3–4 mm long; lip held in upper position oblong, 1.5–2.5 mm long, ± 1 mm wide; lip held in lower position broadly flabellate, 1.2–1.8 mm long, 1.5–3 mm wide, curved around stamens. Staminal filaments 0.8–1.6 mm long, pubescent on upper side; anther thecae 0.2–0.4 mm long, slightly overlapping, highly oblique. Pistil glabrous. Capsule 3–4.5 mm long, puberulous and with short glandular hairs at apex; placental base elastic; seeds 0.8–1.3 mm in diameter, tuberculate, tubercles hooked. Fig. 90: 9/a–d, p. 682.

KENYA. Northern Frontier District: Moyale, foot of scarp, Sept. 1952, *Gillett* 13763! & Katilia forest, 20 km NNE of Kangetet, May 1970, *Mathew* 6391; Machakos District: Nairobi–Emali road, km 111, Feb. 1969, *Napper & Abdallah* 1889!

TANZANIA. Singida District: Iramba Plateau, July 1958, *Hammond* 57!; Njombe District: 65 km on Makumbako–Mbeya road, Apr. 1991, *Bidgood & Vollesen* 2204!; Songea District: Songea Rest House, June 1956, *Milne-Redhead & Taylor* 10920!

DISTR. **K** 1, 3, 4, 6, 7; **T** 1–8; Cape Verde Is. to Sudan and to South Africa; Madagascar, Yemen, India

HAB. *Acacia-Commiphora* bushland and riverine woodland, also a weed of disturbed habitats; 500–1550 m

USES. None recorded on herbarium specimens

CONSERVATION NOTES. Widely distributed and locally abundant. It is tolerant of significant amounts of disturbance, becoming weedy in disturbed open habitats: Least Concern (LC).

SYN. *Dianthera verticillata* Forssk., Fl. Aegypt.-Arab.: 9 (1775)
Justicia cuspidata Vahl, Symb. Bot. 2: 9, 16 (1791), *nom. illegit.* Type as for *D. verticillata*
Justicia umbellata Vahl, Enum. 1: 111 (1807). Type: Senegal, *Jussieu* s.n. (P-JU!, holo.)
Dicliptera umbellata (Vahl) Juss. in Ann. Mus. Par. 9: 268 (1807)
D. micranthes Nees in Wall. Pl. As. Rar. 3: 112 (1832); C.B. Clarke in F.T.A. 5: 258 (1900). Type: Yemen, *Forsskål* 392 (C, microfiche 39: I. 1–2, lecto., selected by J.R.I. Wood, Hillcoat & Brummitt in K.B. 38: 451 (1983), K!, photo.)
D. spinulosa K. Balkwill in K.B. 51: 53 (1996). Type: Sudan, 'Cordofanum Milbeo', *Kotschy* 277 (G-DC, holo.; K!, M!, P!, WAG!, iso.), **syn. nov.**

NOTE. K. Balkwill (l.c.) recognises the widespread form of this taxon, with dense inflorescences, small, narrow bracts and small corollas, as distinct from the type specimen of *D. verticillata* from Yemen. This latter collection has fewer inflorescences per axil, the cymule bracts are broader, obovate and less hairy and the corollas are somewhat larger. It is the former variant, Balkwill's *D. spinulosa*, that occurs in our region. However, analysis of the available Yemen material reveals that *D. verticillata* is rather variable there, and that intermediate plants have been collected. Indeed, one collection cited by Balkwill as referable to *D. verticillata sensu stricto* (*Wood* 2997) differs from *D. spinulosa* only in having slightly larger corollas. Forms with broad bracts are also recorded in West Africa, from Sierra Leone and Ghana. A broader circumscription is therefore adopted here and *D. spinulosa* is considered conspecific with *D. verticillata*.

13. **Dicliptera maculata** *Nees* in DC., Prodr. 11: 485 (1847), excl. var. *b senegambica*; C.B. Clarke in F.T.A. 5: 257 (1900); Friis & Vollesen in Biol. Skr. 51 (2): 441 (2005); Ensermu in Fl. Eth. 5: 444 (2006); Darbyshire in K.B. 63: 374, table 2 (2009). Type: Ethiopia, Tigray, Djeladjeranne [Dscheladscheranne], *Schimper* II.701 (B†, holo.; K!, lecto., BM!, BR!, M!, MPU!, P!, W!, WAG!, isolecto.)

Scrambling, scandent, erect or prostrate perennial herb, 15–150(–240) cm tall; stems (sub-)6-angular, ± ridged, largely glabrous to densely eglandular- and/or glandular-pubescent, sometimes silky white-pilose on older stems. Leaves sometimes immature at flowering, ovate, 1.5–13 cm long, 0.5–7 cm wide, base rounded or shortly attenuate, apex subattenuate to acuminate, surfaces largely glabrous to pubescent; lateral veins (3–)4–8 pairs; petiole 3–55 mm long. Inflorescences axillary and terminal, 1–5 umbels of (2–)3–6(–7) cymules per axil; umbels subsessile or primary peduncle to 40(–85) mm long, peduncles pubescent or pilose, with or without glandular hairs; main axis bracts linear-lanceolate or rarely narrowly ovate, 2–9(–14) mm long, sparsely eglandular- and/or glandular-pubescent, margin often pilose; cymules subsessile or peduncles to 15(–20) mm long; cymule bracts usually "windowed" particularly the smaller of each pair, (oblong-) elliptic, ovate-elliptic or obovate, pairs highly unequal, larger bract 8.5–18 mm long, (1.5–)3–8.5 mm wide, apex rounded to acute, mucronate, surfaces with short glandular and eglandular hairs, margin and midrib often long-pilose, 3-veined; bracteoles lanceolate, 5.5–10 mm long, acuminate, surface shortly glandular-pubescent particularly towards apex, ± ciliate, margin broadly hyaline. Calyx lobes 3–6 mm long, glandular-puberulent and ciliate. Corolla pink, magenta or white, (12.5–)14–25 mm long, palate of lip held in upper position streaked purple, pubescent outside and with scattered glandular hairs; tube (4.5–)6–10.5 mm long; lip held in upper position oblong, 6–14.5 mm long, 2–5 mm wide, palate sparsely puberulous towards mouth or largely glabrous; lip held in lower position (ovate-) elliptic, 5–14 mm long, 3.5–8.5 mm wide. Staminal filaments 5–14.5 mm long, with sparse hairs on upper side; anther thecae 0.5–1 mm long, immediately superposed or slightly overlapping. Ovary glabrous; style sparsely strigulose. Capsule 5.5–8.5(–9.5) mm long, shortly glandular-pubescent, most dense towards apex, with or without scattered eglandular hairs; placental base elastic; seeds 1.2–2.5 mm in diameter, tuberculate, tubercles squat, unhooked.

KEY TO INFRASPECIFIC TAXA

1 Umbels subsessile or shortly pedunculate, often becoming densely clustered at largely leafless nodes towards the stem apices; cymule peduncles with glandular hairs absent or sparse and inconspicuous; calyx lobes sparsely to densely pilose-ciliate . . . . b. subsp **usambarica**
  Umbels usually mainly in the leafy axils and lax, primary peduncle usually 4–15(–20) mm long, rarely subsessile and congested in upper, leafless axils*; cymule peduncles with many, conspicuous glandular hairs; calyx lobes shortly ciliate, rarely sparsely pilose 2
2 Plants slender, prostrate; corolla 14–17 mm long; **K** 7 c. subsp. **A**
  Plants more robust, scrambling, scandent or erect; corolla 18–25 mm long; widespread but absent from **K** 7 . . . . a. subsp. **maculata**

* populations of subsp. *maculata* with subsessile, congested inflorescences are mainly found in **K** 1 and **U** 1 along the geographic boundary between this taxon and subsp. *usambarica* where some intergrading is likely.

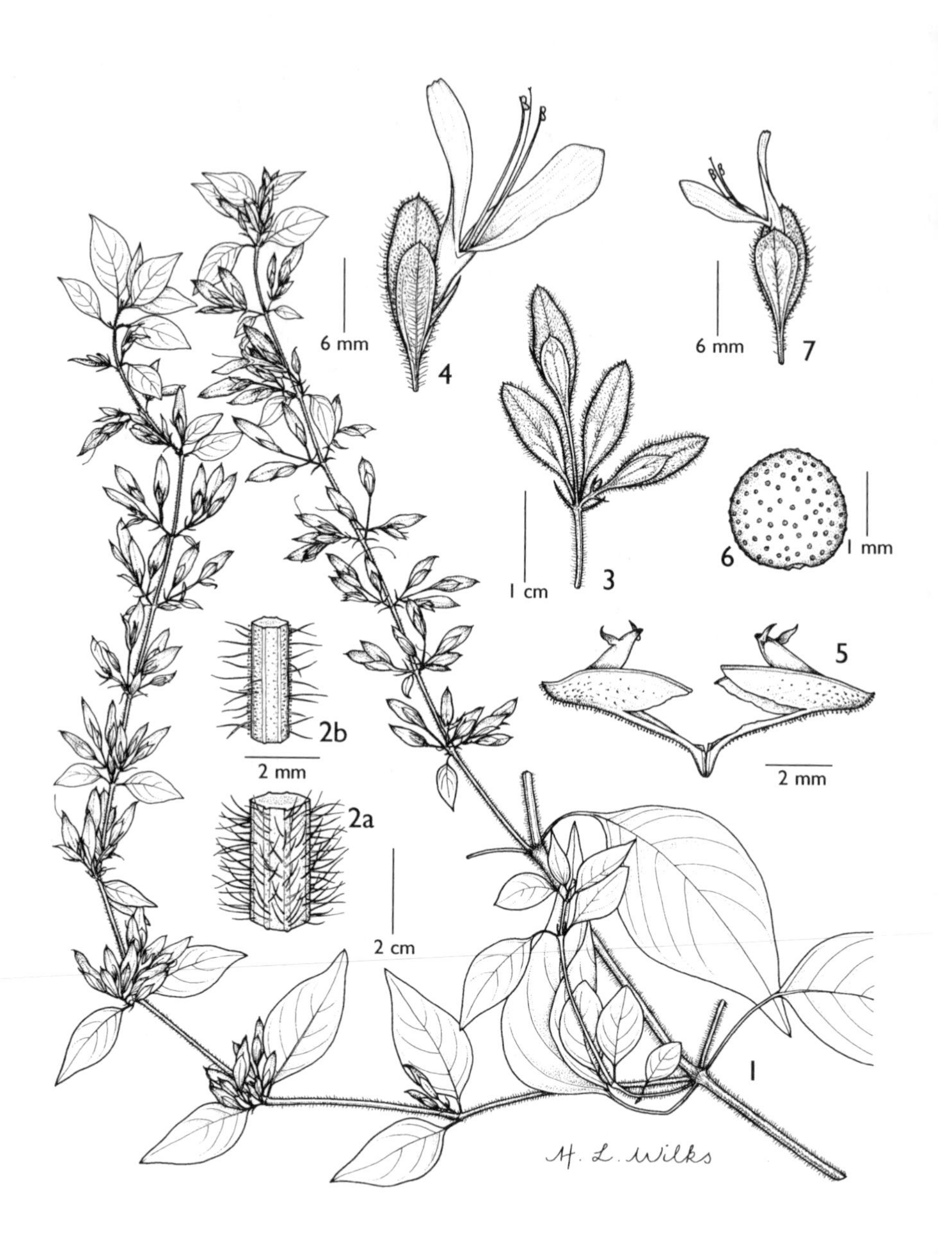

FIG. 92. *DICLIPTERA MACULATA* subsp. *MACULATA* — **1**, habit; **2**, variation in stem indumentum: **a**, lower stem, **b**, upper stem between fertile nodes; **3**, inflorescence; **4**, cymule with flower; **5**, dehisced capsule; **6**, mature seed. *DICLIPTERA HIRTA* (species not recorded in our region but see discussion to *D. maculata*) — **7**, cymule with flower, for comparison. 1, 2, 5 and 6 from *Vollesen* in MRC 4702; **C** from *Haerdi* 620/0; 4 from *Flock* 91; 7 from *Eyles* 3139 ex Zimbabwe. Drawn by Hazel Wilks.

a. subsp. **maculata**, Darbyshire in K.B. 63: 374 (2009)

Scrambling, scandent or erect herb. 1–5 umbels of (2–)3–6(–7) cymules per axil; umbels usually lax, primary peduncle (1–)5–40(–85) mm long, cymule peduncles (1–)4–15(–20) mm long, at least the latter with conspicuous, often dense, glandular hairs; cymule bracts (in our region) not tinged pink. Calyx lobes shortly ciliate, rarely sparsely pilose. Corolla 18–25 mm long. Figs. 90: 10/1–4, p. 682 & 92: 1–6, p. 700.

Uganda. Karamoja District: Lozut, Nov. 1939, *A.S. Thomas* 3198!; Kigezi District: Kachwekano Farm, Dec. 1949, *Purseglove* 3137!; Mengo District: Mawokota, Mar. 1905, *Brown* 208!
Kenya. Northern Frontier District: Mt Nyiru, July 1960, *Kerfoot* 2088! & Ndoto Mts, track up from Ngurunit Mission, June 1979, *Gilbert et al.* 5622!; Turkana District: Lodwar area, Murua Nysigar Peak, Sept. 1963, *Paulo* 1036!
Tanzania. Bukoba District: Minziro Forest Reserve, July 2000, *Bidgood, Leliyo & Vollesen* 4894!; Mpanda District: Ujamba, July 1958, *Mahinde* 74a! & Kungwe Mt, Selimweguru, July 1959, *Newbould & Harley* 4661!
Distr. **U** 1–4; **K** 1, 2; **T** 1, 4, 6–8 (see note); Cameroon to Eritrea and Ethiopia, south to Zambia, Malawi and Mozambique
Hab. Montane and submontane forest, particularly margins and disturbed areas, lower altitude forest, riverine thicket and woodland, roadsides; 100–2750 m
Uses. None recorded on herbarium specimens
Conservation notes. Assessed as of Least Concern (LC) by Darbyshire (l.c.).

Syn. *D. maculata* Nees var. *a panicularis* Nees in DC., Prodr. 11: 485 (1847), *nom. illegit.* Type as for subsp. *maculata*
D. lingulata C.B. Clarke in F.T.A. 5: 257 (1900). Type: Malawi, Nyika Plateau, *Whyte* 192a (K!, holo.)
D. maculata Nees forma *albo-lanata* Lanza in Pl. Erythr.: 70 (1910). Type: Eritrea, Hamasen, Filfil, *Senni* 485 (FT, holo., not traced), synonymy fide Ensermu (l.c.)
D. silvestris Lindau in Z.A.E. 2: 302 (1914). Type: Cameroon, near Muera, *Mildbraed* 2268 (B†, holo.; BR!, iso.)
[*D. clinopodia sensu* Lindau in Z.A.E. 2: 303 (1914) quoad *Mildbraed* 2519 ex Ruwenzori, & *sensu* K. Balkwill *et al.* in K.B. 51: 32 (1996) pro parte quoad *Verdcourt* 1809 ex Kenya, *non* Nees]
D. batesii S. Moore in J.B. 57: 246 (1919). Type: Cameroon, Bitye, *Bates* 608 (BM!, holo.; MO!, NY!, iso.)
D. longipedunculata Mildbr. in B.J.B.B. 14: 357 (1937). Type: Congo-Kinshasa, Lake Kivu, Idjwi Island, *Humbert* 8374 (?B†, holo.; BR!, P, iso.)
D. glanduligera Chiov. in K.B.: 171 (1941). Type: Ethiopia, SE of Yuka, *Gillett* 5407 (K!, holo.)
D. wittei Mildbr. in B.J.B.B. 17: 89 (1943). Type: Congo-Kinshasa, Lake Magera, *de Witte* 1435 (BR!, holo.; see note)
[*D.* sp. *sensu* Milne-Redh. in Brenan *et al.*, Mem. N.Y. Bot. Gard. 9: 28 (1954) quoad *Brass* 17161 ex Malawi]
[*D. umbellata sensu* F.P.S. 3: 173 (1956), & *sensu* Friis & Vollesen in Biol. Skr. 51 (2): 438 (2005) pro parte quoad *Jackson* 1100, *non* (Vahl) Juss.]
[*D.* sp. ? nov. aff. *umbellata* (= *Haerdi* 620/0) *sensu* Vollesen in Opera Bot. 59: 80 (1980)]
[*D.* sp. ? nov. (= *Schlieben* 3999) *sensu* Vollesen in Opera Bot. 59: 80 (1980)]
[*D. colorata sensu* Champluvier in Fl. Rwanda, Sperm. 3: 452 (1985)]
[*D. hirta* K. Balkwill in K.B. 51: 50 (1996), pro part quoad *Torre* 483 ex Mozambique, *non* type]
[*Hypoestes forskahlei* (*forskaolii*) *sensu* Fischer & Killmann, Pl. Nyungwe N.P. Rwanda: fig. p. 593 (2008), *non* (Vahl) R. Br.]

Note. Plants found in low altitude riparian habitats in **T** 6–8 (e.g. *Vollesen* in MRC 4702, *Haerdi* 620/0) differ somewhat from the rest of the material in our region in usually having silky white-pilose hairs on the mature stems, less conspicuously "windowed", oblong cymule bracts and white (-pink) corollas, these being pink or magenta elsewhere. However, some populations from drier mountains in S Sudan and Ethiopia and N Uganda also have silky pilose hairs on the stems. Furthermore, the two specimens seen from the Uluguru Mts (*Schlieben* 3999; *Pócs et al.* 6065/K!) lack the long pilose hairs and have pink corollas, but have oblong cymule bracts with inconspicuous windows and so are largely intermediate between the two forms. I therefore consider the SE Tanzanian material as a lowland form of subsp. *maculata* here. It should be noted that this form is close to *D. hirta* K. Balkwill, recorded from riverine woodland and forest in Zambia, Mozambique, Zimbabwe and Botswana, which again has dense silky hairs on the stems. Indeed, *D. hirta* may well be a further form of *D. maculata* but it usually has smaller flowers (12–15 mm long; see Fig. 92: 7) and the seed tubercles are somewhat longer and minutely hooked; it is therefore maintained as distinct pending further investigation.

I originally (in K.B. 63: 375) listed the BR sheet of *de Witte* 1435 as an isotype of *D. wittei* but it has since been drawn to my attention that Robyns (in B.J.B.B. 17: 65 (1943)) explicitly mentions that Mildbraed examined the BR Acanthaceae material of *de Witte* from Parc National Albert, and the BR sheet of *de Witte* 1435 bears Mildbraed's annotation, hence it should be considered the holotype.

b. subsp. **usambarica** (*Lindau*) *I. Darbysh.* in K.B. 63: 376 (2009). Type: Tanzania, Lushoto District: Usambara, Kwa Mshusa, *Holst* 8914a (B†, holo.; K!, lecto., chosen by Darbyshire, l.c.)

Scrambling, scandent or erect herb. 1–5 umbels of 3–6 cymules per axil, often becoming densely clustered at largely leafless nodes towards apex of stems; umbels and cymules subsessile or shortly pedunculate, if glandular hairs present on the peduncles then usually few and inconspicuous; cymule bracts sometimes tinged pink. Calyx lobes sparsely to densely pilose-ciliate. Corolla (12.5–)16–19(–21) mm long. Fig. 90: 11/a–d, p. 682.

UGANDA. Mt Elgon, Jan. 1918, *Dummer* 3618! & near Mutusyet, Jan. 1969, *Lye* 1559! & Elgon, Mar. 1993, *Naiga* 505A!

KENYA. Laikipia District: Aberdare Forest at Ndaragwa, Mar. 1977, *Hooper & Townsend* 1338!; Kiambu/Machakos District: gorge above Fourteen Falls, Thika, Aug. 1953, *Verdcourt* 1003!; Masai District: ± 5 km SE of Entasekera, Oct. 1977, *Fayad* 267!

TANZANIA. Arusha District: Monduli Forest Reserve near Endepesi Village, July 1999, *Simon & Meliyo* 221!; Same District: Mkomazi Game Reserve, Kinondo Ridge, June 1996, *Abdallah & Vollesen* 96/259!; Lushoto District: W Usambara, Mazumbai Forest Reserve, between Mazumbai and Mqwashi village, Feb. 1984, *Borhidi & Hedrén* 84152!

DISTR. **U** 3; **K** 3–6; **T** 2, 3; not known elsewhere

HAB. Submontane and montane forest, particularly in disturbed areas, forest margins, roadsides, occasionally in wooded grassland; (1150–)1300–2700(–3050) m

USES. "Browsed by sheep and goats" (**K** 6; *Glover, Gwynne & Samuel* 2088)

CONSERVATION NOTES. Assessed as of Least Concern (LC) by Darbyshire (l.c.).

SYN. *D. usambarica* Lindau in E.J. 20: 47 (1894)

[*D. umbellata sensu* C.B. Clarke in F.T.A. 5: 259 (1900) pro parte quoad *Holst* 8914A, *non* (Vahl) Juss.]

[*D. colorata sensu* U.K.W.F. ed. 1: 607, 608 (fig.) (1974) & ed. 2: 276, pl. 120 (1994), *non* C.B. Clarke]

c. subsp. **A** (= *Luke* 2922)

Slender prostrate herb. Umbels solitary in axils or perhaps a second developing later in the season, (1–)2–3(–4) cymules per umbel; umbels lax, primary peduncle 4–13 mm long; cymule peduncles 1.5–7 mm long, with many conspicuous glandular hairs; cymule bracts not tinged pink. Calyx lobes shortly ciliate. Corolla 14–17 mm long.

KENYA. Kwale District: Shimba Hills, Mwele, Oct. 1991, *Luke* 2922! & *idem*, Dec. 2001, *Luke, Stone & Baer* 8243!

DISTR. **K** 7; not known elsewhere

HAB. Lowland moist forest, including roadsides; 300 m

USES. None recorded on herbarium specimens

CONSERVATION NOTES. The status of this taxon is uncertain but if it proves worthy of formal recognition, it is clearly scarce, apparently being restricted to one area of the Shimba Hills where it is threatened by its proximity to a road (W.R.Q. Luke, *pers. comm.*).

SYN. [*D. sp. aff. umbellata* Juss. (*D. sp. 2* of CFS) *sensu* Luke in Journ. E. Afr. Nat. Hist. Soc. 94: 92 (2005) pro parte excl. *Schlieben* 4170]

NOTE. This incompletely known taxon is close to *D. maculata* Nees subsp. *maculata* but differs in its more slender, prostrate habit and in the plants bearing few inflorescences. It may be a depauperate form of that taxon but is notably isolated from other populations. Further collections of the *D. maculata* complex are required from the coastal forests of Kenya.

14. **Dicliptera sp. D** (= *Congdon* 316)

Straggling herb to 100 cm tall, laxly branched; stems (sub-)6-angular, ± ridged, sparsely short-pubescent particularly on ridges below nodes. Leaves immature at flowering, ovate, 1.5–4.5 cm long, 0.5–1.7 cm wide, base shortly attenuate, apex attenuate, surfaces sparsely pubescent, most dense on veins beneath; lateral veins 4–5 pairs, pale and prominent beneath; petiole to 6 mm long. Inflorescences axillary, 1–2 umbels of 2–5 cymules per axil; primary peduncle 1.5–33 mm long, puberulous and/or antrorse-pubescent; main axis bracts linear-lanceolate, 4–7 mm long, puberulous and/or pubescent; cymule peduncles 2–16 mm long, puberulous and with occasional short glandular hairs, sometimes with scattered longer hairs; cymule bracts not or inconspicuously "windowed", elliptic or oblong-elliptic, pairs unequal, larger bract 9.5–14.5 mm long, 3.5–5.5 mm wide, apex rounded to acute, apiculate, surfaces puberulous and with short glandular hairs, sometimes with scattered longer hairs, 3-veined; bracteoles lanceolate, 6–8 mm long, margin hyaline. Calyx lobes 4–5 mm long, glandular-puberulent and with scattered longer glandular hairs, margin with occasional short eglandular hairs. Corolla purple, 16–23 mm long, pubescent and with short glandular hairs outside; tube 5.5–8 mm long; lip held in upper position oblong, 10–16 mm long, 3–4 mm wide, palate sparsely puberulous at mouth; lip held in lower position elliptic, 8.5–15 mm long, 6.5–7.5 mm wide. Staminal filaments 8–12.5 mm long, ± sparsely and shortly hairy mainly on upper side; anther thecae ± 1 mm long, superposed and becoming separated by up to 0.5 mm. Ovary glabrous; style sparsely strigulose. Capsule 6.5–7 mm long, eglandular-pubescent and with occasional glandular hairs near apex; placental base elastic; only immature seeds seen, tuberculate, tubercles most dense towards the rim, sparse elsewhere, hooked. Fig. 90: 12/a–d, p. 682.

TANZANIA. Ufipa District: Mbisi Forest, Aug. 1960, *Richards* 13060! (mixed coll. with *Hypoestes triflora* (Forssk.) Roem. & Schult.); Njombe District: Ikuwo side of Mtorwi Mt, Sept. 1991, *Congdon* 316!

DISTR. **T** 4, 7; not known elsewhere

HAB. Montane forest patches and forest margins; 2100–2400 m

USES. None recorded on herbarium specimens

CONSERVATION NOTES. Apparently localised and scarce in the forests of SW Tanzania, though further populations may come to light following more botanical exploration in this region. Forest over 2000 m remains largely intact in many regions, and that of Mbisi Mt is a forest reserve and so afforded some protection; therefore this taxon may not be threatened.

NOTE. This taxon differs from *D. maculata* in the puberulous inflorescence indumentum, the lanceolate, not lanceolate-acuminate, bracteoles with a narrower hyaline margin, the anther thecae becoming slightly separated at maturity and the predominantly eglandular indumentum of the capsules. These characters unite the two cited specimens, but they otherwise differ in several respects, notably that *Richards* 13060 has larger flowers with more clearly separated anther thecae and almost glabrous filaments, and has more oblong cymule bracts with rounded apices, these being (oblong-) elliptic and with acute apices in *Congdon* 316. Further material is required to delimit this taxon fully and to confirm its status in relation to *D. maculata*.

15. **Dicliptera latibracteata** *I. Darbysh.* in K.B. 63: 366, fig. 3, table 2 (2009). Type: Kenya, Uasin Gishu District, near Moiben R., upper Nzoia, *Brodhurst Hill* 666 (K!, holo.; EA!, iso.)

Scrambling or erect perennial herb or subshrub, 90–150(–300) cm tall; stems 6-angular, ± prominently pale-ridged, sparsely to densely appressed pale-pubescent. Leaves often few at flowering, ovate, 2–6.5 cm long, 1–2.5 cm wide, base rounded to acute, apex acute or shortly attenuate, apiculate, surfaces sparsely to more densely pubescent, veins beneath sometimes pilose; lateral veins 4–6(–7) pairs, conspicuous beneath; petiole 4–20(–32) mm long. Inflorescences axillary, 1–2(–3) umbels of

(2–)3–5(–7) cymules per axil; primary peduncle (2.5–)5–40 mm long, antrorse-pubescent; main axis bracts broadly to narrowly elliptic (-lanceolate) or obovate, 6–19 mm long, 1.5–11 mm wide, apex attenuate, mucronulate, pubescent particularly on margin and midrib, with or without scattered glandular hairs; cymules subsessile or peduncles 1–5 mm long; cymule bracts sometimes tinged purplish or brown towards apex, paler towards base, broadly obovate, pairs highly unequal, larger bract (8–)10.5–21 mm long, (4–)5–10.5 mm wide, apex rounded, obtuse or rarely acute, mucronate, indumentum as main axis bracts but often more densely ciliate towards base, only midrib prominent; bracteoles oblong-lanceolate, (7–)9–13 mm long, apex subattenuate, surfaces with glandular hairs particularly towards apex and with scattered long eglandular hairs mainly on margins. Calyx lobes (5.5–)6.5–8.5 mm long, glandular-puberulent outside, ± sparsely long-ciliate. Corolla magenta, (19–)21–27 mm long, palate of lip held in upper position whitish and purple-streaked, pubescent outside and with sparse short-stalked glands particularly on limb; tube (7–)8.5–11 mm long; lip held in upper position oblong, 10.5–17 mm long, 3–5.5 mm wide, puberulous at mouth; lip held in lower position broadly ovate-elliptic, 9.5–15 mm long, 6–10.5 mm wide. Staminal filaments 8–11.5 mm long, sparsely hairy mainly on upper side; anther thecae (0.7–)1–1.3 mm long, superposed or separated by up to 0.5 mm. Ovary glabrous or sparsely puberulous towards apex; style sparsely strigulose or glabrous. Capsule (6.5–)8–8.5 mm long, eglandular-pubescent and with short glandular hairs towards apex; placental base elastic; seeds ± 2.3 mm in diameter, tuberculate, tubercles squat and unhooked. Fig. 91: 1/a–d, p. 685.

KENYA. Northern Frontier District: Maralal, Lorok Plateau, Nov. 1978, *Hepper & Jaeger* 6701!; Uasin Gishu District: Burnt Forest, Feb. 1933, *Mainwaring* in *Napier* 2500!; Kisumu-Londiani District: Tinderet Forest Station, Nov. 1955, *Irwin* 233!

DISTR. **K** 1–3, 5; S Sudan

HAB. Montane *Podocarpus* and *Podocarpus-Juniperus* forest understorey and forest margins; 1950–2750 m

USES. None recorded on herbarium specimens

CONSERVATION NOTES. Assessed as Near Threatened (NT) by Darbyshire (l.c.).

SYN. [*D. colorata* C.B. Clarke in F.T.A. 5: 260 (1900) pro parte quoad *Scott-Elliot* 6766, *non* lectotype chosen by Darbyshire in K.B. 63: 381 (2009)]

[*D.* sp. B *sensu* U.K.W.F. ed. 1: 609 (1974) & ed. 2: 276 (1994)]

[*D. umbellata sensu* Friis & Vollesen in Biol. Skr. 51 (2): 442 (2005) pro parte quoad *Jackson* 1461, *non* (Vahl) Juss.]

NOTE. Stem and leaf indumentum vary considerably in this taxon, from densely pubescent throughout (e.g. in the type) to sparsely so mainly on the young shoots and leaves (e.g. *Hepper & Jaeger* 6701).

16. **Dicliptera albicaulis** (*S. Moore*) *S. Moore* in J.B. 49: 312 (1911); U.K.W.F. ed. 1: 609 (1974); U.K.W.F. ed. 2: 276 (1994). Type: Kenya, Nairobi, *Kaessner* 975 (BM!, holo.; K!, MO!, iso.)

Procumbent, scrambling or erect perennial herb or subshrub, 25–100 cm tall, much-branched; mature stems woody, whitish; young stems 6-angular, ± ridged, densely glandular-pubescent and/or pale eglandular-pilose, later glabrescent. Leaves often immature at flowering, ovate or elliptic, 0.7–3 cm long, 0.5–2.7 cm wide, base rounded or shortly attenuate, apex obtuse or rounded, apiculate, surfaces densely glandular- and/or eglandular-pubescent when young, sparsely so when mature; lateral veins 3–5 pairs; petiole 1–7 mm long. Inflorescences axillary and subterminal, the latter often on short lateral branches, 1 umbel of (2–)3–4(–5) cymules per axil; primary peduncle of axillary umbels 4–20(–33) mm long, indumentum as stem, subterminal umbels often subsessile; main axis bracts obovate or elliptic, 5.5–14 mm long, 1.5–8.5 mm wide, apiculate, glandular-pubescent and/or eglandular-pilose; cymules subsessile or peduncle to 9 mm long; cymule bracts obovate, pairs unequal, larger bract 7–14 mm long, 2.5–8 mm wide, apex rounded or shallowly emarginate, apiculate, surface glandular-pubescent and usually eglandular-pilose particularly along margins, 3-veined but only midrib conspicuous; bracteoles linear-lanceolate, 5.5–8 mm long, indumentum as cymule bracts, margin hyaline in lower half. Calyx lobes 3–4.5 mm long, glandular-puberulent and sparsely to densely eglandular-pilose. Corolla white to pale mauve with mauve streaking on both lips near mouth, 13–19(–21) mm long, eglandular-pubescent and with scattered glandular hairs outside; tube 6–8.5(–10) mm long; lip held in upper position oblong, 6.5–12 mm long, 2–3.5 mm wide, palate sparsely puberulous at mouth; lip held in lower position elliptic, 7–11.5 mm long, 3–6.5 wide. Staminal filaments 5–8 mm long, sparsely hairy towards base on upper side; anther thecae 1–1.2 mm long, slightly overlapping. Ovary largely glabrous or sparsely puberulent towards apex; style sparsely strigulose. Capsule 8–10 mm long, densely eglandular- and glandular-puberulous; placental base elastic; seeds 2.7–4 mm in diameter, minutely tuberculate. Fig. 91: 2/a–d, p. 685.

KENYA. Machakos District: Kiboko Research Station, June 1971, *Muriithi* 147!; Masai District: Amboseli Game Reserve, Sept. 1954, *Bally* 9869! & Chyulu Plains, June 1991, *Luke* 2853!
TANZANIA. Masai District: E end of Olduwai gorge, Aug. 1956, *Verdcourt* 1556! & Lake Lgarya, Aug. 1962, *Greenway* 10783!; Arusha District: 56 km from Arusha on the road to Nairobi, Sept. 1964, *Leippert* 5002!
DISTR. **K** 4, 6; **T** 1, 2; not known elsewhere
HAB. Open *Acacia* woodland, short grassland, roadsides; 800–1700 m
USES. None recorded on herbarium specimens
CONSERVATION NOTES. Recorded as locally common by several collectors within its limited range. It is found in several protected sites, such as the Ngorongoro Conservation Area and the Amboseli Game Reserve, where human pressures are low and where its favoured habitats are maintained: Least Concern (LC).

SYN. *Diapedium albicaule* S. Moore in J.B. 41: 156 (1903)

NOTE. *Peter* 43420, collected from **T** 2, between Lakes Magadi (Ngorogoro Crater) and Eyasi is unusual in having shortly mucronate cymule bracts. The flowers and bracts of this specimen are at the upper limit of the size range for the species. *Altmann* 137 from **K** 6 Amboseli bears a similar habit to *D. cicatricosa*, having the prominent leaf scars on the old stems typical of that species; the leaves and inflorescence form however clearly place it within *D. albicaulis*.

17. **Dicliptera cicatricosa** *I. Darbysh.* in K.B. 63: 368, fig. 4 (2009). Type: Kenya, Masai District, Olorgesailie, *Verdcourt* 1441 (K!, holo.; BR, EA!, MO, PRE, iso.)

Subshrub, 50–120 cm tall, branches erect; mature stems woody, whitish, lower portions of branches often bare and with many conspicuous leaf scars; young stems subangular, glandular-puberulous with intermixed eglandular hairs, soon glabrescent. Leaves ± dense towards branch apices, ovate to elliptic, 1.5–2.5 cm long, 0.8–2.2 cm wide, base rounded to shortly attenuate, apex acute or obtuse, mucronulate, surfaces glandular- and eglandular-puberulous with occasional longer hairs on margin; lateral veins 3–5 pairs, midrib prominent beneath; petiole 0.5–2 mm long. Inflorescences crowded in upper leaf axils and subterminal, 1 umbel of (1–)3 cymules per axil; primary peduncle 1–7 mm long, glandular-puberulous; main axis bracts elliptic or obovate, 7–12 mm long, 3.5–6 mm wide, apex attenuate, outer surface densely glandular-puberulous, margin pilose; cymules sessile, bracts paler towards base, obovate, pairs slightly unequal, larger bract 8–14 mm long, 3.5–6 mm wide, apex mucronate, indumentum as main axis bracts, 3-veined but often only midrib prominent; bracteoles linear-lanceolate, 7.5–12 mm long, apex attenuate. Calyx lobes 3.5–5 mm long, glandular-puberulous and eglandular-pilose. Corolla white or tinged mauve, 19–27 mm long, pilose and with sparse short glandular hairs outside; tube 8–10.5 mm long; lip held in upper position oblong, 10.5–16 mm long, 3–4 mm wide, palate streaked mauve to crimson, sparsely puberulous at mouth; lip held in lower position elliptic, 10–16 mm long, 6–8.5 mm wide. Staminal filaments

7–9.5 mm long, sparsely and shortly hairy towards base on upper side; anther thecae 1–1.5 mm long, overlapping. Ovary puberulous particularly towards apex; style strigulose except towards apex. Capsule 8.5–11.5 mm long, puberulous on raphes, hairs mainly eglandular; placental base elastic; seeds 3–4.5 mm in diameter, minutely tuberculate or largely smooth.

KENYA. Masai District: Magadi to Nairobi road, km 48, Aug. 1955, *Greenway* 8835! & Nairobi to Magadi road, km 70, Aug. 1961, *Polhill & Greenway* 446! & Olorgesailie prehistory site, Dec. 1965, *Gillett* 16978!

DISTR. **K** 6; not known elsewhere

HAB. Open *Acacia-Commiphora* bushland on lacustrine deposits; 950–1150 m

USES. None recorded on herbarium specimens

CONSERVATION NOTES. Assessed as Vulnerable (VU B1ab(iii)+2ab(iii)) by Darbyshire (l.c.).

SYN. [*D.* sp. C *sensu* U.K.W.F. ed. 1: 609 (1974) & ed. 2: 276 (1994) & *sensu* Beentje, K.T.S.L.: 601 (1994)]

18. **Dicliptera minutifolia** *Ensermu* in K.B. 58: 703, fig. 1 (2003); Ensermu in Fl. Eth. 5: 441, fig. 166.43 (2006). Type: Ethiopia, Sidamo, 63 km from Bokol Mayo towards Filtu, *Sebsebe & Ensermu* 2771 (ETH!, holo.)

Subshrub, 25–100 cm tall, much-branched; stems woody, bark grey, often cracking, young lateral branches densely subappressed pale-pubescent, soon glabrescent. Leaves variously rounded, obovate, elliptic or oblanceolate, 0.5–1.7 cm long, 0.3–0.8 cm wide, base rounded to cuneate, apex obtuse, rounded or emarginate, surfaces pale-pubescent particularly when young; lateral veins 2–3 pairs, inconspicuous; blade sessile or petiole to 3 mm long. Inflorescences axillary on short lateral branches, 1(–2) inflorescences of 1(–2) cymules per axil, sessile or with primary peduncle to 2 mm long, this glandular-puberulent and with scattered longer eglandular hairs; main axis bracts triangular, 0.5–1.5 mm long, pubescent, often caducous; cymule peduncles 0.5–3 mm long; cymule bract pairs unequal, larger bract lanceolate, 2.5–4.5 mm long, 0.5–0.8 mm wide, indumentum as peduncles; bracteoles as cymule bracts but 1.5–3 mm long. Calyx lobes 2.5–4 mm long, glandular-puberulent and with longer eglandular hairs. Corolla white to purple, 13.5–18 mm long, eglandular-pubescent outside and with scattered shorter glandular hairs; tube 6.5–9.5 mm long; lip held in upper position oblong, 7–8.5 mm long, 2.5–5 mm wide, palate streaked purple; lip held in lower position elliptic, 6.5–8 mm long, 3.5–5 mm wide. Staminal filaments 5–7 mm long, shortly hairy on upper side; anther thecae 0.7–0.8 mm long, overlapping and oblique. Ovary with few short glandular and/or eglandular hairs at apex; style strigulose in proximal half. Capsule ± 6 mm long, shortly eglandular-pubescent, with interspersed shorter glandular hairs; placental base ?elastic; only immature seeds seen, tuberculate. Fig. 91: 3/a–d, p. 685.

KENYA. Northern Frontier District: Samburu Game Reserve, July 1976, *Agnew & Timberlake* 11156!

DISTR. **K** 1; S Ethiopia

HAB. *Acacia* woodland, including grazed areas; 450–900 m

USES. None recorded on herbarium specimens

CONSERVATION NOTES. This species is known to the author from only three collections: two in Sidamo, Ethiopia and the third some distance to the south in NE Kenya. The intervening area is, however, undercollected and as this species is rather inconspicuous, particularly when not in flower, it is likely to be overlooked. It appears tolerant of some grazing (*Agnew & Timberlake* 11156) but its ecology is not currently fully understood: Data Deficient (DD).

NOTE. The single collection from our region is rather scant. The flowers are recorded as purple, those in Ethiopia being white to pale blue, and the leaves are narrower; it otherwise closely matches the Ethiopian material.

Only immature capsules have been seen and consequently the dehiscence mechanism remains unconfirmed, although the thickened raphes of each valve are typical of those species which display elastic dehiscence.

19. **Dicliptera melleri** *Rolfe* in Oates, Matabele Land & Victoria Falls ed. 2, App. 5: Botany: 405 (1889); C.B. Clarke in F.T.A. 5: 261, 515 (1900). Types: Zimbabwe, Matabeleland, *Oates* s.n. & Malawi, Manganja Hills, Chiradzulu Mt, *Meller* s.n. (both K!, syn.)

Decumbent to procumbent suffruticose perennial, pyrophytic, producing few to many annual stems 5–30 cm long from a woody base and rootstock; stems ± strongly 6-angular, sulcate, sparsely pubescent. Leaves ? immature at flowering, oblong-elliptic to oblanceolate, 1.5–6 cm long, 0.3–0.6(–1.3) cm wide, base and apex acute to obtuse, the latter apiculate, surfaces largely glabrous; lateral veins 3–4 pairs, inconspicuous; petiole 0–3 mm long. Inflorescences axillary and terminal, crowded in the upper axils where forming dense conical or cylindric heads 1.5–7 cm long, comprisingly many umbellately arranged cymules; primary peduncle of each umbel 2.5–6.5 mm long, sparsely antrorse-pubescent; main axis bracts linear-lanceolate, 5–6.5 mm long, ciliate and with short hairs along midrib; cymules subsessile or peduncles to 2 mm long, these antrorse-pubescent; cymule bract pairs highly unequal, the larger inconspicuously "windowed", linear-oblanceolate, 8.5–10.5(–13) mm long, 0.7–2 mm wide, apex acuminate, the smaller conspicuously 3-veined and "windowed", obovate, apex rounded-apiculate, both bracts with a conspicuously ciliate, hyaline margin, surface with shorter hairs and sometimes short glandular hairs towards apex; bracteoles lanceolate, 6–7 mm long, apex acuminate, surface ciliate and with short hairs on midrib, margin broadly hyaline. Calyx lobes 4–5 mm long, ciliate. Corolla white, 13.5–15 mm long, pubescent outside; tube 6.5–7 mm long; lip held in upper position oblong-ovate, 6.5–8 mm long, 3–5 mm wide, palate often streaked pale purple, pubescent towards mouth; lip held in lower position subrounded, ± 6 mm long, 5.5–7.5 mm wide, margin irregular. Staminal filaments 3.5–4 mm long, shortly hairy mainly on upper side; anther thecae 0.8–1.1 mm long, slightly overlapping. Ovary glabrous; style sparsely strigulose in lower half. Capsule 6.5–7.5 mm long, puberulous towards apex; placental base elastic; seeds discoid, 2.2–3 mm in diameter, smooth, rim somewhat uneven. Fig. 91: 4/a–d, p. 685.

TANZANIA. Ufipa District: Chala to Sitalike road, 24 km from Chala, Sept. 1970, *Richards & Arasululu* 25827A!; Rungwe District: Mlale, Ulambya, Oct. 1971, *Leedal* 681!
DISTR. **T** 4, 7; SE Congo-Kinshasa to Zimbabwe
HAB. Recently burnt grassland and open woodland; ± 1500 m
USES. None recorded on herbarium specimens
CONSERVATION NOTES. Although scarce in our region, this species is locally abundant in recently burnt areas of miombo woodland in Zambia, Malawi and Zimbabwe: Least Concern (LC).

SYN. *Diapedium melleri* (Rolfe) S. Moore in J.B. 38: 205 (1900)
Peristrophe mellerioides Merxm. in Proc. Trans. Rhod. Sci. Assoc. 43: 123 (1951). Type: Zimbabwe, Marandellas, *Dehn* 676/678 (M!, holo.), **syn. nov.**

20. **Dicliptera capitata** *Milne-Redh.* in K.B.: 428 (1937); Darbyshire in K.B. 63: 373, fig. 5J (2009). Type: Zambia, Solwezi District, Solwezi Boma, *Milne-Redhead* 493 (K!, holo.)

Erect annual herb, 5–90 cm tall, unbranched or laxly branched; stems 6-angular, ridged, appressed pale-pubescent mainly on ridges. Leaves oblong-lanceolate, lengthening up the stem, (1.5–)4–7.5 cm long, 0.4–1.5 cm wide, basal leaves sometimes broader, base rounded to subcordate, apex acute, upper and sometimes lower surface ± sparsely pubescent, veins beneath and midrib above with shorter antrorse hairs; lateral veins (3–)5–7 pairs, prominent beneath; blade sessile or petiole to 3 mm long. Inflorescences compounded into a ± globose terminal head, 0.5–2.5 cm in diameter, sessile, immediately subtended by a pair or pseudowhorl of 4 leaves; cymules many or, in small plants, reduced to 3–5; main axis bracts lanceolate, pairs unequal, the larger 5–9.5 mm long, apex acuminate, surface ciliate and with antrorse hairs on midrib, margin hyaline; cymules sessile, bracts lanceolate or the smaller rarely elliptic, pairs somewhat unequal, the larger 7.5–16 mm long, 1.5–3.5 mm wide, apex acute to attenuate, prominently 3-veined, veins parallel, surface pubescent to pilose and with inconspicuous short glandular hairs, margin densely white-pilose; bracteoles lanceolate, 5–8.5 mm long, ciliate, hyaline except for prominent midrib. Calyx lobes 3.5–4 mm long, ciliate. Corolla tube white, limb pink, purple or rarely white, 11–15 mm long, pubescent outside; tube 5–9 mm long; limb lip held in upper position oblong, 5–7.5 mm long, 2–3 mm wide, palate white and streaked dark purple or blackish; lip held in lower position ovate-rhombic, 4–7.5 mm long and wide, margin irregular. Staminal filaments 3–5.5 mm long, shortly hairy mainly on upper side; anther thecae 0.5–0.7 mm long, superposed. Ovary glabrous; style glabrous or sparsely strigulose. Capsule 5–6 mm long, glabrous; placental base elastic; seeds discoid, 2.5–3 mm in diameter, smooth. Fig. 91: 5/a–d, p. 685.

TANZANIA. Ufipa District: 18 km on Namanyere–Kipili road, May 1997, *Bidgood et al.* 3725!; Mbeya District: Magangwe, Rangers Post, Apr. 1970, *Greenway & Kanuri* 14322!; Songea District: 56 km W of Songea just E of Ruanda turnoff, May 1956, *Milne-Redhead & Taylor* 10356!

DISTR. **T** 4, 7, 8; Burundi, SE Congo-Kinshasa, Angola, Zambia and N Mozambique

HAB. Open *Brachystegia* and other miombo woodland and associated grassland; 800–1400 (–1800) m

USES. None recorded on herbarium specimens

CONSERVATION NOTES. This species is locally common in the miombo woodlands of East Africa, particularly in SE Congo-Kinshasa, Zambia and SW Tanzania; only single collections have been seen from Mozambique and Angola, though this may be more a product of under-collection in these countries. As it favours open, grassy areas in woodland, it is likely to tolerate moderate disturbance: Least Concern (LC).

21. **Dicliptera vollesenii** *I. Darbysh.* in K.B. 63: 371, fig. 5A–H (2009). Type: Tanzania, Tabora District, 10 km on Tabora–Itigi road, *Bidgood, Hoenselaar, Leliyo & Vollesen* 5982 (K!, holo.; DSM!, EA!, NHT!, iso.)

Annual herb with lax, erect to decumbent branches, 15–50 cm tall, lateral branches often patent to main stem; stems 6-angular, sulcate, shortly appressed-pubescent on the ridges. Leaves often copper-coloured above, oblong-lanceolate, 3.7–8 cm long, 0.6–1.6 cm wide, base obtuse to rounded, apex acute, margin and veins beneath with inconspicuous short ascending hairs, upper surface with scattered longer hairs or largely glabrous; lateral veins 5 pairs; petiole to 1.5 mm long. Inflorescences compounded into a dense terminal, globose or shortly conical head, 1–3 cm long, sessile, immediately subtended by a pair of leaves; cymules many; main axis bracts lanceolate to oblong, 2.5–4 mm long, apex acuminate, margin hyaline, shortly ciliate; cymules sessile, bract pairs highly unequal, larger bract oblong-lanceolate to narrowly elliptic, 6.5–9 mm long, 1.5–2.5 mm wide, apex acute, smaller bract elliptic or obovate, apex obtuse or rounded, both minutely apiculate, surfaces paler towards base, the smaller bract ± conspicuously "windowed", prominently 3-veined, more rarely only the midrib prominent on the larger bract, puberulous predominanty on veins and margin and with short-stalked glands between the veins; bracteoles linear-lanceolate, 5–6.5 mm long, hyaline except for prominent midrib. Calyx lobes 3–3.7 mm long, minutely ciliate. Corolla white, with red or purple guidelines on lip in upper position, 8.5–10.5 mm long, retrorse-pubescent outside; tube 5–6 mm long; lip held in upper position oblong (-ovate), 3.8–4.7 mm long, 2.5–3 mm wide; lip held in lower position broadly rhombic, 3.5–4 mm long and wide. Staminal filaments 2.5–3 mm long, pubescent particularly on upper side; anther thecae superposed, ± 0.5 mm long. Ovary glabrous; style strigulose towards base. Capsule 5–5.5 mm long, glabrous or with few hairs on the short apical beak, placental base elastic; seeds discoid, 2.3–3 mm in diameter, smooth.

TANZANIA. Tabora District: 22 km on Tabora–Sikonge road, May 2006, *Bidgood et al.* 5878! & 42 km on Ipole–Rungwa road, June 2008, *Bidgood, Leliyo & Vollesen* 7225!; Iringa District: Mufindi, Kwatwanga, May 2008, *Suleiman et al.* 3566!

DISTR. **T** 4, 7; not known elsewhere

HAB. Mature and degraded *Brachystegia* woodland and secondary *Terminalia* bushland, on sandy soils; 1000–1250 m

USES. None recorded on herbarium specimens

CONSERVATION NOTES. Assessed as Data Deficient (DD) by Darbyshire (l.c.). However, during fieldwork in W Tanzania by Vollesen *et al.* in May–June 2008 this species was actively sought for and found to be locally common in the Tabora region. It was also found by the author during fieldwork in the dry miombo woodland a considerable distance to the south in Iringa District. It is therefore now considered unthreatened: Least Concern (LC).

NOTE. The notes to the protologue state that *D. vollesenii* and *D. capitata* have never been found growing together; this was an editing error, and was supposed to read that the two have been seen growing together in central Tanzania (K. Vollesen, *pers. comm.*).

22. **Dicliptera carvalhoi** *Lindau* in P.O.A.: 371 (1895); C.B. Clarke in F.T.A. 5: 257 (1900); Darbyshire in K.B. 63: 376, fig. 6, table 3 (2009). Type: Mozambique, between the lower and middle Zambesi, *de Carvalho* s.n. (COI!, lecto., chosen by Darbyshire, l.c.)

Erect, straggling or decumbent, annual or perennial herb, 30–150(–200) cm tall, laxly branched; stems wiry, 6-angular, sulcate when young, with or without pale ridges, sparsely antrorse- and/or retrorse-pubescent, sometimes also hispid, rarely glabrous. Leaves sometimes immature at flowering, ovate to linear-lanceolate, 1.5–8(–14.5) cm long, 0.3–2.8 cm wide, base obtuse to attenuate, apex acute or shortly attenuate, apiculate, surfaces antrorse-pubescent mainly on midrib and margins, upper surface sometimes hispid, rarely glabrous; lateral veins 3–6 pairs; petiole 1–12(–22) mm long. Inflorescences either of 1–2 pedunculate umbels of (2–)3(–4) cymules in the largely bare upper axils or the umbels congested into a verticillate or dense globose to cylindrical terminal synflorescence; main axis bracts linear-lanceolate, pairs unequal, larger bract (3–)4.5–8(–12.5) mm long, antrorse-pubescent; cymules subsessile or pedunculate for up to 2 mm; cymule bracts darker and often tinged purplish towards apex, (oblong-) oblanceolate or lanceolate, pairs unequal, larger bract 6–16 mm long, 1–3(–4) mm wide, mucronate, surfaces with sparse to dense capitate glandular hairs towards apex particularly within, margin and often also midrib densely pilose, outer surface also with antrorse to patent eglandular hairs; bracteoles linear-lanceolate, 4.5–8 mm long, apex attenuate, margin hyaline towards base. Calyx lobes 2.5–5 mm long, glandular-puberulent and sometimes with few eglandular hairs at least on margin. Corolla white or limb mauve, 9.5–20 mm long, pubescent outside; tube 5–10.5 mm long; lip held in upper position oblong, 5.5–11.5 mm long, 2–4.5 mm wide, palate streaked purple; lip held in lower position obovate (-elliptic), 5–10.5 mm long, 3.5–6 mm wide. Stamens ± long-exserted; filaments 5–13(–16) mm long, sparsely hairy on upper side; anther thecae 0.5–1 mm long, superposed. Pistil glabrous. Capsule 4.5–8 mm long, eglandular-puberulous and/or with short glandular hairs towards apex; placental base elastic; seeds only subflattened, 1–1.7 mm in diameter, smooth or tuberculate, tubercles with or without hooks.

KEY TO INFRASPECIFIC TAXA

1. Cymules held in dense globose to cylindrical terminal heads, sometimes interrupted by reduced leaves and/or becoming verticillate in the lower portion, often with additional subsessile umbels in the uppermost leafy axils 2
   Cymules either held in verticillate heads with distinct bare internodes throughout or in lax, pedunculate umbels in the largely bare upper stem axils 3

2. Seeds smooth except for microscopic verrucae; leaves usually (linear-) lanceolate, rarely ovate .... a. subsp. **carvalhoi**
   Seeds tuberculate; leaves usually (narrowly) ovate, rarely linear-lanceolate .... b. subsp. **nemorum**
3. At least some umbels pedunculate, primary peduncle to 9(–16) mm long, umbels not compounded into a verticillate synflorescence; capsule 5.5–6.5 mm long; seeds smooth or tuberculate, then the tubercles less dense and shorter .... c. subsp. **laxiflora**
   Umbels subsessile, primary peduncle to 2 mm long, umbels compounded into a verticillate synflorescence; capsule (6.5–)7–7.5 mm long; seeds with dense long and slender hooked tubercles .... d. subsp. **erinacea**

a. subsp. **carvalhoi**; Darbyshire in K.B. 63: 376, fig. 6A–B, table 3 (2009)

Leaves lanceolate, linear-lanceolate or rarely ovate. Inflorescences compounded into a terminal, subglobose to cylindrical head, usually comprising many subsessile umbellate cymes (in small plants reduced to 1–2 cymes), sometimes verticillate towards the base of the synflorescence, often with additional subsessile umbels in the upper leafy axils; cymule bracts (6–)9–13.5(–16) mm long. Capsule 4.5–5.5 mm long; seeds smooth. Fig. 91: 6/a–d, p. 685.

TANZANIA. Mpanda District: Kungwe Mt, Kasoje, July 1959, *Newbould & Harley* 4379!; Njombe District: ± 4 km N of Lukumburu, July 1956, *Milne-Redhead & Taylor* 10987!; Songea District: by R. Likondi E of Songea, June 1956, *Milne-Redhead & Taylor* 10907!

DISTR. **T** 4, 6–8; E Congo-Kinshasa, Rwanda, Burundi to Zambia, Zimbabwe and Mozambique

HAB. *Brachystegia* and other miombo woodland, rough grassland, riverine woodland; 500–1600(–2050) m

USES."Leaves eaten as vegetables" (*fide* Schlieben, cited in the protologue of *D. olitoria* Mildbr.)

CONSERVATION NOTES. Assessed as of Least Concern (LC) by Darbyshire (l.c.).

SYN. *D. rogersii* Turrill in K.B.: 314 (1911); Champluvier in Fl. Rwanda, Sperm. 3: 452 (1985). Type: Zambia, Kalomo, *Rogers* 8249 (K!, holo.; GRA!, iso.)

D. cephalantha S. Moore in J.L.S. 40: 162 (1911). Type: Zimbabwe, near Chirinda, *Swynnerton* 514 (BM!, holo.; K!, iso.)

D. olitoria Mildbr. in N.B.G.B. 11: 1085 (1934); Vollesen in Opera Bot. 59: 80 (1980). Type: Tanzania, Ulanga District, Mahenge, *Schlieben* 2295 [number not recorded in protologue] (B†, holo.; BM!, BR!, LISC!, M!, PRE!, iso.; K!, ?iso.)

D. angustifolia Gilli in Ann. Naturhist. Mus. Wien 77: 48 (1973). Type: Tanzania, Njombe District, Lumbila, *Gilli* 520 (W!, holo.; K!, iso.)

NOTE. Specimens from **T** 4, the Mahali Mts (e.g. *Newbould & Harley* 4379) have cymule bracts which narrow more gradually into a longer mucro than typical. However, some collections from **T** 7 are intermediate.

b. subsp. **nemorum** (*Milne-Redh.*) *I. Darbysh.* in K.B. 63: 379, fig. 6C–D, table 3 (2009). Type: Zambia, Solwezi District, Mbulungu Stream, *Milne-Redhead* 712 (K!, lecto. chosen by Darbyshire (l.c.); BR!, K! isolecto.)

Leaves ovate, ovate-lanceolate or rarely linear-lanceolate. Inflorescences compounded into terminal, subglobose to cylindrical heads, comprising many umbellate subsessile cymes, often with additional subsessile umbels in the upper leafy axils; cymule bracts (7–)9.5–14.5 mm long. Capsule 4.5–6.5 mm long; seeds tuberculate.

TANZANIA. Mbeya District: NW part of Muvwa village, June 1999, *Kayombo & Mwangoka* 2453!

DISTR. **T** ?4, 7; Zambia, SE Congo-Kinshasa

HAB. Disturbed riverine forest; 1150 m

USES. None recorded on herbarium specimens

CONSERVATION NOTES. Assessed as of Least Concern (LC) by Darbyshire (l.c.).

SYN. *D. nemorum* Milne-Redh. in K.B. 1937: 429 (1937)

NOTE. This subspecies is largely inseparable from subsp. *carvalhoi* when seeds are absent, although the leaves are often broader. *Bidgood et al.* 3501 (fl.) from Tatanda Mission (**T** 4) has particularly broad, ovate leaves and is likely to be subsp. *nemorum* but seeds are needed from this population for confirmation.

Mwangulango 1257, recently collected from Katavi N.P., **T** 4, resembles *D. carvalhoi* and has unhooked-tuberculate seeds as in subsp. *nemorum* but differs in having larger capsules, 8–9 mm long, and a more verticillate synflorescence with rather large cymule bracts (to 16.5 × 3.5 mm). It is quite possibly an extreme variant of this taxon but more material from this locality is desirable.

c. subsp. **laxiflora** *I. Darbysh.* in K.B. 63: 380. fig. 6E–F, table 3 (2009). Type: Zambia, 16 km on the Mpulungu–Mbala [Abercorn] road, May 1963, *Boaler* 951 (K!, holo.; EA!, K!, iso.)

Leaves narrowly ovate or lanceolate. Inflorescences axillary, 1–2 umbels of (2–)3(–4) cymules in the largely leafless upper axils; umbel peduncles 1.5–9(–16) mm long; cymule bracts 6.5–10(–13) mm long. Capsule 5.5–6.5 mm long; seeds smooth or tuberculate, the latter shortly hooked.

TANZANIA. Mpanda District: Mpanda–Uvinza road, May 2000, *Bidgood, Leliyo & Vollesen* 4469! & *idem, Bidgood, Leliyo & Vollesen* 4577!
DISTR. **T** 4; Zambia
HAB. Open *Brachystegia* and other miombo and riverine woodland, long grassland; 1150–1500 m
USES. None recorded on herbarium specimens
CONSERVATION NOTES. Assessed as of Least Concern (LC) by Darbyshire (l.c.).

d. subsp. **erinacea** *I. Darbysh.* in K.B. 63: 380, fig. 6G–H, table 3 (2009). Type: Tanzania, Mbeya District, Kimani R., Nyengenge waterfall, *P. Lovett & Kayombo* 468 (MO!, holo.; K!, iso.)

Immature leaves only seen, ovate. Inflorescences compounded into a verticillate synflorescence in the largely bare upper axils; umbels subsessile or pedunculate for up to 2 mm; cymule bracts 8–12 mm long. Capsule 6–8 mm long; seeds tuberculate, tubercles long and slender with conspicuous short hooks. Fig. 91: 7/a–d, p. 685.

TANZANIA. Mbeya District: at base of falls of tributary of Kimani R. above Kimani Falls, Iringa–Mbeya Road, June 1990, *Congdon* 270! & Nyengenge waterfall, Kimani R., June 1991, *P. Lovett & Kayombo* 468! (type) & Makete, Igando village, May 2008, *Suleiman et al.* 3571!
DISTR. **T** 7; not known elsewhere
HAB. River and waterfall margins including seasonally flooded areas, open dry *Brachystegia* woodland by roadside; 1350–1500 m
USES. None recorded on herbarium specimens
CONSERVATION NOTES. Assessed as Data Deficient (DD) by Darbyshire (l.c.). This subspecies has recently been found by the author on the edges of extensive miombo woodland in the foothills of the Poroto Mts (*Suleiman et al.* 3571) and it is quite possibly locally common there, so may well prove to be unthreatened.

23. **Dicliptera sp. E** (= *Tanner* 1544)

Erect herb to 45 cm tall, laxly branched; stems 6-angular, ridged, sulcate when young, patent-pubescent throughout, hairs somewhat hispid. Leaves largely absent at flowering; immature blade lanceolate, to 2.5 cm long, pubescent. Inflorescences axillary, 1 umbel of 2–3 cymules per axil; primary peduncle 2–12(–22) mm long, pubescent, hairs spreading to retrorse; main axis bracts linear-lanceolate, 3.5–6 mm long, pubescent; cymules subsessile, bracts linear-lanceolate, pairs unequal, the larger 11.5–16 mm long, ± 1.5 mm wide, apex narrowed gradually into a short mucro, outer surface densely white-pilose particularly on margin and midrib, inner surface with short glandular hairs, particularly towards apex; bracteoles lanceolate, 7.5–8.5 mm long, apex attenuate, pilose on midrib and margin towards base, hairs shorter towards apex, with additional short glandular hairs, margin hyaline towards base. Calyx lobes 4–5 mm long, glandular-puberulent and with sparse eglandular hairs outside. Corolla white, ± 17.5 mm long, pubescent outside; tube ± 9 mm long;

lips 8–8.5 mm long. Stamens long exserted; filaments ± 12 mm long, sparsely hairy on upper side; anther thecae ± 0.8 mm long, superposed. Ovary not seen; style glabrous. Capsule 7.5–8 mm long, puberulous and with scattered short glandular hairs towards apex; placental base elastic; seeds ± 1.7 mm in diameter, tuberculate, tubercles minutely hooked.

TANZANIA. Mwanza District: Nyasaka, Ilemera, June 1953, *Tanner* 1544!
DISTR. **T** 1; not known elsewhere
HAB. Rocky hillside under shade; 1150 m
USES. None recorded on herbarium specimen
CONSERVATION NOTES. This poorly known taxon appears highly localised in the area south of Lake Victoria. It will probably prove to be threatened once further data on its status become available: Data Deficient (DD).

NOTE. The cited specimen is close to *D. carvalhoi* subsp. *laxiflora* but differs in having longer cymule bracts which narrow more gradually into the mucro and are more densely pilose outside, and in having larger capsules. Its location is isolated from the currently known populations of *D. carvalhoi*. Further material, including leaves and further flowering material, is required to delimit this taxon fully and establish its relationship to *D. carvalhoi*.

24. **Dicliptera nilotica** *C.B. Clarke* in F.T.A. 5: 260 (1900) pro parte quoad *Grant* s.n.; Darbyshire in K.B. 63: 380 (2009). Type: Uganda, White Nile District: Madi, *Grant* s.n. (K!, lecto., chosen by Darbyshire, l.c.)

Decumbent or creeping suffruticose perennial, pyrophytic, (few-) much-branched from a woody base and rootstock, shoots to 35 cm tall; stems 6-angular, sulcate, spreading- to retrorse-pubescent to pilose, rarely largely glabrous. Leaves ± immature at flowering, narrowly ovate to elliptic (-obovate), 1–4.5 cm long, 0.5–1.7 cm wide, base cuneate to obtuse, apex acute to rounded, apiculate, surfaces pubescent at least on the margin and veins beneath; lateral veins 3–5 pairs; petiole to 5 mm long. Inflorescences terminal and/or axillary, 1–2 umbels of 2–5 cymules per axil, clearly spaced in the lower axils, often becoming crowded in the largely leafless upper axils, subsessile or pedunculate for up to 5(–15) mm, peduncles shortly pubescent to pilose; main axis bracts linear-lanceolate, 3–6 mm long, pubescent or ciliate; cymules subsessile or pedunculate for up to 3 mm; cymule bracts oblanceolate to linear-lanceolate, pairs somewhat unequal, the larger 6.5–10 mm long, 1.5–4 mm wide, apex apiculate, surfaces puberulous to pilose,hairs longest on margin and midrib, with capitate glandular hairs often dense in the upper half; bracteoles linear-lanceolate, 4–6.5 mm long, margin hyaline. Calyx lobes 3–3.5 mm long, outer surface puberulent and with short glandular hairs, ciliate. Corolla pink to mauve or rarely white, 11–13.5 mm long, pubescent and with occasional glandular hairs outside; tube 5.5–7 mm long; lip held in upper position oblong-obovate, 5.5–7 mm long, 2.5–4 mm wide; lip held in lower position elliptic, 5–7 mm long, 3–5 mm wide. Staminal filaments 5.5–7 mm long, sparsely hairy towards base; anther thecae 0.7–0.9 mm long, superposed. Pistil glabrous. Capsule 4.5–6 mm long, puberulous; placental base elastic; only immature seeds seen, tuberculate. Fig. 91: 8/a–d, p. 685.

UGANDA. West Nile District: Madi, Dec. 1862, *Grant* s.n.! (type); Karamoja District: Kadam [Debasien] Mt, date unknown, *Eggeling* 2646!; Mbale District: NE Elgon, Mar. 1958, *Tweedie* 1524!
KENYA. Trans-Nzoia District: Kitale, Mar. 1953, *Bogdan* 3693! & Milimani area, Apr. 1967, *Tweedie* 3438!; Uasin Gishu District: Turbo, date unknown, *Brodhurst-Hill* 375!
DISTR. **U** 1–3; **K** 2, 3, 5; not known elsewhere
HAB. Short grassland, open woodland and roadsides, particularly in recently burnt areas; 1050–2350 m
USES. None recorded on herbarium specimens
CONSERVATION NOTES. Assessed as of Least Concern (LC) by Darbyshire (l.c.).

SYN. *D. bagshawei* S. Moore in J.B. 49: 311 (1911). Type: Uganda, Bunyoro District, near Masindi, *Bagshawe* 1539 (BM!, holo.), **syn. nov.**

[*D. bupleuroides sensu* Thomson in Speke, Journ. Disc. source Nile. Append.: 644 (1863), *non* Nees]
[*D.* sp. nov. *sensu* Oliver in Trans, Linn. Soc., Bot. 29: 130 (1875)]
[*D.* sp. A *sensu* U.K.W.F. ed. 1: 607 (1974)]
[*D. pumila sensu* U.K.W.F. ed. 2: 276 (1994), *non* (Lindau) Dandy]

NOTE. I initially intended to maintain *D. bagshawei* as separate based on it having exclusively axillary inflorescences with longer peduncles than in *D. nilotica*. However, *Turner* 78.T (EA) from near the type locality of the former is entirely intermediate between the two and so *D. bagshawei* is synonymised here.

D. nilotica differs from the closely related West and Central African species *D. adusta* Lindau in lacking the long-pedunculate inflorescences of that taxon. In addition, the cymule bracts of the latter are often larger (to 13 mm long) and the plants are often, though not always, densely pilose. The variation between the two may, however, prove to be clinal, in which case *D. adusta* would have to be synonymised within *D. nilotica*. In the protologue to *D. nilotica*, Clarke suggested that *D. kamerunensis* Lindau may be conspecific, although he appears not to have seen the type (*Passarge* 128). It is quite possible that this is an older name for *D. adusta* (and the oldest name in the *D. nilotica-adusta* group), but the type appears to have been destroyed and so the placement of *D. kamerunensis* remains uncertain.

25. **Dicliptera pumila** (*Lindau*) *Dandy* in Mem. N. Y. Bot. Gard. 9: 27 (1954); F.P.S. 3: 173 (1956); Friis & Vollesen in Biol. Skr. 51 (2): 441 (2005); Ensermu in Fl. Eth. 5: 445 (2006) pro parte excl. *D. adusta*. Type: Malawi, Shire Highlands, *Buchanan* 1474 (K!, lecto., chosen here)

Suffruticose pyrophyte, usually producing many erect flowering shoots, 5–12 cm tall, later leafy shoots straggling or decumbent, 30–40 cm long, these sometimes with flowers; stems 6-angular, sulcate, flowering shoots densely puberulous, sometimes also pilose and with short capitate glandular hairs; leafy stems glabrous or sparsely puberulous. Flowering shoots often largely leafless. Mature leaves elliptic to obovate, 2–7 cm long, 0.7–2 cm wide, base cuneate, apex acute, shortly ciliate and with short hairs on veins beneath; lateral veins 4–5 pairs; petiole to 5 mm long. Inflorescences axillary and terminal, 1(–2) per axil, cymules usually solitary but appearing umbellately arranged at the apex of lateral branches; primary peduncle 7–30(–95) mm long, puberulous and usually with capitate glandular hairs, sometimes also pilose; main axis bracts absent; cymules dense, subcapitate or shortly spicate to 1–3.5 cm long, 3–8 flowers per cymule, basal flowers sometimes clearly spaced; flowers sessile, each subtended by one bract and a pair of bracteoles except the terminal flower where bracts paired; bracts linear-lanceolate to oblanceolate, (3.5–)5–9 mm long, 1–2.5 mm wide, outermost (cymule) bracts subequal to the other bracts, apex shortly attenuate, apiculate, puberulous and with many capitate glandular hairs, ± ciliate, sometimes with longer hairs along midrib, margin hyaline towards base; bracteoles linear, 3.5–7 mm long. Calyx lobes 3–5 mm long, puberulent and with sparse glandular hairs, ciliate. Corolla white or limb usually pale pink to magenta, 13–17 mm long, pubescent outside; tube 6.5–8 mm long; lip held in upper position oblong-ovate, 6.5–10.5 mm long, 3–5.5 mm wide, palate streaked purple; lip held in lower position elliptic-obovate, 5.5–8 mm long, 3–5 mm wide. Staminal filaments 4.5–9 mm long, sparsely hairy; anther thecae 0.7–1 mm long, superposed. Pistil largely glabrous. Capsule 6–6.5 cm long, puberulous; placental base elastic; seeds 1.5–2 mm in diameter, tuberculate. Fig. 91: 9/a–e, p. 685.

UGANDA. Acholi District, ENE slopes of Lomwaga Mt, Apr. 1945, *Greenway & Hummel* 7298! & 5 km N of Kiten, Dec. 1957, *Langdale-Brown* 2420!; Karamoja District: Lonyili Mt, Apr. 1960, *J. Wilson* 851!

TANZANIA. Kigoma District: Lubalisi village to Ntakatta, June 2000, *Bidgood, Leliyo & Vollesen* 4616!; Njombe District: Kipengere hills, Sept. 1986, *Linder* 3871!; Songea District: 29 km from Songea on Lindi road, Nov. 1966, *Gillett* 17895!

DISTR. **U** 1; **T** 4, 7, 8; S Ethiopia and Sudan to SE Congo-Kinshasa, Zimbabwe and Angola
HAB. Short grassland and open woodland, particularly in recently burnt areas; 950–2850 m
USES. None recorded on herbarium specimens
CONSERVATION NOTES. Widespread and often locally common in suitable habitat. It is probably under-recorded as it usually flowers only after burning, the leafy shoots developing later and usually sterile. Least Concern (LC).

SYN. *Duvernoia pumila* Lindau in E.J. 20: 44 (1894)
Peristrophe usta C.B. Clarke in F.T.A. 5: 244 (1900). Syntypes: Malawi, Zomba Mt, *Whyte* s.n. & idem, *Buchanan* 127 (both K!, syn.) & Tanganyika Plateau, *Carson* s.n. (not traced)
P. pumila (Lindau) Lindau in Wiss. Ergebn. Schwed. Rhod.–Kongo-Exped. 1: 307 (1916) & as *P. pumila* (Lindau) Gilli in Ann. Naturhist. Mus. Wien 77: 53 (1973)

NOTE. Plants from the northern part of the range (S Sudan, Ethiopia and represented in our region by *Langdale-Brown* 2420, **U** 1) sometimes flower from the leafy stems. *Richards* 13071 from **T** 4, Mbisi Forest, has atypical ovate leaves with an acute apex; the inflorescences are again on the leafy stems but are here mainly terminal. This collection matches *Brummitt* 11795 from N Malawi, both being tentatively placed in *D. pumila* on account of their displaying the typical inflorescence form and indumentum of this taxon.

Ensermu (l.c.) places *D. adusta* Lindau in synonymy with *D. pumila*. This species however differs in having umbellately arranged, few-flowered cymules. *D. adusta* is therefore more closely allied to *D. nilotica*, the differences being discussed under that taxon.

26. **Dicliptera brevispicata** *I. Darbysh.* in K.B. 62: 123, fig. 1 (2007). Type: Tanzania, Ufipa District, 2 km NW of Kalaela on Matai–Mwimbi road, *Goyder, Griffiths, Harvey, Kamwela & Paton* 3798 (K!, holo.; C, CAS, DSM!, EA!, WAG, iso.)

Scrambling, prostrate or decumbent suffruticose perennial, few- to much-branched from a woody rootstock, branches 5–90 cm long; stems 6-angular, ± ridged, with pale, somewhat hispid spreading and retrorse hairs, young stems often also with minute glandular and eglandular hairs. Leaves sometimes immature or absent at flowering, ovate, 1.5–8.5 cm long, 0.8–5 cm wide, base rounded to shallowly cordate, apex acute or subacuminate, upper surface and veins beneath hispid; lateral veins 4–7 pairs, pale and conspicuous beneath; petiole 3–17 mm long. Inflorescences axillary and terminal, 1(–2) per axil, cymules often solitary, sometimes in a lax umbel of 2(–3); primary peduncle 4–45(–60) mm long, indumentum as young stems but hairs often antrorse; main axis bracts (if present) lanceolate, (3.5–)5–7(–9) mm long, margin and midrib antrorse-pubescent; cymules dense, subcapitate or shortly spicate; flowers sessile, each subtended by one bract and a pair of bracteoles; bracts initially green, tinged brown with age, (linear-) lanceolate, pairs slightly unequal, larger bract (5.5–)8–13(–16) mm long, 1.5–2(–3) mm wide, the outermost (cymule) bracts subequal to the other bracts, apex acute to subattenuate, outer surface pale antrorse-pubescent particularly on midrib and with inconspicuous minute glandular and eglandular hairs, venation reticulate, conspicuous within, margin narrowly hyaline; bracteoles as bracts but 7–10.5(–12) mm long, 0.7–1.7 mm wide. Calyx lobes 4.5–8 mm long, glandular-puberulent and with occasional eglandular hairs on midrib, margin hyaline, ciliate. Corolla white or limb usually pale pink to mauve, 17–21 mm long, retrorse-pubescent outside; tube 7–9 mm long; lip held in upper position oblong, 9–12 mm long, 3–4.5 mm wide, palate sometimes streaked pink or purple, puberulous particularly towards mouth; lip held in lower position ovate-elliptic, 8.5–11.5 mm long, 6.5–8 mm wide. Staminal filaments 4.5–7 mm long, pubescent; anther thecae ± 1 mm long, superposed and becoming slightly separated. Ovary glabrous or apex with sparse hairs; style sparsely strigulose. Capsule 7.5–9.5 mm long, densely pubescent; placental base inelastic or rising slightly; seeds 1.9–2.3 mm in diameter, sparsely tuberculate, tubercles shortly clavate. Fig. 93, p. 715.

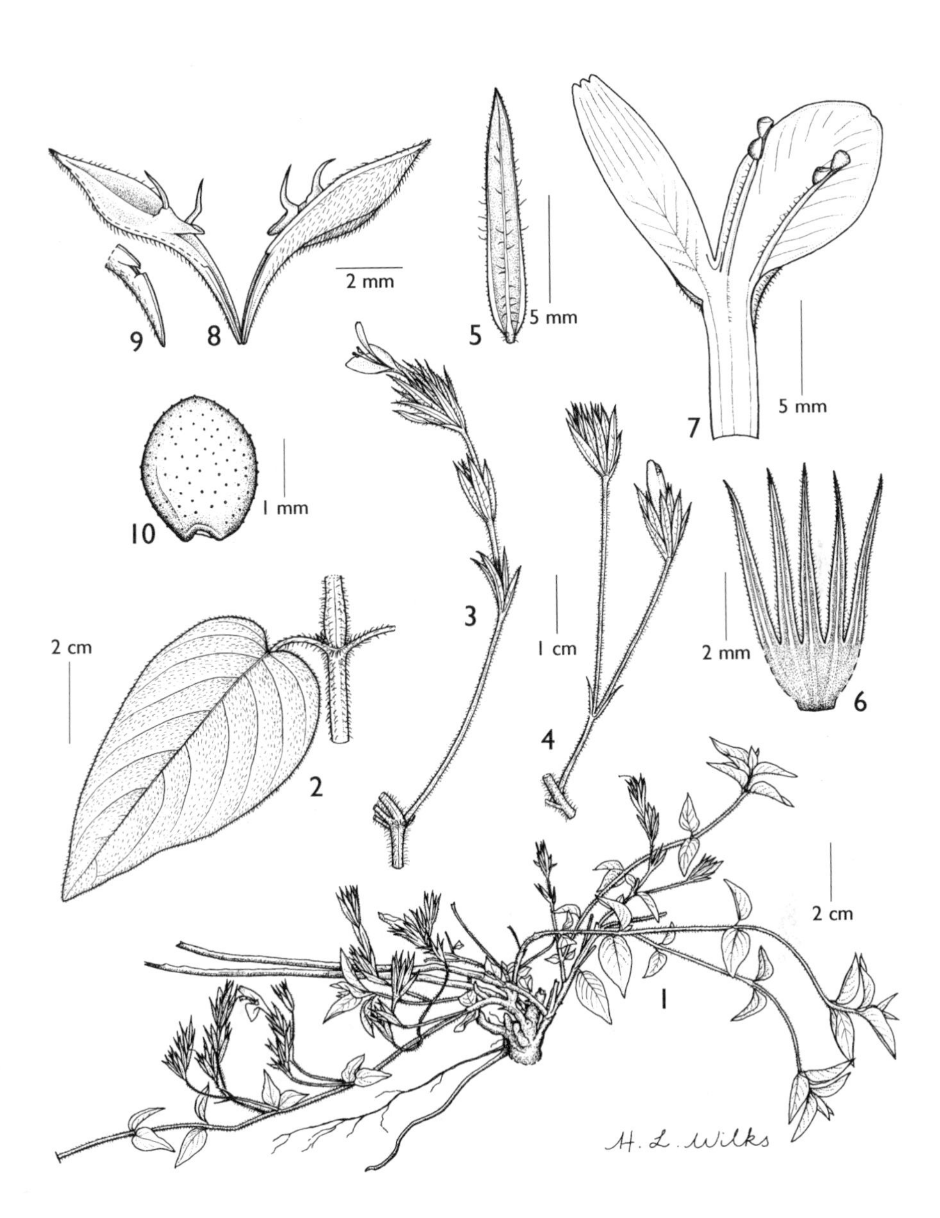

FIG. 93. *DICLIPTERA BREVISPICATA* — **1**, habit, young plant with persistent stems from previous growth season; **2**, mature leaf and stem section; **3**, inflorescence with solitary spicate cymule and mature flower; **4**, inflorescence with two umbellately arranged capitate cymules and flower in bud; **5**, cymule bract, outer surface; **6**, dissected calyx, outer surface; **7**, dissected corolla with attached stamens, internal surface; **8**, mature, dehisced capsule, the right hand valve showing the slight tearing near the apex of the stipe, the left hand side dissected longitudinally to show the slight elasticity of the placenta; **9**, detail of apex of capsule stipe, showing the tendency for the wall to tear; **10**, mature seed. 1, 6, 7, 8 and 10 from *Goyder et al.* 3798; 2 and 4 from *Bidgood et al.* 3864; 3, 5 and 9 from *Richards* 18353. Drawn by Hazel Wilks. Reproduced from Kew Bulletin 62: 125 (2007) with permission of the Trustees of the Royal Botanic Gardens, Kew.

TANZANIA. Mpanda District: 60 km on Chala–Mpanda road, May 1997, *Bidgood et al.* 3864!; Ufipa District: road to Kamsamba village, central Rukwa, Nov. 1963, *Richards* 18353!; Mbeya District: Pungulumo Hills above Mshewe village, Nov. 1989, *J. Lovett & Kayombo* 3380!

DISTR. **T** 4, 7; NE Zambia

HAB. *Brachystegia* and other miombo woodland, particularly in disturbed areas including recently burnt ground, roadsides; 750–1850 m

USES. "Cattle food" (**T** 7; *J. Lovett & Kayombo* 3380)

CONSERVATION NOTES. Assessed as of Least Concern (LC) by Darbyshire (l.c.).

NOTE. *Ngoundai* 157 (EA) is recorded as having been collected from **T** 3, Lushoto District, but this is surely incorrect.

27. **Dicliptera sp. F** (= *P. Lovett & Kayombo* 159)

Erect perennial herb to 100 cm tall, several-branched; stems 6-angular, prominently ridged, pale-hispid on the ridges and with shorter antrorse hairs when young. Leaves ovate, 5.5–9 cm long, 3–4.5 cm wide, base rounded to shallowly cordate, apex attenuate, apiculate, surfaces hispid particularly towards margin above and on veins beneath; lateral veins 6–7 pairs, pale and prominent beneath; petiole 4–22 mm long. Inflorescences axillary and terminal, 1–2 lax umbels of (1–)2 cymules per axil, often compounded; primary peduncle 1–25 mm long, ridged, antrorse-pubescent; main axis bracts lanceolate or narrowly elliptic, 3.5–7 mm long, antrorse-pubescent on margin and veins outside, margin hyaline; cymules several-flowered but only one flower developed at a given time; peduncles highly variable on each cyme, one of a pair much longer than the second, the former 7.5–37 mm long; cymule bracts initially green but soon tinged pinkish-brown, rhombic or elliptic, pairs slightly unequal, larger bract 15–21 mm long, 4.5–7 mm wide, base and apex acute, the latter mucronulate, outer surface antrorse-hispid mainly on margin and midrib, venation pinnate-reticulate, conspicuous within, margin narrowly hyaline towards base; bracteoles as bracts but ovate-lanceolate, 12–14.5(–16) mm long. Calyx lobes 8.5–9 mm long, sparsely puberulent and with longer hairs on midrib and margin, margin hyaline. Corolla "purple and white", 23–25 mm long, retrorse-pubescent outside; tube 9–9.5 mm long, lobes 13.5–17 mm long. Staminal filaments ± 12 mm long, with sparse hairs above towards base; anther thecae ± 1 mm long, superposed and slightly separated. Ovary not seen; style glabrous. Immature capsule only seen, densely pubescent; placental base (?) inelastic; seeds not seen. Fig. 91: 10/a–d, p. 685.

TANZANIA. Mbeya District: near Songwe hotsprings, May 1991, *P. Lovett & Kayombo* 159!

DISTR. **T** 7; not known elsewhere

HAB. Scrub on limestone; 1200 m

USES. None recorded on herbarium specimen

CONSERVATION NOTES. This taxon is currently known from only a single locality and clearly scarce; it is likely to prove threatened once fully delimited and studied further but is currently assessed as Data Deficient (DD).

NOTE. The specimen cited is highly distinctive within the Flora region, appearing most closely related to *D. welwitschii* S. Moore from Angola with which it shares the hispid eglandular indumentum, stiff cymule bracts which are glabrous within and tinged pinkish-brown, calyx lobes with a hyaline margin and broadly ovate, (sub)cordate leaves. *D. welwitschii* however differs in being more conspicuously hairy throughout, in the main axis bracts being broader and ovate in shape and in having more regular umbels. *D. colorata* C.B. Clarke from Malawi falls within the same group but clearly differs in having ± dense axillary umbels, narrowly elliptic cymule bracts, smaller flowers with conspicuous hairs on the palate of the lip held in the upper position and more densely hairy filaments.

28. **Dicliptera paniculata** (*Forssk.*) *I. Darbysh.* in K.B. 62: 122 (2007). Type: Yemen, *Forsskål* 385 (C, microfiche 38: III. 3–4, lecto., chosen by Wood, Hillcoat & Brummitt in K.B. 38: 451 (1983), K!, photo.)

Erect to straggling annual herb, (10–)60–200 cm tall, much-branched; stems 6-angular with prominent pale ridges, these pale-hispid or rarely glabrous, young stems often also with short eglandular and capitate glandular hairs. Leaves often immature at flowering, ovate (-lanceolate), 1.5–10 cm long, 0.8–4.5 cm wide, base rounded to shortly attenuate, apex acute to acuminate, apiculate, surfaces hispid particularly margin and veins beneath, rarely largely glabrous; lateral veins (3–)5–7 pairs; petiole 2–22 mm long. Inflorescences axillary and terminal, compounded into a lax panicle to 20(–40) cm long on the largely bare upper branches; peduncles wiry, with short eglandular hairs and sparse to dense capitate glandular hairs, rarely also pilose; main axis bracts linear-lanceolate, 2–4.5(–5.5) mm long, margin and midrib with sparse antrorse hairs; cymule peduncles (3.5–)8–22(–38) mm long, apex often abruptly bent; cymule bracts linear-lanceolate, pairs highly unequal, larger bract (5.5–)7–14.5(–17) mm long, 0.7–1.7 mm wide, apiculate, outer surface with minute eglandular and glandular hairs, often with additional capitate glandular hairs, midrib only prominent, usually antrorsely hairy, margin sparsely and shortly ciliate, narrowly hyaline towards base; bracteoles lanceolate, 3.5–7 mm long, margin hyaline. Calyx lobes 2–3.5 mm long, ciliate and with occasional minute glandular hairs outside, margin hyaline. Corolla pink or mauve, rarely white, 9–13.5(–17) mm long, pubescent outside; tube 4–6(–7) mm long; lip held in upper position oblong, 4.5–8(–9.5) mm long, 1.5–3 mm wide, palate streaked purple, puberulous at mouth; lip held in lower position elliptic, 4.5–7(–8.5) mm long, 2.5–4 mm wide. Staminal filaments 3.5–7 mm long, sparsely hairy; anther thecae 0.5–0.8 mm long, superposed and somewhat oblique. Pistil largely glabrous. Capsule 8–12.5 mm long, pubescent and with scattered short glandular hairs; placental base inelastic; seeds 1.8–2.5 mm in diameter, tuberculate particularly towards the rim, tubercles short, minutely hooked. Fig. 91: 11/a–d, p. 685.

UGANDA. Acholi District: Agoro, Chua, Nov. 1945, *A.S. Thomas* 4357!; Karamoja District: Moroto township, 1964, *Wilson* 1744!; Teso District: Serere, Kyere, July 1926, *Maitland* 1303!
KENYA. Northern Frontier District: Helu police post ± 7 km E of Moyale, July 1952, *Gillett* 13500!; Machakos District: Athi R. near Kibwezi, May 1959, *Napper* 1254!; Teita District: Voi, May 1931, *Napier* 916!
TANZANIA. Shinyanga District: Shinyanga, Kizumbi Hill, May 1931, *Burtt* 2504!; Mbulu District: near Chem Chem R., Lake Manyara National Park, Feb. 1964, *Greenway & Kanuri* 11234!; Mpwapwa District: Mpwapwa, Aug. 1930, *Greenway* 2483!
DISTR. **U** 1, 3; **K** 1–7; **T** 1–7; widespread in tropical and subtropical Africa; Arabia, India and Thailand
HAB. Open *Acacia-Commiphora* woodland, riverine woodland, grassland, roadsides and as a weed of cultivation; 450–1350 m
USES. "Eaten by stock" (**T** 5; *Hornby* 89)
CONSERVATION NOTES. Widespread and locally abundant, particularly favouring open, disturbed habitats: Least Concern (LC).

SYN. *Dianthera paniculata* Forssk., Fl. Aegypt.-Arab.: 9 (1775)
D. bicalyculata Retz. in Vet. Acad. Handl. Stockholm 36: 297 (1775). Type: India, ? Malabar, *König* s.n. (UPS-LINN)
Justicia bicalyculata (Retz.) Vahl in Symb. Bot. Upsal. 2: 13 (1791)
Peristrophe bicalyculata (Retz.) Nees in Wall., Pl. As. Rar. 3: 113 (1832); Nees in DC., Prodr. 11: 496 (1847); C.B. Clarke in F.T.A. 5: 242 (1900); Heine in F.W.T.A. ed. 2, 2: 424 (1963); U.K.W.F. ed. 1: 609 (1974); U.K.W.F. ed. 2: 275 (1994)
P. kotschyana Nees in DC., Prodr. 11: 497 (1847) pro parte excl. *Schimper* 1095. Type: Sudan, Kordofan, 'Arasch-Cool', *Kotschy* 161 (K!, lecto., chosen here; HBG!, isolecto.)
P. doriae Terrac. in Ann. Ist. Bot. Roma 5: 102 (1894). Type: Eritrea, Ferehan Volcano, *Terracciano* s.n. (FT, holo.)
P. paniculata (Forssk.) Brummitt in K.B. 38: 451 (1983); Friis & Vollesen in Biol. Skr. 51 (2): 452 (2005); Hedrén in Fl. Somalia 3: 425, fig. 286 (2006); Ensermu in Fl. Eth. 5: 447 (2006)

NOTE. *Gilli* 536 from **T** 7, Lupingu displays anomalously large flowers (15–17 mm long) but closely resembles other material of *D. paniculata* in, for example, inflorescence form and indumentum. *Mungai et al.* 310/83, from **K** 7, Kora Game Reserve, has exceptionally small capsules maturing to 5.5–6.5 mm long, each valve containing only a single seed; this specimen is considered aberrant and the measurements are omitted from the description.

29. **Dicliptera aculeata** *C.B. Clarke* in F.T.A. 5: 257 (1900). Type: Malawi, Nyika Mts, *Whyte* s.n. (K!, lecto.), chosen by Brummitt in K.B. 46: 290 (1991)

Erect or straggling suffruticose perennial, (15–)45–200 cm tall, few- to much-branched from a woody rootstock; stems 6-angular, with or without prominent pale ridges, hispid on ridges and nodes or glabrous. Leaves often immature at flowering, (ovate-) lanceolate, 2.5–13.5 cm long, 0.7–3 cm wide, base shortly attenuate, apex acute to acuminate, apiculate, hispid particularly on margin and veins beneath, or largely glabrous; lateral veins 3–7 pairs, pale and prominent beneath; petiole 1–8 mm long. Inflorescences axillary and terminal, lax or more rarely congested, 1–2 umbels of (2–)3–4 cymules per axil, often compounded into panicles; primary peduncle wiry, (1–)4–55 mm long, glabrous or sparsely antrorse-hispid, rarely with scattered subsessile glands; main axis bracts linear-lanceolate, (2–)3.5–8(–14) mm long, glabrous or with sparse antrorse hairs on margin and midrib; cymules subsessile or peduncles to 25(–40) mm long; cymule bracts linear-lanceolate or lanceolate, pairs unequal, larger bract (5.5–)7–13(–24) mm long, 0.8–1.5(–2) mm wide, apex acuminate, apiculate, largely glabrous or with sparse short eglandular and glandular hairs and/or subsessile glands, midrib prominent, this sometimes antrorsely hispid, margin narrowly hyaline towards base; bracteoles lanceolate, 4.5–10.5(–15) mm long, margin hyaline. Calyx lobes 3–8 mm long, ciliate and with occasional minute glandular hairs outside, margin hyaline. Corolla pink, mauve or rarely white, 15–23(–30) mm long, pubescent and with scattered short glandular hairs outside; tube 7–9.5(–11) mm long; lip held in upper position oblong, 8.5–13.5(–19) mm long, 3–5 mm wide, palate streaked purple, puberulous particularly towards mouth; lip held in lower position elliptic, 8–14.5(–20) mm long, 4–7.5 mm wide. Staminal filaments 6.5–12(–14.5) mm long, with short hairs particularly above; anther thecae 0.8–1 mm long, superposed and somewhat oblique. Ovary glabrous or with few apical hairs; style sparsely strigulose. Capsule 9.5–14(–16) mm long, with sparse eglandular hairs mainly towards apex and scattered short glandular hairs elsewhere; placental base inelastic; seeds 1.5–2.5 mm in diameter, tuberculate, tubercles short, somewhat clavate, not hooked. Fig. 91: 12/a–d, p. 685.

TANZANIA. Mbulu District: Babati, Nov. 1930, *Haarer* 1830!; Morogoro District: Turiani, Nov. 1954, *Semsei* 1904!; Songea District: Msena near Litenga Hills, May 1972, *Shabani* 845!

DISTR. **T** 2, 5–8; Malawi, Mozambique and Zambia

HAB. *Brachystegia* and other miombo woodland, grassland including recently burnt areas, riverbanks; 250–1700 m

USES. None recorded on herbarium specimens

CONSERVATION NOTES. Widespread, locally common and adapted to a range of habitats: Least Concern (LC).

SYN. *Peristrophe aculeata* (C.B. Clarke) Brummitt in K.B. 46: 290 (1991)

NOTE. In transferring *Dicliptera aculeata* to *Peristrophe*, Brummitt (in K.B. 46: 290 (1991)) included only specimens from N Malawi with congested inflorescences on reduced side branches, remarking that these were the most distinctive of the perennial plants within the *P. paniculata* complex from East Africa. Perennial specimens with more lax inflorescences remained under *P. paniculata*, but K. Balkwill (in S. Afr. J. Bot. 55: 254–258 (1989)) considered *P. paniculata* to be exclusively annual. This view is followed here, and the laxly flowered perennial plants are considered a variant of *D. aculeata*. The form with congested inflorescences often has the largest, long-acuminate cymule bracts, the longest calyx lobes and markedly ridged stems and peduncles, the ridges beset with hispid hairs. This form is also represented in South-central Tanzania, (e.g. *Shabani* 845). The more widespread form with lax, fewer-flowered inflorescences tends to have smaller bracts (max. 14.5 mm long in our region), shorter calyx lobes (max. 5 mm) and ± inconspicuously ridged, largely glabrous stems and peduncles. However, intermediate specimens are not uncommon, for example lax-flowered specimens with hispid stems are regularly recorded in North and central Malawi and also in our region from northern Tanzania (e.g. *Haarer* 1830). *D. aculeata* is therefore considered a single, highly variable species, although it may be possible to separate out distinct varieties within this complex following further analysis.

Dicliptera aculeata largely replaces *D. paniculata* in miombo woodland and in localised areas of increased moisture availability, for example along river courses. It is one of several closely related perennial species in the *D. paniculata* complex which include *Peristrophe teklei* Ensermu from Arabia and NE Africa and both *Dicliptera decorticans* (K. Balkwill) I. Darbysh. and *Peristrophe cernua* Nees from southern Africa. The combinations in *Dicliptera* are yet to have been made for the two *Peristrophe* names.

50. HYPOESTES

R. Br. in Prodr. Fl. Nov. Holl.: 474 (1810); Nees in DC., Prodr. 11: 501 (1847); C.B. Clarke in F.T.A. 5: 244 (1900); Benoist in Not. Syst., Paris 10: 241–248 (1942); Balkwill & Getliffe-Norris in S. Afr. J. Bot. 51: 133–144 (1984)

Annual or perennial herbs or subshrubs, evergreen; cystoliths present, variable, sometimes conspicuous. Stems (sub-)angular, sulcate or ridged. Leaves opposite-decussate, pairs equal or anisophyllous; margin subentire to crenate-serrate. Inflorescences axillary and/or terminal, comprising a series of monochasial cymules, arrangement of cymules variously fasciculate, umbellate, spiciform, strobiliform or cylindrical, solitary or verticillate; main axis bracts paired, equal to highly unequal, free or basally fused, shape and size variable, rarely foliaceous; cymule bracts paired, equal to unequal, free or fused for up to two thirds of their length; bracteoles paired, equal to somewhat unequal, free, narrower than the cymule bracts, margin hyaline. Calyx divided almost to base or with a distinct tube, 4- or 5-lobed, lobes subequal to somewhat unequal in length, often hyaline. Corolla bilabiate, variously white, pink or purple, usually with darker guidelines; limb resupinate, tube twisted through 180°, cylindrical, somewhat expanded towards mouth, pubescent outside except towards base of tube where glabrous, with two paired lines of pubescence within; lip held in upper position recurved distally, three-lobed or with only the median lobe well developed, palate longer than lobes; lip held in lower position unlobed, apex obtuse to shallowly emarginate. Stamens two; filaments attached at mouth of corolla, exserted, initially straight, becoming divergently reflexed after maturity; anthers monothecous, theca ellipsoid; staminodes absent. Disk cupular, with a V-shaped slit, two-awned. Ovary oblong-ovoid, glabrous, two-locular, two ovules per locule; style filiform; stigma bifid, exserted from the corolla tube, held below the anthers. Capsule clavate, shortly stipitate, placental base inelastic. Seeds held on retinacula, lenticular or compressed-ellipsoid, with a hilar excavation, smooth or tuberculate.

A genus confined to the Old World tropics and subtropics from Africa to Australia with highest diversity on Madagascar. The number of accepted species quoted varies from 40 (Mabberley, Pl. Book, ed. 3: 425 (2008)) to 150 (Dyer, Gen. S. Afr. Fl. Pl. 1 (1975)). In continental Africa, approximately 10 species are recognised. The taxonomy is, however, complicated by the presence of three widespread, polymorphic taxa: *H. triflora*, *H. aristata* and *H. forskaolii*, the delimitations of which have not been fully defined. A revision of the genus in Africa and Arabia would aid such clarification and would probably result in the reduction of several currently recognised species to synonymy or infraspecific rank within one of these complexes. For example, *H. carnulosa* Chiov. from Somalia and *H. pubescens* Balf.f. and *H. sokotrana* Vierh. from Socotra are possibly better treated as infraspecific variants of *H. forskaolii*.

Inflorescence and bract terminology follows that used in *Dicliptera*. Several flowers can develop in a monochasial cyme within each cymule, although in *H. aristata* it is reduced to a single flower.

1. Principal inflorescence terminal, strobiliform or cylindrical; outer bracts conspicuously pale-setose 2
   Inflorescences terminal and axillary, units variously fasciculate, umbellate or spiciform, solitary or verticillate; bracts never pale-setose 3
2. Leaves evenly distributed along the erect stems, inflorescences subtended by a single pair of leaves; blade 2–7 cm long; outer bracts lanceolate-aristate; capsule pubescent at apex . 2. *H. cancellata*
   Leaves tending to cluster towards stem apices, inflorescences often subtended by a pseudowhorl of 4–6 leaves; blade 6.5–14.5 cm long; outer bracts broadly spatulate-aristate; capsule glabrous .... 3. *H. strobilifera*
3. Cymule bract pairs free; cystoliths arcuate, prominent on the leaf blade; inflorescences umbellate, sometimes lax due to extension of the cymule peduncles, cymules (1–)3–4(–6) per inflorescence; staminal filaments glabrous; seeds tuberculate 1. *H. triflora*
   Cymule bract pairs partially fused; cystoliths elliptic or linear, often inconspicuous; cymes fasciculate or unilateral spikes, usually dense, often compounded and verticillate or paniculate; cymules 5–many per inflorescence; staminal filaments glandular-puberulent to pilose; seeds smooth 4
4. Cymule bract pairs fused for up to one third of their length, aristate; calyx lobes 4; corolla lip held in lower position narrowly oblong-elliptic; filaments puberulous; capsule glabrous .... 4. *H. aristata*
   Cymule bract pairs fused for one half to two thirds of their length, not aristate; calyx lobes 5; corolla lip held in lower position trullate-caudate; filaments pilose, usually conspicuously so; capsule puberulous in the upper half or rarely glabrate .... 5. *H. forskaolii*

1. **Hypoestes triflora** (*Forssk.*) *Roem. & Schult.* in Syst. Veg. 1: 141 (1817); Nees in DC., Prodr. 11: 506 (1847); C.B. Clarke in F.T.A. 5: 247 (1900); Benoist in Not. Syst., Paris 10: 243 (1942); F.P.S. 3: 177 (1956); Heine in F.W.T.A. ed. 2, 2: 431 (1963); U.K.W.F. ed. 1: 609, 610 (fig.) (1974); Champluvier in Fl. Rwanda, Sperm. 3: 460, fig. 142: 2 (1985); K. Balkwill & Getliffe-Norris in S. Afr. J. Bot. 51: 139 (1985); Blundell, Wild Fl. E. Afr.: 393, pl. 610 (1987); U.K.W.F. ed. 2: 277, pl. 121 (1994); Friis & Vollesen in Biol. Skr. 51 (2): 444 (2005); Ensermu in Fl. Eth. 5: 450, fig. 167.47: 5–15 (2006); Fischer & Killmann, Pl. Nyungwe N.P. Rwanda: 592 (2008). Syntypes: Yemen, *Forsskål* s.n. (C, microfiche 60: I. 3–4 & 5–6; K!, photo.)

Creeping, straggling or subscandent, annual or perennial herb, 10–100(–300) cm tall, rooting at lower nodes, ± decumbent; stems sulcate, spreading- and/or appressed-pubescent when young, soon glabrescent except for a line of long hairs at nodes. Leaf pairs ± anisophyllous, ovate or rarely narrowly elliptic, 1.5–13 cm long, 0.8–6.5 cm wide, base ± asymmetric, rounded to attenuate, margin crenate-serrate or subentire, apex acuminate, upper surface ± pubescent, most dense on margins, veins beneath pubescent; lateral veins 4–7(–10) pairs; cystoliths arcuate, conspicuous; petiole 0.7–6 cm long, pubescent mainly above, hairs often long, spreading. Inflorescences axillary and terminal, umbellate, solitary or verticillate, (1–)3–4(–6) cymules per umbel; primary peduncle 3–70 mm long, sparsely to densely spreading and/or appressed-pubescent, with or without glandular hairs distally; main axis bracts foliaceous, ovate to oblanceolate, to 20 mm long, one pair per umbel; cymules subsessile or peduncles to 33 mm long; cymule bracts free but clasping at the base, obovate to narrowly elliptic, unequal, the larger 8–17 mm long, 2–7 mm wide, apex rounded, acute or rarely acuminate, variously short-pubescent to long-pilose, sometimes only on margin, with or without glandular hairs; bracteoles linear-lanceolate, unequal, the longer 6–11 mm long, margin hyaline and often ciliate, surfaces with or without short eglandular and/or glandular hairs. Calyx lobes 5, subequal, linear-lanceolate, 2–5(–8) mm long, glabrous or ciliate, hyaline. Corolla white or pink, with purple markings on palate of lip held in upper position, (12–)16–27(–37) mm long, pubescent outside; tube 8–13 mm long; lip held in upper position oblong, (4.5–)7–18(–21) mm long, 2.5–6.5 mm wide, apex 3-lobed, each 1–2.5 mm long; lip held in lower position obovate, (4–)6.5–16(–20) mm long, 3.5–6.5 mm wide, apex rounded or emarginate. Staminal filaments (4–)5.5–11(–14) mm long, glabrous; anthers 0.6–1.7 mm long. Style pubescent towards apex. Capsule (6–)8–11(–13) mm long, glabrous or sparsely puberulous at apex; seeds lenticular, (1–)1.5–2.5 mm in diameter, tuberculate.

Uganda. Karamoja District: Moroto Mt, Sept. 1958, *J. Wilson* 534!; Kigezi District: Luhizha, June 1951, *Purseglove* 3669!; Mbale District: Siti R., N Elgon, Jan. 1953, *Dawkins* 777!

Kenya. Northern Frontier District, Mt Kulal, Nov. 1978, *Hepper & Jaeger* 7081!; Kiambu District: Kikuyu Escarpment Forest, Gatamayu R., Jan. 1969, *Napper & Stewart* 1820!; Kisumu-Londiani District: Londiani Mt, Dec. 1974, *J.G. Williams* 61!

Tanzania. Masai District: Ngorongoro Conservation Area, Embagai Crater, outside East slope, Aug. 1972, *Frame* 31!; Mpanda District: Ntakatta Forest, June 2000, *Bidgood, Leliyo & Vollesen* 4639!; Kilosa District: Ukaguru Mts, track to Trial Plot from Mandege Forest Station, Aug. 1972, *Mabberley* 1431!

Distr. **U** 1–4; **K** 1–6; **T** 1–7; widespread in tropical and southern Africa; southern Arabia, India to China and Thailand

Hab. Forest, forest margins, clearings and grassland, often at high elevations; 900–3200 m

Uses. "Used against cerebral malaria" (**T** 1; *Bidgood et al.* 4783)

Conservation notes. Widespread and common in suitable habitat, often gregarious and dominating the herb layer in montane forest. It also appears tolerant of habitat disturbance. Least Concern (LC).

Syn. *Justicia triflora* Forssk., Fl. Aegypt.-Arab.: 4 (1775)

Hypoestes adoensis A. Rich. in Tent. Fl. Abyss. 2: 162 (1850). Syntypes: Ethiopia, Adoua, *Schimper* I.108 (M!) & *Schimper* II.1111 (not traced); Axum, *Schimper* III.1491 (MPU!, syn.); Adoua, *Quartin-Dillon & Petit* s.n. (K!, P, syn.)

[*H. acuminata* Hochst., *nom. nud.*; T. Anderson in J.L.S. 7: 50 (1864), in syn.]

H. simensis Solms in Beitr. Fl. Aeth.: 111 (1867). Type: Ethiopia, Debre Eski, *Schimper* s.n. (not traced)

H. adoensis A. Rich. var. *andersonii* Engl., Hochgebirgsfl. Trop. Afr.: 395 (1892). Syntypes: Ethiopia, Gajeh-Merki, *Schimper* 600 & Thal Locomdi, *Schimper* 641 (both B†, K!, syn.)

H. phaylopsoides S. Moore in Trans. Linn. Soc., Bot. 4: 34 (1894); C.B. Clarke in F.T.A. 5: 248 (1900) pro parte excl. *Whyte* s.n. ex Misuku Hills [Masuka Plateau]. Type: Malawi, Mlanje, *Whyte* 126 (BM!, holo.)

H. kilimandscharica Lindau in E.J. 19: 47 (1894). Type: Tanzania, Kilimanjaro, Kibosho, W of Sinas Boma, *Volkens* 1663 (B†, holo.; BM!, iso.)

H. inaequalis Lindau in E.J. 20: 50 (1894). Type: Cameroon, between Buea and Mann's Spring, *Preuss* 745 (B†, holo.; K!, iso.)

H. consanguinea Lindau in E.J. 20: 50 (1894); Friis & Vollesen in Biol. Skr. 51 (2): 444 (2005). Syntypes: Togo, Bismarckburg, *Büttner* 315 & *Kling* 189 (both B†, syn.); Cameroon, W of Buea, *Preuss* 599 (B†, K!, syn.), **syn. nov.**

H. ciliata Lindau in E.J. 20: 51 (1894). Type: Cameroon, Buea, Mann's Spring, *Preuss* 732 (B†, holo.)

[*H. rosea sensu* C.B. Clarke in F.T.A. 5: 248 (1900) pro parte quoad *Büttner* 315, *Kling* 189 & *Preuss* 599, *non* P. Beauv.]

H. toroensis S. Moore in J.B. 45: 92 (1907). Type: Uganda, Toro District, Wimi River, *Bagshawe* 1020 (BM!, holo.; MHU!, iso.), **syn. nov.**

H. caloi Chiov. in Ann. Bot. Roma 9: 129 (1911). Syntypes: Ethiopia, Dembia, *Chiovenda* 2385, 2745 & 2818 (all FT!, syn.)

H. sennii Chiov. in Atti R. Accad. Ital., Mem. Cl. Sc. Fis. etc., 9 (Pl. Nov. Aethiop.): 51 (1940). Syntypes: Ethiopia, Shewa, Kabana, *Senni* 110; Addis Ababa, *Senni* 10, 56, 1498 & 1499; between Ambo and Guder, *Senni* 1748 (all FT, syn.), syn. fide Ensermu in Fl. Eth. 5: 450 (2006)

[*H. cf. consanguinea sensu* Fischer & Killmann, Pl. Nyungwe N.P. Rwanda: 592 (2008)]

NOTE. This species is highly variable across its range. In East Africa several forms are discernible, although the widespread occurrence of intermediate specimens prevents their formal recognition. Of particular note is the more robust, often straggling or subscandent form found at forest margins, clearances and in open, disturbed habitats (e.g. *Synnott* 1383, **U** 1). These plants have conspicuously glandular inflorescences, and the cymule peduncles often extend, creating a more lax inflorescence. This form is largely restricted to the F.T.E.A. region, where it is widespread above 2000 m altitude and less commonly down to 900 m alt; it also extends into N Malawi. Glandular forms further south lack the extension of the cymule peduncles and tend towards larger corolla sizes; such forms are represented in East Africa by two specimens from **T** 4 (*Richards* 12964; 13060 *pro parte*). Intermediates between these glandular forms and typical specimens, with less conspicuously glandular inflorescences and with subsessile to shortly pedunculate cymules, are widespread.

Specimens with narrowly elliptic, acute or subacuminate cymule bracts were previously treated as separate species under *H. consanguinea* and *H. toroensis*, the latter additionally having narrowly elliptic leaves. In the F.T.E.A. region, they are mainly restricted to Uganda and W Tanzania in forest at lower altitudes than typical *H. triflora*. However, many intermediate specimens are found throughout the region, particularly at intermediate elevations. Even within individual populations, bract shape can vary widely, e.g. the K sheet of *Archbold* 2471B from **T** 7, Milo, has three plants which display bracts ranging from elliptic, subacuminate and 2 mm wide to obovate, obtuse-rounded and 4 mm wide. In other respects, *H. consanguinea* and *H. toroensis* are within the variation displayed in typical *H. triflora*, and are therefore reduced to synonymy here.

2. **Hypoestes cancellata** *Nees* in DC., Prodr. 11: 505 (1847); C.B. Clarke in F.T.A. 5: 246 (1900); Benoist in Not. Syst., Paris 10: 242 (1942); F.P.S. 3: 176 (1956); Heine in F.W.T.A. ed. 2, 2: 430 (1963); Champluvier in Fl. Rwanda, Sperm. 3: 460, fig. 142: 3 (1985). Type: Sierre Leone, *Thunberg* in herb. Willd. 311 (B-W, holo.)

Slender annual herb, 15–80 cm tall, erect or decumbent; stems 6-angular, ridged, retrorse-pubescent, hairs concentrated on the ridges, sometimes additionally pilose. Leaves ± anisophyllous, oblong-lanceolate, 2–7 cm long, 0.7–1.3 cm wide, basal leaves sometimes oblong-ovate, to 2 cm wide, base ± asymmetric, acute or obtuse, margin subentire, apex acute-apiculate, surfaces ± pubescent with spreading, antrorse or retrorse hairs, midrib densely antrorse-pubescent; lateral veins 4–7 pairs, inconspicuous; cystoliths linear, inconspicuous; petiole 2–5 mm long, antrorse-pubescent with occasional pilose and glandular hairs. Inflorescences principally terminal, sometimes with reduced axillary inflorescences, the former cylindrical, to 4.5 cm long, subsessile, immediately subtended by a pair of leaves; main axis bracts imbricate, pairs highly unequal, inner bract linear, 2–2.5 mm long, glabrous, outer bract linear-lanceolate, aristate, 7–9 mm long, appressed-puberulous towards base, with conspicuous straw-coloured setose hairs along margin below the arista and with short eglandular and glandular hairs on the arista; cymule bracts free, subequal, resembling the outer main axis bract, 7–7.5 mm long, margin hyaline towards base; bracteoles equal, oblanceolate, 5.5–6.5 mm long, mucronate, sparsely appressed-puberulous, margin broadly hyaline. Calyx lobes 5, subequal, lanceolate, 2–2.5 mm long, appressed-puberulous mainly on margins and with scattered subsessile glands, margin or whole lobe hyaline. Corolla pink, red or purple, paler towards base, 9.5–14.5 mm long, retrorse-pubescent outside; tube 6.5–8 mm long; lip held in upper position oblong-elliptic, 3–7.5 mm long, 1–3 mm wide, with one median lobe to 0.35 mm long; lip held in lower position elliptic-obovate, 3.5–7 mm long, 2–3.5 mm wide, apex obtuse to emarginate. Staminal filaments 2–6 mm long, sparsely puberulous towards base; anthers 0.5–1.5 mm long. Style glabrous. Capsule 4–6 mm long, appressed-puberulous towards apex; seeds lenticular, 0.8–1 mm in diameter, tuberculate.

UGANDA. Bunyoro District: Hoima, Oct. 1970, *Katende* 674; Masaka District: Lake Nabugabo, May 1969, *Lye* 3109 & Bukoto county, near Kigo, May 1972, *Lye* 6865!

KENYA. Nandi District: Kiptuiya, Sept. 1960, *Tallantire* 151; North Kavirondo District: Kakamega Forest, Yala R. area, Dec. 1956, *Verdcourt* 1694! & Kakamega Forest, Yala R. valley, Yala Nature Reserve, Sept. 2003, *Mwachala, Fischer & Dumbo* 320!

TANZANIA. Bukoba District: Minziro Forest Reserve, Nyokabanga subvillage near top of Kaiyamba Hill, May 2001, *Festo, Bayona & Wibard* 1474!; Buha District: 20 km on Kasulu–Kibondo road, May 1994, *Bidgood & Vollesen* 3247!; Kigoma District: Gombe Stream Reserve, Peak Ridge, May 1992, *Mbago* 1053!

DISTR. **U** 2, 4; **K** 3, 5; **T** 1, 4; Senegal to Ivory Coast, Nigeria to Sudan, Angola

HAB. Grassland, open bushland, roadsides; 900–1250 m

USES. None recorded on herbarium specimens

CONSERVATION NOTES. Widespread and locally common in West and Central Africa, but becoming more localised at the eastern end of its range. As it favours open habitats, it is likely to tolerate moderate disturbance. Least Concern (LC).

SYN. *H. callicoma* S. Moore in J.B. 18: 41 (1880). Syntypes: Angola, Bembe Mt, *Monteiro* s.n. & Sudan, Bongo, Addai, *Schweinfurth* 2525 (both ?BM, K!, syn.)

NOTE. Specimens from Buha and Kigoma districts, Tanzania, with densely pilose stems in addition to the shorter retrorse hairs, closely resemble material from adjacent Burundi where, however, intermediate forms with sparsely pilose stems are also recorded. Specimens with pilose stems are also recorded from Kenya (*Whyte* s.n.) and Congo-Kinshasa.

3. **Hypoestes strobilifera** *S. Moore* in J.B. 18: 40 (1880); C.B. Clarke in F.T.A. 5: 248 (1900); Benoist in Not. Syst., Paris 10: 242 (1942); F.P.S. 3: 177 (1956); Heine in F.W.T.A. ed. 2, 2: 430 (1963). Syntypes: Sudan, Bongo, Addai, *Schweinfurth* ser. 3, 30 & 2553 (both ?BM, K!)

Erect annual herb, 20–60(–100) cm tall; stems 6-angular, ridged, retrorse-pubescent, hairs concentrated on the ridges, sometimes additionally ± densely pilose on young stems and nodes. Leaves tending to crowd towards stem apex, forming pseudowhorls of 4–6 subtending the inflorescences, pairs subequal, linear-lanceolate, 6.5–14.5 cm long, 0.8–2 cm wide, base obtuse to truncate, margin subentire, apex acute, surfaces sparsely pubescent, most dense on margins and veins beneath, hairs antrorse or spreading; lateral veins 6–10 pairs; cystoliths linear, conspicuous; petiole to 3 mm, upper leaves subsessile. Inflorescences terminal, strobiliform, to 3.5 cm long, subsessile; main axis bracts imbricate, pairs highly unequal, inner bract triangular, 2 mm long, glabrous, outer bract broadly obovate, aristate, 10.5–18 mm long, 3.5–7 mm wide centrally, antrorse-puberulous towards base, with or without conspicuous pale setose hairs centrally particularly along margins, the arista with shorter eglandular hairs; cymule bracts free, unequal, inner bract linear-lanceolate, to 13.5 mm long, outer bract resembling the outer main axis bract but spatulate-aristate, 10.5–14.5 mm long, 2.5–5 mm wide, arista with occasional glandular hairs; bracteoles subequal, oblanceolate, 5.5–6.5 mm long, antrorse-pubescent, margin hyaline. Calyx lobes 5, subequal, linear-lanceolate, 5–6 mm long, ± antrorse-puberulous, margin broadly hyaline. Corolla pink, red or purple, paler towards base, 13.5–16(–20) mm long, retrorse-pubescent outside; tube 8–10.5(–13) mm long; lip held in upper position oblong-ovate, 4.5–7 mm long, 1.5–3 mm wide, with one median lobe to 0.3 mm long; lip held in lower position obovate, 4–7.5 mm long, 2.5–4.5 mm wide, apex rounded or emarginate. Staminal filaments 3–5.5 mm long, sparsely puberulous towards base; anthers 1.3–1.8 mm long. Style glabrous. Capsule 5.5–6.5 mm long, glabrous; seeds lenticular, 1.3–1.6 mm in diameter, tuberculate.

var. **tisserantii** *Benoist* in Not. Syst., Paris 10: 242 (1942); Heine in F.W.T.A. ed. 2, 2: 430 (1963). Type: Central African Republic, Ippy area, *Tisserant* 2681 (P, holo.)

Young stems ± densely pilose. Outer main axis and cymule bracts with conspicuous pale setose hairs below the long arista, particuarly along the margin.

UGANDA. West Nile District: Koboko, Nov. 1940, *Purseglove* 1081!; Acholi District: Gulu, Nov. 1941, *A.S. Thomas* 4020!; Toro District: Toro, Oct. 1932, *Eggeling* 685!

DISTR. U 1, 2; Guinea, Togo to Central African Republic
HAB. Grassland, open bushland, roadsides; 900–1250 m
USES. None recorded on herbarium specimens
CONSERVATION NOTES. *H. strobilifera* is a widespread but apparently local species in West and Central Africa; var. *tisserantii* is the more widespread variety, the typical variety being restricted to Sudan and Congo-Kinshasa. This species can be locally common, and as it favours open grassland, it is likely to tolerate some disturbance. It therefore appears unthreatened at present and both the species as a whole and var. *tisserantii* are assessed here as of Least Concern (LC).

NOTE. The typical variety differs in lacking pilose hairs on the stems and long setose hairs on the bracts, and in the aristae of the bracts being shorter. A somewhat intermediate specimen from N Congo-Kinshasa (*de Saeger* 1475) is sparsely setose-ciliate on the bracts but has short bract aristae and lacks pilose hairs on the stems as in var. *strobilifera*. Heine's (F.W.T.A. ed. 2, 2: 430, 1963) record of var. *strobilifera* from East Africa appears incorrect.

Specimens from Uganda at K were determined in 1960 by Heine as *H. strobilifera* var. *setosa*. This name appears not to have been published and would, in any case, be a later synonym of var. *tisserantii*.

4. **Hypoestes aristata** (*Vahl*) *Roem. & Schult.* in Syst. Veg. 1: 140 (1817); Nees in DC., Prodr. 11: 509 (1847); C.B. Clarke in F.T.A. 5: 245 (1900) as *H. aristata* R. Br.; Benoist in Not. Syst., Paris 10: 243 (1942); F.P.S. 3: 176 (1956); Heine in F.W.T.A. ed. 2, 2: 431 (1963) & in Fl. Gabon 13: 231, pl. 48 (1966); U.K.W.F. ed. 1: 609, 610 (fig.) (1974); K. Balkwill & Getliffe-Norris, S. Afr. J. Bot. 51: 140, figs. 1 g–l, 2, 3, 4 c, d, f, o & q, 5 b, 7 b (1985); Blundell, Wild Fl. E. Afr.: 392, pl. 608 & 800 (1987); U.K.W.F. ed. 2: 277, pl. 121 (1994); Friis & Vollesen in Biol. Skr. 51 (2): 443 (2005); Ensermu in Fl. Eth. 5: 449, fig. 167.47: 1 & 2 (2006). Type: South Africa, Cape of Good Hope, *Bülow* s.n. (C, holo.)

Erect or scrambling perennial herb or subshrub, 40–300 cm tall; stems shallowly sulcate, sparsely to densely puberulous or pubescent, hairs spreading, antrorse or retrorse, older stems often glabrescent except line of persistent hairs at nodes. Leaves broadly to narrowly ovate, 3.5–17 cm long, 1.2–10.5 cm wide, base shortly attenuate, acute or rounded, becoming subcordate in leaves subtending inflorescences, margin entire to obscurely undulate, apex acuminate, surfaces sparsely to densely puberulous or pubescent; lateral veins 6–9 pairs; cystoliths minute, elliptic; petiole to 68 mm long, puberulous to pubescent particularly on the proximal side, uppermost leaves subsessile. Inflorescences terminal and axillary, the former verticillate; cymes fasciculate; main axis bracts lanceolate, 1.5–4(–10) mm long, inconspicuous; cymules many, dense, subsessile, each 1-flowered; cymule bracts subequal, lanceolate-aristate, 5–21 mm long including arista, 0.8–2 mm wide towards base, fused in the basal 1–2.5 mm, antrorse-puberulous, margin hyaline towards base, arista 2–13 mm long, with gland-tipped hairs; bracteoles linear-aristate, 4–8.5 mm long, sparsely puberulous, margin broadly hyaline. Calyx tube 1–2(–4) mm long, lobes 4, subequal, lanceolate or acuminate, 2.8–5 mm long, sparsely ciliate, margins broadly hyaline. Corolla white, pink, purple or red, with darker markings on palate of lip held in upper position, 13–39 mm long, retrorse-pubescent and with short-stalked glands outside; tube 8–18 mm long; lip held in upper position oblong-obovate, 6–18 mm long, 3–9.5 mm wide, lobes 3, 1.2–4.5(–7.5) mm long, ciliate; lip held in lower position narrowly oblong-elliptic, 6–16 mm long, 2–5 mm wide, apex obtuse. Staminal filaments (4.5–)8–14 mm long, puberulous, particularly towards base; anthers 1.2–2 mm long. Ovary 1.5–1.7 mm long; style sparsely pubescent. Capsule 11–17 mm long, glabrous; seeds lenticular, 2–3 mm in diameter, smooth.

UGANDA. West Nile Province: 1.2 km S of Metu rest camp, Sept. 1953, *Chancellor* 237!; Bunyoro District: Budongo Forest, Feb. 1973, *Synnott* 1430!; Mengo District: Namanve Forest, Kiagwe, Apr. 1932, *Eggeling* 675!

KENYA. Ravine District: near Timboroa, Dec. 1986, *Robertson* 4407!; Machakos District: Kibwezi Hill, Nov. 1979, *Gatheri et al.* 79/167!; Masai District: Ol'Pusimoru sawmill ± 15 km from Olokurto, May 1961, *Glover, Gwynne & Samuel* 1361!

TANZANIA. Arusha District: Mt Meru, Mar. 1974, *Richards & Arasululu* 28954!; Mpwapwa District: Rubeho Mts, Aug. 2001, *Mwangoka & Ngombaniza* 2490!; Iringa District: slopes of Nyumbanitu, July 1953, *Carmichael* 219!

DISTR. **U** 1–4; **K** 1, 3–7; **T** 1–7; Nigeria to Ethiopia and to South Africa

HAB. Forest, particularly margins and clearings, more rarely in open bushland or grassland; 400–2900 m

USES. "Masai chew the roots as a cough remedy" (**K** 6; *Glover, Gwynne & Samuel* 1361 & 1777); "used to treat stomach problems" (*Simon & Kuraru* 212)

CONSERVATION NOTES. Widespread and locally abundant across tropical and subtropical Africa, and often dominant in the herbaceous layer of forest margins. It appears tolerant of habitat disturbance and adaptable to a range of environments: Least Concern (LC).

SYN. *Justicia verticillaris* L.f., Suppl. Pl.: 85 (1782). Type: South Africa, Cape of Good Hope, *Herb. Thunberg* 427 (UPS, lecto.); *nom. rejic.*, Brummitt in Taxon 36: 432 (1987)

J. aristata Vahl, Symb. Bot. Upsal. 2: 2 (1794)

Hypoestes verticillaris (L.f.) Roem. & Schult. in Syst. Veg. 1: 140 (1817)

H. aristata (Vahl) Roem. & Schult. var. *macrophylla* Nees in DC., Prodr. 11: 510 (1847); Benoist in Not. Syst., Paris 10: 243 (1942). Type: South Africa, Cape, *Drège* s.n. (B†, holo.; K!, iso.)

H. insularis T. Anderson in J.L.S. 7: 49 (1862); C.B. Clarke in F.T.A. 5: 246 (1900). Type: Equatorial Guinea, Bioko, *Mann* 179 [number not cited in protologue] (K!, holo.)

H. antennifera S. Moore in J.B. 18: 41 (1880); C.B. Clarke in F.T.A. 5: 245 (1900). Type: Kenya, Teita District, N'di, *Hildebrandt* 2563 (BM!, holo.; K!, iso.)

H. staudtii Lindau in E.J. 22: 122 (1897); C.B. Clarke in F.T.A. 5: 246 (1900). Type: Cameroon, Yaoundé [Jaundestation], *Zenker & Staudt* 36 (B†, holo.; K!, iso.)

H. aristata (Vahl) Roem. & Schult. var. *insularis* (T. Anderson) Benoist in Not. Syst., Paris 10: 244 (1942)

H. aristata (Vahl) Roem. & Schult. var. *kikuyuensis* Benoist in Not. Syst., Paris 10: 244 (1942). Type: Kenya, Kiambu District, Kikuyu plateau, *Gronier-Le Petit* s.n. (P, holo.)

H. aristata (Vahl) Roem. & Schult. var. *staudtii* (Lindau) Benoist in Not. Syst., Paris 10: 244 (1942)

NOTE. Two forms of this species can be readily distinguished in Kenya. Plants of forest and forest margins usually at altitudes over 1800 m have small inflorescences with small (13–21 mm long) white corollas, subtended by rather shortly aristate bracts, and have sparsely to densely pubescent stems and leaves. Those of bushland or gallery forest at lower altitudes (600–1500 m) have large, dense fascicles with larger ((20–)25–39 mm) purple to red corollas, subtended by long-aristate bracts, and have densely puberulous stems and leaves. Only the small, white-flowered form is recorded in Uganda, this at lower altitudes (1000–1200 m). These forms broadly correspond to var. *alba* and var. *aristata* respectively, recognised in South Africa by Balkwill & Getliffe-Norris (l.c.). They considered var. *alba* to be restricted to the Afromontane phytochorion, whilst var. *aristata* is more widespread in Africa. In Tanzania, the distinction between the two forms becomes much less clearcut, with a wide range of intermediate specimens recorded. Small-flowered forms do not have exclusively white corollas, but are often pink or purple, and the corolla size is often intermediate (20–26 mm). Furthermore, large-flowered forms are not restricted to lower altitudes. Most notably, plants from Mbeya Peak, **T** 7 at 2400–2600 m altitude (e.g. *Kerfoot* 4160; *Goyder at al.* 3851) have large purple corollas to 27 mm long in dense, many-flowered fascicles. These specimens also have notably long calyx tubes (2–4 mm) and elongate lobes to the corolla lip. Intermediate forms are, again, recorded in neighbouring areas of SW Tanzania. The currently recognised varieties of *H. aristata* cannot therefore be upheld in our region.

H. aristata is unusual in having only four calyx lobes, rather than the five lobes typical in *Hypoestes*. This feature is shared with the closely related West African species, *H. rosea* P.Beauv. which differs from *H. aristata* principally in having spiciform, not fasciculate, cymes; and in having only shortly aristate bracts. *H. barteri* T. Anderson, correctly recognised as a synonym of *H. rosea* by Milne-Redhead in K.B.: 64 (1940), was wrongly placed as a variety of *H. aristata* by Benoist in Not. Syst., Paris 10: 244 (1942).

In East African material the cymules are uniformly reduced to a single flower in the material of *H. aristata* studied. However, in West and Central Africa, a reduced second flower is often found.

5. **Hypoestes forskaolii** (*Vahl*) *R. Br.* in Salt, Voy. Abyss. App.: 63 (1814) as *forskalii*; Nees in DC., Prodr. 11: 507 (1847) as *forskolii*; Lindau in P.O.A.: 371 (1895) as *forskohlii*; C.B. Clarke in F.T.A. 5: 249 (1900) as *forskalei*; Wood, Hillcoat & Brummitt, Kew Bull. 38: 455 (1983) as *forskalei*; K. Balkwill & Getliffe-Norris in S. Afr. J. Bot. 51: 141, figs. 1 m–r, 2, 4 b, g, h, n & p, 5 a, 7 c & d (1984); Champluvier in Fl. Rwanda, Sperm. 3: 458, fig. 142: 1 (1985) as *forskalei*; Blundell, Wild Fl. E. Afr.: 392, pl. 609 (1987) as *forskalei*; U.K.W.F.: 277 (1994) as *forskahlii*; Friis & Vollesen in Biol. Skr. 51 (2): 444 (2005); Hedrén in Fl. Somalia 3: 425, fig. 287 (2006); Ensermu in Fl. Eth. 5: 450, fig. 167.47: 3 & 4 (2006); Pickering & Patzelt, Field Guide Wild Pl. Oman: 17 (2008); Fischer & Killmann, Pl. Nyungwe N.P. Rwanda: 592 (2008) as *forskahlei*, excl. fig. (= *Dicliptera maculata*). Type: Yemen, Jebel Melhan, *Forsskål* 387 (C, lecto., microfiche 60: I. 1–2, selected by Wood, Hillcoat & Brummitt, l.c.; K!, photo.)

Erect or scrambling perennial herb or subshrub, 20–100(–200) cm tall; stems solitary or often much-branched from a woody base, shallowly sulcate, glabrate to densely pubescent or pilose, hairs spreading, antrorse or retrorse, or whole plant appressed-canescent. Leaves ovate, elliptic or lanceolate, 2.2–18 cm long, 0.5–8 cm wide, base shortly attenuate, acute, rounded or subcordate, margin entire to shallowly undulate, apex shortly acuminate, surfaces sparsely to densely pubescent or pilose, particularly on the veins, or glabrate; lateral veins 5–8(–10) pairs; cystoliths elliptic to linear, ± inconspicuous; petiole 2–25(–50) mm long, pubescent and/or pilose, occasionally with glandular hairs. Inflorescences terminal and axillary, the former often verticillate, cymes unilateral, spiciform or fasciculate, to 4 cm long, more rarely lax and paniculate; main axis bracts triangular to lanceolate, 1.5–4(–11) mm long, pairs basally fused, inconspicuous; cymules 5–many, subsessile; cymule bracts equal or somewhat unequal, lanceolate, linear or acuminate, 5–11(–15) mm long, fused in the lower half to two thirds, the connate portion often hyaline, surfaces sparsely to densely puberulous, often additionally pilose and with many gland-tipped hairs on the lobes; bracteoles equal, elliptic-lanceolate, 5–9.5 mm long, indumentum as cymule bracts, margin broadly hyaline. Calyx tube 2.5–4.5 mm long, lobes 5, subequal, hyaline, lanceolate, 1.2–3.5 mm long, with minute spreading hairs outside. Corolla white to purple with purple markings on paler palate of lip held in upper position, 13–20(–25) mm long, pubescent and with ± prominent glandular hairs outside; tube 8–13 mm long; lip held in upper position oblong-obovate, 5.5–8(–11) mm long, 3.5–5(–6) mm wide, lobes 3, 0.8–1.8 mm long; lip held in lower position trullate-caudate, 5–7(–9.5) mm long, 2–4.5 mm wide, apex of acumen rounded to emarginate. Staminal filaments 4–8 mm long, pilose above, hairs inconspicuously glandular; anthers 1.1–1.6 mm long. Ovary 1.2–1.6 mm long; style glabrous to sparsely pubescent. Capsule 6.5–8(–9) mm long, pubescent in upper half or rarely glabrate; seeds broadly ellipsoid, 1.3–2 mm wide, only slightly compressed, smooth. Fig. 94, p. 727.

a. subsp. **forskaolii**

Indumentum highly variable, glabrate to densely pubescent and/or pilose, the hairs spreading, antrorse or retrorse, never appressed canescent-puberulent. Leaves (in our region) ovate to elliptic, rarely lanceolate, 2.2–18 cm long, 0.6–8 cm wide. Fig. 94.

UGANDA. Karamoja District: Mt Moroto, Mar. 1960, *J. Wilson* 1000!; Kigezi District: Mt Muhabura, June 1949, *Purseglove* 2911!; Mbale District: Tororo Hill, Dec. 1968, *Kimani* 148!

KENYA. Northern Frontier District: Lorogi Forest, Dec. 1996, *Bono* 95!; Naivasha District: Lake Naivasha, June 1971, *E. Polhill* 136!; Masai District: Entasekera River, July 1961, *Glover et al.* 2082!

TANZANIA. Mbulu District: Nou Forest Reserve, June 1969, *Carmichael* 1641!; Lushoto District: Sangerawe, Kwamkoro Forest Reserve, Jan. 1987, *Ruffo & Mmari* 2058!; Iringa District: Mufindi, Ngowazi, Apr. 1972, *Paget-Wilkes* 1038!; Zanzibar, Kusini District, Mtende, July 1999, *Fakih* 333!

DISTR. **U** 1–4; **K** 1–7; **T** 1–8; **Z**, **P**; widespread in tropical and southern Africa, extending to the Saharan highlands; Arabia and Madagascar

FIG. 94. *HYPOESTES FORSKAOLII* subsp. *FORSKAOLII* — **1**, habit, × $^2/_3$; **2** & **3**, mature leaves: extreme large and small variants, × $^2/_3$; **4**, paired main axis bracts and cymule bracts, lateral view, × 6; **5**, bracteoles and calyx, × 6; **6**, flower, lateral view, × 4; **7**, detail of corolla lip held in lower position, × 4; **8**, detail of stamens, × 8; **9**, stamens, recurved after maturity, × 12; **10**, capsule with seeds, × 5; **11**, mature seed, × 12. **1** from *Kayombo* 2392; **2** from *Bidgood et al.* 919; **3**, **6–8** from *Greenway* 13902; **4**, **5**, **10** & **11** from *Drummond & Hemsley* 2913; **9** from *Kayombo* 576. Drawn by Juliet Williamson.

HAB. Common in a wide range of habitats including grassland, bushland, forest margins and clearings, cultivated land and coastal foreshores; 0–2500(–3000) m

USES. "Used locally for cooking soda, water is mixed with ashes then filtered" (**K** 3; *Kahuho* 8); "roots rubbed to pulp then eaten for throat and chest complaints" (**K** 6; *Glover et al.* several collections, e.g. 3436); "medicinal" (**K** 6; *Nesbit-Evans* 36); "leaves boiled and then rubbed on limbs for rheumatism" (**T** 1; *Newbould* 5766); "used as a medicine for cows" (**T** 2; *Mlangwa & Masanyika* 461); "to treat swellings: leaves pounded up and put on as a paste" (**T** 3; *Tanner* 2935); "leaves sometimes used as vegetables" (**T** 4; *Semsei* 327); "decoction of roots taken as a contraceptive" (**T** 7; *Wiland & Mboya* 127)

CONSERVATION NOTES. Abundant and often dominant in the herb layer of a range of habitats across Africa; highly tolerant of habitat disturbance: Least Concern (LC).

SYN. *Justicia paniculata* Forssk., Fl. Aegypt.-Arab. 4 (1775) *nom. illegit.*, *non J. paniculata* Burm.f. Type as for *H. forskaolii*

J. forskaolii Vahl, Symb. Bot. 1: 2 (1790) as *forskoalei*

Hypoestes latifolia Nees in DC., Prodr. 11: 509 (1847). Type: Sudan, Kordofan, *Kotschy* 296 (?GZU, holo.; B†, HBG!, K!, M!, P!, iso.)

H. latifolia Nees var. *integrifolia* Nees in DC., Prodr. 11: 509 (1847). Syntypes: Senegal, Casamance, *Heudelot* 552 (?GZU, P!, K!, syn.) & Sudan, Sennaar, *Acerbi* s.n. (G-DC, syn.)

[*H. verticillaris* auctt. plur.: Nees in DC., Prodr. 11: 507 (1847); Lindau in P.O.A.: 371 (1895); Clarke in F.T.A. 5: 250 (1900); F.P.S. 3: 176 (1956); F.P.U.: 140 (1962); Heine in F.W.T.A. ed. 2, 2: 431 (1963) & in Fl. Gabon 13: 227, pl. 47 (1966); U.K.W.F. ed. 1: 610 (fig.), 611 (1974); U.K.W.F. ed. 2: 277 (1994) in syn., *non* (L.f.) Roem. & Schult.]

H. mollis T. Anderson in J.L.S. 7: 49 (1864). Type: Congo-Kinshasa, *Smith* s.n.(K!, holo.; P!, iso.)

H. rothii T. Anderson in J.L.S. 7: 49 (1864). Syntypes: Ethiopia, Shewa, Ankober, *Roth* 485 [number not cited in protologue] (K!, syn.); Ethiopia, without precise locality, *Plowden* s.n. (K!, syn.)

H. verticillaris R. Br. var. *glabra* S. Moore in J.B. 18: 363 (1880). Type: Angola, between Lake Delvantala and Quilangues, *Welwitsch* 5059 (BM, holo.; K!, iso.)

H. preussii Lindau in E.J. 20: 48 (1894); C.B. Clarke in F.T.A. 5: 251 (1900). Type: Cameroon, Buea, *Preuss* 755 (B†, holo.; K!, iso.)

H. echioides Lindau in E.J. 20: 52 (1894). Type: Tanzania, Mpwapwa, *Stuhlmann* 286 (B†, holo.)

H. violaceotincta Lindau in E.J. 24: 323 (1897); C.B. Clarke in F.T.A. 5: 251 (1900). Type: Togo, Misahöhe, *Baumann* 476 (B†, holo.; K!, iso.)

H. mlanjensis C.B. Clarke in F.T.A. 5: 250 (1900). Syntypes: Kenya, Machakos/Kitui District, Ukambani, *Gregory* (BM!, syn.); Malawi, Mt Mlanje, *Whyte* 151 (K!, syn.), **syn. nov.**

H. tanganyikensis C.B. Clarke in F.T.A. 5: 252 (1900). Syntypes: Malawi, Tanganyika Plateau, *Whyte* s.n. & Nyika Plateau, *Whyte* 192 (both K!, syn.), **syn. nov.**

H. verticillaris R. Br. var. *forskaolii* (Vahl) Benoist in Not. Syst., Paris 10: 246 (1942)

H. verticillaris R. Br. var. *mollis* (T. Anderson) Benoist in Not. Syst., Paris 10: 246 (1942)

H. verticillaris R. Br. var. *latifolia* (Nees) Benoist in Not. Syst., Paris 10: 247 (1942)

H. verticillaris R. Br. var. *violaceotincta* (Lindau) Benoist in Not. Syst., Paris 10: 247 (1942)

NOTE. Attempts have previously been made to recognise several species or varieties within this polymorphic taxon, based principally upon indumentum type and inflorescence form. A review of the abundant herbarium material available from across Africa reveals that these taxa, with the exception of the discrete subsp. *hildebrandtii* (see below), merely represent the extremes of a continuum of variation within *H. forskaolii*. Even within populations, variation in leaf and flower size and indumentum can be pronounced, e.g. *Gillett* 13595 from Moyale, **K** 1. However, several trends recognisable in E. Africa are worthy of note:

The most widespread form has mid-size, shortly petiolate leaves, spiciform inflorescences, (oblong-) lanceolate cymule bracts and white flowers; the indumentum is highly variable. Specimens from coastal Kenya and Tanzania, recorded in a range of habitats from coastal forest to bushland and farmbush, develop more linear cymule bracts which in **T** 3, Tanga District, are often strikingly coloured purple towards the apex (e.g. *Kayombo et al.* 3018). The corolla in this form is often pink to purple and the inflorescences are more dense and less clearly spiciform.

In wetter forest margins and clearings, a glabrate form with large long-petiolate leaves, small pale cymule bracts arranged in more lax paniculate inflorescences, and small white flowers is recorded. This form is scattered across the F.T.E.A. region, but it is most abundant and striking in the E Usambara Mts (e.g. *Peter* 15255). Similar plants are recorded in forest areas across Africa, and were previously placed under *H. violaceotincta* and *H. preussii*, the former having pilose bracts. A second form largely restricted to dry forest/woodland, with broadly ovate, long-petiolate leaves with a rounded or subcordate base, long, linear-lanceolate cymule bracts and elongate main axis bracts, is recorded from central and West Tanzania (e.g. *Bidgood et al.* 919). Such plants usually have more fasciculate inflorescences and purple flowers towards the larger end of the size range. The stems and bracts are often densely long-pilose but this indumentum is sometimes absent. This form was originally described from Sudan as *H. latifolia*, and similar plants are recorded from Congo-Kinshasa and Nigeria.

In the Serengeti region of **T** 1/2, stunted, much-branched plants with small, leathery leaves are recorded (e.g. *Greenway* 9916). This growth form is clearly an adaptation to the heavily grazed grasslands of this region, and similarly small-leaved variants are recorded in grazed or disturbed areas elsewhere, for example on Kilimanjaro (*Grimshaw* 93218 & 93219). A small-leaved variant is also recorded in the Ngong area of **K** 6 (e.g. *Greenway* 13902).

Specimens with lax, elongate inflorescences previously described as *H. tanganyikensis* are recorded from several locations in Tanzania. However, such plants appear to occur within populations with more regularly spiciform inflorescences (e.g. *Revell* 92 from Moshi Camp, **T** 2; *Congdon* 93 from Tukuyu, **T** 7) and are regarded here as forms of subsp. *forskaolii*.

b. subsp. **hildebrandtii** (*Lindau*) *I. Darbysh.*, **comb. et stat. nov.** Syntypes: Somalia, Mt Serrut, *Hildebrandt* 1405 & Mt Ahl Yafir [Ahlgebirge (Jafir)], *Hildebrandt* 860d (both B†, syn.)

Whole plant appressed puberulent-canescent, most dense on the stems, petioles and leaf midrib, giving the plants a distinctly pale grey appearance. Leaves lanceolate to narrowly elliptic, 2.2–13.5 cm long, 0.5–2.6 cm wide.

Kenya. Northern Frontier District: Isiolo, ± 1 km W of the township, June 1979, *Gilbert, Kanuri & Mungai* 5664!; Laikipia District: Rumuruti, ± 50 km N near Colcheccio Lodge, Nov. 1978, *Hepper & Jaeger* 6605; Teita District: Voi Gate W, Tsavo National Park (East), Dec. 1966, *Greenway & Kanuri* 12785!

Distr. **K** 1, 3, 4, 7; Somalia, SE Ethiopia

Hab. Dry grassland and open *Acacia-Commiphora* bushland, often on sandy soils; 450–1400(–1800) m

Uses. None recorded on herbarium specimens

Conservation notes. This taxon is much more restricted in range and habitat than the typical subspecies. It appears scarce in Ethiopia, being known from very few collections. In Kenya it is recorded by several collectors as locally common around Isiolo. Few collections have been made further north or east in Kenya, though this is likely a reflection of under-collection here rather than of the taxon's scarcity. In Somalia it is more widespread and often common. Human population density, and so habitat disturbance, is low throughout much of this taxon's range. It does not therefore appear to be threatened: Least Concern (LC).

Syn. *H. hildebrandtii* Lindau in E.J. 20: 48 (1894); C.B. Clarke in F.T.A. 5: 249 (1900); Blundell, Wild Fl. E. Afr.: 392, pl. 702 (1987)

H. paniculata (Forssk.) Schweinf. var. *hildebrandtii* (Lindau) Fiori in Bosch. Pl. Leg. Eritrea: 358 (1912), *nom. illegit.*

H. verticillaris R. Br. var. *hildebrandtii* (Lindau) Benoist in Not. Syst., Paris 10: 246 (1942)

Note. Recent Flora accounts (e.g. Hedrén in Fl. Somalia; Ensermu in Fl. Eth.) have placed this taxon in synonymy with *H. forskaolii sensu stricto*. Although the inflorescence and flower morphology clearly places it close to that taxon, its distinct indumentum renders it clearly separable both in the field and in herbarium material. Subsp. *hildebrandtii* has a non-disjunct geographic distribution unlike other previously recognised varieties within *H. forskaolii*, and although its range overlaps with subsp. *forskaolii*, the latter is usually found in moister habitats than subsp. *hildebrandtii*. For example, in **K** 1, subsp. *forskaolii* is found only on the isolated mountains such as Marsabit and Kulal and in wetter lowland areas such as riparian habitats, whilst subsp. *hildebrandtii* is more widespread in the surrounding dry lowlands. The previous treatment of this taxon as a variety of *H. forskaolii* therefore seems unsatisfactory, and I consider subspecies status more appropriate on both geographical and ecological grounds. Occasional intermediate specimens, with sparser canescent indumentum and the hairs more spreading, have been collected (e.g. *Magogo* 1322, **K** 1), but these are from the ecotonal areas where populations of the two entities are most likely to meet.

H. pubescens Balf. f., endemic to Socotra, shares the canescent stem indumentum of subsp. *hildebrandtii*, but differs in its sparsely puberulent ovate leaves. It is otherwise close to *H. forskaolii* and is best considered a further subspecies of that taxon.

Cultivated Acanthaceae in the *Flora of Tropical East Africa* region

Nearly 100 species of Acanthaceae have been reported, within the literature or in herbarium collections, as having been cultivated in East Africa. Of these, many are native species which have been brought into cultivation locally; these species are treated in the native species accounts, as are those introduced ornamental species which have become naturalised (e.g. *Barleria cristata* L., *Ruellia tuberosa* L.). Of the remaining introduced species, the majority are known only from botanical gardens or a few private gardens and are unlikely to be widely encountered in the Flora region and so are not treated here. In this section we restrict ourselves to treating only the 12 most widely cultivated ornamentals which have not, to our knowledge, become naturalised in East Africa. A maximum of one herbarium specimen per country is cited.

Aphelandra pulcherrima (Jacq.) Kunth

SYN. *A. cristata sensu* Jex-Blake, Gard. E. Afr. ed. 4: 329 (1957), *non* (Jacq.) R.Br.

Shrub; cystoliths absent. Leaves ovate or elliptic, with narrowly cuneate-attenuate base and acuminate apex. Flowers held in dense slender terminal spikes with imbricate ovate bracts 5–7 mm long, woolly-hairy towards the base and along the margin; bracteoles woolly throughout. Calyx slightly longer than bracts, of 5 lanceolate lobes of unequal width. Corolla bright red or red-orange, 5–7 cm long, with a long, slightly curved tube, narrowly and gradually expanded towards the mouth; limb 2-lipped, upper lip erect, 2-lobed, lower lip with a long narrow central lobe declinate or becoming recoiled, lateral pair of lobes very short. Stamens 4, exserted, monothecous.

Native of Costa Rica and West Indies to Bolivia; rather widely cultivated in the tropics.

KENYA: Karen, hort. H.M. Gardner, 25 Apr. 1977, *Gillett* 21079! TANZANIA: State House, Dar es Salaam, 21 Nov. 1972, *Ruffo* 537!

NOTE. Jex-Blake also records *A. tetragona* (Vahl) Nees as occurring in E Africa but no material has been seen; this species is similar to *A. pulcherrima* but has larger bracts and often larger, more numerous corollas.

Eranthemum pulchellum Andrews; Jex-Blake, Gard. E. Afr. ed. 4: 111 (1957), as *E. nervosum* (Vahl) Roem. & Schult.

Shrub; cystoliths present. Leaves ovate or elliptic, base and apex attenuate, lateral veins and scalariform tertiary veins prominent beneath. Flowers held in dense spikes, terminal and in the upper axils; bracts showy, imbricate, white or silverish with green pinnate-reticulate venation, ± elliptic, 14–20 mm long. Calyx of 5 ± equal lanceolate lobes shorter than bracts. Corolla blue, salver-shaped, tube narrowly cylindrical, curved, to ± 20 mm long; limb of 5 subequal spreading lobes to ± 10 mm long. Stamens 2, exserted, bithecous, thecae at an equal height; staminodes 2, minute.

Native of the Indian subcontinent; widely cultivated in the tropics.

KENYA: Kilifi District, 9 Oct. 1945, *Jeffery* K347! TANZANIA: Amani, 4 July 1950, *Verdcourt* 286!

Graptophyllum pictum (L.) Griff.; Jex-Blake, Gard. E. Afr. ed. 4: 342 (1957), as *G. hortense* Nees

Shrub; cystoliths present. Leaves variegated, either green and yellow or dark green and pink, elliptic, base cuneate to attenuate, apex shortly acuminate, glabrous. Flowers held in a short terminal thyrse; bracts and bracteoles inconspicuous, lanceolate, 2–3 mm long. Calyx of 5 ± equal lanceolate lobes c. 4 mm long. Corolla deep reddish-purple, 35–40 mm long, tube gradually expanded to mouth, curved, to ± 25 mm long; limb 2-lipped, upper lip 2-lobed, lower lip divided into 3 oblong lobes. Stamens 2, exserted, bithecous, thecae at an equal height; staminodes 2.

Native of the Indo-Pacific region, though exact native distribution apparently unknown; very widely cultivated in the tropics.

KENYA: Malindi, hort. Robertson, 9 Dec. 2007, *Robertson* 7625B!

NOTE. The genus *Graptophyllum* has an unusual distribution, the species all being from SE Asia and Australasia with the exception of *G. glandulosum* Turrill, native to the Cameroon Highlands.

Justicia adhatoda L.; Jex-Blake, Gard. E. Afr. ed. 4: 111 (1957), as *Duvernoya adhatoda* (*Adhatoda vascia* Nees)

Shrub to 3 m; cystoliths present. Leaves ovate, base attenuate, apex cuspidate. Flowers in pedunculate racemoid cymes from upper axils; bracts imbricate, green, elliptic, to 2 cm long, glabrous; bracteoles oblanceolate or narrowly obovate, to 1.5 cm long. Calyx with 5 lanceolate puberulous lobes to 1 cm long. Corolla white with purple markings on lower lip, to 3 cm long (along upper lip), with a hooded 2-lobed upper lip and flat spreading or deflexed 3-lobed lower lip. Stamens 2, held under upper lip; anthers bithecous, thecae at different height, lower with small white basal appendage.

Native of India; widely cultivated in all tropical and subtropical regions.

KENYA: Mombasa, 6 Jan. 1992, *Luke* 3498! TANZANIA: Dar es Salaam, 23 March 1993, *Mwasumbi* 16772!

Justicia brandegeeana Wassh. & L.B. Sm.; Jex-Blake, Gard. E. Afr. ed. 4: 72 (1957), as *Beloperone guttata* Brandegee. (Mexican Shrimp Plant)

Perennial or shrubby herb to 1 m; cystoliths present. Leaves ovate, base cuneate, apex acute. Flowers in subsessile or shortly pedunculate racemoid cymes from upper axils; bracts imbricate, red to purple or green, ovate, to 2 cm long, puberulous and ciliate; bracteoles similar to bracts. Calyx with 5 lanceolate acute puberulous lobes to 5 mm long. Corolla white with reddish purple stripes on lower lip or whole lower lip red, to 3 cm long (along upper lip), with a hooded 2-lobed upper lip and spreading 3-lobed lower lip. Stamens 2, held under upper lip; anthers bithecous, thecae at different height, both with flat white basal appendage.

Native of tropical America; widely cultivated in all tropical and subtropical regions and as a pot plant in temperate regions.

UGANDA: Kampala, Makerere University Hill, 13 Feb. 1999, *Lye* 23468! KENYA: Nairobi City Council Gardens, 16 Dec. 1971, *Mwangangi* 1902! TANZANIA: Dar es Salaam, State House Grounds, 22 Nov. 1972, *Ruffo* 539!

Justicia spicigera Schlechtend.
SYN. *Jacobinia spicigera* (Schlechtend.) Bailey

Shrubby herb to 1 m; cystoliths present. Leaves ovate to elliptic; apex acute; base attenuate. Flowers in open axillary cymes from upper axils; bracts and bracteoles minute. Calyx with 5 linear glabrous lobes to 2 mm long. Corolla orange or orange-red, to 5 cm long (along upper lip), with a slightly hooded 2-lobed upper lip and recoiled 3-lobed lower lip. Stamens 2, held under upper lip; anthers bithecous with thecae almost at same height, lower theca apiculate.

Native of Central and South America; widely cultivated in the Old World tropics.

KENYA: Nairobi, 29 Oct. 1978, *Gillett* 21691! TANZANIA: Dar es Salaam, 11 Nov. 1978, *Chilongola* 118!

Mackaya bella Harv.; Jex-Blake, Gard. E. Afr. ed. 4: 119 (1957)
SYN. *Asystasia bella* (Harv.) Benth. & Hook. f.

Shrub to 4 m tall; cystoliths present. Leaves elliptic, coarsely toothed; apex acuminate; base cuneate to attenuate. Flowers in elongated terminal racemoid cymes, either 1 or 2 per node; bracts and bracteoles inconspicuous. Calyx with 5 linear cuspidate lobes to 5 mm long. Corolla lilac or pale mauve with darker lines in throat, to 6 cm long, with a narrow basal tube, campanulate throat and 5 erect to spreading subequal lobes. Stamens 2, included in throat; anthers bithecous, thecae linear, at same height, apiculate at base.

Native of South Africa; cultivated in eastern and southern Africa.

KENYA: Nairobi, 13 Nov. 1971, *Gillett* 15011! TANZANIA: Amani, 12 Aug. 1940, *Greenway* 5980!

Odontonema cuspidatum (Nees) Kuntze

Shrub; cystoliths present. Leaves ± elliptic, base cuneate to attenuate, apex shortly acuminate, largely glabrous except along veins beneath. Flowers held in a ± lax terminal thyrse, racemose or often with well-developed lateral branches towards base; bracts and bracteoles inconspicuous, lanceolate or subulate, 1.5–5 mm long; flowers pedicellate. Calyx of 5 ± equal lanceolate or triangular lobes 1–3 mm long. Corolla red, 25–32 mm long, tube slender, 20–27 mm long, basal tube cylindrical, throat narrowly broadened; limb 2-lipped, upper lip shortly 2-lobed, lower lip divided into 3 lobes. Stamens 2, exserted or included (flowers heterostylous), bithecous, thecae at an equal height; staminodes 2, minute.

Native of Mexico and the West Indies; widely cultivated in the tropics.

UGANDA: Kampala, Makerere University, March 2010, *Darbyshire*, sight record; KENYA: Closeburn, Nairobi, 17 May 1958, *G. Bell* in EAH 11407! TANZANIA: Lushoto District, Mang'ula near Pentecostal Church, 20 Aug. 1982, *Kisena* 15!

NOTE. African specimens of this ornamental have often been named *O. strictum* (Nees) Kuntze, a closely related species from Central America, but it appears that *O. cuspidatum* is the correct name for the widely cultivated species; Heine (F.W.T.A. ed. 2, 2: 423) followed this approach.

Sanchezia nobilis Hook. f.; Jex-Blake, Gard. E. Afr. ed. 4: 125 (1957)

Shrub; cystoliths present; stems usually reddish. Leaves with midrib and lateral veins yellow or whitish, contrasting with bright green surface, (oblong-) elliptic, base cuneate-attenuate, margin serrate, apex acuminate, glabrous. Flowers held in a terminal thyrse, spiciform or with a pair of lateral branches towards base, 1–3 sessile flowers per bract; bracts conspicuous, often brightly coloured, yellow or red, ovate, 1.5–4 cm long, apex obtuse. Calyx of 5 narrowly oblong or spathulate lobes ± 20 mm long. Corolla yellow or yellow-orange, to ± 40 mm long, tube gradually expanded upwards; limb ± 4 mm long, equally 5-lobed, lobes revolute at anthesis. Stamens 2, exserted, bithecous, thecae at an equal height, hairy; staminodes 2, linear.

Native of Ecuador; commonly cultivated in the tropics (but see note).

KENYA: Malindi, hort. Robertson, 20 Nov. 2007, *Robertson* 7610! TANZANIA: Muheza, 16 Mar. 1973, *Ruffo* 642!

NOTE. Cultivated African specimens have usually been named *S. nobilis*. However, Leonard & Smith (in Rhodora 66: 313–343 (1964)) say that the true *S. nobilis* has a glabrous corolla and leaf blade broadest near the apex, whilst specimens from E African gardens have a hairy corolla and elliptic leaves. Leonard & Smith (l.c.) cite the type of *S. nobilis* as being held at Kew, but we currently have no material of wild origin under that name with which to compare the cultivated specimens. The other commonly cultivated species is *S. parvibracteata* Sprague & Hutch. which is very similar to *S. nobilis* but has smaller bracts and hairy flowers. The E African material is variable in bract size with some specimens falling between the ranges given for these two species, whilst some large-bracted specimens from elsewhere in Africa also have hairy corollas. It is therefore possible that at least some of the African material is of hybrid origin; we have kept the name *S. nobilis* pending further analysis.

Thunbergia grandiflora (Rottl.) Roxb.; Jex-Blake, Gard. E. Afr. ed. 4: 141 (1957), incl. var. *alba* (Blue Trumpet Vine).

Vigorous herbaceous twiner to 5 m, often totally covering walls, roofs or trees; cystoliths absent. Leaves triangular in outline, to 20 cm long, 7-veined from base, with 2–4 large triangular teeth per side; apex acute to acuminate; base cordate. Flowers in pendulous racemoid cymes to 50 cm long; bracts minute; each flower clasped by two large obovate greenish bracteoles to 3.5 cm long with large black urn-like glands. Calyx an undulate puberulous rim. Corolla white or blue; tube to 2.5 cm long, broadly campanulate; lobes 5, subequal, spreading, to 2.5 × 2.5 cm. Stamens 4, included in throat; anthers bithecous, thecae pubescent, with 5 mm long basal spurs. Native of India and SE Asia; widely cultivated elsewhere in tropical Asia and in Africa.

KENYA: Nairobi, 27 July 2008, *Luke* 12347!

NOTE. The closely related *T. laurifolia* Lindl. (Jex-Blake, Gard. E. Afr. ed. 4: 142 (1957)) is also widely cultivated. It has ovate-elliptic leaves which are 3-5-veined from base and without or with 1–2 basal teeth. This species has occasionally become naturalized (see FTEA, Acanthaceae, part 1: 42 (2008). KENYA: Nairobi, 27 July 2008, *Luke* 12345! TANZANIA: Amani, 28 Sept. 1940, *Greenway* 6029!

Thunbergia holstii Lindau

SYN. *T. affinis* sensu Jex-Blake, Gard. E. Afr. ed. 4: 228 (1957) and *T. erecta* sensu Jex-Blake, Gard. E. Afr. ed. 4: 337 (1957)

An extremely attractive shrub to 1.5 m tall with large bluish purple flowers. Native of East Africa.

UGANDA: Entebbe, 23 Sept. 1932, *Ruck* T1195! KENYA: Nairobi, 7 July 1952, *G.R. Williams* 470! TANZANIA: Amani, 28 Oct. 1933, *Greenway* 3670!

NOTE. This species, which is a native of East Africa, was treated fully in FTEA, Acanthaceae, Part 1: 40 (2008). It is my (K.V.) opinion that the plants which have been grown widely in eastern Africa under the names *T. erecta* T. Anderson and *T. affinis* S. Moore are all this species, and probably all originate from native material. The first confirmed record of cultivation is from 1919 (*Battiscombe* 943) and the label states: "Common at Machakos. Specimen cultivated at Nairobi". It is such an attractive plant that it soon spread and by the early 1930s there were collections from all three countries. It will grow easily from sea level to 1700 m.

Thunbergia mysorensis (Wight) T. Anderson

Vigorous herbaceous twiner to 5 m, often totally covering walls or roofs; cystoliths absent. Leaves ovate, to 15 cm long, base truncate, margin irregularly dentate, apex acuminate to cuspidate, 3-veined from base. Flowers in pendulous racemoid cymes to 50 cm long; bracts minute; each flower clasped by two large ovate purplish brown bracteoles to 2.5 cm long which are partly fused along lower edge, no large urn-shaped glands. Calyx an undulate glabrous rim. Corolla yellow outside, deeper yellow inside and lower lip maroon distally; tube to 3 cm long, campanulate; upper lip of 2 erect and lower of 3 deflexed lobes to 3 cm long. Stamens 4, held under upper lip; anthers bithecous, thecae densely pubescent, with 5 mm long basal spurs.

Native of India. Occasionally cultivated in Africa.

KENYA: Nairobi, Fairview Hotel, May 1995, *Vollesen* s.n.

NOTE. Jex-Blake (1957: 141) mentions *Thunbergia coccinea*. This is vegetatively a very similar species with a similar inflorescence but smaller dull red flowers. I have not seen any material from Africa but there is a possibility that he confused the two species.

INDEX TO ACANTHACEAE

New names validated in part 1

Acanthopale macrocarpa *Vollesen* **sp. nov.**
Brillantaisia richardsiae *Vollesen* **sp. nov.**
Dyschoriste kitongaensis *Vollesen* **sp. nov.**
Dyschoriste sallyae *Vollesen* **sp. nov.**
Dyschoriste tanzaniensis *Vollesen* **sp. nov.**
Dyschoriste trichocalyx (*Oliv.*) *Lindau* subsp. **verticillaris** (*C.B.Clarke*) *Vollesen* **comb. nov.**
Epiclastopelma marroninus *Vollesen* **sp. nov.**
Hygrophila albobracteata *Vollesen* **sp. nov.**
Hygrophila richardsiae *Vollesen* **sp. nov.**
Mellera congdonii *Vollesen* **sp. nov.**
Mellera insignis *Vollesen* **sp. nov.**
Ruellia burttii *Vollesen* **sp. nov.**
Ruellia richardsiae *Vollesen* **sp. nov.**
Thunbergia subgen. **Stellatae** *Vollesen* **subgen. nov.**
Thunbergia austromontana *Vollesen* **sp. nov.**
Thunbergia barbata *Vollesen* **sp. nov.**
Thunbergia citrina *Vollesen* **sp. nov.**
Thunbergia heterochondros (*Mildbr.*) *Vollesen* **comb. nov.**
Thunbergia minziroensis *Vollesen* **sp. nov.**
Thunbergia mufindiensis *Vollesen* **sp. nov.**
Thunbergia napperae *Mwachala, Malombe & Vollesen* **sp. nov.**
Thunbergia racemosa *Vollesen* **sp. nov.**
Thunbergia reniformis *Vollesen* **sp. nov.**
Thunbergia richardsiae *Vollesen* **sp. nov.**
Thunbergia schliebenii *Vollesen* **sp. nov.**
Thunbergia tsavoensis *Vollesen* **sp. nov.**
Thunbergia verdcourtii *Vollesen* **sp. nov.**

New names validated in this part

Anisotes galanae (C. Baden) Vollesen, **comb. et stat. nov.**
Anisotes spectabilis (Mildbr.) Vollesen, **comb. nov.**
Asystasia calcicola *Ensermu & Vollesen,* **sp. nov.**
Asystasia minutiflora *Ensermu & Vollesen,* **sp. nov.**
Asystasia moorei *Ensermu* **nom. nov.**
Asystasia tanzaniensis *Ensermu & Vollesen,* **sp. nov.**
Barleria athiensis *I. Darbysh.* **sp. nov.**
Barleria calophylloides *Lindau* subsp. **pilosa** *I. Darbysh.* **subsp. nov.**
Barleria crassa *C.B. Clarke* subsp. **mbalensis** *I. Darbysh.* **subsp. nov.**
Barleria eranthemoides *C.B. Clarke* var. **agnewii** *I. Darbysh.* **var. nov.**
Barleria faulknerae *I. Darbysh.* **sp. nov.**
Barleria gracilispina (*Fiori*) *I. Darbysh.*, **comb. et stat. nov.**
Barleria granarii *I. Darbysh.* **sp. nov.**
Barleria hirtifructa *I. Darbysh.* **sp. nov.**
Barleria holubii *C.B. Clarke* subsp. **ugandensis** *I. Darbysh.* **subsp. nov.**
Barleria inclusa *I. Darbysh.* **sp. nov.**
Barleria lukei *I. Darbysh.* **sp. nov.**
Barleria lukwangulensis [*Mildbr. ex*] *I. Darbysh.* **sp. nov.**
Barleria maritima *I. Darbysh.* **sp. nov.**
Barleria masaiensis *I. Darbysh.* **sp. nov.**
Barleria paolioides *I. Darbysh.* **sp. nov.**
Barleria penelopeana *I. Darbysh.* **sp. nov.**
Barleria polhillii *I. Darbysh.* **sp. nov.**
Barleria polhillii *I. Darbysh.* subsp. **latiloba** *I. Darbysh.* **subsp. nov.**
Barleria polhillii *I. Darbysh.* subsp. **nidus-avis** *I. Darbysh.* **subsp. nov.**
Barleria polhillii *I. Darbysh.* subsp. **turkanae** *I. Darbysh.* **subsp. nov.**
Barleria pseudosomalia *I. Darbysh.* **sp. nov.**
Barleria ramulosa *C.B. Clarke* var. **dispersa** *I. Darbysh.* **var. nov.**
Barleria robertsoniae *I. Darbysh.* **sp. nov.**
Barleria scandens *I. Darbysh.* **sp. nov.**
Barleria subregularis *I. Darbysh.* **sp. nov.**
Barleria tanzaniana (*Brummitt & J.R.I. Wood*) *I. Darbysh.*, **comb. et stat. nov.**
Barleria trispinosa (*Forssk.*) *Vahl* subsp. **glandulosissima** *I. Darbysh.* **subsp. nov.**
Barleria venenata *I. Darbysh.* **sp. nov.**
Barleria vollesenii *I. Darbysh.* **sp. nov.**
Cephalophis *Vollesen* **gen. nov.**
Cephalopis lukei *Vollesen* **sp. nov.**
Hypoestes forskaolii (Vahl) R.Br. subsp. **hildebrandtii** (*Lindau*) *I. Darbysh.*, **comb. et. stat. nov.**
Isoglossa lactea *Lindau* subsp. **saccata** *I. Darbysh.* **subsp. nov.**
Justicia alterniflora *Vollesen* **sp. nov.**
Justicia attenuifolia *Vollesen* **sp. nov.**
Justicia breviracemosa *Vollesen* **sp. nov.**
Justicia bridsoniana *Vollesen* **sp. nov.**
Justicia callopsoidea *Vollesen* **sp. nov.**
Justicia chalaensis *Vollesen* **sp. nov.**
Justicia drummondii *Vollesen* **sp. nov.**
Justicia faulknerae *Vollesen* **sp. nov.**
Justicia gilbertii *Vollesen* **sp. nov.**
Justicia grandis (*T. Anderson*) *Vollesen*, **comb. nov.**
Justicia kiborianensis (*Hedrén*) *Vollesen*, **comb. nov.**
Justicia kulalensis *Vollesen* **sp. nov.**
Justicia lukei *Vollesen* **sp. nov.**
Justicia microthyrsa *Vollesen* **sp. nov.**
Justicia mkungweensis *Vollesen* **sp. nov.**
Justicia petterssonii (*Hedrén*) *Vollesen*, **comb. nov.**
Justicia rodgersii *Vollesen* **sp. nov.**
Justicia roseobracteata *Vollesen* **nom. nov.**

Justicia schimperiana (*Nees*) *T. Anderson* subsp. **campestris** *Vollesen* **subsp. nov.**
Justicia tricostata *Vollesen* **sp. nov.**
Justicia udzungwaensis *Vollesen* **sp. nov.**
Neuracanthus africanus *S. Moore* subsp. **ruahae** (*Bidgood & Brummitt*) *Vollesen* **comb. et. stat. nov**
Neuracanthus africanus *S. Moore* subsp. **masaicus** (*Bidgood & Brummitt*) *Vollesen* **comb. et. stat. nov.**
Pseuderanthemum usambarensis *Vollesen* **sp. nov.**
Ruspolia australis (*Milne-Redh.*) *Vollesen* **comb. et stat. nov.**

GEOGRAPHICAL DIVISIONS OF THE FLORA

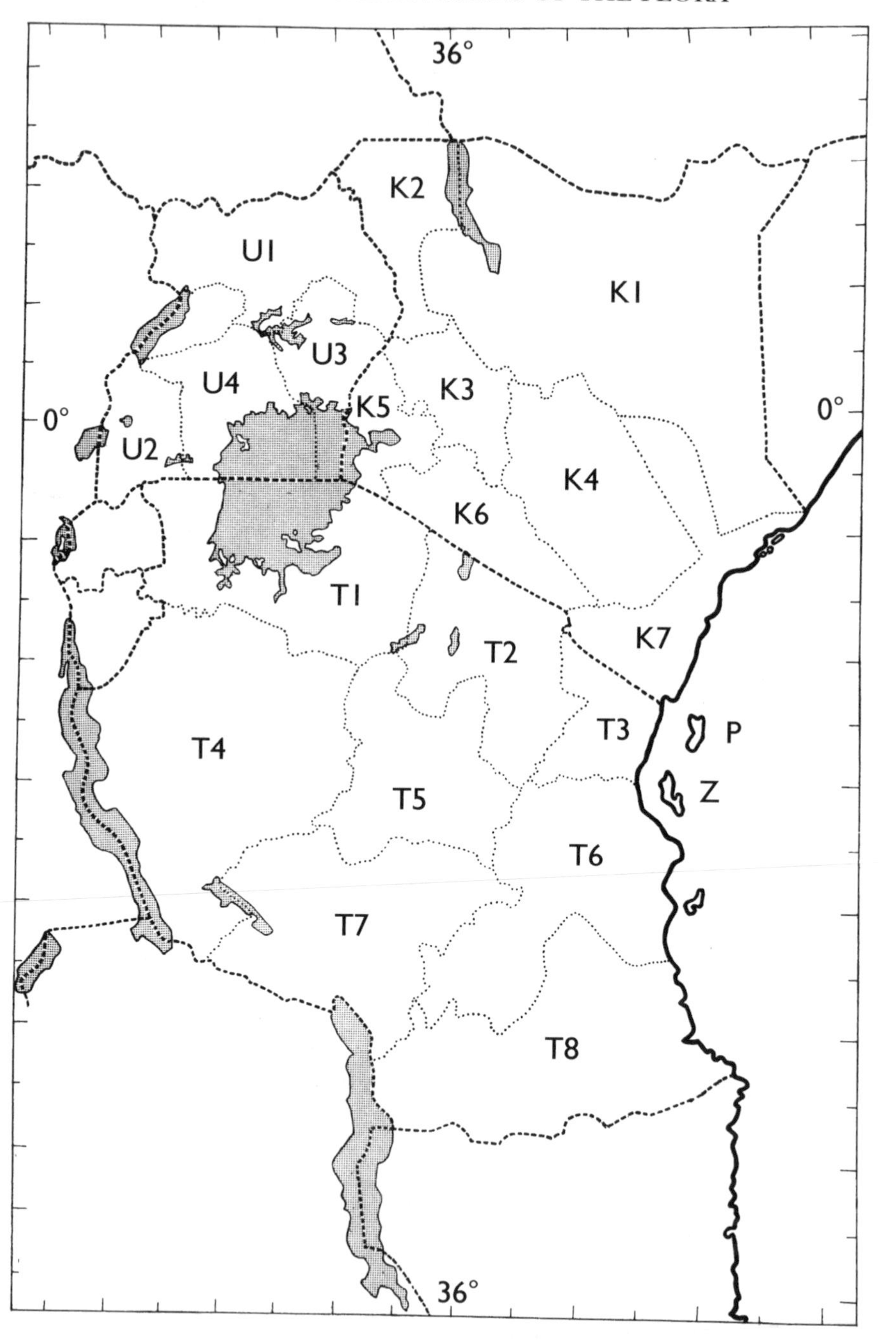

LIST OF ABBREVIATIONS

A.V.P. = O. Hedberg, Afroalpine Vascular Plants; **B.J.B.B.** = Bulletin du Jardin Botanique de l'Etat, Bruxelles; Bulletin du Jardin Botanique Nationale de Belgique; **B.S.B.B.** = Bulletin de la Société Royale de Botanique de Belgique; **C.F.A.** = Conspectus Florae Angolensis; **E.J.** = A. Engler, Botanische Jahrbücher für Systematik, Pflanzengeschichte und Pflanzengeographie; **E.M.** = A. Engler, Monographieen Afrikanischer Pflanzen-Familien und Gattungen; **E.P.** = A. Engler, Das Pflanzenreich; **E.P.A.** = G. Cufodontis, Enumeratio Plantarum Aethiopiae Spermatophyta; in B.J.B.B. 23, Suppl. (1953) et seq.; **E. & P. Pf.** = A. Engler & K. Prantl, Die Natürlichen Pflanzenfamilien; **F.A.C.** = Flore d'Afrique Centrale (*formerly* F.C.B.); **F.C.B.** = Flore du Congo Belge et du Ruanda-Urundi; Flore du Congo, du Rwanda et du Burundi; **F.E.E.** = Flora of Ethiopia & Eritrea; **F.D.-O.A.** = A. Peter, Flora von Deutsch-Ostafrika; **F.F.N.R.** = F. White, Forest Flora of Northern Rhodesia; **F.P.N.A.** = W. Robyns, Flore des Spermatophytes du Parc National Albert; **F.P.S.** = F.W. Andrews, Flowering Plants of the Anglo-Egyptian Sudan *or* Flowering Plants of the Sudan; **F.P.U.** = E. Lind & A. Tallantire, Some Common Flowering Plants of Uganda; **F.R.** = F. Fedde, Repertorium Speciorum Novarum Regni Vegetabilis; **F.S.A.** = Flora of Southern Africa; **F.T.A.** = Flora of Tropical Africa; **F.W.T.A.** = Flora of West Tropical Africa; **F.Z.** = Flora Zambesiaca; **G.F.P.** = J. Hutchinson, The Genera of Flowering Plants; **G.P.** = G. Bentham & J.D. Hooker, Genera Plantarum; **G.T.** = D.M. Napper, Grasses of Tanganyika; **I.G.U.** = K.W. Harker & D.M. Napper, An Illustrated Guide to the Grasses of Uganda; **I.T.U.** = W.J. Eggeling, Indigenous Trees of the Uganda Protectorate; **J.B.** = Journal of Botany; **J.L.S.** = Journal of the Linnean Society of London, Botany; **K.B.** = Kew Bulletin, *or* Bulletin of Miscellaneous Information, Kew; **K.T.S.** = I. Dale & P.J. Greenway, Kenya Trees and Shrubs; **K.T.S.L.** = H.J. Beentje, Kenya Trees, Shrubs and Lianas; **L.T.A.** = E.G. Baker, Leguminosae of Tropical Africa; **N.B.G.B.** = Notizblatt des Botanischen Gartens und Museums zu Berlin-Dahlem; **P.O.A.** = A. Engler, Die Pflanzenwelt Ost-Afrikas und der Nachbargebiete; **R.K.G.** = A.V. Bogdan, A Revised List of Kenya Grasses; **T.S.K.** = E. Battiscombe, Trees and Shrubs of Kenya Colony; **T.T.C.L.** = J.P.M. Brenan, Check-lists of the Forest Trees and Shrubs of the British Empire no. 5, part II, Tanganyika Territory; **U.K.W.F.** = A.D.Q. Agnew (or for ed. 2, A.D.Q. Agnew & S. Agnew), Upland Kenya Wild Flowers; **U.O.P.Z.** = R.O. Williams, Useful and Ornamental Plants in Zanzibar and Pemba; **V.E.** = A. Engler & O. Drude, Die Vegetation der Erde, IX, Pflanzenwelt Afrikas; **W.F.K.** = A.J. Jex-Blake, Some Wild Flowers of Kenya; **Z.A.E.** = Wissenschaftliche Ergebnisse der Deutschen Zentral-Afrika-Expedition 1907–1908, 2 (Botanik).

FAMILIES OF VASCULAR PLANTS REPRESENTED IN THE FLORA OF TROPICAL EAST AFRICA

The family system used in the Flora has diverged in some respects from that now in use at Kew and the herbaria in East Africa. The accepted family name of a synonym or alternative is indicated by the word "see". Included family names are referred to the one used in the Flora by "in" if in accordance with the current system, and "as" if not. Where two families are included in one fascicle the subsidiary family is referred to the main family by "with".

PUBLISHED PARTS

Foreword and preface
*Glossary
Index of Collecting Localities

Acanthaceae
  Part 1
  Part 2
*Actiniopteridaceae
*Adiantaceae
Aizoaceae
Alangiaceae
Alismataceae
*Alliaceae
*Aloaceae
*Amaranthaceae
*Amaryllidaceae
*Anacardiaceae
*Ancistrocladaceae
Anisophyllaceae — as Rhizophoraceae
Annonaceae
*Anthericaceae
Apiaceae — see Umbelliferae
Apocynaceae
  *Part 1
  **Part 2
*Aponogetonaceae
Aquifoliaceae
*Araceae
Araliaceae
Arecaceae — see Palmae
*Aristolochiaceae
**Asclepiadaceae — see Apocynaceae
Asparagaceae
*Asphodelaceae
Aspleniaceae
Asteraceae — see Compositae
Avicenniaceae — as Verbenaceae
*Azollaceae

*Balanitaceae
*Balanophoraceae
*Balsaminaceae
Basellaceae
Begoniaceae
Berberidaceae
Bignoniaceae
Bischofiaceae — in Euphorbiaceae
Bixaceae
Blechnaceae
*Bombacaceae
*Boraginaceae
Brassicaceae — see Cruciferae
Brexiaceae
Buddlejaceae — as Loganiaceae
*Burmanniaceae
*Burseraceae
Butomaceae
Buxaceae

Cabombaceae
Cactaceae
Caesalpiniaceae — in Leguminosae
*Callitrichaceae
Campanulaceae
Canellaceae
Cannabaceae
Cannaceae — with Musaceae
Capparaceae
Caprifoliaceae
Caricaceae
Caryophyllaceae
*Casuarinaceae
Cecropiaceae — with Moraceae
*Celastraceae
*Ceratophyllaceae
Chenopodiaceae
Chrysobalanaceae — as Rosaceae

Clusiaceae — see Guttiferae
Cobaeaceae — with Bignoniaceae
Cochlospermaceae
Colchicaceae
Combretaceae
**Commelinaceae
Compositae
*Part 1
*Part 2
Part 3
Connaraceae
Convolvulaceae
Cornaceae
Costaceae — as Zingiberaceae
*Crassulaceae
*Cruciferae
Cucurbitaceae
Cupressaceae
Cyanastraceae — in Tecophilaeaceae
Cyatheaceae
Cycadaceae
*Cyclocheilaceae
*Cymodoceaceae
Cyperaceae
Cyphiaceae — as Lobeliaceae

*Davalliaceae
*Dennstaedtiaceae
*Dichapetalaceae
Dilleniaceae
Dioscoreaceae
Dipsacaceae
*Dipterocarpaceae
Dracaenaceae
Dryopteridaceae
Droseraceae

*Ebenaceae
Elatinaceae
Equisetaceae
Ericaceae
*Eriocaulaceae
*Eriospermaceae
*Erythroxylaceae
Escalloniaceae
Euphorbiaceae
*Part 1
*Part 2

Fabaceae — see Leguminosae
Flacourtiaceae
Flagellariaceae
Fumariaceae

*Gentianaceae
Geraniaceae
Gesneriaceae
Gisekiaceae — as Aizoaceae
*Gleicheniaceae
Goodeniaceae
Gramineae
Part 1
Part 2
*Part 3
Grammitidaceae
Gunneraceae — as Haloragaceae
Guttiferae

Haloragaceae
Hamamelidaceae
*Hernandiaceae
Hippocrateaceae — in Celastraceae
Hugoniaceae — in Linaceae
*Hyacinthaceae
*Hydnoraceae
*Hydrocharitaceae
*Hydrophyllaceae
*Hydrostachyaceae
Hymenocardiaceae — with Euphorbiaceae
Hymenophyllaceae

Hypericaceae — see also Guttiferae
Hypoxidaceae

Icacinaceae
Illecebraceae — as Caryophyllaceae
*Iridaceae
Irvingiaceae — as Ixonanthaceae
Isoetaceae
*Ixonanthaceae

Juncaceae
Juncaginaceae

Labiatae
Lamiaceae — see Labiatae
*Lauraceae
Lecythidaceae
Leeaceae — with Vitaceae
Leguminosae
Part 1, Mimosoideae
Part 2, Caesalpinioideae
Part 3
Part 4 } Papilionoideae
Lemnaceae
Lentibulariaceae
Liliaceae (*s.s.*)
Limnocharitaceae — as Butomaceae
Linaceae
*Lobeliaceae
Loganiaceae
*Lomariopsidaceae
*Loranthaceae
Lycopodiaceae
*Lythraceae

Malpighiaceae
Malvaceae
Marantaceae
*Marattiaceae
*Marsileaceae
Melastomataceae
*Meliaceae
Melianthaceae
Menispermaceae
*Menyanthaceae
Mimosaceae — in Leguminosae
Molluginaceae — as Aizoaceae
Monimiaceae
Montiniaceae
*Moraceae
*Moringaceae
Muntingiaceae — with Tiliaceae
*Musaceae
*Myricaceae
*Myristicaceae
*Myrothamnaceae
*Myrsinaceae
*Myrtaceae

*Najadaceae
Nectaropetalaceae — in Erythroxylaceae
*Nesogenaceae
*Nyctaginaceae
*Nymphaeaceae

Ochnaceae
Octoknemaceae — in Olacaceae
Olacaceae
Oleaceae
*Oleandraceae
Oliniaceae
Onagraceae
*Ophioglossaceae
Opiliaceae
Orchidaceae
Part 1, Orchideae
*Part 2, Neottieae, Epidendreae
*Part 3, Epidendreae, Vandeae
Orobanchaceae
*Osmundaceae
Oxalidaceae

*Palmae
Pandaceae — with Euphorbiaceae
*Pandanaceae
Papaveraceae
Papilionaceae — in Leguminosae
*Parkeriaceae
Passifloraceae
Pedaliaceae
Periplocaceae — see Apocynaceae (Part 2)
Phytolaccaceae
*Piperaceae
Pittosporaceae
Plantaginaceae
Plumbaginaceae
Poaceae — see Gramineae
Podocarpaceae
Podostemaceae
Polemoniaceae — see Cobaeaceae
Polygalaceae
Polygonaceae
*Polypodiaceae
Pontederiaceae
*Portulacaceae
Potamogetonaceae
Primulaceae
*Proteaceae
*Psilotaceae
*Ptaeroxylaceae
*Pteridaceae

*Rafflesiaceae
Ranunculaceae
Resedaceae
Restionaceae
Rhamnaceae
Rhizophoraceae
Rosaceae
Rubiaceae
Part 1
*Part 2
*Part 3
*Ruppiaceae
*Rutaceae

*Salicaceae
Salvadoraceae
*Salviniaceae
Santalaceae
*Sapindaceae
Sapotaceae
*Schizaeaceae
Scrophulariaceae
Scytopetalaceae
Selaginellaceae
Selaginaceae — in Scrophulariaceae
*Simaroubaceae
*Smilacaceae
**Solanaceae
Sonneratiaceae
Sphenocleaceae
Strychnaceae — in Loganiaceae
*Surianaceae
Sterculiaceae

Taccaceae
Tamaricaceae
Tecophilaeaceae
Ternstroemiaceae — in Theaceae
Tetragoniaceae — in Aizoaceae
Theaceae
Thelypteridaceae
Thismiaceae — in Burmanniaceae
Thymelaeaceae
*Tiliaceae
Trapaceae
Tribulaceae — in Zygophyllaceae
*Triuridaceae
Turneraceae
Typhaceae

Uapacaceae — in Euphorbiaceae
Ulmaceae
*Umbelliferae
*Urticaceae

Vacciniaceae — in Ericaceae
Valerianaceae
Velloziaceae
*Verbenaceae
*Violaceae
*Viscaceae
*Vitaceae
*Vittariaceae

*Woodsiaceae

*Xyridaceae

*Zannichelliaceae
*Zingiberaceae
*Zosteraceae
*Zygophyllaceae

Editorial adviser, National Museums of Kenya: Quentin Luke
Editorial adviser, Makerere University: J. Kalema
Adviser on Linnaean types: C. Jarvis

Parts of this Flora, unless otherwise indicated, are obtainable from:
Royal Botanic Gardens, Kew, Richmond, Surrey TW9 3AB, England. www.kew.org or www.kewbooks.com

*** Only available through CRC Press at:**
UK and Rest of World (except North and South America):
CRS Press/ITPS,
Cheriton House, North Way, Andover, Hants SP10 5BE.
e: uk.tandf@thomsonpublishingservices. co.uk

North and South America:
CRC Press,
2000NW Corporate Blvd, Boco Raton, FL 33431-9868, USA.
e: orders@crcpress.com

**** Forthcoming titles in production**
For availability and expected publication dates please check on our website, www.kew.books.com

Information on current prices can be found at www.kewbooks.com or www.tandf.co.uk/books/

First published in 2010 by
Royal Botanic Gardens, Kew
Richmond, Surrey, TW9 3AB, UK
www.kew.org

ISBN 978 1 84246 386 4

British Library Cataloguing in Publication Data
A catalogue record for this book is available from the British Library

Design and typesetting by Margaret Newman,
Kew Publishing, Royal Botanic Gardens, Kew.

Printed in the the USA by The University of Chicago Press

All proceeds go to support Kew's work in saving the world's plants for life